PRECALCULUS

PRECALCULUS

A VIEW
OF THE WORLD
AROUND US

DAVID WELLS
Penn State University–New Kensington

LYNN TILSON
Penn State University–University Park

Prentice Hall
Upper Saddle River, New Jersey 07458

Library of Congress Cataloging-in-Publication Data

Wells, David
    Precalculus : a view of the world around us / David Wells,
Lynn Tilson.
        p.   cm.
    Includes index.
    ISBN 0-02-425451-7   (hardcover: alk. paper)
    1. Functions.   I. Tilson, Lynn.   II. Title.
QA331.3.W45   1998
515′.25—dc21   97-36136
                                  CIP

Senior Acquisition Editor: Sally Denlow
Editorial Assistant/Supplement Editor: April Thrower
Development Editor: Tony Palermino
Marketing Manager: Patrice Lumumba Jones
Marketing Assistant: Patrick Murphy
AVP, Production and Manufacturing: David W. Riccardi
Production Editor: Elaine W. Wetterau
Senior Managing Editor: Linda Mihatov Behrens
Executive Managing Editor: Kathleen Schiaparelli
Buyer: Alan Fischer
Manufacturing Manager: Trudy Pisciotti
Interior Design: Judy Matz-Coniglio
Asssociate Creative Director: Amy Rosen
Director of Creative Services: Paula Maylahn
Art Manager: Gus Vibal
Art Director, Cover: Jayne Conte

©1998 by David Wells and Lynn Tilson
Published by Prentice-Hall, Inc.
Simon & Schuster/A Viacom Company
Upper Saddle River, New Jersey 07458

Printed in the United States of America
10 9 8 7 6 5 4 3 2 1

ISBN 0-02-425451-7

Prentice-Hall International (UK) Limited, *London*
Prentice-Hall of Australia Pty. Limited, *Sydney*
Prentice-Hall Canada, Inc., *Toronto*
Prentice-Hall Hispanoamericana, S.A., *Mexico*
Prentice-Hall of India Private Limited, *New Delhi*
Prentice-Hall of Japan, Inc., *Tokyo*
Simon & Schuster Asia Pte. Ltd., *Singapore*
Editora Prentice-Hall do Brasil, Ltda., *Rio de Janeiro*

Dave's Dedication

To Alice and Sarah with love—
who kept my life alive while I was writing.

Lynn's Dedications

To Papa with love—
for my father, Dr. Ernest Watts Harvey, Jr., who died February 20, 1994
but who still lives within me and within the pages of this book.

To Mom with love—
for my mother, Harriet Virginia Peaseley Harvey, who died November 7,
1995—This book would not be, had you not been. It is your legacy to me,
through me.

CONTENTS

CHAPTER 3   LINEAR SYSTEMS   101

CHAPTER 5   POLYNOMIAL FUNCTIONS 241

CHAPTER 6   RATIONAL FUNCTIONS 295

CHAPTER 7

EXPONENTIAL AND LOGARITHMIC FUNCTIONS 323

CHAPTER 8

TRIGONOMETRIC FUNCTIONS 391

CHAPTER 9

TRIGONOMETRIC FUNCTIONS AS ANALYTICAL TOOLS   455

A P P E N D I X   B

TIPS FOR GRAPHING FUNCTIONS WITH A CALCULATOR   B1

PREFACE

"The answer must be in the attempt,"

a quote from the character Celine in the movie, *Before Sunrise.*

This idea permeates our view of mathematics
and is found throughout the book.

Precalculus: *A View of the World Around Us,* is the product of our efforts to write something we could both use in our own teaching. Our teaching styles are very different, so the book needed the flexibility to be supporting material for Dave's lectures as well as integral material for Lynn's group work. We developed "hands on" explorations and associated exercises to complement the usual exposition and example approach. This blend of pedagogies created a balance between tradition and innovation that has an appeal to Dave, who always wanted to be a mathematician and a rock musician—and is—and to Lynn, who always wanted to be an actress or veterinarian—and isn't.

This balance created a textbook with both a style and content that closely corresponds to both the AMATYC and NCTM standards. It may be associated with the reform calculus movement for that reason. Although we did not write it with a reform calculus course specifically in mind, it will precede one nicely, as well as precede a traditional course. Instructors who use this book as a prelude to a reform calculus course might place more emphasis on the Mathematical Looking Glasses. Those who use it as a prelude to a traditional calculus course might place more emphasis on algebraic skill. In either case, this book needs only the prerequisite of a traditional intermediate algebra course and gives a solid precalculus foundation.

There are four main defining features of this textbook, all of which are meant to enhance the student's mathematical power.

One of the defining features is the dynamic approach to functions. The chapter on linear functions introduces the average rate of change and increasing and decreasing behavior. We reinforce these topics with polynomial functions, rational functions, exponential functions, logarithmic functions, and finally with trigonometric functions. The connection between the average rate of change and increasing and decreasing behavior is extended to include the idea of instantaneous rate of change. We do this by using the idea of the limit both numerically and graphically.

Another defining feature is the presentation of real physical problems in what we call Mathematical Looking Glasses. They present a rich variety of situations in which students can see mathematics as interesting, timely, and useful. They create a curiosity and understanding of "the world around us," engaging the students, both in action and in interest, instead of allowing them to be passive.

Some examples are

- ***Sonar***, traces the FBI's search for a cause of an airliner crash
- ***The Mad Tea Party***, investigates angular velocity of a ride at Disney World
- ***The Comet's Tale***, combines trigonometry with the history of Halley's comet
- ***Shortstop***, analyzes a critical moment in the 1996 World Series
- ***Crowded Heavens***, connects international politics with satellites
- ***20 Questions***, looks at the game we played as children
- ***A Murder in Ceylon***, discusses the cooling of dead bodies
- ***Fat Cats on Campus***, presents the workings of a T-shirt company
- ***A Case for Algebra***, asks the students to decide someone's guilt or innocence

A third defining feature is George Polya's problem-solving process, which we introduce in Section 1-3 and refer to many times throughout the book. Students use the strategies to solve a number of problems, including Mathematical Looking Glasses. They see how we, as authors, have used them and where it might be appropriate for them to try.

The fourth defining feature is the presentation of concepts numerically, analytically, and graphically, known as the "rule-of-three." This presents "a view of the world around us" from several mathematical perspectives and develops multiple technical skills. Many exercises reinforce these methods. The exercises labeled *Creating Models* require the students to present physical situations as equations, tables, or graphs. The ones called *Interpreting Mathematics* ask for an interpretation of equations, tables, or graphs. The *Writing to Learn* exercises challenge students to communicate mathematical ideas. (The inclusion of communication skills into the rule-of-three is now known in some circles as the "rule-of-four.") Exercises follow directly from examples, giving the book a natural flow from concept to skill. This allows the students to determine whether their understanding is sufficient to attempt the ones at the end of the section.

This textbook's visual appearance and topical sequence give it flexibility and easy accessibility, which allow instructors to adapt it to their own focus. The following topics are individual units that can be moved around or omitted.

- Parametric equations are first presented with linear functions in Section 2-6, then in Section 4-2 with quadratic functions, in Section 7-3 as a supplementary topic with inverse functions, and in Section 10-2 with conic sections.
- Systems of equations are first discussed in Section 3-1 with linear functions and reintroduced in Sections 4-2 and 4-7 with quadratic functions and relations, then in Section 5-4 with polynomial functions, and finally in Section 7-1 with exponential functions.
- Sequences are first presented in Section 2-4 with linear functions and then in Section 7-1 with exponential functions.

- The trigonometry chapters are organized with an emphasis on circular functions. The sine and cosine functions are defined through the unit circle in the first three pages of Section 8-1, and then by their right triangle definitions. This order can be rearranged.
- The appendices are designed for individual reading, leaving classroom discussion to a minimum. Appendix A is a quick, elementary algebra review. The placement after Chapter 10 is a clear message to the student that these skills are prerequisite to the course. Appendix B is a guide to calculator techniques common to most graphing calculators. It is not meant to replace any individual manual.

This book has several more defining features. First, it presents precalculus as part of our historical heritage. Polynomial and trigonometric functions, in particular, have a rich and interesting history, which is reflected in the Mathematical Looking Glasses and the narrative text. Mathematics is also seen as a pattern-seeking device for discovering underlying principles, as in *20 Questions*, for example. Finally, the problem-solving process is complemented by the use of a graphing calculator. It is used as a means to an end, not an end in itself. We guide the students, quite often at first, as to whether the calculator is the most efficient or most comfortable path to a solution. Later, we usually leave that decision up to them.

All of the book's defining features lead students to a deeper understanding of mathematics and a broader view of its applicability. Students will see that the subject is of interest in its own right, not *just* as a course needed for the study of calculus.

Many textbooks have a *letter to the student*. Our first section takes the place of this letter. It details what the students need coming into the book and what they will have going out. It gives them instructions and hints on how to use our book in the most effective way and some personal notes on what we think success in this course means. Most important, it will let them see what this book is all about.

Dave Wells
Lynn Tilson

Please visit our website at:
www.prenhall.com/wells

ACKNOWLEDGMENTS

LYNN'S ACKNOWLEDGMENTS

My new husband James—for his patience and love as he discovers what the world of book writing and publishing are all about.

My daughter Carrie—for her conjured-up enthusiasm about the book and for now being old enough to read it and drink at the same time.

My mother "Boo" Peaseley Harvey—for her faith in me, her role model as teacher, and her financial assistance.

My father Ernest Watts Harvey—for saying, "I didn't know you were so smart," when he first read some of the manuscript, and for keeping me on my intellectual toes with philosophical conversation.

Steve Albert and company (Brian, Mike, & Gene)—for keeping my old Mac running.

Tim Schmitt—for supporting me through graduate school and for his many suggestions that helped form the focus of this book.

The staff (especially Mary Beth Kleist) and patrons at the Hawthorne Club in Hawthorne, Nevada—for believing in me and giving me a second home.

Kathi Arnold—for her warmth and model of strength.

Steve Merrill—for seeing this theatre major through graduate school.

My students—for being themselves and helping me grow.

My co-author Dave—for being there, always.

Barbara Schmitt—for her laughter and love.

Lee Koenig—for her continuous affirmation of friendship.

Ann Crowe—for being my best friend during the last 25 years.

Mary Hembree—for her kindness when I needed it most.

DAVE'S ACKNOWLEDGMENTS

Dave Brown—who showed me what good teaching is.

Roy Myers—for the benefit of his wisdom during the many projects we have worked on together.

My wife Alice—for being a one-person support staff.

My daughter Sarah—for making her life an inspiration to her father.

My mother Jacqueline Gaye Wells—for encouraging me to be a mathematician.

My father Edgar Franklin Wells—for his excitement about the originality of our approach in this project.

My co-author Lynn—for helping me rediscover my sense of wonder.

My long-suffering students—for their class testing the manuscript during 1994–1996.

All my friends at the Unitarian Universalist Church of the North Hills, and at Calliope Songwriter's Night—for keeping me from going over the edge.

OUR ACKNOWLEDGMENTS

Sally Denlow—for being our lovely and bright mathematics editor who in the face of these two inexperienced authors has "put it all together." Thanks Sally—we did it!

Bob Pirtle—for his unwavering support and constant direction for this project.

Elaine Wetterau—for being a unique blend of production editor, stern advisor, and friend.

Tony Palermino—for being our third author, who is disguised by Prentice Hall as a development editor.

Billy Kratsa, Sr.—for keeping us in lunch and cookies.

All our friends and colleagues—for listening to book woes and sagas.

Lee Row, III—for staying in constant touch with us for almost eight years and providing background for several of our Mathematical Looking Glasses.

Fred Rickey—for patiently answering untimely questions about mathematical history.

John Fink, Kalamazoo College, and Nancy Prosenjak, Metropolitan State College of Denver—for exceptionally thorough and valuable reviews.

Steve Monk, University of Washington, and James M. Sullivan, Massachusetts Bay Community College—for making some valuable suggestions in the early developmental stages of this book.

Alden Partridge, Ricks College—for his conscientious efforts and thoughtful recommendations.

Yolanda Davis, Jan Scott, and Mary Stephens, Clayton College and State University—for their many candid comments.

We thank the following reviewers for their comments and suggestions during the development of the manuscript: James E. Daly, University of Colorado at Colorado Springs; Raymond D. McDaniel, Jr., Pembroke State University; Elise Price, Tarrant County Junior College; and James M. Sullivan, Massachusetts Bay Community College.

All those people that we met in our travels over the last eight years who seemed excited about our project—for regenerating our enthusiasm.

The many people whose interests and activities provided material for the Mathematical Looking Glasses—for having interesting lives and letting us in on them. (We generally have not changed names to protect the guilty—but sometimes *did* change their lives.)

David Wells
Lynn Tilson

PRECALCULUS

CHAPTER 1

MODELING AND PROBLEM SOLVING

1–1 A CASE FOR ALGEBRA

PREREQUISITES   MAKE SURE YOU ARE FAMILIAR WITH:
The Coordinate Plane (Section A-3)
Basic Graphing Techniques (Section A-5)
Graphing Linear Equations (Section A-6)

(Prerequisites are indicated at the beginning of many sections in this book. They indicate skills from previous courses that you will need in these sections. In the present section you will need to know the standard methods for graphing linear equations and finding equations of linear graphs. Refer to the sections cited and review them as needed before continuing.)

Mathematics lives in the world around us. For example, mathematics is needed to build a laser, bake a cake, and calculate the Dow Jones average. Because this is so, this book contains descriptions of the world viewed through a "mathematical looking glass," along with mathematical questions that real people might ask. The descriptions are detailed, so that you can see the mathematics embedded in the situation and view the questions in a meaningful context. The accompanying tables, equations, and graphs will help you answer the questions.

Let's use the following situation to illustrate the process of formulating mathematical questions in a physical context, answering the questions, and interpreting the results. Read through it once to understand the context, without worrying about the mathematics.

A MATHEMATICAL LOOKING GLASS
A Case for Algebra

One evening in the summer of 1988, a young man whom we will call Bill was socializing at the Greentree Holiday Inn in Pittsburgh. He drank seven beers between 10:00 P.M. and 2:00 A.M. before leaving with two friends and a young woman they had met. En route to Primanti Brothers Sandwich Shop at 2:10 A.M., Bill's Volkswagen GTI struck a steel pole. The woman was thrown forward and sustained a fatal head injury.

Technicians at Mercy Hospital took a reading of Bill's blood alcohol concentration (BAC) at 2:52 A.M. This reading showed a level of 0.136, indicating that his bloodstream contained 0.136% alcohol. A second reading taken by the police at 4:40 A.M. showed a level of 0.100. Both were at or above Pennsylvania's legal threshold of intoxication, which is 0.100. Bill was therefore charged with vehicular homicide while DUI.

Dr. Charles Winek, chief toxicologist of the Allegheny County Crime Lab, testified at the trial that for a person of Bill's weight, elimination of alcohol would have lowered his BAC at a rate of about 0.020 per hour. Based on this assumption, his BAC at the time of the accident would have been higher than 0.136, as indicated in Figure 1-1.

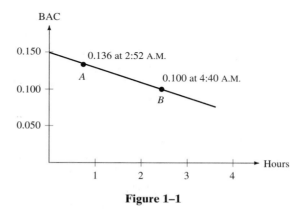

Figure 1–1

During cross-examination Bill's lawyer, Thomas Schuchert, reminded Dr. Winek that Bill had consumed at least two beers between 1:30 A.M. and 2:00 A.M. He then asked how Dr. Winek's conclusions would be affected if neither of those beers had been absorbed into Bill's bloodstream prior to the accident. Dr. Winek replied that in this case, Bill's BAC would have risen by about 0.064 per hour between 2:10 A.M. and 2:52 A.M., so that Figure 1-2 would present a more appropriate picture.

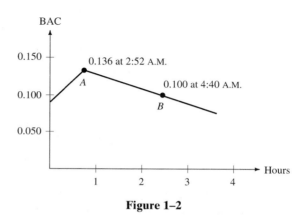

Figure 1–2

The critical problem for both the prosecution and the defense was to determine Bill's BAC at the time of the accident as reliably as possible.   ■

The problem described here certainly has mathematical aspects. Let's look at the situation again and try to formulate it as a mathematical question. Read on, but more slowly now, with pencil in hand.

The graphs in Figures 1-1 and 1-2 each provide a rough estimate of Bill's BAC at 2:10 A.M. How can we improve the accuracy of those estimates? Let's focus on Figure 1-1 first. Since the graph in that figure is a straight line, we should be able to find its equation if we know the coordinates of the points A and B. (See pages A14 and A15 to review this idea.) Let's define the variables

$$t = \text{time in hours after the accident}$$

$$y = \text{Bill's BAC}$$

We have been given some information about these variables. In particular, we were told that Bill's BAC was 0.136 at 2:52 A.M. This was 42 minutes, or 0.7 hours, after the accident. (Don't just take our word for this. Check your understanding of the situation by performing the calculations yourself.) Therefore, point A in Figure 1-1 has coordinates (0.7, 0.136). Similarly, 4:40 A.M. was 2.5 hours after the accident, so point B has coordinates (2.5, 0.100).

The physical question,

"What was Bill's BAC at the time of the accident?"

has now been formulated as a mathematical question,

"What is the equation of the line through the points (0.7, 0.136) and (2.5, 0.100)?"

You can answer this mathematical question in Exercise 1.

EXERCISE   **1.** Find the equation of the line through (0.7, 0.136) and (2.5, 0.100).

Let's interpret our results. In slope-intercept form, your equation in Exercise 1 should be $y = -0.020t + 0.150$. The slope and the y-intercept of this line both have a physical meaning. The slope of -0.020 represents the rate at which alcohol was removed from his bloodstream, and the y-intercept of 0.150 represents Bill's BAC at the time of the accident. Thus, according to the graph in Figure 1-1, Bill was legally intoxicated when the accident occurred.

Now let's focus on the line in the left part of Figure 1-2. It has a slope of 0.064, representing the rate at which Bill's BAC would have increased according to Dr. Winek's testimony. Thus in Figure 1-2, you can calculate Bill's BAC at the time of the accident if you know the equation of the line through (0.7, 0.136) with a slope of 0.064.

EXERCISES   **2.** Find the equation of the line passing through (0.7, 0.136) with a slope of 0.064.

3. (*Interpreting Mathematics*) According to the graph in Figure 1-2, what was Bill's BAC at the time of the accident?

4. (*Writing to Learn*)
   a. Which of the two graphs would Mr. Schuchert be more likely to use in his arguments? Explain your answer.
   b. As a juror, would you accept either graph as evidence? If so, which one?
   c. As a juror, would you need additional information before reaching a decision? If so, what kind of information?

5. (*Creating Models*) Dr. Winek's testimony was based on an assumption that Bill had not eaten. However, one of Bill's friends testified that Bill had some food at the shrimp bar after his last drink.
   a. The presence of food slows the absorption of alcohol into the bloodstream. How would this affect the appearance of the graph in Figure 1-2?
   b. Does Bill's friend's testimony favor the prosecution or the defense?

Many students see no value in a precalculus course except that it fulfills a degree requirement. However, *A Case for Algebra* indicates that mathematics can be useful to you even if you do not use it in your career. To help you understand our reasons for asking you to commit your time to this book, let's discuss a few reasons for studying mathematics.

MATHEMATICS IS AN ANALYTICAL TOOL

A Case for Algebra illustrates dramatically that no matter what your interests or your chosen occupation may be, you sometimes need to process quantitative information to make decisions. Even if you never serve on a jury, you will certainly make personal decisions about mortgages, credit cards, and investments. You will also evaluate the credibility of quantitative information provided by politicians and others, concerning subjects such as the economy, the environment, and national security.

If you are reading as critically as we hope you are, the preceding paragraph has probably not convinced you of the value of mathematics beyond arithmetic. After all, your great aunt Bertha never went beyond third grade and was still able to make shrewd financial investments and intelligent political decisions. We are not claiming that formal mathematical training is the only road to quantitative enlightenment. However, you can analyze the dynamics of quantitative relationships systematically only through algebra and other mathematics based on algebra.

MATHEMATICS IS RICH IN CONCEPTS AND CONTEXTS

Many students believe that mathematics consists only of computation and symbol pushing. By contrast, this book will discuss mathematics in the context of situations that have personally touched your authors, Lynn and Dave. This approach will help you to appreciate the importance of mathematics and to integrate the subject into your life.

Many students also believe that mathematics is a rigid, inflexible discipline. Studying mathematics in context will dissolve this image and reveal the richness of the subject. Behind the operations, there are assumptions that are subject to opinions and judgments, and beyond the operations are results that are open to varying interpretations. For example, in completing Exercises 4 and 5, you formed opinions and judgments based partly on the results of mathematical operations. Several other aspects of Bill's trial also integrate mathematics into a larger context. For example, the actual BAC readings were taken from Bill's blood serum, rather than whole blood. Since Pennsylvania's legal standard for intoxication is based on whole blood samples, Dr. Winek needed to base his testimony on estimates of readings that would have resulted from whole blood samples.

Even within the narrower context of mathematical operations, where questions have only one correct answer, there are usually many paths to that answer. For example, in *A Case for Algebra*, questions about Bill's BAC can be answered by using either the graphs in Figures 1-1 and 1-2 or the equations in Exercises 1 and 2.

MATHEMATICS IS PART OF OUR INTELLECTUAL HERITAGE

Like literature or music, mathematics has been created over the course of centuries by fascinating people who were passionate about their activities as ends in themselves, regardless of the practical applicability of their ideas. You will become acquainted with some of these people as you study their contributions to our mathematical heritage. You will meet, among others, an astronomer who cast astrological charts for kings to avoid starvation while calculating planetary orbits and a political activist who developed an important new body of mathematical theory before he was killed in a duel at the age of 20.

Like literature or music, mathematics has also had a profound effect on the cultures that created it. For example, the computers that affect our everyday lives operate on the principles of Boolean algebra, an area of abstract mathematics developed about 150 years ago by George Boole (1815–1864). The study of mathematics, like that of literature or music, enriches your understanding of your cultural environment.

YOU MAY ACTUALLY USE MATHEMATICS!

We live in an increasingly global economy with changing needs. It is impossible to predict whether the career you choose today will be viable 10 or 20 years from now. Furthermore, our increasing reliance on technology ensures that many career paths will require you to think quantitatively. The more mathematics you know, the more career options you will have.

EXERCISE   **6.** (*Writing to Learn*) Choose one of our statements that describes a benefit to be gained by studying mathematics. Write a paragraph citing your reasons for either agreeing or disagreeing with it. (We won't be offended if you disagree with us. What is important is that you read critically and support your opinions.)

SOME SUGGESTIONS FOR FEELING AT HOME WITH THIS BOOK

By beginning this book, you have given yourself an opportunity to become familiar with the many faces of mathematics. Here are some suggestions for making the most of that opportunity.

• Regard the book as a continuous journey, not a collection of isolated facts. It is best read as you would read a novel, that is, without skipping immediately to the exercises. (Your author Lynn confesses that she reads the last page of a novel first, but swears that she reads every page carefully afterward.)

- Our suggestion to read this book like a novel is not meant to imply that you can read it as quickly as you would read a contemporary romance. A more appropriate comparison would be with Shakespeare. Both reading and doing mathematics take time, and temporary failures are normal. Be patient!

- Exercises are scattered through each section rather than being grouped all at the end. Read with pencil and paper in hand, and participate by completing exercises as you come to them. Feel free to scribble in the wide margins.

- Because of the cumulative nature of the subject, you will sometimes need to refer to your work from an earlier exercise. It will be important to keep all your prior work organized for ready reference.

- The Mathematical Looking Glasses you will encounter fall broadly into two categories, each having its own purpose.

  Some, such as **Ultrasound** (page 221) and **Standing Room Only** (page 323), describe uses of mathematics by mathematicians, scientists, and engineers to establish scientific principles and obtain information. All descriptions and data are accurate to our best knowledge, except where we admit to meddling with facts to enhance an illustration.

  Some, such as **Split-Rail Fence** (page 3) and **Satellite Dish** (page 18), describe quantitative problems encountered in everyday life. Although we have sometimes altered details to make it more natural to use precalculus mathematics, the problems are presented in essence as they actually occurred. In some cases, the mathematical solution is the one used by the person in the situation. Even when this is not the case, applying mathematics to the problem often provides insight into the situation, beyond the answer to the specific questions being asked.

- Exercises are labeled according to their purpose.

  Exercises marked (*Review*) both reinforce skills presented in previous sections and prepare you to use them in other exercises.

  Exercises marked (*Extension*) explore concepts beyond those discussed in the text, or at a depth beyond that already considered. Their purpose is to indicate some possible continuations of the lines of thought developed in the text and to arouse your interest in the further study of mathematics.

  Exercises marked (*Creating Models*) ask you to represent a physical situation by means of an equation, a table, or a graph. Exercises marked (*Interpreting Mathematics*) ask you to interpret an equation (or its solution), a table, or a graph in terms of a physical situation. Exercises marked (*Problem Solving*) ask you to do both. All serve to place mathematical ideas in a physical context.

  Exercises marked (*Writing to Learn*) require responses written in complete sentences. Their purpose is to help you synthesize and express mathematical ideas.

  Exercises marked (*Making Observations*) ask you to make observations about material you have just read or exercises you have just completed. Their purpose is to help you to identify patterns and to use logic to connect mathematical ideas.

  Exercises not otherwise marked are designed to reinforce technical skills developed in the section where the exercise appears.

By the way, Bill was convicted of vehicular homicide while DUI in the fall of 1989. In 1991, he was granted a new trial on the grounds that the court had failed to admit evidence that the steel pole that Bill's car struck had been the location of numerous accidents and had subsequently been removed. His conviction was reversed.

In working with *A Case for Algebra*, you have begun a journey of quantitative problem solving. We hope you remembered to pack your toothbrush and lots of pencils. Bon voyage!

1–2   MODELS OF QUANTITATIVE RELATIONSHIPS

PREREQUISITE   MAKE SURE YOU ARE FAMILIAR WITH:
Accuracy and Precision (Section A-1)

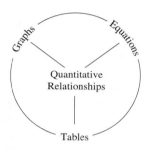

Many questions that affect people's lives involve **quantitative relationships**, that is, relationships among things that can be counted or measured. You have already seen a quantitative relationship in *A Case for Algebra* (page 1). The quantities being related were the time in hours after the accident and Bill's blood alcohol count (BAC). Quantitative relationships are frequently studied by creating mathematical models, such as **tables**, **graphs**, and **equations**, also referred to as **numerical**, **graphical**, and **analytical models**, respectively. Although you will study tables, graphs, and equations for their own sakes, we will use them primarily to study quantitative relationships.

In this section we will look at numerical, graphical, and analytical models of the relationship in *A Case for Algebra*, so that you will have a specific example of each type of model. Afterward we will discuss three other quantitative relationships, focusing on the process of creating one model from another. The wheel at the left will serve as a visual guide through the section.

THREE TYPES OF MODELS

A Numerical Model   The data from the BAC readings that were available to Dr. Winek can be summarized in the form of a table, such as Table 1.

TABLE 1

Time since accident in hours (t)	BAC (y)
0.7	0.136
2.5	0.100

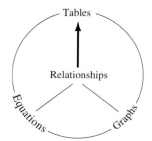

In most relationships you will encounter in this book, it is natural to think of one variable as the **input**, or **independent variable**, and the other as the **output**, or **dependent variable**. In this case, since Bill's BAC *depends on* the elapsed time since the accident, it is natural to think of the variable t as independent and y as dependent.

When a relationship is described by a table, it is traditional to list values of the independent variable in the left column and values of the dependent variable in the right column, as has been done in Table 1.

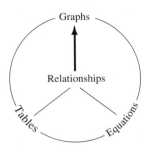

A Graphical Model The graph from Figure 1-1 is reproduced in Figure 1-3.

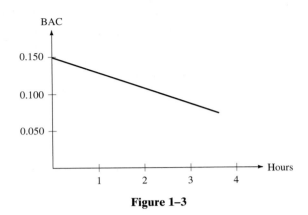

Figure 1–3

When a relationship is described by a graph, it is traditional to measure values of the independent variable along the horizontal axis and the dependent variable along the vertical axis, as has been done in Figure 1-3.

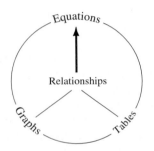

An Analytical Model In Section 1-1 the relationship in Figure 1-3 was described by the equation

$$y = -0.020t + 0.150$$

When a relationship is described by an equation, it is traditional to express the dependent variable in terms of the independent variable, as has been done here.

EXERCISE   1. (*Interpreting Mathematics*) From newspapers, magazines, or your own experience, find numerical, analytical, and graphical models of quantitative relationships. Identify the independent and dependent variables for each relationship.

INFORMATION PROVIDED BY THE THREE TYPES OF MODELS

Now that you have seen examples of tables, equations, and graphs, let's focus on the information each model provides.

Complete Versus Incomplete Information An equation provides complete quantitative information about the relationship it describes. The equation $y = -0.020t + 0.150$ in **A Case for Algebra** allows you to calculate Bill's BAC at any time. By contrast, information provided by tables is often incomplete. Table 1 provides BAC readings only at two specific times.

Graphs provide complete information about values of the input variable within some interval. Figure 1-3 shows Bill's BAC over a time interval of about 4 hours, but does not provide any information outside that interval.

Exact Versus Approximate Information Some equations provide exact information. For example, the equation $A = s^2$ describes the relationship between the side length s and the area A of a square. It allows you to calculate that a square 8 inches on a side has an area of exactly 64 square inches.

Other equations provide only approximate information. In **A Case for Algebra**, the BAC readings of 0.136 and 0.100 are almost certainly approximations. Since the equation $y = -0.020t + 0.150$ is based on those readings, any BAC calculated from the equation is also an approximation.

Tables can also provide either exact or approximate information. For example, the information in Table 1 is approximate. However, a table always provides exact information to the extent that its entries are exact.

Graphs provide only approximate information. In Figure 1-3 you can estimate that Bill's BAC was about 0.090 three hours after the accident, but you cannot calculate it exactly from the graph alone.

Explicit Versus Hidden Information   The entries in a table display explicit information. For example, Table 1 explicitly displays a BAC of 0.100 at a time of 2.5 hours after the accident. By contrast, information provided by equations and graphs is hidden. The BAC after 2.5 hours is not displayed explicitly in the equation $y = -0.020t + 0.150$, but you can calculate it. Similarly, it is not displayed explicitly in the graph in Figure 1-3, but you can estimate it.

CREATING ONE MODEL FROM ANOTHER

Which type of model is the most appropriate way to represent a relationship? This depends on the questions you are trying to answer, or the type of information you wish to convey. Newspapers and magazines abound with quantitative relationships modeled as tables or graphs, because they must present information that readers can absorb quickly and easily. Scientific journals are usually more concerned with precision and, therefore, model many relationships as equations rather than as graphs or tables.

To answer a specific question or gain insight into a relationship, you may need to create a model different from the one presented. Let's look at three different quantitative relationships and ask some questions that are more easily answered by using a different model. The first of our three relationships is initially described by a table.

 A MATHEMATICAL LOOKING GLASS

Pulleys

John Corriere, a friend of your author Dave, owns a woodworking shop. His equipment includes a centrifugal fan, which is part of a dust collection system. The fan is driven by a 0.25-horsepower motor, which turns at 1700 revolutions per minute (rpm) and is connected to the fan by a fan belt of fixed length. The pulleys M and F come in standard diameters ranging from 2 to 6 inches in 1-inch increments, but John can make pulleys of other sizes.

Figure 1–4

Changing the pulley sizes will change the fan speed. In particular, if pulley F (the fan pulley) in Figure 1-4 has a fixed diameter of 5 inches, then increasing the diameter of pulley M (the motor pulley) will cause the fan to turn more rapidly. There is a quantitative relationship between the diameter of the motor pulley and the speed of the fan. Since the pulley diameter controls the fan speed, it is natural to think of the diameter as the input and the speed as the output.

Table 2 contains data from the owner's manual for John's motor and is a numerical model of the quantitative relationship.

TABLE 2

Motor pulley diameter, in inches (input)	Fan speed, in rpm (output)
1.5	510
3	1020
5	1700
8	2720
10	3400

Higher fan speeds collect more wood dust, but speeds above 3600 rpm will damage the motor. John has a pulley 11 inches in diameter and would like to know whether the corresponding fan speed will exceed the motor's limit. Since the numerical model in Table 2 provides incomplete information about the relationship, John cannot answer his question from the table alone. Let's see how he might create an analytical or graphical model of the relationship and how each model will allow him to determine the speed by "filling in the gaps" in the table. ■

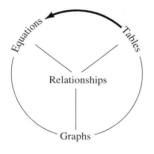

Tables to Equations John would like to know what fan speed is produced by a pulley diameter of 11 inches. One strategy for answering this question is to create an equation by studying the data and searching for patterns. By studying his table John observed that the fan speed divided by the motor pulley diameter is always 340. If he uses the variable names M and S to represent the motor pulley diameter and fan speed, he can write this observation as an equation:

$$S = 340M$$

EXERCISES

2. (*Interpreting Mathematics*) Use John's equation to find the fan speed for a motor pulley diameter of 11 inches. Will this speed exceed the motor's limit of 3600 rpm?

3. (*Interpreting Mathematics*) What motor pulley diameter will produce a fan speed of exactly 3600 rpm?

4. (*Writing to Learn*) Explain why Exercises 2 and 3 can be completed more efficiently by using an equation rather than a table.

Creating an equation from a table is often more difficult than it was in this example. Throughout the book, you will learn techniques for dealing with more complicated cases.

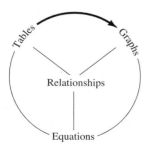

Tables to Graphs A different strategy John can use for answering his question in ***Pulleys*** is to draw a graph of the data, as in Figure 1-5a. The values of the independent variable, pulley diameter, are on the horizontal axis.

So far, the graph allows John to see the correspondence of the five motor pulley diameters in Table 2 with the five fan speeds. To answer his question using his graph, John has sketched a smooth curve through the data points in Figure 1-5b. (Actually, John's graph is a line, but mathematicians often use the word *curve* to include lines.) Drawing a smooth curve "fills in the gaps" in Table 2.

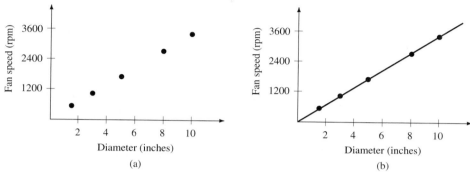

(a)   (b)

Figure 1–5

EXERCISES   **5.** (*Interpreting Mathematics*) Use John's graph to estimate the fan speed for a motor pulley diameter of 11 inches. How does your result compare with that from Exercise 2?

6. (*Interpreting Mathematics*) Use John's graph to estimate the motor pulley diameter that will produce a fan speed of exactly 3600 rpm. How does your result compare with that from Exercise 3?

7. (*Writing to Learn*) Compare John's table, equation, and graph in terms of accuracy and efficiency in answering the questions in Exercises 5 and 6.

The second of our three relationships is initially described by an equation.

 A MATHEMATICAL LOOKING GLASS
Swimming Pool

Skyview Pools, Inc., in Pittsburgh, PA, sells outdoor swimming pools in several standard sizes, with diameters of 15, 18, 21, 24, and 27 feet. The pools are 4 feet deep and are designed to be filled to a depth of 46 inches. See Figure 1-6.

Figure 1–6

To provide information to customers, the salespeople need to know the number of gallons of water required to fill a pool of each standard diameter.

■

EXERCISE   **8.** (*Problem Solving*)

    a. If a pool with a diameter of d feet is filled to a depth of 46 inches, what is the volume of water in cubic feet? Express your answer as an equation $V = (\text{coefficient})d^2$, and round the coefficient to two decimal places. (The volume V of a cylinder with radius r and height h is $V = \pi r^2 h$.)

    b. A cubic foot of water contains about 7.80 gallons. How many gallons of water does the pool contain? Express your answer as an equation $G = (\text{coefficient})d^2$, and round the coefficient to two decimal places.

    c. How many gallons are required to fill a pool of each of the five standard diameters listed in **Swimming Pool**?

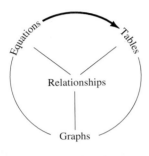

Equations to Tables The information provided by the analytical model in Exercise 8 is hidden. That is, to find the capacity of a pool with a given diameter, you must perform some calculations. It is more convenient for Skyview's salespeople to have immediate access to explicit information, such as a table can provide. Your results from Exercise 8c are reproduced in Table 3.

TABLE 3

Diameter (feet)	Capacity (gallons)
15	5,283
18	7,608
21	10,355
24	13,524
27	17,117

    Table 3 will not give the capacity of pools with every possible diameter. However, a salesperson armed with Table 3 can immediately answer some commonly asked questions about pool capacities without performing any calculations. For example, the table tells immediately that a pool 21 feet in diameter has a capacity of 10,355 gallons.

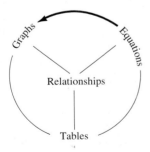

Equations to Graphs Would a graph of the relationship in **Swimming Pool** be helpful to Skyview's salespeople? One way to obtain a graph from the equation is to make a table of values, plot points, and draw a smooth curve through the points. We have used Table 3, obtained from the equation in Exercise 8, to draw the graph in Figure 1-7a. Figure 1-7b shows a graph of the same equation, using a table that extends over diameters from 0 to 30 feet.

    The graphs provide visual estimates without computations, but also without the equation's accuracy. For example, the equation allows us to calculate the capacity of a pool 20 feet in diameter as 9392 gallons. Using the graph in Figure 1-7a, we can only estimate the capacity as a little more than 9000 gallons. An estimate based on the graph in Figure 1-7b would be even less accurate, because its scale is larger.

EXERCISE   **9.** (*Writing to Learn*) As a salesperson for Skyview, would you rather have access to Table 3 or the graphs in Figure 1-7? Explain your answer.

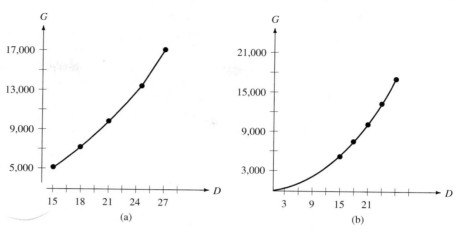

Figure 1–7

It is worth noting that by their visual nature, graphs allow us to observe the dynamic behavior of relationships. For example, in Figure 1-7a we see that larger diameters correspond to larger capacities. In Figure 1-7b we can observe that the capacity grows more rapidly as the diameter increases. Such observations can be used to predict results beyond the scope of the graph.

Several techniques for graphing equations without extensive point plotting will be presented later in this book. In particular, you will learn a more efficient way to graph the equation from *Swimming Pool* in Section 4-1.

The last of our three relationships is initially described by a graph.

A MATHEMATICAL LOOKING GLASS

Greenhouse Effect

Our planet's atmosphere contains nitrogen (75% by weight) and oxygen (23%), along with much smaller concentrations of other gases, such as carbon dioxide. A small change in this balance of elements can have drastic consequences for everything that lives on earth. For example, let's look at some reasons for the recent concern over the buildup of carbon dioxide (CO_2), caused largely by the burning of fossil fuels in automobiles and other vehicles.

Part of the heat produced on earth by sunlight is reflected through the skies and back out into space. Much of this reflected heat could not pass through an atmosphere containing a high concentration of CO_2 and would thus be trapped near ground level. The resulting increase in temperature is often called the **greenhouse effect**, because it would produce the warm, humid conditions found in greenhouses. Increased temperatures would melt the polar icecaps, causing water levels to rise in the oceans and flooding all coastal cities. The greenhouse effect could eventually make the planet too hot to sustain life at all.

The graph in Figure 1-8a shows the level of CO_2 in the atmosphere since 1960. The waviness occurs because the concentration of CO_2 varies with the season. The graph in Figure 1-8b shows the average trend in CO_2 concentration over the same time period.

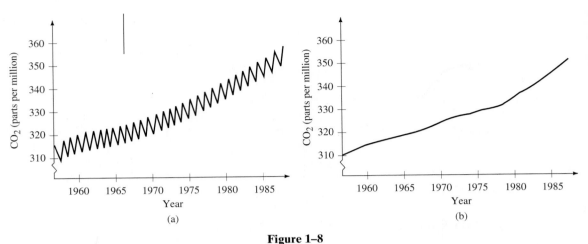

Figure 1–8

Adapted from Michael Seeds, *Foundations of Astronomy*, Wadsworth, 1990.

Although the graphs in Figure 1-8 provide immediate visual information about the greenhouse effect, they are of limited value in predicting the concentration of CO_2 at any time in the future. ∎

EXERCISES

10. (*Writing to Learn*)

    a. What fact about CO_2 concentration can be observed in Figure 1-8a, but not Figure 1-8b?

    b. What fact about CO_2 concentration can be observed more easily in Figure 1-8b?

    c. From which graph can future concentration levels be predicted more easily?

11. (*Interpreting Mathematics*) Use Figure 1-8b to estimate CO_2 concentration in the year 2000, and in 2050. Which estimate do you trust more? Why?

12. (*Writing to Learn*) Scientists who want to estimate future CO_2 levels often try to fit equations to the graphs in Figure 1-8 and use the equations to arrive at their estimates. Why do you think they prefer to work with equations rather than graphs?

Graphs to Tables You can obtain a table from any graph by estimating the coordinates of several points.

EXERCISE

13. (*Problem Solving*)

    a. Use the graph in Figure 1-8b to complete the following table.

Year	CO_2 concentration (parts per million)
1960	
1970	
1980	
1985	

b. Use the table to estimate the concentration in the year 2000, and in 2050.

c. Do you trust these estimates more than the ones you obtained in Exercise 11? Why or why not?

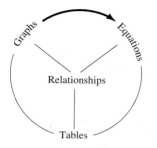

Graphs to Equations There is no all-purpose procedure for constructing equations from *every* graph. However, it is possible to construct equations from *some* graphs. For example, in *A Case for Algebra*, we were able to construct an equation from the graph in Figure 1-3. The process required us to assume that the graph was a straight line, as it appeared to be in the figure. We will return periodically to the problem of constructing equations from graphs, beginning in Chapter 2.

ADDITIONAL EXERCISES   *For each table in Exercises* 14–17,

a. Write an equation that might describe the relationship between *x* and *y*. (*Hint:* Look for a pattern in each table, as John did to construct his equation in ***Pulleys***.)

b. Sketch a possible graph of the relationship.

14.

x	y
1	3
2	4
3	5
4	6
5	7
6	8

15.

x	y
1	9
2	8
3	7
4	6
5	5
6	4

16.

x	y
5	10
10	20
15	30
20	40
25	50
30	60

17.

x	y
1	12
2	6
3	4
4	3
6	2
12	1

For each equation in Exercises 18–21,

a. Complete the following table.

x	y
−4	
−2	
0	
2	
4	

b. Use your table to sketch a possible graph of the relationship between *x* and *y*.

18. $y = 2x + 3$

19. $y = 10 - x$

20. $y = \frac{1}{2}x^2$ **21.** $y = 10 - x^2$

For each graph in Exercises 22–25,

> **a.** Complete a table showing the values of y when $x = -2, -1, 0, 1,$ and 2.
>
> **b.** Make reasonable assumptions about the behavior of the graph outside the region shown, and estimate the value of y when $x = 5$. (There is no single correct answer.)

22.

23.

24.

25.

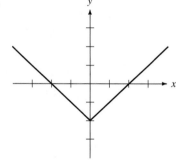

In Exercises 26–29 (Writing to Learn), tell whether the information presented by the model of the relationship is

> **a.** complete or incomplete
>
> **b.** exact or approximate
>
> **c.** explicit or hidden

Explain your answers. (Arguments in favor of both answers can often be given, so your reasons are important.)

26. The table at the left shows how much mortgage can be carried by a family with a given income.

> G = gross annual family income, in dollars
>
> M = maximum monthly PITI (*p*rincipal, *i*nterest, *t*axes, and *i*nsurance), in dollars

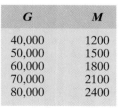

G	M
40,000	1200
50,000	1500
60,000	1800
70,000	2100
80,000	2400

Adapted from Joel Lerner, *Financial Planning for the Utterly Confused*, McGraw-Hill, 1994.

A	E
65	18.4
70	14.8
75	11.6
80	8.7
85	6.4
90	4.7

27. The table at the left shows the life expectancy of older females in the United States.

$$A = \text{present age}$$

$$E = \text{expected years of remaining life}$$

28. If you want to invest a sum of money for a relatively short time, you might buy a Treasury bill, also called a T-bill, from the U.S. Treasury Department. Because the interest on such an investment is paid "up front," the actual rate of interest you earn is higher than the stated rate. For example, suppose you buy a 1-year T-bill for $10,000 (the minimum amount allowed) at 5% interest. After you pay for it, the Treasury Department will send you a check for the interest (5% of $10,000 = $500). In effect, you have paid $9500 and will receive $10,000 at the end of 1 year. As a percentage of your investment, the interest is

$$(100)\left(\frac{500}{9500}\right) \approx 5.3$$

In general, let

$$D = \text{the dollar amount of interest on a \$10,000 1-year T-bill}$$

$$r = \text{the actual interest rate}$$

Then the relationship between D and r is given by the equation

$$r = \frac{100D}{10{,}000 - D}$$

29. Most archaeologists believe that the first natives of North America migrated on foot from Siberia to Alaska. Since the Bering Sea presently separates Siberia from Alaska, the migration could have occurred only during a time when the world's sea levels were much lower than they are today. Therefore, scientists attempting to estimate the date of the earliest migration have an interest in past sea levels. In the following graph,

$$t = \text{time, in years before the present}$$

$$L = \text{sea level, in feet above or below present level}$$

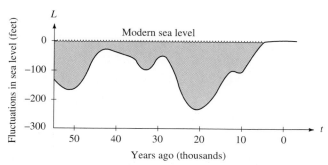

Adapted from Brian Fagan, *The Great Journey*, Thames and Hudson, 1987.

30. (*Problem Solving*) Refer to the table in Exercise 26.

a. Sketch a possible graph of the relationship between G and M.

b. Write an equation that might describe the relationship.

c. Using either your equation or your graph, decide how much mortgage could be carried by a family with a gross annual income of $120,000.

   d. Using either your equation or your graph, decide what gross annual income is needed to carry a mortgage of $1000.

31. (*Creating Models*)

    a. From the table in Exercise 27, sketch a possible graph of the relationship between *A* and *E*.

    b. Estimate the expected years of remaining life for a 100-year-old female.

32. Refer to the table in Exercise 27.

    a. What is the expected age at death for a 65-year-old woman?

    b. What is the expected age at death for a 70-year-old woman?

    c. (*Writing to Learn*) Why do you think your answers in parts (a) and (b) are different?

33. (*Creating Models*) Refer to the equation in Exercise 28.

    a. Complete the table at the left.

    b. Sketch a possible graph of the relationship between *D* and *r*.

34. (*Interpreting Mathematics*) Refer to the graph in Exercise 29. A migration from Siberia to Alaska would have been possible only during periods when the sea level was at least 100 feet below its present level. During what times was this the case?

35. (*Writing to Learn*) Both your purpose in communicating information and your audience determine the most appropriate model to use. Describe actual situations in which quantitative information should be communicated by

    a. using a graph   **b.** using a table   **c.** using an equation

36. (*Writing to Learn*) Compare tables, equations, and graphs as models of quantitative relationships. Include the advantages and limitations of each.

37. (*Writing to Learn*) Refer to one of the models you found in Exercise 1, and make up a question that would be more easily answered using a model different from the one given. Tell what type of model would be appropriate, and why. You do not need to create the model or answer your question.

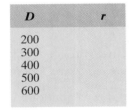

D	r
200	
300	
400	
500	
600	

1–3   PROBLEM-SOLVING STRATEGIES

PREREQUISITES   MAKE SURE YOU ARE FAMILIAR WITH:

    Linear Equations (Section A-2)

    The Pythagorean Theorem (Section A-4)

Throughout this book you will construct mathematical models and interpret the information they provide to answer questions about quantitative relationships. You can do so more effectively by arming yourself with a systematic approach to problem solving. In this section we will develop a general problem-solving method developed by George Polya (*How To Solve It*, Princeton University Press, 1945). We will also consider several specific strategies. These strategies are not rigid rules for solving a narrowly defined type of problem, such as motion or mixture problems. You can think of them as a general and flexible process you can use to solve the wide variety of quantitative problems you encounter in this book, as well as many everyday, nonquantitative problems. The method and the strategies will be developed in the context of the following Mathematical Looking Glass.

A MATHEMATICAL LOOKING GLASS
Split-Rail Fence

The property around your author Lynn's house includes a rectangular pasture enclosed by a split-rail fence, as in Figure 1-9.

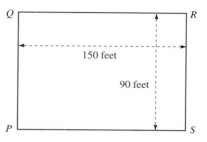

Figure 1–9

A few years ago, Lynn's daughter Carrie wanted to expand the pasture to give her horse, Illiad, more room. She considered moving one side of the fence from QR to TU and installing additional fence along QT and RU, as in Figure 1-10a. She also considered moving one side from SR to VW and installing additional fence along RV and SW, as in Figure 1-10b.

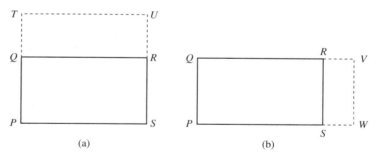

Figure 1–10

Carrie called contractor Steve Ford of Ford Enterprises, Inc. Steve agreed to do the job at a cost of $1.50 per foot to move existing fence and $3.00 per foot to install new fence. Carrie had several hundred dollars set aside for the project but was not sure how much she wanted to spend. To help her decide, she wanted to know which of the alternatives in Figure 1-10 would give her the greater additional area for a given cost. ■

Summarized briefly, Polya's approach to problem solving consists of four steps.

- Understand the problem.
- Devise a plan.
- Carry out the plan.
- Look back.

Let's look at each step in detail as it relates to ***Split-Rail Fence***.

FIRST STEP: UNDERSTAND THE PROBLEM

Here are several questions you can ask yourself to make sure you understand a quantitative problem.

A. Can you state the problem in your own words?

In the last sentence of **Split-Rail Fence**, we have stated Carrie's problem in our own words. You can check your understanding of the problem by doing the same in Exercise 1.

EXERCISE   **1.** (*Writing to Learn*) Reread **Split-Rail Fence**. Then close your book and state Carrie's problem in a sentence or two. (Don't just memorize our statement. Use *your own* words.)

B. What information is given?

The information that is likely to be relevant to Carrie's problem can be summarized in three statements.

- The present dimensions of the pasture are 90 × 150 feet.
- It will cost $1.50 per foot to move existing fence.
- It will cost $3.00 per foot to install new fence.

C. What information, if any, is given but not needed?

Nonquantitative information in physical situations can often be used to decide what quantitative questions are worth asking. Once a quantitative question has been asked, most nonquantitative information is not needed for the solution. For example, the fact that Carrie owns a horse indicates the reason for her concern about the size of the pasture, but does not help her solve the problem.

Some quantitative information may also not be needed. For example, in **Pulleys** (page 9), the fact that John's motor has 0.25 horsepower has no bearing on the solution to his problem.

D. What information, if any, is needed but not given?

In **Split-Rail Fence**, it appears that we have all the information we need to solve Carrie's problem. However, we often encounter situations in which we need to gather more information before solving a problem. For example, suppose that the land around the pasture were heavily wooded. (Fortunately, that is not the case). Then Carrie could not analyze her problem without measuring the distance from each edge of the pasture to the nearest tree. She would also need to decide whether to spend any of her money on tree removal.

E. What are the constants and the variables in the problem? That is, what quantities are fixed, and what quantities can vary?

This book deals with relationships between two varying quantities, under the condition that all other quantities remain fixed. Often, you have the freedom to choose which two quantities are variables. For example, since Carrie wants to know how much additional area she can obtain for a given cost, she might regard additional area and cost as her variables. Alternatively, she might recognize that the area of a rectangle is the product of its length and width. Thus, in Figure 1-10a, where the length is fixed, she could regard width and cost as the variables. Similarly, in Figure 1-10b, she could regard length and cost as the variables.

SECOND STEP: DEVISE A PLAN

According to Polya in *How To Solve It*, "the main achievement in the solution of a problem is to conceive the idea of a plan." After we list some frequently used strategies for doing this, we will see how you might use some of them to solve Carrie's problem. Here is the list.

- *Drawing a Picture or Diagram.* This strategy often takes the form of *Drawing a Graph.*
- *Making a Table.*
- *Looking for a Pattern.* This strategy often takes the form of *Looking for Symmetry.*
- *Guessing and Checking.*
- *Working Backwards.*
- *Examining a Special Case.* This strategy often takes the form of *Examining Extreme Cases.*
- *Examining a Related Problem.* This strategy often takes the form of *Examining a Simpler Problem.*
- *Breaking the Problem into Parts.*
- *Looking for Relationships Among Variables.* This strategy often takes the form of *Writing an Equation* or *Looking for a Formula.*

Of course, you will not use all these strategies on every problem you solve, and you will sometimes use others not listed here. Most problems can be solved in more than one way, and there is no foolproof method for deciding which strategies work best for a given problem. The only way to become a better problem solver is to get plenty of practice!

Now let's return to **Split-Rail Fence**. In Figure 1-9, Carrie has already approached the problem by *Drawing a Picture or Diagram.* Figure 1-10 suggests that she can continue by *Breaking the Problem into Parts*, considering first the plan in Figure 1-10a, then the one in Figure 1-10b.

Next let's recall that in Figure 1-10a, the variables were the width of the pasture and the cost of enlarging it. Carrie needs to know how much additional area she can get for a given cost, so she should try *Looking for Relationships Among the Variables*, probably in the form of *Writing an Equation*.

These strategies provide a starting point for the third step of Polya's process. Other strategies may come into play as we continue.

THIRD STEP: CARRY OUT THE PLAN

Most of the mathematics occurs here. You must often simplify algebraic expressions, solve equations, and draw and interpret graphs. You will need to give some thought to the order of these if the problem is complex. Let's carry out the plan we just devised for the solution of Carrie's problem.

First, let's define the variables

$$A = \text{additional area obtained}$$

$$C = \text{cost of enlarging the pasture}$$

To gain some insight into the relationship between A and C, Carrie might begin by *Examining a Special Case*. Specifically, suppose she expands the pasture as in Figure 1-10a and increases the width by 5 feet. The additional area she obtains is then $(150)(5) = 750$ square feet. What is the cost of the additional area? The length of the new fence to be installed is $(2)(5) = 10$ feet, at a cost of $(3)(10) = 30$ dollars. The length of existing fence to be moved is 150 feet, at a cost of $(1.50)(150) = 225$ dollars. The total cost is $255.

The insight gained from the special case suggests a line of thought for constructing an equation relating A and C. Suppose Carrie expands the pasture as in Figure 1-10a and increases the width by x feet. The additional area she obtains is $(150)(x) = 150x$ square feet. The length of new fence to be installed is $(2)(x) = 2x$ feet, at a cost of $(3)(2x) = 6x$ dollars. The length of existing fence to be moved is 150 feet, at a cost of $(1.50)(150) = 225$ dollars. The total cost is $6x + 225$ dollars. Carrie now knows that if she expands the width by x feet, then

$$A = 150x$$

and

$$C = 6x + 225$$

To calculate the additional area obtainable for a given cost, she would like to express A in terms of C. She can do so by means of the following calculation.

$$C = 6x + 225 \qquad \text{(from our previous work)}$$
$$6x = C - 225$$
$$x = \frac{C - 225}{6}$$
$$A = 150x \qquad \text{(from our previous work)}$$
$$A = 150\left(\frac{C - 225}{6}\right)$$
$$A = 25C - 5625$$

Now Carrie can calculate the additional area she will obtain for any given cost if she increases the width of the pasture. What if she increases the length instead, as in Figure 1-10b? In Exercise 2 you can find out by *Examining a Related Problem*.

EXERCISE   **2.** (*Creating Models*) Suppose Carrie increases the length of the pasture, as in Figure 1-10b. Construct an equation expressing A in terms of C by using a line of reasoning similar to the one just given.

At this point Carrie has two equations relating cost to additional area:

$$A = 25C - 5625, \qquad \text{if she increases the width}$$
$$A = 15C - 2025, \qquad \text{if she increases the length}$$

To solve her problem, she needs to decide which plan gives her more area for her money. She might continue by *Making a Table*, such as Table 4.

TABLE 4

C (cost)	A = 25C − 5625 (width is increased)	A = 15C − 2025 (length is increased)
250	625	1725
300	1875	2475
350	3125	3225
400	4375	3975
450	5625	4725
500	6875	5475

She can now apply the strategy of *Looking for a Pattern*. Table 4 indicates that for a small expansion it is better to increase the length, while for a larger expansion it is better to increase the width.

Although Carrie can probably use Table 4 to decide how much she wants to spend, she might also want to know where the "crossover" occurs. That is, at what cost will the two plans give her the same amount of area? In Exercise 3 you can answer this question by using the strategy of *Guessing and Checking*.

EXERCISE   **3.** By trial and error, find a cost for which the two plans give equal areas. Use Table 4 to make efficient guesses.

FOURTH STEP: LOOK BACK

One purpose of looking back at the problem-solving process is to verify that your solution is correct. Another is to sharpen your problem-solving skills by studying both the problem and the process you used to solve it. Understanding why your plan worked may help you to solve other problems in the future. It is helpful to ask yourself the following questions.

A. Is your solution mathematically accurate?

We hope you have checked the accuracy of all the computations in **Split-Rail Fence**, both yours and ours. Contrary to popular belief, authors are human, and we sometimes make errors.

In general, your checking should include making sure each step leads to an equivalent model. For example, if you are solving an equation, remember that you may lose solutions if you divide both sides by a variable quantity. You should also check the validity of your solution in the original model. In particular, if you use several equations in the process of solving, substitute each solution into the original equation.

B. Is your solution reasonable in the context of the problem?

The pattern that Carrie observed in Table 4 can be supported by a "commonsense" line of reasoning. To increase the width of the pasture by *any* amount, Carrie must pay to have 150 feet of existing fence moved. To increase the length requires moving only 90 feet of existing fence. If the area is to be increased by only a small amount, the cost of new fence will be comparatively small with either plan, so it will be cheaper to increase the length. However, each time the width is increased by a foot, Carrie gains 150 square feet of additional area. Increasing the length by a foot gives her only 90 square feet. Thus, if the additional area is to be sufficiently large, it will be cheaper to increase the width.

Sometimes mathematically correct solutions fail to be reasonable in the context of a problem. In **Swimming Pool**, the equation relating diameter to capacity was $G = 23.48d^2$. If a pool has a capacity of 10,000 gallons, we can find its diameter by solving the equation $10,000 = 23.48d^2$ to obtain $d \approx \pm 20.64$ feet. Both are valid solutions to the equation, but the negative one is not reasonable in the context of the problem.

C. Could you have solved the problem in a different way?

This is an invitation to look at the same problem from several different perspectives. This often gives you greater insight into the problem. Solving the problem in a different way also supports the accuracy of your result more reliably than reworking it in the same way. This is because you are extremely unlikely to arrive at the same wrong answer in several different ways.

After obtaining her equations, Carrie could have continued by solving the system of linear equations

$$A = 25C - 5625$$

$$A = 15C - 2025$$

Methods of solving linear systems are reviewed in Section A-10 and are discussed in greater detail in Chapter 3.

D. Is the problem or solution similar to others you have solved previously? If so, can you generalize the method of solution?

By contrast with (C), this is an invitation to look at several problems from the same perspective. This will help you to understand the common mathematical structure of seemingly diverse problems. As we build a foundation of common experience in the coming chapters, we will often ask you to reflect on the common features of the problems you solve. In particular, you will see that **Health Insurance** (page 80) and **Fat Cats on Campus** (page 28) have mathematical structures similar to those in **Split-Rail Fence** (page 19).

ADDITIONAL EXERCISES   *In Exercises 4–5, use the given information to answer as many of the questions as possible, and tell what additional information you would need to answer the others.*

4. (*Problem Solving*) Three sprinters started a 100-meter dash at the same time. The fastest one averaged 9 meters per second, and the slowest averaged 8 meters per second.
   a. What was the winning time?
   b. What was the winner's margin of victory, in seconds?
   c. Where was the slowest runner when the winner crossed the finish line?
   d. Which runner was ahead at the 50-meter mark?

5. (*Problem Solving*) The following table shows percentage increases from 1990 to 1991 in three categories of health care expenditures in Idaho, Oklahoma, and Wyoming. For example, total expenditures for hospital care in Idaho rose 14.6% between 1990 and 1991.

Category	ID	OK	WY
Hospital care	14.6%	12.1%	10.3%
Physician care	7.0%	1.7%	3.6%
Prescription drugs	12.8%	11.8%	10.1%

Adapted from *Health Care State Rankings*, Morgan Quitno Corp., 1994.

a. In which state was the least money spent on prescription drugs in 1991?

b. Which category of health care experienced the smallest percentage increase in cost between 1990 and 1991?

c. Which category of health care experienced the smallest dollar increase in cost between 1990 and 1991?

d. Which state had the lowest overall percentage increase (in all three categories combined)? (*Hint:* A state's overall percentage increase is *not* the sum of the percentage increases in each category!)

In Exercises 6–20, identify each step in the Polya process as you solve the problem and each strategy being used from the second step. Focus on your thought processes as you solve the problems, not just on the final solution. Even if you can solve the problem with a "flash of insight," identify the strategies that triggered the flash.

6. (*Problem Solving*) A Swimman (like a Walkman, but watertight with submersible headphones) is on sale for $180 after having been marked down 20%. What was its original price?

7. (*Problem Solving*) Fifteen International Master chess players took part in a recent tournament with a round-robin format, so that every participant played every other participant. How many games were played altogether?

8. (*Problem Solving*) If the sponsor of a round-robin chess tournament (see Exercise 7) wants to limit the number of games played to no more than 80, how many players can participate?

9. (*Problem Solving*) A well-known mathematical brain-teaser challenges you to measure out exactly 4 liters of water using only a 3-liter container and a 5-liter container. Describe the procedure for doing so.

10. (*Problem Solving*) Can you solve the brain-teaser in Exercise 9 using only a 3-liter and a 6-liter container? If so, how? If not, why not?

11. (*Problem Solving*) A number of elephants and ostriches are in an enclosure at a zoo. Slats looks over the fence and counts 60 heads. Shorty looks under the fence and counts 194 legs. How many elephants are there?

12. (*Problem Solving*) A few years ago your author Lynn helped Robert Kelly install a ham radio antenna on a trailer owned by Lynn's father, Ernie Harvey. They installed it at the center of the rectangular roof, which was 70 feet long and 40 feet wide. They had no tape measure, but had 165 feet of string with them. Is this enough to locate the center by stretching across the two diagonals, as in Figure 1-11?

Figure 1–11

13. (*Problem Solving*) The antenna in Exercise 12 is 15 feet high and is secured by wires running from the top to each corner of the roof. How long should each wire be?

14. (*Problem Solving*) Your absent-minded author Dave recently had to drive to Atlantic City, NJ, to attend a conference. Thirty minutes after he left, his wife, Alice, noticed that his suitcase was still in the living room. Their daughter, Sarah, grabbed the suitcase and took off in pursuit. She knew that her father would drive at a steady 50 mph, since he had just received a speeding ticket. Risking her license by doing 70, she tried to estimate how far she would need to drive before starting to watch closely for him. How far did she drive before overtaking Dave?

15. (*Problem Solving*) Lynn's friend Robert Kelly is in the habit of feeding the ducks at North Park Lake. He has discovered through trial and error that they adore pumpernickel bread from Periwinkle's Bakery. However, at a cost of $1.75 a loaf, a month's supply of 14 loaves costs more than he is willing to spend. A precise budgeter, Robert has set aside $17 per month as his allowance for duck food. He will therefore buy as much pumpernickel bread as he can, subject to the 14-loaf total, with the remainder being day-old bread at $0.50 a loaf. How much of each will he buy?

16. (*Problem Solving*) When the German mathematician Carl Friedrich Gauss was a child, his class was given the task of finding the sum of the first 100 positive integers as a punishment for some misbehavior. While his classmates labored over the addition, Carl shouted out the answer in a few seconds. Use Figure 1-12 to duplicate his feat.

Figure 1–12

17. (*Problem Solving*) The graphs in Figure 1-13 show the relationship between time and the height of two elevators. Which one would you rather ride, and why?

Figure 1–13

Figure 1–14

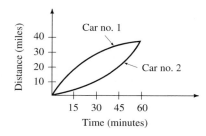

Figure 1–15

18. (This exercise was adapted from a group of problems developed by Steve Monk, of the University of Washington.)

(*Problem Solving*) The graph in Figure 1-14 shows the relationship between time and the distance traveled for each of two cars. Which car has gone farther after an hour? Which is going faster after an hour?

19. (*Problem Solving*) Suppose that the graph in Figure 1-14 shows the relationship between time and the *speed* of the two cars. Which car has gone farther after an hour? Which is going faster after an hour?

20. (*Problem Solving*) Imagine two glasses as in Figure 1-15, one containing water and the other containing red wine. Transfer a teaspoon of wine to the water glass; then transfer a teaspoon of the mixture back to the wine glass. Is there now more water in the wine or more wine in the water?

21. (*Extension*) Find the book *But the Crackling Is Superb*, Nicholas and Giana Kurti, ed. (Adam Hilger, Bristol, 1989), and read the article, "Wine and Water," by H. B. G. Casamir. The conclusion of the article differs from the one you obtained in Exercise 20. Ask a chemist or physicist to explain why.

22. (*Extension*) Think of three problems, mathematical or otherwise, you have encountered in your "real life" (that is, outside of mathematics classrooms). Identify at least three Polya strategies you used in solving these problems, and describe how they were used.

1-4   SOLVING EQUATIONS AND INEQUALITIES GRAPHICALLY

PREREQUISITES   MAKE SURE YOU ARE FAMILIAR WITH:
   Intervals (Section A-7)
   Linear Inequalities (Section A-8)
   The Laws of Exponents (Section A-11)
   Factoring (Section A-12)
   Quadratic Equations (Section A-13)

Although quantitative relationships involve two variables, questions about such relationships can often be answered by solving equations and inequalities in one variable. In the past you have learned analytical methods for solving certain equations and inequalities. (For example, see Sections A-2, A-8, and A-13.) In the present section you will learn a graphical method. Like all graphical methods, it yields only approximate solutions. However, it is powerful in that it can be applied when analytical methods fail. Let's apply the method to answer some questions about a hypothetical business that you will run throughout the book.

SOLVING EQUATIONS GRAPHICALLY

A MATHEMATICAL LOOKING GLASS

Fat Cats on Campus

Your economics professor wants your class to gain some experience in running a business and has agreed to provide up to $1000 of his own money to help you get started. He expects that you and your classmates will reimburse him out of your profits and split any remaining profits. (A similar situation actually occurs at Penn State–New Kensington, where Professor Mark DeHainaut gives his classes $1000 to invest in the stock market.) After doing some marketing research, the class decides to start a T-shirt company. (An account of a real student-run T-shirt company appeared in "Fat Cats on Campus", *USA Weekend*, March 26–28, 1993.)

To run your business profitably, you need to know what it will cost to produce the shirts, and how many you can expect to sell for a given price. You have identified the following costs.

Purchase of heat press	$612.50
White T-shirts in bulk	2.50 each
Transfers to press onto each shirt	1.00 each

In Exercise 1 you can explain why the cost of producing x shirts is $3.50x + 612.50$ dollars.

You have also surveyed students on your campus and operators of other T-shirt companies to help you decide on a reasonable level of production and an appropriate price. For simplicity, let's assume that there are no competing companies nearby. However, to sell more shirts you will need to charge a lower price, as indicated in Table 5.

TABLE 5

Target shirt sales for the semester (x)	Price per shirt (p)
500 shirts	$13.50
900	11.50
1300	9.50
1700	7.50
2100	5.50
2500	3.50

In Exercise 2 you can verify that to sell x shirts during the semester, you must charge a price of $16 - 0.005x$ dollars per shirt.

One of the most fundamental questions you can ask about your company is, "At what sales levels will the company make a profit?" The first step toward an answer is usually to answer the question, "At what sales levels will the company break even, with neither a profit nor a loss?" These sales levels are referred to as **breakeven points**, and the process of finding them is called **breakeven analysis**.

In Exercise 3 you can verify that your net profit from the sale of x shirts is $-0.005x^2 + 12.50x - 612.50$ dollars. Thus, to find your breakeven points, you must solve the equation

$$-0.005x^2 + 12.50x - 612.50 = 0$$
∎

EXERCISES   **1.** (*Creating Models*) Explain why the cost of producing x shirts will be $3.50x + 612.50$ dollars.

2. Verify that in each row of Table 5, you must charge a price of $16 - 0.005x$ dollars per shirt in order to sell x shirts.

3. (*Creating Models*)

  a. Explain why the total revenue from the sale of x shirts at a price of $16 - 0.005x$ dollars per shirt will be $16x - 0.005x^2$ dollars.

  b. Explain why your net profit from the sale of x shirts will be $-0.005x^2 + 12.50x - 612.50$ dollars.

Now let's compare analytical and graphical methods of solving the equation in ***Fat Cats on Campus***.

EXAMPLE 1   *A comparison of analytical and graphical solutions*

  a. To solve

$$-0.005x^2 + 12.50x - 612.50 = 0$$

analytically, you might factor the left side, but it is probably easier to use the quadratic formula.

$$x = \frac{-12.50 \pm \sqrt{12.50^2 - 4(-0.005)(612.50)}}{2(-0.005)}$$

$$x = 50 \quad \text{or} \quad x = 2450$$

  b. To solve

$$-0.005x^2 + 12.50x - 612.50 = 0$$

graphically, begin by graphing

$$y = -0.005x^2 + 12.50x - 612.50$$

The graph is shown in Figure 1-16.

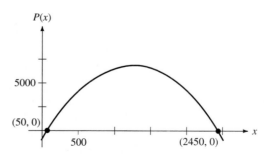

Figure 1–16

On the graph the expressions y and $-0.005x^2 + 12.50x - 612.50$ are identical. Therefore, solving the equation means finding all points on the graph where $y = 0$. This in turn means finding all x-intercepts on the graph. The graph has an x-intercept at about $(50, 0)$, so that the equation has a solution at about $x = 50$. To see whether this is an exact solution, substitute 50 for x in the equation.

$$-0.005(50)^2 + 12.50(50) - 612.50 = 0$$
$$-12.50 + 625 - 612.50 = 0$$
$$0 = 0$$

Similarly, you can verify that the other x-intercept is at $(2450, 0)$. There-fore, the equation has solutions $x = 50$ and $x = 2450$. ∎

EXAMPLE 2   *A comparison of analytical and graphical solutions*

a. To solve $x^3 + 8 = 2x^2 + 10x$ analytically, begin by writing the equation with zero on one side.

$$x^3 - 2x^2 - 10x + 8 = 0$$

Since the left-hand side does not factor easily, the analytical solution involves methods we will not discuss until Section 5-2. For now, we need only observe that it has already become complicated.

b. To solve $x^3 + 8 = 2x^2 + 10x$ graphically, begin by writing the equation with zero on one side.

$$x^3 - 2x^2 - 10x + 8 = 0$$

The graph of $y = x^3 - 2x^2 - 10x + 8$ is shown in Figure 1-17. It has three x-intercepts, one of which appears to occur at about $(4, 0)$. Substitution in the equation verifies that $x = 4$ is a solution. One of the other x-intercepts occurs in the interval $[-3, -2]$. To locate it more precisely, change the viewing window by zooming in on the graph, as in Figure 1-18a.

[−8, 8] by [−20, 20]

Figure 1–17

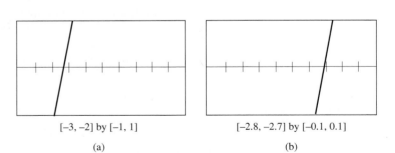

[−3, −2] by [−1, 1]   [−2.8, −2.7] by [−0.1, 0.1]

(a)   (b)

Figure 1–18

Figure 1-18a shows that the intercept is in $[-2.8, -2.7]$. Zooming in again, Figure 1-18b shows that it is in $[-2.74, -2.73]$. You can now say that the intercept occurs at $(-2.73, 0)$, and the solution is $x = -2.73$, with an error of no more than 0.01. Although we can locate the intercept with increas-ing precision by zooming in repeatedly, we may never locate it exactly. It is therefore practical to stop at some point and settle for an estimate. ∎

> **FYI** Unless otherwise indicated, an approximate solution to an equation should have at least two significant digits and should have an error of no more than 0.01.

For the equation in Example 1, the analytical solution is simpler. For the equation in Example 2, the graphical solution is simpler. These examples illustrate that as equations become more complicated, graphical methods become more useful. However, it is still worthwhile to learn some analytical methods for solving complicated equations. As we will see in Chapter 5, a combination of analytical and graphical methods is often more effective than either one alone. Furthermore, you can identify only real solutions graphically. Nonreal solutions must be found analytically. (Nonreal solutions are discussed in Section A-13.)

EXERCISES   **4.** In Example 2, find the third solution of $x^3 + 8 = 2x^2 + 10x$.

Solve the equations in Exercises 5–8 by graphical methods. Graph in the indicated viewing window.

5. $10 - 5x = 2x$ $[-10, 10]$ by $[-10, 10]$
6. $x^2 + x = 10$ $[-10, 10]$ by $[-10, 10]$
7. $x^2 + 400 = 100x - 2000$ $[0, 100]$ by $[-200, 200]$
8. $x^3 + 408x - 1728 = 34x^2$ $[-20, 20]$ by $[-1000, 1000]$

The process of solving equations graphically depends on a hidden assumption. To see what it is, trace along the graphs in Exercises 5–8 near their *x*-intercepts. You will notice that in most cases, your cursor does not show the exact intercept. Instead, it shows two pixels in adjacent columns whose *y*-coordinates are near 0, but opposite in sign. The hidden assumption is that the graph between those points is an unbroken curve, and thus contains an *x*-intercept. Informally, a graph that is an unbroken curve over an interval is said to be **continuous** over that interval. Otherwise, it is **discontinuous**. (You will learn formal definitions of these terms in a calculus course.) The equations you will solve between here and the end of Chapter 5 involve continuous graphs, so you may solve them using the methods of this section. We will return to the idea of continuity in Sections 2-4 and 6-1.

SOLVING INEQUALITIES GRAPHICALLY

To find your profitable levels of production in *Fat Cats on Campus*, you must solve the inequality

$$-0.005x^2 + 12.50x - 612.50 > 0$$

You will see an analytical method for solving this particular inequality in Section 4-2. For now, let's solve it graphically. As Example 3 shows, the first step is to solve the equation

$$-0.005x^2 + 12.50x - 612.50 = 0$$

as we did in Example 1. (This is an instance of *Examining a Simpler Problem*.)

EXAMPLE 3   *Solving an inequality graphically*

To solve

$$-0.005x^2 + 12.50x - 612.50 > 0$$

graphically, graph $y = -0.005x^2 + 12.50x - 612.50$, as in Figure 1-16. The graph is reproduced in Figure 1-19.

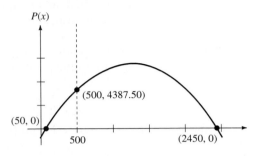

Figure 1–19

The expressions y and $-0.005x^2 + 12.50x - 612.50$ are identical on the graph. Thus solving the inequality means finding all x-values for which $y > 0$ on the graph. To find these values, take an imaginary walk in the positive direction along the x-axis. Look up because $y > 0$ above you. The graph becomes visible above the x-axis when $x > 50$. For example, at $x = 500$ you can see the point (500, 4387.50). Since the graph is visible above any point in (50, 2450), the solution of the inequality is (50, 2450). Thus you will make a profit if you sell between 50 and 2450 T-shirts.   ■

EXAMPLE 4   *Solving an inequality graphically*

To solve

$$x^3 - 2x^2 - 10x + 8 \le 0$$

graphically, graph $y = x^3 - 2x^2 - 10x + 8$, as in Figure 1-17. The graph is reproduced in Figure 1-20.

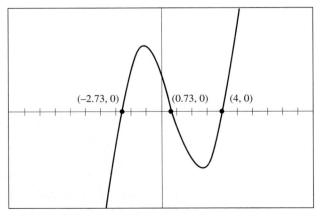

Figure 1–20

This time, as you walk along the x-axis, look directly on and below your path, since $y = 0$ on the x-axis and $y < 0$ below it. The graph intersects your path when $x \approx -2.73, 0.73$, and 4. It is below you in $(-\infty, -2.73) \cup (0.73, 4)$. The solution to the inequality is $(-\infty, -2.73] \cup [0.73, 4]$. ∎

Example 4 shows that to solve inequalities graphically you must find x-intercepts, and to find x-intercepts is to solve equations graphically. Thus, *to solve an inequality, you must first solve an equation.*

EXERCISES   *Solve the inequalities in Exercises 9–12 graphically. Use the viewing windows and the solutions from Exercises 5–8.*

9. $10 - 5x > 2x$ **10.** $x^2 + x \leqslant 10$

11. $x^2 + 400 < 100x - 2000$ **12.** $x^3 + 408x - 1728 \geqslant 34x^2$

13. (*Writing to Learn*) You have just read that to solve an inequality graphically, you must first solve the corresponding equation by locating the x-intercepts. Explain how locating the x-intercepts helps you solve the inequality.

ADDITIONAL EXERCISES   *In each of Exercises* 14–23,

     a. Solve the equation graphically. Use a viewing window of $[-10, 10]$ by $[-10, 10]$ except where another is indicated.

     b. Solve the inequality graphically.

14. $x^2 - 4x = 5$ **15.** $3x - 13 = 2x + 3$
     $x^2 - 4x < 5$ $3x - 13 > 2x + 3$
                               $[-20, 20]$ by $[-20, 20]$

16. $x^3 = 4$ **17.** $x^3 = 4x$
     $x^3 \geqslant 4$ $x^3 \leqslant 4x$

18. $x^2 + 2 = 2x$ **19.** $4 - 3x = 3(x + 7)$
     $x^2 + 2 > 2x$ $4 - 3x \leqslant 3(x + 7)$

20. $25x^2 = x^4$ **21.** $x^2 + 98 = 20x$
     $25x^2 \geqslant x^4$ $x^2 + 98 < 20x$
     $[-10, 10]$ by $[-200, 200]$ $[0, 20]$ by $[-10, 10]$

22. $|2x - 5| = 3$ **23.** $\dfrac{1}{x^2 + 1} = 2$

     $|2x - 5| \leqslant 3$ $\dfrac{1}{x^2 + 1} > 2$

24. Solve the equation in Exercise 14 by the following alternate method.

     a. Graph both $y = x^2 - 4x$ and $y = 5$ in the window $[-10, 10]$ by $[-10, 10]$.

     b. (*Writing to Learn*) Explain why the solutions to the equation are the x-coordinates at the points where the two graphs intersect.

     c. (*Making Observations*) Find the points of intersection, and compare your results with those you obtained in Exercise 14.

25. Solve the inequality in Exercise 14 by the following alternate method.

     a. Graph both $y = x^2 - 4x$ and $y = 5$ in the window $[-10, 10]$ by $[-10, 10]$.

b. (*Writing to Learn*) Explain why the solutions to the inequality are the x-coordinates at the points where the first graph is below the second.

c. (*Making Observations*) Find the solutions to the inequality, and compare your results with those you obtained in Exercise 14.

In Exercises 26–31, solve each equation and each inequality by the methods of Exercises 24–25. Use a viewing window of $[-10, 10]$ by $[-10, 10]$ except where another is indicated.

26. $3x - 13 = 2x + 3$
$3x - 13 > 2x + 3$
$[0, 30]$ by $[0, 60]$

27. $x^3 = 4$
$x^3 \leq 4$

28. $x^3 = 4x$
$x^3 \leq 4x$

29. $\dfrac{4x}{x^2 + 4} = 1$

$\dfrac{4x}{x^2 + 4} \leq 1$

$[-10, 10]$ by $[-2, 2]$

30. $x^2 = x + 2$

$x^2 < x + 2$

31. $x^2 = x - 2$

$x^2 < x - 2$

CHAPTER REVIEW

Complete Exercises 1–17 (Writing to Learn) before referring to the indicated pages.

WORDS AND PHRASES

In Exercises 1–4, explain the meaning of the words or phrases in your own words.

1. (page 7)   **quantitative relationship**

2. (page 7)   **numerical**, **analytical**, and **graphical** models of relationships

3. (page 7)   **independent variable**, **dependent variable**

4. (page 31)   **continuous, discontinuous**

IDEAS

5. (pages 7, 8)   When you look at a table, graph, or equation, how can you tell which variable is the independent variable?

6. (pages 7, 8)   Describe the process for obtaining output values from input values in a table, in a graph, and in an equation.

7. (pages 8, 9)   Discuss the nature of the information provided by a table, a graph, and an equation. In particular discuss whether the information is complete or incomplete, exact or approximate, and explicit or hidden.

8. (pages 10, 11)   Describe the processes of creating a graph or equation from a table. Make up a table and use it to illustrate the processes.

9. (pages 12, 13)   Describe the processes of creating a table or graph from an equation. Make up an equation and use it to illustrate the processes.

10. (page 14)   Describe the process of creating a table from a graph. Make up a graph and use it to illustrate the process. Why do you think it is usually difficult to create an equation from a graph?

11. (pages 9–14)   If a relationship is represented by a table, an equation, or a graph, why might we want to represent it by a different model?

12. (page 18)   Why is it useful to develop a systematic approach to problem solving?

13. (pages 19–24)   What are the four steps in Polya's approach to problem solving? Explain what is involved in each step.

14. (pages 19, 20)   What questions can you ask yourself to make sure you understand a problem? How does each question check your understanding?

15. (pages 20, 21)   Name several frequently used strategies for solving problems, and describe how you have used some of them in solving problems in this chapter.

16. (pages 23, 24)   What questions can you ask yourself to look back at your solution of a problem? How is each question helpful?

17. (pages 28–32)   Describe the processes for solving both an equation and an inequality graphically. Explain the reasoning behind each process.

CHAPTER 2

FUNCTIONS

2–1 AN OVERVIEW OF FUNCTIONS

THE DEFINITION OF FUNCTION

In Chapter 1 you saw many examples of relationships in which it is natural to think of one variable as the independent variable (input) and the other as the dependent variable (output). For example, in **Swimming Pool** (page 11), the independent variable was the pool's diameter and the dependent variable was its capacity in gallons.

All the relationships in Chapter 1 have the property that each input value generates a unique output value. In this chapter you will learn how to tell whether an equation, a graph, or a table represents a relationship with uniqueness of output. You will discover why such relationships are both convenient and useful, and you will explore them in greater depth. Let's begin by noticing how uniqueness of output is exhibited numerically, analytically, and graphically.

- In the numerical model for **Pulleys** (page 9), each entry in the first column of John's table (input) generates a unique entry in the second column (output).

- In the analytical model for **Swimming Pool** (page 11), each value of d in the equation $G = 23.48d^2$ (input) generates a unique value of G (output).

- In the graphical model for **Greenhouse Effect** (page 13), each horizontal coordinate on the graph (input) generates a unique vertical coordinate (output).

We admit that we chose these situations carefully, so that they would clearly illustrate input-output behavior. However, even in the messier quantitative relationships we encounter in our lives, one variable often determines the other. This behavior occurs so frequently that mathematicians have chosen a word to describe it. The word is **function**. Informally, a function is a relationship with uniqueness of output.

Function is an appropriate word, because one of its everyday definitions is, "something closely related to another thing, and dependent upon it

35

for its existence, value, or significance: *growth is a function of nutrition.*" "*(The American Heritage Dictionary)*". One way to state its mathematical definition is as follows.

> If x and y are so related that each value assumed by x determines a unique value for y, then y is a function of x.

EXERCISES   *In each of Exercises 1–4 (Writing to Learn), explain why the given statement is true.*

1. In **A Case for Algebra** (page 1), Bill's blood alcohol count y is a function of the time t, in hours, since the accident.

2. In **Split-Rail Fence** (page 19), the additional area A of pasture Carrie can buy is a function of the cost C of expanding the pasture.

3. In **Fat Cats on Campus** (page 28), your profit P is a function of your sales level x.

4. In **Fat Cats on Campus** (page 28), your company's sales level x is *not* a function of your profit P. (*Hint*: In Example 1 of Section 1-4, you saw that two different sales levels represent breakeven points.)

FUNCTIONAL NOTATION

A function can be viewed as a type of relationship between two variables. It can also be viewed as a process that acts on the variables and is distinct from them. For example, we can think of the function in **Swimming Pool** as the equation

$$G = 23.48d^2$$

relating the variables d and G. We can also think of it as the process of squaring a value of d and multiplying the result by 23.48.

Since we assign symbolic names to the variables, it makes sense to give a symbolic name to the process as well. The symbolic naming of the process is referred to as **functional notation**. The following Mathematical Looking Glass will introduce you to this notation and the rules governing its use.

 A MATHEMATICAL LOOKING GLASS
Big Blue Marble

How much of the earth's surface can be seen from an airplane in flight? From a satellite? From the moon? With the use of trigonometry, a formula can be developed to provide the answers. (The formula is not quite accurate because it assumes that the earth is a perfect sphere with no mountains or valleys.) At an altitude of x kilometers from the surface, the fraction that can be seen is

$$y = \frac{x}{2x + 12{,}800}$$

For example, the fraction that can be seen by a satellite camera at an altitude of 1600 km is $\dfrac{1600}{2(1600) + 12{,}800} = 0.1$, or one-tenth of the surface.

Functions are typically given names such as f, g, and h. Let's call this function g. Then g represents the process of multiplying the independent variable x by 2, adding 12,800 to the result, and dividing x by that result. The value of the dependent variable is written $g(x)$ and is an instruction to apply the process g to the input x. For example,

$$g(1600) = \frac{1600}{2(1600) + 12{,}800} = 0.1$$

More generally,

$$g(x) = \frac{x}{2(x) + 12{,}800}$$ ■

To see how functional notation compares with other ways of making statements about relationships, let's make an English statement about the relationship in **Big Blue Marble**.

At an altitude of 1600 km, you can see $\frac{1}{10}$ of the earth's surface.

Now let's make the same statement using variables x and y.

If $x = 1600$, then $y = \frac{1}{10}$.

Finally, let's make the same statement in functional notation.

$$g(1600) = \frac{1}{10}$$

The functional statement is the shortest of the three. In general, functional notation allows us to make quantitative statements concisely. The notation has other advantages that you will discover later.

   Here are a few important points concerning functional notation and its use.

- The function represents a process, in contrast to the variables, which represent numbers. Thus $g(x)$ does *not* mean a number g multiplied by a number x, but the process g applied to the number x.
- The symbols g and $g(x)$ have different meanings. The symbol g refers to the function, and $g(x)$ refers to the output obtained by applying the function g to the input x. While $g(x)$ has the same meaning as y, g by itself does not.
- Although it is common practice to use the letters f, g, and h to represent functions, we are free to choose any symbol at all to represent either a function or a variable. As authors, it is our responsibility to make clear the meaning of any symbol we use in this book. You have a similar responsibility whenever you communicate a mathematical thought to someone.
- The quantity in parentheses denotes an input to the function. It can be a number or a variable expression representing a number.

EXAMPLE 1   *Using functional notation*

   a. If $f(x) = x^2 + 2x$, then

$$f(-5) = (-5)^2 + 2(-5) = 15$$

   b. If $f(x) = x^2 + 2x$, then

$$f(2z - 5) = (2z - 5)^2 + 2(2z - 5)$$
$$= 4z^2 - 16z + 15$$

The expression $4z^2 - 16z + 15$ represents the result of the process f applied to the number $2z - 5$.   ■

In Exercises 5–8, evaluate $H(7)$, $H(0)$, $H(A)$, *and* $H(A - 1)$.

5. $H(x) = 5x - 4$ **6.** $H(z) = z^2$

7. $H(T) = \dfrac{T}{T + 1}$ **8.** $H(w) = \sqrt{w + 9}$

DOMAIN AND RANGE

You have already observed that certain inputs to functions sometimes fail to yield reasonable results. This can happen for either of the following reasons.

- The *input* or the *output* is not *physically reasonable*.

  In **Big Blue Marble**, we can calculate $g(-9600) = \dfrac{3}{2}$. However, neither the input (-9600 km) nor the output $\left(\dfrac{3}{2}$ of the earth's surface$\right)$ is physically reasonable.

- The *output* is not *mathematically reasonable*.

  If we attempt to evaluate $g(-6400) = \dfrac{-6400}{2(-6400) + 12{,}800}$, we must divide by zero. There is no reasonable output, regardless of whether the function occurs in a physical context.

We will use the word *reasonable* to mean "mathematically reasonable" for an abstract function and "physically reasonable" in a physical context. The collection of all reasonable values for the independent variable is called the **domain** of a function. The collection of all reasonable values of the dependent variable is called its **range**. For example, the function $f(t) = t^2$ is defined for all real values of t, so its domain is the interval $(-\infty, \infty)$. However, the value of $f(t)$ must be nonnegative, so the range of f is $[0, \infty)$.

You can avoid much mathematical and physical nonsense by identifying the domain and range of any function you work with. The domain and range depend on whether you consider the function in the abstract or as a model of a physical situation.

We will spend the remainder of this section answering three questions.

- If a relationship between x and y is modeled by an equation, a graph, or a table, how can you tell whether y is a function of x?
- If a function is represented by an equation, a graph, or a table, how can you determine its domain and range?
- Why is it useful to know whether y is a function of x?

Let's deal with the first two questions simultaneously.

THREE VIEWS OF FUNCTIONS

An Analytical View No procedure always tells you whether an equation in x and y represents y as a function of x. However, if you can solve the equation for y, you can decide whether y is a function of x. If it is, solving for y also allows you to determine its domain.

In **Big Blue Marble**, the relationship between x and y is represented by the equation

$$y = \frac{x}{2x + 12{,}800}$$

Because this equation describes a process that yields a unique value of y for each value of x, y is a function of x.

EXAMPLE 2   *Solving for y to decide whether y is a function of x.*

To decide whether the equation $2x - y^2 = 1$ represents y as a function of x, solve for y.

$$2x - y^2 = 1$$
$$-y^2 = -2x + 1$$
$$y^2 = 2x - 1$$
$$y = \pm\sqrt{2x - 1}$$

The $\pm$ indicates that the same input may lead to two different outputs. For example, the input $x = 5$ leads to the outputs $y = 2$ and $y = -2$. Thus y is not a function of x. ■

FYI   The functions in Chapters 1–6 of this book, called **algebraic functions**, use only the operations of addition, subtraction, multiplication, division, and taking powers and roots. These operations produce undefined results in two instances:

- Dividing by zero (for example, evaluating $\dfrac{1}{x - 2}$ when $x = 2$)
- Taking an even-indexed root of a negative number (for example, evaluating $\sqrt{x - 3}$ when $x = 2$)

The only mathematical restrictions on the domains of algebraic functions result from those two operations.

Example 3 illustrates an analytical method for finding the range of a function.

EXAMPLE 3   *Finding the range of a function analytically*

The range of the function $g(x) = \dfrac{x}{2x + 12{,}800}$ consists of all possible values for $g(x)$, or y. We can more easily see what these values are if we solve the equation for x.

$$y = \frac{x}{2x + 12{,}800}$$
$$(2x + 12{,}800)y = x$$
$$2xy + 12{,}800y = x$$
$$2xy - x = -12{,}800y$$
$$x(2y - 1) = -12{,}800y$$
$$x = -\frac{12{,}800y}{2y - 1}$$

The expression in y is undefined when the denominator is zero. This will happen if and only if $y = \frac{1}{2}$, so the range is $\left(-\infty, \frac{1}{2}\right) \cup \left(\frac{1}{2}, \infty\right)$. ∎

EXERCISES   *In Exercises 9–12,*

       a. Solve for y to decide whether y is a function of x.

       b. If y is a function of x, find its domain and range.

   9. $4x + 2y = 9$ **10.** $0.2x = 3 - 0.1y$

  11. $x^4 + y = 1$ **12.** $x^3 + y^2 = 1$

Many equations, such as $x = y^5 - y + 1$, are impossible to solve for y. Other strategies must therefore be used to decide whether y is a function of x. In Exercise 19, you will apply a graphical strategy to this equation.

By asking whether y is a function of x, we have assigned the role of independent variable to x up until now. It is equally reasonable, although not traditional, to regard y as the independent variable and ask whether x is a function of y.

EXERCISES   **13.** As parts (a) – (d) of this exercise, tell whether x is a function of y in each of Exercises 9–12.

  14. (*Making Observations*) In Example 3, x and y are treated as abstract variables. What are the domain and range of the function g in the physical context of **Big Blue Marble**? (*Hint:* About what fraction of the earth's surface can be seen from a very small height? From a very large height?)

A Graphical View Example 4 shows how to decide whether a relationship has unique outputs for every input by looking at its graph.

EXAMPLE 4   *Looking at a graph to decide whether y is a function of x*

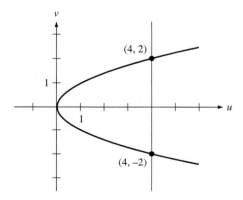

Figure 2–1

Figure 2-1 shows the graph of the equation $u = v^2$. The vertical line $u = 4$ intersects the graph at both $(4, 2)$ and $(4, -2)$. This tells us that in the equation $u = v^2$, the input (u-value) of 4 leads to both of the outputs (v-values) 2 and -2. Thus, v is not a function of u. ∎

Example 4 suggests a test for deciding whether a graph represents a function. If the variables are *x* and *y* and the *x*-axis is horizontal, the test can be stated as follows.

> **The Vertical Line Test**
>
> A graph represents *y* as a function of *x* if and only if no vertical line intersects it more than once.

Example 5 shows how to find the domain and range of a function by looking at its graph.

EXAMPLE 5   *Finding the domain and range of a function graphically*

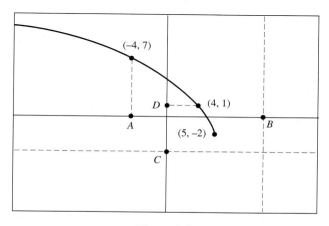

Figure 2–2

Figure 2-2 shows the graph of a function $y = f(x)$. To find the domain of f, let's take an imaginary walk in the positive direction along the *x*-axis. As we walk, we look straight up and down (directly above and below our location on the *x*-axis) for points on the graph. From A, at $x = -4$, we see the point $(-4, 7)$ on the graph, so -4 is in the domain of f. On the other hand, no point on the graph is visible from B, at $x = 10$, so 10 is not in the domain of f. Since the graph is visible from any point in $(-\infty, 5]$, the domain of f is $(-\infty, 5]$.

Similarly, we can find the range of f by climbing up the *y*-axis. If we look directly to our left and right as we climb, the graph is not visible from C, at $y = -4$, so -4 is not in the range of f. On the other hand, the point $(4, 1)$ is visible from D, at $y = 1$, so 1 is in the range of f. Since the graph is visible from any point in $[-2, \infty)$, the range of f is $[-2, \infty)$.   ■

FYI   We cannot read graphs of functions with perfect accuracy, so to "find" the domain and range graphically means to *estimate* them. Furthermore, we can't estimate without knowing what the entire graph looks like. We often need to make reasonable guesses based on what we can see. For example, it is reasonable to guess that the graph in Figure 2-2 extends infinitely far up and to the left.

EXERCISES   *In Exercises 15–18, decide whether y is a function of x. If it is, find the domain and range. Make reasonable guesses about the behavior of the graph outside the viewing window.*

15.

16.

17.

18.

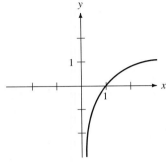

19. (*Writing to Learn*) Figure 2-3a shows the graph of $x = y^5 - y + 1$ (discussed earlier in this section) and Figure 2-3b shows the graph of $x = y^5 + y + 1$. Which equation represents y as a function of x? How do you know?

(a)   (b)

Figure 2–3

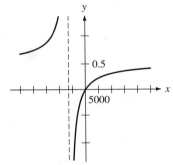

Figure 2–4

20. (*Making Observations*)

Figure 2-4 shows the graph of $g(x) = \dfrac{x}{2x + 12{,}800}$.

a. Use the graph to find its domain and range.

b. Sketch the part of the graph that represents the function g in the physical context of ***Big Blue Marble***, and find the domain and range in context. Make sure your result is the same as the one you obtained in Exercise 14.

A Numerical View Most tables do not contain every data point for a relationship, so we cannot always decide whether a table represents a function unless we know the underlying relationship. If there is a function that could generate it, the table is said to **fit** the function.

EXAMPLE 6   *A table that fits a function*

TABLE 6

x	y
32	58
35	62
41	64
45	62

In Table 6 no number in the first column corresponds to more than one number in the second column. However, we cannot see *all possible* tables that might have been generated by the relationship between *x* and *y*. Without more information, we cannot decide whether Table 6 represents *y* as a function of *x*. To see why, let's look at two situations that might have generated the data.

If *x* represents Dave's age and *y* his resting pulse during his annual checkup, then *y* is a function of *x*, because no matter how the table continues, he will never be the same age twice.

If *x* and *y* represent daily low and high temperatures at Logan International Airport in Boston, then *y* is almost certainly not a function of *x*. For example, some day not listed could have a low temperature of 32 and a high temperature other than 58. ■

If a table is **complete** (that is, if it contains every data point for a relationship), then uniqueness of output can be stated as follows.

> A complete table represents a function if and only if, whenever the first entries in two or more rows are the same, so are the second.

The entries of a complete table provide the domain and range for the function it represents. For example, if Table 6 is complete, it represents a function with domain {32, 35, 41, 45} and range {58, 62, 64}.

EXERCISES   *In Exercises 21 and 22,*

a. Assume that the table is complete, and draw one of the following conclusions.

- *y* is a function of *x*.
- *y* is not a function of *x*.
- More information is needed to decide whether *y* is a function of *x*.

b. If *y* is a function of *x*, find the domain and range.

c. Do your conclusions in part (a) change if the tables are incomplete? What are your new conclusions?

21.

x	y
5	8
9	12
−1	4
2	8
7	0

22.

x	y
8	5
12	9
4	−1
8	2
0	7

x	y
1	
10	
100	
1,000	
10,000	
100,000	
1,000,000	

23. (*Writing to Learn*)

   a. Use the equation $y = \dfrac{x}{2x + 12{,}800}$ to fill in the table at the left.

   b. Explain why the table fits a function.

   c. Without knowing what the variables x and y represent, could you conclude from the table that y is a function of x? Why or why not?

   d. Although the table is incomplete, it supports your conclusions from Exercises 14 and 20 concerning the domain and range of the function in context. Explain how.

REASONS TO STUDY FUNCTIONS AND THEIR MODELS

Now that you have discovered how to tell whether y is a function of x, let's turn to the question of what good it does to know. Functions are useful for two primary reasons.

- Functions reflect a great deal of the physical world. This fact has been recognized at least since the seventeenth century, when mathematicians such as Sir Isaac Newton (1642–1727) were attempting to discover the laws governing gravity and optics. The mathematics of calculus, based heavily on the concept of function, played a fundamental role in developing these laws. For example, the effect of gravity on a falling object is shown by the function $s = 16t^2$, where s is the number of feet the object has fallen after t seconds.

- Functions have uniqueness of output. You have already seen that an equation of a function provides step-by-step instructions for operating on a given input to produce a unique output. Since calculators and computers are designed precisely to execute such step-by-step instructions, they are efficient function-evaluating machines. Thus the more we use technology to perform our calculations, the more we need both an awareness of functions in the physical world and an ability to construct mathematical models for them.

There are also several important reasons to study the connections among analytical, numerical, and graphical models of functions. In particular,

- The connections among the three types of models allow you to change from one to another. For example, as you will begin to see later in this chapter, the equation of a function can provide information about its graph. This information can be used to check the accuracy of calculator-generated graphs and to solve problems graphically.

- The model of a function presented to you may not be the most convenient one for solving your problem. For example, in **Swimming Pool** (page 11), the relationship between the diameter and capacity of a swimming pool was presented as an equation, but it was more convenient for the salespeople to develop a table.

ADDITIONAL EXERCISES   **24.** (*Interpreting Mathematics*) If the sides of a square are s centimeters long, its area in square centimeters is

$$A(s) = s^2$$

A square whose sides are 9 cm long has an area of 81 cm². Make the same statement using functional notation.

25. (*Interpreting Mathematics*) The number of hours required to drive 100 miles at r miles per hour is

$$T(r) = \frac{100}{r}$$

It takes 2 hours to drive 100 miles at 50 miles per hour. Make the same statement using functional notation.

For each function in Exercises 26–31,

    a. (*Writing to Learn*) Describe in words the process represented by the function h.

    b. Find the domain and range of h.

    c. Evaluate $h(4)$, $h(0)$, and $h(-0.5)$.

    d. Write an expression for $h(3Z)$.

26. $h(x) = 20x + 13$ **27.** $h(t) = 100 - 4t^2$

28. $h(W) = \dfrac{W}{W + 1}$ **29.** $h(a) = a + \dfrac{1}{a}$

30. $h(r) = r^3$ **31.** $h(x) = 6$ [*Hint* for part (b): The output is 6 for every input.]

In Exercises 32–37, solve for y, if possible, to decide whether y is a function of x. Identify the equations for which you cannot solve for y.

32. $2x - 3y = 11$ **33.** $y - 2x^5 + 3x = 0$

34. $x^2 + y^2 = 25$ **35.** $x = y^7 + 4y + 17$

36. $y = 6$ **37.** $x = 6$

38. As parts (a)–(f) of this exercise, solve each of the equations in Exercises 32–37 for x, if possible, to decide whether x is a function of y.

In Exercises 39–42, decide whether y is a function of x. If so, estimate the domain and range. Make reasonable guesses about the graph outside the viewing window.

39.

40.

41.

42.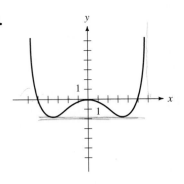

43. In the following graph,

$$t = \text{time in years since 1970}$$

$$M = \text{millions of metric tons of fish caught in the Pacific Ocean}$$

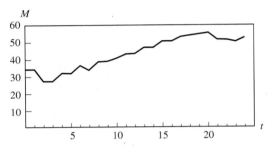

Adapted from *National Geographic*, November 1995.

    a. (*Writing to Learn*) The graph shows that M is a function of t. What does this statement say in physical terms (that is, in the language of time and fish)?

    b. (*Extension*) How does the graph show that t is not a function of M?

In Exercises 44–47,

    a. Assume that the table is complete, and draw one of the following conclusions.

        • y is a function of x.

        • y is not a function of x.

        • More information is needed to decide whether y is a function of x.

    b. If y is a function of x, find its domain and range.

    c. Do your conclusions in part (a) change if the table is incomplete? What are your new conclusions?

44.

x	y
5.3	2.15
0.009	1.645
1.645	0.009
−2.1	7.0

45.

x	y
0	0
−17	−17
$\dfrac{2\pi}{\sqrt{3}}$	$\dfrac{2\pi}{\sqrt{3}}$

46.

x	y
42	2
42	4
42	6
42	8
42	10

47.

x	y
2	42
4	42
6	42
8	42
10	42

48. The following table lists the height and weight of each person in one of Lynn's precalculus classes.

h = height (inches)	w = weight (pounds)
68	150
70	155
67	130
71	155
64	105
70	145

    a. Is w a function of h? Is h a function of w?

    b. (*Interpreting Mathematics*) During the semester, one student withdrew from the class. Afterward, w was a function of h and vice versa. Which student withdrew?

In Exercises 49–52, find the domain and range of the function in the given context.

49. $S = 340M$

    M and S are as in **Pulleys** (page 9).

50. $C = 23.48d^2$

    d and C are as in **Swimming Pool** (page 11).

51. $A = x(2000 - x)$

    x = length, in feet, of a rectangular pasture you are planning to fence off

    A = area of the pasture, in square feet
There are 4000 feet of fencing available.

52. $P = 2x + \dfrac{2{,}000{,}000}{x}$

    x = length, in feet, of a rectangular pasture you are planning to fence off

    P = perimeter of the pasture, in feet

The pasture must have an area of 1,000,000 square feet, and there is an unlimited supply of fencing.

53. (*Interpreting Mathematics*) Find a table that models a physical relationship, different from those discussed in this section, for which y is a function of x. Explain how you know that y is a function of x, and state the domain and range.

54. (*Interpreting Mathematics*) Find a table that models a physical relationship, different from those discussed in this section, for which y is not a function of x. Explain how you know that y is not a function of x.

55. (*Writing to Learn*) Explain why many tables generated by physical relationships do not allow you to decide whether one variable is a function of the other.

56. (*Writing to Learn*) Another student has just said to you, "I don't understand why we just have a *vertical* line test to tell whether *y* is a function of *x*. Why doesn't it matter if a *horizontal* line crosses a graph more than once?" Write a short paragraph to answer your classmate's question.

(Actually, there *is* a horizontal line test for graphs. In Section 7-3 we will see what information it provides.)

SUPPLEMENTARY TOPIC
Nonnumerical Functions

The word *function* has a more general mathematical definition than the one given at the beginning of this section. A **function** from a set *A* of objects to a set *B* is a correspondence with the property that no object in *A* corresponds to more than one object in *B*.

EXAMPLE 7   *Functional and nonfunctional relationships among sets of objects*

a. Define a correspondence between the two sets

$$S = \{\text{Illinois, Indiana, Michigan, Minnesota, New York,}$$
$$\text{Ohio, Pennsylvania, Wisconsin}\}$$

and

$$L = \{\text{Erie, Huron, Michigan, Ontario, Superior}\}$$

by the rule that a state in *S* corresponds to a lake in *L* if and only if the state borders the lake. The correspondence is not a function from *S* to *L*, since Wisconsin borders two lakes and Michigan borders four. (Extra credit if you can tell which lakes.) It is also not a function from *L* to *S*, since Lakes Erie, Michigan, and Superior are all bordered by more than one state.

b. Ludwig von Köchel (1800–1877) devoted much of his life to numbering in chronological order the 626 largely undated musical compositions of Wolfgang Amadeus Mozart (1756–1791). Let *M* denote the set of all musical pieces composed by Mozart, and let *K* denote the set of integers from 1 to 626, inclusive. The correspondence between *M* and *K* that associates each composition with its **Köchel number** is a function both from *M* to *K* and from *K* to *M*. This is because no composition has more than one Köchel number, and no number refers to more than one composition. ■

In Exercises 57–58, decide whether the correspondence is a function from A to B and whether it is a function from B to A.

57. (*Extension*)

$$A = \{\text{lazy, malevolent}\}$$

$$B = \{\text{sluggish, shiftless, evil, idle, wicked}\}$$

An element of *A* corresponds to an element of *B* if the two are synonyms.

58. (*Extension*) A question on an art history quiz asks students to match artists in set *A* with their works in set *B*.

$$A = \{\text{Michelangelo, Picasso, Van Gogh, Jefferson}\}$$

$$B = \{\text{Monticello, Guernica, David, self-portrait}\}$$

59. (*Writing to Learn*) Explain why the equation $y = x^2$ defines y as a function of x according to *both* of the definitions given in this section.

2–2   COMBINATIONS OF FUNCTIONS

ADDING, SUBTRACTING, MULTIPLYING, AND DIVIDING FUNCTIONS

To solve physical problems you often need to define several functions, each of which describes a process of obtaining an output from a given input. Sometimes you need to add, subtract, multiply, or divide the outputs from two or more functions. The sum, difference, product, and quotient of two functions f and g can each be expressed as a single function. These functions are written $f + g$, $f - g$, fg, and $\dfrac{f}{g}$.

EXAMPLE 1   *Combining functions using addition, subtraction, multiplication, and division*

If $f(x) = 2x - 5$ and $g(x) = x^2 + 4x$, then

$$(f + g)(x) = (2x - 5) + (x^2 + 4x)$$
$$= x^2 + 6x - 5$$
$$(f - g)(x) = (2x - 5) - (x^2 + 4x)$$
$$= -x^2 - 2x - 5$$
$$(fg)(x) = (2x - 5)(x^2 + 4x)$$
$$= 2x^3 + 3x^2 - 20x$$
$$\left(\frac{f}{g}\right)(x) = \frac{2x - 5}{x^2 + 4x}$$

■

EXERCISES   *In Exercises 1–4, write expressions to represent* $(f + g)(x)$, $(f - g)(x)$, $(fg)(x)$, *and* $\left(\dfrac{f}{g}\right)(x)$.

1. $f(x) = x + 3$ and $g(x) = x - 3$
2. $f(x) = x^2 - 2$ and $g(x) = 4 - x$
3. $f(x) = \dfrac{3}{x}$ and $g(x) = \dfrac{2}{x}$
4. $f(t) = \sqrt{t - 1}$ and $g(t) = \sqrt{t^2 - 1}$

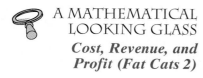

A MATHEMATICAL LOOKING GLASS

Cost, Revenue, and Profit (Fat Cats 2)

Sums, differences, products, and quotients of functions often occur in the study of economics. For example, let's look at several functions of interest to the T-shirt company you started in **Fat Cats on Campus** (page 28). If your target sales level for the semester is x shirts, you will want to know the following quantities.

- The total cost, in dollars, of producing the shirts is your **cost function**, usually named $C(x)$.
- The highest price you can charge and still sell all of the shirts is your **demand function**, usually called $p(x)$.

- The total number of dollars you bring in by selling the shirts is your **revenue function**, usually called $R(x)$.
- Your net profit, in dollars, from the sale of the shirts is your **profit function**, usually called $P(x)$.   ■

EXERCISES   5. (*Making Observations*) In **Fat Cats on Campus**, you verified that for your company,

$$p(x) = 16 - 0.005x$$
$$R(x) = 16x - 0.005x^2$$

a. Verify that $R(x) = xp(x)$.

b. Explain why $R(x) = xp(x)$ for every company.

6. (*Making Observations*) In **Fat Cats on Campus**, you verified that for your company,

$$C(x) = 3.50x + 612.50$$
$$P(x) = -0.005x^2 + 12.50x - 612.50$$

a. Verify that $P(x) = (R - C)(x)$.

b. Explain why $P(x) = (R - C)(x)$ for every company.

COMPOSING FUNCTIONS

Sometimes you need to use the output from one function as the input to another. In the following Mathematical Looking Glass you will learn how to define a single function to describe the entire process of obtaining the final output from the original input.

A MATHEMATICAL LOOKING GLASS

Canadian Crickets

At some time during your life you have undoubtedly found yourself outside on a warm evening, listening to crickets. Have you noticed that they chirp faster on warmer nights? In fact, if a cricket is chirping at the rate of r times per minute, the Fahrenheit temperature $h(r)$ is closely approximated by

$$h(r) = \frac{1}{4}r + 40$$

Thad Rude, a friend of your author Dave, does a lot of camping. He sometimes estimates the temperature by listening to crickets. However, Thad lived in Canada for many years, so it is more natural for him to express the *Celsius* temperature as a function of the number of chirps per minute. He knows that the Celsius temperature corresponding to a Fahrenheit temperature F is

$$g(F) = \frac{5}{9}(F - 32)$$

Thus, if a cricket chirps 112 times in a minute, Thad can evaluate

$$h(112) = \frac{1}{4}(112) + 40 = 68$$

to obtain the Fahrenheit temperature of 68° and then evaluate

$$g(68) = \frac{5}{9}(68 - 32) = 20$$

to obtain the Celsius temperature of 20°.

More generally, if the cricket chirps r times in a minute, Thad can evaluate $h(r) = \frac{1}{4}r + 40$ and use that number as an input to g. The resulting output is

$$g\left(\frac{1}{4}r + 40\right) = \frac{5}{9}\left[\left(\frac{1}{4}r + 40\right) - 32\right]$$

$$= \frac{5}{36}r + \frac{40}{9}$$

Thus the equation

$$C = \frac{5}{36}r + \frac{40}{9}$$

describes Celsius temperature C directly as a function of r.   ■

The function in **Canadian Crickets** is called the **composition** of g and h and is written $g \circ h$. It represents the process g applied to the input $h(r)$, so

$$(g \circ h)(r) = g[h(r)]$$

Figure 2-5 illustrates the relationship of $g \circ h$ to g and h for an input of 112 chirps per minute and for an arbitrary input.

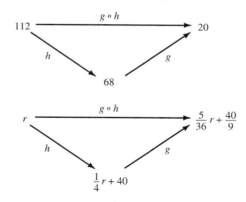

Figure 2-5

FYI   When you are composing functions, order matters! For example, in **Canadian Crickets**, the output from the function $h \circ g$ is

$$(h \circ g)(F) = h[g(F)]$$

$$= h\left[\frac{5}{9}(F - 32)\right]$$

$$= \frac{1}{4}\left[\frac{5}{9}(F - 32)\right] + 40$$

$$= \frac{5}{36}F + \frac{320}{9}$$

which is not the same as the output from $g \circ h$. Also, $h \circ g$ makes no sense in the context of the problem, because the output from g is a Celsius temperature and the input to h must be a frequency of chirps.

Example 2 shows how to find an analytical expression to represent the composition of two functions.

EXAMPLE 2   *Finding an expression for a composed function*

Suppose $U(x) = 3x + 100$ and $f(h) = \dfrac{2h}{h - 100}$. To obtain an expression for $(f \circ U)(x)$, begin by rewriting

$$(f \circ U)(x) = f[U(x)] = f(3x + 100)$$

To evaluate $f(3x + 100)$, substitute $3x + 100$ for every h in the expression for $f(h)$.

$$f(3x + 100) = \frac{2(3x + 100)}{(3x + 100) - 100}$$

$$= \frac{6x + 200}{3x}$$

∎

EXERCISES   **7.** Find an expression for $(U \circ f)(h)$ in Example 2.

In Exercises 8–11, find expressions for $(u \circ v)(x)$ and $(v \circ u)(x)$.

8. $u(x) = 2x + 1$
$v(x) = 5x$

9. $u(x) = 8 - 4x$
$v(x) = x^2$

10. $u(x) = 4x - 5$
$v(x) = \dfrac{x + 5}{4}$

11. $u(x) = x^2 - 3x$
$v(x) = \dfrac{1}{x}$

EXAMPLE 3   *Representing a single function as a composition*

Every function is a sequence of operations performed on a variable, and each operation is itself a function. For example, to evaluate $f(x) = 5x^2 + 3$, we apply first the squaring function, then the multiplying-by-five function, and finally the adding-three function. The sequence of operations can be represented schematically as follows.

$$x \to x^2 \to 5x^2 \to 5x^2 + 3$$

∎

EXERCISES   *In Exercises 12–15, draw a schematic diagram as in Example 4 to represent the sequence of functions applied to x in the given order. Then write an equation in functional notation to represent the composed function f.*

12. the multiplying-by-five function,
the adding-three function, and
the squaring function

13. the squaring function,
the adding-three function, and
the multiplying-by-five function

14. the adding-three function,
the squaring function, and
the multiplying-by-five function

15. the adding-three function,
the multiplying-by-five function, and
the squaring function

The functions in Exercises 16–19 are all compositions of the square root function, the subtracting-two function, and the multiplying-by-six function. In each case, write down the order in which these functions are applied.

16. $g(x) = 6\sqrt{x - 2}$ **17.** $h(x) = \sqrt{6x - 2}$

18. $j(x) = \sqrt{6x} - 2$ **19.** $k(x) = 6\sqrt{x} - 2$

ADDITIONAL EXERCISES *For each function in Exercises 20–29,*

   a. If $g(x) = x$, write expressions for $(h + g)(x)$, $(h - g)(x)$, $(hg)(x)$, and $\left(\dfrac{h}{g}\right)(x)$.

   b. If $f(x) = x + 4$, write expressions to represent $(h \circ f)(x)$ and $(f \circ h)(x)$.

20. $h(x) = 20x + 13$ **21.** $h(y) = y - 0.1$

22. $h(t) = 100t - 4t^2$ **23.** $h(\&) = \&^3 - 5\&^2 - 6$

24. $h(W) = \dfrac{W}{W + 1}$ **25.** $h(a) = a + \dfrac{1}{a}$

26. $h(\partial) = \sqrt{2\partial + 10}$ **27.** $h(p) = \sqrt{4 - p}$

28. $h(r) = r^3$ **29.** $h(x) = 6$

30. If

$$u(P) = (2P - 6)^3 - 5(2P - 6)^2 + 4(2P - 6) - 7$$

and

$$v(Q) = 0.5Q + 3$$

write expressions to represent $(v \circ u)(P)$ and $(u \circ v)(Q)$.

In Exercises 31–34, write an equation for the function that results when the given sequence of functions is applied to x in order.

31. the dividing-by-two function,
the subtracting-five function,
and the cubing function

32. the cubing function,
the subtracting-five function,
and the dividing-by-two function

33. the subtracting-five function,
the cubing function, and
the dividing-by-two function

34. the subtracting-five function,
the dividing-by-two function,
and the cubing function

The functions in Exercises 35–38 are all compositions of the multiplying-by-three function, the adding-one function, and the taking-the-reciprocal function. In each case, write down the order in which these functions are applied.

35. $g(x) = 3\left(\dfrac{1}{x}\right) + 1$ **36.** $h(x) = \dfrac{1}{3x} + 1$

37. $j(x) = 3\left(\dfrac{1}{x + 1}\right)$ **38.** $k(x) = \dfrac{1}{3x + 1}$

In Exercises 39–40, decide which of the given compositions make sense in a physical context. For those that do, describe the physical information provided by the output.

39. (*Interpreting Mathematics*)

$$N(t) = \text{number of sales by a telemarketer in } t \text{ days}$$

$$I(s) = \text{income to the telemarketer from } s \text{ sales}$$

a. $(I \circ N)(t)$ **b.** $(N \circ I)(s)$

40. (*Interpreting Mathematics*)

$$p(A) = \text{air pressure in pounds per square inch at an altitude of } A \text{ feet above sea level}$$

$$q(D) = \text{altitude in feet above sea level at a distance of } D \text{ miles west of St. Louis}$$

a. $(q \circ p)(A)$ **b.** $(p \circ q)(D)$

Use the table at the left to complete Exercises 41–42.

x	f(x)	g(x)
1	3	0
2	5	5
3	1	4
4	2	8
5	4	1

41. Evaluate
a. $(f + g)(2)$ **b.** $(fg)(4)$
c. $(f \circ g)(3)$ **d.** $(g \circ f)(3)$

42. (*Writing to Learn*) Explain why each of the following is undefined.

a. $\left(\dfrac{f}{g}\right)(1)$ **b.** $(f \circ g)(4)$

43. Use the following graph to estimate the value of each expression in parts (a)–(d).

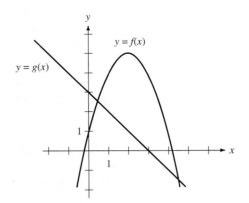

a. $(g - f)(3)$ **b.** $\left(\dfrac{f}{g}\right)(2)$

c. $(f \circ g)(1)$ **d.** $(g \circ f)(1)$

The Kelvin temperature scale was invented by William Thomson (Lord Kelvin, 1824–1907) in 1848, following his discovery that there is a "lowest possible" temperature at which all molecular motion stops. This temperature, which is about $-459.7°$ Fahrenheit and about $-273.2°$ Celsius, is called **absolute zero** and is assigned a value of 0 on the Kelvin scale. In Exercises

44–47, the functions $u(C) = \dfrac{9C}{5} + 32$ and $v(C) = C + 273.2$ express Fahrenheit and Kelvin temperatures, respectively, as functions of Celsius temperature C.

44. (*Problem Solving*) The thermometer on Dave's back porch registers Celsius temperatures, but always reads 5° too low. Suppose his thermometer reading is T, and he wants to determine the actual Fahrenheit temperature. Since the actual Celsius temperature is $T + 5$, the actual Fahrenheit temperature is $u(T + 5)$.

  a. Write an expression in T to represent $u(T + 5)$.

  b. Use this expression to find the Fahrenheit temperature when his thermometer reads 23°.

  c. Express the Kelvin temperature as a function of Dave's thermometer reading T.

  d. Use this expression to find the Kelvin temperature when his thermometer reads 23°.

45. (*Making Observations*) When a kettle of water at room temperature is put onto a hot stove, its Celsius temperature C after t seconds depends on several factors, including how much water is in the kettle and how high the burner is turned up. The Celsius temperature might be approximated for the first few minutes by the equation

$$C = c(t) = 22 + 0.5t$$

  a. Write an expression in t to represent $(u \circ c)(t)$, and evaluate $(u \circ c)(60)$.

  b. Now obtain the answer to part (a) in a different way. Use the function $c(t) = 22 + 0.5t$ to calculate the Celsius temperature of the water after 1 minute, and use the result as an input for the function u.

  c. Write an expression in t to represent the Kelvin temperature of the water after t seconds.

  d. Evaluate this expression when $t = 60$.

46. (*Writing to Learn*) Explain what physical information is given by the calculations in Exercise 45a.

47. (*Making Observations*) Obtain the results from Exercise 45d in a different way. Use the equation $C = 22 + 0.5t$ to calculate the Celsius temperature of the water after 1 minute, and use the result as an input for the function v defined by $v(C) = C + 273.2$.

48. (*Problem Solving*) A car dealer offers a 15% discount on all new cars on his lot. At the same time, the auto manufacturer offers a $1000 rebate.

  a. Let P represent the sticker price of a car. Write a function f to represent its price if only the 15% discount is applied.

  b. Write a function g to represent its price if only the $1000 rebate is applied.

  c. When both the discount and the rebate are applied, the purchase price of the car is either $(f \circ g)(P)$ or $(g \circ f)(P)$, depending on the order in which they are applied. Which would you ask the dealer to apply first? Which composition represents your choice?

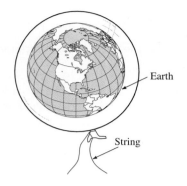

Earth

String

49. A well-known mathematical question states,

Suppose a string is tied around the earth at the equator, so that it fits snugly at all points. If the length of the string is increased by 10 feet, how far off the surface of the earth could it be held at each point? (See the figure at the left.)

Before reading on, make a guess at the answer.

a. (*Writing to Learn*) The function R defined by $R(c) = \dfrac{c}{2\pi}$ gives the radius of a circle whose circumference is c. If c_0 denotes the circumference of the earth, in feet, explain why the answer to the question is given by $R(c_0 + 10) - R(c_0)$.

b. (*Interpreting Mathematics*) Then answer the question by evaluating $R(c_0 + 10) - R(c_0)$. How does the answer compare with what you expected?

50. (*Problem Solving*) Periwinkle's, Lynn's favorite bakery, bakes cakes at a cost of $1.50 each for ingredients. Baker Chris Kozubal devotes half her time to cakes, so that the cost of producing cakes includes half her weekly salary of $425. Because there are several bakeries in the area, Periwinkle's cannot set a price for cakes independently, but must sell them for $10 each. Find Periwinkle's profit function for cakes.

2–3 LINEAR FUNCTIONS AND AVERAGE RATES OF CHANGE

RATE OF CHANGE FOR A LINEAR FUNCTION

Functions vary so widely as to have few global properties in common. However, several useful classes of functions have graphs, equations, and tables with easily identifiable properties. The simplest such class consists of functions whose equations can be written in the form $y = mx + b$. Because the graphs of these functions are straight lines, they are called **linear functions**. (Some basic facts about linear functions and their graphs are discussed in Section A-6.)

Every linear function exhibits a constant **rate of change** between its variables. That is, as the values of the variables change, the ratio

$$\frac{\text{change in the dependent variable}}{\text{change in the independent variable}}$$

remains the same. Conversely, any function with a constant rate of change must be linear. The following Mathematical Looking Glass illustrates how this ratio is reflected in tables, graphs, and equations of linear functions, and what information it can provide in a physical context.

A MATHEMATICAL LOOKING GLASS
Sound Waves

When you sing, sneeze, whisper, whistle, clap, or stomp, you create vibrations that travel through the air around you as sound waves. Their speeds vary depending on conditions such as your elevation above sea level and the temperature and humidity of the air. In particular, the speed of sound S (in meters per second) at sea level is related to the Celsius temperature T by the equation $S = 0.61T + 331.4$.

Table 7 shows the speed of sound at several Celsius temperatures.

TABLE 7

T	S
−10	325.3
0	331.4
10	337.5
20	343.6
30	349.7

The symbol Δ commonly denotes a change in a variable, so in Table 7 we can denote changes in S and T by ΔS and ΔT, respectively. ∎

EXERCISE   **1.** (*Making Observations*) In Table 7 verify that from each row to the next, the ratio $\dfrac{\Delta S}{\Delta T}$ has the same value. What is that value?

Exercise 1 suggests that you have found a property common to all tables from linear functions. It can be stated as follows.

> For any table from a linear function $y = f(x)$, the ratio $\dfrac{\Delta y}{\Delta x}$ is the same between any two rows.

If $\dfrac{\Delta y}{\Delta x}$ is constant for a given table, we say that the table fits a linear function.

The quantities Δx and Δy are called **first differences** in x and y, respectively. If the values of x are equally spaced (that is, if the first differences in x are constant), then it is easy to tell whether the table fits a linear function.

> A table with equally spaced x-values fits a linear function if and only if it has y-values with constant first differences.

EXERCISES   *In Exercises 2–5, decide which tables fit linear functions.*

2.

x	y
13	29
15	26
17	23
19	20
21	17

3.

x	y
−2	−5
−1	−1
0	3
1	5
2	7

4.

x	y
5	7
4	5
2	1
−3	−9
−4	−11

5.

x	y
0.3	17
0.6	11
0.7	9
0.9	5
1.0	3

6. (*Making Observations*) The tables in Exercises 2 and 3 have equally spaced x-values. For which of these are the first differences in y constant? Is this the same table that fits a linear function?

Can a table fit more than one linear function? If it contains more than one point, the answer is no. This is because the graph of every linear function that fits the data is a straight line through all of the data points, and two distinct lines cannot have more than one point in common.

FYI If a table contains only two points, the line through those two points fits the data. This alone is not evidence that the data came from a linear function. On the contrary, it says that *every* two-point table "looks" linear, *no matter what function produced it*! Therefore, be wary about concluding that a two-point table was generated by a linear function. You can explore this idea further in Exercise 73.

EXERCISES   **7.** (*Making Observations*)

    a. Plot the points from Table 7 in **Sound Waves**. How do these points support the conclusion that Table 7 fits a linear function?

    b. What is the slope of the graph of the function $S = 0.61T + 331.4$ from **Sound Waves**?

    c. The slope has the same value as the ratio $\dfrac{\Delta S}{\Delta T}$ that you calculated in Exercise 1. Explain why this is so.

8. For the table in Exercise 2,

    a. Calculate the slope of the line through the points from the table.

    b. Find the equation of the line. Where does the slope appear in the equation?

9. Repeat Exercise 8 for the table in Exercise 4.

10. Repeat Exercise 8 for the table in Exercise 5.

Exercises 1–10 illustrate that for a linear function $f(x) = mx + b$, the following quantities are equal.

- the ratio $\dfrac{\Delta y}{\Delta x}$ between any two points in a table for f

- the slope of the graph of f, calculated as the ratio $\dfrac{y_2 - y_1}{x_2 - x_1}$ for any points (x_1, y_1) and (x_2, y_2) on the graph

- the coefficient m in the equation for f

Each of these quantities is called the **rate of change in y with respect to x**. Figure 2-6 shows the relationships among the three ways to represent rate of change.

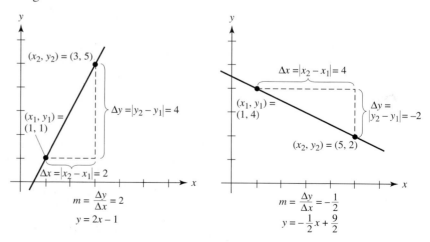

Figure 2–6

To find the equation of a linear function, you need to know both its rate of change and a particular value of the function. In the language of Section A-6, to find the equation of a line, you need to know both its slope and a point on the line.

EXERCISES   *In Exercises* 11–14,

   a. Find the rate of change in *y* with respect to *x*.
   b. Find the equation of the function $y = f(x)$.

11.

12.

13.

14.
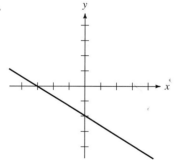

When *x* and *y* represent physical quantities, the rate of change also has physical significance.

EXAMPLE 1   *Interpreting rate of change in a physical context*

   a. In **Sound Waves**, you discovered that the rate of change in *S* with respect to *T* is 0.61. This says that every change of 1 unit in *T* produces a change of 0.61 units in *S*. That is, the speed of sound increases by 0.61 meters per second for each increase of 1 Celsius degree in temperature.
   b. In **Fat Cats on Campus** (page 28), you saw that you can sell *x* T-shirts during the semester if you charge a price of $p = 16 - 0.005x$ dollars per shirt. Thus, the rate of change in *p* with respect to *x* is -0.005. This says that every change of 1 unit in *x* produces a change of -0.005 units in *p*. That is, to increase your target sales level by 1 shirt, you must lower the price per shirt by half a cent ($0.005). ∎

In Exercises 15–18, calculate the rate of change in the dependent variable with respect to the independent variable and explain its physical significance.

15. (*Interpreting Mathematics*) The following table is taken from **Pulleys** (page 9).

Motor pulley diameter, in inches (M)	Fan speed, in rpm (S)
1.5	510
3	1020
5	1700
8	2720
10	3400

16. (*Interpreting Mathematics*) Tony Zarillo, a student in one of Lynn's classes, devised the following problem to illustrate rates of change.

"I am taking a math test about which I know nothing. The following table describes my state of mind during the test."

Minutes since the test began (m)	Percentage of my sanity remaining (s)
8	80
16	60
24	40
32	20

17. (*Interpreting Mathematics*) In one of his frequent failed attempts to budget his money, Dave recently analyzed the monthly cost of driving his 1991 Cavalier. Some costs, such as gasoline and tires, vary with mileage, while others, such as insurance, are fixed. Dave calculated that the total cost in dollars of driving m miles in a month is $C = 0.15m + 100$.

18. (*Interpreting Mathematics*) Lyle Wolf, III sells candy bars in his Lexington, KY neighborhood for $2.50 each. James Tilson is one of his best customers. If James buys b candy bars, then he must pay a total of D dollars.

19. (*Interpreting Mathematics*) Make up your own problem involving a linear table or equation. Calculate the rate of change in the dependent variable with respect to the independent variable, and explain its physical significance.

LINEARIZATION

Linear functions are often used to model relationships that are not linear. To see why it might make sense to do so, try the following experiment. First, graph the function $y = x^2$ in the window $[-10, 10]$ by $[-10, 10]$. Next, zoom in on a small interval around the point $(1, 1)$. Finally, pick any other point on the graph and zoom in on a small interval around it. In each case, what do you see on the screen after zooming in?

As your experiment suggests, any smooth curve can be closely approximated by a line over a small interval. This simple observation has many powerful consequences. In fact, much of calculus is based on it. Its importance to us is suggested by Figure 2-7.

A line such as L, joining two points A and B on the graph of a nonlinear function f, is called a **secant line**. If the secant line is close to the nonlinear graph between A and B, we can use values of the linear function $L(x)$ to esti-

Figure 2-7

mate values of $f(x)$. This process is referred to as **linearization** and is carried out for two primary reasons.

- Linear relationships are simple. Many nonlinear relationships are so complicated that it is impossible to write useful equations to describe them. Even when equations can be written, they are often difficult to analyze.
- In practice, many quantitative questions require only approximate answers. Errors due to linearizing may be acceptable, depending on the question. In addition, it is possible to analyze the size of errors due to linearization. (The analysis requires calculus, so we will not discuss it here.)

In the following situation a nonlinear function f can be closely approximated by a linear function L over an interval.

A MATHEMATICAL LOOKING GLASS
Cold Coffee

As we wrote the preceding paragraphs, Lynn took a sip of her coffee and said, "Yuk! It's cold already." Wondering how long it actually takes a cup of coffee to cool off, we poured another cup and recorded the data in Table 8.

TABLE 8

Time in minutes since coffee was poured	Fahrenheit temperature
0	175
5	147
10	126
15	111
20	100
25	92
30	86

Table 8 does not fit a linear function. In Chapter 7 you will discover a type of function that fits the data. In the meantime, you can provide an approximate fit by linearizing.

Suppose you want to know how long it took the coffee to cool to a temperature of 150°. In Exercises 20–25, you will use a graphical approach to find out. ■

EXERCISES

20. (*Review*) Verify that Table 8 does not fit a linear function.

21. (*Making Observations*) Plot the points from Table 8 on a large piece of graph paper. Sketch a smooth curve through the points, and use your graph to complete the table at the left. Estimate temperatures to the nearest degree.

x	y
0	175
1	
2	
3	
4	
5	147

22. On your graph paper from Exercise 21, sketch the line through (0, 175) and (5, 147).

Exercise 22 illustrates that the line segment through (0, 175) and (5, 147) is "close" to the segment of the curve between the same two points.

23. Find the equation of the line through (0, 175) and (5, 147), and use it to complete the table in Exercise 21.

24. (*Making Observations*) Complete the following table.

x	y (from Exercise 21)	y (from Exercise 23)	Difference
0	175	175	0
1			
2			
3			
4			
5	147	147	0

Exercise 24 illustrates numerically that the line segment is "close" to the curve. That is, the segment of the curve between (0, 175) and (5, 147) is "approximately linear" and can be modeled by a linear function without great loss of accuracy. You can now use this fact to answer the question posed in *Cold Coffee*.

EXERCISE   25. (*Interpreting Mathematics*) Use your linear equation from Exercise 23 to estimate how long it takes the coffee to reach a temperature of 150°.

AVERAGE RATE OF CHANGE

Example 2 shows how a question about the temperature of Lynn's coffee can be answered from the perspective of rates of change.

EXAMPLE 2   *Using rates of change to estimate data*

What was the temperature of the coffee in *Cold Coffee* after 2 minutes? During the first 5 minutes, the change in the temperature was

$$147 - 175 = -28 \text{ degrees}$$

The change in time was

$$5 - 0 = 5 \text{ minutes}$$

Thus the temperature changed at an average rate of

$$\frac{147 - 175}{5 - 0} = -5.6 \text{ degrees per minute}$$

Since its initial temperature was 175 degrees, its temperature after 2 minutes was about $175 + 2(-5.6) = 163.8°$.   ■

EXERCISES   26. (*Interpreting Mathematics*) If the coffee cools at the rate of 5.6° per minute, how long does it take to cool from 175° to 150°? Make sure your answer agrees with the one you obtained in Exercise 25.

27. (*Interpreting Mathematics*) Refer to the data from *Cold Coffee*.

   a. During the time interval between 15 and 20 minutes, the coffee cooled at an average rate of how many degrees per minute?

   b. Use your answer from part (a) to estimate the temperature of the coffee after 18 minutes.

c. Use your answer from part (a) to estimate the time when the temperature was 109°.

Estimate the answers to the questions in Exercises 28–30 by assuming that the relationship is approximately linear. Either construct an equation, as in Exercise 23, or use the idea of rate of change, as in Exercise 26.

28. (*Problem Solving*) A house whose appraised value in 1980 was $52,000 had a value of $97,000 in 1995. What was it worth each year during the 1980s?

29. (*Problem Solving*) The city of Yellowknife in Canada's Northwest Territories receives 20.8 hours of daylight on June 21 and 12.0 hours of daylight on September 21 (92 days later). How many hours of daylight does Yellowknife receive on August 12 (52 days after June 21)?

30. (*Problem Solving*) $\sqrt{16} = 4$ and $\sqrt{36} = 6$. Without using your calculator, estimate the square root of each whole number between 16 and 36.

Let's look more carefully at the idea of average rate of change. Graphs of nonlinear functions, such as the one in *Cold Coffee*, do not have a constant slope. Thus the ratio $\frac{\Delta y}{\Delta x}$ is not constant, but varies according to the two points chosen on the graph. In Figure 2-8 the value of $\frac{\Delta y}{\Delta x}$ is larger between Q and R than between P and Q.

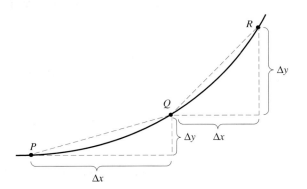

Figure 2–8

The value of $\frac{\Delta y}{\Delta x}$ on the line PQ (its slope) represents the rate of change in y with respect to x along that line. Since P and Q are also on the nonlinear graph, that value of $\frac{\Delta y}{\Delta x}$ also represents the **average rate of change** in $f(x)$ with respect to x between these points. Similarly, the slope of the line QR represents the average rate of change in $f(x)$ with respect to x between Q and R. In general, the average rate of change in $f(x)$ with respect to x over an interval $[a, b]$ is

$$\frac{f(b) - f(a)}{b - a}$$

Example 3 shows how to calculate an average rate of change and illustrates why the word "average" is appropriate in the definition.

EXAMPLE 3   *Calculating and interpreting an average rate of change*

a. If $f(t) = 5t^2 + 45t$, the average rate of change in $f(t)$ with respect to t over $[0, 2]$ is

$$\frac{f(2) - f(0)}{2 - 0} = \frac{110 - 0}{2 - 0} = 55$$

(2, 110)

50

(0, 0)   1   *t*

f(t)

Figure 2–9

Figure 2-9 shows that the average rate of change is the slope of the line connecting the points $(0, 0)$ and $(2, 110)$ on the graph of $y = f(t)$. If $f(t)$ represents the number of miles you have driven during the first t hours of a trip, the average rate of change represents your average speed of 55 mph over the first two hours.

b. If $g(x) = \dfrac{1000}{x}$, the average rate of change in $g(x)$ with respect to x over $[40, 50]$ is

$$\frac{g(50) - g(40)}{50 - 40} = \frac{20 - 25}{50 - 40} = -\frac{1}{2}$$

Figure 2-10 shows that the average rate of change is the slope of the line connecting the points $(40, 25)$ and $(50, 20)$ on the graph of $y = g(x)$.

Figure 2–10

If $f(x)$ represents the number of hours needed to drive 1000 miles at x miles per hour, then the average rate of change has a physical meaning. Specifically, for average speeds between 40 and 50 mph, each increase of 1 mph in your speed reduces the total driving time by about one-half hour.

EXERCISES **31.** (*Making Observations*) In part (a) of Example 3, calculate the average rate of change in $f(t)$ with respect to t over each of the intervals $[0, 1]$ and $[1, 2]$. What do your results tell you about the motion of the car in the example?

32. (*Interpreting Mathematics*) Suppose that the secant line in Figure 2-9 represents the number of miles traveled by a car during the time interval $[0, 2]$.

a. Describe the motion of the car. How does it compare with the motion of the car in Example 3?

b. What is the average speed of the car over [0, 2]? How does it compare with the average speed of the car in Example 3?

Exercises 31 and 32 illustrate that

- An object traveling at a constant velocity has a linear time-distance graph, such as the secant line in Figure 2-9. The slope of the line is equal to the velocity.
- An object traveling at a varying velocity has a nonlinear time-distance graph, such as the curve in Figure 2-9. The slope of a secant line between any two points on the graph is equal to the average velocity between those two points.

To generalize the preceding two statements,

- If the rate of change in y with respect to x is constant, the graph of $y = f(x)$ is linear. The slope of the line is equal to the rate of change.
- If the rate of change in y with respect to x varies, the graph of $y = f(x)$ is nonlinear. The slope of a secant line between any two points on the graph is equal to the average rate of change in y with respect to x between those two points.

EXERCISES

33. (*Interpreting Mathematics*) In **Cold Coffee**, find and interpret the average rate of change between the points (20, 100) and (30, 86).

34. (*Interpreting Mathematics*) In Exercise 28, find and interpret an appropriate average rate of change.

35. (*Interpreting Mathematics*) In Exercise 29, find and interpret an appropriate average rate of change.

36. (*Interpreting Mathematics*) In **Big Blue Marble** (page 36), find and interpret the average rate of change in $g(x)$ with respect to x over the interval [0, 1600].

Economists often linearize nonlinear functions when analyzing cost, revenue, or profit, as in the following Mathematical Looking Glass.

A MATHEMATICAL LOOKING GLASS
Marginal Profit (Fat Cats 3)

Based on the information in **Fat Cats on Campus** (page 28), one of your classmates has constructed a table containing your projected profit for the semester at a variety of sales levels. Table 9 shows some of his projections.

TABLE 9

Number of shirts sold (x)	Profit [$P(x)$]
500	$4387.50
510	4462.00
520	4535.50
530	4608.00

Your economics professor has told you that a company's **marginal profit** is the additional profit resulting from the sale of one additional item.

She has also said that a company's marginal profit varies from one sales level to another. To gain some insight into her statement, you have decided to find your company's marginal profit at sales levels of 505 and 525 shirts.

Although Table 9 does not list the projected profits for every number of shirts between 500 and 530, you can use the idea of average rate of change to estimate your marginal profits.   ■

EXERCISES   **37.** (*Interpreting Mathematics*)

a. Calculate the average rate of change in $P(x)$ with respect to x over the interval $[500, 510]$.

b. What is your projected marginal profit at a sales level of 505 shirts?

c. What do your calculations in parts (a) and (b) tell you in physical terms?

38. (*Interpreting Mathematics*)

a. Calculate the average rate of change in $P(x)$ with respect to x over the interval $[520, 530]$.

b. What is your projected marginal profit at a sales level of 525 shirts?

39. (*Making Observations*) Does your profit increase more if you increase your target sales level from 505 to 506 shirts or if you increase it from 525 to 526 shirts?

DYNAMIC BEHAVIOR: INCREASING AND DECREASING FUNCTIONS

Imagine yourself walking along the x-axis through the domain of a function. As you walk, the values of the independent variable increase. The values of the dependent variable may increase over some intervals and decrease over others, as in Figure 2-11.

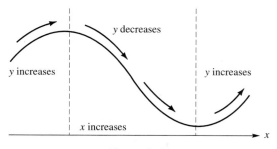

Figure 2–11

If the values of y increase as x increases over an interval, we say that the function is **increasing** over the interval. If the values of y decrease as x increases, the function is **decreasing** over the interval. In Exercise 44, you can discover how the increasing and decreasing behavior of a function on an interval relates to its average rate of change over that interval.

EXAMPLE 4   *Identifying intervals where a function is increasing or decreasing*

Figure 2-12 shows the rate of unemployment in the United States as a function of time between 1982 and 1992. The graph indicates that the unemploy-

Figure 2–12

Adapted from *Time*, July 13, 1992.

ment rate fell from near 10% to just above 5% between 1982 and 1989, then rose to almost 8% by 1992. Therefore, the function is decreasing on the interval (1982, 1989) and increasing on (1989, 1992). ■

EXERCISES   *In Exercises 40–43, identify the intervals on which the function is increasing and those on which it is decreasing.*

40.

41.

42.

43.
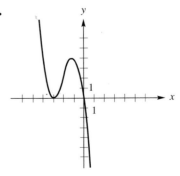

44. (*Making Observations*)

  a. What can you say about the slope of the secant line from $(a, f(a))$ to $(b, f(b))$ if f is increasing on $[a, b]$?

b. If f is increasing on $[a, b]$, must it be true that the average rate of change in $f(x)$ with respect to x over $[a, b]$ is positive? Explain your answer.

c. If the average rate of change in $f(x)$ with respect to x is positive over $[a, b]$, must it be true that f is increasing on $[a, b]$? Explain your answer. (*Hint*: Look at the function $f(x) = x^2$ on $[-1, 2]$.)

ADDITIONAL EXERCISES   *In Exercises 45–50,*

    a. Tell which tables fit linear functions.

If the table fits a linear function, then

    b. State the rate of change in y with respect to x.

    c. Decide whether the function is increasing or decreasing.

    d. Write an equation for the function.

45.

x	y
1	13
2	17
3	21
4	25
5	29

46.

x	y
0	1
0.1	0
0.2	-1
0.3	0
0.4	1

47.

x	y
32	68
37	63
39	61
42	58
43	57

48.

x	y
1	3π
2	6π
5	15π
7	21π
8	24π

49.

x	y
3	1
3	2
3	3
3	4
3	5

50.

x	y
20	2.3
21	2.3
24.5	2.3
$20\sqrt{3}$	2.3
97	2.3

51. (*Problem Solving*) When a large airliner approaches an airport, it begins its descent from about 100 miles away and takes about 20 minutes to descend from an altitude of 36,000 feet.

    a. Express the airliner's altitude A in feet as a linear function of time t in minutes after it begins its descent.

    b. Find and interpret the rate of change in A with respect to t.

    c. Assume that the airliner's altitude A in feet is a linear function of its distance d from the airport in miles. Express A as a function of d.

    d. Find and interpret the rate of change in A with respect to d.

52. (*Making Observations*) (Continuation of Exercise 51) If the cabin of a small aircraft is not pressurized (that is, if the air pressure in the cabin varies with the altitude), it must fly at a lower altitude and descend more gradually to avoid discomfort to the passengers. A typical small aircraft might take 15 minutes to descend from an altitude of 10,500 feet.

a. Express the aircraft's altitude A in feet as a linear function of time t in minutes after it begins its descent.

b. Compare your equation in part (a) with your equation from Exercise 51a. How do the equations reflect the fact that the smaller aircraft must descend more slowly?

c. Graph your equations from Exercises 51a and 52a. How do the graphs reflect the fact that the smaller aircraft must descend more slowly?

In Exercises 53–56,

a. Find a linear function L to approximate the nonlinear function f over the interval $[2, 5]$.

b. Verify that $L(2) = f(2)$ and $L(5) = f(5)$.

c. Calculate the differences $L(3) - f(3)$ and $L(4) - f(4)$.

53. $f(x) = \dfrac{1}{x}$ **54.** $f(x) = \sqrt{x - 1}$

55. $f(x) = x^2 + 6x$ **56.** $f(x) = x^2 - 6x$

In Exercises 57–60, find the average rate of change in y with respect to x over the interval $[0, 10]$.

57. $y = x^3$ **58.** $y = x^3 - 500$

59. $y = \dfrac{1}{x^2 + 100}$ **60.** $y = 7 + 10x - x^2$

61. (*Problem Solving*) In the table at the left m is the dollar amount of each monthly payment on a \$100,000 mortgage at an interest rate of 10%, to be paid back over a period of t years.

a. Find and interpret the average rate of change in m with respect to t over $[25, 30]$.

b. In how many years could you pay off the mortgage at a rate of \$900 per month?

62. (*Interpreting Mathematics*)

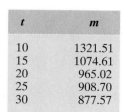

t	m
10	1321.51
15	1074.61
20	965.02
25	908.70
30	877.57

Adapted from *The World Almanac*, Funk and Wagnalls, 1995.

a. Estimate and interpret the rate of change in the number N of farms in the United States with respect to the year y over $[1940, 1980]$.

b. Use your result to estimate the number of farms in the United States in 1993. (The actual number was 2.1 million.)

63. (*Interpreting Mathematics*) The following graph shows the size of the average farm in the United States in each of several years.

Adapted from *The World Almanac*, Funk and Wagnalls, 1995.

a. Estimate and interpret the rate of change in the size *S* of the average farm in the United States with respect to the year *y* over [1940, 1980].

b. Use your result to estimate the size of the average farm in the United States in 1993. (The actual average size was 473 acres.)

64. (*Problem Solving*) In 1983, 26.2% of all persons in the United States between the ages of 18 and 24 were enrolled in college. By 1992 the percentage had risen to 34.4%. (Source: *The Almanac of Higher Education*, University of Chicago Press, 1995.) Let *p* represent the percentage in the year *y*.

a. Find and interpret the average rate of change in *p* with respect to *y* over [1983, 1992].

Use your result from part (a) to estimate

b. the value of *p* in 1987. (The actual value was 29.6%.)

c. the value of *p* in 1993. (The actual value was 34.0%.)

d. the value of *p* in 2063.

e. How accurate do you believe your estimate in part (d) is? Explain your answer.

In Exercises 65–68, identify the intervals on which the function is increasing and those on which it is decreasing.

65.

66.

67.

68.

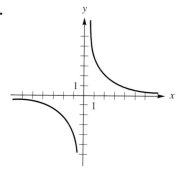

69. (*Writing to Learn*) Let $f(x) = x - 1$ and $g(x) = 10x - 10$.

    a. Which function has the steeper graph?

    b. Graph $y = f(x)$ in the window $[-10, 10]$ by $[-2, 2]$, and graph $y = g(x)$ in the window $[-10, 10]$ by $[-100, 100]$. Which graph appears steeper?

    c. Explain the apparent contradiction in parts (a) and (b).

    Exercise 70 shows that a nonlinear function can generate a table that fits a linear function.

70. (*Extension*) The table at the left was generated by the function $f(x) = x^3$.

x	y
-1	-1
0	0
1	1

    a. Verify that the value of $\dfrac{\Delta y}{\Delta x}$ is the same for each pair of data points.

    b. By generating more points from $f(x) = x^3$, show that $\dfrac{\Delta y}{\Delta x}$ is not the same for all pairs of points from this function.

    Exercise 71 shows that the graph of a nonlinear function can appear linear in some viewing windows.

71. (*Extension*) Suppose $g(x) = \sqrt{x^2 - 1}$.

    a. Graph $y = g(x)$ in the window $[10, 25]$ by $[10, 20]$. Does the graph look linear?

    b. Find a different viewing window to show that g is not linear.

    Exercise 72 illustrates that a table fits a linear function precisely when its points are collinear.

72. (*Making Observations*) Plot the points from Exercise 2 in the coordinate plane. Repeat for Exercises 3–5. Verify that the tables with collinear points are those that fit linear functions.

73. (*Extension*) Is the following statement true or false?

    A graph through $(-1, -1)$, $(0, 0)$, and $(1, 1)$ must be a straight line with a slope of 1 and a y-intercept of 0.

        Explain how you arrived at your conclusion. (*Hint*: Plot the three points, and visualize several graphs that could pass through all three.) (*Second Hint*: See Exercise 70.)

SUPPLEMENTARY TOPIC

Linearizing a Situation

To model a problem with a linear function, we often need to make some simplifying assumptions. In *Cold Coffee* it was assumed that the coffee cooled at a constant rate during the first five minutes. While this was almost certainly not the case, the assumption still yielded reasonable results.

74. (*Writing to Learn*) List at least one simplifying assumption you needed to make to construct a linear model in each exercise.

a. in Exercise 28   b. in Exercise 29
c. in Exercise 30   d. in Exercise 61

75. (*Writing to Learn*) List at least two assumptions you needed to make to construct the linear functions in Exercise 51. Do you think these assumptions closely match the actual behavior of an airliner approaching an airport? Why or why not?

In Exercises 76–78,

a. (*Problem Solving*) First complete the exercise, assuming that the described relationship is linear.

b. (*Writing to Learn*) Explain what the assumption of linearity means in physical terms.

76. If the temperature outside was 45° F at 7:00 A.M. and is 83° F at 2:00 P.M., construct a table giving the approximate temperature at 30-minute intervals between 8:00 A.M. and noon.

77. Driving westward across Kansas on I-70, Dave passed milepost 421 at 9:45 A.M. and milepost 195 at 1:31 P.M. Where was Dave at noon?

78. The air pollution level is 136 parts per million when measured at a distance of 50 feet from a major highway and 84 parts per million when measured at a distance of 180 feet. How far must you go from the highway before the level of pollution drops to 130 parts per million? 120? 110? 100? 90?

SUPPLEMENTARY TOPIC
**Local Linearization:
A "Propinquity Principle"**

The word *propinquity* means "nearness in time or place." Our "Propinquity Principle" states that linearizing a function over an interval usually yields better estimates close to the interval than farther away from it.

79. (*Making Observations*) Recopy your graph from Exercise 21. Obtain two estimates of the time when the coffee reaches a temperature of 90° F as follows.

a. Sketch a straight line through the points (0, 175) and (5, 147), and use the *x*-coordinate when *y* = 90 on this line.

b. Sketch a straight line through the points (20, 100) and (25, 92), and use the *x*-coordinate when *y* = 90 on this line.

c. Decide which estimate is more accurate, and explain your choice.

2–4   SEQUENCES

Sometimes the independent variable in a function represents a quantity that is counted rather than measured. Such a function is usually called a **sequence**. Specifically, a sequence is a function whose domain is a set $\{K, K + 1, K + 2, \ldots\}$ for some integer K. In this section you may assume that $K = 1$, so that the domain is the set of all positive integers. The following Mathematical Looking Glass describes an especially famous sequence.

A MATHEMATICAL
LOOKING GLASS
Fibonacci's Rabbits

Leonardo Fibonacci (c. 1170–1250), also known as Leonardo of Pisa, had a profound influence on the course of mathematical history. At the time of his birth, the world's mathematical expertise was concentrated in northern Africa, the Middle East, and Asia. After being educated in Africa, Fibonacci returned to Italy, where in 1202 he published a free translation of Greek and Arabic mathematics into Latin. His treatise, called *Liber Abaci*, introduced Europeans to Hindu methods of calculating with integers and fractions and extracting square and cube roots. It also helped to popularize the use of modern Arabic numbers in place of Roman numerals. Today, however, Fibonacci is remembered primarily for the function he encountered while studying the following problem.

A particular breed of rabbit grows and reproduces according to a precise schedule. Each female rabbit becomes an adult and produces a pair of babies (one male and one female) at the age of two months and produces exactly one pair every month thereafter. Suppose you begin with one pair of adult rabbits, who produce a pair of babies at the end of the first month. If no rabbits die, what function describes the number of adult female rabbits you will have after a given number of months?

It is likely that Fibonacci first approached this problem by using the strategy of *Making a Table* such as Table 10.

TABLE 10

Month	Newborns (female)	1-Month-olds (female)	Adults (female)
1	1	0	1
2	1	1	1
3	2	1	2
4	3	2	3
5	5	3	5
6	8	5	8

If Table 10 is continued indefinitely, it defines a function called the **Fibonacci sequence**. Its domain is the set of integers $\{1, 2, 3, 4, 5, 6, \ldots\}$ in the left column, and its range is the set of integers $\{1, 2, 3, 5, 8, \ldots\}$ in the right column.

The Fibonacci sequence has many fascinating mathematical properties, some of which you will learn in Exercises 44 and 45. Its numbers also show up unexpectedly often in the physical world. For example, the leaves of many common plants are arranged in patterns involving Fibonacci numbers. Mathematician Rudy Rucker cites several examples in his book, *Mind Tools* (Houghton Mifflin, 1987). In a completely different direction, baseball statistician Bill James' book, *The Politics of Glory* (Macmillan, 1994), gives a fascinating account of his formula, "Fibonacci win points," for evaluating career performances of major league pitchers.

By the way, more than 700 years after Fibonacci's death a group of mathematicians formed the Fibonacci Association and began publication of a quarterly journal dedicated to research on the Fibonacci sequence. ■

EXERCISE   **1.** (*Making Observations*)

a. Continue Table 10 for months 7–10. What pattern do you observe in the last column?

b. Use your observation in part (a) to list the number of adult female rabbits each month for the first two years.

Sequences are usually given functional names using letters at the beginning of the alphabet. Their independent variables are usually denoted by letters in the middle of the alphabet. Furthermore, the values of the dependent variable, called the **terms** of the sequence, are usually written using subscripts such as a_n rather than $a(n)$. This variant of standard functional notation is illustrated in Example 1.

EXAMPLE 1   *Functional notation for sequences*

a. The equation

$$a_k = k^2$$

describes a sequence. For this sequence,

$$a_1 = 1^2 = 1$$
$$a_2 = 2^2 = 4$$
$$a_3 = 3^2 = 9$$
$$a_4 = 4^2 = 16$$

etc.

The same sequence can be described by listing its terms

$$\{1, 4, 9, 16, \ldots\}$$

b. In the Fibonacci sequence, each term beginning with the third is the sum of the preceding two. If the nth term is denoted by c_n, the sequence can be described by the equations

$$c_1 = 1$$
$$c_2 = 1$$
$$c_n = c_{n-1} + c_{n-2}, \qquad \text{for } n \geqslant 3$$

The Fibonacci sequence can also be described by listing its terms

$$\{1, 1, 2, 3, 5, 8, \ldots\}$$ ■

Example 1a illustrates that some sequences can be defined by displaying an explicit formula for a given term. The formula $a_k = k^2$ allows us to calculate a_k for any value of k. For example, $a_{1000} = 1,000,000$. By contrast, the formula $c_n = c_{n-1} + c_{n-2}$ given for the Fibonacci sequence in Example 1b is **recursive**. That is, each term beyond the first few is expressed as a function of the preceding terms. This formula does not allow us to calculate c_{1000} without first calculating c_1 through c_{999}.

EXERCISES   *In Exercises 2–9, write out the first ten terms of the sequence.*

2. $a_n = 2n$

3. $b_k = \dfrac{1}{k}$

4. $c_j = 2 + j - j^2$

5. $A_m = \dfrac{m}{m + 1}$

6. $a_1 = 1$
$a_k = ka_{k-1}$ for $k \geqslant 2$

7. $C_1 = 1$
$C_2 = -1$
$C_n = C_{n-1}C_{n-2}$ for $n \geqslant 3$

8. $b_1 = 1024$
$b_2 = 0$
$b_n = \dfrac{b_{n-1} + b_{n-2}}{2}$ for $n \geqslant 3$

9. $B_1 = 1$
$B_j = B_{j-1} + j$ for $j \geqslant 2$

10. (*Making Observations*)

a. Write out the first four terms of the sequence $b_1 = 2$, $b_k = 2b_{k-1}$ for $k \geqslant 2$.

b. Write an explicit formula for b_k, and use it to calculate b_{20}.

11. (*Making Observations*)

a. Write out the first ten terms of the sequence $a_1 = 3$, $a_n = a_{n-1} + 4$ for $n \geqslant 2$.

b. Write an explicit formula for a_n, and use it to calculate a_{1000}.

ARITHMETIC SEQUENCES

The sequence in Exercise 10 is a **geometric sequence**, in which the ratio of consecutive terms is constant. You will explore geometric sequences in Section 7-1, where we will discuss the type of function that defines them. By contrast, the sequence in Exercise 11 is an **arithmetic sequence**, in which the difference between consecutive terms is constant. Its defining equation is a linear function of n. In the following Mathematical Looking Glass, an arithmetic sequence appears in a physical context.

A MATHEMATICAL
LOOKING GLASS
Class of 2016

Justina Prenatt and Alan Medvick, friends of Dave's daughter Sarah, have two young sons, Lennon and Gareth. To provide for the children's college education, they have started a savings account for each of them. Since Lennon was born on Christmas Eve of 1993, his account was opened at the beginning of 1994. Justina and Alan deposited $1200 into the account during the first year, $1320 during the second year, and so on, increasing the amount by $120 each year. They are now trying to estimate the value of the account on December 31, 2011, the end of the eighteenth year. Before attempting the cumbersome calculation of interest, they have adopted the strategy of *Examining a Simpler Problem* by calculating the total amount they will have deposited into the account by that time. ∎

EXERCISES

12. (*Writing to Learn*)

Let

n = the number of the years since the account was opened

a_n = the number of dollars deposited into the account during the nth year of Lennon's life

a. How does the situation indicate that n and a_n are linearly related?

b. Describe the steps you would use in constructing the equation $a_n = 120n + 1080$.

13. (*Interpreting Mathematics*) How much will Justina and Alan deposit into Lennon's account during the year 2011 (the eighteenth year of his life)?

The equation defining the sequence $a_n = 120n + 1080$ is a linear function of n, so the sequence is arithmetic. An arithmetic sequence differs from other linear functions only in its domain. Because the domain of this sequence consists only of positive integers, its graph is the one in Figure 2-13a, not Figure 2-13b.

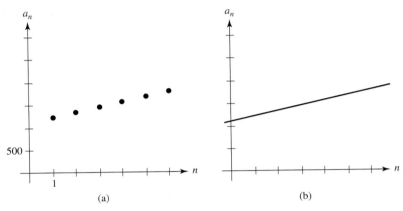

(a)   (b)

Figure 2-13

Arithmetic sequences, like other linear functions, have the property that $\dfrac{\Delta a_n}{\Delta n}$ is constant. Since $\Delta n = 1$ between any two consecutive terms, the difference Δa_n between consecutive terms is constant. That is,

> A sequence a is arithmetic if and only if $a_{n+1} - a_n$ has the same value for each positive integer n.

For any sequence a, the quantities $a_{n+1} - a_n$ are called its **first differences**. Thus, a sequence is arithmetic if and only if its first differences are constant.

An arithmetic sequence can be described by the point-slope equation $a_n - a_1 = d(n - 1)$ or, equivalently,

$$a_n = a_1 + d(n - 1)$$

where d is the first difference and n is any positive integer. If you know a_1 and d, you can use this equation to find a_n for any n.

EXAMPLE 2   *Finding a specified term of an arithmetic sequence*

Suppose we need to find the 247th term in the sequence

$$a = \{3.4, 4.1, 4.8, 5.5, 6.2, \dots\}$$

This sequence is a linear function whose graph consists of the points $(1, 3.4)$, $(2, 4.1)$, $(3, 4.8)$, ... We can calculate the slope of the graph as 0.7 and write its equation in point-slope form as

$$a_n = 3.4 + 0.7(n - 1)$$

Therefore,

$$a_{247} = 3.4 + 0.7(247 - 1) = 175.6$$ ■

EXERCISES

In Exercises 14–17, decide whether the given sequence is arithmetic. If it is, find a formula for the nth term of the sequence, and find the value of the 425th term.

14. $\{4, 7, 10, 13, 16, \ldots\}$

15. $\{90, 100, 110, 120, 130, \ldots\}$

16. $\{7.5, 6, 4.5, 3, 1.5, \ldots\}$

17. $\{1, 3, 6, 10, 15, \ldots\}$

18. (*Problem Solving*) Fausto Obregon runs the Boston Marathon almost every year. A few months before the race, he begins to work on his speed at the Marion High School track in Marion, VA. During the first week of training, Fausto sprints a few miles at a pace of 5:30 per mile. His goal is to decrease his time by 5 seconds per mile each week.

  a. If his pace is a_n seconds per mile during week n, find an expression for a_n.

  b. How many weeks will it take Fausto to decrease his pace to his goal of 4:45 per mile?

ARITHMETIC SERIES

In solving problems you sometimes need to add a number of consecutive terms in an arithmetic sequence. An expression $a_1 + a_2 + \cdots + a_N$ is called an **arithmetic series**. In Exercises 19–26 you will develop and apply a formula to find sums of arithmetic series.

EXERCISES

19. (*Problem Solving*) In **Class of 2016**, how much will Lennon's parents have deposited into his savings account during the first 18 years of his life? (*Hint*: The sequence of deposits has been copied down twice in Figure 2-14. To avoid a lot of tedious addition, refer to Figure 1-12 in Section 1-3 and use the strategy of *Examining a Related Problem*.)

1200	1320	1440	...	...	3000	3120	3240
↑	↑	↑			↑	↑	↑
3240	3120	3000	...	...	1440	1320	1200

Figure 2–14

20. Use the idea in Exercise 19 to find the sum

$$3 + 5 + 7 + \cdots + 2001$$

To apply the idea, you must first find the number of terms in the series. [*Hint*: The equation of the sequence is $a_n = 3 + 2(n - 1)$.]

21. Find the sum $7 + 4 + 1 - 2 - 5 - \cdots - 2987$.

22. Find the sum of the first 73 terms of the sequence

$$\{6, 10, 14, 18, 22, \ldots\}$$

23. Find the sum of the first 824 terms of the sequence

$$\{3.94, 4.03, 4.12, 4.21, 4.30, \ldots\}$$

In general, the sum of the series $a_1 + a_2 + \cdots + a_N$ is

$$S_N = \frac{N(a_1 + a_N)}{2}$$

24. (*Writing to Learn*) Using Exercises 19–23 as examples, explain why this formula is valid.

25. Find the sum $57 + 59 + 61 + \cdots + 123$.

26. Find the sum of the first 51 terms of the sequence

$$\{-15, -10, -5, 0, 5, \ldots\}$$

ADDITIONAL EXERCISES   *In Exercises 27–34, write out the first ten terms of the sequence.*

27. $A_j = 5 - j$ **28.** $c_k = |k - 5|$

29. $b_m = m^3$ **30.** $f_n = 3^n$

31. $C_1 = 1$
$C_i = 0.1 C_{i-1}, \qquad$ for $i \geqslant 2$

32. $d_1 = 1$
$d_2 = 2$
$d_n = d_{n-1} - d_{n-2}, \qquad$ for $n \geqslant 3$

33. $F_1 = 1$
$F_2 = -1$
$F_k = F_{k-2}, \qquad$ for $k \geqslant 3$

34. $a_1 = 1$
$a_2 = 1$
$a_n = a_1 + a_2 + \cdots + a_{n-1}, \qquad$ for $n \geqslant 3$

In Exercises 35–42,

    a. Decide whether the sequence is arithmetic.

    If it is, then

    b. Write an equation to describe it.

    c. Find the value of the 100th term and the value of the 425th term.

    d. Find the sum of the first 100 terms and the sum of the first 425 terms.

35. $\{7, 10, 13, 16, 19, \ldots\}$

36. $\{-8, -3, 2, 7, 12, \ldots\}$

37. $\{3487.6, 3487.5, 3487.4, 3487.3, 3487.2, \ldots\}$

38. $\{\sqrt{6}, \sqrt{6}, \sqrt{6}, \sqrt{6}, \sqrt{6}, \ldots\}$

39. $\{3, 6, 12, 24, 48, \ldots\}$

40. $\{-1, 0, 1, 0, -1, \ldots\}$

41. $\{\pi - 2, 2\pi - 3, 3\pi - 4, 4\pi - 5, 5\pi - 6, \ldots\}$

42. $\{2, 4, 6, 10, 16, \ldots\}$

Time (seconds)	Distance (feet)
1	16
2	48
3	80
4	112

43. (*Interpreting Mathematics*) The table at the left shows the number of feet an object in a vacuum falls each second after it is dropped.

   a. Verify that the table defines an arithmetic sequence.

   b. How far does the object fall during the 60th second?

   c. How far does the object fall altogether during the first 60 seconds?

44. If c_n is the nth term of the Fibonacci sequence, it can be shown that for even values of n,

$$c_n = \frac{(\sqrt{5} + 1)^n - (\sqrt{5} - 1)^n}{2^n \sqrt{5}}$$

Verify that this formula is correct for $n = 2, 4$, and 6.

45. (*Extension*)

   a. If c_n is the nth term of the Fibonacci sequence, show that

$$c_1 + c_2 + \cdots + c_k = c_{k+2} - 1$$

   (*Hint*: First explain why you can write

$$c_1 = c_3 - c_2$$
$$c_2 = c_4 - c_3$$
$$\vdots$$
$$c_k = c_{k+2} - c_{k+1}$$

   b. Use the formula in part (a) to find the sum of the first 20 terms in the Fibonacci sequence. (Use your calculator and the formula in Exercise 44 to evaluate c_{22}.)

46. (*Extension*) Every finite sequence can be continued in more than one way. For example, the fourth term of the sequence $\{1, 2, 4, \ldots\}$ might be 8 (using the rule $a_n = 2a^{n-1}$) or 7 (using the rule $a_n = a_{n-1} + n - 1$). Other continuations are also possible. Describe at least two rules that lead to different continuations of each of the following sequences.

   a. $\{1, 1, 2, \ldots\}$ **b.** $\{1, 2, 3, \ldots\}$

   c. $\{1, 2, 6, \ldots\}$ **d.** $\{1, -1, 2, \ldots\}$

47. (*Extension*) The following sequence was created by John Conway of Princeton University. Can you guess the rule for generating the numbers in each row?

$$1$$
$$11$$
$$21$$
$$1211$$
$$111221$$
$$312211$$
$$13112221$$
$$\vdots$$

48. (*Writing to Learn*) If mathematicians had named baseball's World Series, it would be called the World Sequence. Explain why.

2–5   PIECEWISE DEFINED FUNCTIONS

PREREQUISITE   MAKE SURE YOU ARE FAMILIAR WITH:
Absolute Value Equations and Inequalities (Section A-9)

A MATHEMATICAL
LOOKING GLASS
Health Insurance

After Dave's daughter Sarah graduated from college in 1995, she was employed as a counselor at Bethesda Children's Home. For the first several months her position was part time and included no benefits. In particular, Sarah had to purchase her own health care insurance. She called several providers, including Blue Cross and Blue Shield, who offered her a choice of the three plans summarized in Table 11.

TABLE 11

	Monthly premium	Annual deductible
Plan 1	$73.30	$1000
Plan 2	81.60	500
Plan 3	88.80	250

Under Plan 1 Sarah must pay the first $1000 of any medical expenses each year (the deductible) out of her own pocket. The plans with higher premiums carry smaller deductibles. If medical expenses exceed the deductible for a given plan, Blue Cross and Blue Shield pays 80% of the additional expenses until the total for the year reaches $5000.

Sarah realized that different annual levels of medical expenses might make different plans more desirable. To decide which plan would be best for her, she began by *Making a Table* showing the total cost of each plan at several levels of medical expenses. *Breaking the Problem into Parts*, she began by considering the plan with the lowest premium. Table 12 shows some of her calculations.

TABLE 12

Total annual medical expenses	Total annual cost of Plan 1
$ 0	$ 879.60
250	1129.60
500	1379.60
750	1629.60
1000	1879.60

EXERCISES   **1.** (*Writing to Learn*) Describe how each entry in Table 12 was calculated.

2. (*Writing to Learn*) Explain why Table 12 can be continued as follows.

Total annual medical expenses	Total annual cost of Plan 1
$1000	$1879.60
1250	1929.60
1500	1979.60
1750	2029.60
2000	2079.60

3. (*Writing to Learn*) Let x represent Sarah's annual medical expenses and y represent the total annual cost of Plan 1, both in dollars.

   a. Use Table 12 to show that if Sarah's annual expenses are $1000 or less, the cost of Plan 1 is expressed by the equation

$$y = x + 879.60$$

   b. Use the continuation of Sarah's table in Exercise 2 to show that if her annual expenses are more than $1000, the cost of Plan 1 is expressed by the equation

$$y = 0.20x + 1679.60$$

4. Exercise 3 shows that the cost of Plan 1 is

$$f(x) = \begin{cases} x + 879.60 & \text{if } 0 \leq x \leq 1000 \\ 0.20x + 1679.60 & \text{if } x > 1000 \end{cases}$$

Use these equations and Sarah's table to sketch the graph of f.

THREE VIEWS OF PIECEWISE DEFINED FUNCTIONS

In Exercises 59–60 you can find functions to express the annual cost of Plans 2 and 3 and decide which plan would be best for a given level of medical expenses. In the meantime, let's notice that the function f in Exercise 4 is composed of "pieces" of two separate functions and is therefore called **piecewise defined**. Specifically, a piecewise defined function is one that is described by two or more formulas, each applied over a specified interval.

The usual way of writing such functions analytically was illustrated above in Exercise 4. Let's also look at piecewise defined functions from numerical and graphical perspectives.

A Numerical View Example 1 illustrates how to evaluate a piecewise defined function at a particular value of the independent variable.

EXAMPLE 1   *Evaluating a piecewise defined function*

Suppose $f(x) = \begin{cases} 2x - 1 & \text{if } x \leq 3 \\ 8 - x & \text{if } x > 3 \end{cases}$

The equation defines $f(x)$ by separate rules in the intervals $(-\infty, 3]$ and $(3, \infty)$, respectively. For example,

Since $2 \leq 3$ $f(2) = 2(2) - 1 = 3$

Since $3 \leq 3$ $f(3) = 2(3) - 1 = 5$

Since $4 > 3$ $f(4) = 8 - (4) = 4$ ∎

EXERCISES   **5.** (*Making Observations*)

a. Make a table showing the values of $f(x)$ for $x = 2.5, 2.6, 2.7, \ldots, 3.3,$ 3.4, 3.5.

b. Make a table showing the values of $f(x)$ for $x = 2.95, 2.96, 2.97, \ldots,$ 3.03, 3.04, 3.05.

c. If x is near 3, do your tables indicate that y is near $f(3)$? Explain.

6. (*Making Observations*) Repeat Exercise 5 for

$$g(x) = \begin{cases} x - 2 & \text{if } x < 3 \\ x^2 - 1 & \text{if } x \geqslant 3 \end{cases}$$

A Graphical View   Example 2 illustrates a method for graphing a piecewise defined function.

EXAMPLE 2   *The graph of a piecewise defined function*

Let's sketch the graph of the function

$$f(x) = \begin{cases} 2x - 1 & \text{if } x \leqslant 3 \\ 8 - x & \text{if } x > 3 \end{cases}$$

from Example 1. Since the expression $2x - 1$ defines $f(x)$ for $x \leq 3$, the graph of $y = f(x)$ coincides with the graph of $y = 2x - 1$ on and to the left of its intersection with the line $x = 3$. Similarly, the graph of $y = f(x)$ coincides with the graph of $y = 8 - x$ to the right of $x = 3$. Figure 2-15 indicates how pieces of the two linear graphs combine to form the graph of $f(x)$.

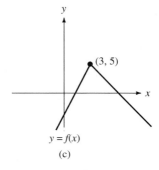

Figure 2–15

EXERCISES   **7.** Use the method of Example 2 to sketch the graph of the function

$$g(x) = \begin{cases} x - 2 & \text{if } x < 3 \\ x^2 - 1 & \text{if } x \geqslant 3 \end{cases}$$

8. (*Writing to Learn*) Describe how the graph in Figure 2-15 and your graph in Exercise 7 support your conclusions in Exercises 5c and 6c.

In Exercises 9–12, sketch the graph of the function. Check your graphs on your calculator.

9. $y = \begin{cases} 3 - x & \text{if } x \leqslant 1 \\ x + 1 & \text{if } x > 1 \end{cases}$

10. $Z = \begin{cases} 0.5W + 4.5 & \text{if } W < -4 \\ 2.5 & \text{if } W \geqslant -4 \end{cases}$

11. $f(t) = \begin{cases} \sqrt{t} & \text{if } 0 \leqslant t \leqslant 4 \\ 7 - t & \text{if } t > 4 \end{cases}$

12. $g(x) = \begin{cases} 1 & \text{if } x < 0 \\ 1 - 0.5x & \text{if } 0 \leqslant x \leqslant 2 \\ x^2 & \text{if } x > 2 \end{cases}$

13. (*Making Observations*) As you have seen, the "pieces" of a piecewise defined graph may join to form a continuous graph as in Figure 2-15, or they may fail to join as in Exercise 7. How can you tell what will happen by looking at the equations of the function? The functions in Exercises 9–12 can help you answer this question.

THREE VIEWS OF LINEAR ABSOLUTE VALUE FUNCTIONS

Since

$$|x| = \begin{cases} -x & \text{if } x < 0 \\ x & \text{if } x \geqslant 0 \end{cases}$$

the function $f(x) = |x|$ is piecewise defined and is an example of a **linear absolute value function**, that is, a function $f(x) = a|x - h| + k$ for some real numbers a, h, and k. Linear absolute value functions occur in physical situations when we are concerned about the *size* of the difference between two quantities, without regard to the *sign*. The following Mathematical Looking Glass provides an example.

A MATHEMATICAL LOOKING GLASS

Salt Container

Morton International, Inc., produces cylindrical 26-ounce salt containers, designed to have a height of 13.90 cm and a radius of 4.280 cm. The heights and radii of the actual containers deviate slightly from these values, so that the actual capacity deviates slightly from the designed capacity of 800 cm³. Morton employees responsible for quality control need to know what deviations in radius and height produce acceptably small deviations in capacity.

We will not attempt to deal with the complexities of considering deviations in radius and height simultaneously. Instead, we will adopt the strategy of *Examining a Simpler Problem.* Let's assume that each container has a radius of exactly 4.280 cm. Then the deviation in capacity results entirely from the deviation between the actual height of the container and the designed height of 13.90 cm.

EXERCISES   **14.** Use the formula $V = \pi r^2 h$ for the volume of a cylinder to verify that a container with the designed height and radius has a capacity of 800 cm³.

15. a. By how much does a container deviate from its designed height if its actual height is 14.00 cm? 13.95 cm? 13.85 cm? 13.80 cm?

b. (*Writing to Learn*) If the actual height of a container is h cm, explain why the deviation in centimeters from its designed height is $|h - 13.90|$.

c. (*Writing to Learn*) Explain why the resulting deviation in cubic centimeters from its designed capacity is

$$\pi(4.280)^2|h - 13.90| \approx 57.55|h - 13.90|$$

The acceptable variations in the heights of Morton salt containers are closely related to the linear absolute value function $E(h) = 57.55\,|h - 13.90|$ in Exercise 15. To prepare ourselves to solve problems involving such functions, let's consider them from numerical, analytical, and graphical perspectives.

A numerical view fails to provide much insight by itself. Without additional information it is usually difficult to tell whether a table was generated by a linear absolute value function. However, a table can help explain the presence of certain features in a linear absolute value graph.

An Analytical View If the equations describing a piecewise defined function are all linear, the function is called **piecewise linear**. Example 3 illustrates that every linear absolute value function is piecewise linear.

EXAMPLE 3   *Writing a linear absolute value function in piecewise linear form*
To write

$$f(x) = 3|x - 2| - 4$$

in piecewise linear form, recall that

$$|x - 2| = \begin{cases} -(x - 2) & \text{if } x - 2 < 0 \text{ (that is, if } x < 2) \\ (x - 2) & \text{if } x - 2 \geq 0 \text{ (that is, if } x \geq 2) \end{cases}$$

Thus

$$f(x) = \begin{cases} -3(x - 2) - 4 & \text{if } x < 2 \\ 3(x - 2) - 4 & \text{if } x \geq 2 \end{cases}$$

In Example 4 you will discover why it is usually better to leave the equations written as they are here, as opposed to multiplying and combining terms in each expression. ■

EXERCISES   *In Exercises 16–19, write the function in piecewise linear form.*

16. $f(x) = 2|x - 1|$ **17.** $g(x) = 3|x| + 4$

18. $F(t) = 0.5|t + 4| - 1$ **19.** $H(z) = 5 - |z|$

A Graphical View As Exercises 16–19 suggest, every linear absolute value graph is V-shaped. The vertical line through its **vertex** (the point of the "V") is an **axis of symmetry** for the graph. That is, the pieces of the graph to the left and right of the line mirror each other as in Figure 2-16.

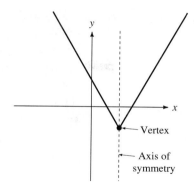

Figure 2–16

The piecewise linear form of the equation provides some clues as to why this is true and allows us to find the vertex, the axis of symmetry, and the range of the function.

EXAMPLE 4   *Graphing a linear absolute value function in piecewise linear form*
To sketch the graph of

$$f(x) = 3|x - 2| - 4$$

write the equation in piecewise linear form, as in Example 3.

$$f(x) = \begin{cases} -3(x - 2) - 4 & \text{if } x < 2 \\ 3(x - 2) - 4 & \text{if } x \geq 2 \end{cases}$$

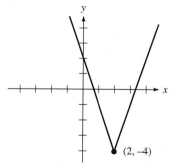

Figure 2–17

Figure 2-17 shows that the graph of $f(x)$ coincides with that of $y = -3(x - 2) - 4$ to the left of $x = 2$ and with that of $y = 3(x - 2) - 4$ to the right of $x = 2$. From their equations, it is easy to verify that the two lines each pass through $(2, -4)$ and that their slopes are -3 and 3, respectively.

The vertex is $(2, -4)$ and the axis of symmetry is the line $x = 2$. The range of the function is $[-4, \infty)$. ■

EXERCISES   *In Exercises 20–23,*

a. Write the function in piecewise linear form.

b. Find the slope of each piece and the coordinates of the vertex.

c. Sketch the graph of the function.

d. Identify the axis of symmetry, and find the range of the function.

20. $y = 2|x - 4|$ **21.** $y = 2|x| - 4$

22. $y = \frac{1}{3}|x + 5| + 1$ **23.** $y = -|x - 6| + 3$

24. (*Making Observations*) On the graph of the function $f(x) = a|x - h| + k$, what are the slopes of the two pieces? Where is the vertex? How can you tell whether the graph opens upward or downward? Your results from Exercises 20–23 can help you answer these questions.

25. (*Making Observations*) The functions in parts (a)–(d) are from Exercises 20–23. Make a table for each, using the given values of x.

a. $y = 2|x - 4|; x = 2, 3, 4, 5, 6$

b. $y = 2|x| - 4; x = -2, -1, 0, 1, 2$

c. $y = \frac{1}{3}|x + 5| + 1; x = -7, -6, -5, -4, -3$

d. $y = -|x - 6| + 3; x = 4, 5, 6, 7, 8$

e. How do the entries in your tables support the appearance of the graphs in Exercises 20–23?

Your observations in Exercise 24 can be stated as follows.

> The graph of $y = a|x - h| + k$ has its vertex at (h, k). It consists of two linear pieces whose slopes are $\pm a$. It opens up if $a > 0$ and down if $a < 0$.

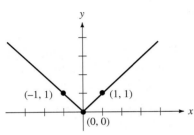

Figure 2–18

A different way to think of the graph of $y = a|x - h| + k$ is outlined in Exercises 26–30. Figure 2-18 shows the graph of the simplest linear absolute value function $y = |x|$.

The graph of every other linear absolute value function can be obtained by transforming (stretching, compressing, reflecting, and shifting) this basic graph in appropriate ways.

EXERCISES

26. (*Making Observations*) Graph the functions $y = |x|$ and $y = -|x|$ on the same set of coordinate axes. Describe how the graph of $y = |x|$ must be transformed to obtain the graph of $y = -|x|$.

27. Graph the functions in parts (a)–(d) on one set of coordinate axes.

a. $y = |x|$ **b.** $y = 2|x|$

c. $y = 3|x|$ **d.** $y = 4|x|$

e. (*Making Observations*) Describe how the graph of $y = |x|$ must be transformed to obtain the graph of $y = a|x|$ if $a > 1$.

28. Graph the functions in parts (a)–(d) on one set of coordinate axes.

a. $y = |x|$ **b.** $y = 0.5|x|$

c. $y = 0.2|x|$ **d.** $y = 0.1|x|$

e. (*Making Observations*) Describe how the graph of $y = |x|$ must be transformed to obtain the graph of $y = a|x|$ if $0 < a < 1$.

29. Graph the functions in parts (a)–(d) on one set of coordinate axes.

a. $y = |x|$ **b.** $y = |x + 1|$

c. $y = |x + 4|$ **d.** $y = |x - 1|$

e. (*Making Observations*) Describe how the graph of $y = |x|$ must be transformed to obtain the graph of $y = |x - h|$.

30. Graph the functions in parts (a)–(d) on one set of coordinate axes.

a. $y = |x|$ **b.** $y = |x| - 1$

c. $y = |x| + 1$ **d.** $y = |x| + 3$

e. (*Making Observations*) Describe how the graph of $y = |x|$ must be transformed to obtain the graph of $y = |x| + k$.

Your observations in Exercises 26–30 are summarized in the following statement and illustrated in Figure 2-19.

To obtain the graph of $y = a|x - h| + k$ by transforming the graph of $y = |x|$,

- Stretch the graph vertically by a factor of $|a|$ if $|a| > 1$, or compress it if $|a| < 1$.
- If $a < 0$, reflect the graph in the x-axis.
- Shift the graph to the right if $h > 0$ or to the left if $h < 0$.
- Shift the graph up if $k > 0$ or down if $k < 0$.

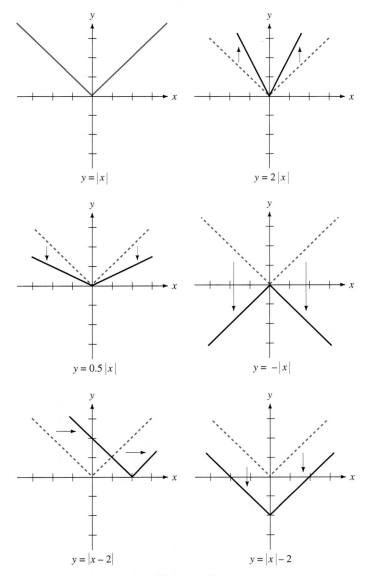

Figure 2–19

EXERCISE   **31.** Describe how the graph of $y = |x|$ must be transformed to obtain the graph of $y = a|x - h| + k$. In particular, describe the transformations that must be used to obtain the graphs of the functions in parts (a)–(d), and sketch each graph.

a. $y = 2|x - 2| + 5$

b. $y = 0.2|x + 5| + 2$

c. $y = -|x + 2| - 5$

d. $y = -0.5|x + 6| - 3$

SOLVING LINEAR ABSOLUTE VALUE INEQUALITIES

In **Salt Container**, Morton employees responsible for quality control need to know what deviations in the height of a container produce acceptably small deviations in its capacity. This question can be answered by solving a linear absolute value inequality. You have already learned how to solve such inequalities both analytically (see Section A-9) and graphically (see Section 1-4). Example 5 reviews the methods.

EXAMPLE 5   *Solving absolute value equations and inequalities*

a. To solve $|3x - 4| < 5$ analytically,

$$-5 < 3x - 4 < 5$$

$$-1 < 3x < 9$$

$$-\frac{1}{3} < x < 3$$

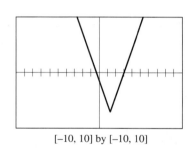

[–10, 10] by [–10, 10]

Figure 2–20

b. To solve $|3x - 4| < 5$ graphically, begin by graphing $y = |3x - 4| - 5$. The graph is shown in Figure 2-20.

The two x-intercepts represent solutions of the equation $|3x - 4| = 5$. By zooming and tracing, you can verify that they occur at $x \approx -0.33$ and $x = 3$. The inequality is true for all values of x where the graph is below the x-axis. The solution is $(-0.33, 3)$. ∎

EXERCISES   *In Exercises 32–35 (Review), solve the equation or inequality both analytically and graphically, and make sure your solutions agree.*

32. $|x + 2| = 9$ **33.** $|2t - 1| = 5$

34. $|Q - 7| < 0.1$ **35.** $|5z + 100| \geqslant 12$

36. (*Problem Solving*) In **Salt Container**, a container whose height is h cm deviates from its designed capacity by $E(h) = 57.55 \, |h - 13.90|$ cm³. What heights will produce a deviation of no more than 10 cm³?

ADDITIONAL EXERCISES   *In Exercises 37–40,*

a. Sketch the graph of the function.

b. Find its domain and range.

c. Find the intervals where the function is increasing, and the intervals where it is decreasing.

37. $f(P) = \begin{cases} \dfrac{2P - 5}{3} & \text{if } P < 2 \\[2mm] \dfrac{P}{3} - 1 & \text{if } P \geqslant 2 \end{cases}$

38. $g(x) = \begin{cases} \sqrt{x + 2} & \text{if } -2 \leqslant x \leqslant 0 \\[2mm] \sqrt{2 - x} & \text{if } 0 < x \leqslant 2 \end{cases}$

39. $H(x) = \begin{cases} -2x - 8 & \text{if } -4 \leqslant x < -1 \\ -6 & \text{if } -1 \leqslant x \leqslant 1 \\ 2x - 8 & \text{if } 1 < x \leqslant 4 \end{cases}$

40. $S = \begin{cases} 5 & \text{if } T < -5 \\ 0.2T^2 & \text{if } -5 \leqslant T \leqslant 5 \\ 5 & \text{if } T > 5 \end{cases}$

In Exercises 41–48,

a. Find the slope of each linear piece of the graph and the coordinates of the vertex.

b. Sketch the graph of the function.

c. Identify the axis of symmetry, and find the range of the function.

d. Identify the intervals where the function is increasing and the intervals where it is decreasing.

e. Describe how the basic graph of $y = |x|$ must be transformed to produce the given graph.

41. $y = 6|x|$ **42.** $y = -0.1|x|$

43. $y = |x - 6|$ **44.** $y = |x| + 4$

45. $y = 3|x + 2|$ **46.** $y = |x - 4| - 5$

47. $y = 5|x + 1| + 7$ **48.** $y = 8 - 2|x|$

In Exercises 49–58, solve the equations and inequalities by any method.

49. $6|x| - 2 = 8$ **50.** $|3 - 0.2x| = 7$

51. $|3 - 0.2x| = 0$ **52.** $|3 - 0.2x| = -7$

53. $|3x + 7| < 19$ **54.** $\dfrac{|x + 4|}{2} \leqslant \dfrac{4}{3}$

55. $2\pi < |\pi x - 2\pi|$ **56.** $\left| \dfrac{x}{3} - 2 \right| \geqslant 0$

57. $\left| \dfrac{x}{3} - 2 \right| + 4 \geqslant 0$ **58.** $\left| \dfrac{x}{3} - 2 \right| - 4 \geqslant 0$

59. (*Creating Models*) Find a piecewise linear function $g(x)$ to describe the cost of Plan 2 in **Health Insurance**, and find a different piecewise linear function $h(x)$ to describe the cost of Plan 3.

60. (*Problem Solving*) For what levels of medical expenses should Sarah choose Plan 1? Plan 2? Plan 3?

61. (*Writing to Learn*) Which of the three plans in **Health Insurance** would you choose for yourself? Explain why.

62. (*Problem Solving*) Dave and his wife, Alice, buy groceries at the Giant Eagle Supermarket, where ground meat costs less per pound if it is purchased in packages of 3 pounds or more.

   a. If the price is \$1.79 per pound for the smaller packages and \$1.39 per pound for the larger ones, express the cost of p pounds as a piecewise linear function $C(p)$.

   b. Sketch the graph of $y = C(p)$.

   c. What quantities of meat weighing less than 3 pounds cost more than a 3-pound package?

63. (*Creating Models*) The 1996 Boston Marathon was won by Kenya's Moses Tanui. Although race officials timed the race with extreme attention to accuracy, Tanui's recorded winning time of 2 hours, 9 minutes, and 16 seconds was almost certainly in error by at least a small fraction of a second. If his actual time was t seconds, express the error as a linear absolute value function $E(t)$.

64. (*Creating Models*) Repeat Exercise 63 using the time of Germany's Uta Pippig, who won the women's race in a recorded time of 2:27:12.

65. (*Problem Solving*) You are driving on Interstate Route I-40, which passes through Arizona's Petrified Forest National Park at milepost 311.

   a. Express your distance from milepost 311 as a function of the milepost number you are currently passing.

   b. At what mileposts might you see a sign saying "Petrified Forest— 15 miles"?

66. (*Problem Solving*) In 1978, Lynn was restoring a Victorian farmhouse in Milwaukee, WI. The baseboard in one bedroom was destroyed by the previous owner's pet rabbit. The baseboard must be replaced around the entire perimeter of the room, except for two doorways, which are each 3 feet wide. Lynn needed to allow herself some slack in ordering the baseboard. However, since it is custom-made and very expensive, she wanted to allow as little slack as possible. The room is square, and she measured the length of one side as 10 feet, 9 inches. If the actual length is a slightly different number of inches, say, L, Lynn will make an error in calculating the total length of baseboard required.

   a. To help Lynn place her order economically, write the size of the error as a function of L.

   b. For what values of L will Lynn have an error of no more than 2 inches in the total length of the baseboard?

67. (*Problem Solving*) Duquesne Light Company provides electric power for customers in the Pittsburgh area. In estimating the cost of cooling and heating a house, the company computes the absolute difference between the average outside temperature and 65°F each day. The result is multiplied by a figure that varies according to the size, shape, and amount of insulation for each house. A typical value for this multiplier might be \$0.25 per day.

   a. Using these figures, express the estimated cost per day as a function of the average outside temperature.

   b. If you can afford up to \$6.00 per day for heating and cooling, what average temperatures will prevent you from paying your entire electric bill?

68. (*Extension*) In **Salt Container**, assume that each container has a height of exactly 13.90 cm, so that the error in capacity results entirely from the deviation in the radius r.

   a. Express the error in capacity as an absolute value function $E(r)$. (It will not be linear.)

   b. What radii produce containers with an error in capacity of no more than 10 cm³? (Use graphical methods to answer.)

69. (*Making Observations*) In parts (a)–(c), solve the equation graphically.

   a. $3|x - 5| - 6 = 0$ **b.** $3|x - 5| = 0$

   c. $3|x - 5| + 6 = 0$

   d. Your experience suggests that every linear absolute value graph is V-shaped. Assuming that this is true, what can you say about the number of solutions to a linear absolute value equation?

   Exercise 70 illustrates that any table can be made to fit a piecewise linear function.

x	y
1	7
2	3
3	5
4	2

70. (*Making Observations*) The table at the left was generated by a function. Find a piecewise linear function to fit it, as follows.

   a. Plot the data on graph paper.

   b. Draw line segments between consecutive data points.

   c. Find the equation of each segment.

   d. Write the equation of a piecewise linear function that fits the data.

71. (*Writing to Learn*) Without additional information, would you believe that the table in Exercise 70 was generated by the function you found? Explain your answer.

2–6   PARAMETRIC EQUATIONS

A moving object often traces out a curve in a plane. Whether the object is a car traveling along a highway, an artillery shell fired at a target, or a satellite orbiting the earth, its path can be described by an equation in x and y. However, such an equation alone cannot tell us the car's speed at a given instant or the time when the satellite will pass over a given point. The following Mathematical Looking Glass shows how functions can be used to provide time-related information about a moving object.

A MATHEMATICAL
LOOKING GLASS

Air Traffic Control

Mark Conaway is an air traffic controller at Pittsburgh International Airport. In a single day he must analyze the flight paths of hundreds of airplanes whose images appear on a screen in front of him. If two airplanes approach each other too closely, Mark must tell one how to alter its course to avoid the risk of a collision. Thus, to ensure that each airplane can take off or land safely, he needs to know both its precise path and the time when it will be at each point along its path.

   Let's look over Mark's shoulder as he monitors the progress of three airplanes. To simplify our discussion, we will assume that his screen shows a square extending 20 miles in each direction from the airport. The viewing window $[-20, 20]$ by $[-20, 20]$ is like a map of that region, with the control tower at the origin, as shown in Figure 2-21.

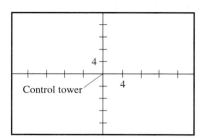

Figure 2–21

We will also assume that each airplane travels at a constant speed in a straight line. (In reality, Mark must deal with changes in both speed and direction.) Table 13 shows the position of each airplane at the time we begin observing ($t = 0$) and one minute later ($t = 1$).

TABLE 13

	Position at $t = 0$	Position at $t = 1$
First airplane	$(-12, -20)$	$(-7, -14)$
Second airplane	$(-9, 20)$	$(-6, 13)$
Third airplane	$(20, -8)$	$(15, -6)$

EXERCISES

1. (*Creating Models*) Write an equation in x and y to describe the path of the first airplane.

Use your equation from Exercise 1 to complete Exercises 2 and 3.

2. (*Interpreting Mathematics*) Will the first airplane pass directly over the control tower? Justify your answer with a calculation.

3. (*Interpreting Mathematics*) What will the first airplane's coordinates be when it leaves the screen?

4. (*Writing to Learn*) We might also want to know where the airplane is after 3.5 seconds or how long it will take to cross the screen. Why can you not answer these questions by using your equation from Exercise 1?

5. (*Creating Models*) Complete the table at the left to show the first airplane's coordinates at each time.

6. (*Creating Models*)

   a. Use your table from Exercise 5 to express x as a linear function of t.

   b. Use your table from Exercise 5 to express y as a linear function of t.

7. (*Interpreting Mathematics*) Use your equations from Exercise 6 to find the airplane's coordinates after 3.5 seconds.

8. (*Interpreting Mathematics*) How long does it take the airplane to cross the screen? (*Hint:* From Exercise 3, you know what the airplane's coordinates are when it leaves the screen.)

t	x	y
0	-12	-20
1	-7	-14
2		
3		
4		
5		

In Exercises 30–33 you can write equations for the paths of the other two airplanes and decide whether any of them should change speed or direction.

The equations $x = -12 + 5t$ and $y = -20 + 6t$ you obtained in Exercise 6 are examples of **parametric equations**. Informally, a set of parametric equations is a pair of functions $x = f(t)$ and $y = g(t)$, with the understanding that their graph consists of all points $(x, y) = (f(t), g(t))$. The variable t is called a **parameter**. In practice, t often represents time. The parametric equations then describe the path of an object moving in a plane by giving its x- and y-coordinates as functions of time. In this section we will focus on motion in a straight line at a constant speed. In that case, the functions $x = f(t)$ and $y = g(t)$ are linear and are called **linear parametric equations**.

Before proceeding any further, make sure you know how to graph parametric equations on your calculator. Consult your owner's manual for details.

EXERCISE   **9.** (*Making Observations*) With your calculator in parametric mode, graph your equations from Exercise 6. Use values of t in $[0, 10]$ and a viewing window of $[-20, 20]$ by $[-20, 20]$.

    a. Trace along your graph to imitate the airplane's motion, and find its coordinates after 3.5 seconds. Does your result agree with your answer to Exercise 7?

    b. Find the time at which the airplane leaves the screen. Does your result agree with your answer to Exercise 8?

The single equation in x and y from Exercise 1 and the parametric equations in x, y, and t from Exercise 6 are two analytical models of the airplane's path. Two differences between them are

- The parametric model involves an extra variable and an extra equation and is thus more complicated than the single-equation model.
- As a trade-off, the parametric model allows you to do time-related calculations, whereas the single-equation model does not.

Since single equations and parametric equations each have unique advantages, it is useful to be able to construct both and to obtain either one from the other.

ELIMINATING THE PARAMETER

Although parametric equations may appear to "lose" some of the information contained in a single equation, this is not the case. In fact, starting from a parametric model of a path, we can construct a single-equation model by eliminating the parameter. Then we have direct access to all information in the single equation, such as the slope of the path.

EXAMPLE 1   *Eliminating the parameter*

Your parametric equations from Exercise 6 are

$$x = -12 + 5t$$
$$y = -20 + 6t$$

To obtain a single equation relating x and y, begin by solving either equation for t. From the first equation,

$$t = \frac{x + 12}{5}$$

Now replace t by $\dfrac{x + 12}{5}$ in the second equation.

$$y = -20 + 6\left(\frac{x + 12}{5}\right)$$

$$y = -5.6 + 1.2x$$

This should agree with your equation from Exercise 1—and should reassure you that the two models are in fact describing the same path. ■

EXERCISES

In Exercises 10–13, eliminate the parameter.

10. $x = t - 1$
      $y = t + 1$

11. $x = 3t - 7$
      $y = 6t$

12. $x = 2t + 5$
      $y = 2t + 5$

13. $x = 3 - 4t$
      $y = 8t + 2$

Each of Exercises 14–17 describes the path of an object in the coordinate plane, starting at time $t = 0$. Complete the following steps for each exercise.

    a. Write an equation in x and y to describe the path.

    b. Write parametric equations to describe the path.

    c. Eliminate the parameter to obtain the equation in part (a).

    d. What are the object's coordinates after 2 seconds?

    e. When does it cross the line $y = 20$?

    f. Graph the parametric equations for $0 \leqslant t \leqslant 20$. Choose a viewing window containing both points given in the exercise.

    g. Use your graph to estimate the object's coordinates after 2 seconds, and the time when it crosses the line $y = 20$. Compare your answers with those in parts (d) and (e).

14. The object moves from $(-4, 2)$ to $(6, 8)$ in 3 seconds.

15. The object moves from $(-4, 22)$ to $(6, 12)$ in 7 seconds.

16. The object moves from $(-5, -3)$ to $(4, -3)$ in 6 seconds.

17. The object moves from $(2, 0)$ to $(8, -3)$ in 1 second.

PARAMETRIZING A PATH

You have just seen that a set of parametric equations specifies a unique path. However, a path does not specify a unique set of parametric equations. Because parametric equations contain information related to time and motion, two objects moving along the same line with different motions (for example, at different speeds) will have different sets of parametric equations.

A MATHEMATICAL LOOKING GLASS

Phoenix

The main streets in Phoenix, AZ, are eight blocks apart and spaced at intervals of one mile. It is about 7 miles along Grande Avenue from the intersection of 59th and Glendale to the intersection of 19th and McDowell. Figure 2-22 shows a map of Phoenix superimposed on a coordinate plane. The origin is at the intersection of McDowell Road and 59th Avenue and each mill represents 1 unit.

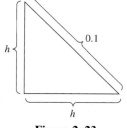

Figure 2–22

Let's look around Grande Avenue on a typical morning. At 7:00 A.M., someone begins jogging northwest at 6 mph, beginning at the Embassy Suites. At 7:10 A.M., someone else begins riding a bicycle at 15 mph from 59th and Glendale toward McDonald's. At 7:20 A.M., a third person leaves McDonald's and drives northwest at 30 mph, proceeding past 59th and Glendale into the desert.

If t denotes time, in minutes past 7:00 A.M., we can represent the path of each person by a set of equations

$$x = a + bt$$

$$y = c + dt$$

Let's focus on the jogger first. Since he is at $(3, 2)$ when $t = 0$, we can write

$$3 = a + b(0)$$

$$2 = c + d(0)$$

so $a = 3$ and $c = 2$. To find b and d, observe that in one minute, the jogger runs $\left(\frac{1}{60} \text{ hours}\right)(6 \text{ miles per hour}) = 0.1$ miles, as in Figure 2-23.

By the Pythagorean theorem, $h^2 + h^2 = (0.1)^2$, so

$$2h^2 = 0.01$$

$$h^2 = 0.005$$

$$h = \sqrt{0.005} \approx 0.07$$

Since he is running northwest, his x-coordinate decreases by 0.07 miles per minute, and his y-coordinate increases by the same amount. Therefore, the equations are

$$x = 3 - 0.07t$$

$$y = 2 + 0.07t$$ ∎

Figure 2–23

EXERCISES

18. (*Creating Models*) Write parametric equations to describe the cyclist's path. Round all coefficients to two decimal places. [*Hint*: Look for equations in the form $x = a + b(t - 10)$, $y = c + d(t - 10)$.]

19. (*Creating Models*) Write parametric equations to describe the driver's path.

20. (*Interpreting Mathematics*) Graph the three sets of parametric equations, using values of t in [20, 40]. Trace along the graphs to estimate the times at which the driver passes the jogger and the cyclist.

21. (*Making Observations*) Eliminate the parameter from all three sets of parametric equations, and verify that they all describe motion along the same line.

DOMAIN AND RANGES OF LINEAR PARAMETRIC FUNCTIONS

Objects typically travel in a straight line at a constant speed only over a finite time interval. The domain of the parametric functions for the motion consists of values of t in that interval. The ranges of the functions (the values of x and y) are also restricted.

EXAMPLE 2   *Finding the domain and the ranges of parametric functions*

If the cyclist in ***Phoenix*** rides for 25 minutes, her parametric functions have the domain $10 \leq t \leq 35$. The ranges are indicated in Figure 2-24.

Figure 2–24

When $t = 10$, she is at $(0, 5)$, and when $t = 35$, she is at $(4.5, 0.5)$. Therefore, the parametric functions have ranges $0 \leq x \leq 4.5$ and $0.5 \leq y \leq 5$, respectively. ■

FYI When x and y are both expressed as functions of t, then t is the independent variable in both functions. Thus the phrase "domain of the parametric functions" refers to values of t, and "ranges of the parametric functions" refers to values of x and y. If we eliminate the parameter, we then regard x as the independent variable, so that "domain" refers to values of x and "range" to values of y. The two situations are summarized in Table 14, using Figure 2-24 as an example.

TABLE 14

Model	Function(s)	Domain	Range(s)
Parametric	$x = 0.18t - 1.8$ $y = 6.8 - 0.18t$	$10 \leq t \leq 35$	$0 \leq x \leq 4.5$ $0.5 \leq y \leq 5$
Single equation	$y = 5 - x$	$0 \leq x \leq 4.5$	$0.5 \leq y \leq 5$

EXERCISES

22. (*Interpreting Mathematics*) If the jogger in **Phoenix** runs for 40 minutes, find the domain and the ranges of his parametric functions.

23. (*Interpreting Mathematics*) If the driver in **Phoenix** drives for an hour, find the domain and the ranges of his parametric functions.

ADDITIONAL EXERCISES

In Exercises 24–29, eliminate the parameter.

24. $x = 3t - 1.5$
$y = 1.5 - 3t$

25. $x = 3 - 2t$
$y = 0.4(5 - t)$

26. $x = t$
$y = 2t + 7$

27. $x = 2t + 7$
$y = t$

28. $x = 100 - t$
$y = 4t + 50$

29. $x = \pi t - 5$
$y = 2\pi t + 9$

*Exercises 30–33 refer to **Air Traffic Control**.*

30. (*Creating Models*)

  a. Write an equation in x and y to describe the path of the second airplane.

  b. Write parametric equations to describe its path.

  c. Eliminate the parameter and verify that you obtain the equation in part (a).

31. (*Creating Models*) Repeat Exercise 30 for the third airplane.

32. (*Interpreting Mathematics*) Graph the parametric equations from Exercises 6, 30, and 31, using the "simultaneous" graphing mode on your calculator.

  a. Does each pair of paths intersect?

  b. Will any of the airplanes collide if they continue on their present courses?

  c. Use your graph to decide which two airplanes come nearest to each other, and estimate the minimum distance between them.

33. (*Writing to Learn*) Use your result from Exercise 32c to decide whether any airplane should change its speed or direction. Give reasons for your answer.

34. (*Writing to Learn*) Suppose that an object moves from $(-2, 3)$ to $(-2, 3)$ in 1 second.

  a. Describe in words the object's motion.

  b. Can you write a linear equation in x and y to describe its path? If so, what is the equation? If not, why not?

  c. Can you write parametric equations to describe its path? If so, what are the equations? If not, why not?

In Exercises 35–38, complete the following steps.

  a. Find the ranges of the parametric functions.

  b. Eliminate the parameter, and find the domain and range of the resulting function.

  c. Support your answers to parts (a) and (b) by graphing the parametric equations.

35. $x = t - 3$
$y = t + 2$ $4 \leqslant t \leqslant 13$

36. $x = 5 - 0.7t$
$y = 5 - 0.7t$ $-3 < t < 0$

37. $x = t$
$y = 3t + 4$ $0 \leqslant t \leqslant 100$

38. $x = 2t + 3$
$y = 3t$ $-\infty < t < \infty$

39. (*Creating Models*) Figure 2-25 shows a coordinate plane superimposed on the solar system.

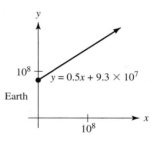

Figure 2–25

Distances are measured in miles, the sun is at the origin, and the earth is at $(0, 9.3 \times 10^7)$. Suppose a rocket is launched from the earth, along the path shown, at 6.92 miles per second (the minimum velocity required to escape the earth's gravity). Write parametric equations for the rocket's path, letting t represent time after the launch, in seconds. (*Hint*: Use the Pythagorean theorem to find the distance traveled in the x- and y-directions each second. See Figure 2-26.)

Figure 2–26

40. (*Interpreting Mathematics*) How long will it take the rocket in Exercise 39 to reach the edge of the solar system, at about

$$(16.86 \times 10^{12}, 8.43 \times 10^{12})?$$

41. (*Writing to Learn*) "An equation in x and y is like a road map. A set of parametric equations is like a bus schedule." Comment on this statement.

42. Think of a situation in your own life in which you or something else moves in a straight line at a constant speed.

a. (*Writing to Learn*) Describe the situation in words.

b. (*Creating Models*) Write a set of parametric equations to describe the motion, and find their domain and ranges.

CHAPTER REVIEW

Complete Exercises 1–57 (Writing to Learn) before refer-ring to the indicated pages.

WORDS AND PHRASES

In Exercises 1–24, explain the meaning of the words or phrases in your own words.

1. (pages 35, 36, 48) **function**
2. (pages 36, 37) **functional notation**
3. (page 38) **domain**, **range**
4. (page 39) **algebraic function**
5. (page 43) **fit** (a table *fits* a function)
6. (page 43) **complete table**
7. (pages 50, 51) **composition** of functions
8. (page 56) **linear function**
9. (page 56) **rate of change**
10. (page 57) **first differences**
11. (pages 60, 61) **secant line**, **linearization**
12. (page 63) **average rate of change**
13. (page 66) **increasing**, **decreasing**
14. (page 72) **sequence**
15. (page 74) **terms** of a sequence
16. (page 74) **recursive** definition of a sequence
17. (pages 75, 76) **arithmetic sequence**, **geometric se-quence**
18. (page 77) **arithmetic series**
19. (page 81) **piecewise defined function**
20. (page 83) **linear absolute value function**
21. (page 84) **piecewise linear function**
22. (page 84) **vertex**, **axis of symmetry** of a linear absolute value graph
23. (pages 92, 93) **parametric equations**, **parameter**
24. (page 93) **linear parametric equations**

IDEAS

25. (pages 36, 37) Using a specific function of your choice, illustrate the meaning of the statement that a function is a process that acts on variables and is dis-tinct from them.
26. (page 37) If $y = f(x)$, explain the meaning of the symbol $f(x)$ and how it differs from f alone.
27. (page 38) What is meant by the phrases *physically reasonable* and *mathematically reasonable* in connec-tion with inputs and outputs of functions? Give exam-ples of a situation in which an output is not physically

reasonable and in which an output is not mathemati-cally reasonable.

28. (pages 38–44) Describe some methods for deciding whether a table, equation, or graph represents a func-tion. Include statements about the conditions under which each method can be applied. Illustrate each method with two examples, one that represents a func-tion and one that does not.
29. (pages 38–44) Describe how to find the domain and range of a function from a table, graph, or equation. Illustrate each process with a specific example.
30. (page 39) What operations can limit the domain of an algebraic function? Illustrate each type of limitation with a specific example.
31. (page 43) Under what conditions do we say that a table fits a function? How does that statement differ from a statement that a table represents a function?
32. (page 44) Why are functions important both in mathematics and as models of physical relationships?
33. (page 44) What do we gain by studying the con-nections among analytical, graphical, and numerical models of functions?
34. (pages 49–52) Describe the processes of adding, subtracting, multiplying, dividing, and composing two functions. Make up two functions and use them to illustrate each process.
35. (page 57) Describe the test for deciding whether a table fits a linear function. Explain how and why the test can be simplified if the values of the independent variable are equally spaced. Illustrate with specific examples.
36. (page 58) If a table contains only two data points, why should you be wary about concluding that it was generated by a linear function?
37. (page 58) Linear functions have a constant rate of change in the dependent variable with respect to the independent variable. How is this fact reflected in a table, graph, or equation of a linear function? Illustrate with specific examples.
38. (pages 59, 60) Describe or make up a specific linear function in which the variables represent physical quantities. What physical information is provided by the rate of change?
39. (pages 60, 61) What do we gain by using a linear function to approximate a nonlinear function over an interval? What do we lose?
40. (pages 60–62) Compare the ideas of rate of change for a linear function and average rate of change for a

nonlinear function. In particular, tell how they are alike and how they are different.

41. (pages 63, 64)   Describe or make up a nonlinear function in which the variables represent physical quantities. What physical information is provided by the average rate of change?

42. (page 66)   What does the graph of an increasing function look like? What can you say about the average rate of change of an increasing function over any interval?

43. (page 73)   Describe a specific physical situation in which a relationship between two variables is represented by a sequence. Use your example to explain how sequences differ from other functions.

44. (page 74)   Compare the processes of calculating a given term of a sequence when it is defined recursively and when it is defined by an explicit formula.

45. (page 75)   How does an arithmetic sequence differ from other linear functions? Illustrate the difference with specific examples.

46. (page 77)   Write down the formula for finding the sum of an arithmetic series, and show why the formula works. Illustrate your argument with a specific series.

47. (page 82)   Explain how to sketch the graph of a piecewise defined function. Use a specific function to illustrate the process and explain the reasoning behind it.

48. (pages 84, 85)   Why can a linear absolute value function be viewed as a piecewise defined function? Illustrate with a specific example. What graphical information is provided by the piecewise linear form of the equation?

49. (pages 83, 84, 88)   Why do linear absolute value functions occur in questions related to errors in measurement?

50. (page 84)   You can graph linear absolute value functions quickly on your calculator. Why do you think it is still helpful to be able to sketch their graphs manually?

51. (page 85)   Tell how to sketch the graph of $y = a|x - h| + k$ quickly. In particular, tell what information is provided by each of the parameters a, h, and k.

52. (page 88)   The solution to a linear absolute value inequality is typically a finite interval or the union of two semi-infinite intervals. The solution may also be empty or contain a single point. Show how the shape of a linear absolute value graph makes it clear that these are the only possibilities.

53. (pages 91–93)   Describe a situation in which it would be natural to use parametric equations to model a relationship, and explain why it would be helpful.

54. (page 93)   What are the advantages and disadvantages of parametric equations, as opposed to a single-equation model of a relationship?

55. (pages 93, 94)   Describe the process of eliminating the parameter from a set of parametric equations, and make up specific equations to illustrate the process.

56. (pages 94, 95)   Give a physical interpretation of the fact that the same line can be parametrized in many different ways.

57. (page 96)   How do the domain and the ranges of a set of parametric functions relate to the domain and range of the function obtained by eliminating the parameter? Illustrate your explanation with a specific example.

CHAPTER 3

LINEAR SYSTEMS

3–1 SYSTEMS OF LINEAR EQUATIONS

PREREQUISITE MAKE SURE YOU ARE FAMILIAR WITH:
Systems of Linear Equations (Section A-10)

According to an old riddle, "A bottle full of wine is worth $50, and the wine alone is worth $45 more than the bottle alone. How much is the bottle worth?" The riddle was designed to lure you into giving a quick (and incorrect) answer of $5. Let's look at two paths to the correct answer.

One path uses the strategy of *Writing an Equation*. If the dollar value of the bottle alone is b, then the dollar value of the wine alone is $b + 45$. Thus

$$b + (b + 45) = 50$$

from which $b = 2.50$, so the bottle is worth $2.50.

A second path uses the strategy of *Looking for Relationships Among Variables*. Let b and w represent the dollar values of the bottle and the wine. Then the riddle tells you that

$$w + b = 50$$

$$w - b = 45$$

These two equations are an example of a **linear system**, that is, a collection of linear equations or inequalities to be solved simultaneously. Constructing a linear system to solve a problem allows you to name all the important unknown quantities quickly and look for relationships among them later. In the first three sections of this chapter you will study several methods of solving systems of linear equations. (In the meantime you can verify that both of the equations above are true if $b = 2.50$ and $w = 47.50$.) The last section will focus on systems of linear inequalities.

101

METHODS OF SOLVING 2 × 2 SYSTEMS

A system of m linear equations in n variables is referred to as an **$m \times n$ system of linear equations**. As we will see, systems involving exactly as many equations as variables are usually the easiest to solve. The following Mathematical Looking Glass involves a 2×2 system.

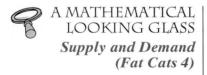

A MATHEMATICAL LOOKING GLASS
Supply and Demand (Fat Cats 4)

The company your economics class formed in **Fat Cats on Campus** (page 28) has held its first board meeting. To increase sales, the class has adopted an incentive program, whereby each student will receive 10% of the net profit on each shirt he or she sells.

When you attempted to agree on a selling price, you discovered that financial rewards were more important to some students than to others.

- All 14 students in the class are willing to produce and sell 100 T-shirts during the semester at a price of $13.50 each.
- Only 4 students are willing to produce and sell 100 T-shirts during the semester at a price equal to the cost of $3.50 each (that is, no net profit per shirt).

At this point you were confronted with what economists refer to as a **supply and demand problem**. Although more students were willing to sell shirts at $13.50 each, fewer customers will purchase them at that price. The class realized that their problem was to find the **equilibrium price**, at which the supply exactly equals the demand for shirts.

After the meeting the class divided into three groups, each of which solved the problem in a different way. ∎

Numerical Methods At your next board meeting, one group presented a numerical solution. From your first meeting they knew that the supply of shirts would be 1400 (14 students supplying 100 shirts each) at a price of $13.50 and 400 at a price of $3.50. They constructed Table 15, whose first two columns duplicate Table 8 on page 61.

TABLE 15

Price per shirt	Demand	Supply
$ 3.50	2500	400
5.50	2100	
7.50	1700	
9.50	1300	
11.50	900	
13.50	500	1400

EXERCISES

1. (*Creating Models*) Assuming, as the group did, that price and supply are linearly related, complete Table 15.

2. (*Interpreting Mathematics*) Generate more entries for Table 15 to find the equilibrium price and demand for T-shirts.

Analytical Methods A second group of students had arrived at an analytical solution. Sharing the first group's assumption of linearity, they had constructed a system of linear equations. Exercises 3–5 will help you duplicate their solution.

EXERCISES

3. (*Creating Models*) Express the demand for T-shirts during the semester as a function of the price per shirt in dollars.

4. (*Creating Models*) Express the available supply of T-shirts during the semester as a function of the price per shirt in dollars.

5. (*Interpreting Mathematics*) Find the equilibrium price and demand by solving your system of equations from Exercises 3 and 4. Make sure your solution agrees with the one you obtained in Exercise 2.

Graphical Methods The third group had solved the problem graphically. They had plotted the data from Table 15 in a coordinate plane. Also making the assumption of linearity, they sketched the graphs in Figure 3-1.

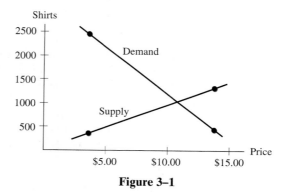

Figure 3–1

The graphs show clearly that supply is an increasing function of price, while demand is a decreasing one. The point of intersection represents the equilibrium price and demand, but they could get only a rough estimate with their hand-drawn graphs. In Exercise 6, you can do better with your calculator.

EXERCISES

6. (*Interpreting Mathematics*) Graph your system of equations, and trace the graphs to estimate the solution. Make sure your solution agrees with those in Exercises 2 and 5.

Solve the systems in Exercises 7–10 graphically.

7. $y = x + 4$
$y = 6 - x$

8. $L = 1.5W - 0.2$
$L = 3.5W + 6.4$

9. $x + 2y = 7$
$3x - y = 7$

10. $x + 2y = 9$
$3x - y = -6$

11. Solve the systems in Exercises 8 and 9 both numerically and analytically.

12. (*Writing to Learn*) Make up a 2×2 system of linear equations, and solve it both numerically and graphically. Then explain how the information in your table is reflected in your graph, and vice versa.

METHODS OF SOLVING 3 × 3 SYSTEMS

The following Mathematical Looking Glass leads to a 3 × 3 system and shows how the methods of solution compare with those used for 2 × 2 systems.

A MATHEMATICAL
LOOKING GLASS
Sunsilk

Sunsilk is a small business run by Paul and Karen Schmitt in Santa Fe, NM. They do three different types of work: landscaping, masonry, and carpentry. Paul likes to schedule their work so that every employee can be fully utilized, but that is not always possible. To analyze Paul's problem, let's make some simplifying assumptions.

- Paul and two helpers do all the manual labor. They can each work 24 days per month.
- Karen manages the paperwork, which includes designing, planning, and billing. She can also work 24 days per month.
- Karen's mother Mary orders and picks up all supplies. Because her grandchildren, Luke and Trevor, keep her busy, she can devote only 12 days per month to the business.

Table 16 shows the division of labor for each type of job they do. For example, on landscaping jobs, 25% of the work is done by Karen, 65% by Paul and the two helpers, and 10% by Mary.

TABLE 16

	Landscaping	Masonry	Carpentry
Karen	25%	15%	20%
Paul and helpers	65%	80%	65%
Mary	10%	5%	15%

Since Paul and the two helpers can be considered together as one "triple person," Paul has to satisfy time requirements for three people. Since they do three types of work, he feels intuitively that he can produce a work schedule which will utilize everyone to full capacity. (We will see later why this intuition is correct.) Let L, M, and C represent the total days per month spent by all employees on landscaping, masonry, and carpentry. If Paul and his helpers are to work at full capacity, then $0.65L + 0.80M + 0.65C = 72$. ∎

EXERCISES
13. (*Interpreting Mathematics*) Explain what the quantities $0.65L$, $0.80M$, and $0.65C$ represent, and explain what the following equation: $0.65L + 0.80M + 0.65C = 72$ says in physical terms.

14. (*Creating Models*) Write a second equation in L, M, and C to describe the use of Karen's time to full capacity. Write a third equation for Mary's time.

Although 2 × 2 systems can be solved by making a table, this method is usually impractical for 3 × 3 systems. (If you don't believe this, you are welcome to try.) Therefore, let's consider analytical and graphical methods of solution.

Analytical Methods Analytical methods for 3×3 systems, such as **Gaussian elimination**, discussed in Section A-10, are extensions of those used for 2×2 systems and are usually the easiest path to the solution.

EXERCISE   **15.** (*Interpreting Mathematics*) Solve the linear system you wrote in Exercises 13–14. How many days of each type of work should be scheduled each month?

Graphical Methods Like 2×2 systems, 3×3 systems can be represented graphically. However, the three variables need a coordinate system with three axes, so the representation is three-dimensional. The graph of a linear equation in two variables is a (one-dimensional) line in two-dimensional space, and the graph of a linear equation in three variables is a (two-dimensional) plane in three-dimensional space. Thus, just as solutions of a 2×2 system correspond to points where two lines intersect, solutions of a 3×3 system correspond to points where three planes intersect. The graphical solution to Paul's problem looks something like Figure 3-2.

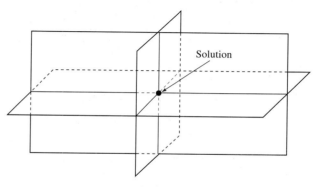

Figure 3–2

We do not expect you to solve 3×3 systems graphically. However, visualizing both 2×2 and 3×3 systems graphically will help you to understand how many solutions each type of system can have.

NUMBER OF SOLUTIONS TO 2 × 2 AND 3 × 3 SYSTEMS

The number of solutions to any system of linear equations is zero, one, or infinitely many. (In Figures 4-3 and 4-4, you will see graphical evidence in support of this fact.) You can say which is most likely to occur if you know how many equations and variables the system has. This knowledge is helpful for at least two reasons.

- If you know in advance how many solutions a physical problem has, then any linear system you construct as a model should have the same number of solutions.

- You can conduct a partial check on the accuracy of your solutions by noticing whether you obtained the expected number of solutions.

2 × 2 Systems Every 2×2 system you have considered so far has a unique solution. Does every 2×2 system have a unique solution? You can discover the answer in Exercises 16–17.

EXERCISES   **16.** (*Making Observations*) For the system

$$x - y = 1$$
$$2x - 2y = 6$$

complete the following table.

x	y (1st equation)	y (2nd equation)
1		
2		
3		
4		

How many solutions does the system have? Explain how the table indicates this.

17. (*Making Observations*) Repeat Exercise 16 for the system

$$x - y = 1$$
$$2x - 2y = 2$$

Exercises 16 and 17 show that a 2×2 system can have zero or infinitely many solutions, as well as one. Are there any other possibilities? The answer to that question is easier to see graphically.

If two people each draw a line in a plane, the two lines nearly always intersect at a unique point, as in Figure 3-3a. These lines represent a linear system with a unique solution. Occasionally, two lines drawn in the plane are parallel, as in Figure 3-3b, and represent a linear system with no solutions. Alternately, they may coincide, as in Figure 3-3c, and represent a linear system with infinitely many solutions.

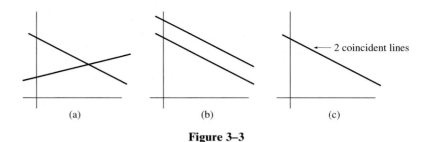

(a)   (b)   (c)

2 coincident lines

Figure 3–3

Figure 3-3 represents all possible types of 2×2 systems of linear equations. In each case, the number of solutions is equal to the number of points where the lines intersect.

A linear system with at least one solution is called **consistent**, while one with no solutions is called **inconsistent**. A 2×2 linear system with at most one solution is called **independent**, while one with infinitely many is called **dependent**. (For larger linear systems, the meaning of independence and dependence becomes more complicated.)

EXERCISES   *Graph the systems in Exercises 18–21. Decide which systems are consistent and independent (have unique solutions), which are inconsistent, and which are dependent.*

18. $x + 5y = 13$
$x = 8 - 5y$

19. $3S = 5T - 4$
$9S - 15T = -12$

20. $2A = 3B$
$2A + 3B = 0$

21. $4P - 3Q = 12$
$3Q - 4P = -12$

22. (*Making Observations*) Choose a system in Exercises 18–21 that you identified as inconsistent, and verify your result analytically.

23. (*Making Observations*) Choose a system in Exercises 18–21 that you identified as dependent, and verify your result analytically.

3×3 Systems   The 3×3 linear system you solved in Exercise 15 had a unique solution. In fact, Figure 3-2 illustrates that almost every 3×3 linear system has a unique solution, because three planes in three-dimensional space almost always have exactly one point in common. This is why, in **Sunsilk**, Paul's intuition that he could schedule three types of work to meet the time requirements for three people was correct.

Occasionally, three planes in three-dimensional space do not intersect at a point. All possible ways in which three planes can intersect are indicated in Figure 3-4. As was the case in two dimensions, the corresponding linear system can have zero, one, or infinitely many solutions.

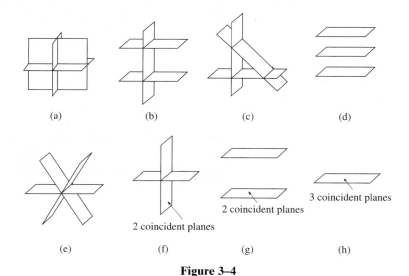

Figure 3–4

EXERCISE   **24.** (*Making Observations*) Decide how many solutions each system in Figure 3-4 has. Remember that a solution corresponds to a point lying on all three planes.

ADDITIONAL EXERCISES   *Solve the 2 × 2 linear systems in Exercises 25–30 both graphically and analytically, and make sure your solutions agree.*

25. $y = 5$
$4x - 3y = 2$

26. $R + 4 = 5S - 3$
$4S + 2R = 0$

27. $2T + 5V = 80$
$3T - 2V = 82$

28. $w + 3z = 0$
$2w - z = 0$

29. $y = 2.5x - 7$
$x = 0.4y + 1$

30. $4x - 6y = 10$
$9y - 6x = -15$

Solve the 3×3 linear systems in Exercises 31–34 (Review) analytically.

31. $x - 2y \qquad = 8$
$2x + 3y - 3z = 8$
$x + 8y - 2z = 8$

32. $3a - 2b - \quad c = 0$
$a + 3b - 5c = 1$
$2a - 7b + 3c = 2$

33. $5p + 2q + \quad r = 0.3$
$2p - \quad q + 3r = 0.1$
$3p - 2q + 6r = 0.3$

34. $x - 0.1y + 30z = 6$
$x + 0.3y - 25z = 3$
$2x + 0.2y - 15z = 5$

Graph the systems in Exercises 35–38. Decide which are consistent and independent, which are inconsistent, and which are dependent.

35. $\alpha - 3\beta = 5$
$3\alpha - \quad \beta = -5$

36. $4(5x + 2y) = 15$
$10x + 4y = 11$

37. $x = -1.5y + 1$
$2x + 3y = 2$

38. $4x = 8 - 2y$
$2x = y - 4$

39. Graph each system, and identify the one that is *not* consistent and independent.

    a. $x + 2y = \quad 6$
        $2x + 3y = 12$

    b. $x + 2y = \quad 6$
        $2x + 4y = 12$

    c. $x + 2y = \quad 6$
        $2x + 5y = 12$

40. (*Making Observations*) The system

$$3x - 4y = \quad 7$$
$$ax - 8y = 14$$

is consistent and independent for every value of a except one. What is the exceptional value of a? (*Hint*: Look at your results from Exercise 39.)

41. Graph each system, and identify the one that is *not* inconsistent.

    a. $x + 2y = 6$
        $2x + 4y = 8$

    b. $x + 2y = \quad 6$
        $2x + 4y = 12$

    c. $x + 2y = \quad 6$
        $2x + 4y = 16$

42. (*Making Observations*) The system

$$3x - 4y = 7$$
$$6x - 8y = a$$

is inconsistent for every value of a except one. What is the exceptional value of a? (*Hint*: Look at your results from Exercise 41.)

43. (*Problem Solving*) Dave contributes to each of two retirement plans through Penn State University. TIAA (Teachers Insurance and Annuity Association) has a more stable rate of return, which was 8.7% during a recent 12-month period. CREF (College Retirement Equities Fund) increased at a rate of 10.5% during the same period, but its rate is subject to greater fluctuations. To obtain an optimum combination of income and security, Dave wants to invest as much as possible in TIAA, as long as the interest rate on the total investment remains at least 9%. Assuming that the above rates remain the same, how much of his $4000 annual contribution should go to TIAA?

44. (*Problem Solving*) Figure 3-5 shows part of the blueprint for a house built for Lynn in 1987. Builder/architect Jack Miller suggested that lengthening the bathroom from 4 feet to 8 feet would provide room for a walk-in closet. Extending into the hallway (shaded area 1) would cost $100 per foot, while extending beyond the outer first-floor wall (shaded area 2) would cost $400 per foot. Lynn wanted to keep the hallway as wide as possible, but could spend no more than $1200 on the bathroom enlargement. How far did she extend the bathroom in each direction?

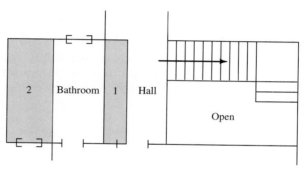

Figure 3–5

45. (*Problem Solving*) On April 4, 1990, millionaire Phil Sokolof placed full-page ads, headlined "The Poisoning of America," in several national newspapers. "McDonald's, Your Hamburgers Have Too Much Fat," proclaimed the ad, which went on to ask McDonald's to reduce the fat in its hamburgers by 10%. Shortly afterward, McDonald's introduced the McLean Deluxe hamburger, a 0.25-pound hamburger containing 10 grams of fat. The company's other hamburgers contain 90 grams of fat per pound. McDonald's uses about 500,000,000 pounds of ground meat each year. To achieve a 10% reduction by weight in total fat content, how many McLean Deluxes must they sell?

46. (*Problem Solving*) A gasoline's **octane** rating reflects both the chemical composition of the gasoline and the nature of its performance in your car. In general, gasolines with higher octane ratings yield higher mileage and smoother engine performance, but at a higher price. One station near Dave's home offers a choice of 87 (regular), 89 (midgrade), or 93 (premium) octane ratings. These were priced at $1.249, $1.449, and $1.549 per gallon in May 1997. Dave can buy 10 gallons of midgrade gasoline for $14.49, but figures he can get a higher octane rating for his money by mixing regular with premium.

  a. How many gallons of each must he buy to get 10 gallons at a cost of $14.49?

b. The octane rating of a mixture of x gallons of regular and y gallons of premium is

$$\frac{87x + 93y}{x + y}$$

What is the octane rating of Dave's mixture?

47. (*Problem Solving*) Allegheny Ludlum Steel Corporation produces steel from scrap metal at two locations in western Pennsylvania. The steel's strength and resistance to rust is determined in part by the amount of chromium and nickel it contains. One of Allegheny Ludlum's customers requires steel containing 18% chromium and 8% nickel by weight. A typical batch of 200,000 pounds of scrap steel might contain 15% chromium and 5% nickel. This can be mixed with an alloy that is 50% iron and 50% chromium by weight and a second alloy that is 50% iron and 50% nickel by weight. How much of each alloy must be used to meet the customer's requirements?

48. (*Problem Solving*) Refer to **Sunsilk**. In the earlier days of the company's operation, Paul and two part-time helpers worked a combined total of 56 days per month. Mary had to spend more time watching the children during those years and could work for the company only 8 days per month. Karen's hours were unchanged. Under those conditions, was it possible to utilize everyone to full capacity? To find out, make a table similar to Table 16, then write and solve a system of equations. Is your result a valid solution to the system of equations? Is it a valid solution to the problem?

49. (*Making Observations*) Identify several Polya strategies you used in completing Exercise 48, and describe when and how they were used.

SUPPLEMENTARY TOPIC

Number of Solutions to $m \times n$ Linear Systems

You have seen that most, although not all, 2×2 and 3×3 linear systems have unique solutions. In Section 4-2 you will solve linear systems having more equations than variables, or vice versa. For now, let's see what we can say about the number of solutions to $m \times n$ linear systems (systems of m linear equations in n variables).

Exercises 50–54 explore $m \times 2$ systems.

50. (*Extension*) Suppose three people each draw a line in two-dimensional space. How many points are usually common to all three lines?

51. (*Extension*) What does your answer to Exercise 50 tell you about the usual number of solutions to a 3×2 system of linear equations?

52. (*Extension*) Suppose one person draws a line in two-dimensional space. How many points lie on the line?

53. (*Extension*) What does your answer to Exercise 52 tell you about the usual number of solutions to a 1×2 system of linear equations?

54. (*Extension*) How many solutions does an $m \times 2$ linear system usually have if $m > 2$? If $m < 2$? If $m = 2$?

Exercises 55–59 explore $m \times 3$ systems.

55. (*Extension*) Suppose four people each draw a plane in three-dimensional space. How many points are usually common to all four planes?

56. (*Extension*) What does your answer to Exercise 55 tell you about the usual number of solutions to a 4 × 3 system of linear equations?

57. (*Extension*) Suppose two people each draw a plane in three-dimensional space. How many points are usually common to both planes?

58. (*Extension*) What does your answer to Exercise 57 tell you about the usual number of solutions to a 2 × 3 system of linear equations?

59. (*Extension*) How many solutions does an $m \times 3$ linear system usually have if $m > 3$? If $m < 3$? If $m = 3$?

60. (*Extension*) How many solutions do you expect an $m \times n$ system of linear equations to have if $m = n$? If $m < n$? If $m > n$?

The following are not exercises, but are simply intended to stimulate thought and discussion. Use your imagination, along with your results from Exercises 50–60.

- What do you think the graph of a linear equation in four variables might look like?

- In a two- or three-dimensional coordinate system, all coordinate axes are mutually perpendicular. Do you think it is possible for four axes in a four-dimensional coordinate system to be mutually perpendicular? Why or why not?

- What might the intersection of the graphs of two linear equations in four variables look like? What about the intersection of three such graphs? Four?

- Give a geometric description of a solution to a system of linear equations in four variables.

If these ideas interest you, continue your study of mathematics long enough to take a course in linear algebra.

3–2 MATRIX SOLUTIONS OF SYSTEMS OF LINEAR EQUATIONS

AUGMENTED MATRIX OF A SYSTEM

You solved the following system in Exercise 15 of Section 3-1, probably by using Gaussian elimination.

$$0.25L + 0.15M + 0.20C = 24$$

$$0.65L + 0.80M + 0.65C = 72$$

$$0.10L + 0.05M + 0.15C = 12$$

In this section you will learn a more efficient method using the same principles. Begin by writing three columns containing the coefficients of L, M, and C and a fourth column containing the constant terms.

$$\begin{bmatrix} 0.25 & 0.15 & 0.20 & 24 \\ 0.65 & 0.80 & 0.65 & 72 \\ 0.10 & 0.05 & 0.15 & 12 \end{bmatrix}$$

The result is a **matrix**, that is, a rectangular array of numbers. The term was first used by James Joseph Sylvester (1814–1897) in 1850, although mathematicians had studied rectangular arrays since about a century earlier. The matrix shown here contains all the information in the original system and is called its **augmented matrix**. (The word *augmented* refers to the fact that the matrix containing the coefficients in the system is augmented by a column containing the constant terms.)

EXERCISES   *In Exercises 1–4, write the augmented matrix of each system of linear equations.*

1. $3x + 2y = 6$
   $x + 7y = 5$

2. $2A + 3B + 4C = 5$
   $6A - 7B + 8C = 9$

3. $2m + 3n = 5$
   $-m - 4n = 0$
   $3m = 12$
   $4n = -4$

4. $\frac{1}{3}U + 3V - 5W = \frac{4}{3}$
   $2U \quad - W = 7$
   $\frac{1}{5}U - 2V \quad = 2$

In Exercises 5–8, write a system of linear equations for each augmented matrix, using any symbols you like as variables.

5. $\begin{bmatrix} 23 & 6 & -2 \\ 7 & 31 & 12 \end{bmatrix}$

6. $\begin{bmatrix} 1 & 4 & -2 & 16 \\ 3 & 0 & 1 & -5 \end{bmatrix}$

7. $\begin{bmatrix} 1 & 3 & 9 & 5 \\ \frac{1}{3} & 4 & -13 & 7 \\ 0 & -2 & 3 & 16 \end{bmatrix}$

8. $\begin{array}{c} 0 \\ \\ 1.5 \\ 11 \end{array} \begin{bmatrix} 0 & 5 & 3 \\ -1 & 4 & 0.60 \\ -7 & 8 & 2.1 \\ -3 & -1 & 0.2 \end{bmatrix}$

MATRIX ROW OPERATIONS

The idea of using a matrix to solve a system of linear equations can be found in the Chinese text *K'ui-ch'ang Fuan-shu* (*Arithmetic in Nine Sections*), dating from the Han dynasty (206 B.C.–A.D. 222). In modern times Arthur Cayley (1821–1895) first used matrices to represent linear systems.

The method of solution you are about to learn, which was developed by the German engineer Wilhelm Jordan (1842–1899), depends on the fact that Gaussian elimination requires only three types of operations on the equations in a system.

- The order of the equations can be changed.
- Any equation can be multiplied by any nonzero constant.
- A multiple of any equation can be added to any other equation.

Each operation produces an equivalent system, that is, one with the same solutions. The corresponding operations on the rows of a matrix can be described as follows.

> The **matrix row operations** equivalent to the operations of Gaussian elimination are
>
> - to interchange two rows,
> - to multiply a row by a nonzero constant, and
> - to add a multiple of a row to another row.

Example 1 illustrates the use of matrix row operations, and shows how they correspond to the operations of Gaussian elimination.

EXAMPLE 1   *Performing row operations on a matrix*

Let's apply each of the three operations of Gaussian elimination to the system at the beginning of this section, and apply the corresponding row operations to its augmented matrix.

1. We can write the third equation first, and vice versa. Equivalently, we can interchange the first and third matrix rows.

System	Matrix
$0.10L + 0.05M + 0.15C = 12$ $0.65L + 0.80M + 0.65C = 72$ $0.25L + 0.15M + 0.20C = 24$	$\begin{bmatrix} 0.10 & 0.05 & 0.15 & 12 \\ 0.65 & 0.80 & 0.65 & 72 \\ 0.25 & 0.15 & 0.20 & 24 \end{bmatrix}$

2. We can multiply the first equation by 10 and the other two by 20. Equivalently, we can multiply the first matrix row by 10 and the other two by 20.

System	Matrix
$L + 0.5M + 1.5C = 120$ $13L + 16M + 13C = 1440$ $5L + 3M + 4C = 480$	$\begin{bmatrix} 1 & 0.5 & 1.5 & 120 \\ 13 & 16 & 13 & 1440 \\ 5 & 3 & 4 & 480 \end{bmatrix}$

3. We can add (-13) times the first equation to the second. Equivalently, we can add (-13) times the first matrix row to the second.

System	Matrix
$L + 0.5M + 1.5C = 120$ $9.5M - 6.5C = -120$ $5L + 3M + 4C = 480$	$\begin{bmatrix} 1 & 0.5 & 1.5 & 120 \\ 0 & 9.5 & -6.5 & -120 \\ 5 & 3 & 4 & 480 \end{bmatrix}$

The large numbers in the last system we obtained may seem to make it more cumbersome than the original one, but the first entry in the second row of the matrix is now zero. Thus we have eliminated the variable L from the second equation, so we have simplified the system.

GAUSS–JORDAN ELIMINATION

In performing row operations as we did in Example 1, our objective is to obtain a system and a matrix like this, with L_0, M_0, and C_0 representing numbers.

System	Matrix
$L \qquad\quad = L_0$ $\quad M \quad\; = M_0$ $\qquad C = C_0$	$\begin{bmatrix} 1 & 0 & 0 & L_0 \\ 0 & 1 & 0 & M_0 \\ 0 & 0 & 1 & C_0 \end{bmatrix}$

There are several ways we might do this. The following sequence of steps, called **Gauss–Jordan elimination**, is based on Gaussian elimination and always works. To understand the steps more clearly, read them along with Example 2.

STEP 1:   Choose a row having a nonzero entry as far to the left as possible. If this is not row 1, interchange it with row 1. Row 1 is now called the **pivot row**, and its first nonzero entry is the **pivot**.

STEP 2:   Multiply the pivot row by a constant so that the value of the pivot becomes 1.

STEP 3:   Add multiples of the pivot row to each of the other rows so that each number in the column of the pivot (except for the pivot itself) becomes 0.

STEP 4:   Choose a row below the previous pivot row having a nonzero entry as far to the left as possible. If this is not row 2, interchange it with row 2. Now row 2 is the pivot row, and its first nonzero entry is the pivot.

STEP 5:   Repeat steps 2–4 until you run out of rows.

EXAMPLE 2   *Solving a linear system by Gauss–Jordan elimination*

In Example 1 we performed row operations on the augmented matrix of the system at the beginning of this section, obtaining

System	Matrix
$L + 0.5M + 1.5C = \quad 120$ $\qquad\quad 9.5M - 6.5C = -120$ $5L + \;\; 3M + \;\; 4C = \quad 480$	$\begin{bmatrix} 1 & 0.5 & 1.5 & 120 \\ 0 & 9.5 & -6.5 & -120 \\ 5 & 3 & 4 & 480 \end{bmatrix}$

Let's continue to solve the system by Gauss–Jordan elimination. Since the first entry in the first row is 1, steps 1 and 2 have been completed. The first row is the pivot row, and the 1 is the pivot.

STEP 3:   The first entry in the second row is already 0. To continue we add (-5) times the first row to the third row.

System	Matrix
$L + 0.5M + 1.5C = 120$ $9.5M - 6.5C = -120$ $0.5M - 3.5C = -120$	$\begin{bmatrix} 1 & 0.5 & 1.5 & 120 \\ 0 & 9.5 & -6.5 & -120 \\ 0 & 0.5 & -3.5 & -120 \end{bmatrix}$

STEP 4:   We interchange the second and third rows. This is not necessary, since both the second and third rows have nonzero entries in the second column. However, the pivot now becomes 0.5 instead of 9.5, making the next step easier.

System	Matrix
$L + 0.5M + 1.5C = 120$ $0.5M - 3.5C = -120$ $9.5M - 6.5C = -120$	$\begin{bmatrix} 1 & 0.5 & 1.5 & 120 \\ 0 & 0.5 & -3.5 & -120 \\ 0 & 9.5 & -6.5 & -120 \end{bmatrix}$

Now we repeat steps 2–4.

STEP 2:   Multiply the second row by 2.

System	Matrix
$L + 0.5M + 1.5C = 120$ $M - 7C = -240$ $9.5M - 6.5C = -120$	$\begin{bmatrix} 1 & 0.5 & 1.5 & 120 \\ 0 & 1 & -7 & -240 \\ 0 & 9.5 & -6.5 & -120 \end{bmatrix}$

STEP 3:   Add (-0.5) times the second row to the first, and (-9.5) times the second row to the third.

System	Matrix
$L + 5C = 240$ $M - 7C = -240$ $60C = 2160$	$\begin{bmatrix} 1 & 0 & 5 & 240 \\ 0 & 1 & -7 & -240 \\ 0 & 0 & 60 & 2160 \end{bmatrix}$

STEP 4:   Since there is only one row below the pivot row, the 60 in the third row becomes the new pivot.

In Exercises 9–10 you can complete the process of Gauss–Jordan elimination for this system.   ■

EXERCISES   **9.** (*Making Observations*) What row operation was performed on the last matrix in Example 2 to produce the following matrix?

System	Matrix

$$\begin{array}{rcr} L + & 5C = & 240 \\ & M - 7C = & -240 \\ & C = & 36 \end{array} \qquad \begin{bmatrix} 1 & 0 & 5 & 240 \\ 0 & 1 & -7 & -240 \\ 0 & 0 & 1 & 36 \end{bmatrix}$$

10. (*Making Observations*) What two row operations were performed on the matrix in Exercise 9 to produce the following matrix?

System	Matrix

$$\begin{array}{rcr} L & = & 60 \\ M & = & 12 \\ C = & 36 \end{array} \qquad \begin{bmatrix} 1 & 0 & 0 & 60 \\ 0 & 1 & 0 & 12 \\ 0 & 0 & 1 & 36 \end{bmatrix}$$

The matrix in Exercise 10 allows us to read the solution to the system directly.

REDUCED ROW-ECHELON MATRICES

The matrix in Exercise 10 is an example of a **reduced row-echelon matrix**, that is, a matrix in which

- the **leading entry** (first nonzero entry) in each row has a value of 1,
- the leading entry in each row is to the right of the leading entry in any row above it,
- in the column containing the leading entry of each row, the other entries are all zeros, and
- any rows consisting entirely of zeros are at the bottom.

We will say that the corresponding linear system is in **reduced form**. Applying Gauss–Jordan elimination always produces a reduced row-echelon matrix and a system in reduced form.

EXAMPLE 3   *Reduced row-echelon matrices*

a. The matrix

$$\begin{bmatrix} \mathbf{1} & 0 & 2 & 0 & 3 \\ 0 & \mathbf{1} & 4 & 0 & 5 \\ 0 & 0 & 0 & \mathbf{1} & 6 \end{bmatrix}$$

is in reduced row-echelon form. The leading entry in each row is indicated in bold color.

b. The matrix

$$\begin{bmatrix} \mathbf{1} & 2 & 2 & 0 & 3 \\ 0 & \mathbf{1} & 4 & 0 & 5 \\ 0 & 0 & 0 & 0 & 0 \\ 0 & 0 & 0 & \mathbf{1} & 6 \end{bmatrix}$$

is not in reduced row-echelon form for two reasons. First, the leading entry in the second row has another nonzero entry above it. Second, the row of zeros is not at the bottom.   ■

EXERCISES   *In Exercises 11–14, decide whether the matrix is in reduced row-echelon form. If it is not, explain why.*

11. $\begin{bmatrix} 1 & 0 & 5 \\ 0 & 1 & 1 \\ 0 & 0 & 1 \end{bmatrix}$

12. $\begin{bmatrix} 1 & 0 & 5 \\ 0 & 1 & 1 \\ 0 & 1 & 0 \end{bmatrix}$

13. $\begin{bmatrix} 1 & 0 & 5 \\ 0 & 1 & 1 \\ 0 & 0 & 0 \end{bmatrix}$

14. $\begin{bmatrix} 5 & 0 & 1 \\ 0 & 1 & 1 \\ 0 & 0 & 0 \end{bmatrix}$

In Exercises 15–18, use Gauss–Jordan elimination to obtain reduced row-echelon matrices.

15. $\begin{bmatrix} 1 & -3 & 5 \\ 4 & 5 & 3 \end{bmatrix}$

16. $\begin{bmatrix} 1 & -1 & 2 & 1 & 3 \\ 4 & -4 & 8 & 6 & 7 \end{bmatrix}$

17. $\begin{bmatrix} 2 & 0 & -1 & -2 \\ 0 & 1 & -2 & 1 \\ 4 & -1 & 1 & 0 \end{bmatrix}$

18. $\begin{bmatrix} 2 & 0 & -1 & -2 \\ 0 & 1 & -2 & 1 \\ 4 & -1 & 0 & 0 \end{bmatrix}$

Use Gauss–Jordan elimination to solve each linear system in Exercises 19–22. The matrices in Exercises 19 and 21 are the ones you worked with in Exercises 15 and 17.

19. $\begin{aligned} x - 3y &= 5 \\ 4x + 5y &= 3 \end{aligned}$

20. $\begin{aligned} -0.5W + 1.4Z &= 11 \\ 2.5W + 3.2Z &= -4 \end{aligned}$

21. $\begin{aligned} 2x \quad\;\; - z &= -2 \\ y - 2z &= 1 \\ 4x - y + z &= 0 \end{aligned}$

22. $\begin{aligned} 3p + q + 4r &= 8 \\ 9p - 3q + 6r &= 12 \\ 2q - 4r &= -2 \end{aligned}$

MATRICES AND NUMBERS OF SOLUTIONS

As you saw in Section 3-1, a system of linear equations can have one, zero, or infinitely many solutions. Its reduced row-echelon matrix can tell us how many solutions it has. Let's return to Paul and Karen's landscaping company from **Sunsilk** (page 104) to see how.

Matrices of Systems with No Solutions   First, let's see what the reduced row-echelon matrix of a system with no solutions looks like.

A MATHEMATICAL
LOOKING GLASS
Sunsilk 2

In Exercise 48 of Section 3-1, you discovered that Paul and Karen could not make full use of everyone's available work time. They responded by making their operations more efficient. Since the lumberyard was directly on the route to most of their work, Paul picked up all carpentry supplies, relieving Mary of that responsibility. In return, Mary assumed Paul's responsibility for loading all plants onto the truck for landscaping jobs. This resulted in the following altered version of Table 16 from *Sunsilk* (Table 17).

TABLE 17

	Landscaping	Masonry	Carpentry
Karen	25%	15%	20%
Paul and helpers	60%	80%	70%
Mary	15%	5%	10%

Paul and his helpers still worked a combined total of 56 days per month, and Karen and Mary worked 24 and 8 days per month, respectively. ■

EXERCISES

23. (*Creating Models*) Let L, M, and C represent, respectively, the total days per month spent by all employees on landscaping, masonry, and carpentry. Write a system of linear equations that Paul must solve to make full use of everyone's time.

24. Use Gauss–Jordan elimination to verify that the reduced row-echelon matrix of your system is

$$\begin{bmatrix} 1 & 0 & 0.5 & 0 \\ 0 & 1 & 0.5 & 0 \\ 0 & 0 & 0 & 1 \end{bmatrix}$$

In the reduced form of the system, the third equation is $0 = 1$. Since no values of the variables can make this equation true, the system has no solution. This result can be generalized as follows.

> If a reduced linear system contains an equation $0 = c$ with $c \neq 0$, the system has no solution.
>
> Equivalently, if its reduced row-echelon matrix contains a row with its leading entry in the last column, the system has no solution.

EXERCISES

25. (*Making Observations*)
   a. Use Gauss–Jordan elimination to solve

   $$2x - 3y = 10$$
   $$4x - 6y = 15$$

   b. Use graphical methods to solve the same system.

   c. Describe how the graph and the reduced row-echelon matrix provide the same information about the system.

26. Use Gauss–Jordan elimination to solve the following system. The matrix is the one you worked with in Exercise 18.

$$2A - \quad C = -2$$
$$B - 2C = \quad 1$$
$$4A - B \quad = \quad 0$$

Matrices of Systems with Infinitely Many Solutions Next, let's see what the reduced row-echelon matrix of a system with infinitely many solutions looks like.

A MATHEMATICAL
LOOKING GLASS
Sunsilk 3

When we left Paul and Karen in *Sunsilk 2*, they had discovered that they could not make full use of everyone's available work time. A few months later, Karen became pregnant and decided to work only 16 days per month. Now their system of equations looked like this.

$$0.25L + 0.15M + 0.20C = 16$$
$$0.60L + 0.80M + 0.70C = 56$$
$$0.15L + 0.05M + 0.10C = \ 8$$ ■

EXERCISE   **27.** For the system in *Sunsilk 3*, use Gauss–Jordan elimination to obtain the reduced row-echelon matrix

$$\begin{bmatrix} 1 & 0 & 0.5 & 40 \\ 0 & 1 & 0.5 & 40 \\ 0 & 0 & 0 & 0 \end{bmatrix}$$

Writing the system in reduced form, we obtain

$$L + \quad 0.5C = 40$$
$$M + 0.5C = 40$$
$$0 \quad = \ 0$$

Each of L and M is a **leading variable** (the first variable in some equation), but C is not. Therefore, we can assign any real value to C and obtain a unique solution for L and M. Thus the system has infinitely many solutions.

To tell which values of L, M, and C are solutions, we can solve for the leading variables L and M in terms of C.

$$L = 40 - 0.5C$$
$$M = 40 - 0.5C$$

Thus if C is assigned any real value, say, c, then

$$L = 40 - \frac{1}{2}c$$

$$M = 40 - \frac{1}{2}c$$
$$C = c$$

is a solution to the system. However, we obtain a solution to the problem only if L, M, and C are all nonnegative. For example, if $c = 20$, then

$$L = 30$$
$$M = 30$$
$$C = 20$$

Thus, one possible way to make full use of everyone's time is to schedule 30 days of landscaping, 30 days of masonry, and 20 days of carpentry each month.

Our observations about the system in **Sunsilk 3** can be generalized as follows.

If a reduced linear system contains no equation $0 = c$ with $c \neq 0$, and has more variables than equations, then the system has infinitely many solutions.

Equivalently, if its reduced row-echelon matrix has no row with its leading entry in the last column, and if (disregarding rows of zeros) it has at least two more columns than rows, the system has infinitely many solutions.

Example 4 further illustrates the process of representing infinitely many solutions to a linear system.

EXAMPLE 4   *Solving a linear system with infinitely many solutions*

Suppose that the reduced row-echelon matrix of a system is

$$\begin{bmatrix} 1 & 0 & 1 & 0 & -1 & 10 \\ 0 & 1 & 1 & 0 & -1 & 1 \\ 0 & 0 & 0 & 1 & 1 & -1 \\ 0 & 0 & 0 & 0 & 0 & 0 \end{bmatrix}$$

The matrix has no row with its leading entry in the last column, and disregarding the row of zeros, it has three rows and six columns. Therefore, the system has infinitely many solutions. To represent them, we can begin by writing the system in reduced form. If the variables are u, v, x, y, and z, we obtain

$$\begin{aligned} u + \quad x - \quad z &= 10 \\ v + x - \quad z &= 1 \\ y + z &= -1 \end{aligned}$$

Since x and z are not leading variables, they can be assigned any real value. The system then has unique solutions for u, v, and y. If x and z are assigned values of x_0 and z_0, then

$$\begin{aligned} u &= 10 - x_0 + z_0 \\ v &= 1 - x_0 + z_0 \\ x &= x_0 \\ y &= -1 \quad\; - z_0 \\ z &= z_0 \end{aligned}$$

When all variables are expressed in terms of the nonleading variables, as they are here, the solution is said to be written in **standard form**. ■

EXERCISES   **28.** (*Making Observations*)

   a. Use Gauss–Jordan elimination to solve

   $$2x - 3y = 10$$
   $$4x - 6y = 20$$

   b. Use graphical methods to solve the system.

   c. Describe how the graph and the row-echelon matrix provide the same information about the system.

In Exercises 29–32, write the reduced linear system and write the solutions in standard form. Use any symbols you like as variables.

29. $\begin{bmatrix} 1 & 0 & 2 & 8 \\ 0 & 1 & 4 & 7 \end{bmatrix}$

30. $\begin{bmatrix} 1 & -1 & 0 & 0 & 12 \\ 0 & 0 & 1 & 3 & 17 \end{bmatrix}$

31. $\begin{bmatrix} 1 & 0 & 10 & 0 & 0.5 \\ 0 & 1 & 20 & 0 & 2.5 \\ 0 & 0 & 0 & 1 & 4 \end{bmatrix}$

32. $\begin{bmatrix} 1 & 0 & 0 & 0 & 3 & 0 \\ 0 & 1 & 0 & 0 & 9 & 0 \\ 0 & 0 & 1 & 0 & 5 & 0 \\ 0 & 0 & 0 & 1 & 7 & 0 \end{bmatrix}$

In Exercises 33–36, solve the system and write the solutions in standard form.

33. $A + B + C + D = 1$
   $\phantom{A + {}}B + C + D = 2$
   $\phantom{A + B + {}}C + D = 3$

34. $x - 3y + 5z = 2$
   $2x \phantom{{}- 3y} - 4z = -3$

35. $p - q + r + s = 0$
   $p + q + r - s = 0$
   $\phantom{p + {}}q - r - s = 0$

36. $p - q + r + s = 0$
   $p + q + r - s = 0$

Matrices of Systems with Unique Solutions The reduced linear system in Exercise 10 has a unique solution because it contains exactly one equation for each variable, assigning a unique value to that variable. This observation can be generalized as follows.

> If a reduced linear system contains no equation $0 = c$ with $c \neq 0$, and has exactly as many variables as equations, then the system has a unique solution.
>
> Equivalently, if its reduced row-echelon matrix has no row with its leading entry in the last column, and if (disregarding rows of zeros) it has exactly one more column than it has rows, the system has a unique solution.

SUMMARY: MATRIX VERSUS NONMATRIX METHODS

Although matrices are more abstract than linear systems, it is useful to learn matrix methods for several reasons.

- With sufficient practice, most people can solve linear systems more quickly with matrices than without.
- Linear systems with infinitely many solutions are usually awkward to solve without matrices.
- Calculators and computers can deal with linear systems in matrix form. In particular, you can do Gauss–Jordan elimination on your calculator.

ADDITIONAL EXERCISES   *In Exercises 37–40, write the augmented matrix of the system of equations.*

37.
$$P + Q + R + S = 0$$
$$3P - Q + 19R + 6S = \frac{1}{7}$$
$$2P = 8$$

38.
$$-4A + 7B = 13$$
$$2A - 15B = 58$$
$$A + B = 91$$

39.
$$2J - 3K - 4L - 5M = 0.001$$
$$2K - 3L - 4M - 5N = -0.001$$

40.
$$w + x - y = 3$$
$$w - x + z = 5$$
$$-w + y + z = 9$$
$$x + y - z = 1$$

In Exercises 41–44, decide whether the matrix is in reduced row-echelon form. If it is not, explain why.

41.
$$\begin{bmatrix} 1 & 2 & 3 & 4 \\ 0 & 0 & 0 & 1 \\ 0 & 0 & 0 & 0 \\ 0 & 0 & 0 & 0 \end{bmatrix}$$

42.
$$\begin{bmatrix} 1 & 0 & 0 & 0 \\ 0 & 0 & 0 & 0 \\ 0 & 1 & 0 & 1 \\ 0 & 0 & 1 & 4 \end{bmatrix}$$

43.
$$\begin{bmatrix} 0 & 1 & 0 & 0 \\ 0 & 0 & 1 & 0 \\ 0 & 0 & 0 & 1 \end{bmatrix}$$

44.
$$\begin{bmatrix} 1 & 1 & 0 & 1 & 1 & 1 \\ 0 & 0 & 1 & 1 & 1 & 1 \end{bmatrix}$$

In Exercises 45–48, obtain reduced row-echelon matrices.

45.
$$\begin{bmatrix} 2 & -7 & 6 \\ 5 & -1 & 4 \end{bmatrix}$$
   46.
$$\begin{bmatrix} 1 & 3 & 4 \\ 10 & 30 & 40 \end{bmatrix}$$

47. $\begin{bmatrix} 1 & 2 & 3 & 0 \\ 1 & 2 & 1 & 0 \\ 2 & 1 & 1 & 0 \end{bmatrix}$

48. $\begin{bmatrix} 1 & 2 & 3 & 0 \\ 1 & 2 & 3 & 1 \\ 1 & 2 & 3 & 2 \end{bmatrix}$

Solve the systems in Exercises 49–56 by Gauss–Jordan elimination.

49. $\begin{aligned} x_1 - 4x_2 &= 6 \\ 4x_1 - 16x_2 &= 18 \end{aligned}$

50. $\begin{aligned} x_1 - 4x_2 &= 6 \\ 4x_1 - 16x_2 &= 24 \end{aligned}$

51. $\begin{aligned} 2x + y + z &= 5 \\ 4x - y + 3z &= 4 \end{aligned}$

52. $\begin{aligned} P - 3Q + 2R &= 4 \\ 3P - 7Q + 5R &= 13 \end{aligned}$

53. $\begin{aligned} R - 2S + T &= 9 \\ R - 2S - T &= 5 \end{aligned}$

54. $\begin{aligned} 3J + 2K - M - N &= 12 \\ J \quad\quad + 2M - N &= 8 \end{aligned}$

55. $\begin{aligned} A + B + C &= 0 \\ 2A + B + C &= 1 \\ A - 2B - 2C &= 0 \\ A + B - C &= 1 \end{aligned}$

56. $\begin{aligned} A + B + C &= 0 \\ A + B - C &= 0 \\ A - B + C &= 0 \\ -A + B + C &= 0 \end{aligned}$

In Exercises 57–60, write the reduced linear system and write the solutions in standard form. Use any symbols you like as variables.

57. $\begin{bmatrix} 1 & 0 & 0 & -2 \\ 0 & 1 & 0.1 & 4 \end{bmatrix}$

58. $\begin{bmatrix} 1 & 0 & 0 & -6 & 1 \\ 0 & 1 & 0 & -6 & 1 \\ 0 & 0 & 1 & -6 & 1 \end{bmatrix}$

59. $\begin{bmatrix} 1 & 0 & 3 & 0 & 0 & 25 \\ 0 & 1 & 1 & 0 & 0 & 0 \\ 0 & 0 & 0 & 1 & 5 & 15 \\ 0 & 0 & 0 & 0 & 0 & 0 \end{bmatrix}$

60. $\begin{bmatrix} 1 & 0 & 0 & -1 & 0 & 0 & 1 \\ 0 & 1 & 0 & 0 & -1 & 0 & 1 \\ 0 & 0 & 1 & 0 & 0 & -1 & 1 \end{bmatrix}$

61. (*Problem Solving*) Nutritionists often need to plan meals for persons with special dietary needs. The process involves analyzing the nutritional content of a large number of combinations of foods. In this simplified example, you can plan a meal using the foods listed in the following table.

	Grams fat	Grams carbohydrates	Grams protein	Milligrams cholesterol	Calories
1 baked potato	0.2	32.8	4.0	0	145
1 stalk broccoli	0.5	8.1	5.6	0	47
4 oz. salmon	6.4	0	30.8	40	192
Small slice apple pie	5.5	22.0	1.1	0	139
1 cup 2% milk	2.6	11.7	8.0	10	102

Adapted from Dean Ornish, *Eat More, Weigh Less*, HarperPerennial, 1993.

Define the variables

P = the number of baked potatoes in the meal

B = the number of stalks of broccoli

S = the number of 4-ounce portions of salmon

A = the number of slices of apple pie

M = the number of cups of 2% milk

a. Write a system of linear equations that must be satisfied if the meal is to contain 15 grams of fat and 50 mg of cholesterol.

b. Solve the system and write the solution in standard form.

c. Verify that two solutions to the system are

$$(P, B, S, A, M) = (0, 1, 1, 1, 1)$$

and

$$(P, B, S, A, M) = (27.5, 1, 1, 0, 1).$$

Describe the meals represented by each solution.

d. By trial and error, find several other solutions with nonnegative values for all the variables. Would any of these solutions make reasonable meals?

e. Identify a few reasons why a nutritionist would probably not plan a meal by solving a system of linear equations.

62. (*Writing to Learn*) In **Sunsilk 3**, C is not a leading variable in the system of equations.

a. Can you rewrite the system in such a way that C is a leading variable? If so, how? If not, why not?

b. Can you rewrite the system in such a way that all the variables are leading variables? If so, how? If not, why not?

63. (*Extension*) Consult the owner's manual for your calculator to learn the proper keystrokes for performing matrix row operations. Then solve each of the systems in Exercises 19–22 on your calculator.

64. (*Extension*) Using the five foods listed in Exercise 61, is it possible to plan a meal having 15 grams of fat, 100 grams of carbohydrates, 50 grams of protein, 50 milligrams of cholesterol, and 600 calories? (*Hint*: Do the matrix row operations on your calculator.)

3-3 DETERMINANTS AND CRAMER'S RULE

In Section 3-2, you learned how to solve a system of linear equations by manipulating the rectangular array of numbers in the system. In this section, you will use the same array of numbers to develop a different method of solution, known as Cramer's Rule. Although it is limited in that it applies only to "square" systems (2×2, 3×3, and so on), Cramer's Rule has at least three advantages.

- You can use it to determine whether a system has a unique solution without actually solving the system.

- You can use it to solve for any one variable in a system without solving the entire system.

- Your calculator can perform the required operations with a few keystrokes.

In the following Mathematical Looking Glass, it is important to know whether a system of linear equations will have a unique solution even before the system is constructed.

A MATHEMATICAL LOOKING GLASS

Rush Hour

Interstate highways in urban areas, such as the Cross-Bronx Expressway in New York or the Santa Monica Freeway in Los Angeles, are designed to carry large volumes of traffic at high speeds. Entrance ramps to such highways are high-risk areas for accidents, as well as a source of congestion during morning and evening rush hours. Figure 3-6a shows a ramp at which entering cars must merge into a lane of faster traffic. The ramp shown in Figure 3-6b allows cars to enter more safely and more quickly by providing a separate lane in which they can continue. However, construction of an extra lane makes the ramp in Figure 3-6b more expensive.

(a)   (b)

Figure 3–6

In designing a highway, engineers often analyze traffic flow on existing roads to decide whether the volume of traffic at a particular ramp will justify the extra expense. The analysis is often carried out by placing devices to count vehicles at strategic points.

Figure 3-7 shows two one-way streets connected by ramps. Eastbound drivers at point A can either continue straight to B or turn north to D. Similarly, northbound drivers at C can either continue straight to D or turn east to B.

Engineers with the state department of transportation are trying to determine the values of the quantities

x_1 = the number of vehicles proceeding from A to B

x_2 = the number of vehicles proceeding from A to D

y_1 = the number of vehicles proceeding from C to B

y_2 = the number of vehicles proceeding from C to D

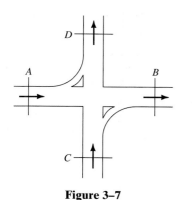

Figure 3–7

during the period between 5:00 P.M. and 6:00 P.M. on a weekday. It has been proposed to place counters at A, B, C, and D. Before placing them, the engineers need to know whether the data they gather will provide them with the information they need. ■

EXERCISES

1. (*Creating Models*) Suppose that the numbers of vehicles passing the counters at A, B, C, and D are a, b, c, and d, respectively. Write a system of linear equations that must be solved to determine the values of x_1, x_2, y_1, and y_2.

2. (*Writing to Learn*) Why is it important to determine whether the system will have a unique solution before placing the counters?

In Exercises 46 and 47, you can decide whether the proposed placement of counters will provide the required information. In the meantime, let's develop the mathematical tools you will need for the job.

SECOND-ORDER DETERMINANTS AND CRAMER'S RULE FOR 2 × 2 SYSTEMS

In Exercises 3–5, you can begin to develop Cramer's Rule by seeing how the solutions to a 2 × 2 linear system are related to the numbers in the system.

EXERCISES   **3.** (*Making Observations*) In the system

$$a_1 x + b_1 y = c_1$$
$$a_2 x + b_2 y = c_2$$

multiply the first equation by b_2 and the second by $-b_1$. Then add the equations and solve for x to verify that $x = \dfrac{c_1 b_2 - b_1 c_2}{a_1 b_2 - b_1 a_2}$.

4. (*Making Observations*) Solve the system in Exercise 3 for y.

5. (*Making Observations*) Explain why the solutions obtained in Exercise 3 are not valid if $a_1 b_2 - b_1 a_2 = 0$.

6. (*Making Observations*) According to your work in Exercises 3–5, the solution to the system

$$3x - 7y = 5$$
$$2x - 4y = 3$$

is

$$x = \frac{(5)(-4) - (3)(-7)}{(3)(-4) - (2)(-7)}, \qquad y = \frac{(3)(3) - (2)(5)}{(3)(-4) - (2)(-7)}$$

Verify that this is actually the solution.

Your results from Exercises 3–5 can be stated as a convenient rule for solving 2 × 2 linear systems. We can begin to develop the rule by *Looking for a Pattern* in Exercises 3 and 4. The denominator of each solution involves the numbers in the **coefficient matrix** of the system, that is, the array

$$\begin{bmatrix} a_1 & b_1 \\ a_2 & b_2 \end{bmatrix}$$

Each denominator is the difference between the diagonal products $a_1 b_2$ and $b_1 a_2$ in the coefficient matrix. This number is called the **determinant** of the matrix. It is also sometimes called a **second-order** (or **2 × 2**) **determinant**. Specifically, the determinant of a 2 × 2 matrix

$$\begin{bmatrix} a & b \\ c & d \end{bmatrix}$$

is the number $ad - bc$ and is written

$$\begin{vmatrix} a & b \\ c & d \end{vmatrix}$$

Determinants have been used to study linear systems since their invention in 1693 by Gottfried Leibniz. Augustin-Louis Cauchy (1789–1857) first arranged the elements of a determinant in a square array, and Arthur Cayley (1821–1895) introduced the vertical line notation.

EXAMPLE 1   *Evaluating a 2 × 2 determinant*

The value of

$$\begin{vmatrix} 8 & 5 \\ 3 & 4 \end{vmatrix}$$

is $(8)(4) - (5)(3) = 17$.   ∎

EXERCISES   *In Exercises 7–10, evaluate the determinant.*

7. $\begin{vmatrix} 2 & -3 \\ 6 & 11 \end{vmatrix}$ **8.** $\begin{vmatrix} 6 & 11 \\ 2 & -3 \end{vmatrix}$

9. $\begin{vmatrix} -4 & 2 \\ 2 & -1 \end{vmatrix}$ **10.** $\begin{vmatrix} 0 & 7 \\ 0 & 1 \end{vmatrix}$

Your results from Exercises 3–5 can now be stated as

Cramer's Rule for 2 × 2 Linear Systems

Given a 2 × 2 system of linear equations

$$a_1 x + b_1 y = c_1$$
$$a_2 x + b_2 y = c_2$$

define the determinants

$$D = \begin{vmatrix} a_1 & b_1 \\ a_2 & b_2 \end{vmatrix}, \qquad D_x = \begin{vmatrix} c_1 & b_1 \\ c_2 & b_2 \end{vmatrix}, \qquad D_y = \begin{vmatrix} a_1 & c_1 \\ a_2 & c_2 \end{vmatrix}$$

The system has a unique solution if and only if $D \neq 0$. In that case, the solution is

$$x = \frac{D_x}{D}, \qquad y = \frac{D_y}{D}$$

The determinant D_x was obtained by substituting the constant terms c_1 and c_2 for the x-coefficients a_1 and a_2 in D. The determinant D_y was obtained similarly.

EXAMPLE 2   *Using Cramer's Rule to solve a 2 × 2 system of linear equations*

You encountered the system of equations

$$3x - 7y = 5$$
$$2x - 4y = 3$$

in Exercise 6. For this system,

$$D = \begin{vmatrix} 3 & -7 \\ 2 & -4 \end{vmatrix} = (3)(-4) - (-7)(2) = 2$$

$$D_x = \begin{vmatrix} 5 & -7 \\ 2 & -4 \end{vmatrix} = (5)(-4) - (-7)(3) = 1$$

$$D_y = \begin{vmatrix} 3 & 5 \\ 2 & 3 \end{vmatrix} = (3)(3) - (5)(2) = -1$$

The solution to the system is $x = \dfrac{D_x}{D} = \dfrac{1}{2}, y = \dfrac{D_y}{D} = -\dfrac{1}{2}.$ ■

It is interesting to note that Cramer's Rule was actually developed by the Scottish mathematician Colin Maclaurin (1698–1746) around 1729 and was included in his *Treatise of Algebra* (1748). However, the rule was also described by Gabriel Cramer (1704–1752) in his widely read *Introduction à l'analyse des lignes courbes algébriques* (1750). Although Cramer did not claim credit for the rule, it came to be named after him. Incidentally, Maclaurin was the beneficiary of a similar historical error. Second-semester calculus students who learn about Maclaurin series are studying an idea published by Maclaurin but developed earlier by Brook Taylor (1685–1731).

EXERCISES   *In Exercises 11–14, solve the system by using Cramer's Rule.*

11. $3x + 4y = 17$
$\quad\ \ 2x + 3y = 12$

12. $5x - 6y = \ \ \ 24$
$\quad\ \ 5x + 4y = -16$

13. $\ \ \ s + 3t = 10$
$\quad\ \ 3s - \ \ t = \ \ 4$

14. $4u - \ \ v = -1$
$\quad\ \ 7u - 3v = \ \ 20$

In Exercises 15 *and* 16,

a. Verify that $D = 0$, so that Cramer's Rule cannot be used to solve the system.

b. Use a method of your choice to verify that the system does not have a unique solution. How many solutions does it have?

15. $8x + 6y = 3$
$\quad\ \ 4x + 3y = 2$

16. $\quad\ \ 2x - \ \ 5y = -8$
$\quad\ -6x + 15y = \ \ 24$

THIRD-ORDER DETERMINANTS AND CRAMER'S RULE FOR 3 × 3 SYSTEMS

Can we extend Cramer's Rule to apply to 3 × 3 linear systems? To do so, we will have to define the determinant of a 3 × 3 matrix in a suitable way. Over the next few pages, you can discover the appropriate definition and develop Cramer's Rule for 3 × 3 systems.

Let's use the symbol a_{ij} to denote the number in row i and column j of a 3 × 3 matrix. Then we can write the matrix

$$A = \begin{bmatrix} a_{11} & a_{12} & a_{13} \\ a_{21} & a_{22} & a_{23} \\ a_{31} & a_{32} & a_{33} \end{bmatrix}$$

The **minor** of the element a_{ij} is the determinant M_{ij} of the matrix obtained from A by deleting the row and column containing a_{ij}. For example, if

$$A = \begin{bmatrix} 7 & -3 & 1 \\ 4 & 0 & 5 \\ -1 & 2 & 6 \end{bmatrix}$$

the minor M_{11} of a_{11} is the determinant of the matrix obtained from A by deleting row 1 and column 1, as shown below.

$$\begin{vmatrix} 7 & 3 & 1 \\ 4 & 0 & 5 \\ -1 & 2 & 6 \end{vmatrix}$$

Thus,

$$M_{32} = \begin{vmatrix} 0 & 5 \\ 2 & 6 \end{vmatrix} = (0)(5) - (2)(6) = -12$$

The **cofactor** of a_{ij} is defined to be $C_{ij} = (-1)^{i+j} M_{ij}$. Thus, for the matrix A just given, the cofactor of a_{11} is $C_{11} = (-1)^{1+1} M_{11} = (1)(10) = 10$. You can think of the cofactor C_{ij} as the minor M_{ij} multiplied by ± 1, according to the checkerboard pattern shown below.

$$\begin{vmatrix} + & - & + \\ - & + & - \\ + & - & + \end{vmatrix}$$

EXAMPLE 3   *Minors and cofactors in a 3 × 3 determinant*

Some other minors and cofactors in the determinant

$$\begin{vmatrix} 7 & -3 & 1 \\ 4 & 0 & 5 \\ -1 & 2 & 6 \end{vmatrix}$$

are

$$M_{12} = \begin{vmatrix} 4 & 5 \\ -1 & 6 \end{vmatrix} = (4)(6) - (5)(-1) = 29$$

$$C_{12} = (-1)^{1+2} M_{12} = (-1)(29) = -29$$

$$M_{13} = \begin{vmatrix} 4 & 0 \\ -1 & 2 \end{vmatrix} = (4)(2) - (0)(-1) = 8$$

$$C_{13} = (-1)^{1+3} M_{13} = (1)(8) = 8$$

EXERCISES   *In Exercises 17–20, refer to the determinant in Example 3.*

> **a.** Write the given minor M_{ij} as a determinant.
> **b.** Find its value.
> **c.** Find the value of the corresponding cofactor C_{ij}.

17. M_{21} **18.** M_{22}

19. M_{31} **20.** M_{32}

In Exercises 21 and 22, you can see how cofactors of 3×3 matrices are related to solutions of 3×3 linear systems.

In the system

$$a_{11}x + a_{12}y + a_{13}z = c_1$$

$$a_{21}x + a_{22}y + a_{23}z = c_2$$

$$a_{31}x + a_{32}y + a_{33}z = c_3$$

it is possible to solve for x by combining the equations as you did for the 2×2 system in Exercise 3. It can be shown that the value of x is

$$\frac{c_1 a_{22} a_{33} - c_1 a_{23} a_{32} - a_{12} c_2 a_{33} + a_{12} c_3 a_{23} + a_{13} c_2 a_{32} - a_{13} c_3 a_{22}}{a_{11} a_{22} a_{33} - a_{11} a_{23} a_{32} - a_{12} a_{21} a_{33} + a_{12} a_{31} a_{23} + a_{13} a_{21} a_{32} - a_{13} a_{31} a_{22}}$$

21. (*Making Observations*) Show that the denominator of the solution for x is equal to $a_{11}C_{11} + a_{12}C_{12} + a_{13}C_{13}$ in the coefficient matrix

$$A = \begin{bmatrix} a_{11} & a_{12} & a_{13} \\ a_{21} & a_{22} & a_{23} \\ a_{31} & a_{32} & a_{33} \end{bmatrix}$$

22. (*Making Observations*) Show that the numerator of the solution for x is equal to $c_1 C_{11} + a_{12}C_{12} + a_{13}C_{13}$ in the matrix

$$A_x = \begin{bmatrix} c_1 & a_{12} & a_{13} \\ c_2 & a_{22} & a_{23} \\ c_3 & a_{32} & a_{33} \end{bmatrix}$$

Exercises 21 and 22 suggest an appropriate way to define a 3×3 determinant and extend Cramer's Rule to apply to 3×3 linear systems. The **third-order** (or **3 × 3**) **determinant** of the matrix

$$A = \begin{bmatrix} a_{11} & a_{12} & a_{13} \\ a_{21} & a_{22} & a_{23} \\ a_{31} & a_{32} & a_{33} \end{bmatrix}$$

is defined as the number $a_{11}C_{11} + a_{12}C_{12} + a_{13}C_{13}$. In words, the determinant is equal to the sum of each element in the first row multiplied by its cofactor.

EXAMPLE 4   *Evaluating a 3 × 3 determinant*

The value of

$$\begin{vmatrix} 7 & -3 & 1 \\ 4 & 0 & 5 \\ -1 & 2 & 6 \end{vmatrix}$$

is

$$(7)(-1)^{1+1}\begin{vmatrix} 0 & 5 \\ 2 & 6 \end{vmatrix} + (-3)(-1)^{1+2}\begin{vmatrix} 4 & 5 \\ -1 & 6 \end{vmatrix} + (1)(-1)^{1+3}\begin{vmatrix} 4 & 0 \\ -1 & 2 \end{vmatrix}$$

$$= (7)(1)(0 - 10) + (-3)(-1)[24 - (-5)] + (1)(1)(8 - 0) = 25 \quad \blacksquare$$

EXERCISES   *In Exercises 23–26, evaluate the determinant.*

23. $\begin{vmatrix} 1 & 4 & 3 \\ 2 & -2 & 2 \\ 1 & -2 & -1 \end{vmatrix}$ **24.** $\begin{vmatrix} -5 & 1 & 1 \\ 6 & 2 & 0 \\ 1 & 2 & 0 \end{vmatrix}$

25. $\begin{vmatrix} 4 & 17 & -4 \\ 0 & 3 & 0 \\ 2 & 23 & -1 \end{vmatrix}$ **26.** $\begin{vmatrix} 2 & -1 & 3 \\ 5 & 3 & 2 \\ 3 & 4 & -1 \end{vmatrix}$

The definition of a 3×3 determinant, coupled with your results from Exercises 21 and 22, allow us to state

---

Cramer's Rule for 3×3 Linear Systems

Given a 3×3 system of linear equations

$$a_{11}x + a_{12}y + a_{13}z = c_1$$
$$a_{21}x + a_{22}y + a_{23}z = c_2$$
$$a_{31}x + a_{32}y + a_{33}z = c_3$$

let D be the determinant of the coefficient matrix A. Let D_x be the determinant of the matrix obtained from A by substituting c_1, c_2, and c_3 for the coefficients of x, and let D_y and D_z be defined similarly. Then the system has a unique solution if and only if $D \neq 0$. In that case, the solution is

$$x = \frac{D_x}{D}, \qquad y = \frac{D_y}{D}, \qquad z = \frac{D_z}{D}$$

EXERCISES   *In Exercises 27–30, solve the system by using Cramer's Rule. (You solved the same systems by other methods in Exercises 31–34 of Section 3-1.)*

27. $\begin{aligned} x - 2y &= 8 \\ 2x + 3y - 3z &= 8 \\ x + 8y - 2z &= 8 \end{aligned}$ **28.** $\begin{aligned} 3a - 2b - c &= 0 \\ a + 3b - 5c &= 1 \\ 2a - 7b + 3c &= 2 \end{aligned}$

29. $\begin{aligned} 5p + 2q + r &= 0.3 \\ 2p - q + 3r &= 0.1 \\ 3p - 2q + 6r &= 0.3 \end{aligned}$ **30.** $\begin{aligned} x - 0.1y + 30z &= 6 \\ x + 0.3y - 25z &= 3 \\ 2x + 0.2y - 15z &= 5 \end{aligned}$

ALTERNATE METHODS OF EVALUATING A 3 × 3 DETERMINANT

So far, you have been evaluating 3×3 determinants by using the process specified by the definition. You can sometimes simplify the calculation involved by using either of two alternate methods of evaluation. Let's look at each in turn.

Expanding Along Any Row or Column It can be shown that the value of a 3×3 determinant is equal to the sum of each element in *any row or column* multiplied by its cofactor. This process is referred to as **expanding** the determinant **along a row or column**. For example, the determinant in Example 4 could be expanded along the second column by evaluating

$$a_{12}C_{12} + a_{22}C_{22} + a_{32}C_{32}.$$

EXERCISES

31. (*Making Observations*) Expand the determinant in Example 4 along the second column, and verify that the value is the same as that obtained in Example 4.

32. (*Making Observations*) Expand the determinant in Exercise 23 along the third row, and verify that the value is the same as that obtained in Exercise 23.

33. (*Making Observations*) Expand the determinant in Exercise 24 along the third column. How did the choice of this column simplify the calculation?

34. (*Making Observations*) Evaluate the determinant in Exercise 25 by choosing a row or column to make the calculation as simple as possible.

The Criss-Cross Method In Exercise 35, you can verify that the following procedure produces the correct value of a 3×3 determinant.

- Beginning with a 3×3 matrix, recopy the first two columns to the right of the matrix, as shown below.

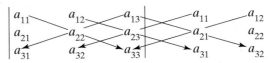

- Add together the products of the elements along each diagonal indicated by the arrows pointing toward the right to obtain

$$a_{11}a_{22}a_{33} + a_{12}a_{23}a_{31} + a_{13}a_{21}a_{32}.$$

- Add together the products of the elements along each diagonal indicated by the arrows pointing toward the left to obtain

$$a_{13}a_{22}a_{31} + a_{11}a_{23}a_{32} + a_{12}a_{21}a_{33}.$$

- Subtract the second sum from the first to obtain

$$(a_{11}a_{22}a_{33} + a_{12}a_{23}a_{31} + a_{13}a_{21}a_{32}) - (a_{13}a_{22}a_{31} + a_{11}a_{23}a_{32} + a_{12}a_{21}a_{33})$$

EXAMPLE 5   *Evaluating a 3×3 determinant by the criss-cross method*

To evaluate

$$\begin{vmatrix} 2 & 4 & 1 \\ -3 & 6 & 2 \\ 8 & 0 & 5 \end{vmatrix}$$

begin by recopying the first two columns as shown.

The value of the determinant is

$$[(2)(6)(5) + (4)(2)(8) + (1)(-3)(0)] - [(1)(6)(8) + (2)(2)(0) + (4)(-3)(5)]$$
$$= (60 + 64 + 0) - [48 + 0 + (-60)] = 136 \quad \blacksquare$$

EXERCISES   **35.** (*Making Observations*) Show that the criss-cross method produces the correct value for a 3×3 determinant by showing that the expression

$$(a_{11}a_{22}a_{33} + a_{12}a_{23}a_{31} + a_{13}a_{21}a_{32}) - (a_{13}a_{22}a_{31} + a_{11}a_{23}a_{32} + a_{12}a_{21}a_{33})$$

obtained by the criss-cross method is equal to $a_{11}C_{11} + a_{12}C_{12} + a_{13}C_{13}$.

36. (*Making Observations*) Evaluate the determinant in Example 5 by expanding along a row or column of your choice, and verify that your result is equal to that obtained in Example 5.

In Exercises 37–40, evaluate the determinant by the criss-cross method. You evaluated them by expanding along the first row in Exercises 23–26.

37. $\begin{vmatrix} 1 & 4 & 3 \\ 2 & -2 & 2 \\ 1 & -2 & -1 \end{vmatrix}$ **38.** $\begin{vmatrix} -5 & 1 & 1 \\ 6 & 2 & 0 \\ 1 & 2 & 0 \end{vmatrix}$

39. $\begin{vmatrix} 4 & 17 & -4 \\ 0 & 3 & 0 \\ 2 & 23 & -1 \end{vmatrix}$ **40.** $\begin{vmatrix} 2 & -1 & 3 \\ 5 & 3 & 2 \\ 3 & 4 & -1 \end{vmatrix}$

HIGHER-ORDER DETERMINANTS AND CRAMER'S RULE FOR LARGER SYSTEMS

For an $n \times n$ matrix with $n \geqslant 4$, the minor and the cofactor of an element are defined in exactly the same way as in the 3×3 case. The **determinant** of an $n \times n$ matrix is defined to be $a_{11}C_{11} + a_{12}C_{12} + \cdots + a_{1n}C_{1n}$, the sum of each element in the first row multiplied by its cofactor.

FYI   The criss-cross method used for 3×3 determinants does *not* produce correct results for determinants of larger size. However, an $n \times n$ determinant can always be evaluated by expanding along any row or column. Example 6 shows that a thoughtful choice can often simplify the calculation.

EXAMPLE 6   *Evaluating a 4×4 determinant*

To evaluate

$$\begin{vmatrix} 3 & 4 & -2 & 0 \\ 0 & 1 & 1 & 5 \\ 0 & -3 & 1 & 1 \\ 1 & 2 & 6 & -1 \end{vmatrix}$$

let's expand along the first column. Since that column contains two zeros, we have to calculate only the cofactors C_{11} and C_{41}. The value of the determinant is

$$(3)(-1)^{1+1}\begin{vmatrix} 1 & 1 & 5 \\ -3 & 1 & 1 \\ 2 & 6 & -1 \end{vmatrix} + (1)(-1)^{4+1}\begin{vmatrix} 4 & -2 & 0 \\ 1 & 1 & 5 \\ -3 & 1 & 1 \end{vmatrix}$$

$$= (3)(1)(-108) + (1)(-1)(16) = -340 \quad \blacksquare$$

EXERCISES   **41.** (*Making Observations*) How would the calculation in Example 6 be made more difficult if you expanded the determinant along the second column?

In Exercises 42 and 43, evaluate the determinant by expanding along a row or column of your choice.

42. $\begin{vmatrix} 1 & 3 & 3 & 0 \\ 0 & 2 & 0 & 1 \\ -1 & 0 & 0 & 2 \\ 1 & 6 & 4 & 1 \end{vmatrix}$

43. $\begin{vmatrix} 3 & 5 & 2 & 1 \\ 1 & 0 & 1 & 4 \\ 4 & 5 & 3 & 5 \\ 2 & 0 & 1 & 0 \end{vmatrix}$

It can be shown that Cramer's Rule can be extended in a natural way to apply to $n \times n$ linear systems for $n \geqslant 4$. Specifically, if D is the determinant of the coefficient matrix A, then the system has a unique solution if and only if $D \neq 0$. In that case, if x is any variable in the system, let D_x be the determinant of the matrix obtained from A by substituting each constant term in the system for the corresponding coefficient of x. Then $x = \dfrac{D_x}{D}$.

EXERCISES   In Exercises 44 and 45, use Cramer's Rule to decide whether the system has a unique solution. If it does, find the solution for y. (You evaluated the determinant of the coefficient matrices in Exercises 42 and 43.)

44. $\begin{aligned} w + 3x + 3y \qquad\quad &= 16 \\ 2x \qquad\; + z &= 8 \\ -w \qquad\qquad + 2z &= 7 \\ w + 6x + 4y + \;\; z &= 29 \end{aligned}$

45. $\begin{aligned} 3w + 5x + 2y + \;\; z &= 9 \\ w \qquad\; + y + 4z &= 5 \\ 4w + 5x + 3y + 5z &= 0 \\ 2w \qquad\; + y \qquad\quad &= 6 \end{aligned}$

46. In Exercise 2, related to **Rush Hour**, you obtained the system

$$\begin{aligned} x_1 + x_2 \qquad\qquad\quad &= a \\ x_1 \qquad + y_1 \qquad &= b \\ y_1 + y_2 &= c \\ x_2 \qquad\quad + y_2 &= d \end{aligned}$$

Although the values of a, b, c, and d are not known, you can use Cramer's Rule to decide whether the system will have a unique solution. Will it?

47. (*Problem Solving*) Could you obtain the required information in **Rush Hour** by moving one of the counters in Figure 3-7 to a different location? If so, where? Use Cramer's Rule to justify your answer.

ADDITIONAL EXERCISES   *In Exercises 48–57, evaluate the determinant.*

48. $\begin{vmatrix} 9 & 3 \\ 5 & 2 \end{vmatrix}$ **49.** $\begin{vmatrix} -4 & -6 \\ -1 & -7 \end{vmatrix}$

50. $\begin{vmatrix} 1 & 1 \\ 0 & 1 \end{vmatrix}$ **51.** $\begin{vmatrix} 0.01 & 1 \\ 0 & 100 \end{vmatrix}$

52. $\begin{vmatrix} 7 & 0 & 3 \\ -4 & 1 & 0 \\ 12 & -2 & 3 \end{vmatrix}$ **53.** $\begin{vmatrix} 1 & 2 & 3 \\ 4 & 5 & 6 \\ 7 & 8 & 9 \end{vmatrix}$

54. $\begin{vmatrix} -2 & 0 & 8 \\ 5 & 0 & -1 \\ -1 & 0 & 4 \end{vmatrix}$ **55.** $\begin{vmatrix} 2 & -1 & 3 \\ 2 & -1 & 3 \\ 4 & 11 & 9 \end{vmatrix}$

56. $\begin{vmatrix} 1 & -1 & 2 & 2 \\ 0 & 3 & 4 & -2 \\ 0 & -3 & 0 & 6 \\ 1 & 0 & 5 & -5 \end{vmatrix}$

57. $\begin{vmatrix} 5 & 23 & -41 & 60 \\ 0 & 2 & 153 & 98 \\ 0 & 0 & 1 & 15 \\ 0 & 0 & 0 & 3 \end{vmatrix}$

In Exercises 58–73, use Cramer's Rule to decide whether the system has a unique solution. If it does, use Cramer's Rule to find the solution.

58. $4x + 3y = 6$
    $5x + 4y = 2$

59. $3x - 7y = 0.3$
    $2x - 8y = 1.5$

60. $5x - 3y = 21$
    $y = x - 5$

61. $q = 6p - 9$
    $q = 4p + 11$

62. $x = 8 - 3y$
    $y = 8 - 3x$

63. $42x - 79y = 0$
    $35x + 27y = 0$

64. $y = 3 - 2x$
    $x = 5 - 0.5y$

65. $2s + 6t = 5$
    $2s - 6t = 1$

66. $x + y + z = 0$
    $2x + y - z = -4$
    $-x + 4y + z = 7$

67. $3x - 2y + 4z = 7$
    $x + 5y - 2z = 24$
    $4x + 3y + z = 29$

68. $2x - 3y + z = 1$
    $2x + y - 6z = 2$
    $2y + z = 0$

69. $2x + y - 3z = 17$
    $2x + y + z = 21$
    $2x - y - 5z = -5$

70. $x - y = 1$
    $y - z = 2$
    $x - z = 5$

71. $x - y = 1$
    $y - z = 2$
    $x + z = 5$

72. $a + b - 3c = 0$
    $a - 2b + c = 0$
    $2a - b - c = 0$

73. $a + b - 3c = 0$
    $a - 2b + c = 0$
    $2a - b - c = 3$

In Exercises 74 and 75, use Cramer's Rule to decide whether the system has a unique solution. If it does, use Cramer's Rule to solve for w.

74. $w + x + y + z = 21$
$w + x + y - z = 17$
$w + x - y - z = 11$
$w - x - y - z = 21$

75. $w + x + y + z = 21$
$w + x \qquad = 17$
$w \qquad + y \qquad = 11$
$w \qquad\qquad + z = 21$

76. (*Writing to Learn*) Of the methods for solving linear systems described in this chapter and in Section A-10, which do you prefer to use? Give reasons for your answer. Does your preference depend on the size of the system? If so, describe how.

77. (*Making Observations*) To evaluate a 4×4 determinant, you need to evaluate four 3×3 cofactors. To evaluate each of these, you need to evaluate three 2×2 cofactors. Therefore, you need to evaluate twelve 2×2 cofactors altogether. How many 2×2 cofactors do you evaluate in the process of evaluating a 5×5 determinant? A 6×6 determinant? A 10×10 determinant? An $n \times n$ determinant?

3–4   SYSTEMS OF LINEAR INEQUALITIES AND LINEAR PROGRAMMING

GRAPHS OF LINEAR INEQUALITIES IN TWO VARIABLES

In this section you will encounter some relationships that can be more accurately modeled by inequalities in two variables rather than equations. Just as the solution to an inequality in one variable can be conveniently represented by a graph on a one-dimensional number line, the solution to an inequality in two variables can be conveniently represented by a graph in a two-dimensional coordinate plane. The following Mathematical Looking Glass explores both the use of an inequality to represent a relationship and the use of a graph to represent its solution.

A MATHEMATICAL LOOKING GLASS
Horse Feed

Lynn's daughter Carrie owns a 16-hand Lippizan-Percheron cross named Illiad. His diet includes a mix of timothy hay and alfalfa. During the fall of 1991, Carrie could buy timothy for $1.50 per bale. Alfalfa was harder to obtain, and cost $2.50 per bale. Carrie had set aside $165 to spend on Illiad and wanted to know what combinations of timothy and alfalfa she could buy. However, she realized that she might want to spend only part of the $165 on feed and buy other horse supplies with the rest. ■

EXERCISES

1. (*Writing to Learn*) The combinations of timothy and alfalfa Carrie can buy are the solutions to the equation $1.50x + 2.50y = 165$, where

$x =$ the number of bales of timothy she buys

and

$y =$ the number of bales of alfalfa she buys

Explain how this equation was obtained.

2. (*Creating Models*) Write an inequality to express the combinations of timothy and alfalfa Carrie can buy for no more than $165.

The equation in Exercise 1 and the inequality in Exercise 2 each have infinitely many solutions. In both cases, we can represent the solutions graphically. The equation can be solved for y to obtain $y = 66 - 0.6x$. Its graph is shown in Figure 3-8a.

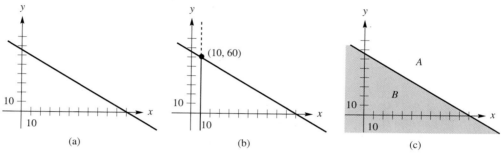

Figure 3–8

The inequality in Exercise 2 can be written $y \leq 66 - 0.6x$. To see what its graph looks like, let's focus on one value of x, say $x = 10$. The inequality then becomes $y \leq 60$. Figure 3-8b shows the points on the line $x = 10$ with $y \leq 60$. Similar calculations show that for every value of x, the inequality is true at points on or below the line $y = 66 - 0.6x$. The graph of the inequality is shown in the shaded portion of the plane in Figure 3-8c. The boundary line $y = 66 - 0.6x$ is included in the graph. It is common practice to draw a boundary as a solid line if it is included in the graph—and as a dashed line if it is not.

EXERCISE   **3.** (*Writing to Learn*) The graph of a linear equation in two variables is always a line that divides the plane into two regions. The solution to every linear inequality in two variables can be represented as one such region. Explain why.

EXAMPLE 1   *Graphing a linear inequality*

To graph the inequality $2x + y < 6$, begin by graphing the following equation: $2x + y = 6$, as in Figure 3-9a.

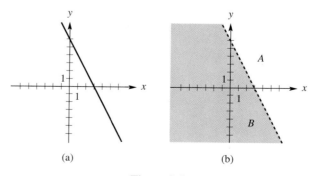

Figure 3–9

The solution to the inequality is one of the two regions *A* and *B*. To see which region is the solution, choose a test value in either region. For example, (0, 0) is in region *B*, and the inequality is true there. Figure 3-9b shows how to indicate that region *B* is the solution.   ■

EXERCISES   *Graph the inequalities in Exercises 4–7.*

4. $3x + 2y < 5$ **5.** $3x - 2y < 5$

6. $y - 7x \geqslant 3$ **7.** $x + 3y \geqslant 0$

GRAPHS OF SYSTEMS OF LINEAR INEQUALITIES IN TWO VARIABLES

In Section 3-1 you saw that some problems can be modeled by a system of equations in two or more variables. Graphically, the solutions to the system are those points common to the graphs of all the equations in the system. Example 2 describes a problem that can be modeled by a system of linear inequalities in two variables. Graphically, the solutions to the system are those points common to the graphs of all the inequalities in the system.

EXAMPLE 2   *Graphing a system of linear inequalities*

Carrie's problem in ***Horse Feed*** involves the following system of linear inequalities.

$$1.50x + 2.50y \leqslant 165$$

$$x \geqslant 0$$

$$y \geqslant 0$$

The first inequality represents her budget constraint, and the other two represent the constraints that quantities of timothy and alfalfa must be nonnegative. Figure 3-10 shows the graph of each inequality in the system.

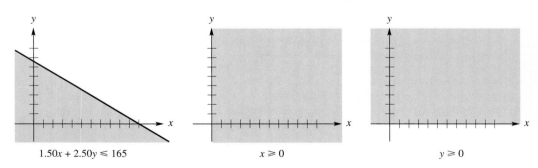

$1.50x + 2.50y \leqslant 165$ $x \geqslant 0$ $y \geqslant 0$

Figure 3–10

A point (x, y) represents a solution to the system if and only if it makes all three inequalities true. Therefore, the graph of the system is the region common to the graphs of the inequalities in the system. Figure 3-11 shows the solution.

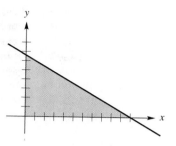

Figure 3–11

The solution to a system of inequalities is called its **feasible region**. The feasible regions of larger systems can be obtained in a similar manner. For example, Carrie must buy at least twice as much timothy as alfalfa, because alfalfa is too rich for Illiad to digest unless it is mixed with large quantities of timothy. This additional constraint is expressed by the inequality $x \geqslant 2y$.

EXERCISES

8. Graph the inequality $x \geqslant 2y$.

9. Find the feasible region for the system of inequalities

$$1.50x + 2.50y \leqslant 165$$

$$x \geqslant 0$$

$$y \geqslant 0$$

$$x \geqslant 2y$$

10. (*Interpreting Mathematics*) What does the feasible region in Exercise 9 represent in physical terms?

11. An additional constraint in **Horse Feed** is that Carrie's barn holds only 100 bales of hay. Graph the resulting inequality $x + y \leqslant 100$.

12. Find the feasible region for the system of inequalities

$$1.50x + 2.50y \leqslant 165$$

$$x \geqslant 0$$

$$y \geqslant 0$$

$$x \geqslant 2y$$

$$x + y \leqslant 100$$

13. (*Interpreting Mathematics*) What does the feasible region in Exercise 12 represent in physical terms?

Graph the solution to each system of inequalities in Exercises 14–17.

14. $y < 2x + 2$
    $2y \geqslant x - 2$

15. $y \leqslant x + 1$
    $2x + y \leqslant -2$

16. $2x + 3y < 12$
    $2x + 3y > 6$
    $x - y < 3$

17. $2x + 3y < 12$
    $2x + 3y > 6$
    $x - y < 3$
    $x - y > 0$

LINEAR PROGRAMMING

As you have seen, systems of linear inequalities in two variables typically have infinitely many solutions. In most situations it is natural to look for one solution that is optimal in some respect. One method of finding optimal solutions, called **linear programming**, was developed during World War II to determine the optimal use of resources in military operations. Linear programming is also used today by businesses and government agencies to solve problems in such diverse areas as economics, personnel, and transportation. Read on to see how Carrie can use linear programming to find an optimal solution to her system of inequalities in *Horse Feed*.

A MATHEMATICAL LOOKING GLASS

Horse Feed 2

Carrie's system of inequalities from Exercise 12 is

$$1.50x + 2.50y \leq 165$$
$$x \geq 0$$
$$y \geq 0$$
$$x \geq 2y$$
$$x + y \leq 100$$

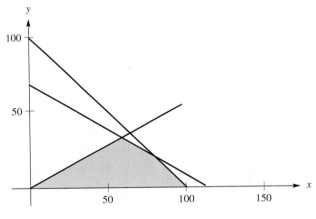

Figure 3–12

Her feasible region, shown in Figure 3-12, contains infinitely many solutions. Her problem is to decide on a "best" solution. She will probably do this by seeking an optimal combination of many factors, such as nutritive value and resistance to spoilage. Let's keep life simple and assume that she needs to consider only one factor. Suppose that Carrie plans to ride Illiad in a strenuous cross-country event and wants to feed him a protein-rich diet in preparation. Her problem is to choose the solution with the most digestible protein.

Carrie knows that timothy is 6.1% digestible protein and alfalfa is 12.9% digestible protein and that an average bale weighs 50 pounds. Therefore, the total amount of digestible protein in the feed shipment is

$$(0.061)(50x) + (0.129)(50y)$$

or

$$3.05x + 6.45y$$

This expression is Carrie's **objective function**. She must now find the point (x, y) in the feasible region where the value of the objective function is largest.

Let's consider several values for the objective function. Any equation $3.05x + 6.45y = k$ represents a feed shipment containing k pounds of digestible protein. The dashed lines in Figure 3-13 represent feed shipments with $k = 100, 200,$ and 300 pounds of digestible protein. Their equations are

$$3.05x + 6.45y = 100, 3.05x + 6.45y = 200$$

and

$$3.05x + 6.45y = 300$$

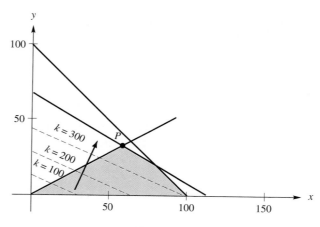

Figure 3–13

The graph of the objective function moves in the direction of the arrow as k increases. Carrie needs to find the largest value of k for which the line $3.05x + 6.45y = k$ intersects the feasible region. This happens when the graph of the objective function passes through P. Since P is the intersection of the lines $x = 2y$ and $1.50x + 2.50y = 165$, $P = (60, 30)$. Thus, the optimal line passes through $(60, 30)$ and has equation $3.05x + 6.45y = k$, so

$$(3.05)(60) + (6.45)(30) = k$$

$$k = 376.5$$

Carrie will buy a shipment of 60 bales of timothy and 30 bales of alfalfa, having 376.5 pounds of digestible protein.   ∎

As Figure 3-13 illustrates, the maximum (or minimum) value of the objective function in a linear programming problem always occurs at an **extreme point** (a corner point) of the system's feasible region. This observation leads to the following method of solution.

To solve a linear programming problem,

- identify the feasible region for the system of inequalities representing the constraints,
- solve systems of linear equations to identify the extreme points of the feasible region, and
- evaluate the objective function at each extreme point.

18. (*Making Observations*) For the linear programming problem in *Horse Feed 2*,

    a. Find the coordinates of the three other extreme points of the feasible region.

    b. Evaluate the objective function at each extreme point. Do your results support the conclusion that the maximum value occurs at $(60, 30)$?

Complete Exercises 19–21 to find the minimum value of $50 - 2x - 3y$ subject to the constraints

$$x \geqslant 0$$
$$y \geqslant 0$$
$$x + y \leqslant 12$$
$$x + 2y \leqslant 17$$

19. Find the feasible region of the system of inequalities.

20. The feasible region in Exercise 19 has four extreme points, representing the solutions to the systems of equations shown. Find their coordinates.

    a. $x = 0$ **b.** $y = 0$
         $y = 0$ $x + 2y = 17$

    c. $x = 0$ **d.** $x + y = 12$
        $x + y = 12$ $x + 2y = 17$

21. Evaluate the objective function $50 - 2x - 3y$ at each extreme point. Which value is the solution?

Solve the linear programming problems in Exercises 22–23.

22. Maximize $x + 3y$,   **23.** Minimize $2x + 3y$,
     subject to: subject to:

          $y \leqslant 2x + 2$ $y \leqslant x + 1$
          $2y \geqslant x - 2$ $2x + y \leqslant -2$
          $x + y \leqslant 14$ $y \geqslant -5$

Graph the inequalities in Exercises 24–31.

24. $y - 2x \leqslant 10$ **25.** $y - 2x \leqslant -10$

26. $x + y < 2$ **27.** $5x - 3y \geqslant 0$

28. $x > 2y$ **29.** $2x + 3y \geqslant -12$

30. $0.1x + 0.3y < -3$ **31.** $0.4x - 0.2y > -1$

Graph the systems of inequalities in Exercises 32–37.

32. $2x + 3y > 12$ **33.** $x > -2 - y$
      $2x + 3y < 6$ $y < x + 2$
                            $x \geqslant 2$

34. $4y \leqslant 12 + 3x$ **35.** $4y < 12 + 3x$
       $x + y \leqslant 5$ $x + y < 5$
         $x < 4y - 4$ $x \leqslant 4y - 4$

36. $y < 3x - 3$
  $y > x - 7$
  $y > -3$
  $y < 0$

37. $x > y$
  $x \leqslant 3y$
  $y \geqslant 5 - x$

Solve the linear programming problems in Exercises 38–41.

38. Maximize $2S - T$,
  subject to:

$$S \geqslant -2 - T$$
$$T \leqslant S + 2$$
$$S \geqslant 2$$
$$S \leqslant 10$$

39. Minimize $3Z_1 - Z_2$,
  subject to:

$$4Z_2 \leqslant 12 + 3Z_1$$
$$Z_1 + Z_2 \leqslant 5$$
$$Z_1 + 2Z_2 \leqslant 20$$
$$Z_1 \leqslant 4Z_2 - 4$$
$$Z_1 \geqslant -1$$

40. Maximize $5x + 3y$,
  subject to:

$$y \leqslant 3x - 3$$
$$y \geqslant x - 7$$
$$y \leqslant 6 - x$$
$$y \geqslant -3$$
$$y \leqslant 0$$

41. Maximize $4x_1 + 3x_2$,
  subject to:

$$2x_1 + 3x_2 \leqslant 6$$
$$-3x_1 + 2x_2 \leqslant 3$$
$$2x_2 \leqslant 5$$
$$2x_1 + x_2 \leqslant 4$$
$$x_1 \geqslant 0$$
$$x_2 \geqslant 0$$

42. (*Creating Models*) You are running for governor of Delaware and would like to get a majority of the votes in each of the state's three counties to demonstrate broad support. Current polls show that you need an additional 9000 votes in New Castle county, 10,000 in Kent County, and 3000 in Sussex County. Your strategy is to buy newspaper ads in the *Delaware State News* and spots on TV station KYW in nearby Philadelphia. You make an educated guess that each newspaper ad will gain you 20 additional votes in New Castle County, 70 in Kent County, and 10 in Sussex County, while each TV spot will gain 400, 100, and 50 votes, respectively. Write and graph a system of inequalities to show all combinations of newspaper ads and TV spots that will accomplish your goal.

43. (*Problem Solving*) Refer to Exercise 42. If TV spots cost 10 times as much as newspaper ads, how many of each should you purchase to obtain a majority of the votes in each county at minimum expense?

44. (*Problem Solving*) (Continuation of Exercise 61 of Section 3-2)

	Grams fat	Grams carbohydrates	Grams protein	Milligrams cholesterol	Calories
1 baked potato	0.2	32.8	4.0	0	145
4 oz. salmon	6.4	0	30.8	40	192

Adapted from Dean Ornish, *Eat More, Weigh Less*, HarperPerennial, 1993.

a. Write and graph a system of inequalities to show all possible combinations of potatoes and salmon having at most 400 calories and at most 8 grams of fat.

b. Among all such combinations, which has the highest protein content?

c. In Exercise 61 of Section 3-2 you saw that a system of linear equations is not a practical model for most dietary problems. Why can a linear programming problem provide a more realistic model?

CHAPTER REVIEW

Complete Exercises 1–4 (Writing to Learn) before referring to the indicated pages.

WORDS AND PHRASES

In Exercises 1–20, explain the meaning of the words or phrases in your own words.

1. (pages 101, 102)   **linear system, $m \times n$ system of linear equations**

2. (page 106)   **consistent, inconsistent** system of linear equations

3. (page 106)   **dependent, independent** system of linear equations

4. (page 112)   **matrix, augmented matrix**

5. (page 113)   **matrix row operations**

6. (page 114)   **Gauss–Jordan elimination**

7. (page 114)   **pivot row, pivot**

8. (page 116)   **reduced row-echelon matrix, leading entry, reduced form**

9. (page 119)   **leading variable** in a system of linear equations

10. (page 120)   **standard form** of the solution to a system of linear equations

11. (page 126)   **coefficient matrix**

12. (page 126)   **second-order (2×2) determinant**

13. (page 129)   **minor**

14. (page 129)   **cofactor**

15. (page 130)   **third-order (3×3) determinant**

16. (page 132)   **expanding along a row or column** of a determinant

17. (page 133)   **determinant** of an $n \times n$ matrix

18. (page 139)   **feasible region** of a system of linear inequalities

19. (page 140)   **linear programming**

20. (page 141)   **objective function**

IDEAS

21. (page 101)   Any problem that can be modeled by a system of linear equations can also be modeled by a single linear equation. Discuss some of the advantages and disadvantages of each type of model.

22. (pages 102, 103)   Compare the ease, speed, and accuracy of numerical, analytical, and graphical methods of solving systems of linear equations. Illustrate your points with a particular system.

23. (pages 104, 105)   Why is it usually impractical to solve a 3×3 system of linear equations by numerical or graphical methods?

24. (pages 106, 107)   Most 2×2 and 3×3 systems of linear equations have exactly one solution. How do the graphs of such systems support this statement?

25. (pages 106, 107)   Explain why it is impossible for a system of linear equations in two or three variables to have more than one solution unless it has infinitely many.

26. (pages 111, 112)   In what respect is it more efficient to represent a system of linear equations in matrix form?

27. (pages 112, 113)   Describe the operations on a system of linear equations corresponding to the matrix row operations, and explain why none of these operations change the solutions to the system.

28. (pages 114–117)   Explain why the process of Gauss–Jordan elimination always leads to a reduced row-echelon matrix.

29. (pages 116, 117)   Why is the reduced row-echelon form of the augmented matrix especially helpful in solving a system of linear equations?

30. (page 118)   If a system of linear equations has no solutions, what does its reduced row-echelon matrix look like? Why does the form of the matrix guarantee that the system has no solution? Give an example of such a matrix, and write down the corresponding system.

31. (page 119)   If a system of linear equations has infinitely many solutions, what does its reduced row-echelon matrix look like? Why does the form of the matrix guarantee that the system has infinitely many solutions? Give an example of such a matrix, write down the corresponding system, and list several solutions.

32. (page 121)   If a system of linear equations has a unique solution, what does its reduced row-echelon matrix look like? Why does the form of the matrix guarantee that the system has a unique solution? Give an example of such a matrix, and write down the corresponding system.

33. (page 122)   What are some of the advantages of using matrix methods to solve a system of linear equations? When might you prefer to use nonmatrix methods?

34. (pages 124, 125)   What are some advantages of Cramer's Rule in comparison to other methods of solving linear systems? What limitations does it have?

35. (page 126)   Describe the process of evaluating a 2×2 determinant, and illustrate your explanation with a specific example.

36. (page 127)   Explain why Cramer's Rule gives correct results for a 2×2 system and why it cannot be used if the determinant of the coefficient matrix is 0.

37. (page 127)   What is the difference between a cofactor and a minor? Use a specific 3×3 matrix to give an example of each.

38. (pages 131–133)   Describe the different processes that can be used to evaluate a 3×3 determinant. Make up a specific determinant, and use each process on it.

39. (pages 133, 134)   Describe the process of evaluating an $n \times n$ determinant and the process of solving an $n \times n$ linear system using Cramer's Rule. How do they compare with the corresponding processes for 3×3 systems?

40. (pages 136–138)   Describe the process of solving a linear inequality in two variables graphically, and illustrate with a particular inequality.

41. (pages 138, 139)   The solution of a system of linear inequalities in two variables is usually represented graphically, rather than analytically. Explain why.

42. (pages 140–142)   Describe the process of solving a linear programming problem in two variables. In particular, explain why the optimum value of the objective function always occurs at a corner point of the feasible region.

CHAPTER 4

QUADRATIC FUNCTIONS AND RELATIONS

4–1 QUADRATIC FUNCTIONS

PREREQUISITE MAKE SURE YOU ARE FAMILIAR WITH:
Operations with Complex Numbers (Section A-14)

In Chapters 2 and 3 you viewed linear functions from several perspectives and learned how to use them in modeling and problem solving. Along the way, you learned that their graphs have a distinctive shape, their equations have a distinctive form, and their tables have a distinctive property by which they can be identified.

In the first two sections of this chapter you will study quadratic functions, which are the simplest nonlinear functions. You will see how they appear in situations related to calculating areas, analyzing the motion of a falling object, and maximizing a company's profit. Although quadratic functions are radically different from linear functions, their graphs, equations, and tables also have distinctive shapes, forms, and properties.

The following Mathematical Looking Glass illustrates how a quadratic function can be of interest in a physical context.

A MATHEMATICAL
LOOKING GLASS
*Maximum Profit
(Fat Cats 5)*

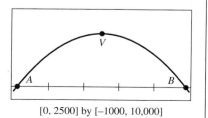
[0, 2500] by [−1000, 10,000]

Figure 4–1

The T-shirt company your class started in ***Fat Cats on Campus*** (page 28) is about to begin production. A group of students from the class has pointed out that without an intensive marketing campaign, higher sales will not necessarily mean higher profits. Using the profit function

$$P(x) = -0.005x^2 + 12.50x - 612.50$$

that you developed in Exercise 3 of Section 1-4, they present the graph in Figure 4-1 at your next board meeting.

The horizontal axis measures the number of shirts sold, and the vertical axis measures profit in dollars. The group explains that if your level of production is between A and V on the graph, increasing production will cause profits to rise. If your level of production is between V and B, increasing production will cause profits to fall. This is because to sell a large number of shirts, you will have to lower the price.

147

The class must now identify the target sales level that will generate the largest profit. ∎

1. (*Interpreting Mathematics*)

   a. Explain why the sales level that will generate the maximum profit is represented by the point *V* in Figure 4-1.

   b. Use Figure 4-1 to estimate the number of shirts you must sell to generate the maximum profit.

2. (*Review*) In **Fat Cats on Campus** (page 28) you discovered that your company's breakeven sales levels were 50 and 2450 shirts.

   a. Explain why the breakeven sales levels are of interest to your company.

   b. Explain why the breakeven sales levels are represented by the *x*-intercepts *A* and *B* in Figure 4-1.

3. (*Making Observations*) Explain how identifying the *x*-intercepts in Figure 4-1 can help you estimate the sales level that will generate the maximum profit.

The profit function in **Maximum Profit** is an example of a **quadratic function**, that is, one that can be written

$$f(x) = ax^2 + bx + c$$

for some real numbers *a*, *b*, and *c* with $a \neq 0$. By studying these functions you will learn systematic methods of solving your company's problem and other similar problems.

A GRAPHICAL VIEW OF QUADRATIC FUNCTIONS

Features of Quadratic Graphs Figure 4-2a shows the graph of the simplest quadratic function $y = x^2$. We will refer to it as the *basic* quadratic graph. It shares several striking features with the graph of $y = P(x)$ from **Maximum Profit**, which is reproduced in Figure 4-2b.

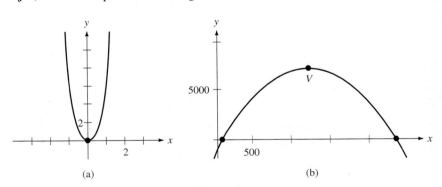

(a)   (b)

Figure 4–2

Every quadratic graph has the following features.

- It is a **parabola**, a U-shaped curve opening upward as in Figure 4-2a or downward as in Figure 4-2b. In contrast to linear graphs, quadratic graphs do not exhibit a constant rate of change. You can verify this in Exercise 63.

- It has a **vertex**, that is, its extreme point (the origin in Figure 4-2a and the point V in Figure 4-2b). Therefore, the range of a quadratic function does not include all values of the dependent variable, but is an interval extending infinitely in only one direction. From their graphs, the functions in Figure 4-2 appear to have ranges of $[0, \infty)$ and approximately $(-\infty, 7200]$. Later in this section you will learn how to identify the range of a quadratic function analytically.

- The vertical line through the vertex is an **axis of symmetry**. The parts of the graph to the left and right of this line mirror each other. You will learn how to identify the axis of symmetry later in this section.

- It has 0, 1, or 2 **x-intercepts**. You can explain why in Exercises 21–23.

FYI *Parabola* is not a generic name for any U-shaped curve. Parabolas have some distinctive properties that you will study in Section 4-6.

Graphing Quadratic Functions Exercises 1–3 indicate that the vertex and the upward or downward orientation of a parabola are often of physical importance. Let's see if we can read these graphical features quickly from the equation of a quadratic function. We can proceed by *Examining a Special Case* (actually several special cases).

EXERCISES

4. Graph all of the following functions on one set of coordinate axes.

   a. $y = x^2$ **b.** $y = 4x^2$

   c. $y = 0.5x^2$ **d.** $y = -3x^2$

5. (*Making Observations*) Based on your graphs from Exercise 4, guess the answers to the following questions about the graph of $y = ax^2$ for an arbitrary real number a not equal to zero.

   a. Where is the vertex on the graph of $y = ax^2$?

   b. How does the steepness of the graph compare with that of the basic graph of $y = x^2$ if $|a| > 1$? If $|a| < 1$?

   c. What is the orientation (up or down) of the graph if $a > 0$? If $a < 0$?

6. (*Making Observations*) Use your answers from Exercise 5 to sketch each of the following graphs. Use a calculator graph to support your results.

   a. $y = 2x^2$ **b.** $y = 0.2x^2$

   c. $y = -x^2$ **d.** $y = -0.5x^2$

7. Graph all of the following functions on one set of coordinate axes.

   a. $y = x^2$ **b.** $y = (x + 1)^2$

   c. $y = (x + 4)^2$ **d.** $y = (x - 1)^2$

8. (*Making Observations*) Based on your graphs from Exercise 7, guess the location of the vertex on the graph of $y = (x - h)^2$ for any real value of h. Then sketch each of the following graphs, and use a calculator graph to support your results.

   a. $y = (x - 2)^2$ **b.** $y = (x - 5)^2$

   c. $y = (x + 2)^2$ **d.** $y = (x + 6)^2$

9. Graph all of the following functions on one set of coordinate axes.

   a. $y = x^2$ **b.** $y = x^2 - 1$

   c. $y = x^2 + 1$ **d.** $y = x^2 + 3$

10. (*Making Observations*) Based on your graphs from Exercise 9, guess the location of the vertex on the graph of $y = x^2 + k$ for any real value of k. Then sketch each of the following graphs, and use a calculator graph to support your results.

a. $y = x^2 + 5$ **b.** $y = x^2 + 2$
c. $y = x^2 - 5$ **d.** $y = x^2 - 3$

Your observations in Exercises 4–10 can be generalized as follows.

- The graph of $y = a(x - h)^2 + k$ is a parabola with its vertex at (h, k).
- The graph opens up if $a > 0$ and down if $a < 0$.
- The graph is steeper than the basic graph of $y = x^2$ if $|a| > 1$ and flatter if $|a| < 1$.

EXAMPLE 1   *Sketching the graph of a function $y = a(x - h)^2 + k$*

The graph of

$$y = -0.5(x + 4)^2 + 2$$

is a parabola with its vertex at $(-4, 2)$. It opens down since $a = -0.5 < 0$. Since $|a| = 0.5 < 1$, the graph is flatter than the basic graph of $y = x^2$. Specifically, it is half as steep as the basic graph, so the points 1 unit to the left and right of the vertex have coordinates $(-5, 1.5)$ and $(-3, 1.5)$. See Figure 4-3.

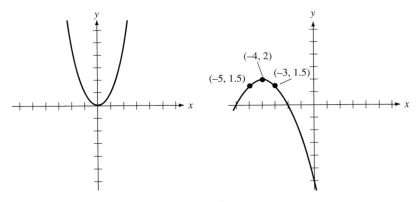

Figure 4–3   ■

EXERCISE   **11.** (*Making Observations*) Sketch each of the following graphs, and use a calculator graph to support your results.

a. $y = 2(x - 2)^2 + 5$
b. $y = 0.2(x - 5)^2 + 2$
c. $y = -(x + 2)^2 - 5$
d. $y = -0.5(x + 6)^2 - 3$

Your observations in Exercises 4–10 are important because, as you will soon see, you can write the equation of every quadratic function in the form $y = a(x - h)^2 + k$. When the equation of a quadratic function is written in this form, you can sketch its graph quickly, as you saw in Exercise 11.

(We hope you noticed the resemblance to the standard graphing form $y = a|x - h| + k$ for the linear absolute value functions you encountered in Section 2-5.)

EXERCISE   **12.** (*Making Observations*) Sketch the graph of $y = -0.5|x + 4| + 2$. In what respects is it like the graph of $y = -0.5(x + 4)^2 + 2$ in Example 1? In what respects is it different?

The graph of a quadratic function, like that of a linear absolute value function, can be described in terms of transformations applied to a basic graph.

To obtain the graph of $y = a(x - h)^2 + k$ from the basic graph of $y = x^2$:

- Stretch the graph vertically by a factor of $|a|$ if $|a| > 1$, or compress it if $|a| < 1$.
- If $a < 0$, reflect the graph in the x-axis.
- Shift the graph to the right if $h > 0$ or to the left if $h < 0$.
- Shift the graph up if $k > 0$ or down if $k < 0$.

EXAMPLE 2   *Sketching the graph of a quadratic function in the form $y = a(x - h)^2 + k$*

The function

$$g(x) = 2(x - 3)^2 - 5$$

has the form $y = a(x - h)^2 + k$, with $a = 2$, $h = 3$, $k = -5$. Its graph is that of $y = x^2$ stretched vertically by a factor of 2, then shifted 3 units to the right and 5 units down. As shown in Figure 4-4, its vertex is $(3, -5)$. The points 1 unit to the left and right of the vertex are 2 units above the vertex, at $(2, -3)$ and $(4, -3)$.

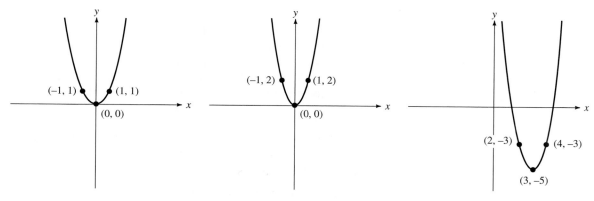

Figure 4–4

EXERCISES   *In Exercises 13–16,*

a. Describe the ways in which the basic graph of $y = x^2$ must be stretched, compressed, reflected, and shifted to produce the graph of the given function.

b. Sketch the graph. Identify the vertex and the points 1 unit to the left and right of the vertex.

13. $g(x) = 2(x - 4)^2 + 1$ **14.** $g(x) = -3(x + 1)^2 + 3$

15. $g(x) = -\dfrac{3}{4}x^2 - 3$ **16.** $g(x) = 0.1(x - 2)^2$

Examples 1 and 2 and Exercises 4–16 indicate some of the strong connections between the graphs and the equations of quadratic functions. To establish these connections more clearly, let's look at quadratic functions analytically.

AN ANALYTICAL VIEW OF QUADRATIC FUNCTIONS

The equation of every quadratic function can be written in the following three forms.

$$f(x) = ax^2 + bx + c \qquad \textbf{(expanded form)}$$

$$f(x) = a(x - r_1)(x - r_2) \qquad \textbf{(factored form)}$$

$$f(x) = a(x - h)^2 + k \qquad \textbf{(standard graphing form)}$$

EXAMPLE 3   *The three analytical forms of a quadratic function*

The function

$$f(x) = -0.5(x + 4)^2 + 2$$

is written in standard graphing form. Perform the multiplication and combine like terms to obtain the expanded form

$$f(x) = -0.5x^2 - 4x - 6$$

Factor the expanded form to obtain the factored form

$$f(x) = -0.5(x + 6)(x + 2) \qquad \blacksquare$$

Each form provides useful information about the function it represents. Example 3 shows that the value of the parameter a is the same for all three forms, so that it always provides information about the direction and steepness of the graph. You can use each form to find the x-intercepts, vertex, and axis of symmetry, as well as the range of the function.

Expanded Form   The x-intercepts, vertex, and axis of symmetry cannot be read directly from the expanded form $f(x) = ax^2 + bx + c$. However, we can identify them with the help of a few formulas.

In Exercise 93 you will show that the vertex on the graph of $f(x) = ax^2 + bx + c$ has an x-coordinate of $-\dfrac{b}{2a}$. For now you can use this fact to find the vertex. Knowing the vertex, you can also identify the axis of symmetry and the range of $f(x)$.

EXAMPLE 4   *Finding the vertex on the graph of $f(x) = ax^2 + bx + c$*

The vertex on the graph of $f(x) = 2x^2 - 6x + 5$ occurs when

$$x = -\frac{-6}{2(2)} = \frac{3}{2}$$

$$y = 2\left(\frac{3}{2}\right)^2 - 6\left(\frac{3}{2}\right) + 5 = \frac{1}{2}$$

As shown in Figure 4-5, the axis of symmetry is the vertical line through the vertex and thus has the equation $x = \frac{3}{2}$. The range of $f(x)$ extends upward from the y-coordinate of the vertex, so the range is $\left[\frac{1}{2}, \infty\right)$.

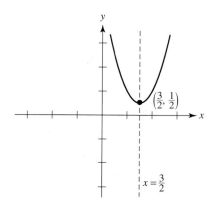

Figure 4–5

EXERCISES   *In Exercises 17–20, identify the vertex and the axis of symmetry on the graph of $y = H(x)$ and the range of $H(x)$.*

17. $H(x) = x^2 + 8x + 9$ **18.** $H(x) = 3x^2 - 6x + 2$
19. $H(x) = 9 - x^2$ **20.** $H(x) = 9x - x^2$

The quadratic formula allows you to locate the x-intercepts and also shows how the intercepts are related to the vertex.

EXERCISES   **21.** (*Review*) Use the quadratic formula

$$x = \frac{-b \pm \sqrt{b^2 - 4ac}}{2a}$$

to solve each equation.
  a. $x^2 - 5x + 2.25 = 0$
  b. $x^2 - 5x + 6.25 = 0$
  c. $x^2 - 5x + 10.25 = 0$

22. (*Making Observations*)
  a. How many real solutions does each equation in Exercise 21 have?
  b. Graph the functions $y = x^2 - 5x + 2.25$, $y = x^2 - 5x + 6.25$, and $y = x^2 - 5x + 10.25$ on the same set of coordinate axes. How many x-intercepts does each graph have?
  c. How can you decide whether a solution to a quadratic equation $f(x) = 0$ corresponds to an x-intercept on the graph of $y = f(x)$?

23. (*Making Observations*) How do your observations in Exercise 22 support the conclusion that a quadratic graph can have 0, 1, or 2 x-intercepts?

24. (*Making Observations*) The quadratic formula can be written

$$x = -\frac{b}{2a} \pm \frac{\sqrt{b^2 - 4ac}}{2a}$$

How does this observation support the conclusion that when a quadratic graph has two x-intercepts, they are at equal distances to the left and right of the vertex?

The discriminant $\sqrt{b^2 - 4ac}$ of a quadratic expression

$$f(x) = ax^2 + bx + c$$

can be used to determine the number of real solutions to the equation $f(x) = 0$. (See Section A-13.)

EXERCISES   *In Exercises 25–28, use the discriminant to determine the number of x-intercepts on the graph of $y = f(x)$. Use a calculator graph to support your results.*

25. $f(x) = x^2 + 4$ **26.** $f(x) = x^2 + 4x$

27. $f(x) = x^2 + 4x + 4$ **28.** $f(x) = -x^2 + 4x + 4$

Factored Form You may have learned in an algebra course that the quadratic expression $2x^2 - 5x + 2$ is factorable, while $x^2 - 2$ and $x^2 + 2x + 5$ are not. It is more accurate to say that $x^2 - 2$ and $x^2 + 2x + 5$ cannot be written in the form $a(x - r_1)(x - r_2)$ *if r_1 and r_2 are required to be rational numbers. Every* quadratic expression can be written in that form if r_1 and r_2 can take on any complex value. When r_1 and r_2 are real, as in Example 5, the factored form of the function is closely connected with the x-intercepts on its graph.

EXAMPLE 5   *Writing quadratic functions in factored form*

a. The function

$$h(x) = 2x^2 - 5x + 2$$

can be written

$$h(x) = (2x - 1)(x - 2)$$

We can then continue by writing $2x - 1$ as $2\left(x - \frac{1}{2}\right)$ to obtain the factored form

$$h(x) = 2\left(x - \frac{1}{2}\right)(x - 2)$$

The factored form makes it clear that $h\left(\frac{1}{2}\right) = 0$ and $h(2) = 0$, so the x-intercepts on the graph of $y = h(x)$ are $\left(\frac{1}{2}, 0\right)$ and $(2, 0)$. The graph in Figure 4-6a supports these conclusions.

Figure 4–6

b. To write the factored form of

$$g(x) = x^2 - 2$$

factor $g(x)$ as a difference of squares.

$$g(x) = \left(x + \sqrt{2}\right)\left(x - \sqrt{2}\right)$$

The factored form makes it clear that $g\left(\sqrt{2}\right) = 0$ and $g\left(-\sqrt{2}\right) = 0$, so the x-intercepts on the graph of $y = g(x)$ are $\left(-\sqrt{2}, 0\right)$ and $\left(\sqrt{2}, 0\right)$. The graph in Figure 4-6b supports this conclusion. ■

EXERCISES *In Exercises 29–32, write the function in factored form.*

29. $f(x) = 3x^2 + 8x$ **30.** $Q(x) = -x^2 - 14x - 45$

31. $P(t) = 6x^2 + 7x - 5$ **32.** $G(v) = 13 - v^2$

33. The equation $15x^2 + 41x + 12 = 0$ has solutions $x = -\dfrac{12}{5}$ and $x = -\dfrac{1}{3}$.

Use this fact to help you write the function $f(x) = 15x^2 + 41x + 12$ in factored form.

In Exercises 34–37, obtain the factored form of $Q(x)$ by solving the equation $Q(x) = 0$ first.

34. $Q(x) = x^2 + 2x - 10$ **35.** $Q(x) = 2x^2 - 4x - 1$

36. $Q(x) = 3 + x - x^2$ **37.** $Q(x) = 0.5x^2 + 4x + 5$

38. (*Making Observations*) Solve $x^2 + 2x + 5 = 0$. How does the solution show that the graph of $y = x^2 + 2x + 5$ has no x-intercepts? Support this conclusion by graphing the function.

The observations on the preceding few pages can be generalized into a statement that allows you to read the x-intercepts on a quadratic graph directly from the factored form of the function.

> If r_1 and r_2 are real, the graph of $f(x) = a(x - r_1)(x - r_2)$ has x-intercepts at $(r_1, 0)$ and $(r_2, 0)$. [If $r_1 = r_2$, there is a single x-intercept at $(r_1, 0)$.]
>
> If r_1 and r_2 are not real, the graph of $f(x) = a(x - r_1)(x - r_2)$ has no x-intercepts.

EXERCISE   **39.** (*Interpreting Mathematics*) Write the factored form of the function $P(x) = -0.005x^2 + 12.50x - 612.50$ from **Maximum Profit**. Use the factored form to find the breakeven points A and B in Figure 4-1.

Can we also read the vertex of a quadratic graph directly from the factored form of the function? You can explore this question in Exercises 40–41.

EXERCISES   **40.** (*Making Observations*) Find the vertex on the graph of each function in Exercises 29–32. How is the vertex related to the x-intercepts in each case?
41. Suppose that the graph of a quadratic function $y = f(x)$ has x-intercepts at $(-2, 0)$ and $(8, 0)$. Use your observations from Exercise 40 to find the x-coordinate of the vertex.

Your observations from Exercise 40 can be stated as a formula that relates the vertex of a quadratic graph to the factored form of its equation. The formula is valid even when the factors involve complex numbers. You can learn why in Exercise 94.

> If $f(x) = a(x - r_1)(x - r_2)$, the vertex on the graph of $f(x)$ occurs when
> $$x = \frac{r_1 + r_2}{2}.$$

EXAMPLE 6   *Finding the vertex on the graph of $f(x) = a(x - r_1)(x - r_2)$*
The vertex on the graph of

$$f(x) = -4\big[x - (5 + 2i)\big]\big[x - (5 - 2i)\big]$$

occurs when

$$x = \frac{(5 + 2i) + (5 - 2i)}{2} = \frac{10}{2} = 5$$

To find the y-coordinate of the vertex, substitute 5 for x in the equation for $f(x)$.

$$y = -4\big[5 - (5 + 2i)\big]\big[5 - (5 - 2i)\big]$$
$$= -4(-2i)(2i)$$
$$= 16i^2$$
$$= -16$$

The vertex is $(5, -16)$. As usual, the location of the vertex determines the axis of symmetry and the range of f. The axis of symmetry is the line $x = 5$, and the range of f is $(-\infty, -16]$. See Figure 4-7. ■

Figure 4–7

EXERCISES   *In Exercises 42–45, find the vertex and axis of symmetry on the graph of $y = f(x)$, and the range of f.*

42. $f(x) = 3\left(x - \dfrac{5}{3}\right)(x - 8)$

43. $f(x) = -\left(x + \sqrt{7}\right)\left(x - \sqrt{7}\right)$

44. $f(x) = -5\left[x - \left(10 + \sqrt{13}\right)\right]\left[x - \left(10 - \sqrt{13}\right)\right]$

45. $f(x) = 2\left[x - (1 + 3i)\right]\left[x - (1 - 3i)\right]$

Standard Graphing Form By completing the square you can write every quadratic function in the form $f(x) = a(x - h)^2 + k$. The technique of completing the square is discussed in Section A-13. Let's review it in Example 7.

EXAMPLE 7   *Writing a quadratic function in standard graphing form*
To write

$$g(x) = 2x^2 - 12x + 13$$

in standard graphing form, first group the terms containing x^2 and x, and factor out the coefficient of x^2.

$$g(x) = 2(x^2 - 6x) + 13$$

Square half of the x-coefficient [half of -6 is -3, and $(-3)^2 = 9$]. Then add and subtract that quantity inside the parentheses.

$$g(x) = 2(x^2 - 6x + 9 - 9) + 13$$

Remove the -9 from the parentheses, remembering to multiply it by the 2 in front.

$$g(x) = 2(x^2 - 6x + 9) - 18 + 13$$

Finally, write the expression inside the parentheses as the square of a binomial, and combine the constants outside.

$$g(x) = 2(x - 3)^2 - 5$$ ■

When a quadratic function is written in standard graphing form, you can easily find its vertex, its axis of symmetry, and its range. For the function $g(x)$ in Example 7, these features were found in Example 2.

EXERCISES   *In Exercises 46–49,*

    a. Write the equation in standard graphing form.

    b. Sketch the graph of the function. Identify the vertex and the axis of symmetry, and find the range of the function.

46. $y = 4x^2 + 8x - 3$ **47.** $y = 0.5x^2 + 4x + 9$

48. $s = -t^2 - 2t - 4$ **49.** $w = -3z^2 + 6z$

50. (*Interpreting Mathematics*) First write the standard graphing form of $P(x) = -0.005x^2 + 12.50x - 612.50$ from **Maximum Profit**. Then find the maximum profit and the target sales level that will generate it.

It is possible to read the x-intercepts on the graph of $f(x) = a(x - h)^2 + k$ directly from the equation (see Exercise 56). However, it is more practical to find them by solving the equation $a(x - h)^2 + k = 0$.

EXERCISES   *In Exercises 51–54, find the t-intercepts on the graph of $y = F(t)$.*

51. $F(t) = (t + 6)^2 - 4$ **52.** $F(t) = 9(t - 1)^2 - 16$

53. $F(t) = -2(t + 3)^2 + 1$ **54.** $F(t) = 2(t + 3)^2 + 1$

55. (*Writing to Learn*) Explain why the graph of $y = a(x - h)^2 + k$ has no x-intercepts if a and k have the same sign.

56. (*Making Observations*) Solve $a(x - h)^2 + k = 0$ for x.

    a. How does the result verify that the x-intercepts on the graph of $y = a(x - h)^2 + k$ are $\left(h \pm \sqrt{-\dfrac{k}{a}}, 0\right)$ if a and k have opposite signs?

    b. How does the result show that the vertex is halfway between the x-intercepts?

Comparisons Among the Three Forms Each analytical form of a quadratic function is useful in its own way. The vertex of the graph is obtained most easily from the standard graphing form, while the x-intercepts are obtained most easily from the factored form. In Section 4-2 you will discover that a quadratic function to fit a table can be constructed most easily in expanded form.

A NUMERICAL VIEW OF QUADRATIC FUNCTIONS

The following Mathematical Looking Glass illustrates that it can be useful to know that a table was generated by a quadratic function.

A MATHEMATICAL LOOKING GLASS

Stopping Distance

The data in Table 18 was gathered by General Motors. The left column represents the speed of a car at the instant the driver recognizes a need to stop. The right column represents the distance traveled from that instant until the car comes to a stop. The data assumes that the pavement is dry].

TABLE 18

Initial speed, in mph (*v*)	Stopping distance, in feet [*S*(*v*)]
30	77
40	128
50	190
60	263

The relationship of the car's initial speed v to the required stopping distance $S(v)$ is important in many contexts.

- A Department of Transportation employee responsible for timing traffic lights needs to know the stopping distance required by a driver traveling toward an intersection.

- A police officer investigating an accident may need to calculate a driver's speed based on evidence such as the distance of skid marks from the point of impact. ∎

To fit a function to Table 18, let's adopt the strategy of *Looking for a Pattern*. In Table 19 we have added a third column for the first differences in $S(v)$, and a fourth column for the **second differences** (the first differences of the first differences).

TABLE 19

v	$S(v)$	First differences	Second differences
30	77		
		51	
40	128		11
		62	
50	190		11
		73	
60	263		

Are the constant second differences significant? Let's explore a little further.

EXERCISE **57.** (*Making Observations*) Construct a table for each function, using the indicated x-values. Calculate the first and second differences for the y-values.

 a. $y = 3x^2$; $x = 1.0, 1.5, 2.0, 2.5, 3.0$

 b. $y = ax^2$; $x = -1, 0, 1, 2, 3$

 c. $y = ax^2$; $x = x_0, x_0 + h, x_0 + 2h, x_0 + 3h, x_0 + 4h$

In Exercise 57 you have shown that for any function $y = ax^2$, any table with equally spaced x-values has y-values with constant second differences. Let's see whether the same is true of any function $y = ax^2 + bx + c$.

EXERCISE **58.** (*Making Observations*) Construct a table for each function, using the indicated x-values. Calculate the first and second differences for the y-values.

 a. $y = 3x^2 + 2x - 5$

 $x = 1.0, \ 1.5, \ 2.0, \ 2.5, \ 3.0$

 b. $y = ax^2 + bx + c$

 $x = -1, \ 0, \ 1, \ 2, \ 3$

 c. $y = ax^2 + bx + c$

 $x = x_0, \ x_0 + h, \ x_0 + 2h, \ x_0 + 3h, \ x_0 + 4h$

In Exercise 58 you have verified the first of the following statements. The second will be illustrated in Section 4-2, where you will find a quadratic function to fit the table in ***Stopping Distance***.

> For a quadratic function $y = f(x)$, a table with equally spaced x-values has y-values with constant second differences.
>
> Conversely, if a table with equally spaced x-values has y-values with constant second differences, it fits a quadratic function.

FYI As with linear functions, we cannot say that a function is quadratic on the basis of a table without additional information. We can only say that the table fits a quadratic function. However, if the second differences are not constant, then the function is not quadratic.

EXERCISES   *In Exercises 59–62, decide whether the table fits a linear function, a quadratic function, or neither.*

59.

x	y
3	23
4	16
5	11
6	8
7	7

60.

x	y
−2	−1
−1.5	7
−1	25
−0.5	62
0	123

61.

x	y
−1	−4
1	−6
3	−6
5	−4
7	0
9	6

62.

x	y
−1	−4
1	0
3	4
5	8
7	12
9	16

63. (*Making Observations*)

a. Use Table 18 to find the average rate of change in $S(v)$ with respect to v over the interval [30, 40]. What physical information is provided by this result?

b. Verify that the average rate of change in $S(v)$ with respect to v is larger over the interval [40, 50] than over [30, 40]. What physical information is provided by this result?

ADDITIONAL EXERCISES   *In Exercises 64–69,*

a. Describe how the basic graph of $y = x^2$ must be transformed to obtain the graph of the given function.

b. Identify the vertex and the points 1 unit to the left and right of the vertex.

c. Sketch the graph.

d. Find the axis of symmetry and the range of the function.

e. Identify the intervals where the function increases and decreases.

64. $y = (x − 6)^2 + 1$ **65.** $f(x) = (x − 2.5)^2 − 6.8$

66. $z = 2(x + 10)^2$ **67.** $A(r) = 2\pi(r + 1)^2 − 2\pi$

68. $P(x) = −4(x − 200)^2 + 9000$

69. $q = 100(p + 3)^2 − 5$

In Exercises 70–81,

a. Find the x-intercepts if there are any, and identify the vertex. If there are fewer than two x-intercepts, identify the points 1 unit to the left and right of the vertex.

b. Sketch the graph.

c. Find the axis of symmetry and the range of the function.

d. Identify the intervals where the function increases and decreases.

70. $y = 3x^2 + 4x + 1$ **71.** $y = 6 + x - x^2$

72. $g(x) = -0.3x^2 + 7x$ **73.** $s(t) = 400 - 16t^2$

74. $y = x^2 + 4x + 5$ **75.** $y = x^2 - 3x + 1$

76. $y = (x + 7)(x - 4)$ **77.** $V(u) = 2u(u + 6)$

78. $w = -(s + i)(s - i)$ **79.** $z = \left(x + \sqrt{5}\right)\left(x - \sqrt{5}\right)$

80. $y = -2\left[x - \left(4 + \sqrt{3}\right)\right]\left[x - \left(4 - \sqrt{3}\right)\right]$

81. $\beta = 5\left[\alpha - (2 + 4i)\right]\left[\alpha - (2 - 4i)\right]$

In Exercises 82–85, find an equation to fit the graph. (Hint: Write the equation in factored form, and use the point that is not an x-intercept to find the value of a.)

82.

83.

84.

85.
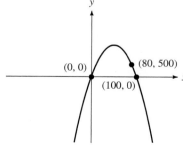

In Exercises 86–89, find an equation to fit the graph. (Hint: Write the equation in standard graphing form, and use the point that is not the vertex to find the value of a.)

86.

87.

88.

89.

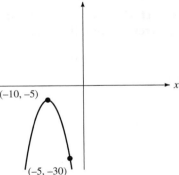

90. (*Interpreting Mathematics*) Assume that each relationship is quadratic. Decide whether the function first increases and then decreases or vice versa.

    a. T = elapsed time after a ball is thrown upward
        H = height of the ball

    b. L = length of a rectangle whose perimeter is 20
        A = area of the rectangle

    c. D = distance traveled as you ride up, then down a hill on the American Eagle roller coaster at Great America, IL
        S = speed of the roller coaster

    d. T = elapsed time from the onset of a recession until the economy recovers
        P = price of a typical share of stock

    e. On May 30, 1986, several creeks northeast of Pittsburgh overflowed, flooding our favorite ice cream stand to a depth of 3 feet.
        T = elapsed time from the beginning of the flood
        L = water level around Ice Cream World

91. (*Making Observations*) What name is given to the point at which a quadratic function stops increasing and starts to decrease or vice versa?

92. (*Making Observations*) Compare the process of graphing the equation $y = a|x - h| + k$ (see Section 2-5) with that of graphing $y = a(x - h)^2 + k$. In particular, describe the information provided by a, h, and k in each case.

93. (*Making Observations*)

    a. Complete the square to verify that $f(x) = ax^2 + bx + c$ has the standard graphing form $f(x) = a\left(x + \dfrac{b}{2a}\right)^2 - \dfrac{b^2 - 4ac}{4a}$.

    b. Explain why this shows that the vertex on the graph of $y = f(x)$ occurs when $x = -\dfrac{b}{2a}$.

94. (*Making Observations*) Write $g(x) = a(x - r_1)(x - r_2)$ in expanded form. Explain why the result shows that the vertex on the graph of $y = g(x)$ occurs when $x = \dfrac{r_1 + r_2}{2}$.

SUPPLEMENTARY TOPICS

Effect of Viewing Window on Apparent Steepness of Graphs

Exercises 95 and 96 illustrate that the steepness of a parabola can be distorted by the viewing window on your calculator. (Your calculator may have a key or menu item, such as "square," that automatically defines viewing windows in which steepness is not distorted. Consult your owner's manual for details.)

95. (*Making Observations*)

    a. Which of the functions $f(x) = x^2$, $g(x) = 4x^2$, or $h(x) = 0.25x^2$ has the steepest graph? The flattest graph?

    b. Graph $f(x)$ in the window $[-10, 10]$ by $[-100, 100]$, graph $g(x)$ in $[-5, 5]$ by $[-100, 100]$, and graph $h(x)$ in $[-20, 20]$ by $[-100, 100]$. Do your graphs support what you said in part (a)? Explain any apparent contradiction.

    c. Graph all three functions in the window $[-10, 10]$ by $[-100, 100]$, and indicate which graph is which. Do these graphs support what you said in part (a)?

96. (*Making Observations*)

    a. Graph $f(x) = 6x - x^2$ in the window $[-10, 10]$ by $[-10, 10]$.

    b. By the trial and error method, find a viewing window in which the graph of $g(x) = 3f(x) = 18x - 3x^2$ appears identical to your graph in part (a).

    c. Graph both functions in a window that allows you to compare the graphs accurately.

Complex Factors

The factored form of a quadratic function $f(x)$ may involve numbers with imaginary components. In this case the factors of $f(x)$ do not correspond to x-intercepts on the graph, but they still correspond to solutions of the equation $f(x) = 0$.

EXAMPLE 8

A quadratic function with complex factors

To write the factored form of

$$P(x) = x^2 + 2x + 5$$

it is easiest to solve the equation $P(x) = 0$ first.

$$x^2 + 2x + 5 = 0$$

$$x = \frac{-2 \pm \sqrt{2^2 - 4(1)(5)}}{2(1)}$$

$$= \frac{-2 \pm \sqrt{-16}}{2}$$

$$= \frac{-2 \pm 4i}{2}$$

$$= -1 \pm 2i$$

The solutions are $-1 + 2i$ and $-1 - 2i$. Therefore, in factored form,

$$P(x) = \left[x - (-1 + 2i)\right]\left[x - (-1 - 2i)\right]$$

■

EXERCISES   *In Exercises 97–100 obtain the factored form of $F(x)$ by solving $F(x) = 0$ first.*

97. $F(x) = x^2 - 4x + 13$ **98.** $F(x) = 4x^2 + 4x + 5$

99. $F(x) = 2x^2 + 8x + 9$ **100.** $F(x) = -x^2 + 2x - 2$

4-2 MODELING AND PROBLEM SOLVING WITH QUADRATIC FUNCTIONS

The Mathematical Looking Glasses you encountered in Section 4-1 illustrated that being familiar with quadratic functions can help you solve physical problems. In this section we will focus on methods for solving some common types of problems. Specifically, we will look at problems involving maximum and minimum values of quadratic functions, the solution of quadratic inequalities, and the construction of a quadratic function to fit a table.

OPTIMIZATION PROBLEMS

An **optimization problem** asks for the **extreme values** of a function, that is, its largest or smallest values. We encounter optimization problems when we want to know a company's most economical production schedule or the point of a comet's closest approach to the earth.

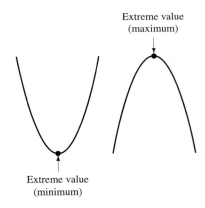

Extreme value
(maximum)

Extreme value
(minimum)

Figure 4–8

Figure 4-8 illustrates that every quadratic function has an extreme value, which occurs at the vertex on its graph. Thus in most optimization problems involving quadratic functions, the most difficult work in the solution is to find an appropriate function. Once that is done, you need only to find the vertex on its graph.

You have already solved an optimization problem. In *Maximum Profit* (page 147), your company needed to find the largest value of its profit function. You solved that problem in Exercise 50 of Section 4-1. Here is another.

A MATHEMATICAL LOOKING GLASS

Gravity

If you fire a toy rocket into the air, you might wonder how high it goes before falling back to earth. By using both physics and calculus, it can be shown that if you launch the rocket from a height of s_0 feet at v_0 feet per second, its height in feet after t seconds later is

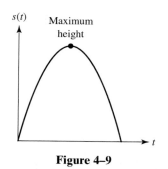

Figure 4–9

$$s(t) = -16t^2 + v_0 t + s_0$$

See Figure 4-9.

   This formula considers only the effect of gravity on the rocket and ignores other forces such as thrust and air resistance. If you continue your study of mathematics long enough to learn differential equations, you can construct a more accurate model.

   If you launch the rocket from ground level, then $s_0 = 0$. You probably don't know the value of v_0 offhand, but you might observe that the rocket's flight lasts 10 seconds. This is enough information for you to construct an equation from which you can calculate the rocket's maximum height. ■

EXERCISE   **1.** (*Problem Solving*)

   a. Explain why $s(10) = 0$.

   b. Use this fact to construct the equation for $s(t)$ and find the value of v_0.

   c. Find the vertex on the graph of $y = s(t)$. How high does the rocket go?

   To solve a quadratic optimization problem is simply to find the extreme value (the y-coordinate of the vertex) on its graph. You have learned how to do this in Section 4-1.

EXAMPLE 1   *Finding the extreme value of a quadratic function*

   The vertex on the graph of $f(x) = -0.5(x + 4)^2 + 2$ is $(-4, 2)$, as shown in Figure 4-10, so the extreme value of f is 2. Since the graph opens downward, the extreme value is a maximum.

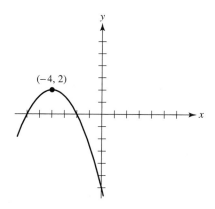

Figure 4–10

   If $f(x)$ is not written in standard graphing form, you can still find its extreme points.

   If $f(x)$ is written in its expanded form $f(x) = -0.5x^2 - 4x - 6$, we can calculate the x-coordinate of the vertex as $-\dfrac{b}{2a} = -4$. The y-coordinate is then $f(-4) = 2$.

   If $f(x)$ is written in its factored form $f(x) = -0.5(x + 6)(x + 2)$, we can read its x-intercepts $(-6, 0)$ and $(-2, 0)$, so the x-coordinate of the vertex is halfway between, at -4. The y-coordinate is then $f(-4) = 2$. ■

EXERCISES *In Exercises 2–5, find the extreme value of the function and state whether it is a maximum or a minimum.*

2. $f(x) = 7 + 4x - x^2$ **3.** $g(x) = 2x^2 + 4x + 5$

4. $F(t) = 0.1t(t - 100)$ **5.** $G(t) = -6(t + 20)(t - 8)$

SOLVING QUADRATIC INEQUALITIES

Sometimes you can solve a problem by solving a quadratic inequality. In such a case you can use the method you learned in Section 1-4 to solve quadratic inequalities graphically. Let's use the following Mathematical Looking Glass to learn an analytical method of solution.

A MATHEMATICAL LOOKING GLASS
Dance Floor

In March of 1992 Pittsburgh's Community Dance Network (CDN) sponsored a Vintage Dance Weekend featuring Richard Powers, founder of the Flying Cloud Academy of Vintage Dance in Cincinnati. The weekend included an evening ball, preceded by instruction by Mr. Powers in dances from the era 1890–1920.

Prior to the weekend, Lisa Tamres of CDN arranged to rent space at the Fox Chapel Yacht Club. The rented space was to be a rectangular area bounded on two sides by a permanent glass wall, and on the remaining sides by a number of movable panels, as shown in Figure 4-11.

Glass wall — Movable panels

Figure 4–11

Mr. Powers required 30 square feet of floor space per person for the daytime instruction and 20 square feet per person for the evening ball. CDN expected 50 to 70 people for instruction and 100 to 130 people for the ball. Lisa wanted to allow enough floor space for each event. ■

EXERCISE **6.** (*Problem Solving*) The yacht club had 105 linear feet of movable panels, and the room was large enough to arrange them in any rectangular shape Lisa might want. Let W represent the width of the rented space in feet.

    a. Express the number of square feet of rented floor space as a function of W.

    b. To satisfy Mr. Powers' requirements, explain why Lisa needs to solve the inequality $105W - W^2 \geq 2600$.

    c. Use graphical methods to solve the inequality, and describe all acceptable arrangements of paneling.

To solve the inequality in Exercise 6, we needed to

- find the W-intercepts on the graph of $S(W) = -W^2 + 105W - 2600$,
- observe that the intercepts divide the W-axis into the intervals $(-\infty, 40)$, $(40, 65)$, and $(65, \infty)$, and

• observe that the function is positive (so that the inequality is true) only in the interval (40, 65).

Let's see how we can obtain the same information by analytical methods. The process can be carried out in three steps.

EXAMPLE 2   *Solving a quadratic inequality analytically*

STEP 1:   To solve the inequality

$$-W^2 + 105W - 2600 \geqslant 0$$

begin by solving the equation

$$-W^2 + 105W - 2600 = 0$$
$$W^2 - 105W + 2600 = 0$$
$$(W - 40)(W - 65) = 0$$
$$W = 40, 65$$

This step finds the W-intercepts on the graph of $S(W) = -W^2 + 105W - 2600$.

STEP 2:   Next indicate the W-intercepts 40 and 65 on a number line.

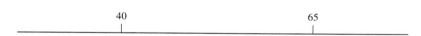

This step identifies the intervals $(-\infty, 40)$, $(40, 65)$, and $(65, \infty)$, where the sign of $S(W)$ cannot change.

STEP 3:   Choose a "test value" of W in $(-\infty, 40)$ and evaluate $S(W)$ there to see whether the inequality is true or false. For example, since $S(0) = -2600$, the inequality $S(W) \geqslant 0$ is false there. Since the sign of $S(W)$ cannot change in $(-\infty, 40)$, the inequality is false for *every* value of W in that interval.

Continue by choosing a test value in $(40, 65)$ and one in $(65, \infty)$. In each case the inequality is true or false over the entire interval according to whether it is true or false at the test value.

Since the inequality is true in $(40, 65)$ and also at the endpoints 40 and 65, the solution is $[40, 65]$.   ∎

EXAMPLE 3   *Solving a quadratic inequality analytically*

STEP 1:   To solve the inequality

$$2x^2 + 6x + 5 > 0$$

begin by solving the equation

$$2x^2 + 6x + 5 = 0$$

$$x = \frac{-6 \pm \sqrt{6^2 - 4(2)(5)}}{2(2)}$$

$$= \frac{-6 \pm \sqrt{-4}}{4}$$

Since there are no real solutions, the graph of $f(x) = 2x^2 + 6x + 5$ has no x-intercepts, so the function never changes sign.

STEP 2:   There is no need to draw a number line. The only interval to consider is $(-\infty, \infty)$.

STEP 3:   Any real number can be used as a test value. For example, $f(0) = 5$. Since the inequality is true there, it must be true on the entire interval $(-\infty, \infty)$.   ■

EXERCISES   **7.** (*Writing to Learn*) Graph the function $f(x)$ from Example 3. How does the graph support the conclusion of the example?

In Exercises 8–11, solve the inequality analytically.

8. $x^2 + x - 12 < 0$ **9.** $2x^2 + x - 1 > 0$

10. $5x - x^2 \leq 0$ **11.** $100 + 15x - x^2 \geq 0$

12. (*Interpreting Mathematics*) In **Maximum Profit** (page 147) your company's profit function is $P(x) = -0.005x^2 + 12.50x - 612.50$.

  a. Use graphical methods to solve the inequality $P(x) \geq 0$.

  b. Use analytical methods to solve $P(x) \geq 0$. Make sure your solution agrees with the one you obtained in part (a).

  c. What useful information does the solution provide?

FITTING A QUADRATIC FUNCTION TO A TABLE

Table 18 from **Stopping Distance** (page 158) suggests that a car's stopping distance is a quadratic function of its initial velocity. If that is the case, can we use the table to construct an equation for the function? Let's find out.

EXAMPLE 4   *Fitting a quadratic function to a table*

Table 18 is reproduced here.

Initial speed, in mph (v)	Stopping distance, in feet [$S(v)$]
30	77
40	128
50	190
60	263

If $S(v)$ is a quadratic function of v, then

$$S(v) = av^2 + bv + c$$

for some choice of real numbers a, b, and c. Furthermore,

$$S(30) = \ 77, \text{ so } \quad a(30)^2 + b(30) + c = \ 77$$

$$S(40) = 128, \text{ so } \quad a(40)^2 + b(40) + c = 128$$

$$S(50) = 190, \text{ so } \quad a(50)^2 + b(50) + c = 190$$

$$S(60) = 263, \text{ so } \quad a(60)^2 + b(60) + c = 263$$

Aha! We have constructed a system of linear equations in the variables a, b, and c. In Chapter 3 you learned that a system of three linear equations in three variables usually has a unique solution. The first three equations in the system can be rewritten as

$$900a + 30b + c = \ 77$$

$$1600a + 40b + c = 128$$

$$2500a + 50b + c = 190$$

In Exercise 13 you can solve this system and verify that the solution is $a = 0.055$, $b = 1.25$, $c = -10$. Thus the function is

$$S(v) = 0.055v^2 + 1.25v - 10$$

In Exercises 60–62 you can decide how accurately this function models stopping distances.   ∎

EXERCISES   **13.** (*Review*) Solve the 3×3 system of linear equations in Example 4, and confirm that your solution fits all four rows of Table 18.

In Exercises 14–17, find the equation of a quadratic function that fits the table.

14.

x	y
0	19
1	12
2	7
3	4

15.

x	y
2	-6
4	6
6	14
8	18

16.

x	y
300	520
400	530
500	520
600	490

17.

x	y
0.5	4
1.0	1
1.5	0
2.0	1

18. (*Writing to Learn*)

a. Why does the technique of Example 4 suggest that nearly every table containing exactly three data points fits a quadratic function?

b. Why does your answer to part (a) suggest that you should look for additional evidence before concluding that a table with three data points was generated by a quadratic function?

In the following Mathematical Looking Glass you can put the skills you just developed to good use.

A MATHEMATICAL
LOOKING GLASS
Gravity 2

Imagine that you are an astronaut on the moon, performing an experiment to confirm accepted information about the strength of the moon's gravitational field. You launch a ball upward from ground level, record its flight with a videocamera, and use the images to determine its height at several instants during its flight. Using the strategy of *Making a Table*, you put the information in tabular form.

TABLE 20

Time, in seconds	Height, in feet
5	383.5
10	634
15	751.5
20	736
25	587.5
30	306

You would like to know the total time of the ball's flight, and the maximum height it attains. As is often the case with tables, Table 20 allows only crude estimates. The total time of the flight is slightly more than 30 seconds and the maximum height is slightly more than 751.5 feet, but you need an equation for more precise answers. The formula in *Gravity* (page 164) does not apply on the moon. ∎

EXERCISES

19. (*Creating Models*) Verify that Table 20 fits a quadratic function, and find an equation for the function.

20. (*Interpreting Mathematics*) Find the ball's total flight time and its maximum height.

PARAMETRIC MODELS FOR QUADRATIC RELATIONSHIPS

The flight path of an object launched upward at an angle is parabolic in the absence of air, and approximately parabolic under more usual conditions. In Section 2-6 you learned that the motion of an object in a plane can often be described by a pair of parametric equations. In the following Mathematical Looking Glass, you can discover how to write parametric equations to express the object's horizontal and vertical movement as separate functions of time.

A MATHEMATICAL
LOOKING GLASS
Volleyball

Volleyball was invented in 1895 by William G. Morgan, physical director of the YMCA in Holyoke, MA. The game was designed as an indoor sport for businessmen who found basketball too strenuous. It grew rapidly in popularity, and was introduced into the Olympic games in 1964. The pace of today's Olympic volleyball games would probably surprise Mr. Morgan.

When serving a volleyball, Olympic players usually use an overhand serve, producing a trajectory like the one in Figure 4-12a. By contrast, a player at an informal picnic game might use an underhand serve, producing a trajectory like the one in Figure 4-12b.

A successful serve must clear the net without going beyond the end line. The path of the serve depends on both its initial velocity and its initial position and direction. For simplicity, let's assume that the overhand serve is

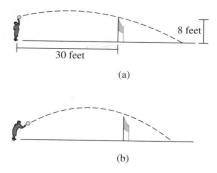

Figure 4–12

from a height of 8 feet at an angle of 14° with the horizontal, while the underhand serve is from a height of 3 feet at an angle of 45° with the horizontal. The path is then determined solely by the initial velocity. In Exercises 21–24, you will find the velocities needed for a successful serve of each style. ∎

Let

v_0 = the initial velocity of the serve,

t = time after the ball is served, in seconds,

x = horizontal distance traveled by the ball, and

y = height of the ball.

In Section 9-4, you will discover why the equations for the two types of serves are

overhand: $x = 0.97v_0t$

$y = 8 + 0.24v_0t - 16t^2$

underhand: $x = 0.71v_0t$

$y = 3 + 0.71v_0t - 16t^2$

In each case, we can eliminate the parameter to obtain an equation for the flight path in x and y. For the overhand serve, we can solve the first equation for t.

$$t = \frac{x}{0.97v_0}$$

Then we can substitute for t in the second equation.

$$y = 8 + 0.24v_0\left(\frac{x}{0.97v_0}\right) - 16\left(\frac{x}{0.97v_0}\right)^2$$

$$y = 8 + 0.25x - \frac{17}{v_0^2}x^2$$

As the equation shows, the path depends on the value of v_0.

EXERCISES   **21.** Eliminate the parameter and express y as a function of x for the underhand serve.

22. (*Interpreting Mathematics*) What is the initial velocity of an overhand serve that just clears the net? (*Hint*: For such a serve, $y = 8$ when $x = 30$.)

23. (*Interpreting Mathematics*) What is the initial velocity of an overhand serve that just touches the end line?

24. (*Interpreting Mathematics*) Repeat Exercises 22 and 23 for an underhand serve.

25. (*Making Observations*) Use your results from Exercises 21–24 to answer the following questions.

    a. Which style of serve must be hit harder in order to clear the net?

    b. For which style of serve is it more difficult to clear the net without going beyond the end line?

For each set of parametric functions in Exercises 26–29, eliminate the parameter to express y as a function of x.

26. $x = t - 3$
    $y = t^2$

27. $x = 3t$
    $y = 6t - 9t^2$

28. $x = 2t - 1$
    $y = -4t^2 + 8t - 3$

29. $x = 5 - t$
    $y = t^2 + 2t - 3$

ADDITIONAL EXERCISES

In Exercises 30–37, find the extreme value of the function, and state whether it is a maximum or a minimum.

30. $y = 3(x - 1)^2$

31. $y = -3(x - 1)^2$

32. $y = x^2 + 4x + 5$

33. $y = x^2 + 4x + 3$

34. $y = -x^2 - 12x + 2$

35. $y = 0.4x^2 + 0.8$

36. $y = -3(x + 3)(x - 9)$

37. $y = (1 - x)(2 - x)$

In Exercises 38–45, solve the inequality analytically. Show a calculator graph to support your answer.

38. $(x + 5)(x - 13) \leq 0$

39. $-3(x + 7)(x - 7) \geq 0$

40. $2x^2 + 7x - 9 > 0$

41. $14 - x - 3x^2 < 0$

42. $x^2 + 16 \geq 8x$

43. $4x^2 + 4x \leq -1$

44. $x^2 + 5x + 8 < 0$

45. $2x^2 + 3x + 2 > 0$

In Exercises 46–51, decide whether the table fits a quadratic function. If so, find its equation.

46.

x	y
-3	11
-2	6
-1	3
0	2

47.

x	y
4	16
5	18
6	18
7	16

48.

x	y
4	18
6	21
8	23
10	26

49.

x	y
2	37
5	35
8	32
11	28

50.

x	y
-6	100
-2	94
2	92
6	94

51.

x	y
0	3
1	4
2	6
3	10

52. (*Problem Solving*) In **Dance Floor**, what dimensions should the rented space have in order to provide the maximum floor space?

53. (*Problem Solving*) In **Fat Cats on Campus** (page 28) you saw that to sell x T-shirts during the semester, your company would need to charge a price of $16 - 0.005x$ dollars per shirt.

    a. What target sales level will generate the maximum revenue?

    b. Is this the same sales level that will generate the maximum profit (see Exercise 50 of Section 4-1)?

54. (*Problem Solving*) In Exercise 53, what sales levels will generate more than $10,000 revenue for your company?

55. (*Problem Solving*) A charter bus service has offered a deluxe 7-day tour from Atlanta to Florida's Disney World at a special rate of $800 per person for groups of 50 people. The company will also give a discount of $15 per person for every additional two people in the group. What number of people will generate the greatest revenue for the company?

56. (*Problem Solving*) In Exercise 55, what numbers of people will generate at least $45,000 revenue for the company?

57. (*Problem Solving*) A stretch of Interstate Highway I-35 in Oklahoma consists of straight downhill and uphill segments connected by a parabolic transitional segment. Figure 4-13 shows the required elevations in feet above sea level at three points on the transitional segment.

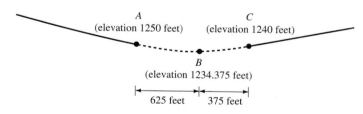

Figure 4–13

In a coordinate system with the x-axis at sea level and the y-axis passing through point A of Figure 4-13, find the equation of the transitional segment. (Adapted from Jeanne Agnew and Marvin Keener, *Industry Related Problems for Mathematics Students*, Oklahoma State University.)

58. (*Problem Solving*) A drain is located at point B in Figure 4-13. If a heavy rain causes the roadway to flood to a depth of 6 inches above the drain, how many horizontal feet of the road will be under water?

59. Your car probably gets its best gas mileage at a speed of about 45 miles per hour. Gas mileage is close to optimum at speeds between about 40 and 50 mph, but gets noticeably worse at lower or higher speeds. Of course at a speed of 0 mph you get 0 miles per gallon. Let v and M represent your speed in miles per hour and your gas mileage in miles per gallon.

a. (*Writing to Learn*) Explain why the relationship between v and M might be closely approximated by a quadratic function over an interval of normal driving speeds.

b. (*Creating Models*) For a car whose maximum gas mileage is 30 miles per gallon, express M as a quadratic function of v. (*Hint*: Write the function as $M = a(v - h)^2 + k$ and use the given information to find a, h, and k.)

c. (*Writing to Learn*) Explain why your equation in part (b) is not a good model for gas mileage over all possible driving speeds. For what speeds do you think your model is reasonably accurate?

SUPPLEMENTARY TOPIC
Validity of a Quadratic Model

In Example 4 we constructed $S(v) = 0.055v^2 + 1.25v - 10$ to fit Table 18 in *Stopping Distance*.

60. (*Interpreting Mathematics*) Evaluate $S(5)$. What does $S(5)$ mean in physical terms? What does your result say about the validity of $S(v)$ as a model for stopping distances from slow initial speeds?

61. (*Writing to Learn*) Explain why a good model $Q(v)$ for stopping distance should have $Q(0) = 0$ and $Q(v) > 0$ whenever $v > 0$.

62. (*Making Observations*) By *Guessing and Checking*, you might choose $Q(v) = 0.06v^2 + 0.8v$ as a model. It does not quite fit Table 18, but let's make some observations about it.

a. Evaluate $Q(30)$, $Q(40)$, $Q(50)$, and $Q(60)$. How do the values compare with the entries in Table 18?

b. Graph both $y = S(v)$ and $y = Q(v)$ in the window $[25, 65]$ by $[50, 300]$. How do the stopping distances predicted by Q compare with those predicted by S?

c. Graph both $y = S(v)$ and $y = Q(v)$ in the window $[0, 10]$ by $[-10, 20]$. How do the stopping distances predicted by Q compare with those predicted by S?

d. Based on your observations in parts (a) through (c), which function do you think is a better model for predicting stopping distance? Explain why.

4–3   RELATIONS

PREREQUISITES   MAKE SURE YOU ARE FAMILIAR WITH:
Radicals and Rational Exponents (Section A-18)
Equations with Radicals (Section A-19)

By now you are aware that relationships between two variables often occur in everyday life, as well as in academic fields as diverse as physics, environmental science, economics, and law. Because so many useful relationships are functions, most of this book focuses on functions. The remainder of this chapter is an exception. Here you will encounter relationships of a more general type. Read on for a simple but important example of a relationship in which neither variable is a function of the other.

A MATHEMATICAL LOOKING GLASS
Fermilab

According to the current theory of atomic structure, every atom of every substance in the universe is made up of two kinds of elementary particles, called **quarks** and **leptons**. The theory holds that there are six types of each. With the recent discovery of the so-called **top quark**, the existence of each type has now been confirmed. Prior to its discovery, the importance of the search for the top quark was summarized by James Trefil in the December 1989 issue of *Discover*.

> With the top quark, the most fundamental theory in all of science will be complete. Without the top quark, the theory is in deep trouble.

The search was made difficult by the extreme elusiveness of quarks. They are normally housed inside larger subatomic particles called **protons** and become observable only when a proton is broken apart by a collision with an **antiproton** at a speed near the speed of light. Thus an important piece of equipment in the search is the **synchrotron**, a circular tunnel in which magnetic and electrical fields are used to accelerate protons to the required speeds.

The first synchrotron capable of accelerating protons to near light speed was the Tevatron, built in 1983 at the Fermi National Accelerator Laboratory in Batavia, IL. Four smaller accelerators first take the protons to a speed of about 45,000 km/sec. The Tevatron then accelerates them to 99.99% of light speed (300,000 km/sec).

A diagram of the Tevatron is shown in Figure 4-14a. Its radius is 1 kilometer. Figure 4-14b shows a graphical model of the Tevatron centered at $(0, 0)$.

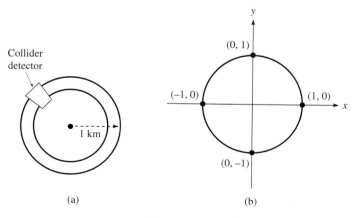

Figure 4–14

The process of accelerating protons and antiprotons to produce a collision is referred to as a "shot." Quarks produced by a shot can be detected only if the collision takes place within the collider detector in Figure 4-14a. To ensure that this occurs, the scientist in charge (the "shotmaster") must be able to calculate the position of each particle at any time. The correspondence between a particle's two coordinates is thus of great interest. ■

The circle in Figure 4-14b is referred to as the **unit circle**. It plays an important role in the development of trigonometry, and in a variety of physical problems. In Exercises 1–5 you can verify that neither variable is a function of the other.

EXERCISES   **1.** (*Review*) How does the graph in Figure 4-14b show that y is not a function of x?

2. (*Review*) Use Figure 4-14b to complete the following table by using two different values of y.

x	y
-1	0
0	
0	
1	0

How does the table show that y is not a function of x?

3. (*Writing to Learn*) Explain how the diagram at the left shows that the equation of the unit circle is $x^2 + y^2 = 1$.

4. (*Review*) Solve the equation in Exercise 3 for y. How does your result show that y is not a function of x?

5. (*Review*) Use each of the following to show that x is not a function of y.

   a. the graph in Figure 4-14b

   b. the table in Exercise 2

   c. the equation in Exercise 3

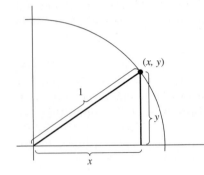

THE DEFINITION OF RELATION

Although the correspondence between x and y in **Fermilab** is not a function, it is a **relation**. A relation is *any* correspondence between two variables, so that every function is a relation, although not all relations are functions. The correspondence in **Fermilab** is a **quadratic relation**, one whose equation is either linear or quadratic in both variables. Most of the relations in this chapter are quadratic.

Let's define relations more precisely from numerical, analytical, and graphical perspectives and contrast the definition of a relation with that of a function.

- Any table giving values of two variables defines a relation. The table fits a function if and only if no two rows have the same first entry and different second entries. (See pages 43 and 44 to review this idea.)
- Any equation in two variables defines a relation. The relation is a function if and only if there is at most one value of the dependent variable for each value of the independent variable. There is no general procedure for deciding whether an equation represents y as a function of x, but solving for y sometimes works. (See pages 38–40 to review this idea.)

- Any curve or set of points in a coordinate plane defines a relation. The relation defines y as a function of x if and only if no vertical line intersects the graph more than once. (See pages 40–42 to review this idea.)

IMPLICIT FUNCTIONS

Relations do not always have unique outputs for each input. However, in a physical situation one particular output, among the many possible, is often of interest to us. Furthermore, additional information about the situation often allows us to express the output of interest as a function of the input. In Exercises 6 and 7 you can explore these ideas in the context of **Fermilab**.

EXERCISES

6. (*Writing to Learn*) The unit circle can be viewed as the union of an upper semicircle and a lower semicircle. Explain why each semicircle is the graph of a function.

7. (*Making Observations*) If you solve the equation $x^2 + y^2 = 1$ for y, you obtain $y = \pm\sqrt{1 - x^2}$. Use this observation to find equations for the upper and lower semicircles of the unit circle. How can you tell which equation corresponds to which semicircle?

In the context of **Fermilab**, Exercises 6 and 7 show that a proton's y-coordinate is $f_1(x) = \sqrt{1 - x^2}$ or $f_2(x) = -\sqrt{1 - x^2}$. We can decide which is the case if we know whether the proton is above or below the x-axis. In general, when a quadratic relation is not a function, its equation can be solved for y to obtain two functions of x. The graph of the relation is then the union of the graphs of the functions.

The equation $x^2 + y^2 = 1$ is true whenever $y = f_1(x)$. That is,

$$x^2 + [f_1(x)]^2 = 1$$

for all x in the domain of f_1, as you will see in Example 1. A similar statement applies to f_2. The functions f_1 and f_2 are therefore said to be defined **implicitly** by the relation $x^2 + y^2 = 1$. (*The American Heritage Dictionary* defines *implicit* to mean "contained in the nature of something although not readily apparent.") This is in contrast to the equation $y = \sqrt{1 - x^2}$, which defines $f_1(x)$ **explicitly**.

EXAMPLE 1

Verifying that a function is defined implicitly by a relation

One way to decide whether the function $f_1(x) = \sqrt{1 - x^2}$ is defined implicitly by the relation $x^2 + y^2 = 1$ is to solve the equation of the relation for y. Another way is to substitute $f_1(x)$ for y in the equation of the relation and reduce the equation to an identity (an equation that is true for all values of x).

$$x^2 + [f_1(x)]^2 = 1$$
$$x^2 + \left(\sqrt{1 - x^2}\right)^2 = 1$$
$$x^2 + (1 - x^2) = 1$$
$$1 = 1$$

Since the equation reduces to an identity, the calculation verifies that $f_1(x)$ is defined implicitly by the relation. ∎

EXERCISES   **8.** Use the method discussed in Example 1 to verify that the function $f_2(x) = -\sqrt{1 - x^2}$ is defined implicitly by the relation $x^2 + y^2 = 1$.

In Exercises 9–12, use the method of Example 1 to verify that the function $f(x)$ is defined implicitly by the given relation.

9. $x = y^2$; $f(x) = \sqrt{x}$

10. $x^2 + 4y^2 = 16$; $f(x) = -\dfrac{1}{2}\sqrt{16 - x^2}$

11. $9x^2 + (y - 5)^2 = 36$; $f(x) = 5 - 3\sqrt{4 - x^2}$

12. $x^2 - 25(y + 2)^2 = 1$; $f(x) = \dfrac{1}{5}\sqrt{x^2 - 1} - 2$

Example 2 shows how to find implicitly defined functions whose equations describe the graph of a quadratic relation.

EXAMPLE 2   *Finding functions to describe the graph of a relation*
Figure 4-15 shows the graph of the relation $(y - 1)^2 - x^2 = 1$.

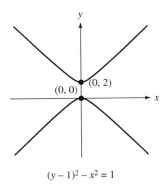

$(y - 1)^2 - x^2 = 1$

Figure 4–15

To find functions to describe the graph, begin by solving for y.

$$(y - 1)^2 = 1 + x^2$$
$$y - 1 = \pm\sqrt{1 + x^2}$$
$$y = 1 \pm \sqrt{1 + x^2}$$

The relation can thus be described by the two functions

$$f_1(x) = 1 + \sqrt{1 + x^2} \quad \text{and} \quad f_2(x) = 1 - \sqrt{1 + x^2}$$

If (x, y) is any point on the graph in Figure 4-15, then either $y = f_1(x)$ or $y = f_2(x)$. To decide what piece of the graph is described by f_1, observe that $f_1(x) \geq 1$ throughout its domain while $f_2(x) \leq 1$ throughout its domain. The graphs of $y = f_1(x)$ and $y = f_2(x)$ are shown in Figure 4-16.

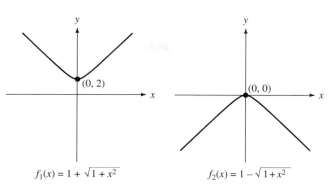

$$f_1(x) = 1 + \sqrt{1 + x^2} \qquad f_2(x) = 1 - \sqrt{1 + x^2}$$

Figure 4–16   ■

EXERCISES   *Each of Exercises* 13–16 *shows the graph of a quadratic relation. Solve for y to obtain two functions of x, and sketch the portion of the graph represented by each function.*

13. $x + 4y^2 = 4$

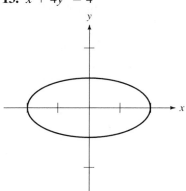

14. $x^2 + y^2 = 2y$

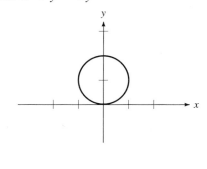

15. $4y^2 - x^2 = 4$

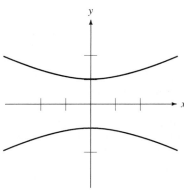

16. $x^2 - 4y^2 = 4$

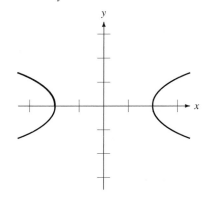

17. (*Writing to Learn*) Why can no vertical line intersect the graph of a quadratic relation more than twice?

In Exercises 18–21,

   a. Find a pair of functions defined implicitly by the relation.

    b. Graph each function, and combine the graphs to obtain the graph of the relation.

18. $4x^2 - 9y^2 = 36$ **19.** $4x^2 + 9y^2 = 36$

20. $x = 2y^2 - 6$ **21.** $x - x^2 = 2y^2 - 6$

Obtaining equations for functions defined implicitly by a relation can be useful for several reasons. Specifically,

- Often, as in **Fermilab**, an implicitly defined function identifies the output of interest from among the many possible outputs.
- Many graphing calculators cannot graph relations unless they are expressed as functions.

Finding equations for implicitly defined functions is often difficult or impossible. However, you can use the quadratic formula to obtain equations for functions defined implicitly by quadratic relations. The technique is discussed in Section B-4. You can apply it in Exercises 57–60.

In Sections 4-4 and 4-6 you will explore equations and graphs of quadratic relations. Analyzing tables generated by quadratic relations usually requires more advanced mathematics than you will learn in this book. As evidence to support this statement, we can point to the German astronomer and mathematician Johannes Kepler (1571–1630). Kepler spent years poring over data from observations made by Danish astronomer Tycho Brahe (1546–1601) before discovering that the relations describing planetary orbits are quadratic.

SQUARE ROOT FUNCTIONS

It often happens that a relationship is best modeled not by a quadratic relation, but by one of its implicitly defined functions. As you have seen, those functions usually involve square roots. Let's look at a relationship between two physical variables that produces a square root function.

A MATHEMATICAL LOOKING GLASS
Tsunami

Tsunamis are commonly called seismic sea waves or, incorrectly, tidal waves. The National Oceanic and Atmospheric Administration describes them in their pamphlet *Tsunami Watch and Warning* (1975).

The phenomenon we call "tsunami" is a series of traveling ocean waves of extremely long length and period, generated by disturbances associated with earthquakes occurring below or near the ocean floor. As the tsunami crosses the deep ocean, its length from crest to crest may be 100 miles or more, its height from trough to crest only a few feet. It cannot be felt aboard ships in deep water, and cannot be seen from the air. But in deep water, tsunami waves may reach forward speeds exceeding 600 miles per hour.

As the tsunami enters the shoaling water of coastlines in its path, the velocity of its waves diminishes and wave height increases. It is in these shallow waters that tsunamis become a threat to life and property, for they can crest to heights of more than 100 feet, and strike with devastating force.

If you live in Hawaii or on the Pacific Coast of the United States or Canada, you may receive an occasional warning that a tsunami is approaching. To tell you how much time you have to evacuate the area, the National Weather Service must predict the tsunami's speed. *Tsunami Watch and Warning* says,

> Tsunami speed is determined solely by water depth, and this fixed relationship makes it possible to forecast tsunami arrival times for distant locations. The tsunami illustrated here, although somewhat exaggerated in the vertical dimension, is characteristic.

Speed (mph)	670	600	519	424	299	212	94	30
Depth (feet)	30,000	24,000	18,000	12,000	6,000	3,000	600	60

Figure 4–17

Adapted from *Tsunami Watch and Warning*, National Oceanic and Atmospheric Administration, 1975.   ■

EXERCISES

22. (*Making Observations*) Show that, in Figure 4-17, allowing for rounding in the data, the ratio $\dfrac{d}{s^2}$ is constant (where d is depth in feet and s is mph).

23. (*Writing to Learn*) Explain why the data in Figure 4-17 fits an equation $d = ks^2$. What is the value of k?

24. (*Writing to Learn*) Tell why you could not use the method of Section 4-1 to discover that the data fits a quadratic function.

25. The National Weather Service needs to use ocean depths as inputs and obtain predicted tsunami speeds as outputs. For this purpose, it is more convenient to express s as a function of d.

   a. (*Creating Models*) Use your equation from Exercise 23 to express the speed of the tsunami as a function of water depth.

   b. (*Writing to Learn*) Explain why s is not a function of d if s and d are abstract variables and $d = ks^2$.

26. (*Interpreting Mathematics*) How does your equation reflect the fact that a tsunami slows down as the water becomes shallower?

27. (*Interpreting Mathematics*) Use the data in Figure 4-17 to decide which statement best describes the speed of a tsunami as it goes from deep to shallow water. Explain your conclusions.

   a. It slows down at a constant rate.

   b. It slows down gradually in deep water and more sharply as it enters shallow water.

   c. It slows down rapidly in deep water and more gradually as it enters shallow water.

28. (*Making Observations*) Graph your function from Exercise 25a on your calculator. How does the graph support your conclusion from Exercise 27?

Your function in Exercise 25a involves a square root. To see how the equations of such functions relate to their graphs, let's proceed by *Examining a Related Problem*. In Sections 2-5 and 4-1 you saw how the graphs of the basic functions $y = |x|$ and $y = x^2$ can be transformed to produce the graph of any function $y = a|x - h| + k$ or $y = a(x - h)^2 + k$. Therefore, let's begin to study **square root functions** by looking at the graph of the basic function $y = \sqrt{x}$ in Figure 4-18.

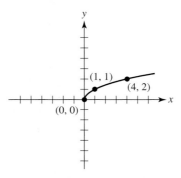

Figure 4–18

EXERCISES

29. (*Writing to Learn*) The graph in Figure 4-18 appears to be half of a parabola. Is it really? How do you know?

30. (*Review*) Both the domain and range of the function $y = \sqrt{x}$ are $[0, \infty)$.
   a. How does the graph of $y = \sqrt{x}$ show this?
   b. How does the equation show this?

You can use your knowledge of the basic graph of $y = \sqrt{x}$ and your experience with linear absolute value functions and quadratic functions to sketch the graphs of a variety of square root functions.

EXERCISES

31. (*Making Observations*) Refer to the boxed statements on pages 87 and 151 and use the strategy of *Looking for a Pattern* to make a conjecture about how to obtain the graph of a function $y = a\sqrt{x - h} + k$ from the basic graph of $y = \sqrt{x}$.

32. (*Making Observations*) Use your conjecture from Exercise 31 to sketch the graph of each function below. Then check the validity of your conjecture by graphing each function on your calculator.
   a. $y = 3\sqrt{x}$ **b.** $y = -3\sqrt{x}$
   c. $y = \sqrt{x} - 4$ **d.** $y = \sqrt{x - 4}$

33. Show how the graph of the function $f(d) = \sqrt{15d}$ in **Tsunami** can be obtained by transforming the basic square root graph.
   (*Hint:* $\sqrt{15d} = \sqrt{15}\sqrt{d}$.)

In Exercises 34–37,

    a. Sketch the graph of the equation by transforming the basic graph of $y = \sqrt{x}$. Identify the points obtained by moving the points $(0, 0)$ and $(1, 1)$ on the basic graph.

    b. Find the domain and range of the function.

34. $y = 1 + \sqrt{x + 3}$ **35.** $y = 1 - \sqrt{x + 3}$

36. $y = 3\sqrt{x} - 2$ **37.** $y = \sqrt{3x} - 2$

ADDITIONAL EXERCISES   *In Exercises 38–41, break the graph of the relation into two pieces, each of which represents y as a function of x.*

38.

39.

40.

41.

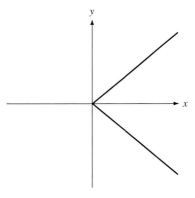

In Exercises 42–45, use the method of Example 1 to verify that the function $f(x)$ is defined implicitly by the given relation.

42. $3x - y^2 = 6;\ f(x) = \sqrt{3x - 6}$

43. $9x^2 + 25y^2 = 225;\ f(x) = -\dfrac{3}{5}\sqrt{25 - x^2}$

44. $x^2 - 4(y - 1) = 4x;\ f(x) = \dfrac{1}{4}(x - 2)^2$

45. $x^2 - 4(y + 1)^2 = 4;\ f(x) = \dfrac{1}{2}\sqrt{x^2 - 4} - 1$

In Exercises 46–53,

    a. Sketch the graph of the equation by transforming the basic graph $y = \sqrt{x}$. Identify the points obtained by moving the points $(0, 0)$ and $(1, 1)$ on the basic graph.

b. Find the domain and range of the function.

46. $y = -2\sqrt{x}$ **47.** $y = \sqrt{2x}$

48. $y = -2\sqrt{2x}$ **49.** $y = 4 + \sqrt{x - 1}$

50. $y = 4 + \sqrt{x + 1}$ **51.** $y = 4 - \sqrt{x + 1}$

52. $y = \sqrt{2x + 8} - 1$ [*Hint*: Think of $2x + 8$ as $2(x + 4)$.]

53. $y = 3\sqrt{0.5x - 2}$

Exercises 54 and 55 refer to ***Tsunami***. If a tsunami's point of origin and time of origin are known, its arrival time at any coastal point can be predicted. In these exercises let's assume that the ocean's depth x miles east of Honolulu is $15x + 135$ feet, as in Figure 4-19.

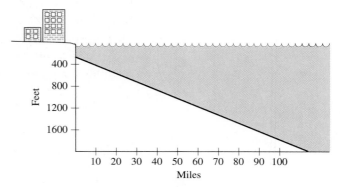

Figure 4–19

If a tsunami is generated x miles east of Honolulu, calculus can be used to show that it will take $h(x) = \dfrac{\sqrt{x + 9} - 3}{15}$ hours to strike the coast.

54. (*Review*)

  a. What are the domain and range of $h(x)$ if x is regarded as an abstract variable?

  b. What are the domain and range of $h(x)$ if x is given its physical meaning?

55. (*Interpreting Mathematics*)

  a. Calculate the average rate of change in $h(x)$ with respect to x over $[160, 280]$.

  b. What physical information does your answer provide?

  c. Calculate the average rate of change in $h(x)$ with respect to x over $[0, 16]$.

  d. Which of your answers to parts (a) and (c) is larger? What physical information does that observation provide?

56. (*Writing to Learn*) Why does the method of Example 1 guarantee that a given function is defined implicitly by a given relation?

In Exercises 57–60,

  a. Use the quadratic formula to find two functions defined implicitly by the relation.

   b. Graph each function, and combine the two graphs to obtain the graph of the relation.

57. (*Extension*) $\frac{1}{2}x^2 + xy + \frac{1}{2}y^2 + 4x + 5y + 10 = 0$

58. (*Extension*) $x^2 + xy + y^2 + 4x + 5y + 10 = 0$

59. (*Extension*) $x^2 - xy + y^2 - 12 = 0$

60. (*Extension*) $x^2 + 10\sqrt{3}\,xy + 11y^2 - 64 = 0$

Exercise 61 illustrates that there is more than one way to choose a pair of implicitly defined functions to express a quadratic relationship.

61. (*Making Observations*)

   a. Verify that the functions $f_1(x) = x$, $f_2(x) = -x$, $g_1(x) = |x|$, and $g_2(x) = -|x|$ are all defined implicitly by the relation $x^2 - y^2 = 0$.

   b. Sketch the graphs of $y = f_1(x)$ and $y = f_2(x)$, and combine the graphs to obtain the graph of the relation.

   c. Sketch the graphs of $y = g_1(x)$ and $y = g_2(x)$, and combine the graphs to obtain the graph of the relation.

4–4 A GRAPHICAL VIEW OF CONIC SECTIONS

To begin exploring graphs of quadratic relations, take a break and perform a simple experiment.

Get a flashlight, a marker, and three sheets of poster paper, at least 2 × 3 feet each. Tape one sheet of paper to a vertical wall, and shine the flashlight so that the beam falls entirely within the paper, as in Figure 4-20a. Hold the flashlight steady, and have a friend trace around the edge of the beam. The curve, shown in Figure 4-20b, is an **ellipse**.

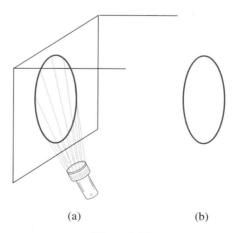

(a)   (b)

Figure 4–20

Next, tape a second sheet of paper to the wall, hold the flashlight a few inches from the wall near the bottom of the paper, and shine it so that one edge of the beam hits a point on the ceiling directly above the flashlight, as in

Figure 4-21a. Again, have a friend trace around the edge of the beam. The curve on the wall, shown in Figure 4-21b, is a **parabola**.

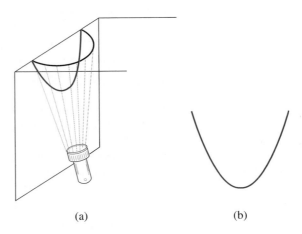

(a)   (b)

Figure 4–21

Finally, tape the third sheet of paper to the wall, hold the flashlight a few inches from the wall near the bottom of the paper, and shine the flashlight straight up, as in Figure 4-22a. Once more have a friend trace around the edge of the beam. The curve on the wall, shown in Figure 4-22b, is one branch of a **hyperbola**. The entire hyperbola, shown in Figure 4-22c, would be traced by the beam of a flashlight with a bulb at each end.

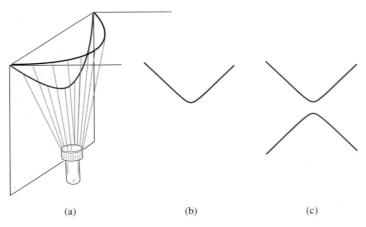

(a)   (b)   (c)

Figure 4–22

The curves you have traced are called **conic sections**, because each is the intersection of a cone (the light beam) with a plane (the wall). We will develop formal definitions for all three types in Section 4-6, but for now you can think of them as describing the curves traced out by your flashlight. In this section you will discover that the graphs of many second degree equations in two variables resemble the curves you traced with the flashlight beam.

The equations in this chapter have the form

$$Ax^2 + Cy^2 + Dx + Ey + F = 0$$

A quadratic relation can also contain a term Bxy, but a systematic study of such equations and their graphs requires trigonometry. In this chapter it will always be the case that both x and y appear in the equation and that the equation is actually second degree (A and C are not both 0).

We can explore the graphs of such equations most easily by completing the square in each variable, if possible. This leads to one of several standard graphing forms, depending on whether the product AC is zero, positive, or negative. Let's explore each case in turn.

A GRAPHICAL VIEW OF PARABOLAS

The condition $AC = 0$ means that either $A = 0$ or $C = 0$. When $C = 0$ the equation contains no term involving y^2, so y is a quadratic function of x. In this case completing the square produces the **standard graphing form** of a quadratic function

$$y = a(x - h)^2 + k$$

In Section 4-1 you learned a few distinctive graphical features of such a function. They are restated here.

The graph of $y = a(x - h)^2 + k$ is a parabola.

- Its vertex is (h, k).
- Its axis of symmetry is the line $x = h$.
- It opens up if $a > 0$ and down if $a < 0$.
- The points 1 unit to the left and right of the vertex on the parabola have a y-coordinate of $k + a$.

Similarly, when $A = 0$, x is a quadratic function of y. In this case completing the square produces the standard graphing form

$$x = a(y - k)^2 + h$$

In Section 4-5 you will learn the graphing principles needed to establish the following graphical features of this equation.

The graph of $x = a(y - k)^2 + h$ is a parabola.

- Its vertex is (h, k).
- Its axis of symmetry is the line $y = k$.
- It opens right if $a > 0$ and left if $a < 0$.
- The points 1 unit above and below the vertex on the parabola have an x-coordinate of $h + a$.

Figure 4-23 illustrates the preceding statements.

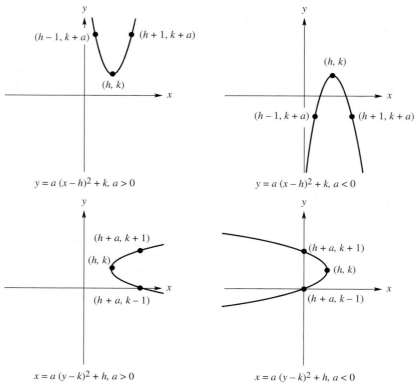

Figure 4–23

EXAMPLE 1   *Sketching the graph of an equation $x = a(y - k)^2 + h$*

To sketch the graph of

$$y^2 + 2x - 4y + 6 = 0$$

first write it in the standard graphing form

$$x = -\frac{1}{2}(y - 2)^2 - 1$$

Since the quadratic term involves y, the graph is a parabola opening horizontally. Its vertex is $(-1, 2)$, and it opens to the left since $a = -\frac{1}{2} < 0$. The axis of symmetry is the line $y = 2$. The points 1 unit above and below the vertex have an x-coordinate of $-1 - \frac{1}{2} = -\frac{3}{2}$. The graph is shown in Figure 4-24. ■

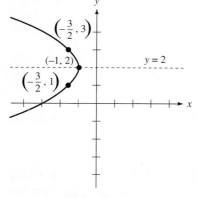

Figure 4–24

EXERCISES   *In Exercises 1–4, sketch the graph of the equation. Identify the vertex and two other points on the graph and the axis of symmetry.*

1. $x^2 + 8x - y + 12 = 0$

2. $x^2 - 2x + 2y + 3 = 0$

3. $y^2 + x - 10y + 18 = 0$

4. $2y^2 - x + 8y + 7 = 0$

A GRAPHICAL VIEW OF ELLIPSES

The condition $AC > 0$ means that A and C have the same sign. By completing the square in both x and y you produce an equation of the form $A(x - h)^2 + C(y - k)^2 = G$. Dividing by G on both sides then produces the **standard graphing form**

$$\left(\frac{x - h}{\text{number}}\right)^2 + \left(\frac{y - k}{\text{number}}\right)^2 = 1$$

as long as $G > 0$. Later in this section we will see what happens when $G \leqslant 0$.

To explore the graphs of these equations, let's begin with a simple example. Figure 4-25 shows the graph of

$$\left(\frac{x}{3}\right)^2 + \left(\frac{y}{2}\right)^2 = 1$$

The graph has the shape of an ellipse. You can explain some of its distinctive graphical features in Exercises 5 and 6.

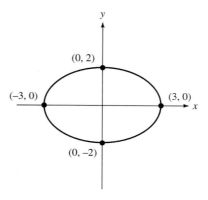

Figure 4–25

EXERCISES

5. (*Writing to Learn*)

    a. Show that if $|x| > 3$, the equation has no real solutions for y. Why does this imply that the graph lies on or between the lines $x = -3$ and $x = 3$?

    b. Explain why the graph also lies on or between lines $y = -2$ and $y = 2$.

    c. Show that the points $(\pm 3, 0)$ and $(0, \pm 2)$ are on the graph.

6. (*Writing to Learn*)

    a. If a point (a, b) is on the graph, show that $(-a, b)$ is also on the graph. Why does this imply that the y-axis is an axis of symmetry for the graph?

    b. If a point (a, b) is on the graph, show that $(a, -b)$ is also on the graph. Why does this imply that the x-axis is an axis of symmetry for the graph?

The features described in Exercises 5 and 6 can be found in every ellipse. Specifically, an ellipse has two **axes of symmetry**, one horizontal and one vertical, as indicated in Figure 4-26. The two axes intersect at the **center** of the ellipse. The longer of the two chords along the axes of symmetry is the

major axis of the ellipse, and the shorter is its **minor axis**. (The terms *major axis* and *minor axis* can also be used to refer to the entire lines containing the two chords.) The endpoints of the major axis are the **vertices** of the ellipse, and the endpoints of the minor axis are its **covertices**.

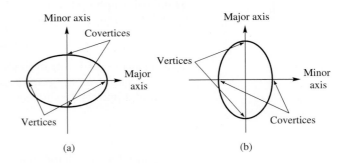

(a)   (b)

Figure 4–26

Let's see how to sketch the graph of a general equation

$$Ax^2 + Cy^2 + Dx + Ey + F = 0 \quad \text{with } AC > 0.$$

Example 2 shows how to obtain a standard graphing form.

EXAMPLE 2   *Standard graphing form for $Ax^2 + Cy^2 + Dx + Ey + F = 0$ when $AC > 0$*
To write $9x^2 + 4y^2 - 180x - 56y + 1060 = 0$ in standard graphing form, first complete the square in x and y.

$$9(x^2 - 20x) + 4(y^2 - 14y) + 1060 = 0$$

$$9(x^2 - 20x + 100 - 100) + 4(y^2 - 14y + 49 - 49) + 1060 = 0$$

$$9(x^2 - 20x + 100) - 900 + 4(y^2 - 14y + 49) - 196 + 1060 = 0$$

$$9(x - 10)^2 + 4(y - 7)^2 = 36$$

Then divide by 36 on both sides.

$$\frac{(x - 10)^2}{4} + \frac{(y - 7)^2}{9} = 1$$

$$\left(\frac{x - 10}{2}\right)^2 + \left(\frac{y - 7}{3}\right)^2 = 1$$

EXERCISE   **7.** (*Writing to Learn*)

a. Show that if $\left|\dfrac{x - 10}{2}\right| > 1$, the equation in Example 2 has no real solutions for y. Why does this imply that the graph lies on or between the lines $x = 8$ and $x = 12$?

b. Explain why the graph also lies on or between lines $y = 4$ and $y = 10$.

c. Show that the points $(8, 7)$, $(12, 7)$, $(10, 4)$, and $(10, 10)$ are on the graph.

Figure 4–27

In Section 4-5 you will see why the lines $x = 10$ and $y = 7$ are the axes of symmetry for the ellipse in Example 2. The graph is shown in Figure 4-27.

It is traditional to denote the two denominators in the standard graphing form by a and b, with $a > b$. Thus the graphical features of ellipses can be summarized in the following statements.

If $a > b$, the graph of $\left(\dfrac{x-h}{a}\right)^2 + \left(\dfrac{y-k}{b}\right)^2 = 1$ is an ellipse.

- Its center is (h, k).
- Its vertices are a units to the left and right of the center, at $(h \pm a, k)$.
- Its covertices are b units above and below the center, at $(h, k \pm b)$.
- Its major axis is the line $y = k$, and its minor axis is the line $x = h$.

If $a > b$, the graph of $\left(\dfrac{x-h}{b}\right)^2 + \left(\dfrac{y-k}{a}\right)^2 = 1$ is an ellipse.

- Its center is (h, k).
- Its vertices are a units above and below the center, at $(h, k \pm a)$.
- Its covertices are b units to the left and right of the center, at $(h \pm b, k)$.
- Its major axis is the line $x = h$, and its minor axis is the line $y = k$.

Figure 4-28 illustrates the preceding statements.

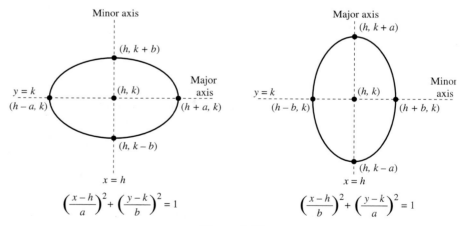

Figure 4–28

Example 3 shows how to sketch an ellipse quickly from its standard graphing form.

EXAMPLE 3   *Sketching an ellipse quickly from the standard graphing form of its equation*
To sketch the graph of

$$\left(\frac{x-1}{5}\right)^2 + \left(\frac{y-3}{2}\right)^2 = 1$$

begin by plotting the center at $(1, 3)$. The major axis is the horizontal line $y = 3$. The vertices are 5 units to the left and right of the center, at $(-4, 3)$

and $(6, 3)$. The minor axis is the vertical line $x = 1$. The covertices are 2 units above and below the center, at $(1, 1)$ and $(1, 5)$. The graph is shown in Figure 4-29.

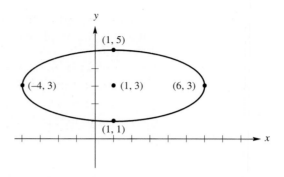

Figure 4–29

EXERCISES   *In Exercises 8–13, write the equation in standard graphing form and sketch the graph. Identify the center, vertices, covertices, and major and minor axes.*

8. $9x^2 + 25y^2 = 225$ **9.** $4x^2 + y^2 = 64$

10. $(x - 3)^2 + 5y^2 = 20$

11. $(x + 4)^2 + 4(y - 2)^2 = 16$

12. $49x^2 + 9y^2 + 18y - 432 = 0$

13. $x^2 + 6y^2 - 60x + 300y + 4644 = 0$

A GRAPHICAL VIEW OF CIRCLES

If $A = C$ in an equation

$$Ax^2 + Cy^2 + Dx + Ey + F = 0$$

then also $a = b$ in its standard graphing form. In this case, the major and minor axes have equal lengths, and the ellipse is a circle. Equations of circles are usually written in the form

$$(x - h)^2 + (y - k)^2 = r^2$$

In Exercise 14 you can explain why the value of r is the radius of the circle.

EXERCISE   **14.** (*Writing to Learn*) Use the following figure and the distance formula to explain why a point (x, y) is on a circle of radius r centered at (h, k) if and only if $(x - h)^2 + (y - k)^2 = r^2$.

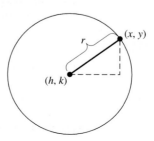

EXAMPLE 4   *Sketching a circle*

To obtain the standard graphing form of the equation

$$4x^2 + 4y^2 + 4x - 8y - 31 = 0$$

begin by completing the square in both x and y to obtain

$$4(x + 0.5)^2 + 4(y - 1)^2 = 36$$

Since $A = C$, do not divide both sides of the equation by 36. Instead, divide both sides by 4 to obtain

$$(x + 0.5)^2 + (y - 1)^2 = 9$$

The graph is a circle with radius 3 centered at $(-0.5, 1)$. Its graph is shown in Figure 4-30.

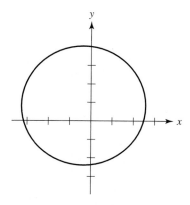

Figure 4–30

EXERCISES   *In Exercises 15–20, sketch the circle and identify its center and radius.*

15. $x^2 + y^2 = 25$ **16.** $x^2 + y^2 - 4x = 0$

17. $(x + 6)^2 + (y - 2)^2 = 4$

18. $9(x + 1)^2 + 9(y + 3)^2 = 1$

19. $x^2 + y^2 - 8x - 6y + 23 = 0$

20. $4x^2 + 4y^2 - 12x + 4y - 15 = 0$

A GRAPHICAL VIEW OF HYPERBOLAS

The condition $AC < 0$ means that A and C have opposite signs. By completing the square in both x and y you produce an equation of the form $A(x - h)^2 + C(y - k)^2 = G$. Dividing by G on both sides then produces a **standard graphing form**

$$\left(\frac{x - h}{a}\right)^2 - \left(\frac{y - k}{b}\right)^2 = 1$$

or

$$\left(\frac{y - k}{a}\right)^2 - \left(\frac{x - h}{b}\right)^2 = 1$$

as long as $G \neq 0$. Later in this section we will see what happens when $G = 0$.

FYI Although these standard graphing forms closely resemble that of an ellipse, there are two important differences.

- The terms on the left side have opposite signs.
- The denominator of the positive term is denoted by a, regardless of which denominator is larger.

EXAMPLE 5   *Standard graphing form for $Ax^2 + Cy^2 + Dx + Ey + F = 0$ when $AC < 0$*

To write $9x^2 - 4y^2 - 180x + 56y + 668 = 0$ in standard graphing form, first complete the square in x and y.

$$9(x^2 - 20x) - 4(y^2 - 14y) + 668 = 0$$

$$9(x^2 - 20x + 100 - 100) - 4(y^2 - 14y + 49 - 49) + 668 = 0$$

$$9(x^2 - 20x + 100) - 900 - 4(y^2 - 14y + 49) + 196 + 668 = 0$$

$$9(x - 10)^2 - 4(y - 7)^2 = 36$$

Then divide by 36 on both sides.

$$\frac{(x - 10)^2}{4} - \frac{(y - 7)^2}{9} = 1$$

$$\left(\frac{x - 10}{2}\right)^2 - \left(\frac{y - 7}{3}\right)^2 = 1$$

∎

The simplest equations having the standard graphing form just discussed are $x^2 - y^2 = 1$ and $y^2 - x^2 = 1$. Exercises 21–25 will give you some insight into their graphs.

EXERCISES   21. (*Writing to Learn*) In the first quadrant, show that the equation $x^2 - y^2 = 1$ is equivalent to $y = \sqrt{x^2 - 1}$. Why are the two equations not equivalent in general?

22. (*Review*) What positive values of x are in the domain of the function $y = \sqrt{x^2 - 1}$?

23. Complete the following table for $y = \sqrt{x^2 - 1}$.

x	y
1	
2	
10	
100	

24. (*Writing to Learn*) How does Exercise 23 suggest that if x is large, the graph of $y = \sqrt{x^2 - 1}$ is near the graph of $y = x$?

In the first quadrant the graph of $x^2 - y^2 = 1$ coincides with that of $y = \sqrt{x^2 - 1}$, shown in Figure 4-31a.

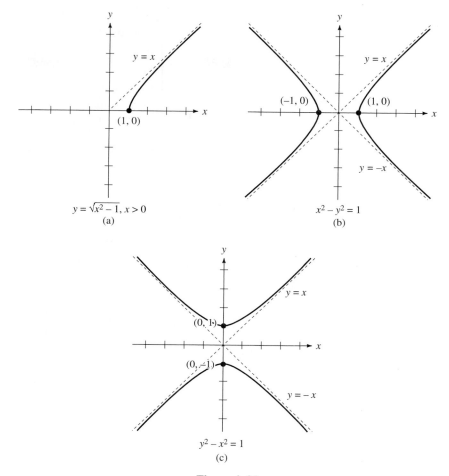

Figure 4–31

EXERCISE **25.** (*Making Observations*) Show that if a point (a, b) is on the graph of $x^2 - y^2 = 1$, so are both the points $(-a, b)$ and $(a, -b)$.

Exercise 25 shows that both the x-axis and the y-axis are axes of symmetry for the graph of $x^2 - y^2 = 1$. The entire graph is shown in Figure 4-31b. The lines $y = x$ and $y = -x$, indicated by the dashed lines, are **asymptotes**. In general an asymptote is a line approached arbitrarily closely by a graph as one of the variables becomes arbitrarily large. In Exercise 64 you can show analytically why the lines $y = \pm x$ must be asymptotes for the graph of $x^2 - y^2 = 1$. Figure 4-31c shows the graph of $y^2 - x^2 = 1$, whose shape can be established by observations similar to those in Exercises 21–25. The graphs of both equations are hyperbolas.

The hyperbolas in Figures 4-31b and 4-31c each have two axes of symmetry, intersecting at the **center** of the hyperbola. One axis of symmetry, the **transverse axis**, intersects the hyperbola in two points, called the **vertices**. The other axis of symmetry is the **conjugate axis**.

Observations like those in Exercises 21–25 can be used to establish the following facts.

The graph of $\left(\dfrac{x-h}{a}\right)^2 - \left(\dfrac{y-k}{b}\right)^2 = 1$ is a hyperbola.

- Its center is (h, k).
- Its vertices are a units to the left and right of the vertex, at $(h \pm a, k)$.
- Its asymptotes are the lines

$$\frac{x-h}{a} = \pm \frac{y-k}{b} \qquad \text{or, equivalently,} \qquad y = \pm \frac{b}{a}(x-h) + k$$

The graph of $\left(\dfrac{y-k}{a}\right)^2 - \left(\dfrac{x-h}{b}\right)^2 = 1$ is a hyperbola.

- Its center is (h, k).
- Its vertices are a units above and below the vertex, at $(h, k \pm a)$.
- Its asymptotes are the lines

$$\frac{y-k}{a} = \pm \frac{x-h}{b} \qquad \text{or, equivalently,} \qquad y = \pm \frac{a}{b}(x-h) + k$$

Figure 4-32 illustrates the preceding statements.

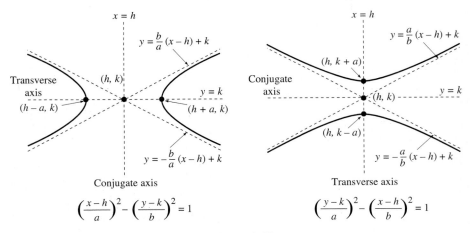

Figure 4–32

EXERCISE   **26.** (*Writing to Learn*) Adapt the argument in Exercises 21–25 to show that

the hyperbola $\left(\dfrac{x-h}{a}\right)^2 - \left(\dfrac{y-k}{b}\right)^2 = 1$ has asymptotes $\dfrac{x-h}{a} = \pm \dfrac{y-k}{b}$

In the standard graphing form of a hyperbola, the positive denominator is the distance from the center to each vertex. To sketch the asymptotes, locate the vertices and the two points b units from the center along the conjugate axis. Sketch a box through these points, as shown in Figure 4-33. The asymptotes are the diagonals of this box.

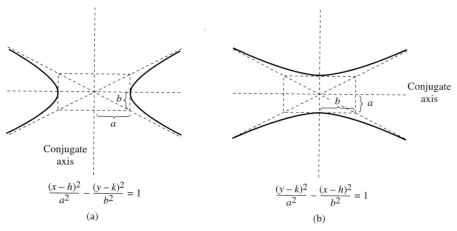

$$\frac{(x-h)^2}{a^2} - \frac{(y-k)^2}{b^2} = 1$$

(a)

$$\frac{(y-k)^2}{a^2} - \frac{(x-h)^2}{b^2} = 1$$

(b)

Figure 4–33

To find the equations of the asymptotes, observe that if x and y are large, the terms on the left of an equation

$$\left(\frac{x-h}{a}\right)^2 - \left(\frac{y-k}{b}\right)^2 = 1$$

are large in comparison to the 1 on the right. Thus the asymptotes have equations

$$\left(\frac{x-h}{a}\right)^2 - \left(\frac{y-k}{b}\right)^2 = 0$$

Simplifying,

$$\left(\frac{x-h}{a}\right)^2 = \left(\frac{y-k}{b}\right)^2$$

$$\frac{x-h}{a} = \pm\frac{y-k}{b}$$

$$y - k = \pm\frac{b}{a}(x - h)$$

Similarly, a hyperbola

$$\left(\frac{y-k}{a}\right)^2 - \left(\frac{x-h}{b}\right)^2 = 1$$

has asymptotes $y - k = \pm\frac{a}{b}(x - h)$.

You can now sketch any hyperbola quickly from the standard graphing form of its equation.

EXAMPLE 6   *Sketching a hyperbola quickly from the standard graphing form of its equation*

To sketch the graph of

$$\left(\frac{y-3}{2}\right)^2 - \left(\frac{x-1}{5}\right)^2 = 1$$

begin by plotting the center at $(1, 3)$. The transverse axis is the vertical line $x = 1$, and the conjugate axis is the horizontal line $y = 3$. The vertices are 2 units above and below the center, at $(1, 1)$ and $(1, 5)$. The equations of the asymptotes are

$$\left(\frac{y-3}{2}\right)^2 = \left(\frac{x-1}{5}\right)^2$$

which can be simplified to

$$y - 3 = \pm\frac{2}{5}(x - 1)$$

The graph is shown in Figure 4-34.

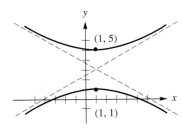

Figure 4–34

EXERCISES   *In Exercises 27–30, sketch the graph of the hyperbola. Identify the center and vertices, and the equations of the transverse axis, conjugate axis, and asymptotes.*

27. $\dfrac{(x+5)^2}{9} - \dfrac{(y+6)^2}{16} = 1$

28. $\dfrac{(x-2)^2}{7} - \dfrac{(y+1.5)^2}{4} = 1$

29. $\dfrac{(y-85)^2}{1225} - \dfrac{(x+129)^2}{2500} = 1$

30. $\dfrac{(x-15)^2}{144} - \dfrac{(y+13)^2}{144} = 1$

EXCEPTIONAL GRAPHS

You know now that the graph of an equation

$$A(x - h)^2 + C(y - k)^2 = G$$

is an ellipse when $AC > 0$ *as long as* $G > 0$, and a hyperbola when $AC < 0$, *as long as* $G \neq 0$. In Exercises 31–34 you can see what happens when the conditions on G are not met.

EXERCISES

31. (*Writing to Learn*) Explain why the graph of $x^2 + y^2 = 0$ is a single point. What is that point?

32. (*Writing to Learn*) Explain why the graph of $x^2 + y^2 = -1$ is empty (contains no points).

33. (*Writing to Learn*) Explain why the graph of $x^2 - y^2 = 0$ consists of two intersecting lines. What are the equations of the two lines?

34. (*Writing to Learn*) Based on your observations in Exercises 31–33, explain why each of the following statements is true about the graph of an equation $A(x - h)^2 + C(y - k)^2 = G$.

a. If $AC > 0$ and $G = 0$, the graph is the single point (h, k).

b. If $A > 0$, $C > 0$, and $G < 0$, the graph is empty.

c. If $AC < 0$ and $G = 0$, the graph consists of the two lines

$$y - k = \pm \sqrt{\frac{A}{C}} (x - h)$$

In Exercises 35–38, describe the graph of the equation.

35. $x^2 + 4y^2 - 6x + 10 = 0$ **36.** $x^2 + 4y^2 - 6x + 9 = 0$

37. $x^2 - 4y^2 - 6x + 9 = 0$ **38.** $4x^2 - y^2 + 12x + 9 = 0$

ADDITIONAL EXERCISES

In Exercises 39–62, sketch the graph of each equation and identify the following graphical features.

For a parabola: *coordinates of the vertex, and equation of the axis of symmetry*

For an ellipse: *coordinates of the center, vertices, and covertices, and equations of the major and minor axes*

For a hyperbola: *coordinates of the center and vertex, and equations of the transverse axis, conjugate axis, and asymptotes*

If the graph is one of the exceptional cases discussed in Exercises 31–34, describe the graph.

39. $x = 3y^2$ **40.** $x^2 - 4y^2 + 16 = 0$

41. $2x^2 + 7y^2 = 28$ **42.** $4x^2 + 9y^2 = 1$

43. $4(x + 1)^2 + 4(y - 2)^2 = 9$

44. $4(x + 1)^2 - 4(y - 2)^2 = 9$

45. $4(x + 1)^2 - 4(y - 2) = 9$

46. $100x^2 - 4(y - 1)^2 = -400$

47. $5x^2 + y^2 - 70x + 225 = 0$

48. $9x^2 - 36y^2 - 90x + 360y - 999 = 0$

49. $2x^2 - 12x - y + 10 = 0$

50. $4x^2 + y^2 + 24x + 8y + 52 = 0$

51. $4x^2 - y^2 + 24x - 8y + 20 = 0$

52. $y^2 - 2x + 8y + 12 = 0$

53. $100x^2 + 20x + 400y^2 - 480y - 9855 = 0$

54. $9x^2 + y^2 - 72x + 4y + 157 = 0$

55. $4x^2 + 25y^2 + 24x - 50y - 39 = 0$

56. $-x^2 - 4y^2 + 8x - 8y - 4 = 0$

57. $x^2 + 5y^2 + 6x + 20y + 34 = 0$

58. $3x^2 + 2y^2 - 4y - 22 = 0$

59. $4x^2 - 25y^2 + 24x + 50y - 89 = 0$

60. $-x^2 + 4y^2 + 8x + 8y - 12 = 0$

61. $x^2 - 5y^2 + 6x - 20y - 6 = 0$

62. $3x^2 - 2y^2 + 4y - 26 = 0$

63. (*Creating Models*) Imagine that you are an automotive engineer and your company is designing a new model of sports car. One of your responsibilities is to design the headlamp unit. You have been given the specifications shown in Figure 4-35.

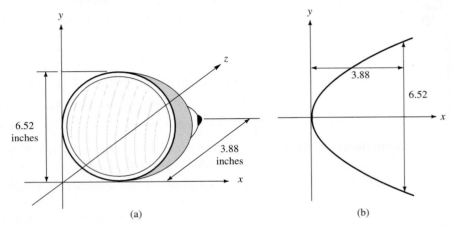

(a)   (b)

Figure 4–35

In Section 4-6 you will discover why the reflective surface should have parabolic cross sections. For now, find the equation of the cross section shown.

64. (*Extension*) Follow these steps to explain why the portion of the hyperbola $x^2 - y^2 = 1$ in the first quadrant becomes and remains arbitrarily close to the line $y = x$.

a. For a fixed value of x, why is the difference of the y-coordinates on the line and hyperbola equal to $x - \sqrt{x^2 - 1}$?

b. Rationalize the numerator to show that the difference can also be expressed as $\dfrac{1}{x + \sqrt{x^2 - 1}}$. (The process of rationalizing the numerator is discussed in Section A-18.)

c. Why is the difference always less than $\dfrac{1}{x}$ and, therefore, as small as you please if x is sufficiently large?

4–5   GRAPHICAL TRANSFORMATIONS

In Section 4-4 you learned how to sketch the graph of a quadratic relation from the standard graphing form of its equation. In this section you will learn an alternate method that uses the idea of transforming a basic graph. You have seen this idea before in Sections 2-5, 4-1, and 4-3, where you sketched the graphs of linear absolute value functions, quadratic functions, and square root functions by transforming the basic graphs of $y = |x|$, $y = x^2$, and $y = \sqrt{x}$. Now we will take a closer look at how changes in an equation produce changes in a graph. Our objective is to develop a unified set of principles that we can use to make quick and accurate sketches of a wide variety of graphs. Let's begin our investigation in the context of a situation in which both the equation and the graph of a quadratic relation are of interest.

A MATHEMATICAL
LOOKING GLASS
The Blue Planets

Our knowledge of the solar system that is our home has increased greatly over the last two decades, largely because of the development of unmanned spacecraft capable of transmitting visual images back to earth. The images can often be quite dramatic as well as informative. In *The Planets: Portraits of New Worlds*, Nigel Henbest describes humanity's first close glimpse of Neptune:

> When the spacecraft *Voyager 2* swept past its final planet in 1989, astronomers viewing the pictures on Earth were in for a nostalgic surprise. *Voyager*'s cameras showed Neptune to be a blue planet with wispy white clouds, accompanied by a large, rocky-looking moon. No one had expected to find, in the outer fringes of the solar system, a world so reminiscent of our own "Blue Planet." Up till then, we knew virtually nothing about Neptune. The planet is so far away that we need a telescope to see it at all. Astronomers did not know of its existence until the middle of the last century, when they noticed that some unexpected force was pulling Uranus as it moved round its orbit. The best explanation was the gravitational tug of an unknown planet, further from the Sun. Mathematicians in England and France calculated where this planet should lie, and in 1846 observers at Berlin Observatory picked it out as a faint moving "star."

Voyager's extraordinary success was a triumph for mathematics as much as for astronomy. To observe details such as clouds on Neptune, the spacecraft traveled nearly three billion miles along an extremely precise path. To calculate that path mathematicians needed the equations and graphs of the orbits of both earth and Neptune.

Since the time of Johannes Kepler, whom you met in Section 4-3, it has been known that the orbit of each planet is an ellipse, a type of curve closely related to a circle. Before returning to planetary orbits in Exercise 38, let's proceed by *Examining a Related Problem*. Over the next few pages you can discover the changes in the unit circle (see **Fermilab**, page 175) that are produced by various changes in its equation. ■

STRETCHES AND COMPRESSIONS

First, let's see what happens to the unit circle if we replace x by $\dfrac{x}{a}$ or y by $\dfrac{y}{b}$ in its equation. Table 21 was generated by the equation $x^2 + y^2 = 1$ of the unit circle.

TABLE 21

x	y
−1	0
0	1
0	−1
1	0

EXERCISES

x	y
−2	0
0	1
0	−1
2	0

1. (*Making Observations*)

a. Verify that the table at the left fits the relation $\left(\dfrac{x}{2}\right)^2 + y^2 = 1$.

b. We have *divided* x by 2 in the equation of the unit circle, but to obtain the table in part (a) from Table 21 we have *multiplied* each x-coordinate by 2. Explain why this occurs.

c. Sketch the graph of $\left(\dfrac{x}{2}\right)^2 + y^2 = 1$. How does the graph support your observations in part (b)?

d. Describe how the unit circle was changed to produce the graph in part (c).

e. Support your result in part (c) by finding and graphing two functions defined implicitly by the relation.

2. Use your results from Exercise 1 to sketch the graph of each relation.

a. $\left(\dfrac{x}{3}\right)^2 + y^2 = 1$ b. $\left(\dfrac{x}{5}\right)^2 + y^2 = 1$

c. $(2x)^2 + y^2 = 1$ d. $(4x)^2 + y^2 = 1$

3. (*Making Observations*)

a. Sketch the graph of $x^2 + \left(\dfrac{y}{2}\right)^2 = 1$.

b. Describe how the unit circle was changed to produce the graph in part (a).

c. Support your result in part (a) by finding and graphing two functions defined implicitly by the relation.

4. Use your results from Exercise 3 to sketch the graph of each relation.

a. $x^2 + \left(\dfrac{y}{3}\right)^2 = 1$ b. $x^2 + \left(\dfrac{y}{5}\right)^2 = 1$

c. $x^2 + (2y)^2 = 1$ d. $x^2 + (4y)^2 = 1$

Your observations in Exercises 1−4 can be generalized as follows.

If x is replaced by $\dfrac{x}{a}$ with $a > 0$ in the equation of a relation, each x-coordinate on the graph is multiplied by a. The graph is stretched horizontally if $a > 1$ and compressed horizontally if $0 < a < 1$.

Similarly, if y is replaced by $\dfrac{y}{b}$ with $b > 0$ in the equation, then each y-coordinate on the graph is multiplied by b. The graph is stretched vertically if $b > 1$ and compressed vertically if $0 < b < 1$.

EXERCISES   *In Exercises 5–10, describe how the first graph must be transformed to produce the second, and sketch both graphs on the same set of axes.*

5. $y = x^2$; $y = (3x)^2$

6. $x = y^2$; $x = \left(\dfrac{y}{4}\right)^2$

7. $x^2 - y^2 = 1$; $\left(\dfrac{x}{2}\right)^2 - y^2 = 1$

8. $y^2 - x^2 = 1$; $(2y)^2 - x^2 = 1$

9. $x^2 + y^2 = 1$; $\left(\dfrac{x}{3}\right)^2 + \left(\dfrac{y}{4}\right)^2 = 1$

10. $x^2 - y^2 = 1$; $\left(\dfrac{x}{3}\right)^2 - \left(\dfrac{y}{4}\right)^2 = 1$

SHIFTS

Next let's see what happens to the unit circle if we replace x by $x - h$ or y by $y - k$ in its equation.

EXERCISES   **11.** (*Making Observations*)

   a. Verify that the following table fits the relation $(x - 3)^2 + y^2 = 1$.

x	y
2	0
3	1
3	−1
4	0

   b. We have *subtracted* 3 from x in the equation of the unit circle, but to obtain the table in part (a) from Table 21 we have *added* 3 to each x-coordinate. Explain why this occurs.

   c. Sketch the graph of $(x - 3)^2 + y^2 = 1$. How does the graph support your observations in part (b)?

   d. Describe how the unit circle was changed to produce the graph in part (c).

   e. Support your result in part (c) by finding and graphing two functions defined implicitly by the relation.

   12. Use your results from Exercise 11 to sketch the graph of each relation.

   a. $(x - 4)^2 + y^2 = 1$

   b. $(x - 1)^2 + y^2 = 1$

   c. $(x + 2)^2 + y^2 = 1$

   d. $(x + 5)^2 + y^2 = 1$

   13. (*Making Observations*)

   a. Sketch the graph of $x^2 + (y - 3)^2 = 1$.

   b. Describe how the unit circle was changed to produce the graph in part (a).

   c. Support your result in part (a) by finding and graphing two functions defined implicitly by the relation.

14. Use your results from Exercise 13 to sketch the graph of each relation.

  a. $x^2 + (y - 4)^2 = 1$

  b. $x^2 + (y - 1)^2 = 1$

  c. $x^2 + (y + 2)^2 = 1$

  d. $x^2 + (y + 5)^2 = 1$

Your observations in Exercises 11–14 can be generalized as follows.

> If x is replaced by $x - h$ in the equation of a relation, h is added to each x-coordinate on the graph. The graph is shifted to the right if $h > 0$ and to the left if $h < 0$.
>
> Similarly, if y is replaced by $y - k$ in the equation, k is added to each y-coordinate on the graph. The graph is shifted up if $k > 0$ and down if $k < 0$.

EXERCISES   *In Exercises 15–20, describe how the first graph must be transformed to produce the second, and sketch both graphs on the same set of axes.*

15. $y = x^2$; $y = (x - 3)^2$

16. $x = y^2$; $x = (y + 4)^2$

17. $x^2 - y^2 = 1$; $(x + 2)^2 - y^2 = 1$

18. $y^2 - x^2 = 1$; $(y - 2)^2 - x^2 = 1$

19. $x^2 + y^2 = 1$; $(x + 3)^2 + (y + 4)^2 = 1$

20. $x^2 - y^2 = 1$; $(x + 3)^2 - (y + 4)^2 = 1$

REFLECTIONS

Finally, let's see what happens to the unit circle if we replace x by $-x$ or y by $-y$ in its equation.

EXERCISE   **21.** (*Making Observations*) In the equation $x^2 + y^2 = 1$, replace x by $-x$ and simplify. Then replace y by $-y$ and simplify. What do your results say about the graphs of the resulting equations?

Surprise! For this particular relation, there is no effect at all on either the equation or the graph. Let's consider some different relations to see how the replacements of x by $-x$ and y by $-y$ affect graphs in general, and why they do not affect the unit circle.

22. (*Making Observations*)

  a. In the equation $(x - 2)^2 + y^2 = 1$, replace x by $-x$ and verify that the resulting equation can be simplified to $(x + 2)^2 + y^2 = 1$.

  b. Sketch the graphs of both equations in part (a). How do the two graphs relate to each other?

23. (*Making Observations*)

  a. In the equation $x^2 + (y - 2)^2 = 1$, replace y by $-y$ and verify that the resulting equation can be simplified to $x^2 + (y + 2)^2 = 1$.

b. Sketch the graphs of both equations in part (a). How do the graphs relate to each other?

24. (*Writing to Learn*) Use your results from Exercises 22 and 23 to explain why the unit circle is not affected when we replace x by $-x$ or y by $-y$ in its equation.

Your observations in Exercises 22 and 23 can be generalized as follows.

> If x is multiplied by -1 in the equation of a relation, each x-coordinate on the graph is multiplied by -1, and the graph is reflected in the y-axis.
>
> Similarly, if y is multiplied by -1 in the equation, each y-coordinate on the graph is multiplied by -1, and the graph is reflected in the x-axis.

If the replacement of x by $-x$ produces an equation equivalent to the original one, as with the unit circle, the graph of the relation is its own reflection in the y-axis. Such a graph is said to be **symmetric about the y-axis**. Similarly, if the replacement of y by $-y$ produces an equation equivalent to the original one, the graph of the relation is **symmetric about the x-axis**. We will return to the idea of symmetry in Section 5-3.

EXAMPLE 1   *Detecting symmetry analytically*

In the equation $\left(\dfrac{x-6}{2}\right)^2 + \left(\dfrac{y}{3}\right)^2 = 1$, if y is replaced by $-y$, the equation becomes

$$\left(\frac{x-6}{2}\right)^2 + \left(\frac{-y}{3}\right)^2 = 1$$

or, equivalently,

$$\left(\frac{x-6}{2}\right)^2 + \left(\frac{y}{3}\right)^2 = 1$$

Thus, its graph is symmetric about the x-axis.

If x is replaced by $-x$, the equation becomes

$$\left(\frac{-x-6}{2}\right)^2 + \left(\frac{y}{3}\right)^2 = 1$$

or, equivalently,

$$\left(\frac{x+6}{2}\right)^2 + \left(\frac{y}{3}\right)^2 = 1$$

Thus, the graph is not symmetric about the y-axis. The graph in Figure 4-36 supports these conclusions. ∎

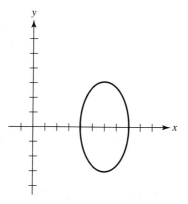

Figure 4–36

EXERCISES   *In Exercises 25–28, decide whether the graph is symmetric about the x-axis, the y-axis, both, or neither. Then use the equation to confirm your conclusion.*

25. $x - y^2 = 1$

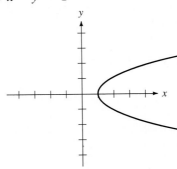

26. $x^2 - y^2 = 1$

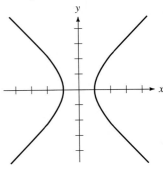

27. $x^2 + \left(\dfrac{y}{4}\right)^2 = 1$

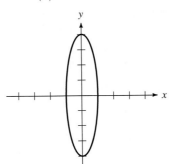

28. $(x + 1)^2 + \left(\dfrac{y}{4}\right)^2 = 1$

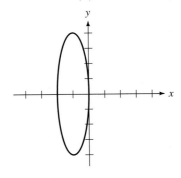

APPLYING A SEQUENCE OF TRANSFORMATIONS

The value of the graphical transformation principles you have just learned is that they allow you to sketch a variety of graphs quickly and accurately by memorizing just a few basic graphs and equations, such as the unit circle and its equation. If a basic equation undergoes several changes of the types just described, you can determine the graphical effects by applying the appropriate transformations in sequence.

EXAMPLE 2   *Graphing a relation by transforming a basic graph*

To sketch the graph of

$$x^2 - \left(\frac{y - 3}{2}\right)^2 = 1$$

begin with the basic equation

$$x^2 - y^2 = 1$$

whose graph is shown in Figure 4-37a. Then divide y by 2 to obtain

$$x^2 - \left(\frac{y}{2}\right)^2 = 1$$

whose graph is shown in Figure 4-37b. Finally, subtract 3 from y to obtain

$$x^2 - \left(\frac{y - 3}{2}\right)^2 = 1$$

whose graph is shown in Figure 4-37c.

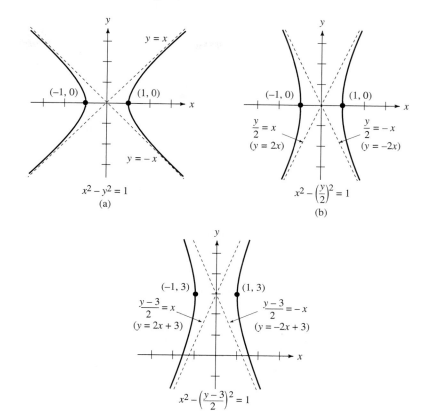

Figure 4–37

Figure 4-37 shows that the basic graph is first stretched vertically by a factor of 2, then shifted 3 units up.   ■

FYI When you are applying a sequence of graphical transformations, order matters! In Exercise 29 you can demonstrate that applying the two transformations in Example 2 in the opposite order yields a different graph with a different equation.

EXERCISES   **29.** (*Making Observations*)

    a. Beginning with the unit circle, shift the graph 3 units up, then stretch it vertically by a factor of 2. Sketch each graph in the sequence and confirm that the final graph is different from the one in Figure 4-37c.

    b. Beginning with the equation of the unit circle, first subtract 3 from y, then divide y by 2. Confirm that the resulting equation is different from the last equation in Example 2.

In Exercises 30–33, sketch the ellipse. Identify the coordinates of the center, vertices, and covertices, and the equations of the major and minor axes.

30. $\left(\dfrac{x+5}{3}\right)^2 + \left(\dfrac{y+6}{4}\right)^2 = 1$

31. $\left(\dfrac{x + 129}{50}\right)^2 + \left(\dfrac{y - 85}{35}\right)^2 = 1$

32. $\left(\dfrac{x - 2}{\sqrt{7}}\right)^2 + \left(\dfrac{y + 1.5}{2}\right)^2 = 1$

33. $(x - 0.5)^2 + \left(\dfrac{y - 1.5}{2}\right)^2 = 1$

In Exercises 34–37, sketch the hyperbola. Identify the coordinates of the center and vertices and the equations of the transverse axis, conjugate axis, and asymptotes.

34. $\left(\dfrac{x + 5}{3}\right)^2 - \left(\dfrac{y + 6}{4}\right)^2 = 1$

35. $\left(\dfrac{x + 129}{50}\right)^2 - \left(\dfrac{y - 85}{35}\right)^2 = 1$

36. $\left(\dfrac{y + 1.5}{2}\right)^2 - \left(\dfrac{x - 2}{\sqrt{7}}\right)^2 = 1$

37. $\left(\dfrac{y - 1.5}{2}\right)^2 - (x - 0.5)^2 = 1$

38. Sketch the graph of each planet's orbit. Distances are measured in astronomical units (1 astronomical unit = 92.9 million miles).

a. Earth: $(x - 0.02)^2 + \left(\dfrac{y}{0.9998}\right)^2 = 1$

b. Mercury: $\left(\dfrac{x - 0.08}{0.39}\right)^2 + \left(\dfrac{y}{0.38}\right)^2 = 1$

SUMMARY

The graphical transformation principles of this section are summarized in Table 22.

TABLE 22

Change applied to equation	Effect on coordinates	Effect on graph
$x \to \dfrac{x}{a}$	Multiply x by a	Horizontal stretch $(a > 1)$ or compression $(0 < a < 1)$
$y \to \dfrac{y}{b}$	Multiply y by b	Vertical stretch $(b > 1)$ or compression $(0 < b < 1)$
$x \to x - h$	Add h to x	Horizontal shift
$y \to y - k$	Add k to y	Vertical shift
$x \to -x$	Multiply x by -1	Reflection in y-axis
$y \to -y$	Multiply y by -1	Reflection in x-axis

ADDITIONAL EXERCISES   *In Exercises 39–50, sketch the graph of each equation by transforming one of the basic graphs $y = x^2$, $x = y^2$, $x^2 + y^2 = 1$, $x^2 - y^2 = 1$, or $y^2 - x^2 = 1$. In each case describe how the basic graph is transformed, and identify the following graphical features.*

For a parabola:	*coordinates of the vertex, and equation of the axis of symmetry*
For an ellipse:	*coordinates of the center, vertices, and covertices, and equations of the major and minor axes*
For a circle:	*radius and coordinates of the center*
For a hyperbola:	*coordinates of the center and vertex, and equations of the transverse axis, conjugate axis, and asymptotes*

39. $x = -5y^2$

40. $4x^2 - y^2 + 64 = 0$

41. $5x^2 + 6y^2 = 30$

42. $9x^2 + 4y^2 = 1$

43. $(x + 10)^2 + (y - 20)^2 = 25$

44. $(x + 10)^2 - (y - 20)^2 = 25$

45. $(x + 10)^2 - (y - 20) = 25$

46. $36x^2 - (y + 7)^2 = -144$

47. $\left(\dfrac{x + 2}{4}\right)^2 + \left(\dfrac{y + 3}{6}\right)^2 = 1$

48. $\left(\dfrac{x - 4}{7}\right)^2 + \left(\dfrac{y - 1}{3}\right)^2 = 1$

49. $\left(\dfrac{x + 2}{4}\right)^2 - \left(\dfrac{y + 3}{6}\right)^2 = 1$

50. $\left(\dfrac{y - 1}{3}\right)^2 - \left(\dfrac{x - 4}{7}\right)^2 = 1$

In Exercises 51–56, decide whether the graph is symmetric about the x-axis, the y-axis, both, or neither. Then use the equation to confirm your conclusion.

51. $y = x^2 - 4$

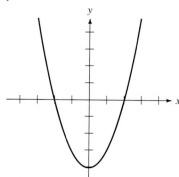

52. $y = x^3 - 4x$

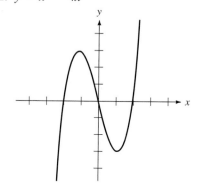

53. $y = x^3 - 4x^2$

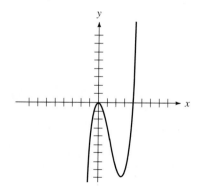

54. $y = \dfrac{x^2}{x^2 + 1}$

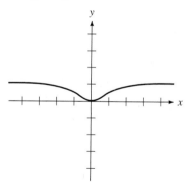

55. $x^3 - 3x^2 - y^2 = 0$ **56.** $x^6 + y^6 = 1$

                    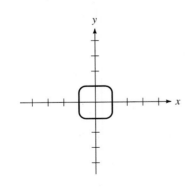

57. What equation results from each sequence of transformations?

  a. The unit circle is shifted 5 units to the right and 2 units down.

  b. The unit circle is stretched horizontally by a factor of 2 and compressed vertically by a factor of 2.

  c. The unit circle is stretched horizontally by a factor of 3, then shifted 1 unit to the left.

  d. The unit circle is shifted 1 unit to the left, then stretched horizontally by a factor of 3.

58. What equation results from each sequence of transformations?

  a. The graph of $x = y^2$ is compressed horizontally by a factor of 2 and then reflected in the y-axis.

  b. The graph of $y^2 - x^2 = 1$ is shifted 4 units left and then stretched vertically by a factor of 5.

  c. The graph of $y = x^2$ is reflected in the x-axis and then shifted 3 units down.

  d. The graph of $x^2 - y^2 = 1$ is stretched horizontally by a factor of 3 and then shifted 1 unit right.

The graph of the relation $x^3 - 4x^2 + y^2 = 0$ is shown below. In Exercises 59–62, sketch the graph of the given relation, and identify three points on the graph.

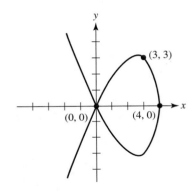

59. $x^3 - 4x^2 + \left(\dfrac{y + 1}{3}\right)^2 = 0$

60. $(-4x)^3 - 4(-4x)^2 + y^2 = 0$

61. $(x + 3)^3 - 4(x + 3)^2 + (y - 4)^2 = 0$

62. $\left(\dfrac{x + 3}{2}\right)^3 - 4\left(\dfrac{x + 3}{2}\right)^2 + [2(y - 1)]^2 = 0$

63. (*Creating Models*) Imagine that you are a civil engineer working on a highway construction project. A portion of the roadbed is shown in Figure 4-38a with the coordinate system that one of your colleagues has imposed on it.

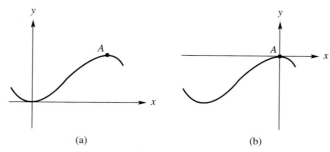

Figure 4–38

You would rather work with the coordinate system in Figure 4-38b since you will be working near the crest of the hill at A. In your colleague's coordinate system, the coordinates of A are (5000, 200), and the equation of the roadbed is

$$y = (2.4 \times 10^{-5})x^2 - (3.2 \times 10^{-9})x^3$$

with both x and y given in feet. What is the equation of the roadbed in your coordinate system?

64. (*Writing to Learn*) The information in the first four lines of Table 22 can be characterized loosely by the statement, "Whatever you do in the equation, do the opposite on the graph." Explain why the information in the last two lines can be characterized in the same way.

SUPPLEMENTARY TOPIC

Graphical Transformations in the Context of Functions

In Sections 2-5, 4-1, and 4-3, you saw how the graphs of $y = a|x - h| + k$, $y = a(x - h)^2 + k$, and $y = a\sqrt{x - h} + k$ could be obtained by transforming the basic graphs of $y = |x|$, $y = x^2$, and $y = \sqrt{x}$. Are those transformations compatible with the principles we just developed? They should be, since the equations of those basic graphs define relations as well as functions. Let's find out.

65. a. (*Review*) Use the methods of Section 2-5 to sketch the graph of $y = 3|x - 4| - 2$.

b. Use the methods of this section to sketch the graph of

$$\dfrac{y + 2}{3} = |x - 4|$$

c. Your graphs should appear to be identical. Verify that their equations are equivalent.

Your observations in Exercise 65 indicate that if $y = f(x)$, the transformations affecting y can be summarized as in Table 23.

TABLE 23

New function	Effect on coordinates	Effect on graph
$y = bf(x)$	Multiply y by b	Vertical stretch $(b > 1)$ or compression $(0 < b < 1)$
$y = f(x) + k$	Add k to y	Vertical shift
$y = -f(x)$	Multiply y by -1	Reflection in x-axis

66. (*Writing to Learn*) If $y = f(x)$, explain why the information in Table 23 is consistent with that in Table 22.

In Exercises 67–70, sketch the graph of the first function by transforming the graph of the second, shown below the equations. Identify three points on each graph.

67. $y = 2\sqrt{x} + 5$
$y = \sqrt{x}$

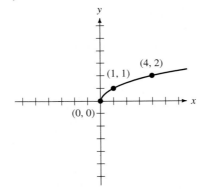

68. $y = (x - 4)^4 + 1$
$y = x^4$

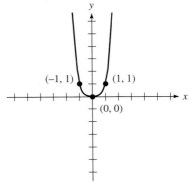

69. $y = 4 - 2x^3$
$y = x^3$

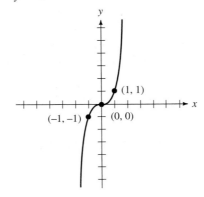

70. $y = 3^{2x} - 5$
$y = 3^x$

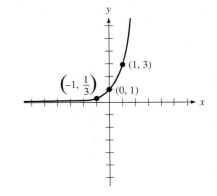

In Exercises 71–74, sketch the graph of each function by transforming the graph of the basic function shown. Identify the points obtained by moving the three identified points on the basic graph.

71. **a.** $y = -3x^5$
   b. $y = (x + 3)^5 + 4$
   c. $y = 0.1(x - 2)^5$
   d. $y = 2(x - 1)^5 + 3$

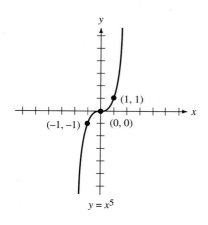

72. **a.** $y = -\sqrt{x}$
   b. $y = \sqrt{-x}$
   c. $y = -\sqrt{-x}$
   d. $y = 2\sqrt{x - 1} + 3$

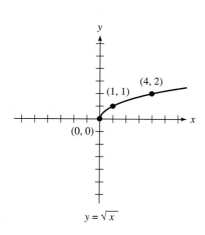

73. **a.** $y = 2^{x+3} - 5$
   b. $y = 0.5(2^{3x})$
   c. $y = 7 + 2^{-x}$
   d. $y = 2(2^{x-1}) + 3$

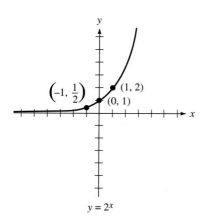

74. **a.** $y = \dfrac{4}{x} - 2$

b. $y = -\dfrac{1}{x + 5}$

c. $y = 6 - \dfrac{1}{x}$

d. $y = \dfrac{2}{x - 1} + 3$

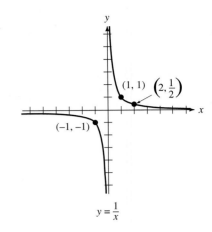

4-6   AN ANALYTICAL VIEW OF CONIC SECTIONS

The universe abounds with conic sections. The orbits of planets and comets and the cross sections of optical lenses are parabolic, elliptical, or hyperbolic. In this section you will learn that conic sections appear so often in the physical world because they possess some interesting geometric properties related to their equations. You will use the methods of analytic geometry (geometry in the framework of a coordinate system) to discover that the equations of the curves with the required properties are quadratic relations.

AN ANALYTICAL VIEW OF PARABOLAS

Try this experiment. You will need a sheet of wax paper about a foot square and a marker. Make a mark about 1 inch up from the middle of the bottom edge, as in Figure 4-39a. Next, fold up part of the bottom edge so that it just touches your mark, as in Figure 4-39b.

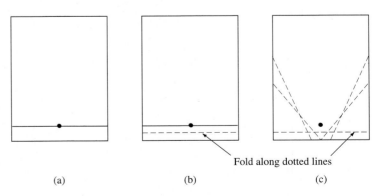

Fold along dotted lines

(a)   (b)   (c)

Figure 4–39

Continue as in Figure 4-39c to make about 15 to 20 such folds at different angles. Now spread the paper flat. Do you see a curve on the paper? What type of curve does it look like?

Let's find an equation to describe the wax paper curve. To do this, we need to place the paper into a convenient coordinate system. We choose one in which the y-axis passes through the mark labeled F in Figure 4-40a, and the origin is midway between the mark and the bottom edge. Thus F has coordinates $(0, p)$, and the bottom edge has the equation $y = -p$ for some $p > 0$. Every point $P = (x, y)$ on the curve was produced by a fold that superimposed a point $D = (x, -p)$ on F, as in Figure 4-40b.

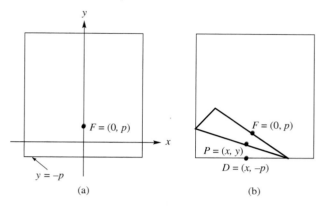

(a)   (b)

Figure 4–40

EXERCISES   **1.** (*Writing to Learn*) Explain why $\overline{PF} = \overline{PD}$ in Figure 4-40.

2. (*Extension*) Use the distance formula to show that x and y must satisfy the equation $\sqrt{x^2 + (y - p)^2} = y + p$.

3. (*Extension*) Show that the equation in Exercise 2 can be simplified to $y = \dfrac{1}{4p} x^2$.

So the wax paper curve is a parabola! In fact, your experiment provides a formal definition of the word *parabola* at last. A **parabola** is the set of all points in the plane that are equidistant from a fixed line and a fixed point not on the line. The fixed line is called the **directrix** of the parabola (the bottom edge of the wax paper), and the fixed point is its **focus** (the mark on the paper). See Figure 4-41.

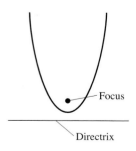

Focus

Directrix

Figure 4–41

Your work in Exercises 1–3 shows that every equation $y = ax^2$ describes a parabola according to our new definition. More generally, in Exercise 45 you will establish the following facts.

- Every equation $y = a(x - h)^2 + k$ or $x = a(y - k)^2 + h$ describes a parabola, so that our new definition coincides with the one given in Section 4-4.
- If $p = \dfrac{1}{4a}$ $\left(\text{equivalently, if } a = \dfrac{1}{4p}\right)$, the focus and directrix are each located p units from the vertex.
- The focus is always on the "inside" of the parabola, and the directrix is always on the "outside." The possible arrangements of focus and directrix are shown in Figure 4-42.

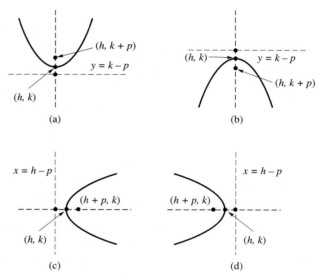

Figure 4–42

Example 1 shows how to find the focus and directrix of a parabola whose equation is known.

EXAMPLE 1   *Using the equation of a parabola to locate its focus and directrix*

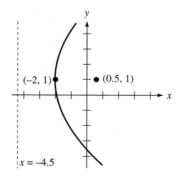

Figure 4–43

Figure 4-43 shows the graph of $x = 0.1(y - 1)^2 - 2$. Its vertex is $(-2, 1)$. Since the quadratic term involves y and has a positive coefficient, the graph opens toward the right. To locate its focus and directrix, first find the value of p.

$$p = \frac{1}{4(0.1)} = 2.5$$

The focus is (0.5, 1), 2.5 units to the right of the vertex. The directrix is $x = -4.5$, the vertical line 2.5 units to the left of the vertex. ∎

EXERCISES *In Exercises 4–7, find the focus and directrix of the parabola, and include them in a sketch of the graph.*

4. $y = 8 - x^2$ **5.** $y^2 + 8x - 2y = 15$

6. $x = 0.1(y + 5)^2 - 10$ **7.** $y = 5(x - 0.4)^2 + 0.03$

Your wax paper experiment illustrated another geometric property of parabolas. The line formed by each of your folds was tangent to the parabola, as indicated in Figure 4-44. (If you take calculus, you will learn the mathematical definition of a line tangent to an arbitrary curve. For now, let's just use the intuitive idea that the line and the curve have the same direction at the point of tangency.)

(a)   (b)

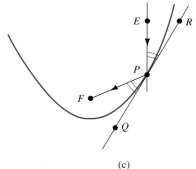

(c)

Figure 4–44

In the act of folding as in Figure 4-44a, you verified that the angles *FPQ* and *DPQ* are equal. Since *DPQ* and *EPR* are vertical angles, as shown in Figure 4-44b, they are also equal. The resulting equality of angles *FPQ* and *EPR* has some physical consequences. If you imagine the parabola as a mirror, any light ray traveling along *EP* will strike the mirror and be reflected toward *F* as illustrated in Figure 4-44c. In general, rays traveling parallel to the axis of symmetry and striking the parabola will be reflected toward the focus. Conversely, rays directed toward the parabola from the focus will be reflected parallel to the axis of symmetry.

A MATHEMATICAL
LOOKING GLASS
Satellite Dish

Lynn's father, Ernie Harvey, lives in Hawthorne, NV, a remote area that had no access to cable television services until 1985. To improve television reception in his home, Dr. Harvey decided in 1983 to buy a satellite dish. The dish, which has parabolic cross sections, collects signals beamed in from a satellite and concentrates them by reflecting them to a receiver at the focus, as shown in Figure 4-45.

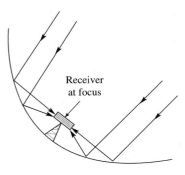

Receiver
at focus

Figure 4–45

After learning that a dish would cost $9000, Dr. Harvey decided to construct his own. He knew that its power would be proportional to the square of the diameter at the edge, so he decided to devote his entire back yard to a dish with a diameter of 20 feet. To minimize the amount of material used in construction, he wanted the dish to be as shallow as possible. However, for a shallow dish he must place the receiver farther from the vertex than for a deep dish. ■

EXERCISES   **8.** (*Writing to Learn*) Explain why the receiver for a shallow dish must be farther from the vertex than the receiver for a deeper dish of the same diameter.

9. (*Problem Solving*) Dr. Harvey wanted his receiver to be only 6 feet from the vertex of his dish. How deep should the dish be?

AN ANALYTICAL VIEW OF ELLIPSES

Try this experiment. You will need a one-foot-square piece of corkboard, enough construction paper to cover it, two thumbtacks, a loop of string at least 15 inches long, and a pencil. Secure the paper on top of the corkboard with the thumbtacks, placed about 2 inches in from opposite edges. Tie the string to each thumbtack, leaving about 10 inches of string between the knots. See Figure 4-46a on the following page.

(a)   (b)

Figure 4–46

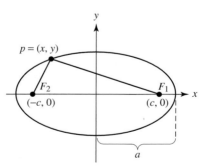

Figure 4–47

Place the pencil below the string so that the string is taut, and trace an arc above the tacks by moving the pencil so the string remains taut, as in Figure 4-46b. Then place the pencil above the string and trace an arc below the tacks. Your two arcs should join to form a single closed curve. What type of curve does it look like?

Let's find an equation to describe the pencil-and-string curve. To do this, let's place the paper into a coordinate system with the thumbtacks on the x-axis and equidistant from the origin. The tacks, at F_1 and F_2 in Figure 4-47, have coordinates $(c, 0)$ and $(-c, 0)$ for some $c > 0$.

EXERCISES

10. (*Writing to Learn*) If the length of the string between the knots is $2a$ and $P = (x, y)$ is any point on the curve, explain why $\overline{PF_1} + \overline{PF_2} = 2a$.

11. (*Extension*) Use the distance formula to show that x and y must satisfy the equation

$$\sqrt{(x + c)^2 + y^2} + \sqrt{(x - c)^2 + y^2} = 2a$$

12. (*Extension*) Show that the equation in Exercise 11 can be simplified to $(a^2 - c^2)x^2 + a^2y^2 = a^2(a^2 - c^2)$.

Figure 4–48

Figure 4-48 illustrates that $a > c$. Therefore, $a^2 - c^2 > 0$, and the equation in Exercise 12 describes an ellipse. In fact, your experiment provides a formal definition of the word *ellipse*. An **ellipse** is the set of all points in the plane, the sum of whose distances from two fixed points is constant. The fixed points are called the **foci** of the ellipse.

If we denote the number $a^2 - c^2$ by b^2, the standard graphing form of the equation in Exercise 12 is

$$\frac{x^2}{a^2} + \frac{y^2}{b^2} = 1$$

and every such equation describes an ellipse according to our new definition. More generally, in Exercise 46 you will establish the following facts.

- Every equation $\dfrac{(x - h)^2}{a^2} + \dfrac{(y - k)^2}{b^2} = 1$ or $\dfrac{(x - h)^2}{b^2} + \dfrac{(y - k)^2}{a^2} = 1$ describes an ellipse, so that our new definition coincides with the one given in Section 4-4.

• If $c^2 = a^2 - b^2$ (equivalently, $b^2 = a^2 - c^2$), the foci are located c units on each side of the center along the major axis. The possible arrangements of the foci with respect to the vertices and covertices are shown in Figure 4-49.

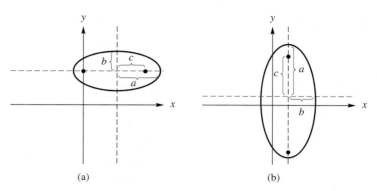

(a)   (b)

Figure 4–49

Example 2 shows how to find the foci of an ellipse whose equation is known.

EXAMPLE 2   *Finding the foci of an ellipse*

To find the foci of the ellipse $\dfrac{(x-2)^2}{9} + \dfrac{(y+5)^2}{16} = 1$, first locate its center at $(2, -5)$. Its major axis is the vertical line $x = 2$ since the y-term has the larger denominator. The value of c can be calculated from the equation $c^2 = 16 - 9$, so $c = \sqrt{7}$. Thus the foci are $\sqrt{7}$ units above and below the center at $(2, -5 + \sqrt{7})$ and $(2, -5 - \sqrt{7})$. From Section 4-4 you also know that the vertices are 4 units above and below the center at $(2, -1)$ and $(2, -9)$ and the covertices are 3 units to the left and right of the center at $(-1, -5)$ and $(5, -5)$. See Figure 4-50.

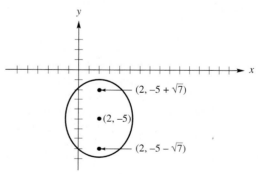

Figure 4–50   ■

Exercise 13 suggests that the relative size of a and c determines the shape of an ellipse.

EXERCISE   **13.** (*Making Observations*) Repeat the pencil-and-string experiment shown in Figure 4-46, but place the thumbtacks about 2 inches on either side of the origin, and leave about 12 inches of string between the knots.

a. How does the shape of your new ellipse compare with the one you constructed before?

b. What would the curve look like if you placed both thumbtacks at the origin?

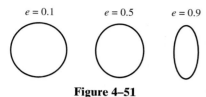

$e = 0.1$ $e = 0.5$ $e = 0.9$

Figure 4–51

Exercise 13 supports the conclusion that a circle is a special case of an ellipse. It also illustrates that an ellipse whose foci are near the vertices is very elongated, while one whose foci are near the center is more nearly circular. The degree of elongation of an ellipse is called its **eccentricity**, denoted by e and is measured by the quantity c/a. See Figure 4-51.

EXERCISES

14. (*Writing to Learn*)

a. Explain why the eccentricity of a circle is 0.

b. Explain why the eccentricity of a very elongated ellipse is close to 1. Is it ever 1?

c. Explain why $0 \leq e < 1$ for any ellipse.

In Exercises 15–18, *find the foci of the ellipse and calculate its eccentricity. Then sketch the graph and include the foci.*

15. $16x^2 + 25y^2 = 400$

16. $x^2 + \dfrac{(y - 10)^2}{36} = 1$

17. $9x^2 + 4y^2 + 18x + 24y - 99 = 0$

18. $3x^2 + 6y^2 - 24x - 24y = 0$

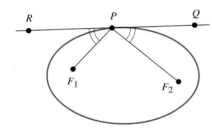

Figure 4–52

Like parabolas, ellipses have a reflective property, illustrated in Figure 4-52. Using calculus, it can be shown that the angles F_1PR and F_2PQ are equal. If you imagine the ellipse as a mirror, any light ray traveling along F_1P will strike the mirror and be reflected toward F_2, and vice versa. That is, it will follow the path of the string in your experiment. Furthermore, since $\overline{PF_1} + \overline{PF_2} = 2a$ no matter where P is on the ellipse, all rays leaving F_1 at the same time will arrive at F_2 at the same time because they travel the same distance.

A MATHEMATICAL LOOKING GLASS

Ultrasound

The reflective property of ellipses provides physicians with a noninvasive procedure (that is, one not requiring surgery) for removing kidney stones. The patient is placed within a device with elliptical walls, so that the kidney stone is at one focus. A source of high-frequency sound waves, called ultrasound, is at the other focus, as in Figure 4-53.

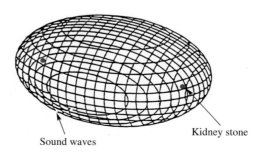

Kidney stone

Sound waves

Figure 4–53

The ultrasound waves converge at the kidney stone, causing vibrations that shatter it. The small pieces can then be passed painlessly in the patient's urine. ■

EXERCISE   19. (*Problem Solving*) If the ellipse is 15 feet long and 5 feet wide, where should the patient and the ultrasound source be placed?

AN ANALYTICAL VIEW OF HYPERBOLAS

We know you are expecting to perform an activity that will generate a hyperbola, but first you will need some information about how radar works in air traffic control systems. A transmitter at an airport sends out electromagnetic signals to an aircraft. These signals travel at the speed of light, about 328 yards every microsecond (1/1,000,000 of a second). When two or more transmitters send out signals simultaneously, the pilot can identify the aircraft's location by comparing arrival times of the signals.

Now try this experiment! You will need a commercial pilot's license and a Boeing 747. Take off from a major airport near your home, and fly in the general direction of Washington, DC. When you are sufficiently close, you will pick up radar signals from both Washington National Airport and Baltimore–Washington International Airport, about 30 miles apart. Adjust your course so that you always receive the signals from Washington National 100 microseconds sooner than those from Baltimore–Washington. This means that you are always (100)(328) = 32,800 yards, or about 18.6 miles, closer to Washington National, as in Figure 4-54a. What kind of curve does your airplane appear to trace?

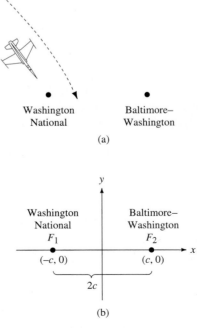

Washington
National

Baltimore–
Washington

(a)

y

Washington
National
F_1

Baltimore–
Washington
F_2

$(-c, 0)$ $(c, 0)$

x

$2c$

(b)

Figure 4–54

Let's find an equation to describe the airplane's curve. To do this, let's place the Washington area into a coordinate system with the two airports on the *x*-axis and equidistant from the origin. In Figure 4-54b, they are at $F_1 = (-c, 0)$ and $F_2 = (c, 0)$. Denote the difference in your distances from the two airports as $2a$.

EXERCISES

20. (*Writing to Learn*) If $P = (x, y)$ is any point on the curve, explain why $\overline{PF_2} - \overline{PF_1} = 2a$.

21. (*Extension*) Use the distance formula to show that *x* and *y* must satisfy the equation

$$\sqrt{(x - c)^2 + y^2} - \sqrt{(x + c)^2 + y^2} = 2a$$

22. (*Extension*) Show that the equation in Exercise 21 can be simplified to $(c^2 - a^2)x^2 - a^2y^2 = a^2(c^2 - a^2)$.

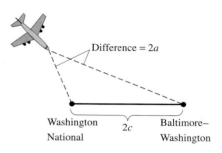

Difference = $2a$

Washington
National

$2c$

Baltimore–
Washington

Figure 4–55

Figure 4-55 shows that the difference in your distances from the airports ($2a$) must be less than their distance from each other ($2c$). Therefore, $c > a$, so $c^2 - a^2 > 0$, and the curve is a hyperbola. In fact, your experiment provides a formal definition of the word *hyperbola*. A **hyperbola** is the set of all points in the plane, the difference of whose distances from two fixed points is constant. The fixed points are called the **foci** of the hyperbola.

As you may have noticed, your airplane traces only the left half of the hyperbola. You would have traced the right half if you had adjusted your course to receive the signals from Washington National 100 microseconds *later* than those from Baltimore–Washington.

If we denote the number $c^2 - a^2$ by b^2, the standard graphing form of the equation in Exercise 22 is

$$\frac{x^2}{a^2} - \frac{y^2}{b^2} = 1$$

and every such equation describes a hyperbola according to our new definition. More generally, in Exercise 47 you will establish the following facts.

- Every equation $\dfrac{(x - h)^2}{a^2} - \dfrac{(y - k)^2}{b^2} = 1$ or $\dfrac{(y - k)^2}{a^2} - \dfrac{(x - h)^2}{b^2} = 1$ describes a hyperbola, so that our new definition coincides with the one given in Section 4-4.

- If $c^2 = a^2 + b^2$ (equivalently, $b^2 = c^2 - a^2$), the foci are located *c* units on each side of the center along the transverse axis. The possible arrangements of the foci with respect to the vertices are shown in Figure 4-56.

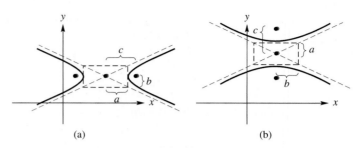

(a)   (b)

Figure 4–56

Example 3 shows how to find the foci of a hyperbola whose equation is known.

EXAMPLE 3   *Locating the foci of a hyperbola*

To find the foci of the hyperbola $\dfrac{(x-2)^2}{9} - \dfrac{(y-5)^2}{16} = 1$, first locate its center at $(2, 5)$. Its transverse axis is the horizontal line $y = 5$ since the x-term has a positive coefficient. The value of c can be calculated from the equation $c^2 = 16 + 9$, so $c = 5$. Thus the foci are 5 units to the left and right of the center, at $(-3, 5)$ and $(7, 5)$. From Section 4-4 you know that the vertices are 3 units to the left and right of the center at $(-1, 5)$ and $(5, 5)$. The asymptotes are the lines $\dfrac{y-5}{4} = \pm\dfrac{x-2}{3}$. See Figure 4-57.

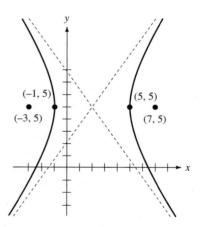

Figure 4–57

FYI   The equations relating the parameters a, b, and c for ellipses and hyperbolas are similar, but different.

For an ellipse: $c^2 = a^2 - b^2$

For a hyperbola: $c^2 = a^2 + b^2$

EXERCISES   *In Exercises 23–26, find the foci of the hyperbola, and include them in a sketch of the graph.*

23. $25y^2 - 16x^2 = 400$

24. $x^2 - \dfrac{(y-10)^2}{36} = 1$

25. $9x^2 - 4y^2 + 18x - 24y + 171 = 0$

26. $3x^2 - 3y^2 - 24x + 24y + 96 = 0$

Like parabolas and ellipses, hyperbolas have a reflective property, illustrated in Figure 4-58.

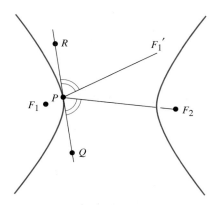

Figure 4–58

Using calculus, it can be shown that the angles $F_1'PR$ and F_2PQ are equal. If you imagine the hyperbola as a mirror, any light ray traveling along $F_1'P$ will strike the mirror and be reflected toward F_2, and vice versa.

A MATHEMATICAL LOOKING GLASS

Earthquakes

When an earthquake occurs, seismologists immediately try to locate its epicenter, or point of origin. Let's see how data from different seismograph stations might have been used to locate the epicenter of the Loma Prieta earthquake of October 17, 1989, also called the San Francisco World Series earthquake because it occurred during a world series game in that city. This earthquake registered 7.1 on the Richter scale and was responsible for $5.6 billion in damage and 62 deaths. According to Lee Row III of the National Oceanic and Atmospheric Administration in Boulder, CO, "The most deadly structural failure of the earthquake occurred when the upper deck of the Interstate 880 (Nimitz Freeway) in Oakland fell onto the lower roadway, causing an official death toll of 41." The report continues, "It is interesting to compare this earthquake with the slightly smaller (magnitude 6.9) Armenian earthquake of December, 1988, that killed 25,000 people and destroyed entire towns. Good construction and engineering practices in the Loma Prieta area obviously contributed to the preservation of property and human lives."

For our discussion let's make the following assumptions.

• The epicenter is located near the surface of the earth. (It was actually 19 kilometers below the surface.)

- The primary shock wave travels from the epicenter at a uniform rate of 25 kilometers per second. (In reality, the shock wave's speed varies as it passes through different types of rock.)
- The portion of the earth's surface containing the seismograph positions can be modeled by a portion of a coordinate plane. (A more accurate model requires spherical geometry.)

The primary wave from the World Series earthquake was detected by a seismograph near Sacramento at 22.3 seconds after 3:04 P.M., Pacific Daylight Time. It was detected in Petaluma at 22.8 seconds after 3:04 P.M. Based on the difference between these times, the epicenter must be located somewhere on one branch of a hyperbola with foci at Sacramento and Petaluma. To describe the hyperbola by means of an equation, let's introduce the coordinate system in Figure 4-59.

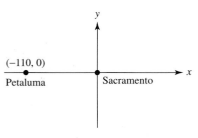

Figure 4–59

Units are measured in kilometers, Sacramento is at the origin, and Petaluma is 110 kilometers west at $(-110, 0)$. ∎

EXERCISES

27. (*Writing to Learn*) Explain why the value of c for this hyperbola is half the distance between the seismograph positions and the value of a is 12.5 times the difference in the arrival times.

28. (*Problem Solving*) Find an equation for the hyperbola, sketch its graph, and identify the branch on which the epicenter lies.

In Section 4-7, you will learn how data from a third seismograph location can be used to locate the epicenter.

ADDITIONAL EXERCISES

In Exercises 29–38, *sketch the graph of each conic section, and identify the following features.*

For each parabola:	vertex, focus, directrix
For each ellipse:	center, vertices, covertices, foci
For each hyperbola:	center, vertices, asymptotes, foci

If an equation leads to one of the exceptional graphs discussed in Section 4-4, these features will not be present, so just sketch the graph.

29. $(x - 5)^2 - 7y^2 = 14$ **30.** $(x - 5)^2 - 7y = 14$

31. $(x - 5)^2 + 7y^2 = 14$

32. $4(x + 0.05)^2 + (y - 0.1)^2 = 0.02$

33. $9(x - 500)^2 = 4(y - 300)^2$

34. $9(x - 500)^2 = -4(y - 300)^2$

35. $x^2 + 9y^2 - x + 9y - 38 = 0$

36. $3y^2 + 4x - 12y + 168 = 0$

37. $4x^2 + 4y^2 - 556x - 372y = 0$

38. $8x^2 - y^2 + 16x + 4y = 0$

39. (*Problem Solving*) In Exercise 63 of Section 4-4 you analyzed the automobile headlamp unit shown in Figure 4-60.

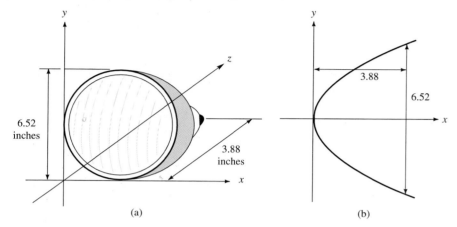

Figure 4–60

  a. Explain why the headlamp works most efficiently if the cross section of the reflective surface has the shape of a parabola.

  b. The equation you found in Exercise 63 of Section 4-4 was approximately $x = 0.37y^2$. Where should the bulb be placed?

40. (*Writing to Learn*) Satellites and space shuttles travel in elliptical orbits, with the earth at one focus. For such orbits, the distances a and c are inconvenient to measure. It is easier to measure the maximum and minimum distances M and m from the space vehicle to the earth. The orbit's eccentricity can be expressed as

$$e = \frac{M - m}{M + m}$$

  Explain why this formula is valid.

41. (*Creating Models*) The formula in Exercise 40 can also be used to calculate the eccentricity of planetary orbits if M and m represent the planet's maximum and minimum distance from the sun. For the planet Neptune, $M = 30.33$ and $m = 29.79$ if distances are measured in astronomical units. (One astronomical unit is the mean distance from the earth to the sun, about 149.6 million km.)

  a. Calculate the eccentricity of Neptune's orbit.

  b. Use the given information to find the values of a and c for Neptune's orbit. (*Hint*: The sun is at one focus.)

  c. Find the equation of Neptune's orbit in a coordinate system with the sun at the origin, the orbit's major axis on the x-axis, and the sun at the left-hand focus.

42. (*Creating Models*) Repeat Exercise 41 for the planet Pluto ($M = 49.3$, $m = 29.6$).

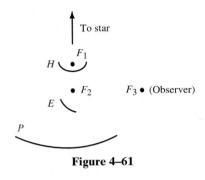

Figure 4–61

43. (*Interpreting Mathematics*) Graph your equations from Exercises 41 and 42 on the same screen. By zooming in on appropriate portions of the graphs, estimate the coordinates of the points where the orbits cross. About what fraction of Pluto's orbit is inside Neptune's orbit?

44. (*Writing to Learn*) Explain how the telescope in Figure 4-61 works. P is an arc of a parabola whose axis of symmetry points directly toward the star, and whose focus is F_1. H is an arc of a hyperbola with foci F_1 and F_2. E is an arc of an ellipse with foci F_2 and F_3.

45. (*Extension*) Show that the parabola in Figure 4-62 has the equation $y = \dfrac{1}{4p}(x - h)^2 + k$. (*Hint:* Apply the strategy of *Looking for Relationships Among Variables* by using the distance formula and the fact that $\overline{PF} = \overline{PD}$.)

Figure 4–62

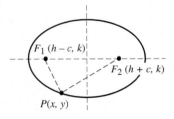

Figure 4–63

46. (*Extension*) Show that the ellipse in Figure 4-63 has the equation $\dfrac{(x - h)^2}{a^2} + \dfrac{(y - k)^2}{b^2} = 1$. (*Hint:* Use the distance formula and the fact that $\overline{PF_1} + \overline{PF_2} = 2a$.)

47. (*Extension*) Show that the hyperbola in Figure 4-64 has the equation $\dfrac{(x - h)^2}{a^2} - \dfrac{(y - k)^2}{b^2} = 1$. (*Hint:* Apply the strategy of *Examining a Related Problem* by adapting your solution from Exercise 46.)

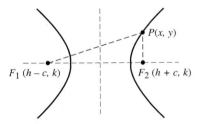

Figure 4–64

4–7   SYSTEMS OF QUADRATIC EQUATIONS AND INEQUALITIES

Earlier in this chapter you have seen physical situations in which two variables satisfy a quadratic relation. It sometimes happens that two variables must satisfy two quadratic relations simultaneously. The relations then form a **system of quadratic equations** that can be solved either analytically or graphically, using methods similar to those used for linear systems in Chapter 3.

A MATHEMATICAL
LOOKING GLASS
Earthquakes 2

In *Earthquakes* (page 225) you used data from two seismographs to place the epicenter of the San Francisco World Series earthquake on one branch of a hyperbola with foci at Sacramento and Petaluma, as in Figure 4-65a.

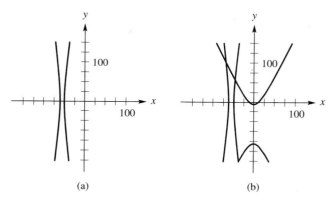

(a)   (b)

Figure 4–65

Now let's see how a third seismograph might have been used to locate the epicenter exactly. Recall our assumptions that the primary shock wave traveled at 25 kilometers per second and was detected in Sacramento at 22.3 seconds after 3:04 P.M. Pacific Daylight Time. Suppose that a seismograph in Modesto detected it at 18.3 seconds after 3:04 P.M. Once again, the difference in arrival times can be used to place the epicenter on one branch of a hyperbola. We have sketched its graph for you in Figure 4-65b. ■

EXERCISE

1. (*Problem Solving*) Modesto is located at $(0, -110)$ in the coordinate system of Figure 4-65. Find an equation for the hyperbola, and identify the branch on which the epicenter lies. (Refer to Exercises 27 and 28 of Section 4-6 for help.)

Since the earthquake's epicenter lies on both hyperbolas in Figure 4-65, it is located at a point where they intersect. At such a point the equations of both hyperbolas are true. Thus to find the points of intersection you must solve a system of equations. This system differs from those in Chapter 3 in that its equations are quadratic rather than linear. Before solving it, you will need to learn some commonly used techniques for solving quadratic systems.

It is usually impractical to solve systems of quadratic equations in two variables numerically, so we will consider analytical and graphical methods in turn.

SOLVING SYSTEMS OF QUADRATIC EQUATIONS ANALYTICALLY

No simple analytical method will solve every system of quadratic equations. Let's look at some methods of solving the types of systems that occur most often.

Systems with One Quadratic and One Linear Equation If one of the equations in the system is linear, rather than quadratic, you can solve the system by substitution. Since you have used substitution to solve linear systems, we are applying the strategy of *Examining a Related Problem*.

EXAMPLE 1   *A system with one quadratic and one linear equation*

To solve the system

$$x^2 + y^2 = 25$$
$$x - 2y = -5$$

first solve the linear equation for either x or y. In this case it is easier to solve for x.

$$x = 2y - 5$$

Now replace x by $2y - 5$ in the quadratic equation.

$$(2y - 5)^2 + y^2 = 25$$
$$4y^2 - 20y + 25 + y^2 = 25$$
$$5y^2 - 20y = 0$$
$$5y(y - 4) = 0$$
$$y = 0 \quad \text{or} \quad y = 4$$

You can now use the equation $x = 2y - 5$ to find x.

If $y = 0$, then $x = 2(0) - 5 = -5$.

If $y = 4$, then $x = 2(4) - 5 = 3$.

The solutions are $(-5, 0)$ and $(3, 4)$.   ■

EXERCISES   *Solve the systems in Exercises 2–5 analytically.*

2. $2x^2 - 3y^2 = -3$ **3.** $x + y^2 = 1$
   $\quad\ x - y = 1$ $\quad\ x + 2y = 2$

4. $x^2 + 3xy = 13$ **5.** $4x^2 + y^2 = 100$
   $\quad\ y = x + 5$ $\quad\ 3x - 2y = 0$

Systems Involving Only x^2 and y^2 If the variable terms in a quadratic system contain only x^2 or y^2, you can solve the system by treating it as linear in the variables x^2 and y^2.

EXAMPLE 2   *A quadratic system involving only x^2 and y^2*

To solve the system

$$3x^2 - y^2 = 6$$
$$2x^2 + 3y^2 = 37$$

you can proceed by elimination, as in Section 3-1, regarding x^2 and y^2 as the variables. Multiply both sides of the first equation by 3.

$$9x^2 - 3y^2 = 18$$
$$2x^2 + 3y^2 = 37$$

Then add the equations.

$$11x^2 = 55$$

$$x^2 = 5$$

$$x = \pm\sqrt{5}$$

Replace x^2 by 5 in the first equation in the system.

$$3(5) - y^2 = 6$$

$$y^2 = 9$$

$$y = \pm 3$$

The solutions are $\left(\sqrt{5}, 3\right), \left(-\sqrt{5}, 3\right), \left(\sqrt{5}, -3\right),$ and $\left(-\sqrt{5}, -3\right).$ ■

EXERCISES   *Solve the systems in Exercises 6–9 analytically.*

6. $7x^2 + 2y^2 = 13$
    $2x^2 - 3y^2 = -16$

7. $x^2 = 2y^2 - 14$
    $y^2 = 2x^2 - 5$

8. $5x^2 + 4y^2 = 2$
    $6x^2 + 7y^2 = 9$

9. $x^2 + 2y^2 = 11$
    $2x^2 - y^2 = 7$

Systems of the Form $Ax^2 + Cy^2 = G$, $xy = K$ An equation $xy = K$ in a system can be solved for either variable. Substitution into the other equation then produces a single equation in one variable. If the other equation has the form $Ax^2 + Cy^2 = G$, the resulting equation can be solved.

EXAMPLE 3   *A quadratic system of the form $Ax^2 + Cy^2 = G$, $xy = K$*

To solve the system

$$4x^2 + y^2 = 26$$

$$xy = 6$$

you can begin by solving the second equation for either variable. Solving for y,

$$y = \frac{6}{x}$$

Substitution into the first equation yields

$$4x^2 + \left(\frac{6}{x}\right)^2 = 26$$

$$4x^2 + \frac{36}{x^2} = 26$$

$$4x^4 + 36 = 26x^2$$

$$4x^4 - 26x^2 + 36 = 0$$

$$2(x^2 - 2)(2x^2 - 9) = 0$$

$$x^2 = 2 \quad \text{or} \quad x^2 = \frac{9}{2}$$

$$x = \pm\sqrt{2} \quad \text{or} \quad x = \pm\sqrt{\frac{9}{2}} = \pm\frac{3\sqrt{2}}{2}$$

Since $y = \dfrac{6}{x}$, the corresponding values of y are

$$y = \pm\frac{6}{\sqrt{2}} = \pm 3\sqrt{2} \quad \text{or} \quad y = \pm\frac{12}{3\sqrt{2}} = \pm 2\sqrt{2}$$

The solutions to the system are

$$\left(\sqrt{2}, 3\sqrt{2}\right) \quad \left(-\sqrt{2}, -3\sqrt{2}\right) \quad \left(\frac{3\sqrt{2}}{2}, 2\sqrt{2}\right) \quad \left(-\frac{3\sqrt{2}}{2}, -2\sqrt{2}\right) \quad \blacksquare$$

EXERCISES   *Solve the systems in Exercises 10–13 analytically.*

10. $x^2 + y^2 = 25$
    $xy = -12$

11. $x^2 + y^2 = 25$
    $xy = 2\sqrt{39}$

12. $2x^2 + 3y^2 = 10$
    $xy = -2$

13. $x^2 + 4y^2 = 40$
    $xy = -6$

SOLVING SYSTEMS OF QUADRATIC EQUATIONS GRAPHICALLY

The analytical methods you have just learned apply only to a few special classes of systems of quadratic equations in two variables. By contrast, you can solve any such system graphically. Furthermore, although a graph does not provide exact solutions, it can indicate the number of solutions and their approximate values at a glance.

Number of Solutions Graphical methods depend on the idea that solutions to a system are represented by points where the graphs of all the equations in the system intersect. You encountered this idea before in connection with linear systems. However, in contrast to linear systems, quadratic systems can have more than one solution without having infinitely many.

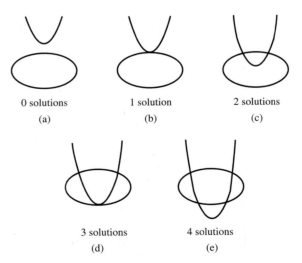

0 solutions   1 solution   2 solutions
  (a)   (b)   (c)

3 solutions   4 solutions
  (d)   (e)

Figure 4–66

Figure 4-66 suggests that two distinct conic sections cannot intersect at more than four points, so that a system of quadratic equations cannot have any more than four solutions without having infinitely many. You can provide additional evidence in Exercises 14 and 15.

EXERCISES   **14.** (*Making Observations*) Draw a five-part figure, similar to Figure 4-66, to illustrate that two ellipses can intersect at 0, 1, 2, 3, or 4 points.

15. (*Making Observations*) Repeat Exercise 14 for

    a. an ellipse and a hyperbola

    b. a parabola and a hyperbola

Finding Points of Intersection You can solve a system of quadratic equations graphically by the method described for linear systems in Section 3-1, but you need to make sure you have found *all* points of intersection on the graphs or, equivalently, all solutions to the system of equations.

EXAMPLE 4   *Solving a quadratic system graphically*

Figure 4-67 shows the graph of the system

$$x^2 - y^2 = 4$$
$$x = y^2 - 2$$

(To graph the system, solve each equation for y to obtain $y = \pm\sqrt{x^2 - 4}$ and $y = \pm\sqrt{x + 2}$.) It appears to show three solutions, labeled A, B, and C.

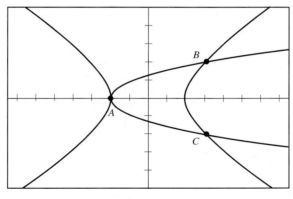

Figure 4–67

You can't be sure from the figure whether the two graphs cross, touch, or do not quite touch in the vicinity of A. However, you might guess that $(-2, 0)$ is a solution and substitution into each equation confirms the guess. Furthermore, by looking at the equations in the system you can verify that $(-2, 0)$ is the vertex of the parabola and also a vertex of the hyperbola. Therefore, the graphs simply touch at that point and do not cross, so there is no other solution close to $(-2, 0)$.

The solutions at B and C, estimated by zooming in, are $(3.00, \pm2.24)$ with an error of no more than 0.01. $\left[\left(\text{The exact solutions, obtained analytically, are} \left(3, \pm\sqrt{5}\right).\right]\right.$ ∎

    Since a quadratic system can have more than one solution, you may need to decide which solution is of interest when the system models a physical problem. The situation often provides clues, as in ***Earthquakes 2***. In Exercise 16, you will use those clues to attempt to locate the epicenter of the San Francisco World Series earthquake.

EXERCISES   **16.** (*Problem Solving*) In Exercises 27 and 28 of Section 4-6, and Exercise 1 of this section, you placed the earthquake's epicenter on a particular branch of each of two hyperbolas. Find all points of intersection of the hyperbolas, and tell which point is the epicenter.

In Exercises 17–20, *solve the system in the indicated exercise graphically.*

17. Exercise 4   **18.** Exercise 6
19. Exercise 8   **20.** Exercise 12

SOLVING SYSTEMS OF QUADRATIC INEQUALITIES

The following Mathematical Looking Glass leads to a **system of quadratic inequalities**, that is, a collection of quadratic inequalities in two variables to be solved simultaneously. Over the next few pages you will discover that the solution of a quadratic inequality, like that of a linear inequality, is represented most conveniently as a region of the plane. The solution of a system of quadratic inequalities is then the region common to the solutions of all inequalities in the system.

A MATHEMATICAL LOOKING GLASS
Bermuda Triangle

The Bermuda Triangle, a region of ocean bounded by Bermuda, Miami, and Puerto Rico, is a notorious area where ships and airplanes disappear mysteriously. In fact, the number of ships lost in the Bermuda Triangle is due largely to its unusually large volume of boating traffic. As Charles J. Cazeau and Stuart D. Scott point out in *Exploring the Unknown* (Da Capo Press, 1980),

> During 1975 the U.S. Coast Guard came to the rescue of 140,000 people. The Coast Guard aided 5600 boats that sank or capsized, 1300 with fires or explosions, and 1000 boats that did not know where they were. . . . We hasten to point out that these statistics apply to *all* of the United States coastline and not just the Bermuda Triangle. However, of the 37,000 U.S. Coast Guardsmen, the busiest are those rescuing people in the Bermuda Triangle, one of the great resort areas of the world . . . . We think that it is a remarkable tribute to the Coast Guard that they let only 2.5 to 6(?) ships and/or planes slip away from them each year.

To assist in their rescue work, the Coast Guard maintains helicopters at several stations in the Caribbean. Each can operate within a 100-mile radius of its base. That is, it can fly to a specific location up to 100 miles away, spend up to a half hour there, and fly back without refueling. Let's follow a hypothetical helicopter rescue mission. Figure 4-68 shows a portion of the Bermuda Triangle between Miami and Nassau. It is placed in a coordinate system with units measured in miles, so that Miami is at $(-80, 0)$ and Nassau is at $(100, 0)$.

A ship is lost in a storm shortly after leaving Orange Cay, at $(0, -40)$. Although it may have been blown in any direction, the Coast Guard is reasonably certain that the ship is still within 20 miles of Orange Cay.

Figure 4–68

■

EXERCISES 21. (*Creating Models*)

a. Copy Figure 4-68 onto a sheet of graph paper, and shade in the region where the ship might be.

b. Write the equation of the boundary of the region.

22. **a.** Use a different style of shading to indicate the region that could be covered by a helicopter based in Miami.

b. Write the equation of the boundary of the region.

23. (*Interpreting Mathematics*) Can the helicopter in Exercise 22 search the entire area where the ship might be located? How can you tell?

Your shaded areas in Exercises 21 and 22 are solutions of quadratic inequalities in two variables. Such inequalities often model physical problems. Like linear inequalities, they are solved most conveniently by graphical methods. After a general discussion of systems of quadratic inequalities, we will return to **Bermuda Triangle** to see how the Coast Guard might improve their chances of finding the lost ship.

Figure 4-69 indicates how the graph of a quadratic relation divides the coordinate plane into two regions. (For our present purpose, consider the regions "outside" the hyperbola in Figure 4-69c as one region.)

(a)   (b)   (c)

Figure 4–69

Any quadratic inequality in two variables is true at all points in one region and false at all points in the other. The solution of the inequality can be found by evaluating the inequality at a test value.

EXAMPLE 5   *Solving a quadratic inequality in two variables*

To solve the inequality

$$4x^2 - 9y^2 \geq 36$$

first graph $4x^2 - 9y^2 = 36$, as in Figure 4-70a.

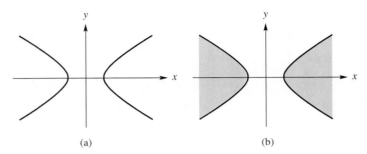

(a)   (b)

Figure 4–70

You can use any point not on the hyperbola as a test value. We have chosen $(0, 0)$. At that point, the inequality asserts

$$4(0)^2 - 9(0)^2 \geqslant 36$$

which is false. The solution of the inequality is therefore the region not containing $(0, 0)$, indicated in Figure 4-70b. The hyperbola itself is included in the solution, since the inequality is true there. ∎

The solution of a system of quadratic inequalities is the region common to the solutions of the inequalities in the system. For example, the solution of the system

$$4x^2 - 9y^2 \geqslant 36$$
$$4x^2 + 25y^2 < 100$$

is shown in Figure 4-71.

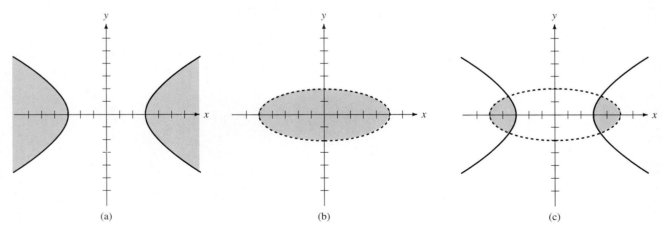

(a)   (b)   (c)

Figure 4–71

EXERCISES   *Solve the inequalities in Exercises 24–27.*

24. $x^2 + y^2 > 64$ **25.** $3x - y^2 \leqslant 6$

26. $y^2 - 16x^2 < 96$ **27.** $y^2 - 16x^2 \geqslant 96$

Solve the systems of inequalities in Exercises 28–31.

28. $x^2 + y^2 > 64$
$3x - y^2 \leqslant 6$

29. $3x - y^2 \leqslant 6$
$y^2 - 16x^2 < 96$

30. $x^2 + y^2 > 64$
$3x - y^2 \leqslant 6$
$y^2 - 16x^2 < 96$

31. $x^2 + y^2 > 64$
$x^2 + y^2 < 100$
$y - x^2 \geqslant 8$

32. (*Creating Models*) Write inequalities to describe the regions you shaded in Exercises 21 and 22.

33. (*Problem Solving*) Refer to **Bermuda Triangle**. Write and solve graphically a system of inequalities to represent the overlap of the lost ship's possible locations with the area that can be covered by a helicopter based in Nassau.

34. (*Problem Solving*)

a. Suppose that a helicopter takes off from Nassau, conducts a search, and lands in Miami, flying a total of 200 miles. Explain why the region it can cover lies inside an ellipse with foci at Nassau and Miami.

b. Write an inequality to represent the area that can be covered by the helicopter in part (a).

c. Write and solve graphically a system of inequalities to represent the overlap of the lost ship's possible locations with the area that can be covered by this helicopter.

35. (*Interpreting Mathematics*) Suppose that you are in charge of helicopter operations out of Nassau, and you have received a call from Miami, saying that a helicopter is already searching for the lost ship, as in Exercise 22. You are to assist them by sending a second helicopter, which will either return to Nassau, as in Exercise 33, or fly on to Miami, as in Exercise 34. Which option do you choose? Why?

ADDITIONAL EXERCISES *Solve the systems of equations in Exercises 36–43 analytically.*

36. $y = x^2 - 1$
$y = x + 1$

37. $x^2 + 5y^2 = 24$
$xy = -4$

38. $3x^2 + 8y^2 = 139$
$2x^2 - 3y^2 = 1$

39. $x^2 + y^2 = 4$
$9x^2 + 16y^2 = 144$

40. $x^2 - y^2 = 9$
$5x - 4y = 9$

41. $x^2 + y^2 = 36$
$y = x^2 - 6$

42. $3(x - 1)^2 + 2(y + 2)^2 = 216$
$x + 4y = 7$

43. $x^2 + 4y^2 - 8x + 16y + 16 = 0$
$xy = 0$

44. Use graphical methods to find the real solutions to the system in the indicated exercise.

a. Exercise 36   **b.** Exercise 37

c. Exercise 38   **d.** Exercise 39

e. Exercise 40   **f.** Exercise 41

Solve each system of inequalities in Exercises 45–48.

45. $3x^2 + 5y^2 \geqslant 15$
$3x^2 + 5y^2 < 90$

46. $x^2 - y^2 < 4$
$y^2 - x^2 < 4$

47. $25x^2 + 4y^2 \leqslant 100$
$y^2 - 9x^2 \geqslant 9$
$xy < 4$

48. $y > x^2 - 16$
$y > 4x - x^2$
$y < 2x + 1$

49. (*Problem Solving*) In an appropriate coordinate system, the orbit of the earth around the sun has the equation

$$\frac{x^2}{1.03} + \frac{y^2}{0.97} = 1$$

Distances are measured in astronomical units (an astronomical unit is the mean distance from the earth to the sun, about 92.9 million miles). In the same coordinate system, a newly discovered comet might have an orbit whose equation is

$$xy + x + y = 0$$

a. Comets with elliptic orbits return periodically to the same position, while comets with parabolic or hyperbolic orbits visit the solar system only once. Will this comet ever be seen again? (*Hint*: Solve the comet's equation for *y*, and graph it.)

b. How many times will the comet cross the earth's orbit? At what points? (*Hint*: Solve the earth's equation for *y*, and answer the question by graphical methods.)

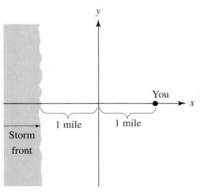

Figure 4–72

50. (*Problem Solving*) Imagine that as you are taking your morning jog, you hear over your portable radio that a storm front is approaching at 8 mph. Looking up and seeing the front about 2 miles ahead, you run for shelter at 8 mph.

a. Explain why the points you can reach without getting wet lie on one side of a parabola with its focus at your present position and its directrix along the storm front.

b. In the coordinate system of Figure 4-72, write and solve graphically an inequality to show all such points.

51. (*Problem Solving*) Figure 4-73 depicts a critical moment in a baseball game. The batter at *A* has just hit a ground ball to the fielder at *B*. The fielder must throw the ball to first base (*C*) before the batter gets there. Assume that the batted ball and the thrown ball both travel in a straight line at 120 feet per second, that the batter can run to first base in 4 seconds, and that the fielder takes 1 second after catching the ball before making the throw.

a. Explain why the fielder's throw will reach first base before the batter arrives if the fielder is positioned inside an ellipse with foci at *A* and *C*.

b. Write and solve graphically a system of inequalities to show all points in fair territory (above the *x*-axis and to the right of the *y*-axis) where the fielder can play.

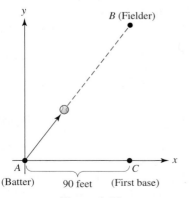

Figure 4–73

CHAPTER REVIEW

Complete Exercises 1–68 (Writing to Learn) before referring to the indicated pages.

WORDS AND PHRASES

In Exercises 1–26, explain the meaning of the words or phrases in your own words.

1. (page 148) **quadratic function**
2. (page 149) **vertex**
3. (page 149) **axis of symmetry**
4. (page 152) **expanded form, factored form, standard graphing form**
5. (page 158) **second differences**
6. (page 164) **optimization problem**
7. (page 174) **unit circle**
8. (page 176) **relation, quadratic relation**
9. (page 177) **explicitly defined function**
10. (page 177) **implicitly defined function**
11. (page 186) **conic section**
12. (page 189) **standard graphing form** of an ellipse
13. (page 190) **center, major axis, minor axis, vertices, covertices** of an ellipse
14. (page 193) **standard graphing form** of a hyperbola
15. (page 195) **asymptote**
16. (page 195) **center, transverse axis, vertices, conjugate axis** of a hyperbola
17. (page 205) **symmetric about the y-axis, symmetric about the x-axis**
18. (pages 106, 215) **parabola**
19. (page 215) **directrix, focus** of a parabola
20. (pages 185, 219) **ellipse**
21. (page 219) **foci** of an ellipse
22. (page 221) **eccentricity** of an ellipse
23. (pages 186, 223) **hyperbola**
24. (page 223) **foci** of a hyperbola
25. (page 229) **system of quadratic equations**
26. (page 234) **system of quadratic inequalities**

IDEAS

27. (page 149) In what ways are all quadratic graphs alike?
28. (page 150) What information do the values of a, h, and k provide about the graph of $y = a(x - h)^2 + k$?
29. (page 151) Describe how the basic graph $y = x^2$ is transformed to obtain the graph of $y = a(x - h)^2 + k$.

In particular, describe how the graph is affected by the values of a, h, and k.

30. (pages 152, 153) Describe the process of finding the vertex and the x-intercepts on a quadratic graph from the expanded form of its equation. Illustrate with a specific example.
31. (page 153) If you know the coordinates of the vertex on a quadratic graph, how can you identify its axis of symmetry and find the range of the function?
32. (pages 155, 156) Describe the process of finding the vertex and the x-intercepts on a quadratic graph from the factored form of its equation. Illustrate with a specific example.
33. (page 157) Describe the process of finding the vertex and the x-intercepts on a quadratic graph from the standard graphing form of its equation. Illustrate with a specific example.
34. (page 158) Name at least one way in which each of the three analytical forms of a quadratic function is more useful than the other two.
35. (page 159) How can you decide whether a table fits a quadratic function? Make up a table and use it to illustrate.
36. (page 164) Why do optimization problems often arise in a physical context?
37. (pages 166, 167) Describe how each step in the analytical solution of a quadratic inequality corresponds to a step in the graphical solution. Illustrate your description by solving a specific inequality both ways.
38. (pages 168, 169) Describe the process of fitting a quadratic function to a table. To illustrate your description, make up a table that fits a quadratic function and find the equation of the function.
39. (page 169) If a table fits a quadratic function, why are three data points enough to find its equation? Why are two data points not enough?
40. (pages 176, 177) In what respects are the concepts of function and relation alike? In what respects are they different? Give an example of a correspondence that is both a function and a relation and an example of a correspondence that is a relation but not a function.
41. (pages 176, 177) A relation may be represented by a table, by a graph, or by an equation. In each case, explain how to decide whether the relation is a function.
42. (page 180) Why might we want to obtain equations for functions defined implicitly by a relation?
43. (pages 177, 178) Describe a method for verifying that a given function is defined implicitly by a given relation.

44. (pages 180, 181)   Even when a quadratic relation does not represent a function in the abstract, it may represent a function if the variables represent physical quantities. Explain how this can occur.

45. (page 187)   In an equation

$$Ax^2 + Cy^2 + Dx + Ey + F = 0$$

what do the values of A and C tell you about the graph?

46. (page 187)   In an equation

$$y = a(x - h)^2 + k \quad \text{or} \quad x = a(y - k)^2 + h$$

what do the values of a, h, and k tell you about the graph?

47. (page 191)   In an equation $\left(\dfrac{x - h}{a}\right)^2 + \left(\dfrac{y - k}{b}\right)^2 = 1$ or $\left(\dfrac{x - h}{b}\right)^2 + \left(\dfrac{y - k}{a}\right)^2 = 1$, what do the values of a, b, h, and k tell you about the graph?

48. (page 192)   What condition on an equation

$$Ax^2 + Cy^2 + Dx + Ey + F = 0$$

distinguishes a circle from other ellipses?

49. (page 192)   In an equation

$$(x - h)^2 + (y - k)^2 = r^2$$

what do the values of h, k, and r tell you about the graph?

50. (page 196)   In an equation $\left(\dfrac{x - h}{a}\right)^2 - \left(\dfrac{y - k}{b}\right)^2 = 1$ or $\left(\dfrac{y - k}{a}\right)^2 - \left(\dfrac{x - h}{b}\right)^2 = 1$, what do the values of a, b, h, and k tell you about the graph?

51. (pages 196, 197)   How does the equation of a hyperbola relate to the equations of its asymptotes?

52. (page 198)   The graph of a quadratic relation is not always a parabola, ellipse, or hyperbola. Describe all the exceptional graphs. How can you identify each exceptional graph from its equation?

53. (page 202)   How does the graph of a relation change if x is replaced by $-x$, by $\dfrac{x}{a}$, or by $x - h$ in its equation? How does the graph change if y is replaced by $-y$, by $\dfrac{y}{b}$, or by $y - k$?

54. (page 207)   When you are applying a sequence of graphical transformations, explain why order is important. Specifically, compare the effects of applying a reflection in the x-axis before or after a vertical shift.

55. (pages 215–217)   How are the points on a parabola determined by its focus and directrix? In an equation $y = a(x - h)^2 + k$ or $x = a(y - k)^2 + h$, what do the values of a, h, and k tell you about the location of the focus and directrix?

56. (page 217)   Describe the reflective property of parabolas, and give an example of its use in a physical context.

57. (pages 219, 220)   In an equation

$$\left(\dfrac{x - h}{a}\right)^2 + \left(\dfrac{y - k}{b}\right)^2 = 1 \quad \text{or} \quad \left(\dfrac{x - h}{b}\right)^2 + \left(\dfrac{y - k}{a}\right)^2 = 1$$

what do the values of a, b, h, and k tell you about the location of the foci?

58. (page 221)   What are the possible values for the eccentricity of an ellipse? What does the eccentricity tell you about the shape of the ellipse?

59. (page 221)   Describe the reflective property of ellipses, and give an example of its use in a physical context.

60. (pages 223, 224)   In an equation

$$\left(\dfrac{x - h}{a}\right)^2 - \left(\dfrac{y - k}{b}\right)^2 = 1 \quad \text{or} \quad \left(\dfrac{y - k}{a}\right)^2 - \left(\dfrac{x - h}{b}\right)^2 = 1$$

what do the values of a, b, h, and k tell you about the location of the foci?

61. (page 225)   Describe the reflective property of hyperbolas, and give an example of its use in a physical context.

62. (page 230)   Describe an analytical method of solution for a system consisting of one quadratic equation and one linear equation. Illustrate your description with a specific system.

63. (pages 230, 231)   Describe an analytical method of solution for a system of quadratic equations involving only x^2 and y^2. Illustrate your description with a specific system.

64. (pages 231, 232)   Describe an analytical method of solution for a system of quadratic equations of the form $Ax^2 + Cy^2 = G$, $xy = K$. Illustrate your description with a specific system.

65. (page 232)   Why is it useful to know both analytical and graphical methods for solving systems of quadratic equations?

66. (page 232)   How many solutions can a system of quadratic equations have? How do the graphs of conic sections support your answer?

67. (page 232)   In what respects is the process of solving a system of quadratic equations like that of solving a system of linear equations? In what respects is it different?

68. (pages 235, 236)   Describe the process of solving a system of quadratic inequalities. In what respects is it like that of solving a system of linear inequalities? In what respects is it different?

CHAPTER 5

POLYNOMIAL FUNCTIONS

5–1 POWER FUNCTIONS

BASIC POWER FUNCTIONS AND THEIR GRAPHS

In Chapter 2 you studied linear functions, whose equations have the form $f(x) = mx + b$. In Chapter 4 you studied quadratic functions, whose equations have the form $f(x) = ax^2 + bx + c$. Both are special cases of **polynomial functions**, whose equations have the form

$$f(x) = a_n x^n + a_{n-1} x^{n-1} + \cdots + a_1 x + a_0$$

for some nonnegative integer n and the choice of real numbers $a_n, a_{n-1}, \ldots, a_1, a_0$, with $a_n \neq 0$. The value of n is the **degree** of the function. Thus the equations

$$f(x) = 3x - 4$$

$$f(x) = x^2 + \pi x + \sqrt{2}$$

$$f(x) = x^7 - \frac{2}{3} x^4 - 97x$$

$$f(x) = 3$$

define polynomial functions of degrees 1, 2, 7, and 0. The equation

$$f(x) = (x - 1)(x + 1)(x - 4)$$

defines a polynomial of degree 3 since it can be rewritten in the required form as

$$f(x) = x^3 - 4x^2 - x + 4$$

The equations

$$f(x) = 2\sqrt{x} - 5$$

$$f(x) = 1 + \frac{1}{x}$$

do not define polynomial functions, since they cannot be rewritten

$$f(x) = a_n x^n + a_{n-1} x^{n-1} + \cdots + a_1 x + a_0$$

You already know that the graph of every linear function is a straight line, and the graph of every quadratic function is a parabola. By contrast, the graphs of polynomial functions of higher degree appear at first to have little in common with each other. Figure 5-1 shows some typical polynomial graphs.

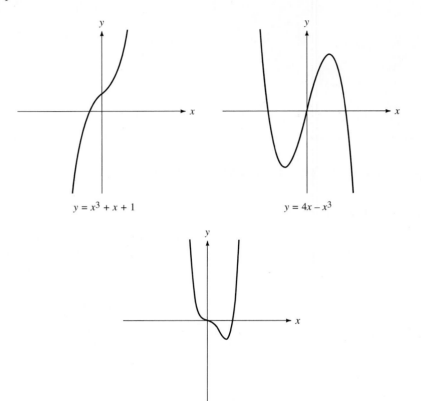

$y = x^3 + x + 1$

$y = 4x - x^3$

$y = 3x^4 - 4x^3$

Figure 5–1

In Sections 5-2 to 5-5 you will discover some of the properties shared by all polynomial functions. In this section we will adopt the strategy of *Examining a Simpler Problem* and confine our attention to a simple but important type of polynomial function. These are the **power functions** $y = x^n$, and their graphs can be regarded as basic graphs. You are already familiar with the graphs of $y = x$ and $y = x^2$. Figure 5-2 shows the next few basic graphs.

EXERCISES

1. (*Making Observations*) Graph the power functions $y = x^3$, $y = x^5$, $y = x^7$, and $y = x^9$ together, first in the window $[-2, 2]$ by $[-10, 10]$, then in the window $[-1, 1]$ by $[-1, 1]$. Describe the similarities and differences among the graphs.

2. (*Making Observations*) Repeat Exercise 1 for the power functions $y = x^4$, $y = x^6$, $y = x^8$, and $y = x^{10}$.

3. (*Making Observations*) Guess what the graphs of $y = x^{49}$ and $y = x^{50}$ look like, and sketch them. Use calculator graphs to check your guesses.

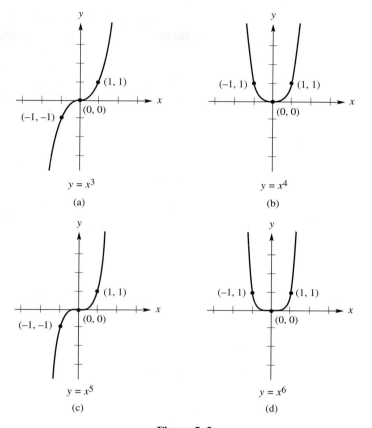

Figure 5–2

Your observations in Exercises 1–3 can be summarized as follows.

> If n is odd, the function $f(x) = x^n$
>
> - has domain $(-\infty, \infty)$ and range $(-\infty, \infty)$,
> - is increasing on $(-\infty, \infty)$, and
> - passes through $(-1, -1), (0, 0)$, and $(1, 1)$.
>
> If n is even, the function $f(x) = x^n$
>
> - has domain $(-\infty, \infty)$ and range $[0, \infty)$,
> - is decreasing on $(-\infty, 0)$ and increasing on $(0, \infty)$, and
> - passes through $(-1, 1), (0, 0)$, and $(1, 1)$.

EXERCISE   **4.** (*Making Observations*) Explain why each of the preceding statements in the box is true.

GRAPHICAL TRANSFORMATIONS

In Section 4-1, you discovered that you could sketch the graph of every quadratic polynomial function by transforming the basic graph $y = x^2$. No similar statement can be made for polynomials of higher degree. However, many polynomial graphs can be obtained by transforming the graphs of power functions, as in Example 1.

EXAMPLE 1   *Using transformations to graph a polynomial function*

To graph $y = 2(x + 3)^6$, begin with the basic graph of $y = x^6$, and make the changes

$$y = x^6 \rightarrow y = 2x^6 \rightarrow y = 2(x + 3)^6$$

These changes stretch the graph of $y = x^6$ vertically by a factor of 2 and then shift it 3 units left, as in Figure 5-3.

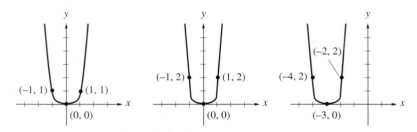

Figure 5–3   ■

EXERCISES   *In Exercises 5–8, sketch each graph. Support your results with a calculator graph.*

5. $y = (x - 6)^5$ **6.** $y = -x^6$

7. $y = \dfrac{1}{2}(x - 5)^4$ **8.** $y = 6 - 2x^3$

In Exercises 9–12,

    a. Sketch the graph of the function whose graph is that of $f(x) = x^3$ transformed as described.

    b. Write the equation of the function

9. reflected in the y-axis.

10. reflected in the x-axis and compressed horizontally by a factor of 2.

11. stretched vertically by a factor of 4 and then shifted 2 units up.

12. shifted 2 units up and then stretched vertically by a factor of 4.

DIRECT VARIATION

If a relationship is described by an equation $y = ax^n$, we say that **y varies directly with the nth power of x**. For example, the volume V of a sphere varies directly with the cube of its radius r, since $V = \dfrac{4}{3}\pi r^3$. In particular, if $y = ax$, we say that y **varies directly** with x, **varies linearly** with x, or is **directly proportional** to x. For example, the weight W, in pounds, of an object on earth is related to its mass M, in kilograms, by the equation $W = 2.20M$, so that W varies directly with M. The coefficients $\dfrac{4}{3}\pi$ and 2.20 in the preceding equations are called **constants of proportionality**.

Direct variation has the following property.

If $y = ax^n$, then whenever x is multiplied by a constant k, y is multiplied by k^n.

Informally,

For nth power variations, doubling x multiplies y by 2^n.

For example, if V varies directly with r^3, then doubling r multiplies V by $2^3 = 8$. In particular, if y varies directly with x, then whenever x is doubled, y is also doubled.

It is easy to tell whether a table fits a variation equation of a given degree and to find its equation when it does.

> In any table generated by $y = ax^n$, the ratio $\dfrac{y}{x^n} = a$.
>
> *Exception*: A row with $x = 0$ fits if and only if $y = 0$ also.

EXAMPLE 2   *Deciding whether a table fits a variation equation of a given degree*

To see whether the table

x	y
1	3
2	48
3	243
4	768

fits a variation equation of degree 4, evaluate the ratio $\dfrac{y}{x^4}$ for each row of the table.

$$\frac{3}{1^4} = 3, \qquad \frac{48}{2^4} = 3, \qquad \frac{243}{3^4} = 3, \qquad \frac{768}{4^4} = 3$$

The table fits the equation $y = 3x^4$.   ■

EXERCISES   *In Exercises 13–16, decide whether the table fits a variation equation $y = ax^n$ for the given value of n. If so, write an equation relating x and y.*

13.

x	y ($n = 5$)
1	0.1
2	3.2
3	24.3
5	312.5
6	777.6

14.

x	y ($n = 3$)
3	54
4	128
6	432
9	1458
12	3456

15.

x	y ($n = 2$)
2	4
4	16
5	32
7	128
9	512

16.

x	y ($n = 4$)
1	0.5
2	8
3	40.5
4	128
5	312.5

17. If y varies linearly with x and $y = 35$ when $x = 21$, find the value of y when $x = 39$.

18. If W varies directly with Z and $W = 0.004$ when $Z = 0.001$, find the value of Z when $W = 30,000$.

19. In Exercises 13–16, if the table fits a polynomial variation, find the value of y when $x = 20$.

In the following Mathematical Looking Glass, direct variation provides insight into a biological question.

A MATHEMATICAL
LOOKING GLASS
Weighing Dinosaurs

How large were the largest dinosaurs? We may never be able to answer this question with more than rough accuracy. The largest for which a nearly complete skeleton has been found is *Brachiosaurus*, which lived over 100 million years ago and despite its immense size was probably a vegetarian. From its skeleton we can be reasonably sure that *Brachiosaurus* stood about 43 feet high. Its weight is more open to question, since we do not have access to the flesh that once surrounded the skeleton. In *Dynamics of Dinosaurs and Other Extinct Giants* (Columbia University Press, 1989), R. McNeill Alexander describes several methods for estimating dinosaur weights. In his preface he states,

> I have used the methods of physics and engineering to try to discover how extinct animals could have lived and moved. I like to think about animals in the kinds of ways that engineers think about machines and vehicles.

Let's estimate *Brachiosaurus'* weight by a method that is not highly accurate, but is easily carried out. The plan is to estimate the amount of weight that its skeleton could have supported. The strategy for implementing the plan is to compare the *Brachiosaurus* skeleton to that of a modern large animal, say, a hippopotamus, whose weight is known.

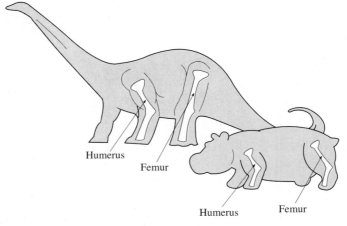

Figure 5–4

When either animal is standing, most of its weight is supported by its upper leg bones, the femur and humerus, as shown in Figure 5-4. In *Dynam-*

ics of Dinosaurs, Professor Alexander describes the work of an international team of scientists who measured and added together the circumference of the two *Brachiosaurus* leg bones. Repeating this for a hippopotamus, they found that the sum was about 400 mm for the hippopotamus, which has a weight of about 5500 pounds, and about 1400 mm for *Brachiosaurus*. In Exercise 20 you can use these data to estimate *Brachiosaurus'* weight. You will need to make the following assumptions.

- For each animal, the femur and humerus bones have circular cross sections of equal size.
- The amount of weight any animal bone can support is proportional to its cross-sectional area.
- The weight of each animal is equal to the greatest weight its bones could support. ∎

EXERCISES   **20.** (*Problem Solving*)

    a. Use the given data to express an animal's weight W in pounds as a function of the circumference C of its femur or humerus bone in millimeters.

    b. Use your result from part (a) to estimate *Brachiosaurus'* weight. (Researchers' estimates quoted in *Dynamics of Dinosaurs* range from about 69,500 to 191,400 pounds. How does your estimate compare with those?)

21. (*Problem Solving*) Professor Alexander's own estimate of the weight of *Brachiosaurus* is based on an experiment he performed using a model *Brachiosaurus* from London's Natural History Museum. By submerging the model in water, he determined that its volume was 728 cm³.

    a. If the model was $\frac{1}{40}$ life size, explain why the dinosaur's volume was $(40)^3(728)$ cm³.

    b. The bodies of most large land animals have about the same density as water (1 kg for each 1000 cm³). Use this fact and your result from part (a) to estimate *Brachiosaurus'* weight in pounds. How does this estimate compare with the one in Exercise 20?

ADDITIONAL EXERCISES   *In Exercises 22–25,*

    a. Use graphical transformations to sketch each graph.

    b. State the domain and range of the function.

22. $y = (x - 1)^3 - 5$ **23.** $y = 10 - 0.2x^7$

24. $y = \frac{1}{3}x^4 + 12$ **25.** $y = \frac{1}{3}(x^4 + 12)$

26. (*Writing to Learn*) Describe how the domain and range of a power function $y = x^n$ are affected by each of the following transformations.

    a. a vertical stretch or compression

    b. a reflection in the *x*-axis

    c. a horizontal shift

    d. a vertical shift

In Exercises 27–32,

a. Sketch the graph of the function whose graph is that of $f(x) = x^4$ transformed as described.

b. Write the equation of the function

27. reflected in the *y*-axis.

28. reflected in the *y*-axis and then shifted 1 unit right.

29. shifted 1 unit right and then reflected in the *y*-axis.

30. stretched vertically by a factor of 3.

31. stretched vertically by a factor of 3 and then shifted 2 units up.

32. shifted 2 units up and then stretched vertically by a factor of 3.

In Exercises 33–36, decide whether the table fits a polynomial variation $y = ax^n$ for the given value of n. If it does, write an equation relating x and y.

33.

x	$y \ (n = 4)$
0.2	48
0.3	243
0.5	1,875
0.8	12,288
1.0	30,000

34.

x	$y \ (n = 1)$
3	45
5	75
10	150
12	180
18	270

35.

x	$y \ (n = 2)$
-20	-8
10	-2
30	-18
40	-32
100	-200

36.

x	$y \ (n = 3)$
2	1
4	16
6	54
8	128
10	250

37. In Exercises 33–36, if the table fits a polynomial variation, find the value of *y* when $x = 1$.

38. If *t* is directly proportional to *s* and $t = 57$ when $s = 126$, find *t* when $s = 168$.

39. If *N* varies directly with *M*, and $N = 1/3$ when $M = 4/5$, find *M* when $N = 3/4$.

40. If *A* varies linearly with *B*, and the constant of proportionality is 0.7, find the value of *A* when $B = 62$.

41. If *q* varies directly with *p*, and the constant of proportionality is 250, find the value of *q* when $p = 0.004$.

42. (*Writing to Learn*) Decide whether each of the following is a relationship of linear variation, and explain your answer.

a. H = the number of hours a 100-watt light bulb is burned

C = the cost, in cents, of burning the light bulb

b. N = the number of people using a reservoir

G = the number of gallons of water taken from the reservoir each day

c. P = the number of people living in a house

C = the amount of closet space per person

d. N = the number of pages in a telephone book

W = the weight of the book in pounds

43. (*Creating Models*) If the relationship referred to is one of linear variation, use the given information to write its equation. If the relationship is not one of linear variation, do not attempt to write an equation.

a. (Exercise 42a) It costs $0.06375 to burn a 100-watt bulb for 5 hours.

b. (Exercise 42b) 30,000 people use a combined total of 2,400,000 gallons of water from the reservoir each day.

c. (Exercise 42c) When 5 people share the house, each one has 23 cubic feet of closet space.

d. (Exercise 42d) A telephone book with 1568 pages weighs 4.8 pounds.

44. (*Problem Solving*) In the absence of air, the distance an object falls in t seconds after being dropped is proportional to t^2. If an object takes 0.1 seconds to fall 4.9 cm, how long will it take to fall 10 cm?

45. (*Problem Solving*) Under certain conditions, the volume occupied by a gas varies directly with its Kelvin temperature. If its volume was 500 cm^3 when the temperature was 0° Celsius and is now 550 cm^3, what is the present Celsius temperature? [*Hint*: If the temperature is C° Celsius, it is $(C + 273.2)$ Kelvin.]

46. (*Problem Solving*) Imagine you are an aeronautical engineer, employed by NASA. You have been assigned to a project whose objective is to put a satellite into orbit around Mars. A similar satellite is currently orbiting the earth at an altitude of 230 km, making one revolution every 89 minutes. You are to calculate an altitude for the Mars satellite so that its period of revolution will also be 89 minutes.

You know that the radius r of a satellite's orbit is related to the mass M of the central body by the equation

$$M = \left(\frac{4\pi}{GT^2}\right) r^3$$

where T is the period of revolution, and G is a universal gravitational constant. The value of r is the distance of the satellite from the center of the central body. For example, since the earth has a radius of 6370 km, the value of r for the earth satellite is $6370 + 230 = 6600$ km. The radius of Mars is 3398 km, and its mass is 0.11 times that of the earth. What should you tell your colleagues at NASA about the altitude of the Mars satellite's orbit?

47. (*Problem Solving*) Physiologist J. L. Poiseuille (1779–1869) discovered that the volume of blood flowing through a small artery (arteriole) per unit time varies directly as the fourth power of the arteriole's radius. If a heart attack victim is given a medicine, such as Cardizem, to dilate his blood vessels by 10%, by what factor will the flow of blood be increased?

5–2 AN ANALYTICAL VIEW OF POLYNOMIAL FUNCTIONS

PREREQUISITE   MAKE SURE YOU ARE FAMILIAR WITH:
Division and Synthetic Division of Polynomials
(Section A-15)

You have seen that polynomial functions of higher degree are studied as models of physical problems, although they occur less often than linear or quadratic functions. Another reason to study them is simply for their own sake. This reason is not as flimsy as it might seem. Great people often do things out of sheer passion and curiosity. Sir George Mallory, whose disappearance on the slopes of Mount Everest in 1924 made him a legend among climbers, was once asked why he was so determined to conquer the mountain. His reply was, "Because it is there!"

Mathematicians of the sixteenth century were climbing mountains of their own as they attempted to find solutions to polynomial equations. But sometimes even they were overwhelmed by its difficulty, as Howard Eves noted in *An Introduction to the History of Mathematics* (Holt, Rinehart, and Winston, 1976). He refers to François Viete's laborious method for solving equations of higher degree by noting that ". . . one seventeenth century mathematician described it as 'work unfit for a Christian.'" Eves presents a delightful account of this struggle.

Probably the most spectacular mathematical achievement of the sixteenth century was the discovery, by Italian mathematicians, of the algebraic solution of cubic and quartic equations. The story of this discovery, when told in its most colorful version, rivals any page written by Benvenuto Cellini. Briefly told, the facts seem to be these. About 1515, Scipione del Ferro (1465–1526), a professor of mathematics at the University of Bologna, solved algebraically the cubic equation $x^3 + mx = n$, probably basing his work on earlier Arabic sources. He did not publish his result but revealed the secret to his pupil Antonio Fior. Now about 1535, Nicolo Fontana of Brescia, commonly referred to as Tartaglia (the stammerer) because of a childhood injury that affected his speech, claimed to have discovered an algebraic solution of the cubic equation $x^3 + px^2 = n$. Believing this claim was a bluff, Fior challenged Tartaglia to a public contest of solving cubic equations, whereupon the latter exerted himself and only a few days before the contest found an algebraic solution for cubics lacking a quadratic term. Entering the contest equipped to solve two types of cubic equations, whereas Fior could solve but one type, Tartaglia triumphed completely. Later Girolamo Cardano, an unprincipled genius who taught mathematics and practiced medicine in Milan, upon giving a solemn pledge of secrecy, wheedled the key to the cubic from Tartaglia. In 1545, Cardano published his *Ars magna*, a great Latin treatise on algebra, at Neuremberg, Germany, and in it appeared Tartaglia's solution of the cubic. Tartaglia's vehement protests were met by Ludovico Ferrari, Cardano's most capable pupil, who argued that Cardano had received his information from del Ferro through a third party and accused Tartaglia of plagiarism from the same source. There ensued an acrimonious dispute from which Tartaglia was perhaps lucky to escape alive.

Since the actors in the above drama seem not always to have had the highest regard for truth, one finds a number of variations in the details of the plot.

Benvenuto Cellini (1500–1571), mentioned in Eves' account above, was an artist, craftsman, and sculptor of great significance. His fame, however, is due mainly to his autobiography, which despite its manifest exaggerations and boastful tone is surprisingly frank and authentic.

In this section we will study polynomial functions from an historical perspective, focusing on three questions that occupied mathematicians until the early part of the nineteenth century.

- What is the largest number of zeros that a polynomial function can have? [A complex number r is a **zero** of a polynomial function P if and only if $P(r) = 0$. Our discussion will focus primarily on real zeros.]
- What is the smallest number of zeros that a polynomial function can have?
- Can we find all zeros of a polynomial function?

FACTORS AND ZEROS

Let's focus on the first of the preceding questions. Exercises 1–4 suggest that the number of zeros of a polynomial function is related to its degree.

EXERCISES   **1.** (*Review*) Find all the zeros of each function P by solving the equation $P(x) = 0$.

   a. $P(x) = x^2 + x - 6$
   b. $P(x) = x^3 - 5x^2 + 6x$
   c. $P(x) = 2x^3 + 4x^2 - 2x - 4$
   d. $P(x) = x^4 - 10x^2 + 9$

2. (*Making Observations*) How is the number of zeros for each function in Exercise 1 related to the degree of the function?

3. (*Review*) Find all the zeros of each function Q by solving the equation $Q(x) = 0$.

   a. $Q(x) = x^2 - 8x + 16$
   b. $Q(x) = x^5 - 9x^3$
   c. $Q(x) = x^3 - x^2 + x - 1$
   d. $Q(x) = x^4 - 4x^2 + 4$

4. (*Making Observations*) Exercise 3 illustrates that the number of zeros of a polynomial function can be less than its degree. Do you think the number of zeros can ever exceed the degree? Why or why not?

By completing Exercises 1–4 you have duplicated the reasoning outlined by René Descartes in his *Discours de la méthode pour bien conduire sa raison et chercher la vérité dans les sciences* (Discourse of the method for conducting reason well and seeking the truth in the sciences), published in 1637. His conclusion, stated here in more modern language, has come to be known as the Factor Theorem.

The Factor Theorem

A polynomial function P has a zero at a complex number r if and only if $P(x)$ has a factor of $x - r$.

Example 1 illustrates the computational power of the Factor Theorem.

EXAMPLE 1   *Using the Factor Theorem to find the zeros of a polynomial function*

Figure 5-5 shows a graph of $P(x) = x^3 - 4x^2 + 2x + 3$. The real zeros of P correspond to the x-intercepts on the graph. One x-intercept appears to be at about $(3, 0)$. A quick check shows that

$$P(3) = (3)^3 - 4(3)^2 + 2(3) + 3 = 0$$

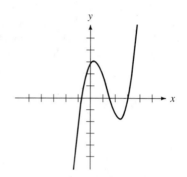

Because $x = 3$ is a zero of P, $x - 3$ must be a factor of $P(x)$. Dividing $P(x)$ by $x - 3$ yields

$$P(x) = (x - 3)(x^2 - x - 1)$$

The other zeros must be solutions of

$$x^2 - x - 1 = 0$$

Figure 5–5

By the quadratic formula, they are

$$x = \frac{1 \pm \sqrt{5}}{2} \approx -0.62, \ 1.62$$ ∎

EXERCISES   *In Exercises 5–8, graph $y = f(x)$ in the indicated window. Use the graph to guess the value of one of the zeros of f, then use division to find the others, as in Example 1.*

5. $f(x) = x^3 + 2x^2 - 5x - 6;\ [-10, 10]$ by $[-10, 10]$

6. $f(x) = 2x^3 - 7x^2 - 5x + 4;\ [-10, 10]$ by $[-30, 30]$

7. $f(x) = -x^3 - x^2 + 3x + 2;\ [-10, 10]$ by $[-10, 10]$

8. $f(x) = 3x^3 + x^2 - 7x - 5;\ [-10, 10]$ by $[-10, 10]$

The Factor Theorem answers the first of our questions at the beginning of this section. Since a polynomial of degree n has at most n linear factors, the Factor Theorem implies that a polynomial function of degree n has at most n zeros. This fact was stated as early as 1600 by the German mathematician Peter Rothe in *Arithmetica philosophic* and later by Descartes and Albert Girard (1593–1632), both Frenchmen working in Holland. Girard's statement appeared in *L'invention novelle en l'algebra* (New results in algebra), a 1629 publication based on a 1585 treatise by Simon Stevin. Girard and Descartes recognized the nature of polynomial zeros because they were among the first mathematicians to take imaginary numbers seriously.

Multiplicity of Zeros Let's turn now to the second of our three questions from the beginning of the section. What is the smallest number of zeros that a polynomial function can have? The Factor Theorem suggests that we begin by asking whether a polynomial function of degree n must always have n zeros. The answer to this question depends on how we count the zeros. If we confine our attention to real zeros, Exercise 9 shows that the answer is no.

EXERCISE **9.** (*Review*)

    a. Sketch the graph of $f(x) = x^2 + 1$, and verify that it does not have any x-intercepts.

    b. Explain why the graph suggests that f has no real zeros.

    c. Confirm that f has no real zeros by solving the equation $x^2 + 1 = 0$.

Although the function in Exercise 9 has no real zeros, it has two complex zeros. This suggests that we ask whether a polynomial function of degree n must always have n complex zeros. The answer still depends on how we count the zeros. If a number can only be counted once, Example 2 shows that the answer is still no.

EXAMPLE 2 *A cubic polynomial function with fewer than three distinct zeros*

In Descartes' *Discours*, he discusses the function

$$P(y) = y^3 - 10y^2 + 17y - 8$$

which can be written in the factored form

$$P(y) = (y - 1)^2(y - 8)$$

Its only zeros are, therefore, $y = 1$ and $y = 8$. ∎

Although the function in Example 2 has only two distinct zeros, it has three linear factors (two factors of $y - 1$ and one factor of $y - 8$). Descartes' *Discours* describes this situation by saying that the polynomial $P(y)$ "... has only three roots, two of them, 1 and 8, being true roots, and the third, also 1, being false; ...". A more modern view is to say that $P(y)$ has a **double zero** at $y = 1$, corresponding to the factor $(y - 1)^2$. More generally, if k is a positive integer, a zero $x = r$ corresponding to a factor $(x - r)^k$ is called a **simple zero** if $k = 1$ and a **multiple zero** if $k > 1$. The number k is called the **multiplicity** of the zero. In our last example, the zero $y = 1$ has multiplicity 2, and the zero $y = 8$ has multiplicity 1.

EXERCISE **10.** As parts (a)–(d) of this exercise, state the multiplicity of each zero for each function in Exercise 3.

THE FUNDAMENTAL THEOREM OF ALGEBRA

Now let's reconsider the question of whether a polynomial function of degree n must always have n zeros. If we mean complex zeros, and if we count them according to their multiplicity, the answer is yes. This statement is referred to as the Fundamental Theorem of Algebra.

> **The Fundamental Theorem of Algebra**
>
> Every polynomial function P of degree n can be written in factored form as
>
> $$P(x) = a_n(x - r_1)(x - r_2) \cdots (x - r_{n-1})(x - r_n)$$
>
> for some choice of complex numbers $r_1, r_2, \ldots, r_{n-1}, r_n$, not necessarily distinct.
>
> Equivalently, every polynomial function of degree n has exactly n complex zeros, counted according to their multiplicity.

Both Girard and Descartes stated the Fundamental Theorem. Its proof, however, was the Mount Everest of seventeenth-century mathematics. Such mathematical giants as Newton, Euler, D'Alembert, and Lagrange all attempted unsuccessfully to establish its truth. The Fundamental Theorem was not conquered until 1799, when Carl Friedrich Gauss proved it in his doctoral dissertation, written at the age of 22 at the University of Helmstädt. (You have seen another example of Gauss' ingenuity in Section 1-3.) Gauss thus answered our second question from the beginning of this section.

EXERCISE   **11.** (*Making Observations*) Show that the Fundamental Theorem of Algebra is true for each of the functions in Exercise 3 by verifying that the number of complex zeros is equal to its degree.

Now let's turn to our third question from the beginning of this section. Can we find all the zeros of a polynomial function? The Fundamental Theorem of Algebra tells how many zeros there are, but does not tell how to find them. The task of finding zeros becomes increasingly more complex as the degree of the polynomial increases.

Your expertise in solving linear and quadratic equations may have been shared by your (great)^{200}-grandparents. Methods for solving linear equations were known in Egypt as far back as 3500 B.C. The quadratic formula was known to the Babylonians of 1700 B.C. A similar formula for cubic equations was first published by Girolamo Cardano (1501–1576) in his *Ars magna* in 1545. Cardano himself rejected negative and imaginary solutions, because he saw no physical meaning in them. The acceptance of such solutions was delayed until 1732, when Leonhard Euler (1707–1783), writing in the journal *Commentarii Academiae Scientiarum Petropolitanae*, showed that Cardano's formula can be used to obtain three complex zeros of every cubic function. *Ars magna* also contained a formula of Ludovico Ferrari (1522–1565) for solving a general quartic equation. You can find the details of Cardano's and Ferrari's formulas in Howard Eves' *An Introduction to the History of Mathematics*.

For over 250 years following the appearance of *Ars magna*, European mathematicians searched in vain for a formula for the zeros of general quintic functions. François Viète (1540–1603), James Gregory (1638–1675), Ehrenfried Walter von Tschirnhausen (1651–1708), Joseph Louis Lagrange (1736–1813), and Leonhard Euler all tried and failed. Gregory, through his work on calculus, was led to believe that no such formula existed. In 1824 Norwegian Niels Henrik Abel (1802–1829), building on the work of

Lagrange, proved Gregory right. There is no general formula for finding zeros of polynomials of degree five or greater. Abel thus answered our third question from the beginning of this section.

Which equations of degree five or greater can be solved by algebraic formulas? This question was answered in 1831 by the Frenchman Evariste Galois. Any discussion of his work is far too ambitious for this book, but it is interesting to note how it came to the attention of the rest of the world. Galois was jailed as a political activist several times while still in his teens. At twenty he was challenged to a duel, probably provoked for political reasons, over a woman of dubious reputation. The only surviving original record of his work was hastily jotted down the night before the duel, in which he was mortally wounded. In Leopold Infeld's biography of Galois, *Whom the Gods Love*, it is told how the manuscript was entrusted to his brother Alfred as he lay dying.

> "I must tell you something. Important. Mathematical manuscripts are in my room. On the table. A letter to Chevalier. It is for you, too. Both. Alfred and Auguste. Take care of my work. Make it known. Important."

He felt relieved and listened to Alfred's violent words:

> "I swear to you that I will do it. I shall do everything. I swear to you that they will be published and recognized. I swear to you that if necessary I shall devote my whole life to it."

> Now he broke out with a new, more violent torrent of tears. Evariste looked at him with pity and said very slowly, tearing the words out of his body with increasing pain, "Don't cry, Alfred. I need all my courage—to die at twenty."

Alfred did indeed devote his life to obtaining proper recognition for his brother's work. A copy of the manuscripts eventually found its way into the hands of Joseph Liouville, who published them in *Journal de mathématiques pures et appliquées* in 1846.

[−10, 10] by [−20, 20]

Figure 5–6

Galois' work shows that no algebraic formula will find the zeros of $P(x) = x^5 - x - 1$. Yet the graph in Figure 5-6 reveals a zero in $[1, 2]$. Although we cannot find this zero analytically, we can approximate it graphically. In Section 5-4 you will also learn how to approximate it numerically.

ADDITIONAL EXERCISES

In Exercises 12–15, find all the zeros of the polynomial function.

12. (*Review*) $f(x) = 16x - x^3$

13. (*Review*) $h(t) = t^4 + t^3 - 42t^2$

14. (*Review*) $Q(z) = 2z^5 - 10z^3 + 8z$

15. (*Review*) $r(v) = v^3 - 3v^2 - 2v + 6$

In Exercises 16–19,

    a. Decide whether $x = -1$ is a zero of g by evaluating $g(-1)$.

    b. Write $g(x)$ in its factored form, and list all of the zeros of g. Make sure the result agrees with your conclusion from part (a).

16. $g(x) = x^3 + 2x^2$ **17.** $g(x) = x^7 + x^5 - 2x^3$

18. $g(x) = x^6 - 1$ **19.** $g(x) = 2x^4 + 5x^2 + 2$

In Exercises 20–25, identify the zeros of the function and state the multiplicity of each zero.

20. $f(x) = (x + 10)(x - 20)^6$

21. $A = (b - 4)^3(b + \pi)^5$

22. $S(x) = x^4(x^2 - 16)^4$

23. $h(r) = (r^2 - 2r - 8)^3(r^2 + 2r - 15)^2$

24. $y = x^3 + 2x^2 - 4x - 8$

25. $z = x^6 - 2x^4 - 4x^2 + 8$

26. (*Making Observations*) Verify that the Fundamental Theorem of Algebra is true for each function in Exercises 20–25.

The functions in Exercises 27–30 each have some nonreal zeros. Find the zeros of each function by factoring and verify that the Fundamental Theorem of Algebra is satisfied.

27. $y = 2x^2 + 4x + 10$ **28.** $C(x) = x^4 + 7x^2 + 12$

29. $F(r) = r^5 + 6r^3 + 9r$

30. $w = z^5 - 2z^4 + 4z^3 - 8z^2$

31. In Exercise 8 of Section 1-4 you solved the following equation $x^3 - 34x^2 + 408x - 1728 = 0$ graphically, obtaining $x = 12$.

  a. (*Writing to Learn*) Why does the Fundamental Theorem of Algebra guarantee that the equation has two other solutions?

  b. Use division to find the other solutions, and thus verify that there are no other real solutions.

32. (*Writing to Learn*) Suppose that f and g are polynomials, with $f(3) = 7$ and $g(3) = 0$. What does the Factor Theorem tell you about f and g?

33. (*Making Observations*) You can use the Factor Theorem to find polynomial functions having specified zeros with specified multiplicities. For example, a function having zeros $x = \dfrac{1}{2}$ with multiplicity 2 and $x = \pm\sqrt{3}$ each with multiplicity 1 is

$$P(x) = \left(x - \frac{1}{2}\right)^2\left(x + \sqrt{3}\right)\left(x - \sqrt{3}\right)$$

  a. Verify that in expanded form,

$$P(x) = x^4 - x^3 - \frac{11}{4}x^2 + 3x - \frac{3}{4}$$

  b. Graph $y = P(x)$ in the window $[-3, 3]$ by $[-10, 10]$. Then graph

$$y = 4P(x) = 4x^4 - 4x^3 - 11x^2 + 12x - 3$$

  on the same screen. How do the graphs suggest that the two functions have the same zeros?

  c. How could you draw the same conclusion without seeing the graphs?

  d. If the zeros of a polynomial function Q are the same as those of P with the same multiplicities, explain why Q is a constant multiple of P.

In Exercises 34–39, use the idea in Exercise 33 to find a polynomial function with integer coefficients having the given zeros with the indicated multiplicities.

34. $x = 1$, multiplicity 3

35. $x = 3$, multiplicity 2; $x = \dfrac{1}{2}$, multiplicity 1

36. $x = \pm\sqrt{5}$, each with multiplicity 1; $x = 0$, multiplicity 3

37. $x = -5$, multiplicity 1; $x = -\dfrac{3}{2}$, multiplicity 2

38. $x = 0$, multiplicity 4; $x = \dfrac{1}{3}$, multiplicity 2

39. $x = \pm\sqrt{2}$, each with multiplicity 2

40. (*Writing to Learn*) Write a few pages on the life of Tartaglia, Cardano, Descartes, Gauss, or Galois. You will need biographical information beyond what we have provided. One of the sources we cited might be a good starting point.

SUPPLEMENTARY TOPIC
The Rational Root Theorem

As you saw in Example 1, one strategy for finding the zeros of a polynomial function is to graph the function, use the graph to guess the value of one or more zeros, then use division to find the remaining zeros. You can often reduce the amount of guesswork in the process by using the following theorem, developed by Descartes.

> **The Rational Root Theorem**
>
> If a polynomial function
>
> $$f(x) = a_n x^n + a_{n-1}x^{n-1} + a_{n-2}x^{n-2} + \cdots + a_1 x + a_0$$
>
> with integer coefficients $a_n, a_{n-1}, \ldots, a_1, a_0$ has a zero $x = \dfrac{p}{q}$ with p and q integers and $\dfrac{p}{q}$ in lowest terms, then a_0 is evenly divisible by p, and a_n is evenly divisible by q.

The Rational Root Theorem allows you to make a list of all possible rational zeros of a given polynomial function with integer coefficients. For example, suppose

$$P(x) = 3x^4 + 8x^3 - 72x^2 - 127x + 50$$

The possible values of p are the integer factors of 50, which are $\pm 1, 2, 5, 10, 25, 50$. The possible values of q are the integer factors of 3, which are $\pm 1, 3$. The possible rational zeros $x = \dfrac{p}{q}$ are

$$x = \pm\, 1,\, 2,\, 5,\, 10,\, 25,\, 50,\, \frac{1}{3},\, \frac{2}{3},\, \frac{5}{3},\, \frac{10}{3},\, \frac{25}{3},\, \frac{50}{3}$$

Example 3 illustrates the use of the Rational Root Theorem in identifying the zeros of this polynomial function.

EXAMPLE 3

Using the Rational Root Theorem to find the zeros of a polynomial function

The graph of $P(x) = 3x^4 + 8x^3 - 72x^2 - 127x + 50$ is shown in Figure 5-7. We know that the window has captured all the zeros of P, because it shows four x-intercepts and the degree of P is 4. The intercepts in the intervals $[-6, -5]$ and $[4, 5]$ must be irrational, because none of our listed possible rational zeros lies in either of these intervals. The other two intercepts appear to be near $(-2, 0)$ and $\left(\frac{1}{3}, 0\right)$.

You can use synthetic division to verify that $x = -2$ is a zero of P, so

$$P(x) = (x + 2)(3x^3 + 2x^2 - 76x + 25)$$

Similarly, you can verify that $x = \frac{1}{3}$ is a zero of P, so

$$P(x) = (x + 2)\left(x - \frac{1}{3}\right)(3x^2 + 3x - 75)$$

The remaining zeros must satisfy the quadratic equation

$$3x^2 + 3x - 75 = 0$$

You can find them through the quadratic formula, so that altogether the zeros are

$$x = -2, \ \frac{1}{3}, \ \frac{-1 \pm \sqrt{101}}{2}$$ ∎

[−10, 10] by [−500, 500]

Figure 5–7

In Exercises 41–48, use the Rational Root Theorem together with a calculator graph to find all rational zeros of the polynomial. If possible, use synthetic division and the quadratic formula to find the remaining zeros. Otherwise, estimate them graphically.

41. $P(x) = x^3 - x^2 - 17x + 20$

42. $G(x) = 2x^4 - 3x^3 + 2x^2 - x - 3$

43. $s(t) = 6t^3 - 7t^2 - 40t + 21$

44. $w = 3x^3 - 10x + 6$

45. $f(x) = x^3 - 2x^2 - 12x + 16$

46. $r = x^3 + x^2 - 20x + 12$

47. $P(w) = w^4 + 2w^3 + w - 24$

48. $Y(x) = 4x^3 + 12x^2 - \frac{7}{2}$

5-3   A GRAPHICAL VIEW OF POLYNOMIAL FUNCTIONS

Sixteenth-century mathematicians could not view polynomial functions graphically, because the coordinate plane had not yet been invented. The first steps in its development were taken by Descartes, who assigned equations to curves in a plane, and by Pierre de Fermat (1601?–1665), who began with equations and studied the corresponding curves. According to Howard Eves, in *An Introduction to the History of Mathematics*,

> There are a couple of legends describing the initial flash that led Descartes to the contemplation of analytic geometry. According to one story, it came to him in a dream. On St. Martin's Eve, November 10, 1616, while encamped in the army's winter quarters on the banks of the Danube, Descartes experienced three singularly vivid and coherent dreams that, he claimed, changed the whole course of his life. The dreams, he said, clarified his purpose in life and determined his future endeavors by revealing to him "a marvelous science" and "a wonderful discovery." Descartes never explicitly disclosed just what were the marvelous science and the wonderful discovery, but some believe them to have been analytical geometry, or the application of algebra to geometry, and then the reduction of all science to geometry. It was eighteen years later that he expounded some of his ideas in his *Discours*.

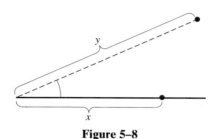

Figure 5–8

The coordinate plane is often called the Cartesian plane after Cartesius (the Latinized spelling of Descartes), but he did not in fact develop it as we know it. In particular, he did not use negative coordinates or mutually perpendicular axes. As Eves describes, he simply ". . . marked off *x* on a given axis and then a length *y* at a fixed angle to this axis and endeavored to construct points whose *x*'s and *y*'s satisfy a given relation," as in Figure 5-8. The coordinate plane as we now know it evolved gradually, and it was not until the eighteenth century that mutually perpendicular axes became standard.

With the coordinate plane and graphing technology, you can gain graphical insights into polynomial functions more easily than the best mathematicians of Descartes' day. Let's begin by building on our analytical work from Section 5-2.

POINTS OF INTEREST

x-intercepts Of all the points on the graph of a polynomial function P, the *x*-intercepts have a special significance. The graph has an *x*-intercept at $x = r$ if and only if $P(r) = 0$. According to the Factor Theorem, this happens if and only if $x - r$ is a factor of $P(x)$. Since a polynomial of degree n has at most n real linear factors, we can draw a graphical conclusion from our observations.

> The graph of a polynomial function of degree n has at most n *x*-intercepts.

EXERCISES

1. (*Review*) List all real zeros of P according to multiplicity, and list all *x*-intercepts on the graph of $y = P(x)$.

   a. $P(x) = (x + 4)(x + 1)(x - 3)^3$

   b. $P(x) = x^3 + 4x$

   c. $P(x) = x^2(x + 2\sqrt{7})(x - 2\sqrt{7})$

   d. $P(x) = x^4 - 16$

2. (*Review*) Write a polynomial function with integer coefficients having the given graph. Assume that all zeros are real and simple (have multiplicity 1).

a.

b.

c.

d.
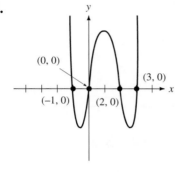

3. (*Review*) Write a polynomial function with integer coefficients having the given zeros with the indicated multiplicities, and list the *x*-intercepts on its graph.

a. $-2, 1, 3$ (each multiplicity 2)

b. $0, \sqrt{6}, -\sqrt{6}$ (each multiplicity 1)

c. $\dfrac{3}{2}$ (multiplicity 2), $-\dfrac{1}{3}$ (multiplicity 1)

d. $\pm\sqrt{2}$ (each multiplicity 2), $\pm\sqrt{3}$ (each multiplicity 1)

x-intercepts at Multiple Zeros   The shape of a polynomial graph around an *x*-intercept is determined by the multiplicity of the corresponding zero. You can use this relationship to sketch graphs more accurately and quickly. In Section 5-5 you will also use it to solve polynomial inequalities.

EXAMPLE 1   *The shape of a polynomial graph around an x-intercept*

Figure 5-9 shows the graph of $P(x) = (x + 4)^2(x - 1)$.

One striking feature of this graph is that it crosses the *x*-axis at $(1, 0)$, but not at $(-4, 0)$. Stated differently, the sign of $P(x)$ changes at $x = 1$, but not at $x = -4$.

What could account for this difference? Since $P(x)$ is the product of the factors $(x + 4)^2$ and $(x - 1)$, the sign of $P(x)$ at each *x*-value depends on the sign of each factor. Let's follow the graph from left to right, and analyze the sign of each factor along the way.

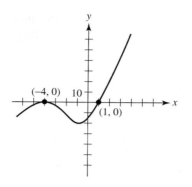

Figure 5–9

To the left of $x = -4$, $(x + 4)^2$ is positive while $(x - 1)$ is negative. As we pass $x = -4$ neither factor changes sign, so that $P(x)$ remains negative and the graph does not cross the x-axis. As we pass $x = 1$ the sign of $(x + 4)^2$ does not change, but $(x - 1)$ changes from negative to positive. Thus the sign of $P(x)$ changes and the graph crosses the x-axis. ■

Example 1 illustrates a general principle.

If a polynomial function P has a zero of multiplicity k at $x = a$, then the graph of $y = P(x)$:

• crosses the x-axis at $(a, 0)$ if k is odd.
• touches, but does not cross, the x-axis at $(a, 0)$ if k is even.

Using calculus, it can be shown that at an x-intercept corresponding to a zero of multiplicity $k > 1$ a polynomial graph flattens out as in Figures 5-10b and 5-10c. Therefore, the three possible shapes of a polynomial graph around an x-intercept of multiplicity k are as shown in Figure 5-10.

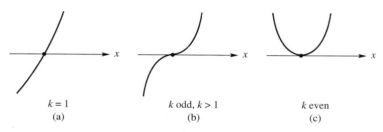

$k = 1$	k odd, $k > 1$	k even
(a)	(b)	(c)

Figure 5–10

EXERCISES *In Exercises 4–7, write an equation in factored form for a sixth-degree polynomial whose graph resembles the one shown.*

4.

5.

6.

7.

Extreme Values Most polynomial graphs have one or more **turning points** (points where the function changes from increasing to decreasing, or vice versa). Using calculus, it can be shown that a polynomial function of degree n has at most $n - 1$ turning points. Their y-coordinates are called **extreme values** of the function. The following terms describe specific types of extreme values shown in Figure 5-11.

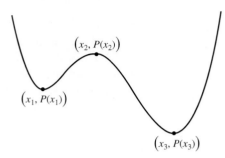

Figure 5–11

- A **local minimum** is the smallest value of the function within some open interval.
  Example: $P(x_1)$, $P(x_3)$.
- A **local maximum** is the largest value of the function within some open interval.
  Example: $P(x_2)$.
- An **absolute minimum** is the smallest value of the function over its entire domain. An absolute minimum is also a local minimum.
  Example: $P(x_3)$.
- An **absolute maximum** is the largest value of the function over its entire domain. An absolute maximum is also a local maximum.
  None is present in Figure 5-11.

Johannes Kepler, whom you met in Section 4-3, was among the first mathematicians to study the problem of locating extreme values of polynomial functions. His concern came from the practical problem of finding an optimal shape for a wine cask. Kepler calculated y-values for a large number of x-values, and estimated the extreme value by noticing where the function increased and decreased. Since Kepler did not have a graphing calculator, or even the concept of a graph, he computed each y-value by hand. In 1629 Fermat devised a method of locating extreme values exactly, using the just-emerging ideas of analytic geometry. Fermat's work was criticized by Descartes and other mathematicians who questioned the logical soundness of his methods. Always reluctant to publish his work, Fermat waited until 1637 to describe his method in a manuscript, which was published by his son in 1679, fourteen years after his death. The theory to justify his method was finally developed about 150 years later, by Augustin Louis Cauchy (1789–1857).

If you take calculus, you will learn Fermat's method of finding exact extreme values, with the benefit of a few centuries of refinements. For now, you can follow the spirit of Kepler by finding approximate extreme values, but with the aid of your graphing calculator you will be able to accomplish your task in minutes, rather than months.

FYI In abstract problems, estimates of extreme values should have an error of less than 0.01 in both coordinates. Calculator techniques for estimating extreme values are discussed in Section B-5.

EXAMPLE 2 *Finding extreme values of a polynomial function*

Figure 5-12 shows the graph of $f(x) = 2x^3 - 12x + 6$, and indicates that f has two turning points. By zooming and tracing, you can estimate that they occur at $(-1.41, 17.31)$ and $(1.41, -5.31)$. See Figures 5-13a and 5-13b, respectively.

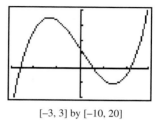

[–3, 3] by [–10, 20]

Figure 5–12

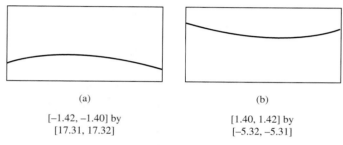

(a)

[–1.42, –1.40] by
[17.31, 17.32]

(b)

[1.40, 1.42] by
[–5.32, –5.31]

Figure 5–13   ■

From the turning points of a polynomial graph you can identify the intervals where it is increasing or decreasing. For example, the function in Example 2 is increasing on $(-\infty, -1.41)$ and $(1.41, \infty)$ and decreasing on $(-1.41, 1.41)$.

EXERCISES   *In Exercises 8–11,*

a. Find all extreme values of the polynomial function, and identify any which are absolute.

b. Identify the intervals where the function is increasing and decreasing.

c. Use the graph of the function to estimate its range.

8. $f(x) = x^2 - 6x + 8$ **9.** $G(x) = 12x - x^3$

10. $y = x^4 - 4x^3$ **11.** $w = 3x^3 - 10x + 6$

A MATHEMATICAL
LOOKING GLASS
Sonar

On July 17, 1996, a TWA airliner on a flight from New York to Paris crashed into the Atlantic Ocean off Long Island after an apparent midair explosion. A team of FBI agents was sent to the site to determine whether the explosion was caused by mechanical failure or by a bomb. Their investigation depended on their ability to locate and retrieve pieces of the wreckage, including two flight recorders, from the ocean floor.

A key piece of equipment in the FBI's search was a side-scanning sonar, which emits sound waves underwater and detects reflected waves from distant objects. To locate objects on the basis of sonar data, it is necessary to know how fast the sound waves travel. The speed of sound underwater depends on the depth, salinity, and temperature of the water. Specifically, at a depth of d meters, a salinity of s grams per liter of water, and a temperature of T degrees Celsius, the speed of sound in meters per second is

$$V = 0.0003T^3 - 0.055T^2 + (8.8 - 0.12s)T + (1400 + 0.017d)$$

The relationship is approximately valid for $0 \leq d \leq 1000$, $0 \leq s \leq 45$, and $0 \leq T \leq 35$. ∎

EXERCISES

12. The Atlantic Ocean near Long Island has a salinity of about 35 grams per liter. On the same screen, in the window [0, 35] by [1400, 1500] graph V as a function of T for each value of d.

    a. $d = 0$ **b.** $d = 500$

    c. $d = 1000$

13. (*Making Observations*)

    a. How do the graphs in Exercise 12 relate to each other?

    b. What graphing transformation principle could you have used to predict that relationship?

14. (*Making Observations*)

    a. Sound always travels faster in warmer water. How do your graphs in Exercise 12 support this statement?

    b. In the window [0, 100] by [1400, 1500], graph V as a function of T with $d = 100$ and $s = 50$. How does your graph tell you that the equation given in **Sonar** does not model the speed of sound accurately for water with a salinity of 50 grams per liter?

END BEHAVIOR

The x-intercepts and extreme values on a polynomial graph occur in the graph's "midsection," referred to by textbook author John Saxon as its **region of interest**. The graph's behavior outside the region of interest is referred to as its **end behavior**. For an example of physical information provided by a function's end behavior, see **Robert Arnold** (page 312).

    The simplest question about end behavior is whether the "tails" of the graph point up or down. Let's look at an example to see how to answer that question.

EXAMPLE 3   *End behavior of a polynomial graph*

Figure 5-14 shows a graph of $P(x) = x^4 - 100x^3$.

Does Figure 5-14 present an accurate picture of the graph's end behavior? We can check by looking at the factored form $P(x) = x^3(x - 100)$.

If $x < 0$, both x^3 and $x - 100$ are negative, so their product is positive. Thus the left tail of the graph actually does point up. When $x > 100$, both x^3 and $x - 100$ are positive, so the right tail of the graph also points up! Figure 5-15 shows a more accurate picture.

[–5, 5] by [–100, 100]

Figure 5–14

[–200, 200] by
[–10 000 000, 10 000 000]

Figure 5–15   ∎

Exercises 15 and 16 illustrate a way to analyze the end behavior of P by looking at the terms of $P(x)$, rather than the factors.

EXERCISES   **15.** Complete the following table for the function P in Example 3.

Value of x	Value of x^4	Value of $-100x^3$	Value of $P(x)$
$-10{,}000$			
$-1{,}000$			
-100			
0			
100			
$1{,}000$			
$10{,}000$			

16. (*Making Observations*) How does the table in Exercise 15 suggest that when $|x|$ is large, the sign of $P(x)$ is determined by the sign of x^4?

Your observation in Exercise 16 can be generalized. If we go far enough to the left or right along the graph of any polynomial function

$$P(x) = a_n x^n + a_{n-1}x^{n-1} + a_{n-2}x^{n-2} + \cdots + a_1 x + a_0$$

the size (absolute value) of the term $a_n x^n$ of largest degree eventually exceeds that of all the other terms combined. This means that on the tails of the graph, the sign of $P(x)$ will agree with that of $a_n x^n$. This conclusion can be stated as

**The Highest Degree Theorem
(Polynomial Version)**

The end behavior of the graph of any polynomial function

$$P(x) = a_n x^n + a_{n-1}x^{n-1} + a_{n-2}x^{n-2} + \cdots + a_1 x + a_0$$

is the same as that of $y = a_n x^n$. (In end behavior, only the term of highest degree counts.)

The graph of $y = a_n x^n$ is always that of a basic power function, possibly stretched or compressed vertically, and reflected in the x-axis if $a_n < 0$. Thus the end behavior of all polynomial functions is summarized in Table 24.

TABLE 24

Conditions on n and a_n	End behavior	Graph	Example
n odd, $a_n > 0$	Left tail down, right tail up		$y = x^3$
n odd, $a_n < 0$	Left tail up, right tail down		$y = -x^3$
n even, $a_n > 0$	Left tail up, right tail up		$y = x^2$
n even, $a_n < 0$	Left tail down, right tail down		$y = -x^2$

You do not need to memorize Table 24. It is easier to apply the principles that generated it, as in Example 4.

EXAMPLE 4   *Describing the end behavior of a polynomial function*

The end behavior of

$$P(x) = -4x^5 + 6x^4 + x^3 - 10x^2 + 5x - 1$$

is the same as that of $y = -4x^5$. The graph of $y = -4x^5$ is that of $y = x^5$ stretched vertically by a factor of 4 and reflected in the x-axis. On the graph of $y = x^5$, the left tail points down and the right tail points up as in Figure 5-16a. The reflection reverses the direction of the tails so that the left tail on the graph of $y = -4x^5$ points up and the right tail points down as in Figure 5-16b. Figure 5-16c shows that the graph of $y = P(x)$ exhibits the same end behavior.

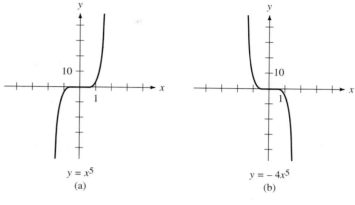

$$y = x^5$$
(a)

$$y = -4x^5$$
(b)

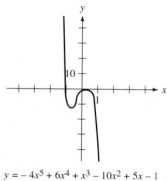

$$y = -4x^5 + 6x^4 + x^3 - 10x^2 + 5x - 1$$
(c)

Figure 5–16

EXERCISES   *In each of Exercises 17–20, describe the end behavior of the graph of the function.*

17. $g(t) = t^4 - 5t^3 + 7$

18. $P(x) = 2x^5 - 9x^4 + 12x^3$

19. $f(x) = 20 - x^2 - 5x^4$

20. $y = 1{,}000{,}000x - x^5$

Match the equations in Exercises 21–24 with the graphs in Figure 5-17 on page 268.

21. $y = x^3 - 4x$ **22.** $y = 4x^3 - 3x^4$

23. $y = -2x^5 + 2x + 3$ **24.** $y = x^6 + 3x^4 - 4x^2$

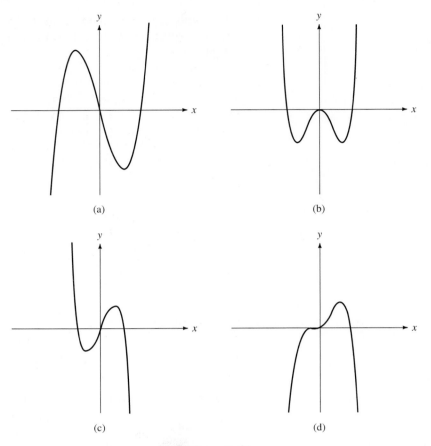

Figure 5–17

25. (*Writing to Learn*) Explain how to generate Table 24 from your knowledge of power function graphs and transformation principles.

SYMMETRY

You may have noticed the symmetry of the graphs in Figures 5-17a, b, and c. You can detect symmetry analytically, using ideas similar to those in Section 4-5.

One type of symmetry is illustrated in Figure 5-18a, which shows that $f(a) = b$ and $f(-a) = b$.

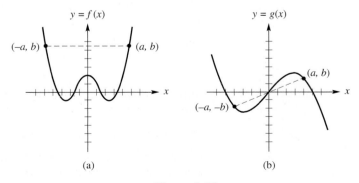

Figure 5–18

More generally in Figure 5-18a, $f(-x) = f(x)$ for any x, so that replacing x by $-x$ leaves the equation unchanged. This means that reflecting the graph about the y-axis leaves the graph unchanged, so the graph is symmetric about the y-axis.

The graph in Figure 5-18b is not symmetric about any axis, but is **symmetric about the origin**. The figure shows that $g(a) = b$ and $g(-a) = -b$. The symmetry results from the more general fact that $g(-x) = -g(x)$ for any x. Thus the combination of replacing x by $-x$ and y by $-y$ leaves the equation unchanged. [Remember that on the graph y and $g(x)$ are the same.]

These results can be summarized in the following analytical tests for symmetry.

> If $f(-x) = f(x)$ for each x in the domain of f, then the graph of $y = f(x)$ is symmetric about the y-axis.
>
> If $f(-x) = -f(x)$ for each x in the domain of f, then the graph of $y = f(x)$ is symmetric about the origin.

EXERCISES *In Exercises 26–31, decide whether the graph of $y = f(x)$ is symmetric about the y-axis, the origin, or neither by evaluating $f(-x)$.*

26. $f(x) = 4 - x^2$ **27.** $f(x) = x^3 + x^2$

28. $f(x) = 3x$ **29.** $f(x) = x^7 - 4x^5 + 10x^3$

30. $f(x) = 2x^6 - x^4 + 3$ **31.** $f(x) = 2x^5 - x^3 + 3$

32. (*Writing to Learn*) An analytical method of detecting symmetry about the y-axis was described in Section 4-5. In what respects is it equivalent to the one in this section? In what respects is it different?

33. (*Writing to Learn*) Explain why you never need to test the graph of a function for symmetry about the x-axis.

It is no coincidence that those functions in Exercises 26–31 exhibiting symmetry about the y-axis involve only even powers of x, and that those exhibiting symmetry about the origin involve only odd powers. (A constant term may be regarded as a coefficient of x^0.) A function whose graph is symmetric about the y-axis is often called **even**, while one whose graph is symmetric about the origin is called **odd**. Most polynomial functions involve both even and odd powers of x and are, therefore, neither even nor odd.

FYI The terms "even" and "odd" can be applied to functions other than polynomials. In this wider context, functions with odd powers of x can be even, and vice versa. For example, $f(x) = |x|$ is even, and $g(x) = \dfrac{x^4 + 1}{x}$ is odd.

Symmetry in a graph that models a physical situation often results from symmetry in the situation. For example, in **Satellite Dish** (page 218), the path of a signal to the receiver from point A in Figure 5-19 mirrors the path from point B. The graph of the resulting equation of the dish, $y = \dfrac{x^2}{24}$, is symmetric about the y-axis. Here, symmetry provides a partial check on the accuracy of our model. If we see symmetry in the physical situation, but not in the model, we need to modify the model.

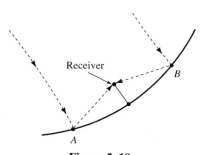

Figure 5–19

Conversely, symmetry in a mathematical model often leads to the discovery of symmetry in a physical situation. For example, during the 1860s, James Clerk Maxwell (1831–1879) rewrote the equations of electromagnetism to make them symmetric. Although no experimental evidence suggested that his altered equations had any physical meaning, Maxwell used them to predict, among other things, the existence of radio waves. Some years later, Heinrich Hertz (1857–1894) confirmed Maxwell's predictions by detecting radio waves experimentally. Perhaps it is for this reason that, according to A. Zee in *Fearful Symmetry*, physicists today "...search their own minds for that which constitutes symmetry and beauty. In the silence of the night, they listen for voices telling them about yet-undreamed-of symmetries."

ADDITIONAL EXERCISES

In Exercises 34–37, write the equation of a polynomial function with only simple real zeros, having the indicated x-intercepts.

34. $(1, 0), (4, 0)$

35. $(-5, 0), (0, 0), (3, 0)$

36. $(0.1, 0), (1, 0), (10, 0), (100, 0)$

37. $(-\sqrt{2}, 0), (\sqrt{2}, 0), (4, 0)$

In Exercises 38–41, write an equation in factored form for a sixth-degree polynomial whose graph resembles the one shown.

38.

39.

40.

41.

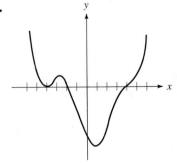

In Exercises 42–47,

a. Find the exact or approximate value of each real zero of the function.

b. Find all extreme values, and identify any that are absolute.

c. Describe the end behavior of the graph.

d. Identify a viewing window that accurately depicts the intercepts, turning points, and end behavior of the graph. State the information you needed in order to choose the window.

    e. Determine the intervals over which the function is increasing and decreasing.

42. $y = x^4 - 6x^2 + 4$ **43.** $y = 4x^5 - 5x^4$

44. $y = 49 - 97x + 23x^2 - x^3$

45. $y = 6 + x + 0.3x^2 - 0.01x^5$

46. $y = -x^4 - x^3 + 3x^2 - 4x + 28$

47. $y = x^5 - 42x^3 + 6x^2 + 185x - 30$

Classify the functions in Exercises 48–51 as even, odd, or neither, using an analytical test. Support your answer with a calculator graph.

48. $U(x) = x^4 - 1$ **49.** $V(x) = x(x^4 - 1)$

50. $W(x) = (x - 1)^4$ **51.** $Z(x) = (x - 1)^4(x + 1)^4$

52. (*Making Observations*) Carefully sketch the graph of $y = \frac{1}{3}x^3 - 3x$ over the interval $[-4, 4]$, making each unit about half an inch.

    a. Calculate the average rate of change in y with respect to x over each of the intervals $[-2.5, -2]$ and $[3, 3.5]$. Over which interval does the average rate of change have the larger absolute value?

    b. Draw secant lines (lines joining points on the graph) over each of the intervals $[-2.5, -2]$ and $[3, 3.5]$. Over which interval is the secant line steeper?

    c. Calculate the average rate of change in y with respect to x over each of the intervals $[-1.5, -0.5]$ and $[-0.5, 0.5]$. Over which interval does the average rate of change have the larger absolute value?

    d. Draw secant lines on the graph over each of the intervals $[-1.5, -0.5]$ and $[-0.5, 0.5]$. Over which interval is the secant line steeper?

53. (*Writing to Learn*) The range of a polynomial function is $(-\infty, \infty)$ if and only if its degree is odd. Explain how your knowledge of end behavior leads to this conclusion.

Find the domain and range of each function in Exercises 54–59.

54. (*Review*) $g(x) = 400x - x^3$

55. (*Review*) $f(x) = x^6 - 7x^4 + 9x^2 - 12$

56. (*Review*) $f(x) = -2x^6 + 24x^4 - 96x^2$

57. (*Review*) $f(x) = x^4 - 14x^3 + 24x^2$

58. (*Review*) $f(x) = x^6 - 12x^3 + 100$

59. (*Review*) $f(x) = x^7 - 1000x$

60. "From a distance, the world looks blue and green,
    And the snow-capped mountains white."

                                  From a Distance
                                    Julie Gold

    (*Writing to Learn*) Graph each of the following functions: $f(x) = x^7$ and $g(x) = x^7 - 3x^6 + x^5$, first in the window $[-2, 4]$ by $[-50, 50]$ and then in the window $[-100, 100]$ by $[-10^{14}, 10^{14}]$. These graphs provide visual support for the Highest Degree Theorem. Explain why. (What does the line from the song have to do with the exercise? You may want to use it in your explanation.)

61. (*Writing to Learn*) In examining extreme values of polynomial functions, Johannes Kepler adopted the strategies of *Making a Table* and *Looking for a Pattern*. In *Nova stereometria doliorum vinariorum*, published in 1615, he stated that "near a maximum the decrements on both sides are in the beginning only imperceptible." What do you think he meant by this? (Before attempting to answer, graph a polynomial function that has a maximum. Trace along the graph, and notice how the *y*-coordinates change as you trace both near the maximum and farther away from it.)

5-4 A NUMERICAL VIEW OF POLYNOMIAL FUNCTIONS

*n*TH-ORDER DIFFERENCES

What tables fit polynomial functions? This question is of considerable practical importance. In this section you will see how scientists can check the accuracy of data transmitted from a space probe by fitting a polynomial function to the data.

Let's use the strategy of *Guessing and Checking* to explore the question. Since tables from linear and quadratic functions have constant first and second differences, respectively, we might guess that tables from polynomial functions of degree n have constant nth-order differences. We can begin to check our guess by *Examining a Special Case*. Table 25 was generated by the simplest cubic function $f(x) = x^3$.

TABLE 25

x	$f(x) = x^3$	First differences	Second differences	Third differences
-3	-27			
		19		
-2	-8		-12	
		7		6
-1	-1		-6	
		1		6
0	0		0	
		1		6
1	1		6	
		7		
2	8			

EXERCISES   **1.** (*Making Observations*) Construct a table for each cubic function using the indicated *x*-values. Calculate the first, second, and third differences for the *y*-values.

a. $y = 3x^3$; $x = 1.0, 1.5, 2.0, 2.5, 3.0, 3.5$

b. $y = ax^3$; $x = -1, 0, 1, 2, 3, 4$

c. $y = ax^3$; $x = x_0, x_0 + h, x_0 + 2h, x_0 + 3h, x_0 + 4h, x_0 + 5h$

In Exercise 1 you have shown that for any function $y = ax^3$, any table with equally spaced *x*-values has *y*-values with constant third differences.

Now let's see whether the same thing is true of any function

$$y = ax^3 + bx^2 + cx + d$$

2. (*Making Observations*) Construct a table for each cubic function using the indicated x-values. Calculate the first, second, and third differences for the y-values.

  a. $y = 3x^3 + x^2 + 2x - 5$
    $x = 1.0, 1.5, 2.0, 2.5, 3.0, 3.5$
  b. $y = ax^3 + bx^2 + cx + d$
    $x = -1, 0, 1, 2, 3, 4$

Exercise 2 suggests that for any cubic function, a table with equally spaced x-values always has y-values with constant third differences. Conversely, if a table with equally spaced x-values has y-values with constant third differences, it fits a cubic function. It can be shown that similar statements apply to polynomial functions of any degree.

> Any table with equally spaced x-values fits a polynomial function of degree n if and only if it has y-values with constant nth-order differences.

FYI As with linear and quadratic functions, we cannot say that a function is a polynomial on the basis of a table without additional information. We can only say that the table fits a polynomial function. However, if the nth-order differences are not constant, then the function is not a polynomial of degree n.

EXERCISES *In Exercises 3–6, find the smallest degree of a polynomial function that fits the table.*

3.

x	y
2	0
3	15
4	48
5	105
6	192
7	315
8	480

4.

x	y
-2	-45
-1.5	35.5
-1	17
-0.5	-1.5
0	-5
0.5	-2.5
1.0	33

5.

x	y
-3	-1
-1	-3
1	-3
3	-1
5	3
7	9
9	17

6.

x	y
0	1
1	0
2	-1
3	-2
4	-1
5	0
6	1

FYI A table with only $n + 1$ points always fits a polynomial function of degree n. For this reason, you should look for additional evidence before concluding that a polynomial function of degree n actually generated the table.

FITTING A POLYNOMIAL FUNCTION TO A TABLE

The following Mathematical Looking Glass illustrates that it is sometimes useful to find the equation of a polynomial function that fits a given table, even if that function did not generate the table.

A MATHEMATICAL LOOKING GLASS

Error-Correcting Codes

Recent advances in technology have enabled us to produce photographlike images of objects that are inaccessible to standard photographs. We see a CT scan of a human brain or a detailed view of a distant planet so often that we may not stop to think about how the images are generated. The first step in "photographing" such an object is to gather data and store it in a computer. For example, a CT scan sends x-rays through a person's head from many directions. The intensity of each ray decreases sharply as it passes through the skull and less sharply as it passes through the less dense material of the brain. As it exits the head, its intensity is recorded as a number that indicates the average density of the material it has passed through. A visual image is then reconstructed from the collection of numbers, using techniques that depend heavily on calculus. Any area of abnormal density in the brain, such as a tumor, appears on the reconstructed image.

To obtain images of distant planets, a space probe has sensors that record the intensity of light at many points. The intensities are beamed back to Earth as numbers from which a visual image can be mathematically reconstructed. A portion of the data sent by the probe might be

$$(0, 6), (1, 5), (2, 4), (3, 9)$$

indicating that the intensity level was 6 at position 0, 5 at position 1, 4 at position 2, and 9 at position 3. However, the data may become garbled as it travels through space, and may arrive on Earth as

$$(0, 6), (1, 5), (2, 9), (3, 4)$$

The accuracy of the reconstructed visual image can be affected by such errors. Fitting the data to a polynomial function is used to check the accuracy of the transmissions. In Exercise 8 you can find a cubic function to fit the four ordered pairs sent from the probe. The next step is to generate and transmit many more ordered pairs that fit the same function. These "extra" ordered pairs, which are not used to reconstruct the visual image, are referred to as **error-correcting codes**. Even if a few of the many ordered pairs become altered in transmission, it is generally easy to identify the function that generated the collection. The process is described in more detail by Linda Kurz in "Error-Correcting Codes," *The AMATYC Review*, Vol. 17, No. 1, 1995.

By the way, scientists at NASA are thankful that they have computers to do their number-crunching for them. The data from *Voyager 2*, whose images of Neptune were described in ***The Blue Planets*** (page 201), had to be fit to a polynomial of degree 222. ∎

EXERCISES

7. (*Writing to Learn*) Why is it reasonable to look for a cubic function to fit the data in ***Error-Correcting Codes***, rather than a polynomial function of some other degree?

8. (*Creating Models*) Find a cubic function to fit the ordered pairs $(0, 6)$, $(1, 5)$, $(2, 4)$, $(3, 9)$. [*Hint:* The function has the form $f(x) = ax^3 + bx^2 + cx + d$. You know $f(0) = 6$, so $a(0)^3 + b(0)^2 + c(0) + d = 6$. Use the other three ordered pairs to create a system of four equations in the variables a, b, c, and d, and solve the system.]

9. Find six other appropriate ordered pairs to transmit back to earth along with the original four.

In Exercises 10–13, *find a polynomial function of the smallest possible degree to fit the given ordered pairs.*

10. $(2, 1)$, $(3, 4)$, $(4, 5)$

11. $(0, 1)$, $(1, 3)$, $(2, 1)$, $(3, 3)$

12. $(1, 7)$, $(2, 1)$, $(3, -1)$, $(4, -1)$

13. $(0, 1)$, $(1, -3)$, $(2, -5)$, $(3, 19)$, $(4, 117)$

A NUMERICAL METHOD FOR FINDING ZEROS

Analytical and graphical methods for finding zeros of polynomial functions are discussed in Sections 5-2 and 1-4. You are about to see a numerical method, called the **bisection method**, that approximates zeros by repeatedly refining an initial guess. The method evolved over the course of centuries, rather than being developed by a single person.

To use the method, you must first find an interval containing a zero. You might find the interval by using a graph or a table. Example 1 applies the method to $P(x) = x^5 - x - 1$.

EXAMPLE 1   *Using the bisection method to approximate a zero*

The following table was generated by $P(x) = x^5 - x - 1$.

$$P(0) = -1$$

$$P(1) = -1$$

$$P(2) = 29$$

$$P(3) = 239$$

$$P(4) = 1019$$

$$P(5) = 3119$$

Since $P(1)$ and $P(2)$ have opposite signs, it is reasonable to expect that P has a zero in the interval $(1, 2)$.

To find a smaller interval around the zero, evaluate P at the midpoint of $(1, 2)$. Now your table becomes

$$P(1) = -1$$

$$P(1.5) \approx 5.09$$

$$P(2) = 29$$

Since $P(1)$ and $P(1.5)$ have opposite signs, P has a zero in $(1, 1.5)$. The point 1.5 bisects the interval $(1, 2)$ and thus cuts the size of the interval containing the zero in half. This explains why the technique is called the bisection method.

You can bisect the interval again by evaluating P at the midpoint of $(1, 1.5)$, so that your table becomes

$$P(1) = -1$$
$$P(1.25) \approx 0.80$$
$$P(1.5) \approx 5.09$$

and there is a zero in $(1, 1.25)$.

At the next step, the table becomes

$$P(1) = -1$$
$$P(1.125) \approx -0.32$$
$$P(1.25) \approx 0.80$$

and there is a zero in $(1.125, 1.25)$.

To approximate the zero with an error of less than 0.01, continue until the width of the interval around the zero is less than 0.01. The work is summarized in Table 26. Since the width of the last interval is less than 0.01, you can choose any number within it as your estimate. One possible choice is $x = 1.17$. ∎

TABLE 26

Step	Interval	Midpoint	Value of P at left endpoint	Value of P at midpoint	Value of P at right endpoint
1	$(1, 2)$	1.5	-1	5.09	29
2	$(1, 1.5)$	1.25	-1	0.80	5.09
3	$(1, 1.25)$	1.125	-1	-0.32	0.80
4	$(1.125, 1.25)$	1.1875	-0.32	0.17	0.80
5	$(1.125, 1.1875)$	1.15625	-0.32	-0.09	0.17
6	$(1.15625, 1.1875)$	1.171875	-0.09	0.04	0.17
7	$(1.15625, 1.171875)$	1.1640625	-0.09	-0.03	0.04
8	$(1.1640625, 1.171875)$				

FYI If $P(x_1)$ and $P(x_2)$ have opposite signs, then P has at least one zero in (x_1, x_2). However, if $P(x_1)$ and $P(x_2)$ have the same sign, there still may be zeros in (x_1, x_2), as in Figure 5-20. A graph can help you decide whether this is the case.

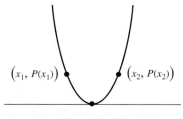

Figure 5–20

EXERCISES   *In Exercises 14–17, use the bisection method to approximate the largest real zero of the function with an error of no more than 0.01. Use a graph to establish initial intervals in which the zero can be found.*

14. $f(x) = x^3 + 4x - 8$ **15.** $w = 3x^3 - 10x + 6$

16. $G(x) = 2x^4 - 3x^3 + 2x^2 - x - 3$

17. $y = x^7 - 3x^5 + 2x^2 - 1$

Now that you understand how the bisection method works, you might program it into your calculator for future use.

A COMPARISON OF METHODS FOR FINDING ZEROS

The method you choose to find the zeros of a polynomial function depends on the nature of the function and the required degree of precision of your answer. A comparison of analytical, graphical, and numerical methods reveals several differences in effectiveness and convenience.

- The analytical method of factoring is convenient only for polynomials that factor easily.
- The graphical method of "zooming-in" can be used with any polynomial function, but several iterations are often required to produce the desired accuracy. Using the Rational Root Theorem can enhance the efficiency of these methods.
- The numerical bisection method can be programmed into your calculator, and can easily provide a high degree of accuracy. However, it requires prior knowledge of an interval containing a zero. A graph can reveal such an interval.
- Analytical methods yield exact results. Graphical and numerical methods usually yield only approximations.
- Although neither graphical nor numerical methods can identify multiple zeros, a graph indicates when multiple zeros may be present.

As you have seen, it is impractical to rely exclusively on any one method for all cases. The most painless "route to the roots" usually involves a combination of two or more. Combining methods is an application of the Polya strategy of *Breaking the Problem into Parts*.

EXERCISES   *In Exercises 18–21, find all real zeros of the function, exact if possible, by any combination of methods.*

18. $P(x) = 3x^3 + 8x^2 - 3x$

19. $Q(t) = t^4 - 40t + 39$

20. $H(z) = z^4 - 20z^3 + 126z^2 - 260z + 25$

21. $y = x^7 - 3x^5 + 2x^2 + 1$ (Compare the methods needed here to those needed for Exercise 17.)

ADDITIONAL EXERCISES   *In Exercises 22–25, determine the smallest degree of a polynomial function that fits the table.*

22.

x	y
-2	18
-1	-15
0	-20
1	-21
2	-30
3	-47
4	-60

23.

x	y
1	0
2	9
3	13
4	14
5	14
6	15
7	18

24.

x	y
-3	-41
-2	-5
-1	5
0	1
1	-5
2	-1
3	25
4	85

25.

x	y
-0.9	-3.12
-0.8	-2.28
-0.7	-1.48
-0.6	-0.72
-0.5	0
-0.4	0.68
-0.3	1.32
-0.2	1.92

In Exercises 26–29, find a polynomial function of the smallest possible degree to fit the data.

26. $(2, -16), (4, -8), (6, 48), (8, 176)$

27. $(0, 0), (1, 10), (2, 4), (3, 7)$

28. $(1, 2), (2, 17), (3, 82), (4, 257), (5, 626)$

29. $(-1, 25), (0, 18), (1, 12), (2, 7), (3, 3)$

In Exercises 30–33, use the bisection method to approximate the largest real zero of the polynomial. Use a graph to establish an initial interval in which the zero can be found.

30. $y = 2 - x^5$ **31.** $y = x^4 - 6x^2 + 4$

32. $y = 98 - 194x + 46x^2 - 2x^3$

33. $y = 6 + x + 0.3x^2 - 0.01x^5$

In Exercises 34–37, find all real zeros of the function, exact if possible, by any combination of methods.

34. $y = x^5 + x^3 - 3x^2$ **35.** $H(z) = z^5 + z^3 - 2z - 4$

36. $S(a) = a^6 + a^3 - 2$ **37.** $P(x) = 3x^4 - 4x^3 - 16$

38. (*Problem Solving*) The following table shows the population of the United States according to censuses taken in the given years.

Year	Population, in millions
1950	151
1960	179
1970	203
1980	227

a. Find a cubic function to fit the data.

b. Use your function to predict the population in 1990, and compare it to the actual population of 249 million.

c. Use your function to predict the population in each of the years 2000, 2050, and 2100. Which of these predictions, if any, do you trust to be accurate to a reasonable degree?

d. Do you trust the method used to obtain the estimates in parts (b) and (c)? Why or why not?

5-5 SOLVING POLYNOMIAL INEQUALITIES

The graphical and analytical methods you learned in Chapter 4 for solving quadratic inequalities can be applied to polynomials of higher degree. As with quadratic inequalities, the first step in solving a polynomial inequality is to solve a polynomial equation. For most polynomials of higher degree, this first step is the most difficult part of the solution. Because of this, mathematicians of past centuries concentrated on solving polynomial equations. The precise historical origins of the methods used to solve polynomial inequalities are unknown.

SOLVING POLYNOMIAL INEQUALITIES GRAPHICALLY

In the following Mathematical Looking Glass, your T-shirt company from *Fat Cats on Campus* (page 28) encounters a polynomial inequality.

A MATHEMATICAL
LOOKING GLASS
*Growth of Cost
(Fat Cats 6)*

As a special assignment, your economics professor has asked your class to see how an increase in the size of your company would affect your pricing strategies and your profit. The cost and revenue functions you calculated in *Fat Cats on Campus* are

$$C(x) = 3.50x + 612.50$$

$$R(x) = -0.005x^2 + 16x$$

If your company grows, it will be more convenient to express sales in units of 100 shirts. In Exercise 5 you can verify that if x represents hundreds of shirts, the cost and revenue functions should be

$$C(x) = 350x + 612.50$$

$$R(x) = -50x^2 + 1600x$$

Your professor has suggested that a more accurate cost function might be

$$C(x) = 0.08x^3 - 6x^2 + 350x + 612.50$$

In Exercise 6 you can discover why costs for a larger company might be modeled accurately by a cubic function. Because your profit is the difference between revenue and cost, your profit function will also be cubic. To discover your profitable target sales levels you need to solve the cubic inequality $P(x) \geq 0$. ■

The inequality in *Growth of Cost* can be solved either graphically or analytically. You already know how to solve polynomial inequalities graphically. Let's look at Example 1 to review the method.

EXAMPLE 1   *Solving a polynomial inequality graphically*

To solve $x^3 + 2x^2 - 9x < 0$, begin by graphing $f(x) = x^3 + 2x^2 - 9x$, as in Figure 5-21.

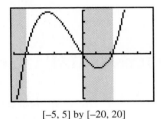

[–5, 5] by [–20, 20]

Figure 5–21

The inequality is true at those values of x where $f(x) < 0$ or, equivalently, where the graph is below the x-axis. The solution is shaded in Figure 5-21. You can identify it by finding the x-intercepts or, equivalently, by solving $f(x) = 0$. Then you can verify that the intercepts are at $x = 0$ and $x = -1 \pm \sqrt{10} \approx -4.16, 2.16$. The inequality is false at each intercept, but true to the left of $-1 - \sqrt{10}$ and between 0 and $-1 + \sqrt{10}$. The solution is $(-\infty, -1 - \sqrt{10}) \cup (0, -1 + \sqrt{10}) \approx (-\infty, -4.16) \cup (0, 2.16)$.   ■

EXERCISES   *In Exercises 1–4, solve the inequality graphically.*

1. $x^3 - 8x^2 + 19x - 13 > 0$

2. $3 + 2x^2 - x^4 > 0$

3. $x^4 - 4x^3 - 2x^2 + 12x + 9 \leq 0$

4. $x^5 + x^4 + 2x^3 \geq 0$

5. (*Writing to Learn*) In *Growth of Cost*, if x represents hundreds of shirts, explain why your former cost and revenue functions should be modified to

$$C(x) = 350x + 612.50$$

$$R(x) = -50x^2 + 1600x$$

   [*Hint*: Begin with the former functions

$$C(x) = 3.50x + 612.50$$

$$R(x) = -0.005x^2 + 16x$$

   and write expressions for $C(100x)$ and $R(100x)$.]

6. (*Interpreting Mathematics*) As your company grows, the average cost of producing additional shirts may decrease for a while. For example, you may be able to buy larger quantities of shirts at a cheaper rate. Economists refer to such savings as **economies of scale**. Eventually, however, the average cost of producing additional shirts may increase again due to size-related costs, called **diseconomies of scale**. For example, you may need to hire new employees and rent additional space.

a. For your professor's suggested cost function

$$C(x) = 0.08x^3 - 6x^2 + 350x + 612.50$$

calculate the average rate of change in $C(x)$ with respect to x over each of the intervals $[0, 10]$, $[20, 30]$, and $[50, 60]$.

b. Interpret the results from part (a) in physical language. (Remember that x is expressed in units of 100 shirts.)

c. Over which interval in part (a) do you see evidence of economies of scale? Over which interval do you see evidence of diseconomies of scale? Explain your answers.

7. (*Problem Solving*)

a. Use the cost and revenue functions

$$C(x) = 0.08x^3 - 6x^2 + 350x + 612.50$$

$$R(x) = -50x^2 + 1600x$$

to find your company's profit function P.

b. Graph $y = P(x)$ in the window $[0, 50]$ by $[-10\,000, 10\,000]$ and solve $P(x) \geq 0$ to find your profitable target sales levels.

c. Graph $y = P(x)$ in the window $[-1000, 1000]$ by $[-3\,000\,000, 3\,000\,000]$ and solve $P(x) \geq 0$ as an abstract inequality. (*Hint:* The graph has three x-intercepts.) How do your solutions in parts (b) and (c) differ? Why?

SOLVING POLYNOMIAL INEQUALITIES ANALYTICALLY

The Test-Value Method Polynomial inequalities can be solved analytically by using the test-value method you learned in Section 4-2. Example 2 reviews the method.

EXAMPLE 2   *Solving a polynomial inequality analytically*

STEP 1:   To solve $x^3 + 2x^2 - 9x < 0$, begin by solving $x^3 + 2x^2 - 9x = 0$ to obtain $x = 0$ and $x = -1 \pm \sqrt{10} \cong -4.16, 2.16$.

STEP 2:   Indicate $-4.16, 0$, and 2.16 on a number line.

STEP 3:   Evaluate $f(x) = x^3 + 2x^2 - 9x$ at a test value in each of the intervals $(-\infty, -4.16)$, $(-4.16, 0)$, $(0, 2.16)$, and $(2.16, \infty)$.

$f(-10) = -710$	$f(-1) = 10$	$f(1) = -6$	$f(10) = 1110$
$f(x) < 0$ is true	$f(x) < 0$ is false	$f(x) < 0$ is true	$f(x) < 0$ is false

(with markers: $\uparrow$ −4.16 $\uparrow$ 0 $\uparrow$ 2.16 $\uparrow$)

The solution is

$$(-\infty, -1 - \sqrt{10}) \cup (0, -1 + \sqrt{10}) \approx (-\infty, -4.16) \cup (0, 2.16)$$ ∎

EXERCISES   *In Exercises 8–11, solve the inequality analytically.*

8. $x^3 - 9x > 0$ **9.** $8x^2 - x^4 \leqslant 0$

10. $x^5 + x^4 - 2x^3 \geqslant 0$ **11.** $x^5 + x^4 + 2x^3 \geqslant 0$

12. (*Writing to Learn*) Compare your graphical solution to Exercise 4 with your analytical solution to Exercise 11. In what respects, if any, was each solution easier?

The Scan Method The following method is especially useful for inequalities involving factored polynomials. It uses some graphical observations from Section 5-3 to create a streamlined analytical procedure we will call the **scan** method (our acronym for Sign Change ANalysis).

EXAMPLE 3   *Solving a polynomial inequality by the scan method*

Let $P(x) = (x + 5)(2x + 1)^2(x - 3)^3$. To solve $P(x) \geqslant 0$,

STEP 1: Solve $P(x) = 0$ to obtain $x = -5, -\dfrac{1}{2}$, and 3.

STEP 2: Indicate these solutions on a number line.

STEP 3: To find the sign of $P(x)$ on $(-\infty, -5)$, we need to determine the left-end behavior of P. The leading term of $P(x)$ in its expanded form is $(x)(2x)^2(x)^3 = 4x^6$. [This is the product of the terms of highest degree in each factor. You do not need to multiply $P(x)$ out.] The end behavior of $P(x)$ agrees with that of $4x^6$, so $P(x) > 0$ on $(-\infty, -5)$.

STEP 4: To find the sign of $P(x)$ in the intervals $\left(-5, -\dfrac{1}{2}\right), \left(-\dfrac{1}{2}, 3\right)$, and $(3, \infty)$, *scan* the zeros, remembering that the graph of P crosses the x-axis only at a zero of odd multiplicity. That is, the sign of $P(x)$ changes only at a zero of odd multiplicity.

The zero at $x = 5$ has multiplicity 1, so the sign of $P(x)$ changes there. Thus $P(x)$ becomes negative in $\left(-5, -\dfrac{1}{2}\right)$.

The zero at $x = -\dfrac{1}{2}$ has multiplicity 2, so the sign of $P(x)$ does not change there. Thus $P(x)$ remains negative in $\left(-\dfrac{1}{2}, 3\right)$.

The zero at $x = 3$ has multiplicity 3, so the sign of $P(x)$ changes there. Thus $P(x)$ becomes positive in $(3, \infty)$.

We can now see that $P(x) > 0$ in $(-\infty, -5) \cup (3, \infty)$. Since $P(x) = 0$ at each of the zeros, the solution to $P(x) \geqslant 0$ is

$$(-\infty, -5] \cup \left\{-\frac{1}{2}\right\} \cup [3, \infty) \qquad \blacksquare$$

EXERCISES

In Exercises 13–16, solve the inequality using the scan method.

13. $(x + 7)(3x + 4) \geqslant 0$

14. $(x + 7)^2(3x + 4)^3 \geqslant 0$

15. $(1 - x)(7 - 2x)^2(5 - x)^2 < 0$

16. $(x + 2)^3(x - 5)^2(x^2 + 4) > 0$ (*Hint:* The factor of $x^2 + 4$ does not correspond to any real zeros.)

17. (*Writing to Learn*) Compare the convenience of the test-value and scan methods of solving polynomial inequalities. How might the form of the inequality influence your choice of methods?

18. (*Writing to Learn*) Would any of the inequalities in this section be easier to solve graphically? If so, which ones? Explain your answer.

ADDITIONAL EXERCISES

In Exercises 19–22, solve the inequality graphically.

19. $x^3 + 2x^2 - 15x < 0$ 20. $4x^3 - x^4 > 0$

21. $x^6 - 12x^4 + 48x^2 - 64 \leqslant 0$

22. $x^4 - 25x^2 + 2x + 3 \geqslant 0$

In Exercises 23–30, solve the inequality analytically, using either the test-value method or the scan method.

23. $x^3 + 2x^2 - 15x < 0$ 24. $4x^3 - x^4 > 0$

25. $x^6 - x^4 - 16x^2 + 16 \leqslant 0$

26. $4x - 3x^3 - x^5 \geqslant 0$

27. $(2x + 5)(3x - 1)(x - 7) < 0$

28. $(x + 4)^2(x - 12)^2 > 0$

29. $(x + 4)^3(x - 12)^3 > 0$

30. $(x + 79)^3(x + 0.3)^4(x - 140) \leqslant 0$

SUPPLEMENTARY TOPIC

Using the Scan Method to Graph Polynomial Functions

The scan method relies on knowledge of the end behavior of polynomial graphs, as well as the behavior around multiple zeros. The processes of solving an inequality by the scan method, and of graphing a polynomial function, thus complement each other. In particular, the reasoning of the scan method can be used to obtain a quick rough sketch of a polynomial graph.

EXAMPLE 4   *Using the scan method to graph a polynomial function*

The information contained in Example 3 can help you to sketch a graph of $P(x) = (x + 5)(2x + 1)^2(x - 3)^3$. Since $P(x) > 0$ in $(-\infty, -5) \cup (3, \infty)$, the graph is above the x-axis there and below the x-axis in $\left(-5, -\frac{1}{2}\right) \cup \left(-\frac{1}{2}, 3\right)$.

Furthermore, the graph flattens out at $x = -\frac{1}{2}$ and 3, since the multiplicity of each zero is greater than 1. Figure 5-22 shows a rough sketch of the graph. (You cannot show extreme values accurately without plotting some additional points.)

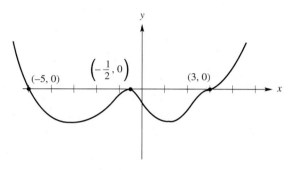

Figure 5–22

EXERCISES   *In Exercises 31–34, sketch a graph of the function by using information from Exercises 13–16. Your sketch should accurately depict end behavior and behavior around the zeros. It does not need to show extreme values to scale. Support your sketch with a calculator graph.*

31. $y = (x + 7)(3x + 4)$

32. $y = (x + 7)^2(3x + 4)^3$

33. $y = (1 - x)(7 - 2x)^2(5 - x)^2$

34. $y = (x + 2)^3(x - 5)^2(x^2 + 4)$

35. (*Writing to Learn*) Describe how your graph in Exercise 33 supports your solution in Exercise 15, and vice versa.

In Exercises 36–39, sketch a graph of the function using information from Exercises 27–30. Your sketch should accurately depict end behavior and behavior around the zeros. It does not need to show extreme values to scale. Support your sketch with a calculator graph.

36. $y = (2x + 5)(3x - 1)(x - 7)$

37. $y = (x + 4)^2(x - 12)^2$

38. $y = (x + 4)^3(x - 12)^3$

39. $y = (x + 79)^3(x + 0.3)^4(x - 140)$

5-6 THE BINOMIAL THEOREM

Mathematicians of the seventeenth century often looked at polynomials as algebraic expressions, rather than functions. They sought to develop formulas for rewriting polynomial expressions in a variety of forms. In Section 5-2, you learned about some of their efforts to express polynomials in factored form. They were also concerned about expressing certain polynomials in expanded form. In this section, you will learn a formula that will allow you to write a binomial power $(p + q)^n$ in expanded form without performing the actual multiplication. The statement of this formula is called the Binomial Theorem. Although its development was primarily the result of intellectual curiosity, you will discover that the Binomial Theorem can also be used to solve physical problems.

PASCAL'S TRIANGLE

To write the expanded form of $(p + q)^n$, you need to know how many terms it contains, what powers of p and q are involved in each term, and what the numerical coefficient of each term is. Let's begin to answer these questions by *Examining a Special Case*. The two simplest binomial powers are

$$(p + q)^0 = 1$$
$$(p + q)^1 = p + q$$

Next, let's recall how the expanded form of $(p + q)^2$ is obtained.

$$(p + q)^2 = (p + q)(p + q)$$
$$= p^2 + pq$$
$$\underline{ + pq + q^2}$$
$$= p^2 + 2pq + q^2$$

Now let's see how the expanded form of $(p + q)^3$ can be obtained from that of $(p + q)^2$.

$$(p + q)^3 = (p + q)(p^2 + 2pq + q^2)$$
$$= p^3 + 2p^2q + pq^2$$
$$\underline{ + p^2q + 2pq^2 + q^3}$$
$$= p^3 + 3p^2q + 3pq^2 + q^3$$

Continuing, we obtain

$$(p + q)^4 = (p + q)(p^3 + 3p^2q + 3pq^2 + q^3)$$
$$= p^4 + 3p^3q + 3p^2q^2 + pq^3$$
$$\underline{ + p^3q + 3p^2q^2 + 3pq^3 + q^4}$$
$$= p^4 + 4p^3q + 6p^2q^2 + 4pq^3 + q^4$$

EXERCISES   1. (*Making Observations*) Write a single paragraph to answer each question below. To answer the questions, look at the expanded forms of $(p + q)^2$, $(p + q)^3$, and $(p + q)^4$ and use the strategy of *Looking for a Pattern*. Do not expand $(p + q)^5$.

    a. How many terms are involved in the expanded form of $(p + q)^5$?

    b. What is the sum of the exponents in each term? That is, what is the degree of each term?

    c. What powers of p are involved in each term? What powers of q are involved in each term?

2. (*Making Observations*) To answer each question, look at the expanded forms of $(p + q)^2$, $(p + q)^3$, and $(p + q)^4$ and use the strategy of *Looking for a Pattern*. Do not expand $(p + q)^5$.

   a. In the expanded form of $(p + q)^5$, what is the coefficient of p^5? Of q^5?

   b. What is the coefficient of p^4q? Of pq^4?

   c. How do the coefficients of p^3q^2 and p^2q^3 relate to each other?

3. **a.** (*Writing to Learn*) Explain why the following arrangement of numbers gives the coefficients in the expanded form of $(p + q)^5$.

$$
\begin{array}{ccccccc}
1 & 4 & 6 & 4 & 1 & & \\
 & 1 & 4 & 6 & 4 & 1 & \\
\hline
1 & 5 & 10 & 10 & 5 & 1 &
\end{array}
$$

   b. Write out the expanded form of $(p + q)^5$ without performing the multiplication.

4. Check your answers to Exercises 1–3 by using the expanded form of $(p + q)^4$ to obtain the expanded form of $(p + q)^5$.

5. (*Making Observations*) Answer the following questions by generalizing your answers from Exercise 1.

   a. How many terms are involved in the expanded form of $(p + q)^n$?

   b. What is the degree of each term?

   c. What powers of p are involved in each term? What powers of q are involved in each term?

6. (*Making Observations*) Answer the following questions by generalizing your answers from Exercise 2.

   a. In the expanded form of $(p + q)^n$, what is the coefficient of p^n? Of q^n?

   b. What is the coefficient of $p^{n-1}q$? Of pq^{n-1}?

   c. Describe the symmetry exhibited by the coefficients.

7. (*Writing to Learn*) By generalizing your response in Exercise 3, explain how to obtain the coefficients in the expanded form of $(p + q)^n$ from those of $(p + q)^{n-1}$.

   Your work in Exercises 1–7 demonstrates that for every integer $n \geqslant 0$, the coefficients of $(p + q)^n$ are generated by **Pascal's triangle**, shown here.

$$
\begin{array}{ccccccccc}
 & & & & 1 & & & & \\
 & & & 1 & & 1 & & & \\
 & & 1 & & 2 & & 1 & & \\
 & 1 & & 3 & & 3 & & 1 & \\
1 & & 4 & & 6 & & 4 & & 1 \\
 & & & & \cdot & & & & \\
 & & & & \cdot & & & & \\
 & & & & \cdot & & & &
\end{array}
$$

   In this triangle, the outside entries in each row are 1's. Each other entry is the sum of the two nearest entries in the preceding row. For example, the 6 in the last row shown is the sum of the two 3's in the preceding row.

The top row of Pascal's triangle (row 0) contains the coefficients of $(p + q)^0 = 1$. The next row (row 1) contains the coefficients of $(p + q)^1 = p + q$. In general, row n contains the coefficients of $(p + q)^n$.

Pascal's triangle was known in China at least since the thirteenth century, when it appeared in a book written by Yang Hui. According to Morris Kline in *Mathematical Thought from Ancient to Modern Times* (Oxford University Press, 1972), it was also known to the Arabs of the thirteenth century and to many European mathematicians in the sixteenth century. However, many of its properties were first discovered by Blaise Pascal (1623–1662). A French mathematician of exceptional genius, Pascal was also, in the words of historian Howard Eves, "a religious neurotic" who often saw divine messages in commonplace events. After narrowly escaping death at the age of 24 when his runaway horses plunged over a bridge, he believed that he had been given a sign from God that he should abandon his mathematical activities and spent most of the remainder of his life in religious contemplation.

Example 1 shows how to use Pascal's triangle to expand a power of a binomial.

EXAMPLE 1 *Using Pascal's triangle to expand a binomial power*

The expanded form of

$$(3x - y^2)^4$$

is

$$(\underline{\hspace{0.3cm}})(3x)^4 + (\underline{\hspace{0.3cm}})(3x)^3(-y^2) + (\underline{\hspace{0.3cm}})(3x)^2(-y^2)^2 + (\underline{\hspace{0.3cm}})(3x)(-y^2)^3 + (\underline{\hspace{0.3cm}})(-y^2)^4$$

Reading the coefficients from row 4 of Pascal's triangle, we obtain

$$(1)(3x)^4 + (4)(3x)^3(-y^2) + (6)(3x)^2(-y^2)^2 + (4)(3x)(-y^2)^3 + (1)(-y^2)^4$$

$$= 81x^4 - 108x^3y^2 + 54x^2y^4 - 12xy^6 + y^8 \qquad \blacksquare$$

EXERCISES *In Exercises 8–11, use Pascal's triangle to write the expression in expanded form.*

8. $(p + q)^6$ **9.** $(p + q)^7$

10. $(p + q)^8$ **11.** $(p + q)^9$

In Exercises 12–15, write the expression in expanded form.

12. $(5w + 2x)^3$ **13.** $(2a - 3b)^4$

14. $(p^3 + q^2)^6$ **15.** $(2a^2 - b^5)^5$

BINOMIAL COEFFICIENTS

Although Pascal's triangle generates the coefficients of every binomial power, it is impractical to use if the power is large. For example, to find the coefficient of p^6q^{27} in the expansion of $(p + q)^{33}$, you must write down the first 32 rows of the triangle. (Later in this section you will see how that coefficient is of practical interest in designing a state lottery game.)

In 1665 Isaac Newton discovered how to find a given coefficient in the expansion of $(p + q)^n$ without expanding $(p + q)^{n-1}$. To discover how to do so, let's begin by looking at the expanded form of $(p + q)^4$ in a different way.

If we think of $(p + q)^4$ as $(p + q)(p + q)(p + q)(p + q)$ and carry out the multiplication without using exponents, we obtain

$$pppp + pppq + ppqp + ppqq$$
$$+ pqpp + pqpq + pqqp + pqqq$$
$$+ qppp + qppq + qpqp + qpqq$$
$$+ qqpp + qqpq + qqqp + qqqq$$

The terms in the expanded form correspond to all possible four-letter strings consisting of p's and q's. The coefficient of a given term, say, p^2q^2, is equal to the number of strings consisting of two p's and two q's. Similarly, the terms in the expanded form of $(p + q)^n$ correspond to all possible n-letter strings consisting of p's and q's. The coefficient of p^rq^{n-r} is equal to the number of strings in which the number of p's is $n - r$ and the number of q's is r. This observation suggests a way to calculate the coefficient of a given term p^rq^{n-r} directly. As an example, let's use this observation to calculate the coefficient of p^3q^2 in $(p + q)^5$.

EXERCISE **16.** (*Writing to Learn*) Explain why the coefficient of p^3q^2 is equal to the number of all possible five-letter strings containing three p's and two q's.

To form all five-letter strings containing three p's and two q's, let's begin by choosing a position for the first p. There are five possibilities, shown in Table 27.

TABLE 27

p_ _ _ _   _ p _ _ _   _ _ p _ _   _ _ _ p _   _ _ _ _ p

Each choice leads to four possible positions for the second p, as shown in Table 28.

TABLE 28

pp_ _ _	p p_ _ _	p_ p_ _	p_ _ p_	p_ _ _ p
p_p_ _	_ p p_ _	_p p_ _	_ p_ p_	_p_ _ p
p_ _p_	_ p_p_	_ _ p p_	_ _p p_	_ _p_ p
p_ _ _p	_ p_ _p	_ _ p_p	_ _ _ p p	_ _ _p p

At this stage, notice that of the twenty strings in Table 28 only ten are different, since each string appears twice. In Exercise 17, you can show how to account for this duplication in counting the total number of five-letter strings containing three p's and two q's.

EXERCISES **17.** (*Making Observations*) Each string in Table 28 has three possible positions for the third p.

  a. If we were to create another table to show all of these positions, explain why the new table would contain 60 strings.

  b. Show that the string ppp_ _ , representing $pppqq$, would appear six times in the new table. (*Hint*: Use Table 28 as a starting point. Using pencils of three different colors may make the demonstration easier.)

    c. Explain why every string appears six times in the new table.

    d. Use your observations in parts (a) and (c) to calculate the total number of five-letter strings containing three p's and two q's. Check your calculation by listing all such strings.

    e. How do your observations allow you to conclude that the coefficient of p^3q^2 in $(p + q)^5$ is 10?

18. (*Writing to Learn*) To determine the coefficient of p^4q^5 in $(p + q)^9$, you could construct a sequence of tables similar to Tables 27 and 28 and the table described in Exercise 17.

    a. Explain why you would construct a total of four tables.

    b. Explain why there would be nine string entries in your first table, $(9)(8) = 72$ in your second table, $(9)(8)(7) = 504$ in your third table, and $(9)(8)(7)(6) = 3024$ in your fourth table.

    c. Explain why each string in your second table would appear 2 times, each string in your third table would appear $(1)(2)(3) = 6$ times, and each string that is shown in your fourth table would appear $(1)(2)(3)(4) = 24$ times.

    d. Explain why the coefficient of p^4q^5 in $(p + q)^9$ is $\dfrac{(9)(8)(7)(6)}{(1)(2)(3)(4)} = 126$.

Your observations in Exercises 17 and 18 can be generalized. In the expansion of $(p + q)^n$, the coefficient of p^rq^{n-r} is

$$\frac{(n)(n - 1)(n - 2) \cdots (n - r + 1)}{(1)(2)(3) \cdots (r)}.$$

You can write this expression more easily by using a notation introduced in 1808 by Christian Kramp (1760–1826). For any positive integer k, Kramp used the symbol $k!$ (read k **factorial**) to represent the product $(1)(2)(3) \cdots (k)$. Thus, for example, $5! = (1)(2)(3)(4)(5) = 120$. By convention, $0!$ is defined to be 1. With this notation the coefficient of p^4q^5 in $(p + q)^9$, which you obtained in Exercise 18d, can be written

$$\frac{(9)(8)(7)(6)}{(1)(2)(3)(4)} \cdot \frac{(5)(4)(3)(2)(1)}{(1)(2)(3)(4)(5)} = \frac{9!}{(4!)(5!)}$$

The generalization of this statement, called the **Binomial Theorem**, provides a formula for every coefficient in the expansion of every binomial power.

The Binomial Theorem

$$(p + q)^n = \frac{n!}{n!\,0!}\,p^n + \frac{n!}{(n - 1)!\,1!}\,p^{n-1}q + \frac{n!}{(n - 2)!\,2!}\,p^{n-2}q^2$$

$$+ \cdots + \frac{n!}{2!\,(n - 2)!}\,p^2q^{n-2} + \frac{n!}{1!(n - 1)}\,pq^{n-1} + \frac{n!}{0!\,n!}\,q^n$$

The expression $\dfrac{n!}{r!\,(n - r)!}$, representing the coefficient of p^rq^{n-r}, is called a **binomial coefficient** and is sometimes written

$$\binom{n}{r} \quad \text{or} \quad {}_nC_r$$

EXERCISES

19. Evaluate each factorial.

    a. 6!   **b.** 7!

    c. 10!   **d.** 25!

20. Find the value of each binomial coefficient.

    a. $_7C_2$ **b.** $_6C_5$

    c. $_9C_3$ **d.** $_9C_6$

In Exercises 21–24, use the Binomial Theorem to find the coefficient of each term described.

21. p^2q^4 in $(p + q)^6$

22. p^7q^3 in $(p + q)^{10}$

23. p^3q^5 in $(p + q)^8$

24. p^5q^3 in $(p + q)^8$

In Exercises 25–28, use the Binomial Theorem to expand the expression. Make sure your result agrees with the one you obtained in the indicated exercise.

25. $(p + q)^6$ (Exercise 8)

26. $(p + q)^9$ (Exercise 11)

27. $(5w + 2x)^3$ (Exercise 12)

28. $(2a^2 - b^5)^5$ (Exercise 15)

Many physical uses of the Binomial Theorem are based on a surprising connection between binomial powers and the theory of probability, discovered by James (Jakob) Bernoulli (1654–1705). This connection is illustrated by the following Mathematical Looking Glass.

A MATHEMATICAL LOOKING GLASS
Key Ball

During the past decade, government-run lotteries have become increasingly widespread. Pennsylvania, like many other states, runs a number of different lottery games in order to increase interest and participation. One of the most successful has been Keystone Jackpot, popularly called Key Ball, introduced in November, 1995. According to Cris Oblack of the lottery's Public Relations Department, Key Ball generated about $290,000,000 in sales during its first 18 months of existence. Half of that amount has been paid out as prizes, 40% has been used to fund projects that benefit senior citizens, and 7% has been paid as commissions to retailers who sell game tickets. The lottery's budget accounts for the remaining 3%.

To conduct a game of Key Ball, officials of the Pennsylvania State Lottery use two containers, each containing balls numbered from 1 to 33. Six white balls are drawn from the first container and one blue ball (the Key Ball) from the second. To win the top prize, a player must correctly mark (in advance, of course!) the first six numbers on one section of his or her ticket and the Key Ball number on another section.

Key Ball is played weekly. The value of the top prize is always at least $3,000,000 but can vary, depending on how many tickets are sold and whether anyone won the top prize during the previous week. As a result, the prize is sometimes in excess of $30,000,000.

To determine appropriate prize levels, lottery officials must know the probability that a given ticket will win. To gain insight into the calculations that are needed, let's adopt the strategy of *Examining a Simpler Problem* by considering only the top prize. We can also use the strategy of *Breaking the Problem into Parts* by considering the first six numbers and the Key Ball number separately. Thus, the first question we will ask is, "What is the probability that the first six numbers on a lottery ticket will match those on the six white balls drawn from the first container?"   ■

The first section of a Key Ball ticket contains 33 numbered spaces, of which a player must mark six. To see how the question in **Key Ball** relates to the Binomial Theorem, imagine that you are to write p in the space for each number you select and q in the remaining spaces. For example, if you select 1, 5, 6, 13, 21, and 32, your ticket would look like this.

$$\mathbf{p}\,q\,q\,q\,\mathbf{p}\,\mathbf{p}\,q\,q\,q\,q\,q\,q\,q\,\mathbf{p}\,q\,q\,q\,q\,q\,q\,q\,q\,\mathbf{p}\,q\,q\,q\,q\,q\,q\,q\,q\,q\,q\,\mathbf{p}\,q$$

EXERCISE   **29.** (*Writing to Learn*)

    a. Explain why the number of different ways to complete the first section of the ticket is equal to the number of different strings containing 6 p's and 27 q's.

    b. In how many different ways can you complete the first section of your ticket?

The answer to the question in **Key Ball** relies on the following principle of probability theory.

If all outcomes in an experiment are equally likely to occur, the probability of a favorable outcome is

$$\frac{\text{number of favorable outcomes}}{\text{number of possible outcomes}}$$

In this case, the experiment is the drawing of the six white balls from the container. As you discovered in Exercise 29, there are 1,107,568 possible combinations (outcomes). The balls are selected at random, so that every possible outcome is equally likely to occur. Since there is only one winning combination (favorable outcome), the probability that the first six numbers selected on your ticket will match those drawn from the first container is

$$\frac{1}{1,107,568}$$

EXERCISE   **30.** (*Extension*) You will win the top prize only if you also match the Key Ball number correctly. What is the probability that your ticket will win the top prize?

ADDITIONAL EXERCISES   **31.** Find the value of each binomial coefficient.

    a. $_5C_5$ **b.** $_6C_2$

    c. $_{10}C_7$ **d.** $_{12}C_1$

32. Use the Binomial Theorem to find the coefficient of each term described.

    a. p^2q^6 in $(p + q)^8$

    b. p^4q^3 in $(p + q)^7$

    c. p^7q^2 in $(p + q)^9$

    d. p^5q^5 in $(p + q)^{10}$

In Exercises 33–36,

    a. Use Pascal's triangle to expand the expression.

    b. Use the Binomial Theorem to expand the expression, and make sure your result agrees with the one you obtained in part (a).

33. $(p + q)^5$ **34.** $(p - q)^{10}$

35. $(6y + 5z)^3$ **36.** $(m^4 - n^3)^4$

In Exercises 37–42, use the Binomial Theorem to expand the expression.

37. $(a - 3b)^6$ **38.** $(ab + cd)^7$

39. $(10u + v^2)^4$ **40.** $(p^3 - q^3)^3$

41. $(2x^2 + 3y^3)^5$ **42.** $[(2x)^2 + (3y)^3]^5$

43. (*Writing to Learn*) In the expansion of $(p + q)^n$, explain why the sum of the coefficients is 2^n. (*Hint:* In every n-letter string of p's and q's, each letter must be either p or q. Given those choices, how many different strings can you construct?)

In Exercises 44–48, imagine that you have just flipped a balanced coin ten times. In such a situation, every possible sequence of heads and tails is equally likely to occur.

44. (*Writing to Learn*) Explain why there are $2^{10} = 1024$ possible sequences of heads and tails. (*Hint:* How is this exercise related to Exercise 43?)

45. (*Problem Solving*)

    a. How many sequences contain exactly 4 heads and 6 tails?

    b. What is the probability that you have obtained exactly 4 heads and 6 tails?

46. (*Problem Solving*)

    a. What is the probability that you have obtained exactly two heads?

    b. What is the probability that you have obtained at most two heads?

47. (*Making Observations*) What is the probability that your first four flips were heads and the rest were tails? How do this question and your answer differ from those in Exercise 45b?

48. (*Extension*)

    a. How many sequences of flips begin with 8 consecutive heads?

    b. How many of the sequences in part (a) have tails on the ninth flip?

    c. Many people believe that after a long string of heads, a tail is more likely to occur because the coin is "due" to come up tails. Do your observations in parts (a) and (b) support this belief? Explain your answer.

49. (*Extension*) In professional baseball, basketball, and hockey, league championships are decided by a series of games played between two teams, ending when one team has won four games. If the teams are evenly matched, what is the probability that a series will last seven games? (*Hint:* The series will last seven games if and only if each team wins three of the first six.)

CHAPTER REVIEW

Complete Exercises 1–42 (Writing to Learn) before referring to the indicated pages.

WORDS AND PHRASES

In Exercises 1–15, explain the meaning of the words or phrases in your own words.

1. (page 241)   **polynomial function**

2. (page 241)   **degree** of a polynomial function

3. (page 242)   **power function**

4. (page 244)   **y varies directly with the nth power of** x

5. (page 244)   **varies directly, varies linearly, directly proportional**

6. (pages 251, 252)   **zero** of a polynomial function

7. (page 253)   **simple zero, multiple zero, multiplicity** of a zero

8. (page 262)   **turning point, extreme value**

9. (page 262)   **local minimum, local maximum, absolute minimum, absolute maximum**

10. (page 264)   **end behavior** of a polynomial graph

11. (pages 268, 269)   **symmetry about the origin**

12. (page 269)   **even** function, **odd** function

13. (page 286)   **Pascal's triangle**

14. (page 289)   **factorial**

15. (page 289)   **binomial coefficient**

IDEAS

16. (pages 243, 244)   Describe the graph of $y = x^n$ for all integers $n \geq 2$, and describe how to obtain the graph of $y = a(x - h)^n + k$ from the basic graph $y = x^n$. Illustrate with a specific example.

17. (page 245)   How can you tell whether a table fits a polynomial variation of a given degree? Make up a table and use it to illustrate the process.

18. (page 245)   Why can you regard linear variation as a special case of polynomial variation?

19. (page 252)   What does the Factor Theorem say? How can it be helpful in finding the zeros of a polynomial function?

20. (page 253)   How does the Factor Theorem guarantee that a polynomial function of degree n has at most n zeros?

21. (page 253)   Why is it sometimes reasonable to refer to a zero of a polynomial function as a double zero?

22. (pages 253, 254)   What does the Fundamental Theorem of Algebra tell us about the number of zeros of a polynomial function? In the statement of the theorem,

are we to count all complex zeros or only the real ones? Are we to count multiple zeros more than once?

23. (page 255)   For what values of n are there formulas for solving all polynomial equations of degree n? When were these formulas discovered? When was it discovered that there are degrees for which no formula exists?

24. (page 259)   What is the maximum number of x-intercepts on the graph of a polynomial function of degree n? How can this result be established analytically?

25. (pages 260, 261)   Describe the behavior of a polynomial graph around a zero of multiplicity k for all integers $k \geq 1$.

26. (page 262)   What are the similarities and differences between the phrases *local minimum* and *absolute minimum*? Use a sketch to illustrate your explanation.

27. (page 264)   Why is the end behavior of a polynomial graph determined by the term of highest degree in its equation?

28. (page 266)   Describe the four possible end behaviors for a polynomial graph. Under what conditions does each occur?

29. (pages 268, 269)   How can you use the equation of a function to decide whether its graph is symmetric about the y-axis? About the origin? Use specific polynomial functions to illustrate.

30. (page 269)   What is the origin of the terms *even function* and *odd function*? Must every polynomial function be either even or odd? Exhibit specific functions in support of your answer.

31. (pages 269, 270)   Why is it important to consider symmetry in constructing mathematical models of physical situations?

32. (page 272)   How can you tell whether a table fits a polynomial function of a given degree? Make up a table and use it to illustrate.

33. (page 273)   Why is it true that every finite table fits a polynomial function of some degree?

34. (pages 273)   Describe the process of fitting a polynomial function to data. In particular, explain why we can fit a polynomial function of at most degree n to any collection of $n + 1$ data points.

35. (page 275)   Explain how the bisection method works. Make up a polynomial equation and use it to illustrate the method.

36. (page 275)   What are some of the advantages and limitations of the various methods of finding zeros of poly-

nomial functions? How can the effectiveness of the methods be improved by using them in combination?

37. (page 281)   Describe the steps in the test value method of solving polynomial inequalities, and explain why the method works.

38. (page 282)   Describe the steps in the scan method of solving polynomial inequalities, and explain why the method works.

39. (page 286)   How is each row of Pascal's triangle generated from the previous row? Why does the process always produce the coefficients in the expansion of a binomial power?

40. (page 289)   In the expansion of $(p + q)^n$, explain why the coefficient of $p^r q^{n-r}$ is $\dfrac{n!}{r! \, (n - r)!}$.

41. (page 289)   Make up a binomial, and show how to expand it by using the Binomial Theorem.

42. (page 291)   What does the Binomial Theorem have to do with the calculation of probabilities in an experiment with equally likely outcomes?

CHAPTER 6

RATIONAL FUNCTIONS

6–1 RECIPROCAL POWER FUNCTIONS

PREREQUISITES MAKE SURE YOU ARE FAMILIAR WITH:
Algebraic Fractions (Section A-16)
Equations with Algebraic Fractions (Section A-17)

The category of polynomial functions (Chapter 5) includes all linear functions (Chapter 2) and quadratic functions (Chapter 4). The present chapter explores a still more inclusive category. **A rational function** is one that can be written as a ratio (quotient) of two polynomial functions. That is, f is rational if and only if $f(x) = \dfrac{N(x)}{D(x)}$ for some polynomial functions N and D.

Every polynomial function is a rational function. For example, the function $f(x) = x^3 + 5$ can be thought of as $f(x) = \dfrac{x^3 + 5}{1}$. However, as you will see in this chapter, rational functions do not always behave like polynomial functions. In particular,

- Rational functions are not always continuous.
- The end behavior of some rational graphs is different from that of polynomial graphs.

These differences will be explored in depth in Section 6-2. However, they are illustrated by the simplest rational functions, the **reciprocal power functions** $y = \dfrac{1}{x^n}$. These, and other rational functions closely related to them, will be the focus of the present section.

A MATHEMATICAL
LOOKING GLASS
Speed Limits

Shortly after Dave graduated from college in 1966, he and his friend Dick Sills drove to Everglades National Park. At that time the speed limit on many interstate highways in southern states was 75 mph, so that Dave and Dick were able to drive a stretch of 450 miles from Atlanta, GA, to Orlando, FL, in 6 hours without breaking any laws. Along the way they were passed by many drivers who were less observant of the speed limit. As one car whizzed by, Dick wondered aloud why anyone would feel the need to exceed a limit of 75 mph. Dave observed that people probably speed in a 75-mph zone for the same reasons that they speed in a 45-mph zone. Speeding cuts time off the trip, and it gives the speeder a feeling of "getting away with something."

Recently, Dave recalled this incident and decided to look more critically at his observation. Specifically, he asked whether speeding saves the same amount of time in a 45-mph zone as in a 75-mph zone. You can duplicate his calculations in Exercises 1–3. ∎

EXERCISES

1. (*Creating Models*) Let t represent the time, in hours, required to drive from Atlanta to Orlando at a speed of s miles per hour. Express t as a function $g(s)$.

2. (*Interpreting Mathematics*)

   a. Calculate the average rate of change in $g(s)$ with respect to s in each of the intervals [45, 60] and [75, 90].

   b. What does the average rate of change in $g(s)$ with respect to s represent physically?

   c. How does your result from part (a) enable you to answer Dave's question in **Speed Limits**?

3. (*Problem Solving*) Today the speed limit between Atlanta and Orlando is 70 mph.

   a. If you drive 85 mph, how much time will you save?

   b. If you drive 85 mph and are stopped for speeding, you will be delayed for about 20 minutes and the ticket will cost about $150. What is your net savings in time? What is the cost of that time in dollars per minute?

Although your function in Exercise 1 is not a reciprocal power function, it is closely related to $f(x) = \dfrac{1}{x}$. Over the next few pages you will discover several distinctive properties shared by reciprocal power functions and certain related functions. You will also see what information those properties can provide in a physical context like that of **Speed Limits**.

THREE VIEWS OF RECIPROCAL POWER FUNCTIONS

An Analytical View A reciprocal power function is described analytically by an equation $f(x) = \dfrac{1}{x^n}$ for some positive integer n. We can identify several properties of f by looking at its equation. In Exercise 4 you can explain why each of the following statements is true.

- The domain of f is $(-\infty, 0) \cup (0, \infty)$.
- The value of $f(x)$ is never 0, but can be made arbitrarily close to 0 by making $|x|$ sufficiently large. Similarly, $|f(x)|$ can be made arbitrarily large by making x sufficiently close to 0.
- The range of f is $(-\infty, 0) \cup (0, \infty)$ if n is odd, and $(0, \infty)$ if n is even.

EXERCISE   4. (*Making Observations*)

    a. Explain how each of the three preceding statements can be derived from the equation $f(x) = \dfrac{1}{x^n}$.

    b. Describe how each of the statements is reflected in the graphs of $y = \dfrac{1}{x}$ and $y = \dfrac{1}{x^2}$, shown in Figure 6-1.

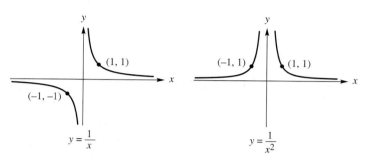

Figure 6–1

A Numerical View The preceding analytical observations can be supported by appropriate tables.

EXERCISES   5. For $f(x) = \dfrac{1}{x}$, complete each table.

a.

x	$f(x)$
1	
10	
100	
1000	

b.

x	$f(x)$
-1	
-10	
-100	
-1000	

c.

x	$f(x)$
1	
0.1	
0.01	
0.001	

d.

x	$f(x)$
-1	
-0.1	
-0.01	
-0.001	

6. Repeat Exercise 5 for $g(x) = \dfrac{1}{x^2}$.

7. (*Writing to Learn*) How do your tables in Exercises 5 and 6 support the statements preceding Exercise 4?

A Graphical View Figure 6-2 shows the graphs of the next few reciprocal power functions after $y = \dfrac{1}{x}$ and $y = \dfrac{1}{x^2}$.

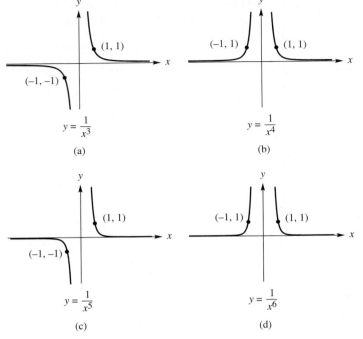

Figure 6–2

EXERCISE   **8.** (*Writing to Learn*) How do the graphs in Figure 6-2 support the statements preceding Exercise 4? How do they support your tables in Exercises 5 and 6?

Recall from Section 4-2 that an **asymptote** is a line that a graph approaches as one of the variables becomes arbitrarily large. As $|x|$ becomes large, the graph of $f(x) = \dfrac{1}{x^n}$ comes arbitrarily close to the x-axis, so that the x-axis is a **horizontal asymptote** for the graph. Similarly, the y-axis is a **vertical asymptote**.

INVERSE VARIATION

Now let's see what properties of reciprocal power functions are shared by functions of the form $y = \dfrac{a}{x^n}$. If a relationship is described by such an equation, we say that y **varies inversely with the nth power of x**. For example, Newton's law of gravitation states that the gravitational force between any two bodies in the universe is described by an equation $F = \dfrac{k}{d^2}$, where k is a constant depending on the mass of the two bodies and d is the distance between them. Because the force varies inversely with the square of the distance, Newton's law of gravitation is sometimes called an inverse square law. The attractive force between two magnets and the intensity of light from a fixed source are also described by inverse square laws.

In Exercise 1 you showed that the number of hours required to drive from Atlanta to Orlando at a speed of s miles per hour is $g(s) = \dfrac{450}{s}$. Thus the required time varies inversely with your speed. In Exercises 9–12 you can verify that this function has the properties just described for reciprocal power functions.

EXERCISES

9. How many hours would it take each of the following to go from Atlanta to Orlando?

    a. a jet at 600 mph

    b. a spaceship at 10,000 mph

    c. a light ray at 186,000 miles per *second*

10. (*Writing to Learn*) What feature on the graph of $g(s) = \dfrac{450}{s}$ is suggested by your result from Exercise 9?

11. How many hours would it take each of the following to go from Atlanta to Orlando?

    a. a pedestrian at 2.5 mph

    b. a snail at 0.01 mph

    c. a glacier at 0.1 miles per *year*

12. (*Writing to Learn*) What feature on the graph of $g(s) = \dfrac{450}{s}$ is suggested by your result from Exercise 11?

Inverse variations behave in a manner similar to other types of variations. In particular,

> If $y = \dfrac{a}{x^n}$, then whenever x is multiplied by a constant k, y is divided by k^n.

Informally,

> For inverse nth power variations, doubling x divides y by 2^n.

For example, if y varies inversely with x^3, then doubling x divides y by $2^3 = 8$. It is easy to tell whether a table fits an inverse nth power variation, and to find its equation when it does.

> In any table generated by $y = \dfrac{a}{x^n}$, the product $x^n y = a$ for every row.

EXAMPLE 1   *Deciding whether a table fits an inverse variation of a given degree*

To see whether the table

x	y
1	8100
2	506.25
3	100
5	12.96

fits an inverse variation of degree 4, evaluate the product $x^4 y$ for each row of the table.

$$(1)^4(8100) = 8100, \quad (2)^4(506.25) = 8100,$$

$$(3)^4(100) \ = 8100, \quad (5)^4(12.96) \ = 8100$$

The table fits the equation $y = \dfrac{8100}{x^4}$. ■

EXERCISES In Exercises 13–16, decide whether the table fits an inverse variation $y = \dfrac{a}{x^n}$ for the given value of n. If it does, write an equation relating x and y.

13.

x	y (n = 1)
3	28
4	21
6	14
7	12
12	7

14.

x	y (n = 1)
10	240
50	200
100	150
125	125
150	100

15.

x	y (n = 2)
2	225
3	100
5	36
6	25
10	9

16.

x	y (n = 3)
2	25
4	3.125
5	1.6
10	0.2
20	0.025

17. If P varies inversely with V and $P = 20$ when $V = 6$, what is the value of P when $V = 10$?

18. If N varies inversely with M and $N = 7.2$ when $M = 3$, what is the value of N when $M = 10.8$?

19. If s varies inversely with r^3 and $s = 250$ when $r = 1$, what is the value of s when $r = 5$?

20. If y varies inversely with x^4 and $y = 13$ when $x = 2$, what is the value of y when $x = 3$?

GRAPHICAL TRANSFORMATIONS

Not all rational graphs have shapes similar to those in Figure 6-2. In particular, not all have vertical or horizontal asymptotes. In Section 6-2 you will learn how to decide whether a rational graph has an asymptote of either type. For now, let's consider a class of rational functions whose graphs always have exactly one asymptote of each type. These are the **linear rational functions**. Every linear rational function has the form

$$f(x) = \frac{Ax + B}{Cx + D}$$

for some real numbers A, B, C, and D with $C \neq 0$. You can always obtain the graph of a linear rational function by transforming the basic graph of $y = \dfrac{1}{x}$ as in Example 2.

EXAMPLE 2 *Using transformations to graph a rational function*

Before sketching the graph of $f(x) = \dfrac{2x + 7}{x + 1}$, perform the division.

$$\begin{array}{r} 2 \\ x + 1 \overline{\smash{\big)}\ 2x + 7} \\ \underline{2x + 2} \\ 5 \end{array}$$

Thus $f(x) = \dfrac{5}{x + 1} + 2$. To sketch its graph, begin with the basic graph of

$y = \dfrac{1}{x}$, and make the changes

$$y = \frac{1}{x} \rightarrow y = \frac{5}{x} \rightarrow y = \frac{5}{x + 1} \rightarrow y = \frac{5}{x + 1} + 2$$

These changes stretch the graph vertically by a factor of 5 and then shift it 1 unit left and 2 units up, as in Figure 6-3.

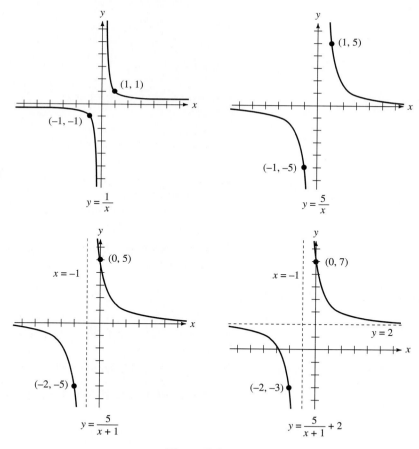

Figure 6–3

EXERCISES **21.** (*Writing to Learn*) In Example 2, what tells us that the graph of f has a vertical asymptote at $x = -1$ and a horizontal asymptote at $y = 2$?

In Exercises 22–25, sketch the graph of each function by transforming the basic graph of $y = \dfrac{1}{x}$. Identify the points obtained by moving (1, 1) and (−1, −1,), the vertical and horizontal asymptotes, and the domain and range of the function.

22. $B(x) = -\dfrac{1}{x-4}$ **23.** $q = \dfrac{1}{x+3} - 2$

24. $y = \dfrac{5}{x} + 1$ **25.** $R(x) = 7 - \dfrac{3}{x}$

It is possible to sketch the graph of a linear rational function and identify important graphical features without performing the division. In Exercise 26 you can verify the following facts.

The graph of $f(x) = \dfrac{Ax + B}{Cx + D}$ has

- a y-intercept at $\left(0, \dfrac{B}{D}\right)$ if $D \neq 0$

- an x-intercept at $\left(-\dfrac{B}{A}, 0\right)$ if $A \neq 0$

- a vertical asymptote at $x = -\dfrac{D}{C}$

- a horizontal asymptote at $y = \dfrac{A}{C}$

EXERCISES

26. (*Writing to Learn*) Explain why the graph of $f(x) = \dfrac{Ax + B}{Cx + D}$ must have the features described above. (*Hint*: Perform the division to locate the horizontal asymptote.)

In Exercises 27–30, sketch the graph of the function by locating its intercepts and asymptotes.

27. $y = \dfrac{3x + 7}{x + 1}$ **28.** $y = \dfrac{2x}{x - 4}$

29. $y = \dfrac{3 - x}{x - 2}$ **30.** $y = \dfrac{x}{3x + 6}$

ADDITIONAL EXERCISES

In Exercises 31–34, decide whether the table fits an inverse variation $y = \dfrac{a}{x^n}$ for the given value of n. If it does, write an equation relating x and y.

31.

x	$y\,(n = 2)$
0.01	1000
0.1	10
1	0.1
10	0.001

32.

x	$y\,(n = 1)$
2	105
3	70
5	42
6	35

33.

x	$y\ (n = 2)$
1	4096
2	1024
4	256
8	64

34.

x	$y\ (n = 3)$
1	1728
2	216
3	64
4	27

35. If b varies inversely with a and $b = 54$ when $a = 7$, what is the value of b when $a = 21$?

36. If H varies inversely with G and $H = 2$ when $G = \pi$, what is the value of H when $G = 2$?

37. If v varies inversely with t^3 and $v = 4$ when $t = 3$, what is the value of v when $t = 6$?

38. If y varies inversely with x^2 and $y = 25$ when $x = \sqrt{3}$, what is the value of y when $x = \sqrt{5}$?

39. (*Problem Solving*) Under certain conditions, the pressure exerted by a gas in a closed container varies inversely with the volume of the container. Suppose that the air in a balloon exerts an outward pressure of 20 lb/in^2.

   a. If the volume of the balloon is reduced to one-third of its present volume, how much outward pressure will be exerted by the air?

   b. If the balloon is designed to withstand an outward pressure of 100 lb/in^2, to what fraction of its present volume can it be reduced without breaking?

40. (*Problem Solving*) Your weight varies inversely with the square of your distance from the center of the earth. Suppose that you weigh 150 lb at sea level, 3960 miles from the center of the earth.

   a. How much would you weigh on the top of Mt. Everest, about 5.5 miles above sea level?

   b. At what altitude above sea level would you weigh 1 pound?

41. (*Problem Solving*) The attractive force between two magnets varies inversely with the square of the distance between them. If the attractive force between two particular magnets is 5 newtons at a distance of 3 meters, what is the force at a distance of 10 meters?

42. (*Problem Solving*) Tides on the earth's oceans are caused by the gravitational pull of external bodies, primarily the moon and sun. The tide-raising force of an external body varies directly with its mass and inversely with the cube of its distance from the earth. That is, the tide-raising force of a body of mass m at a distance d from the earth is $F = \dfrac{km}{d^3}$ for some constant k.

   a. If the moon were twice as far from the earth as it actually is, by what factor would its mass have to be increased to exert the same tide-raising force?

   b. The sun is 391 times as far from the earth as the moon, and its tide-raising force is 0.45 times that of the moon. What is the ratio of the mass of the sun to the mass of the moon?

In Exercises 43–46, sketch the graph of each function by transforming the basic graph of $y = \dfrac{1}{x}$. Identify the points obtained by moving $(1, 1)$ and $(-1, -1)$, $(-1, -1)$, the vertical and horizontal asymptotes, and the domain and range of the function.

43. $f(x) = \dfrac{10}{x - 2}$ **44.** $g(x) = -\dfrac{3}{x}$

45. $y = \dfrac{2}{x + 7}$ **46.** $z = 4 - \dfrac{1}{x + 3}$

In Exercises 47–50, sketch the graph of the function by locating its intercepts and asymptotes.

47. $y = \dfrac{x + 7}{x + 5}$ **48.** $y = \dfrac{4x - 9}{x - 3}$

49. $y = \dfrac{2x + 4}{x}$ **50.** $y = \dfrac{x}{2x + 4}$

In Exercises 51–54, sketch the graph of each function by transforming a reciprocal power graph $y = \dfrac{1}{x^n}$. Identify the points obtained by moving $(1, 1)$ and either $(-1, 1)$ or $(-1, -1)$, identify the vertical and horizontal asymptotes, and find the domain and range of the function.

51. $f(x) = \dfrac{6}{x^3}$ **52.** $T(x) = \dfrac{1}{x^4} + 5$

53. $u = 3 - \dfrac{1}{x^2}$ **54.** $K = \dfrac{5}{(x + 10)^4}$

6–2 DISCONTINUITIES AND END BEHAVIOR OF RATIONAL FUNCTIONS

In Section 6-1 you observed that the graphs of reciprocal power functions have two striking features not found in polynomial graphs.

- The graph consists of two pieces separated by a vertical asymptote. Thus there is a break, or **discontinuity**, in the graph, and the function is said to be **discontinuous**.
- The "tails" of the graph approach a horizontal asymptote.

In this section you will discover that not all rational functions have these graphical features. As we begin to explore the graphs of rational functions, it will be helpful to frame a few questions to guide our exploration.

- Is every rational function discontinuous at some point?
- Is every discontinuity a vertical asymptote?
- How does a table or an equation indicate the behavior of a rational function near a discontinuity?
- When a rational function models a physical problem, what physical information is provided by a discontinuity?
- Does the graph of every rational function have a horizontal asymptote?

- How does a table or an equation indicate the end behavior of a rational graph?
- When a rational function models a physical problem, what physical information is provided by the end behavior of its graph?

DISCONTINUITIES AND VERTICAL ASYMPTOTES

Let's focus on discontinuities first. Your experience in Section 6-1 suggests that a rational function is discontinuous only where its denominator is zero. Using calculus, it can be shown that this is actually the case. You can use this observation to answer our first two questions, and in each case the answer is no. The function in Exercise 1 is not discontinuous at any point, and the function in Exercise 2 has a discontinuity that is not a vertical asymptote.

EXERCISES   **1.** (*Making Observations*) For $f(x) = \dfrac{10}{x^2 + 1}$,

    a. Verify that the denominator is never 0, so that f is continuous for all values of x.

    b. Graph $y = f(x)$ in the window $[-10, 10]$ by $[-10, 10]$. How does the graph support your conclusion in part (a)?

    2. (*Making Observations*) For $g(x) = \dfrac{x^2 - 1}{x - 1}$,

    a. Use the equation to verify that g is discontinuous at $x = 1$.

    b. Graph $y = g(x)$ in the window $[-10, 10]$ by $[-10, 10]$. Does the graph have a vertical asymptote at $x = 1$?

You may have been surprised by the graph in Exercise 2. The function g is discontinuous at $x = 1$, and yet the graph appears to be an unbroken line. To resolve this apparent paradox, let's consider our third question from the beginning of the section. How does an equation or a table indicate the behavior of a rational function near a discontinuity? In Exercises 3–6 you can explore this question for $h(x) = \dfrac{1}{x - 1}$ and $g(x) = \dfrac{x^2 - 1}{x - 1}$. From Section 6-1 you know that $h(x)$ has a vertical asymptote at $x = 1$, and from Exercise 2 you know that $g(x)$ does not.

EXERCISES   **3.** (*Review*)

    a. Complete the following table for $h(x) = \dfrac{1}{x - 1}$.

x	$h(x)$
0	
0.9	
0.99	
0.999	
1	
1.001	
1.01	
1.1	
2	

b. How does the table indicate that the graph of $y = h(x)$ has a vertical asymptote at $x = 1$?

4. (*Making Observations*)

    a. Repeat Exercise 3a for $g(x) = \dfrac{x^2 - 1}{x - 1}$.

    b. How does the table indicate that the graph of $y = g(x)$ has no vertical asymptote at $x = 1$?

Your tables in Exercises 3 and 4 reveal something about the behavior of h and g near their discontinuities. When x is near 1, $|h(x)|$ is large, while $g(x)$ is near 2. In Exercises 5 and 6 you can see how the equations of the two functions support this observation.

5. (*Writing to Learn*) For $h(x) = \dfrac{1}{x - 1}$, explain without using any specific values of x why $|h(x)|$ must be large if x is near 1.

6. (*Making Observations*)

    a. Simplify the expression $g(x) = \dfrac{x^2 - 1}{x - 1}$, and verify that $g(x) = x + 1$ if $x \neq 1$.

    b. Why is $g(x)$ not equal to $x + 1$ when $x = 1$?

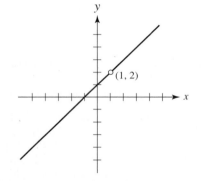

Figure 6–4

Both the table in Exercise 4 and the equation in Exercise 6 indicate that the graph of $y = g(x)$ coincides with that of $y = x + 1$ except that the point $(1, 2)$ is missing. Thus your calculator graph in Exercise 2 was not quite accurate. A more accurate graph is shown in Figure 6-4. A discontinuity such as the one in Figure 6-4 is called a **missing-point discontinuity**.

Your observations in Exercises 3–6 can be generalized.

A rational function $f(x) = \dfrac{N(x)}{D(x)}$ has a discontinuity at $x = a$ if $D(a) = 0$.

The discontinuity is

- a vertical asymptote as in Figure 6-5a if $\dfrac{N(x)}{D(x)}$ is in lowest terms and

- a missing-point discontinuity as in Figure 6-5b if $f(x)$ can be reduced to an expression whose denominator is not 0 at $x = a$.

EXERCISES   *In Exercises 7–10,*

    a. List all discontinuities for each function.

    b. Decide whether each discontinuity is a vertical asymptote or a missing point.

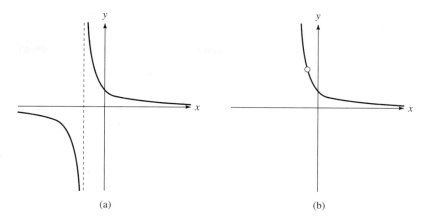

Figure 6–5

7. $y = \dfrac{x}{x + 4}$ **8.** $y = \dfrac{x}{x^2 - 16}$

9. $y = \dfrac{2x - 10}{x - 5}$ **10.** $y = \dfrac{2x - 10}{x^2 - 5x}$

Let's turn now to our fourth question from the beginning of the section. When a rational function models a physical problem, what physical information is provided by a discontinuity?

A MATHEMATICAL LOOKING GLASS

Average Cost (Fat Cats 7)

The T-shirt company started by your economics class in **Fat Cats on Campus** (page 28) has done so well that the class has decided to expand its operations. You plan to open a store near your school, in competition with several other companies. You must therefore sell your shirts at or near the prevailing local price of $9 each. You can make a profit as long as the selling price exceeds the average cost of producing each shirt. To determine profitable production levels you need to know your anticipated average cost, which will be a function of your production level.

Your profits from the semester were used to pay off the heat press you purchased to produce the shirts, so it is no longer considered a cost. However, you need to rent production and office space at $700 per month. Materials are still $3.50 per shirt. You estimate the advertising cost of selling x shirts per month to be $0.01x^2$. The average cost per shirt is the total cost of producing and selling the shirts, divided by the number of shirts sold. ■

EXERCISES

11. (*Interpreting Mathematics*) Near your campus you can advertise inexpensively through posters and word of mouth. To reach a wider market, you must use more expensive forms of advertising, such as newspaper ads or radio commercials. Why does this indicate that a quadratic expression is a better model for your advertising costs than a linear expression?

12. (*Writing to Learn*)

   a. Explain why the cost in dollars of producing x shirts per month is

$$C(x) = 700 + 3.50x + 0.01x^2$$

b. Explain why the average cost per shirt is

$$A(x) = \frac{700 + 3.50x + 0.01x^2}{x} \text{ dollars}$$

c. At what value of x is A discontinuous? Is the discontinuity a vertical asymptote or a missing point?

13. (*Making Observations*)

   a. Complete the following table for $y = A(x)$.

x	y
100	
10	
1	
0.1	

(Think of 0.1 shirts per month as 1 shirt every 10 months.) What physical information is provided by each row in the table?

   b. What does the table in part (a) tell you about the average cost at low production levels? How does your answer reflect the type of discontinuity in the average cost function?

14. Suppose that you are currently producing 300 shirts per month, and you are thinking about increasing your level of production to s shirts per month.

   a. (*Writing to Learn*) Explain why the average cost of producing each additional shirt will be

$$f(s) = \frac{C(s) - C(300)}{s - 300}$$

   b. Show that $f(s)$ can be written

$$f(s) = \frac{0.01s^2 + 3.50s - 1950}{s - 300}$$

   c. At what value of s is f discontinuous? Is the discontinuity a vertical asymptote or a missing point?

15. (*Interpreting Mathematics*) If you increase production by a small amount, the value of s will be near 300.

   a. The value of $f(s)$ will be near what number? [*Hint:* Simplify the expression for $f(s)$ by factoring the numerator.]

   b. If you are currently selling shirts for $9 each, does your result in part (a) indicate that you should increase production? Explain.

END BEHAVIOR AND HORIZONTAL ASYMPTOTES

Now let's explore the end behavior of rational graphs, starting with our fifth question from the beginning of the section. Does the graph of every rational function have a horizontal asymptote? In Exercises 16–18 you can confirm that the answer is no.

EXERCISES   **16.** (*Review*) Use the methods of Section 6-1 to sketch the graph of $f(x) = \dfrac{-6}{x-1}$ and identify its horizontal asymptote.

17. (*Review*)

    a. For $g(x) = \dfrac{x-6}{x-1}$, perform the division to verify that

$$g(x) = 1 - \frac{5}{x-1}$$

    b. Sketch the graph of $y = g(x)$ and identify its horizontal asymptote.

18. (*Making Observations*)

    a. For $h(x) = \dfrac{x^2 + x - 6}{x-1}$, perform the division to verify that

$$h(x) = x + 2 - \frac{4}{x-1}$$

    b. Graph $y = x + 2$ along with $y = h(x)$ in the window $[-10, 10]$ by $[-10, 10]$. What happens when $|x|$ is large?

19. (*Writing to Learn*)

    a. As $|x|$ becomes large, explain why any expression $\dfrac{\text{constant}}{x-1}$ becomes small.

    b. Why does this observation guarantee that when $|x|$ is large, the graphs of the functions in Exercises 16–18 approach the graphs $y = 0$, $y = 1$, and $y = x + 2$, respectively?

In Exercises 20–22 you can see how tables support your conclusions in Exercises 16–19.

20. (*Making Observations*) Complete the following table for $f(x) = \dfrac{-6}{x-1}$.

x	$f(x)$
10	
100	
1,000	
10,000	

    How does the table indicate the end behavior of the graph of $y = f(x)$?

21. (*Making Observations*) Repeat Exercise 20 for $g(x) = \dfrac{x-6}{x-1}$.

22. (*Making Observations*) Complete the following table for

$$h(x) = \frac{x^2 + x - 6}{x-1}$$

x	$x+2$	$h(x)$
10		
100		
1,000		
10,000		

How does the table indicate the end behavior of the graph of $y = h(x)$?

Exercises 16–22 suggest the answer to our sixth question from the beginning of the section. How does a table or an equation indicate the end behavior of a rational graph? By performing division, you can express every rational function as the sum of a polynomial function and a remainder. The polynomial part is 0 if the degree of the numerator in the rational function is less than that of the denominator, as in Exercise 16. This case is illustrated by Figure 6-6a. The polynomial part is a nonzero constant if the degrees are equal, as in Exercise 17, and has degree at least 1 if the degree of the numerator is larger than that of the denominator, as in Exercise 18. These cases are illustrated by Figures 6-6b and 6-6c. In each case, the remainder is a **proper rational function**. That is, its numerator has a smaller degree than its denominator.

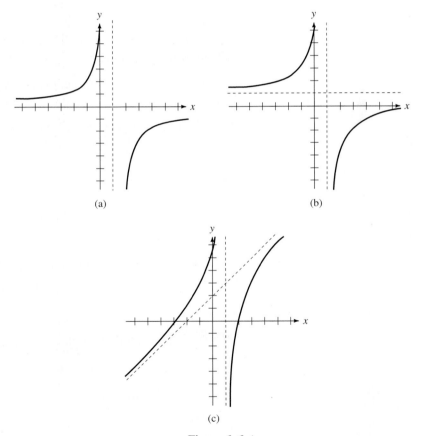

(a)

(b)

(c)

Figure 6–6

In Exercises 48–49 you can show that every proper rational function becomes small as $|x|$ becomes large, so that the end behavior of a rational function depends only on its polynomial part. Your observations are summarized in the following statement.

**The Highest Degree Theorem
(Rational Function Version)**

The end behavior of the graph of any rational function

$$f(x) = \frac{N(x)}{D(x)} = \frac{a_n x^n + a_{n-1} x^{n-1} + \cdots + a_1 x + a_0}{b_d x^d + b_{d-1} x^{d-1} + \cdots + b_1 x + b_0}$$

is the same as that of $y = \dfrac{a_n x^n}{b_d x^d}$. (In end behavior, only the terms of highest degree count.) In particular, the graph has

- a horizontal asymptote at $y = 0$ if $n < d$,

- a horizontal asymptote at $y = \dfrac{a_n}{b_d}$ if $n = d$, and

- no horizontal asymptote if $n > d$.

EXAMPLE 1 *Identifying a horizontal asymptote by using the Highest Degree Theorem*

a. The Highest Degree Theorem implies that the end behavior of

$$g(x) = \frac{8x^2 + 3}{4x^2 - 2x + 1}$$

is the same as that of $y = \dfrac{8x^2}{4x^2} = 2$. Thus the graph of $y = g(x)$ has a horizontal asymptote at $y = 2$.

b. The end behavior of

$$h(x) = \frac{8x^3 + 3}{4x^2 - 2x + 1}$$

is the same as that of $y = \dfrac{8x^3}{4x^2} = 2x$. Thus the left tail of the graph points down, and the right tail points up. In particular, there is no horizontal asymptote. ∎

EXERCISES *For each function in Exercises 23–26, use the Highest Degree Theorem to decide whether the graph has a horizontal asymptote. If there is one, find its equation. If not, decide whether each tail of the graph points up or down.*

23. $f(x) = \dfrac{x^2 + 4x}{3x^2 - 5}$ 　　　　　　**24.** $y = \dfrac{2x^3 - 16}{x^3 + 3x^2}$

25. $A(t) = \dfrac{4t - t^5}{6t^3 - 4t + 10}$

26. $Z = \dfrac{100s^2 + 37s + 2751}{s^4 - s - 1}$

Now let's turn to our seventh and final question from the beginning of the section. When a rational function models a physical problem, what physical information is provided by the end behavior of its graph?

A MATHEMATICAL
LOOKING GLASS
Robert Arnold

Taos, NM, is a famous ski resort and artist's colony. One of Taos' nationally known resident artists is Robert Arnold, a painter of southwestern scenes. To make his works accessible to a wide audience, he has concentrated on marketing his art since 1987. By 1990, he had produced about 500 original and limited-edition pieces. Since then his reputation has grown, and he has expanded his market by producing affordable open-edition items such as note cards, prints, and images on ceramic tiles, which are sold by several distributors. Since 1990 he has produced about 50 original and limited-edition pieces and about 2000 open-edition pieces each year. ∎

EXERCISES

27. (*Creating Models*) Let *t* denote time in years since 1990.

   a. Write a function $P(t)$ to express the total number of open-edition pieces produced prior to time *t*.

   b. Write a function $Q(t)$ to express the total number of original and limited-edition pieces produced by Robert Arnold prior to time *t*, including those produced before 1990.

   c. Write an expression for the rational function $f(t) = \dfrac{P(t)}{Q(t)}$. What does $f(t)$ represent in physical terms?

28. (*Interpreting Mathematics*) The graph of the function *f* in Exercise 27c has a horizontal asymptote.

   a. Find its equation.

   b. What physical information does it provide?

ADDITIONAL EXERCISES   *In Exercises* 29–44,

   a. List the values of *x* where each function is discontinuous.

   b. Decide whether each discontinuity is a vertical asymptote or a missing point.

   c. Decide whether the graph has a horizontal asymptote. If there is one, find its equation. If not, decide whether each tail of the graph points up or down.

29. $y = \dfrac{x + 8}{x - 4}$

30. $y = \dfrac{2x + 3}{3x + 2}$

31. $y = \dfrac{x^2 - 4x}{x}$

32. $y = \dfrac{x^2 - 4x}{x - 4}$

33. $y = \dfrac{6 - 2x}{x + 3} + 1$

34. $y = \dfrac{6 - 2x}{x - 3} - 1$

35. $y = \dfrac{1}{x^2 - 3}$

36. $y = \dfrac{1}{x^2 + 3}$

37. $y = \dfrac{x - 1}{x^2 - 8x + 7}$

38. $y = \dfrac{x - 2}{x^2 - 8x + 7}$

39. $y = \dfrac{x^2 + 4x + 4}{x + 2}$

40. $y = \dfrac{x + 2}{x^2 + 4x + 4}$

41. $f(x) = \dfrac{x^4 - x^3}{2 - x^4}$

42. $J(x) = \dfrac{x^2 - 1}{x}$

43. $R(x) = \dfrac{10x^2}{x^4 + 16}$

44. $Q(x) = \dfrac{6x^3 + 4x - 3}{2x^3 - x^2 - x}$

45. (*Interpreting Mathematics*) The function

$$f(t) = \frac{2000t}{50t + 500}$$

in Exercise 27 has a discontinuity. Where does it occur, and what information, if any, does it provide?

46. (*Interpreting Mathematics*) Determine the end behavior of the average cost function

$$A(x) = \frac{700 + 3.50x + 0.01x^2}{x}$$

in Exercise 12. What physical information, if any, does it provide?

47. A cylindrical 12-ounce soft drink can must have a volume of 21.7 in³. Let r denote its radius in inches.

    a. (*Creating Models*) Express the height of the container in terms of r.

    b. (*Creating Models*) Express the surface area of the container as a function $S(r)$.

    c. (*Interpreting Mathematics*) Where is S discontinuous? What physical information, if any, does the discontinuity provide?

    d. (*Interpreting Mathematics*) Describe the end behavior of the graph of S. What physical information, if any, does it provide?

48. (*Writing to Learn*)

    a. Verify that the function

$$f(x) = \frac{2x^4 + x^3 + 5x - 10}{2x^2 + x - 4}$$

    can be written

$$f(x) = x^2 + 2 + \frac{3x - 2}{2x^2 + x - 4}$$

    b. Explain why the remainder can be written

$$\frac{\dfrac{3}{x} - \dfrac{2}{x^2}}{2 + \dfrac{1}{x} - \dfrac{4}{x^2}}$$

    c. Explain why the expression in part (b) is small if $|x|$ is large. (*Hint:* Which terms in the numerator and denominator become small as $|x|$ becomes large?)

49. (*Writing to Learn*)

    a. Graph $y = f(x)$ from Exercise 48 and $y = x^2 + 2$ in a window that shows the end behavior of both graphs. Why does Exercise 48 suggest that the two graphs are close when $|x|$ is large?

    b. Let $P(x)$ denote the polynomial obtained when the numerator of the rational expression $f(x)$ is divided by its denominator. Explain why the graph of $y = f(x)$ approaches the graph of $y = P(x)$ when $|x|$ is large.

SUPPLEMENTARY TOPIC

Slant Asymptotes

In Exercise 18 you showed that the graph of $h(x) = \dfrac{x^2 + x - 6}{x - 1}$ approaches that of $y = x + 2$ when $|x|$ is large. The line $y = x + 2$ is called a **slant asymptote** for the graph of $y = h(x)$, shown in Figure 6-7.

50. (*Writing to Learn*) If $f(x) = \dfrac{N(x)}{D(x)}$ and the degree of N exceeds the degree of D by 1, explain why the graph of $y = f(x)$ has a slant asymptote.

In Exercises 51–58, find the slant asymptote for each function. Graph the function and its asymptote together in the window $[-10, 10]$ by $[-10, 10]$.

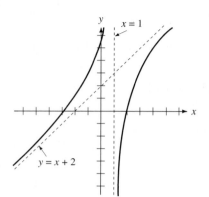

51. $y = \dfrac{x^2 + 1}{x}$

52. $y = \dfrac{x^2 + 2x}{x - 2}$

53. $y = \dfrac{6x^2}{2x - 3}$

54. $y = \dfrac{6x^2}{3 - 2x}$

55. $y = \dfrac{2x^3}{x^2 - x + 1}$

56. $y = \dfrac{x^3 + x^2 - 4x}{4 - x^2}$

57. $y = \dfrac{x^4 - 2x^2 + 1}{2x^3 + 8x}$

58. $y = \dfrac{0.1x^6}{1 - x^5}$

Figure 6–7

6-3   SOLVING RATIONAL INEQUALITIES

In **Average Cost** (page 307) you discovered that it costs your T-shirt company an average of

$$A(x) = \frac{700 + 3.50x + 0.01x^2}{x}$$

dollars per shirt to produce x shirts per month. You can make a profit as long as you can sell the shirts for more than it costs to produce them. Since your selling price is fixed at \$9, your profitable levels of production are the solutions of the inequality

$$\frac{700 + 3.50x + 0.01x^2}{x} < 9$$

In this section you will learn how to solve inequalities involving rational functions. Since the first step in solving any inequality is to solve an equation, let's begin by recalling that rational equations can be solved graphically as in Section 1-4 or analytically as in Section A-17. Example 1 reviews the analytical method.

EXAMPLE 1   *Solving a rational equation*

To solve

$$\frac{5x + 1}{x^2 - 1} = 2$$

begin by multiplying both sides by $x^2 - 1$.

$$5x + 1 = 2(x^2 - 1)$$

$$2x^2 - 5x - 3 = 0$$

$$(2x + 1)(x - 3) = 0$$

$$x = -\frac{1}{2}, 3$$

In solving the equation, you multiplied both sides by the variable quantity $x^2 - 1$. Therefore, your solutions are valid if and only if they do not produce a value of 0 in the denominator of the original equation. A quick check shows that both solutions are valid. ■

EXERCISES   *In Exercises 1–4, solve the equation.*

1. (*Review*) $\dfrac{2x + 7}{x - 1} = 5$

2. (*Review*) $\dfrac{x - 2}{x} = \dfrac{x + 4}{2x + 3}$

3. (*Review*) $\dfrac{2x - 1}{x - 5} - 3 = \dfrac{9}{x - 5}$

4. (*Review*) $\dfrac{2x - 1}{x - 5} - x = \dfrac{9}{x - 5}$

5. a. Solve the equation $\dfrac{700 + 3.50x + 0.01x^2}{x} = 9$ from ***Average Cost***.

b. (*Interpreting Mathematics*) What physical information is provided by the solution?

As with polynomial inequalities, the next step in solving a rational inequality is to rewrite it in one of the forms $f(x) \geqslant 0, f(x) > 0, f(x) \leqslant 0, f(x) < 0$. When f is a rational function, however, you also need to find the discontinuities of f. This is done in Example 2. In Example 3 you will see why the extra step is needed.

EXAMPLE 2   *Preparing to solve a rational inequality*

To prepare to solve

$$\frac{5x + 1}{x^2 - 1} \geqslant 2$$

begin by rewriting it as

$$\frac{5x + 1}{x^2 - 1} - 2 \geqslant 0$$

The function

$$f(x) = \frac{5x + 1}{x^2 - 1} - 2$$

is discontinuous when the denominator $x^2 - 1$ is zero, that is, when $x = \pm 1$. ■

The information from Examples 1 and 2 can be used to solve the inequality either graphically or analytically. Example 3 shows the graphical solution, and Examples 4 and 5 show two analytical methods of solution.

A GRAPHICAL METHOD

EXAMPLE 3   *Solving a rational inequality graphically*

To solve

$$\frac{5x + 1}{x^2 - 1} \geq 2$$

[–5, 5] by [–5, 5]

Figure 6–8

begin by rewriting it as in Example 2. Then graph $f(x) = \dfrac{5x + 1}{x^2 - 1} - 2$, as in Figure 6-8.

The inequality is true at all x-values where the graph is on or above the x-axis. Figure 6-8 indicates that the endpoints of the solution intervals correspond to the x-intercepts and the discontinuities on the graph. The x-intercepts are the solutions to $f(x) = 0$. From Example 1, these are $x = -\dfrac{1}{2}$ and $x = 3$. From Example 2, the discontinuities occur at $x = \pm 1$. The inequality is true at both x-intercepts, but false at both discontinuities. The solution is $\left(-1, -\dfrac{1}{2} \right] \cup (1, 3]$. ∎

EXERCISES   *In Exercises 6–9, solve the inequality graphically.*

6. $\dfrac{x + 1}{x^2 - 9} \leq 0$ **7.** $\dfrac{(x + 3)(x - 7)}{x - 2} \geq 0$

8. $\dfrac{1}{(x + 4)^2} < 1$ **9.** $\dfrac{1}{x} > \dfrac{1}{x^2}$

10.   **a.** Solve the inequality $\dfrac{700 + 3.50x + 0.01x^2}{x} < 9$ from *Average Cost*.

      b. (*Interpreting Mathematics*) What physical information is provided by the solution?

THE TEST-VALUE METHOD

Like polynomial inequalities, rational inequalities can be solved by the test-value method, but because of the possible presence of discontinuities, the method must be modified as shown in Example 4.

EXAMPLE 4   *Solving a rational inequality by the test-value method*

To solve

$$\frac{5x + 1}{x^2 - 1} \geq 2$$

define $f(x) = \dfrac{5x + 1}{x^2 - 1} - 2$, so that the inequality is $f(x) \geq 0$.

STEP 1:   Solve $f(x) = 0$ as in Example 1. The solutions are $x = -\dfrac{1}{2}$ and $x = 3$.

STEP 2: Determine where f is discontinuous as in Example 2. The discontinuities occur when $x = \pm 1$.

STEP 3: As Figure 6-8 showed, the sign of $f(x)$ can change at a value of x where $f(x)$ is either zero or discontinuous. Indicate these values on a number line.

STEP 4: Evaluate $f(x)$ at a test value in each interval.

As in Example 3, the inequality is true when $f(x) = 0$ and false when $f(x)$ is discontinuous. The solution is $\left(-1, -\dfrac{1}{2}\right] \cup (1, 3]$. ∎

EXERCISES

11. As parts (a)–(d) of this exercise, solve the inequalities in Exercises 6–9 by the test-value method.

12. (*Writing to Learn*) Compare the graphical and test-value methods of solving rational inequalities. Which method do you prefer to use, and why? Does your choice depend on the inequality you are solving? If so, how?

THE SCAN METHOD

Knowing the end behavior of rational functions, you can use the scan method to solve rational inequalities. Example 5 indicates how the method must be modified.

EXAMPLE 5 *Solving a rational inequality by the scan method*

Suppose $f(x) = \dfrac{x - 4}{x^2(x - 6)}$. To solve $f(x) \le 0$,

STEP 1: Solve $f(x) = 0$ to obtain $x = 4$.

STEP 2: Determine that f is discontinuous when $x = 0$ or 6.

STEP 3: Indicate these x-values on a number line.

STEP 4: Now you must find the sign of $f(x)$ in each of the four intervals shown. Find the sign of $f(x)$ on $(-\infty, 0)$ by looking at the behavior

of the left side of its graph. If the denominator of $f(x)$ is multiplied out, the terms of highest degree in its numerator and denominator are x and x^3, respectively. Thus the graph of $f(x)$ has the same left-end behavior as that of $\dfrac{x}{x^3} = \dfrac{1}{x^2}$, so that $f(x) > 0$ on $(-\infty, 0)$.

STEP 5:   Next, find the sign of $f(x)$ in the intervals $(0, 4)$, $(4, 6)$, and $(6, \infty)$ by *scan*ning the zeros and discontinuities.

The discontinuity at $x = 0$ corresponds to the factor x^2 in the denominator of f. This factor has multiplicity 2, so the sign of $f(x)$ does not change there and remains positive in $(0, 4)$.

Since the zero at $x = 4$ has multiplicity 1, the sign of $f(x)$ changes there and becomes negative in $(4, 6)$.

The discontinuity at $x = 6$ corresponds to the factor of $x - 6$ in the denominator of f. This factor has multiplicity 1, so the sign of $f(x)$ changes there and becomes positive in $(6, \infty)$.

Thus $f(x) < 0$ in $(4, 6)$. Since $f(x) = 0$ at the zero $x = 4$, the solution to $f(x) \leq 0$ is $[4, 6)$.

The solution is supported by the graph of $y = f(x)$, shown in Figure 6-9.

Figure 6–9

EXERCISES   **13.** As parts (a)–(d) of this exercise, solve the inequalities in Exercises 6–9 by the scan method. (You also solved them by the test-value method in Exercise 11.)

14. (*Writing to Learn*) Compare the test-value and scan methods of solving rational inequalities. Which method do you prefer to use, and why? Does your choice depend on the inequality you are solving? If so, how?

ADDITIONAL EXERCISES   *In Exercises 15–32, solve the inequality by any method.*

15. $\dfrac{x^2 + 3x}{x - 4} > 0$

16. $\dfrac{(x - 1)^2}{x^2 - 25} < 0$

17. $\dfrac{x^2 + 3x - 1}{x + 2} \geq 1$

18. $\dfrac{x^2 + 3x - 1}{x + 2} \geq x$

19. $\dfrac{x}{x - 1} < 0$

20. $\dfrac{x}{(x - 1)^2} < 0$

21. $\dfrac{x^2 + x - 6}{x} \geq 1$

22. $\dfrac{x^2 + x - 6}{x} \geq 2$

23. $\dfrac{(x + 20)^2(x - 10)^3}{(x + 10)(x - 30)^4} > 0$

24. $\dfrac{1}{x} \leq 0$

25. $\dfrac{x}{x^2 + 1} \geq 0$

26. $\dfrac{x}{x^2 + 1} \geq 2$

27. $\dfrac{x^3 + 2x - 1}{x^3} \geq 1$

28. $\dfrac{x^3 + x^2 - 2x}{x^2 - 16} \leq 0$

29. $\dfrac{(x - 1)(x - 7)}{(x - 3)^4} > 0$

30. $\dfrac{x^3}{9 - x^2} < 0$

31. $\dfrac{x^2(x + 5)^2}{x + 2} \leq 0$

32. $\dfrac{(x + 7)(x - 4)}{x(x - 8)^3} \geq 0$

33. (*Problem Solving*) In Exercise 47 of Section 6-2 you found that a cylindrical 12-ounce soft drink can of radius r inches has a surface area of

$$S(r) = \pi r^2 + \frac{43.4\pi}{r}$$

square inches.

a. For what values of r is the surface area of the can less than 100 in^2?

b. For what value of r is the surface area the smallest?

c. The can in part (b) costs the least to produce, since it uses the least material. How does its shape differ from that of most soft drink cans? List several reasons why soft drink companies do not use cans of this optimal shape.

34. (*Problem Solving*) The intensity I of light from a fixed source at a distance d from the source is $\dfrac{I_0}{d^2}$, where I_0 is a constant reference intensity.

In this problem, assume that units of measurement have been chosen so that $I_0 = 1$. Figure 6-10 shows part of a line of lamps designed to illuminate a stretch of highway near an interchange.

Figure 6–10

a. Show that the intensity of light reaching the car from the two lamps in Figure 6-10 is

$$f(x) = \frac{1}{x^2 + h^2} + \frac{1}{(d - x)^2 + h^2}$$

(For simplicity, ignore the light reaching the car from the more distant lamps.)

b. Suppose the lamp poles are planned to be 40 feet high and 200 feet apart. Graph $y = f(x)$ in the window $[0, 200]$ by $[0, 0.001]$. Your graph suggests that the light intensity is lowest halfway between the poles. Show that in general the intensity halfway between the poles is

$$\frac{8}{4h^2 + d^2}$$

c. Suppose $d = 200$ and $h = 40$ as in part (b). At what points between the two poles is the intensity of light reaching the car less than 0.0002? (*Hint*: Solve the inequality graphically.)

d. The intensity of light along the road can be increased by lowering the poles and/or placing them closer together. If the intensity must be at least 0.0002 at each point for safety, explain why lowering the poles alone will not be satisfactory.

CHAPTER REVIEW

Complete Exercises 1–21 (Writing to Learn) before referring to the indicated pages.

WORDS AND PHRASES

In Exercises 1–8, explain the meaning of the words or phrases in your own words.

1. (page 295)   **rational function**

2. (page 295)   **reciprocal power function**

3. (page 298)   **horizontal asymptote, vertical asymptote**

4. (page 298)   ***y* varies inversely with the *n*th power of *x***

5. (page 300)   **linear rational function**

6. (page 304)   **discontinuity, discontinuous** function

7. (page 306)   **missing-point discontinuity**

8. (page 310)   **proper rational function**

IDEAS

9. (page 295)   How do the equations and graphs of polynomial functions differ from those of other rational functions?

10. (pages 296, 297)   Describe the graph of $y = $ for $\frac{1}{x^n}$ for all integers $n \geq 1$. What are the domain and range of the function in each case?

11. (page 298)   Why are certain laws of physics referred to as inverse square laws?

12. (page 299)   How can you tell whether a table fits an inverse variation of a given degree? Make up a table and use it to illustrate the process.

13. (page 301)   Describe the process of obtaining the graph of a linear rational function by transforming the basic graph $y = \dfrac{1}{x}$. Illustrate the process with a specific function.

14. (page 302)   Describe the process of sketching the graph of a linear rational function by locating its intercepts and asymptotes. Illustrate the process with a specific function.

15. (page 304)   Under what conditions is a rational function discontinuous? Give an example of a rational function $f(x)$ that is continuous for all values of x.

16. (pages 305, 306)   What are the two possible types of discontinuity for a rational function? How do tables, equations, and graphs indicate the presence of each type? Make up specific functions to illustrate your explanation.

17. (pages 308–311)   What does the Highest Degree Theorem say about the end behavior of a rational function? Make up three functions to illustrate the three cases described by the theorem.

18. (pages 308–311, 313)   Explain informally why the Highest Degree Theorem is true.

19. (pages 314–316)   To solve the rational inequality $f(x) \geqslant 0$, why is it necessary to find the zeros of f? Why is it necessary to find the discontinuities of f?

20. (pages 316, 317)   Describe the process of solving a rational inequality by the test-value method. Make up an inequality to illustrate your explanation. What is the difference between the test-value method for polynomials and for rational functions?

21. (pages 317, 318)   Describe the process of solving a rational inequality by the scan method. Make up an inequality to illustrate your explanation. What steps in the scan method are the same as in the test-value method? What steps are different?

CHAPTER 7

EXPONENTIAL AND LOGARITHMIC FUNCTIONS

7–1 EXPONENTIAL FUNCTIONS

Exponential functions are arguably the most useful mathematical functions. They are used to describe phenomena that grow at an increasingly rapid rate. (You may have heard the phrase *exponential growth*.) Exponential functions are used to calculate the interest on a bond, estimate the age of an artifact, and place the frets on the neck of a guitar. In this chapter you will learn how they are used to model a variety of problems. Let's begin with the problem of global overpopulation.

A MATHEMATICAL LOOKING GLASS
Standing Room Only

Rudy Rucker observes in *Mind Tools*, Houghton Mifflin, 1987.

> Overpopulation can only lead to famine, war, and plague. . . . As we seem unable to change our patterns of reproduction, shouldn't we just resign ourselves to a lot more plagues, wars, and famines in the future? Why get so upset every time a lot of people die? . . . no one is upset if a tenth of the crabs in Chesapeake Bay are steamed to death every year. . . . Why don't we feel this way about famine in Ethiopia, typhoons in Bangladesh, or an earthquake in Mexico? Most people are about the same, so there's no great global information loss if a few million disappear. . . . The argument seems perfectly logical, but if I accept it, I'm put in a position of having to accept the unacceptable statement: "It makes no real difference if all my friends and relatives drop dead tomorrow." The difference for me between people and crabs seems to be that *I* am a person, and I care about people as individuals. Perhaps a crab sees things differently.

Professor Rucker points out a serious moral dilemma that continued population growth forces upon us. Since we do not want to steam people like crabs, we must learn to control the population and/or generate new sources of food and air. A first step toward solving the population problem is to understand both its scope and its urgency. A mathematical model can provide the needed insight. Let's see if we can construct one, and use it to figure out how long it will take the present population to double. You will need to know that for the last several decades world population has grown at the rate of about 2% per year, reaching 6 billion in 1992. ∎

To answer the question about doubling time, it will be convenient to express the population P in billions as a function of time t in years since 1992. As a first step in constructing the function we can make a table, such as Table 29. An annual increase of 2% means that each year the population is multiplied by 1.02. Thus each value of P in Table 29 is 1.02 times the previous value.

TABLE 29

t	P
0	6
1	6.12
2	6.2424
3	6.367248
4	6.49459296

Intuition might tell you that a 2% increase each year means a 100% increase (doubling the population) in 50 years. However, Table 29 shows that each successive year's increase is 2% *of an increasingly larger number*, so that the doubling time should be less than 50 years. Since it would be tedious to continue the table until the value of P reached 12, let's construct an equation to represent the population function.

EXERCISE   **1.** (*Writing to Learn*)

   a. Explain why the entry opposite $t = 4$ in Table 29 can be written as $6(1.02)^4$.

   b. Explain why the relationship of time to population can be described by the equation $P = 6(1.02)^t$.

To find the doubling time, we need to solve the equation $6(1.02)^t = 12$. You will learn how to solve this equation analytically in Section 7-4. For now, let's solve it graphically.

EXERCISE   **2.** (*Interpreting Mathematics*) Graph the function $y = 6(1.02)^t$ in the window $[0, 50]$ by $[0, 18]$.

   a. How does the graph support the observation that each successive year's increase is 2% of an increasingly larger number?

   b. Estimate, to the nearest year, the number of years required for the population to double.

The function $f(t) = 6(1.02)^t$ is an example of an **exponential function**, that is, a function for which increases of equal *size* in the independent variable produce increases (or decreases) of equal *percentage* in the dependent variable. Let's explore exponential functions from numerical, graphical, and analytical perspectives.

A NUMERICAL VIEW OF EXPONENTIAL FUNCTIONS

Tables of Exponential Functions The definition just given tells you how to decide whether a table with equally spaced x-values fits an exponential function $y = f(x)$.

A table with equally spaced x-values fits an exponential function if and only if the ratio of consecutive y-values is constant.

EXERCISES *In Exercises 3–6, decide whether the table fits an exponential function. If it does, specify the ratio between consecutive values of y.*

3.

x	y
0	$\dfrac{2}{3}$
1	2
2	6
3	18

4.

x	y
0	64
2	16
4	4
6	1

5.

x	y
-3	0
0	16
3	256
6	1296

6.

x	y
1	8
2	-12
3	18
4	-27

FYI A table with only two rows always fits an exponential function. Therefore, look for more evidence before concluding that it was generated by an exponential function.

Predicting Graphical Features from Tables A table generated by an exponential function can suggest some graphical features that you can confirm later in this section. Table 30 was generated by an exponential function $y = f(x)$ for which $f(0) = 1$ and y triples (increases by 200%) whenever x increases by 1.

TABLE 30

x	y
0	1
1	3
2	9
3	27
4	81

EXERCISES

7. (*Making Observations*)

a. Extend Table 30 to show values of y for $x = 5, 6, 7, 8, 9$, and 10. What does your table indicate about values of y when x is large?

b. Extend Table 30 to show values of y for $x = -1, -2, \ldots, -10$. What does your table indicate about values of y when $x < 0$ and $|x|$ is large?

8. (*Making Observations*) Suppose that $y = g(x)$ is a function for which $g(0) = 1$ and y is cut in half (decreases by 50%) whenever x increases by 1. Make a table to show values of y for $x = -10, -9, -8, \ldots, 8, 9, 10$.

(*Hint*: Start at $x = 0$ and work in both directions.) What does your table indicate about values of y when x is large? When $x < 0$ and $|x|$ is large?

9. (*Making Observations*) Based on your results from Exercises 7 and 8, make guesses about

   a. the range of an exponential function

   b. the end behavior on the graph of a decreasing exponential function

   c. the end behavior on the graph of an increasing exponential function

Over the next few pages you can see how accurate your guesses are.

AN ANALYTICAL VIEW OF EXPONENTIAL FUNCTIONS

The Equation of an Exponential Function Exercise 1 suggests that every exponential function can be described by an equation $y = Ab^x$. To determine if that is the case, let's see if we can find such an equation for a table that fits an exponential function. Example 1 illustrates the method.

EXAMPLE 1   *Fitting an exponential function to a table*

The following table fits an exponential function $y = f(x)$.

x	y
2	4
4	36
6	324
8	2916

To find an equation $f(x) = Ab^x$ that fits the table, use the first two rows of the table to obtain

$$Ab^2 = 4$$

$$Ab^4 = 36$$

You now have a system of two equations in the two variables A and b. If the equations were linear, you could solve the system by the method of substitution. Although this system is not linear, the method still works. (By looking at linear systems to gain insight into nonlinear systems, you are applying the strategy of *Examining a Related Problem*.)

Solve the first equation for A.

$$A = \frac{4}{b^2}$$

Replace A by $\frac{4}{b^2}$ in the second equation.

$$\left(\frac{4}{b^2}\right)(b^4) = 36$$

$$b^2 = 9$$

$$b = 3$$

Now $A = \frac{4}{b^2} = \frac{4}{9}$, so $f(x) = \frac{4}{9}(3^x)$.   ∎

EXERCISES

10. Verify that the table in Example 1 fits an exponential function.

11. In Exercises 3–6, if the table fits an exponential function, find its equation.

The method of Example 1 can be shown to produce an equation $y = Ab^x$ for every exponential table. Thus another definition of an **exponential function** is one that can be written $f(x) = Ab^x$ for some constants A and b with $A \neq 0$, $b \neq 1$, and $b > 0$.

These restrictions are imposed on A and b because

- If $A = 0$ or $b = 0$, then $f(x) = 0$ for all x, so that f is a constant function.
- If $b = 1$, then $f(x) = A \cdot 1^x = A$ for all x, so that f is again a constant function.
- If $b < 0$, then $f(x)$ is not always real. For example, if $A = 1$ and $b = -4$, then $f\left(\dfrac{1}{2}\right) = (-4)^{1/2} = \sqrt{-4} = 2i$, which is not a real number.

EXERCISES

12. (*Making Observations*) For each function, evaluate $f(0)$, $f(1)$, $f(2)$, and $f(3)$, and confirm that the ratio of consecutive values of f is constant.
   a. $f(x) = 3(4^x)$ **b.** $f(x) = -2(0.4)^x$

13. (*Writing to Learn*) How does an equation $f(x) = Ab^x$ indicate that increases of equal size in x produce increases of equal percentage in $f(x)$? If x increases by 1, what is the ratio of the new value of f to the previous value?

Irrational Exponents By now it may have occurred to you to ask what b^x means if x is irrational. For example, your calculator tells you that $2^\pi \approx 8.82497782708. . .$, but where does that value come from? Let's explore.

Figure 7-1a shows the points obtained from a table of the function $f(x) = 2^x$ for the values $x = 0, 1, 2,$ and 3.

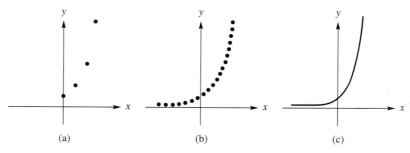

(a)   (b)   (c)

Figure 7–1

You already know that $f(x)$ can be defined for all rational values of x, since, for example, $2^{-3} = \dfrac{1}{8}$ and $2^{1/2} = \sqrt{2}$. Figure 7-1b suggests the graph of $f(x) = 2^x$ for all rational values of x. To obtain the unbroken curve in Figure 7-1c, we must define 2^x for irrational values of x. Let's see how to define 2^π so that the graph of $f(x) = 2^x$ is unbroken when $x = \pi$.

A graphing calculator can illustrate the definition. As Figure 7-2a shows, π is in the interval $(3, 4)$, so we should define 2^π to be in the interval $(2^3, 2^4) = (8, 16)$. Zooming in, Figure 7-2b shows that π is in $(3.1, 3.2)$, so 2^π should be in $(2^{3.1}, 2^{3.2}) \approx (8.57, 9.19)$. Similarly, Figure 7-2c shows that π is in $(3.14, 3.15)$, so 2^π should be in $(2^{3.14}, 2^{3.15}) \approx (8.82, 8.88)$.

Figure 7–2

Repeated zooming in makes it appear that there is a number (approximately 8.82497782708) smaller than 2^x for any rational $x > \pi$ and larger than 2^x for any rational $x < \pi$. In a more advanced course, it can be proved that exactly one such number exists and can, therefore, be defined as 2^π. Values of b^x for other bases and other irrational exponents are defined in a similar manner.

EXERCISES

14. (*Extension*) Estimate the value of $5^{\sqrt{3}}$ by continuing the following table, using $\sqrt{3} = 1.732050\ldots$

$$5^1 = \qquad 5^2 =$$

$$5^{1.7} = \qquad 5^{1.8} =$$

$$5^{1.73} = \qquad 5^{1.74} =$$

Continue until you can estimate $5^{\sqrt{3}}$ with an error of no more than 0.01. Confirm your estimate by evaluating $5^{\sqrt{3}}$ on your calculator.

15. (*Extension*) Use tables similar to the one in Exercise 14 to estimate each of the following with an error of no more than 0.01.
    a. $3^{\sqrt{6}}$ **b.** π^π

Predicting Graphical Features from Equations In Exercise 9 you made guesses about several graphical features of functions $f(x) = b^x$. Let's see if we can predict the same features analytically. Specifically, let's see if the equation supports the following statements about an exponential function $f(x) = b^x$.

- The domain of f is $(-\infty, \infty)$.
- The range of f is $(0, \infty)$.
- If $b > 1$, f is increasing. The value of f becomes arbitrarily large when x is large and becomes arbitrarily close to 0 when $x < 0$ and $|x|$ is large.
- If $0 < b < 1$, f is decreasing. The value of f becomes arbitrarily close to 0 when x is large and becomes arbitrarily large when $x < 0$ and $|x|$ is large.

Based on our discussion of irrational exponents, the expression b^x is defined for all real values of x. Therefore, the first statement in the preceding bulleted list is true. You can explore the rest in Exercises 16 and 17.

EXERCISES   16. (*Writing to Learn*) Suppose $f(x) = 2^x$.

  a. What can you say about the value of $f(x)$ if x is large? Explain.

  b. What can you say about the value of $f(x)$ if $x < 0$ and $|x|$ is large? Explain.

  c. Which of the statements preceding this exercise are supported by your observations in parts (a) and (b)?

  17. (*Writing to Learn*) Repeat Exercise 16 for $g(x) = \left(\dfrac{1}{2}\right)^x$.

A GRAPHICAL VIEW OF EXPONENTIAL FUNCTIONS

Basic Graphs Our preceding discussion suggests that the graphical features of a function $f(x) = b^x$ depend on whether $b > 1$ or $0 < b < 1$. Figures 7-3a and 7-3b show the graphs of $y = 2^x$ and $y = \left(\dfrac{1}{2}\right)^x$, respectively.

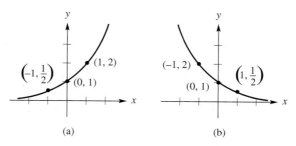

(a)   (b)

Figure 7-3

Exercises 18 and 19 show that these graphs illustrate behavior typical of exponential functions.

EXERCISES   18. (*Making Observations*)

  a. Graph the following functions together in the window $[-2, 2]$ by $[-1, 5]$: $y = 2^x$, $y = 5^x$, $y = 20^x$.

  b. Graph the following functions together in the window $[-2, 2]$ by $[-1, 5]$: $y = 2^x$, $y = 1.5^x$, $y = 1.1^x$.
  Based on your graphs, guess the answers to the following questions about a function $f(x) = b^x$ with $b > 1$.

  c. What is the domain of f? What is its range?

  d. Does the graph have any x-intercepts? y-intercepts? If so, where are they?

  e. In terms of b, what is the value of $f(1)$? $f(-1)$?

f. Does the graph have any vertical asymptotes? Horizontal asymptotes? If so, where are they?

g. Is f increasing or decreasing?

19. (*Making Observations*)

a. Repeat Exercise 18a for the functions $y = \left(\dfrac{1}{2}\right)^x$, $y = \left(\dfrac{1}{5}\right)^x$, $y = \left(\dfrac{1}{20}\right)^x$.

b. Repeat Exercise 18b for the functions $y = \left(\dfrac{1}{2}\right)^x$, $y = \left(\dfrac{2}{3}\right)^x$, $y = \left(\dfrac{9}{10}\right)^x$.

c–g. Repeat Exercise 18c–18g for functions $f(x) = b^x$ with $0 < b < 1$.

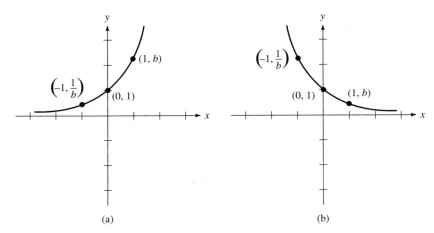

(a)   (b)

Figure 7–4

Exercises 18 and 19 illustrate that the graph of $y = b^x$ resembles Figure 7-4a if $b > 1$, and Figure 7-4b if $0 < b < 1$. To sketch it quickly:

- Begin by plotting the y-intercept at $(0, 1)$.

- Then plot $(1, b)$ and $\left(-1, \dfrac{1}{b}\right)$, which are always on the graph.

- Finally, draw a smooth curve through those points with one of the two basic shapes.

Graphical Transformations   The graphs of many exponential functions in the standard graphing form, $y = Ab^{x-h} + k$, can be obtained by transforming the graph of $y = b^x$.

EXAMPLE 2   *Transforming the graph of $y = b^x$*

To sketch the graph of $y = 3 \cdot 5^{-x}$, begin with the basic graph $y = 5^x$, and make the changes

$$y = 5^x \rightarrow y = 5^{-x} \rightarrow y = 3 \cdot 5^{-x}$$

Replacing x by $-x$ reflects the graph in the y-axis, and multiplying the function by 3 stretches the graph vertically by a factor of 3. Figure 7-5 shows how the points $\left(-1, \dfrac{1}{5}\right)$, $(0, 1)$, and $(1, 5)$ on the basic graph move as the transformations are applied.

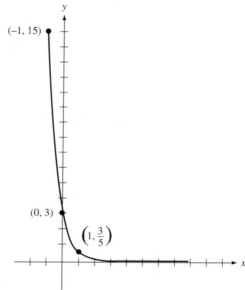

Figure 7–5

EXERCISES *In Exercises 20–23, sketch the graph of each function by transforming an appropriate graph $y = b^x$. In each case,*

    a. State the domain and range of the function.

    b. Decide whether the function is increasing or decreasing.

    c. Plot the points obtained by moving $\left(-1, \dfrac{1}{b}\right)$, $(0, 1)$, and $(1, b)$.

20. $y = 4^{x+3}$ **21.** $y = 0.1^{2x} - 2$

22. $y = 7\left(\dfrac{1}{2}\right)^{x-1}$ **23.** $y = \left(-\dfrac{1}{2}\right)(3^{-x})$

24. (*Writing to Learn*) Describe how each transformation in parts (a)–(f) affects the following graphical features of $y = b^x$.

    • domain and range

    • asymptotes

- intercepts
- increasing/decreasing behavior

   a. a horizontal stretch or compression

   b. a vertical stretch or compression

   c. a reflection in the x-axis

   d. a reflection in the y-axis

   e. a horizontal shift

   f. a vertical shift

Exercises 25–28 illustrate that every exponential function $f(x) = Ab^x$ has a constant **doubling time**. That is, for each such function there is a number h such that $f(x + h) = 2f(x)$ for every value of x. In each exercise, estimate the doubling time by graphing $f(x)$ and estimating the values of x for which $f(x) = 1, f(x) = 2, f(x) = 4,$ and $f(x) = 8$.

25. $f(x) = 1.5^x$ **26.** $f(x) = 4(3^x)$

27. $f(x) = 5(2^x)$ **28.** $f(x) = 0.8^x$ (*Hint*: The doubling time is negative.)

GEOMETRIC SEQUENCES AND SERIES

Many problems can be modeled by an exponential function whose domain is a set $\{K, K + 1, K + 2, \ldots\}$ for some integer K. Such a function is called a **geometric sequence**. (In Section 2-4 you learned that a sequence is a function whose domain is such a set.) The following situation can be modeled by a geometric sequence.

 A MATHEMATICAL LOOKING GLASS
20 Questions

As a child, you probably played the game of "20 Questions," in which your opponent thinks of an object, and you try to identify it by asking up to twenty yes/no questions. Did it ever surprise you that among the millions of objects your opponent could choose, you were often able to identify the right one? You are about to discover that your ability is less amazing than it might appear.

Let's play a simplified version of "20 Questions" in which we will think of a symbol, such as a letter of the alphabet, and you must guess it. To identify it with certainty, the number of questions you need to ask depends on the size of our pool of symbols.

- If our pool contains only one symbol, then you can identify it without asking any questions.

- If our pool contains two symbols, say, A and B, then you can identify the chosen letter by asking the single question, "Is it A?" Our answer will tell you whether it is A or B.

- If our pool contains the four symbols A, B, C, and D, you can identify the chosen letter by asking at most two questions. For example, you can first ask, "Is it one of the symbols A or B?" If our answer is yes, you know that the symbol is either A or B. If the answer is no, you know it is either C or D. In either case, you have reduced the size of the pool to two symbols, and you can guess the correct one with one more question.

• If our pool contains the eight symbols A, B, C, D, E, F, G, and H, you can identify the chosen letter by asking at most three questions. For example, you can first ask, "Is it one of the symbols A, B, C, or D?" Regardless of our answer, you have reduced the size of the pool to four symbols, and you can guess the correct one with two more questions.

With this strategy, each additional question allows you to identify the correct symbol from an original pool twice as large. For example, by asking four questions, you can identify the correct symbol from a pool of 16.

This strategy, called a **binary search**, is illustrated in Figure 7-6. Starting at the top vertex, ask a question about one-half of the pool of symbols. If the answer is "yes," proceed on the path toward that half. If the answer is "no," proceed toward the other half. When you arrive at the next vertex, repeat the procedure. ∎

Figure 7–6

EXERCISES

29. (*Creating Models*) If n represents the number of questions needed to guess the correct symbol, write a function of n to represent the maximum size of the pool of symbols.

30. (*Interpreting Mathematics*) If you can ask 20 questions, how large can the pool of symbols be?

31. (*Writing to Learn*) Why do you think children playing "20 Questions" succeed so often in identifying the correct object? In particular, how does your answer to Exercise 30 provide insight? What other factors might be involved?

The function in Exercise 30 is a geometric sequence, and its values 1, 2, 4, 8, . . . are the terms of the sequence. We can give the sequence a functional name, such as a, and denote its terms by $a_0 = 1$, $a_1 = 2$, $a_2 = 4$, $a_3 = 8$, ... Geometric sequences can also be described by the set of values of the dependent variable, so that this sequence may be written

$$a = \{1, 2, 4, 8, \ldots\}$$

The geometric sequence in Exercise 30 can also be described by the equation $f(n) = 2^n$. An equation has the advantage of giving a formula for each term in the sequence.

EXERCISES

In Exercises 32–35, write out the terms a_0, a_1, a_2, a_3, and a_4, and decide whether the sequence is geometric.

32. $a_n = 4^n$ **33.** $a_n = \dfrac{2}{(-3)^n}$

34. $a_n = \dfrac{1}{5} n^2$ **35.** $a_n = \dfrac{1}{5} 2^n$

In Exercises 36–39, decide whether the given sequence is geometric. If it is, find a formula for a_n, and use your formula to find a_{10}.

36. $a = \{1, 3, 9, 27, \ldots\}$

37. $a = \{4, 8, 12, 16, \ldots\}$

38. $a = \{1000, 200, 40, 8, \ldots\}$

39. $a = \{1, 1.1, 1.21, 1.331, \ldots\}$

In solving problems we sometimes need to add a large number of consecutive terms in a geometric sequence. Such a sum is referred to as a **geometric series**. The addition is usually cumbersome, but in Exercises 40 and 41 you can develop a formula for the sum.

40. For the sequence $a_n = \left(\dfrac{1}{4}\right)^n$, let

$$S_4 = 1 + \frac{1}{4} + \left(\frac{1}{4}\right)^2 + \left(\frac{1}{4}\right)^3 + \left(\frac{1}{4}\right)^4$$

a. Multiply each term by $\dfrac{1}{4}$ to obtain an expression for $\dfrac{1}{4} S_4$.

b. Subtract this expression from the one for S_4, and simplify to obtain a shorter expression for $\dfrac{3}{4} S_4$.

c. Divide by $\dfrac{3}{4}$ to obtain the value of S_4.

Now, in Exercise 41, generalize your work from Exercise 40 to obtain the formula

$$S_n = \frac{A(1 - r^{n+1})}{1 - r}$$

41. For the sequence $a_n = Ar^n$, let

$$S_n = A + Ar + Ar^2 + \cdots + Ar^{n-1} + Ar^n$$

a. Write an expression for rS_n in terms of A, r, and n.
b. Subtract this from the expression for S_n, and simplify to obtain a shorter expression for $(1 - r)S_n$.
c. Divide by $1 - r$ to obtain a formula for S_n.

Use the formula in Exercise 41 to find S_5, S_{10}, and S_{20} for each sequence in Exercises 42–45.

42. $a_n = 6\left(\dfrac{1}{3}\right)^n$ **43.** $\left\{16, -4, 1, -\dfrac{1}{4}, \ldots\right\}$

44. $a_n = (-2)^n$ **45.** $\{2, 2.2, 2.42, 2.662, \ldots\}$

46. (*Making Observations*) In two of the sequences in Exercises 42–45, the values of S_n appear to approach a particular value as n becomes large. Which two sequences are they, and what is the value approached by S_n in each case?

If S_n approaches a particular value as n becomes large, that value is called the **sum of the infinite geometric series** $a_1 + a_2 + a_3 + \cdots$. Every geometric sequence $a_n = Ar^n$ with $|r| < 1$ has a sum, such as the ones you observed in Exercise 46. By using the formula

$$S_n = \frac{A(1 - r^{n+1})}{1 - r}$$

you can derive an expression for the sum in terms of A and r. If $|r| < 1$ and n becomes sufficiently large, r^{n+1} becomes arbitrarily small, so that S_n becomes arbitrarily close to $\dfrac{A}{1-r}$. This number is the sum of the series whenever $|r| < 1$. (If $|r| > 1$ and n becomes sufficiently large, $|r^{n+1}|$ and $|S_n|$ become arbitrarily large as well, so that the series has no sum. In Exercise 74, you can discover what happens when $|r| = 1$.)

47. (*Making Observations*) Check your conclusions about the sums in Exercise 46 by evaluating $\dfrac{A}{1-r}$.

ADDITIONAL EXERCISES   *In Exercises 48–51, decide whether the table fits an exponential function. If it does, find its equation.*

48.

x	y
1	100,000
2	10,000
3	1,000
4	100

49.

x	y
−0.5	0.007
0	0.021
0.5	0.063
1	0.189

50.

x	y
1.5	8
2.5	32
3.5	128
4.5	512

51.

x	y
2	90
3	99
4	99.9
5	99.99

In Exercises 52–55, find an equation for an exponential function $f(x) = A \cdot b^x$ passing through the given points.

52. $(0, 6)$ and $(2, 24)$ **53.** $(0, 7)$ and $(2, 49)$

54. $(3, 1)$ and $(4, 10)$ **55.** $(3, 40)$ and $(8, 30)$

In Exercises 56–59, sketch the graph of each function by transforming an appropriate graph $y = b^x$. In each case,

   a. State the domain and range of the function.

   b. Decide whether the function is increasing or decreasing.

   c. Plot the points obtained by moving $\left(-1, \dfrac{1}{b}\right)$, $(0, 1)$, and $(1, b)$.

56. $y = (0.1)(3^{2x})$ **57.** $y = 5^{x-2} + 1$

58. $y = -4^{-x}$ **59.** $y = 8 - 2^x$

60. (*Making Observations*)

   a. Use transformations to sketch the graphs of $y = 3^{x+2}$ and $y = 9 \cdot 3^x$. The two graphs should appear to be identical. Support this observation with calculator graphs.

   b. Part (a) suggests that any function $y = A \cdot b^{x-h}$ can be written $y = A_1 \cdot b^x$ for some value of A_1. Now find a value of A_1 for which $y = 3 \cdot 4^{x+1}$ is equal to $y = A_1 \cdot 4^x$. (*Hint:* Use the laws of exponents to rewrite 4^{x+1}.)

61. (*Making Observations*)

    a. Use transformations to sketch the graphs of $y = 3^{2x}$ and $y = 9^x$. The two graphs should appear to be identical. Support this observation with calculator graphs.

    b. Part (a) suggests that any function $y = A \cdot b^{cx}$ can be written $y = A \cdot (b_1)^x$ for some value of b_1. Now find a value of b_1 for which $y = 4^{3x}$ is equal to $y = (b_1)^x$. (*Hint:* Use the laws of exponents to rewrite 4^{3x}.)

In Exercises 62–69,

    a. Decide whether the sequence is geometric.

    b. If the sequence is geometric, find a formula for a_n, and use your formula to find a_8.

    c. If the sequence is geometric, decide whether the infinite series $a_1 + a_2 + a_3 + \cdots$ has a sum. If the sum exists, find it.

62. $a = \left\{ 1, \dfrac{1}{3}, \dfrac{1}{9}, \dfrac{1}{27}, \ldots \right\}$

63. $a = \{-1, -0.9, -0.81, -0.729, \ldots\}$

64. $a = \left\{ \dfrac{1}{2}, \dfrac{1}{4}, \dfrac{1}{6}, \dfrac{1}{8}, \ldots \right\}$

65. $a = \{10, 50, 250, 1250, \ldots\}$

66. $a = \left\{ 1, -\dfrac{4}{3}, \dfrac{16}{9}, -\dfrac{64}{27}, \ldots \right\}$

67. $a = \left\{ 1, -\dfrac{3}{4}, \dfrac{9}{16}, -\dfrac{27}{64}, \ldots \right\}$

68. $a = \left\{ 9, 6, 4, \dfrac{8}{3}, \ldots \right\}$

69. $a = \{2, 8, 18, 32, \ldots\}$

70. (*Problem Solving*) The Occupational Health and Safety Act of 1971 established permissible levels of occupational noise in the United States. In the table below, t hours is the maximum acceptable duration of a noise level of A decibels.

A	t
95	4
100	2
105	1
110	0.5
115	0.25

    a. Write an equation for t as a function of A.

    b. Graph the function in part (a) and estimate the maximum acceptable duration for a noise level of 103 decibels.

    c. What is the maximum noise level that can be tolerated through an 8-hour workday?

71. (*Creating Models*) The **pitch**, or frequency, of a musical note determines whether we hear it as a high sound like that produced by a flute or violin or a low sound like that produced by a tuba or bass viol.

Figure 7–7

Figure 7-7 shows the neck of a guitar. It must be constructed so that moving your finger from one fret to the next higher fret always produces the same percentage increase in the frequency of the note. The frequency produced at the 12th fret is twice the frequency produced at the nut (the 0th fret). For the fifth string, the frequency at the nut is 220 cycles per second. Find an equation to express the frequency F of the fifth string at the nth fret as a function of n.

72. (*Problem Solving*) The frequency of the note produced by a guitar string at a given fret is inversely proportional to the distance D from the fret to the bridge (see Figure 7-7). On Dave's Gibson ES-335 guitar the distance from the nut to the bridge is 24.75 inches.

    a. If *F* is the frequency of the note produced by the fifth string when the distance from fret to bridge is *D* inches, write an equation to express *D* as a function of *F*.

    b. Combine your functions in Exercises 71 and 72a to express *D* as a function of *n*.

    c. At what distance from the bridge should each of the first 12 frets be placed on a Gibson ES-335 guitar?

73. (*Problem Solving*) A diabolical mathematician has imprisoned a pregnant fruit fly in a jar. She notices that the number of flies doubles each day. At the end of the 30th day, the jar is full. After how many days was it half full?

74. (*Writing to Learn*) The sequences {1, 1, 1, . . .} and {1, −1, 1, −1, . . .} are each geometric.

    a. What is the value of *r* in each case?

    b. Does either infinite series

$$1 + 1 + 1 + \cdots \text{ or } 1 + (-1) + 1 + (-1) + \cdots$$

have a sum? Explain your answer.

7-2   THE SPECIAL NUMBER *e*

THE DEFINITION OF *e*

Exponential functions are used to calculate interest on an investment or a loan. In this section you will discover that continuously compounded interest is calculated by using an exponential function with a particular irrational base called *e*. You will also see how the same exponential function is used to model other physical phenomena. In the following Mathematical Looking Glass you will learn how the number *e* is defined, and you will find its approximate value.

A MATHEMATICAL LOOKING GLASS
Dealing with e

In September 1997, the Ford Motor Company was running a TV ad offering Ford Explorers at 5.9%. Suppose that your local Ford dealer offers you 5.9% annual interest over a period of 5 years. This offer appears to define a specific interest rate. However, the amount of interest you pay depends on whether it is compounded once a year, once a month, once a day, or continuously.

    To demonstrate the effect of compounding in a simple setting, suppose that you are going to borrow $10,000 at 5.9% annual interest, keep it for 5 years, and pay it back all at once. How much will you pay altogether? (You would pay back a real loan by making monthly payments. Determining these is much more difficult, and we leave the problem to students of accounting and/or differential equations.) ■

EXERCISES   **1.** Suppose that the interest is compounded annually (once a year). Then the amount you owe is increased by 5.9% once a year.

    a. (*Writing to Learn*) Explain why after one year of interest you owe $10,000(1 + 0.059).

    b. Complete the following table to find the amount you owe after five years.

Interest periods	Amount owed
0	$10,000
1	$10,000(1 + 0.059) = $10,590
2	$10,590(1 + 0.059) = $11,214.81
3	
4	
5	

c. (*Creating Models*) The table in part (b) defines a geometric sequence. Write an expression Ab^t for the amount owed after t years.

2. Suppose that the interest is compounded semiannually (twice a year). Then the amount you owe is increased by half of 5.9% twice a year.

   a. (*Writing to Learn*) Explain why after half a year of interest you owe $10,000\left(1 + \dfrac{0.059}{2}\right)$.

   b. Complete the following table to find the amount you owe after five years.

Interest periods	Amount owed
0	$10,000
1	$10,000\left(1 + \dfrac{0.059}{2}\right) = \$10,295$
2	$\$10,295\left(1 + \dfrac{0.059}{2}\right) = \$10,598.70$
3	
4	
.	
.	
.	
10	

   c. (*Creating Models*) The table in part (b) defines a geometric sequence. Write an expression Ab^k for the amount owed after k interest periods. Then rewrite the expression to denote the amount owed after t years. (*Hint*: There are $2t$ interest periods in t years.)

3. (*Creating Models*) Suppose that the interest is compounded three times a year. Then the amount you owe is increased by one-third of 5.9% three times a year.

   a. Find the amount you owe after one-third of a year.

   b. Complete a table to find the amount you owe after 5 years (15 interest periods).

   c. Write an expression Ab^k for the amount owed after k interest periods. Then rewrite the expression to denote the amount owed after t years.

Exercises 1–3 illustrate that if $P is borrowed for t years at an interest rate of r, compounded m times per year, the total amount owed is

$$A = P\left(1 + \frac{r}{m}\right)^{mt}$$

As stated in **Dealing with e**, the amount you owe depends on how often the interest is compounded. However, this is not the end of the story. If the interest is compounded ridiculously often, you should now be asking whether your debt could be ridiculously large. You can answer this question in Exercises 4 and 5.

EXERCISES   **4.** Using the formula just developed, compute the amount you owe after 5 years if you borrow $10,000 at 5.9% interest compounded

    a. once a month   **b.** once a day

    c. once an hour   **d.** once a minute

5. (*Making Observations*) Exercise 4 suggests the presence of a number above which the amount will not rise, no matter how often the interest is compounded. To the nearest dollar, what is this number?

You might now guess that if you borrow any amount $P at any interest rate r for any period of t years, there is a ceiling on the amount you owe, no matter how often the interest is compounded. In Exercises 6–8 you can describe that ceiling in terms of P, r, and t only.

EXERCISE   **6.** (*Writing to Learn*) Explain why the formula

$$A = P\left(1 + \frac{r}{m}\right)^{mt}$$

can be written in each of the following ways.

$$A = P\left(1 + \frac{r}{m}\right)^{(m4r)rt}$$

$$A = P\left[\left(1 + \frac{r}{m}\right)^{(m4r)}\right]^{rt}$$

Exercises 4 and 5 suggest that we should look for the ceiling by exploring the behavior of $\left(1 + \frac{r}{m}\right)^{(m/r)}$ as the number of compoundings m becomes large. If $n = \frac{m}{r}$, then the above expression becomes $\left(1 + \frac{1}{n}\right)^{n}$. Since $n = \frac{m}{r}$, if m is large, then n will be large also, so let's ask whether the value of $\left(1 + \frac{1}{n}\right)^{n}$ approaches a ceiling as n becomes large.

EXERCISES   **7.** (*Making Observations*) Complete the following table to estimate the ceiling on the values of $\left(1 + \frac{1}{n}\right)^{n}$ as n becomes large.

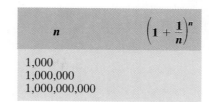

n	$\left(1 + \dfrac{1}{n}\right)^n$
1,000	
1,000,000	
1,000,000,000	

8. (*Making Observations*) Estimate the ceiling on the values of $\left(1 + \dfrac{1}{n}\right)^n$ as n becomes large by looking at the end behavior of the graph of $y = \left(1 + \dfrac{1}{x}\right)^x$.

Using calculus, it can be shown conclusively that if n is large, the value of $\left(1 + \dfrac{1}{n}\right)^n$ approaches 2.71828182846. . . . This special number is called e to honor Swiss mathematician Leonhard Euler, whom you met in Section 5-2.

In summary, if interest is compounded many times each year, our interest formula becomes approximately

$$A = Pe^{rt}$$

This equation is used to define what is called **continuous compounding**. (This does not mean that the interest is actually compounded in infinitesimally small time periods. It simply means that the amount owed is calculated according to this formula.)

Don't be confused by the number e. Like π, e is an irrational number that has been given a letter name because it is used in so many situations. Its decimal expansion neither terminates nor repeats, but for many purposes, such as sketching graphs, you may approximate e by 2.7. For most other purposes in this book, you may approximate it by 2.72. If greater accuracy is desired, your calculator can provide it. (Actually, you can do better than your calculator. The first 15 decimal places of e can be grouped conveniently as $e = 2.7\ 1828\ 1828\ 45\ 90\ 45$ . . . . This memory device is due to George Simmons, *Calculus with Analytic Geometry*, McGraw-Hill, 1985, p. 214.)

EXERCISES   **9.** Find the amount owed when P is borrowed for t years at an annual interest rate r compounded annually. Repeat the calculation for interest compounded monthly, then continuously.

    a. $P = 4000, t = 3, r = 14.9\%$

    b. $P = 50{,}000, t = 30, r = 8.6\%$

    c. $P = 500, t = 0.5, r = 200\%$

10. Find the amount obtained when $P is invested for t years at an annual interest rate r compounded annually. Repeat the calculation for interest compounded monthly, then continuously.

   a. $P = 2000$, $t = 2$, $r = 2.5\%$

   b. $P = 8200$, $t = 1$, $r = 20.85\%$

   c. $P = 5$, $t = 0.25$, $r = 7.5\%$

11. Sketch the graph of $y = 2^x$, $y = e^x$, and $y = 3^x$ on the same set of axes, using $e \approx 2.7$. Support your results with calculator graphs.

USING THE BASE e TO EXPRESS EXPONENTIAL FUNCTIONS

The number e is often used as a base for exponential functions to model physical relationships. Exercises 12 and 13 suggest one reason why.

EXERCISES

12. (*Making Observations*) In each case, calculate the average rate of change in $f(x) = e^x$ over the interval $[a, b]$, and verify that your result is a number between e^a and e^b.

   a. $a = 1, b = 3$ **b.** $a = 1.9, b = 2.1$

   c. $a = 3, b = 5$ **d.** $a = 3.99, b = 4.01$

13. (*Writing to Learn*) Your results in Exercise 12 suggest that the average rate of change in $f(x) = e^x$ over any interval is always equal to the value of $f(x)$ at some point within the interval. Explain why this in turn suggests that $f(x)$ is an appropriate model for physical quantities, such as compound interest and world population, that grow at a rate proportional to their size.

    If you take a course in calculus, you will expand on the ideas in Exercises 12 and 13, and you will discover that the operations of calculus are performed more easily on exponential functions when the base is e. The following Mathematical Looking Glass illustrates that every exponential function can be written with a base of e.

 A MATHEMATICAL LOOKING GLASS

Crater Lake

Crater Lake is a circular lake in southwestern Oregon, about 6 miles in diameter and almost 2000 feet deep. It is contained in a crater formed about 6700 years ago when the volcano Mount Mazama erupted with such violence that its top was blown off (much like Washington's Mount St. Helens in 1980). The lake is remarkable for its dramatic origin and its deep blue color. Its level remains surprisingly constant, even though it is fed entirely by snowfall and rainfall and drained entirely by evaporation.

    Have you ever wondered how scientists can determine when a prehistoric event, such as the formation of Crater Lake, occurred? One method that can determine age up to 50,000 years with reasonable accuracy is known as radiocarbon dating. It is based on two facts.

• Among the carbon atoms contained in all living organisms, the portion of radioactive carbon 14 atoms is constant. When an organism dies, its carbon 14 begins to decay into nonradioactive carbon 12, so that the portion of carbon 14 atoms among all carbon atoms decreases.

- Every radioactive substance has a **half-life**, a fixed length of time during which half of the atoms in any sample of the substance will decay. The half-life of carbon 14 is about 5730 years.

This means that a sample of bone from an animal or wood from a tree 5730 years old contains 50% of the carbon 14 it contained when the animal or tree was alive. A sample 11,460 (twice 5730) years old contains 25% (50% of 50%) of the original portion of carbon 14.   ∎

If the half-life of a radioactive substance is known, we can write a function $f(t) = 2^{-kt}$ for the portion of the substance present after t years. The base 2 occurs naturally, because when the value of t is equal to the half-life of the substance, the value of f is $\frac{1}{2} = 2^{-1}$.

EXERCISES   14. (*Creating Models*) Write a function $f(t) = 2^{-kt}$ for the portion of carbon 14 in a sample of material from an organism that died t years ago. [*Hint:* To find k, use the fact that $f(5730) = 2^{-1}$.]

15. (*Interpreting Mathematics*) The estimate of Crater Lake's age is based in part on a sample of charcoal from a tree that burned in the eruption of Mount Mazama when the lake was formed. Given that the lake is 6700 years old, what percentage of the carbon 14 found in living wood was in the sample?

Scientists often write exponential functions in the form $f(t) = e^{kt}$. In Exercise 34 of Section 7-4, you will learn how to find k analytically. In the meantime, you can use a graphical approach in Exercise 16.

EXERCISES   16. **a.** Zoom in on the graph of $y = e^x$ to find a value $x = c$ with $e^c = 2$.

b. Rewrite your function from Exercise 14 as $f(t) = e^{kt}$ by substituting e^c for 2.

c. Use your function from part (b) to answer the question in Exercise 15. Make sure this answer agrees with the one you obtained there.

In Exercises 17–20, use the method of Exercise 16 to write each function in the form $y = Ae^{kt}$.

17. $y = 3^t$ 18. $y = 5^{2t}$

19. $y = 3\left(\frac{1}{2}\right)^t$ 20. $y = 7(0.8)^{3t}$

ADDITIONAL EXERCISES   *In Exercises 21–24, sketch the graph of each function by transforming the graph of $y = e^x$. In each case, plot the points obtained by moving $\left(-1, \frac{1}{e}\right)$, $(0, 1)$, and $(1, e)$.*

21. (*Review*) $y = e^{0.5x}$ 22. (*Review*) $y = e^{-0.5x}$

23. (*Review*) $y = 2e^{-0.5x}$

24. (*Review*) $y = 2e^{-0.5x} - 3$

25. (*Writing to Learn*) Describe how each graph in Exercises 21–24 differs from the preceding one.

26. (*Problem Solving*) A few years ago Lynn bought a 6-month CD with Atlantic Financial Savings and Loan in the amount of $10,000 at an annual interest rate of 6.2%.

   a. Explain why the amount of money in her account after *t* months is

   $$A(t) = 10,000e^{0.062t/12}$$

   b. After 2.2 months Lynn's bank was bought by Ellwood Federal Savings. How much was her account worth at that time?

27. (*Problem Solving*) (continuation of Exercise 26) Ellwood Federal Savings lowered the annual interest rate on Lynn's CD to 4.85%.

   a. Write a function for the amount of money in Lynn's account after *t* months at 4.85% annual interest. (*Hint*: When *t* = 0, the amount is your answer to Exercise 26b.)

   b. If she leaves her money in the account at 4.85% for the remaining 3.8 months of its term, what will her account be worth?

28. (*Problem Solving*) Lynn wanted to know whether to leave her money in the account, as in Exercise 27, or pay a penalty for early withdrawal and reinvest elsewhere at a higher rate. Head teller Madeline Kelley told her that the penalty would be $119.59. (She also said that if Lynn had acted more promptly after the takeover, there would have been no penalty.)

   a. If she pays the penalty, how much will she have left to reinvest?

   b. If she reinvests at an annual interest rate of *r*, write a function of *r* for the amount of money in her account after 3.8 months.

   c. Graph your function from part (a). Use the graph and your answer from Exercise 27 to determine what interest rates would make it profitable to pay the penalty.

29. (*Problem Solving*) Six months ago, you received a high school graduation gift of $10,000 from your grandfather. Since then you have had it in a CD while making long-term plans for it. Your grandfather has cautioned you that your return on a long-term investment will be very sensitive to the interest rate.

   a. Six months ago, you could have invested the money at 6.2% at a bank 20 miles away. You chose instead to put it into the bank around the corner at 6%. How much did you lose?

   You point out that taking a lower interest rate for the sake of convenience didn't cost you very much. Your grandfather smiles knowingly and says, "Yes, but that was only for six months, and you will keep this investment until you retire."

   b. Assuming that you will keep the money for 50 years before passing it on to your grandchildren, what will be the difference between your returns at 6.2% and at 6%?

30. (*Problem Solving*) How old is a sequoia tree if its (nonliving) heartwood contains 65% of its original carbon 14? (*Hint*: Solve graphically.)

31. (*Problem Solving*) Radiocarbon dating, as described in **Crater Lake**, can be used reliably to date only organic objects up to about 50,000 years old. Dating older objects requires other radioactive elements with longer half-lives. For example, rocks are sometimes dated by analyzing their uranium 238, which has a half-life of 4.5 billion years and decays

into lead 206. The ratio of uranium to lead indicates the portion of uranium that has decayed since the rock was formed.

  a. Write a function $f(t) = 2^{-kt}$ for the portion of uranium 238 remaining in a rock t years after it is formed.

  b. What portion of the original uranium 238 is left in a rock whose age is 200 million years? (*Hint*: Write your function from part (a) so that the base is e.)

  c. How old is a rock that contains 73% of its original uranium 238?

32. (*Problem Solving*) The following elements are among the by-products of a nuclear explosion. How long does it take for 99% of each one to decay?

  a. polonium 212 (half-life 3×10^{-7} seconds)

  b. krypton 91 (half-life 10 seconds)

  c. strontium 90 (half-life 28 years)

In calculus, you will learn another way to construct the number e. Exercises 33 and 34 demonstrate the construction, while leaving the theoretical justification to calculus.

33. (*Extension*) If n is any positive integer, then the symbol $n!$ denotes $1 \cdot 2 \cdot 3 \cdot \cdots \cdot (n - 1) \cdot n$. Thus

$$1! = 1$$

$$2! = 1 \cdot 2 = 2$$

$$3! = 1 \cdot 2 \cdot 3 = 6$$

and so on. By convention, $0! = 1$.

Evaluate each of the numbers $6!$, $7!$, $10!$, $25!$.

34. (*Extension*) Define a sequence $\{c_n\}$ by

$$c_0 = \frac{1}{0!}$$

$$c_1 = \frac{1}{0!} + \frac{1}{1!}$$

$$c_2 = \frac{1}{0!} + \frac{1}{1!} + \frac{1}{2!}$$

and so on.

Find the values of the terms c_0 through c_{10}. Do the values of c_n appear to approach a ceiling as n becomes large? What does that ceiling appear to be? You're right!

7–3   INVERSE FUNCTIONS

In **Standing Room Only** (page 323), we asked how long it would take the world's population to double. To answer the question you constructed the equation

$$P = 6(1.02)^t$$

to describe population as a function of time. This equation is essentially of the form $y = b^x$. Up to this point, we have regarded x as the input and y as the output in such equations. By contrast, the question in **Standing Room Only** provides a value of y and asks you to solve for x. In Section 7-1 you solved the equation by graphical methods. In Section 7-4 you will learn an analytical method utilizing the concept of an inverse function, which you will learn in the present section.

You have seen the word *inverse* before in mathematics. For example, the additive inverse of 2 is -2, and the multiplicative inverse of 2 is $\frac{1}{2}$. The word can also be applied to mathematical operations. For example, the inverse of adding 2 is subtracting 2, and the inverse of multiplying by 2 is dividing by 2.

When a pair of inverse operations is applied in succession, the net result is no change. For example, if we multiply 16 by 2, then divide the result by 2, our final result is our starting point 16. When we speak of inverse functions, we have the same idea in mind. To illustrate, suppose that f is the "multiplying by 2" function defined by $f(x) = 2x$ and g is the "dividing by 2" function defined by $g(x) = \frac{1}{2}x$. If these two functions are applied in succession in either order, the net result is no change. Recall from Section 2-2 that applying f and g in succession produces one of the composed functions $f \circ g$ or $g \circ f$.

$$(f \circ g)(x) = f[g(x)] = 2[g(x)] = 2\left[\frac{1}{2}x\right] = x$$

$$(g \circ f)(x) = g[f(x)] = \frac{1}{2}[f(x)] = \frac{1}{2}[2x] = x$$

Similarly, the equations $F(x) = x + 2$ and $G(x) = x - 2$ define a pair of inverse functions. If they are applied in succession in either order, our final output is our original input x.

EXERCISE

1. (*Review*) For the functions F and G just mentioned, verify that $(F \circ G)(x) = x$ and $(G \circ F)(x) = x$ for all real values of x.

The following definition is made with these examples in mind.

> Two functions f and g are called **inverse functions** if and only if $(f \circ g)(x) = x$ for all x in the domain of g, and $(g \circ f)(x) = x$ for all x in the domain of f. In this case it is customary to write $g = f^{-1}$ and $f = g^{-1}$.

FYI The notation f^{-1} denotes a compositional inverse, and not a multiplicative inverse. In general, $f^{-1}(x)$ and $\frac{1}{f(x)}$ are not the same.

ONE-TO-ONE FUNCTIONS

Not all functions have inverses. A prime example is the squaring function $f(x) = x^2$. For this function, $f(2)$ and $f(-2)$ are both equal to 4. If f^{-1} existed,

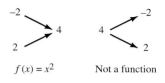

$f(x) = x^2$ Not a function

Figure 7–8

$f^{-1}(4)$ would have to be both 2 and -2, violating the requirement that a function must have only one output for each input. See Figure 7-8.

Which functions have inverses? We can begin to answer this question by focusing on what prevents the squaring function from having an inverse. It is the fact that two input values (2 and -2) produce the same output value (4). Equivalently, the squaring function $y = x^2$ does not have an inverse because x is not a function of y. Conversely, if only one input produces each output, the function has an inverse. Such functions are called **one-to-one**. Since only one-to-one functions have inverses (and only one-to-one functions can *be* inverses), our study of inverse functions begins with a study of one-to-one functions.

The following observation is useful, both in deciding whether a function is one-to-one and in finding the inverse of a one-to-one function.

A function $y = f(x)$ is one-to-one if and only if x is also a function of y.

A Numerical View of One-to-One Functions   It is easy to tell whether a function described by a (complete) table is one-to-one. A function is one-to-one if no two distinct inputs lead to the same output. Equivalently, it is one-to-one if no two rows of the table with different first entries have the same second entry. More generally,

A function is one-to-one if and only if no two rows in any table of values have different first entries and the same second entry.

EXERCISES   *In Exercises 2–5, decide whether the table represents a one-to-one function. Assume that each is complete.*

2.

x	y
0	0
1	1
2	4
3	9
4	16

3.

x	y
-4	16
-2	4
0	0
2	4
4	16

4.

x	y
2	35
5	7
8.1	-2
19	100
$-\sqrt{3}$	0.05

5.

x	y
10	3
20	3
30	3
40	3
50	3

A Graphical View of One-to-One Functions   It is also easy to tell whether a function described by a graph is one-to-one. Figure 7-9a shows the graph of $y = x^2$, which is not one-to-one.

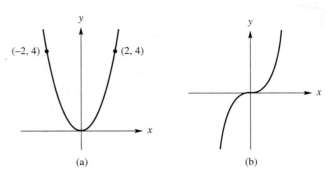

Figure 7–9

The points $(-2, 4)$ and $(2, 4)$ on the graph have different x-coordinates, but the same y-coordinate. Therefore they both lie on the horizontal line $y = 4$. In general, if two distinct values of x produce the same value of y, the points on the graph lie on the same horizontal line.

By contrast, Figure 7-9b shows a graph of $y = x^3$, which is one-to-one, since no two distinct real numbers have the same cube. Equivalently, no two points on the graph of $y = x^3$ lie on the same horizontal line.

The Horizontal Line Test

A function is one-to-one if and only if no horizontal line intersects its graph more than once.

EXERCISES   *In Exercises 6–9, decide whether the function is one-to-one.*

6.

7.

8.

9.
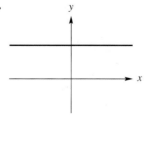

In Exercises 10–13, graph the function and decide whether it is one-to-one.

10. $f(x) = 4 - x^2$ **11.** $f(x) = 2x + 3$

12. $f(x) = \sqrt{x - 3}$ **13.** $f(x) = 2^x$

14. (*Making Observations*)

   a. If $f(x) = mx + b$ with $m \neq 0$, does the graph of $y = f(x)$ pass the horizontal line test?

   b. If $f(x) = ax^2 + bx + c$ with $a \neq 0$, does the graph of $y = f(x)$ pass the horizontal line test?

   c. If $f(x) = Ab^x$ with $A \neq 0$, $b \neq 1$, and $b > 0$, does the graph of $y = f(x)$ pass the horizontal line test?

Exercise 14 allows you to conclude that nonconstant linear and exponential functions are *always* one-to-one, while quadratic functions *never* are.

Except for a few classes of functions such as those in Exercise 14, it is generally difficult or impossible to tell whether a function described by an equation is one-to-one. One method that sometimes works is to attempt to find the equation of the function's inverse. You will learn this method later in this section.

FINDING INVERSES OF ONE-TO-ONE FUNCTIONS

Once you know that a function is one-to-one, how do you find its inverse? To describe the process, let's consider a pair of inverse functions you have seen earlier in this book.

In ***Canadian Crickets*** (page 50) you encountered the following function $C = \dfrac{5}{9}(F - 32)$, which converts Fahrenheit temperatures to Celsius.

Similarly, the function $F = \dfrac{9}{5}C + 32$ converts Celsius to Fahrenheit. If the two are applied in succession, the net result is no change, so that the functions are inverses. You can verify this in Exercise 15.

EXERCISE   **15.** Define functions f and g by $f(x) = \dfrac{9}{5}x + 32$ and $g(x) = \dfrac{5}{9}(x - 32)$, and simplify the expressions for $f[g(x)]$ and $g[f(x)]$. How does your result confirm that f and g are inverses?

The two functions from Exercise 15 provide another way of thinking about inverse functions. A pair of inverse functions looks at the same relationship from two different points of view, by reversing the roles of the independent and dependent variables. This insight suggests the following process for finding the inverse of a given one-to-one function.

- To reverse the roles of the independent and dependent variables, physically interchange them. For example, if the variables are x and y, relabel each x as y and vice versa.
- Display the new dependent variable explicitly as the output.

Over the next few pages you will apply this process to functions described numerically, graphically, and analytically.

Numerical Methods If a one-to-one function is described by a (complete) table, then to find the inverse function is to describe it by a table.

EXAMPLE 1   *Finding the inverse of a one-to-one function numerically*

Suppose the function f is given by the following table.

x	y
−1	10
2	7
3	4
6	9
9	1

To obtain a table for f^{-1}, first interchange the labels "x" and "y" to reverse the roles of independent and dependent variables. Then interchange the positions of the columns to display the new y column as the output.

x	y
10	−1
7	2
4	3
9	6
1	9

■

EXERCISE **16.** In Exercises 2–5, if the function is one-to-one, write a table to describe the inverse function.

Graphical Methods If a one-to-one function is described by a graph, then to find the inverse is to describe it by a graph.

EXAMPLE 2 *Finding the inverse of a one-to-one function graphically*

The function $f(x) = \sqrt{x}$ is one-to-one. The graph of $y = f(x)$ is shown in Figure 7-10a.

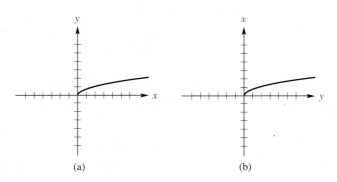

(a) (b)

Figure 7–10

To reverse the roles of independent and dependent variables, relabel the axes as in Figure 7-10b.

To display the new y as output, move the plane so that the positive x-axis points to the right and the positive y-axis points upward, as usual. The required movement is a reflection of the plane across the diagonal line (the graph of $y = x$) shown in Figure 7-11.

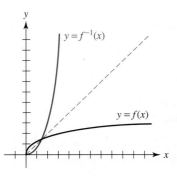

Figure 7–11

Later in this section you will find an equation for the graph in Figure 7-11 by showing analytically that $f^{-1}(x) = x^2$ for $x \geq 0$. ■

The process described in Example 2 can be summarized as follows.

> If f is a one-to-one function, the graph of $y = f^{-1}(x)$ is the reflection of the graph of $y = f(x)$ in the line $y = x$.

EXERCISES   **17.** (*Making Observations*)

a. Sketch the graph of $f(x) = 2^x$ and label the points $\left(-1, \frac{1}{2}\right)$, $(0, 1)$, $(1, 2)$, and $(2, 4)$.

b. Explain why the points $\left(\frac{1}{2}, -1\right)$, $(1, 0)$, $(2, 1)$, and $(4, 2)$ must be on the graph of $y = f^{-1}(x)$.

c. Plot the points in part (b). How do the results suggest that the graph of $y = f^{-1}(x)$ is the reflection of the graph of $y = f(x)$ in the line $y = x$?

d. Sketch the graph of $y = f^{-1}(x)$. (You will find its equation in Section 7-4.)

18. In Exercises 6–13, if the function is one-to-one, sketch the graph of its inverse.

Analytical Methods If a one-to-one function is described by an equation, then to find the inverse function is to describe it by an equation. The analytical process requires an extra step, compared to the numerical and graphical processes. Example 3 shows why.

EXAMPLE 3   *Finding the inverse of a one-to-one function analytically*

Suppose $f(x) = \dfrac{x}{x + 2}$. The graph in Figure 7-12 indicates that f is one-to-one and, therefore, has an inverse.

To find an equation for $f^{-1}(x)$, first write f using the variables x and y.

$$y = \frac{x}{x + 2}$$

Then interchange x and y.

$$x = \frac{y}{y + 2}$$

Displaying the new y as output means solving for y.

$$x(y + 2) = y$$
$$xy + 2x = y$$
$$xy - y = -2x$$
$$y(x - 1) = -2x$$
$$y = -\frac{2x}{x - 1}$$

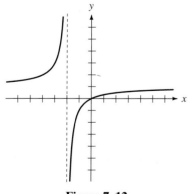

Figure 7–12

We now have a new function $g(x) = -\dfrac{2x}{x - 1}$, but we can't be too hasty in concluding that $g = f^{-1}$. In solving for y we have performed two operations that may not lead to equivalent equations. We have multiplied both sides of the equation by a variable quantity ($y + 2$) and divided both sides by a variable quantity ($x - 1$). As a result, we do not know whether g is the same as f with the roles of x and y reversed.

To see whether $g = f^{-1}$, we need to apply the definition of inverse function. If the composition $(g \circ f)(x) = x$ for all x in the domain of f, and $(f \circ g)(x) = x$ for all x in the domain of g, then we will know that $g = f^{-1}$.

First, let's look at $(g \circ f)(x)$, bearing in mind that the domain of f includes all real numbers $x \neq -2$.

$$(g \circ f)(x) = g[f(x)] = -\frac{2f(x)}{f(x) - 1}$$

$$= -\frac{2\left(\dfrac{x}{x + 2}\right)}{\dfrac{x}{x + 2} - 1}$$

$$= -\frac{2x}{x - (x + 2)} \qquad (\text{if } x \neq -2)$$

$$= -\frac{2x}{-2}$$

$$= x$$

Similarly, you can verify that $(f \circ g)(x) = x$ for all x in the domain of g, so $g = f^{-1}$.   ■

EXERCISES   **19.** For the functions f and g in Example 3:

    a. Find the domain of g.

    b. Verify that $(f \circ g)(x) = x$ for all x in the domain of g.

In Exercises 20–23,

    a. Sketch the graph of $y = f(x)$ to verify that f is one-to-one, and sketch the graph of $y = f^{-1}(x)$.

    b. Find f^{-1} analytically, and verify your result by simplifying the compositions $(f \circ f^{-1})(x)$ and $(f^{-1} \circ f)(x)$.

    c. Support your results by graphing $y = f(x)$ and $y = f^{-1}(x)$ on your calculator.

20. $f(x) = 3x + 5$ **21.** $f(x) = 6 - 4x$

22. $f(x) = 2x^3 - 7$ **23.** $f(x) = \dfrac{1}{2x - 1}$

    Finding the inverse of a function analytically can be difficult for two reasons.

• After interchanging x and y, you may not be able to solve the new equation for y. In this case you can describe the inverse function only numerically or graphically.

• Even if you can solve the new equation explicitly for y, the new function g is not guaranteed to be the inverse of the original function f. That is, it may be that $g[f(x)] \neq x$ for some x in the domain of f, or that $f[g(x)] \neq x$

for some x in the domain of g. When this occurs, you can often obtain f^{-1} by restricting the domain of g. You have seen such a restriction in Example 2, where we found graphically that the inverse of $f(x) = \sqrt{x}$ is $g(x) = x^2$ for $x \geq 0$. Without the restriction on the domain, $g(x) = x^2$ is not one-to-one, and so could not be f^{-1}. Example 4 provides an analytical illustration.

EXAMPLE 4   *Finding the inverse of a one-to-one function analytically*

Suppose that $f(x) = \sqrt{1 - 2x}$. The graph in Figure 7-13 indicates that f is one-to-one.

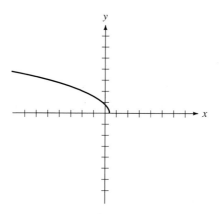

Figure 7–13

To find f^{-1}, first write f using the variables x and y.

$$y = \sqrt{1 - 2x}$$

Then interchange x and y.

$$x = \sqrt{1 - 2y}$$

Next, solve for y to obtain

$$y = \frac{1}{2}(1 - x^2)$$

We now have a new function $g(x) = \frac{1}{2}(1 - x^2)$. To see whether $g = f^{-1}$, let's compose the two. First, let's look at $(g \circ f)(x)$, bearing in mind that the domain of f is $\left(-\infty, \frac{1}{2}\right]$.

$$(g \circ f)(x) = g[f(x)] = \frac{1}{2}\left[1 - f(x)^2\right]$$

$$= \frac{1}{2}\left[1 - \left(\sqrt{1 - 2x}\right)^2\right]$$

$$= \frac{1}{2}[1 - (1 - 2x)]$$

$$= \frac{1}{2}(2x)$$

$$= x \text{ for all } x \text{ in } \left(-\infty, \frac{1}{2}\right]$$

Next, let's look at $(f \circ g)(x)$, bearing in mind that the domain of g is $(-\infty, \infty)$.

$$(f \circ g)(x) = f[g(x)] = \sqrt{1 - 2g(x)}$$

$$= \sqrt{1 - 2\left[\frac{1}{2}(1 - x^2)\right]}$$

$$= \sqrt{1 - (1 - x^2)}$$

$$= \sqrt{x^2}$$

$$= |x|$$

Thus, $(f \circ g)(x)$ is not equal to x on $(-\infty, \infty)$, but is equal to x on $[0, \infty)$. Therefore, the inverse of f is $f^{-1}(x) = \frac{1}{2}(1 - x^2)$ for $x \geqslant 0$. The restriction on the domain of f^{-1} is critical. As Figure 7-14 shows, the function $y = \frac{1}{2}(1 - x^2)$ is not one-to-one on $(-\infty, \infty)$, but is one-to-one on $[0, \infty)$.

Figure 7–14

EXERCISES   *In Exercises* 24–27,

    a. Sketch the graph of $y = f(x)$ to verify that f is one-to-one, and sketch the graph of $y = f^{-1}(x)$.

    b. Find f^{-1} analytically, and verify your result by simplifying the compositions $(f \circ f^{-1})(x)$ and $(f^{-1} \circ f)(x)$.

    c. Support your results by graphing $y = f(x)$ and $y = f^{-1}(x)$ on your calculator.

24. $f(x) = \sqrt{x - 4}$ **25.** $f(x) = \sqrt{5 - x}$

26. $f(x) = \sqrt{3x + 2}$ **27.** $f(x) = \sqrt[4]{x}$

Sometimes a physical problem is modeled by a function that is not one-to-one over its entire domain, but is one-to-one when its domain is suitably restricted. In the following Mathematical Looking Glass, it is useful to find the inverse of such a restricted function.

A MATHEMATICAL LOOKING GLASS

Safe Speeds

The distance d in feet required to stop a car traveling at an initial speed of v miles per hour is approximately

$$d = 0.06v^2 + 0.8v$$

This function was discussed in Exercise 62 of Section 4-2. It is a convenient model of stopping distance as long as speed is the independent variable.

- A member of the National Safety Council might want to construct a table of stopping distances associated with selected initial velocities.
- A Department of Transportation employee who is setting the timing of traffic lights needs to know the stopping distance for a vehicle traveling at the speed limit.

However, there are situations in which it is more convenient to use stopping distance as the independent variable. For example,

- A police officer investigating an accident may need to calculate a driver's initial speed, based on evidence such as the distance from some skid marks to the point of impact.
- A Department of Transportation employee who is setting speed limits needs to know the normal range of visibility on each road so that drivers traveling at the speed limit can stop safely in that distance.

The function $d = 0.06v^2 + 0.8v$ is not one-to-one on $(-\infty, \infty)$, but its domain in a physical context is $[0, \infty)$. In Exercises 28 and 29 you can verify that the function is one-to-one on this domain, and you can find its inverse. ■

EXERCISES

28. (*Making Observations*) When you find the vertex on the graph of $d = 0.06v^2 + 0.8v$, how does your result show that the function is one-to-one on each of the intervals $\left(-\infty, -\dfrac{20}{3}\right]$ and $\left[-\dfrac{20}{3}, \infty\right)$?

29. **a.** To solve the equation $d = 0.06v^2 + 0.8v$ for v, you can rewrite it as $0.06v^2 + 0.8v - d = 0$ and use the quadratic formula.

   b. How does your result indicate that the functions $d = 0.06v^2 + 0.8v$ and $v = \dfrac{5\sqrt{6d + 16} - 20}{3}$ are inverses when v and d are both non-negative?

In Exercises 30–33, break the domain of f into two intervals so that the restriction of f to each interval is one-to-one. Then find the inverse of each restricted function.

30. $f(x) = (x - 3)^2$ 31. $f(x) = 10 - 2x^2$

32. $f(x) = (x + 4)^2 - 5$ 33. $f(x) = 3(x - 1)^2 + 2$

34. Lynn's daughter, Carrie, often rides her horse near her home on Old Mill Road. Because many other residents also ride horses, she feels that the speed limit should be no more than 25 mph. She went to a township meeting armed with the two functions from **Safe Speeds** and the fact that because of curves, drivers must be able to stop within 50 feet.

   a. (*Interpreting Mathematics*) Use the function $v = \dfrac{5\sqrt{6d + 16} - 20}{3}$ to argue her point that the speed limit should be no more than 25 mph.

   b. (*Interpreting Mathematics*) Use the function $d = 0.06v^2 + 0.8v$ to argue her point.

c. (*Writing to Learn*) Which function makes it easier for Carrie to argue her point? Why?

35. (*Writing to Learn*) When you don't use material you have learned, you tend to forget much of it. For example, suppose you memorize 50 new vocabulary words for a final exam in Spanish, and thereafter never speak Spanish again. A reasonable model for the number of those words you retain over time might be

$$n = 10 + \frac{30}{t}$$

where

t = number of days after the exam

n = number of words retained

Describe at least one situation in which it is convenient to use the equation in its present form and at least one situation in which you would want to solve for t and use the inverse function.

ADDITIONAL EXERCISES   *In Exercises 36–39, assume that each table is complete, and decide whether it describes a one-to-one function. If it does, write a table to describe its inverse.*

36.

x	y
-1	-3
0	-2
8	0
9	-1
23	1

37.

x	y
1	1
2	2
3	3
4	4
5	5

38.

x	y
3	6
5	6
2π	6
7.96	6
$\sqrt{139}$	6

39.

x	y
0	1
1	2
2	4
3	8
4	16

In Exercises 40 and 41, write a table to represent f^{-1}, and demonstrate that $f^{-1}[f(x)] = x$ for each x in the domain of f and $f[f^{-1}(x)] = x$ for each x in the domain of f^{-1}.

40.

x	$f(x)$
-7	28
-4	13
0	-2
5	25

41.

x	$f(x)$
1	$\sqrt{3}$
2.8	12,946
7π	0
$\dfrac{2}{3}$	e

In Exercises 42–45, decide whether the graph represents a one-to-one function. If it does, sketch the graph of the inverse function.

42.

43.

44.

45.

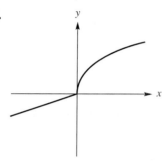

In Exercises 46–53, graph the function and decide whether it is one-to-one. If it is, sketch the graph of its inverse.

46. $f(x) = x^3 - 4x$

47. $f(x) = x^3 + 4x$

48. $f(x) = \dfrac{1}{x^3 + 4x}$

49. $f(x) = 2 \cdot 3^{-x}$

50. $f(x) = \sqrt{1 + x^3}$

51. $f(x) = \sqrt{1 + x^2}$

52. $f(x) = x - \dfrac{1}{x}$ on $(0, \infty)$

53. $f(x) = \dfrac{1}{1 + x^2}$ on $[0, \infty)$

In Exercises 54–61, decide whether f and g are inverses by simplifying the compositions $f[g(x)]$ and $g[f(x)]$.

54. $f(x) = 8x + 16, \; g(x) = \dfrac{1}{8}x - 2$

55. $f(x) = 3 - x, \; g(x) = 3 - x$

56. $f(x) = \dfrac{1}{x}, \; g(x) = \dfrac{1}{x}$

57. $f(x) = \dfrac{x}{x + 6}, \; g(x) = \dfrac{6x}{1 - x}$

58. $f(x) = (x - 7)^5, \; g(x) = \sqrt[5]{x} + 7$

59. $f(x) = (x - 7)^4, \; g(x) = \sqrt[4]{x} + 7$

60. $f(x) = 2x^3 + 1, \; g(x) = \sqrt[3]{\dfrac{x - 1}{2}}$

61. $f(x) = 4, \; g(x) = \dfrac{1}{4}$

In Exercises 62–69, find f^{-1} analytically, and graph f and f^{-1} on the same set of axes.

62. $f(x) = 0.2x$ **63.** $f(x) = 0.2x - 0.8$

64. $f(x) = x^7$ **65.** $f(x) = \dfrac{x^3 + 8}{3}$

66. $f(x) = \dfrac{x}{x - 5}$ **67.** $f(x) = \dfrac{5}{x - 5}$

68. $f(x) = \sqrt{x}$ **69.** $f(x) = \sqrt{10 - x}$

In Exercises 70–73, break the domain of f into two intervals so that the restriction of f to each interval is one-to-one. Then find the inverse of each restricted function.

70. $f(x) = (2x - 1)^2$ **71.** $f(x) = 7 - 3x^2$

72. $f(x) = 7 - 3x^4$ **73.** $f(x) = x^2 + 4x + 5$

In Exercises 74–77, graph the function, and find at least one interval on which it has an inverse.

74. $h(x) = x^4 - 8x^2$ **75.** $P(t) = \dfrac{1}{t^2 - 1}$

76. $y = x^3 - 12x$ **77.** $A(z) = |z^2 - 9|$

SUPPLEMENTARY TOPIC

Parametric Representations of Inverse Functions

Example 5 shows how to use parametric equations to graph the inverse of a one-to-one function on your calculator. Since the method does not require an equation for the inverse function, it is especially useful when an equation is inconvenient or impossible to find.

EXAMPLE 5   *Using parametric functions to graph the inverse of a one-to-one function*

A calculator graph indicates that $f(x) = x^3 + x + 1$ is one-to-one and, therefore, has an inverse. To find the inverse graphically, first represent f parametrically.

$$x = t$$
$$y = t^3 + t + 1$$

The function f^{-1} is represented by interchanging x and y.

$$x = t^3 + t + 1$$
$$y = t$$

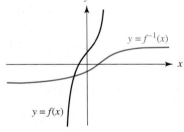

Figure 7–15

Figure 7-15 shows graphs of f and f^{-1} on the same axes.   ■

EXERCISES   **78.** (*Writing to Learn*) Explain why the method in Example 5 is guaranteed to produce the graph of f^{-1} when f is one-to-one.

In Exercises 79–82, use parametric functions to graph f and f^{-1} in the same window.

79. $f(x) = 4 - x^3$ **80.** $f(x) = \sqrt{x - 4}$

81. $f(x) = 2^x$ **82.** $f(x) = \dfrac{1}{x}$

7–4 LOGARITHMIC FUNCTIONS

Section 7-3 began with a promise that our discussion of inverse functions would produce an analytical method for solving exponential equations. Here is a situation in which the solution of such an equation was crucial.

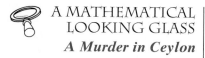

A MATHEMATICAL LOOKING GLASS
A Murder in Ceylon

Newton's law of cooling states that when a hot object cools, the difference between its temperature and that of the surrounding medium decreases by equal percentages over time intervals of equal length. The temperature difference is thus an exponential function of time. In *Clouds in a Glass of Beer* (John Wiley & Sons, 1987), Craig Bohren tells how Newton's law of cooling played an important role in a murder trial.

> Judged by his superbly written autobiography, *Mostly Murder*, the late Sir Sydney Smith must have been a charming and delightful man. But the field in which he achieved eminence, forensic pathology, could hardly be so described; morbidly fascinating might be more apt. The function of a forensic pathologist is to determine how the dead met their end: murder, suicide, or accident. Sir Sydney chose to recount mostly the murders he investigated during his long, distinguished career, and one of them gives the title to this chapter.
>
> About thirty years ago Sir Sydney was involved in a case of murder in Ceylon (now Sri Lanka). A man was accused of murdering his wife. His whereabouts were known reliably after a certain time, and the victim was known to have been alive before a certain time. Whether or not the accused was to be hanged therefore crucially depended on determining the time of death as accurately as possible. After death a body cools slowly at a rate depending on its size and clothing and its surroundings (still or windy, for example). From a body's *cooling curve* the time of death can be estimated by extrapolating backwards from when the temperature is known to when it may be assumed to have been normal (about 37° C). . . .

Here are Sir Sydney's words on what happened:

> The experiments on the cooling of recently dead bodies that I had suggested had been carried out. Professor deSaram [a former pupil of Sir Sydney's] had been fortunate in obtaining three bodies of young murderers immediately after they had been executed. Dressed in petticoats and sarees, the bodies were kept under conditions as close as possible to those in which the body of the murdered woman had been placed, and their temperatures were taken every half-hour. Approximately seven hours were required for these bodies to lose 5.2 degrees. ■

EXERCISES

1. (*Creating Models*) It is reasonable to assume that the victim's temperature at the time of her death was 37° C, that the temperature of her surroundings was 20° C, and that her body, like those of the executed murderers, cooled 5.2° in the first seven hours after death.

a. Find a function $y = 20 + Ab^t$ to express her body's temperature t hours after death.

b. Use the graphical method of Section 7-2 to rewrite your function in the form $y = 20 + Ae^{-kt}$.

2. (*Interpreting Mathematics*)

a. The victim's body was discovered at 6:00 A.M., at which time its temperature was 27°. Use your graph to estimate the time of the murder.

b. Her husband's whereabouts were known at all times since 10:30 A.M. the previous day. What does your graph suggest about his guilt or innocence?

c. Write an equation in t that must be solved to determine the time of the murder more precisely.

We will return to this murder mystery in Exercise 34, where you can solve the equation in Exercise 2. To see how the concept of an inverse function will help, let's consider a simple equation with variable exponents, say, $2^x = 7$.

EXERCISE   3. (*Writing to Learn*)

a. The function $g(x) = 2^x$ has an inverse. How can you tell?

b. Explain why solving the equation $2^x = 7$ is the same as finding $g^{-1}(7)$.

If we could find an equation for the inverse of $g(x) = 2^x$, it would be a simple matter to evaluate $g^{-1}(7)$. Unfortunately, no algebraic operation will solve $y = 2^x$ for x. We are thus in the awkward position of needing to work with the function g^{-1} without having an equation for it. John Napier (1550–1617), the first person to study inverses of exponential functions, provided the name **logarithm** to describe their outputs, such as $g^{-1}(7)$. (The name, meaning "proportionate numbers," was suggested by the computational processes he used.) Mathematicians today, following Napier, call the inverse of the exponential function $y = b^x$ the **logarithmic function** to the base b, and write it as $y = \log_b x$.

Simply naming the inverses of exponential functions provides neither insight into their nature nor help in using them to solve equations. However, let's see what we can discover about logarithmic functions from numerical, graphical, and analytical perspectives.

A NUMERICAL VIEW OF LOGARITHMIC FUNCTIONS

Table 31 was generated by $y = 2^x$.

TABLE 31

x	$y = 2^x$
0	$1 = 2^0$
1	$2 = 2^1$
2	$4 = 2^2$
3	$8 = 2^3$
4	$16 = 2^4$

By interchanging x and y we obtain Table 32, which fits $y = \log_2 x$.

TABLE 32

x	$y = \log_2 x$
$1 = 2^0$	0
$2 = 2^1$	1
$4 = 2^2$	2
$8 = 2^3$	3
$16 = 2^4$	4

So logarithms are just disguised exponents!

> The logarithm to the base b of a number x is the power to which b must be raised in order to obtain x. That is,
>
> $$y = \log_b x \quad \text{means} \quad x = b^y$$

We can use this observation to obtain certain values of logarithmic functions.

EXAMPLE 1   *Evaluating a logarithm (a disguised exponent)*

To evaluate $\log_8 64$, ask yourself to what power 8 must be raised to obtain 64. Recognizing that $8^2 = 64$, you can write $\log_8 64 = 2$.   ∎

We can't use the same process to find all values of logarithmic functions. For example, to find $\log_2 7$, we need to find the power to which 2 must be raised in order to obtain 7. Equivalently, we have to solve $2^x = 7$ for x, which we cannot do algebraically. The evaluation of $\log_2 7$ will have to wait until Exercise 46.

EXERCISES   **4.** (*Making Observations*) Use the x-values 0, 1, 2, 3, and 4 to make a table for $y = 3^x$. Then make a second table for $y = \log_3 x$. Verify that each value of $\log_3 x$ is the power to which 3 must be raised to obtain x.

5. Make a table of values for $y = \log_5 x$, using the x-values 5, 25, 125, and 625.

6. Use the observation that logarithms are disguised exponents to evaluate each of the following.

   a. $\log_4 64$ **b.** $\log_2 128$

   c. $\log_7 7$ **d.** $\log_{12} 1$

   e. $\log_6 \dfrac{1}{6}$ **f.** $\log_9 3$

7. (*Writing to Learn*) Explain why it must be true that $\log_b 1 = 0$ and $\log_b b = 1$ for any base b.

In Section 7-1 you learned that a table with equally spaced x-values fits an exponential function $y = Ab^x$ if and only if the ratio of the consecutive y-values is constant. Interchanging x and y, we can discover how to tell whether a table fits the inverse of a function $y = Ab^x$. In Exercise 41, you will verify that such a function has the form $y = \log_b x + k$. Therefore,

A table with equally spaced y-values fits a function $y = \log_b x + k$ if and only if the ratio of consecutive x-values is constant.

EXERCISES   *In Exercises 8–11, decide whether the table fits a function $y = \log_b x + k$.*

8.

x	y
2	-4
8	-3
32	-2
128	-1
512	0

9.

x	y
162	1
108	3
72	5
48	7
32	9

10.

x	y
1	4
1.1	5
1.11	6
1.111	7
1.1111	8

11.

x	y
1	4
1.1	5
1.21	6
1.331	7
1.4641	8

Until we can verify analytically that functions $y = \log_b x + k$ are inverses of functions $y = Ab^x$, Exercises 12 and 13 will provide some numerical evidence.

12. (*Making Observations*)

   a. Use the x-values 0, 1, 2, 3, and 4 to make a table for $y = 3^x$.

   b. Make a second table for $y = \log_3 x$.

   c. Use your table in part (b) to make a table for $y = \log_3 x - 2$.

13. a. (*Making Observations*) Use the x-values $-2, -1, 0, 1,$ and 2 to make a table for $h(x) = 9 \cdot 3^x$.

   b. Make a second table for the function h^{-1}.

   c. (*Writing to Learn*) Explain why your tables in Exercises 12c and 13b suggest that $y = 9 \cdot 3^x$ and $y = \log_3 x - 2$ are inverses.

Evaluating Logarithms on Your Calculator   Until the advent of computers, the only convenient way to evaluate logarithmic functions accurately was to use lengthy tables first compiled by Napier. (Napier also invented the decimal point.) According to Isaac Asimov's *Biographical Encyclopedia of Science and Technology* (Doubleday & Co., 1982),

Napier spent twenty years working out rather complicated formulas for obtaining exponential expressions for various numbers. . . . Finally, in 1614, Napier published his tables of logarithms, which were not improved on for a century, and they were seized on with avidity. Their impact on the science of the day was something like that of computers on the science of our own time.

Your calculator almost certainly has two keys labeled "log" and "ln." Asimov's *Biographical Encyclopedia of Science and Technology* describes the historical reasons for the presence of those two keys, dating back to John Napier and his contemporary, Henry Briggs (1561–1630):

> [Briggs] is remembered chiefly for his reaction to Napier's publication of logarithms. He was lost in admiration for the beauty of the system and its simplicity (and aghast at his own stupidity in not seeing it until it was shown him).

> He went to the considerable trouble of making a trip to Edinburgh to see Napier and talk to him. Napier had written his exponential numbers as e^2, $e^{2.32}$, $e^{3.97}$, and so on, where e is an unending decimal fraction that starts 2.7182818284 . . . . There are good mathematical reasons for doing this and such Napierian or "natural" logarithms are still used in calculus. However, Briggs pointed out during his conversation with Napier the convenience of using exponential numbers such as 10^2, $10^{2.32}$, $10^{3.97}$, and so on. Logarithms in this fashion are called Briggsian or "common" logarithms and are almost invariably used for ordinary calculations.

> Briggs worked out the first logarithm tables for numbers from 1 to 20,000 and from 90,000 to 100,000 (to fourteen places!) in 1624. Briggs also invented the modern method of long division.

Thus 10 and e have always been, and continue to be, the most common and useful bases for logarithms. The notations $\log_{10} x$ and $\log_e x$ are usually abbreviated as $\log x$ and $\ln x$, respectively. Your two calculator keys will give values of logarithms to the base 10 (**common** logarithm) and base e (**natural** logarithm).

EXERCISE   **14.** (*Making Observations*)

   a. Guess the value of log 3, using the fact that the value is the power to which 10 must be raised to obtain 3.

   b. Use your calculator to evaluate log 3. Confirm your result on your calculator by raising 10 to that power. How does your result compare with your guess in part (a)?

   c. Use your calculator to evaluate ln 3. Confirm your result on your calculator by raising e to that power.

A GRAPHICAL VIEW OF LOGARITHMIC FUNCTIONS

Basic Graphs Since the functions $y = b^x$ and $y = \log_b x$ are inverses, their graphs are reflections of each other in the line $y = x$. Furthermore, each point on the graph of $y = \log_b x$ is a point on the graph of $y = b^x$ with its coordinates reversed. For example, Figure 7-16 shows graphs of $g(x) = 2^x$ and $g^{-1}(x) = \log_2 x$.

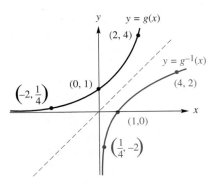

Figure 7–16

Since any function $f(x) = b^x$ with $b > 1$ has a graph similar to that of $g(x) = 2^x$, any function $f^{-1}(x) = \log_b x$ with $b > 1$ has a graph similar to that of $g^{-1}(x) = \log_2 x$.

EXERCISES

15. Sketch the graph of $y = \log_3 x$ by reflecting the graph of $y = 3^x$ in the line $y = x$. Identify the points obtained by moving $\left(-1, \frac{1}{3}\right)$, $(0, 1)$, and $(1, 3)$.

16. Repeat Exercise 15 for base 5, using the points $\left(-1, \frac{1}{5}\right)$, $(0, 1)$, and $(1, 5)$.

17. a. (*Making Observations*) For any $b > 0$ and any x, it is true that $\left(\frac{1}{b}\right)^x = b^{-x}$. Use this fact to demonstrate that $\log_{1/b} x = -\log_b x$. $\left[\textit{Hint:}\right.$
If $y = \left(\frac{1}{b}\right)^x$, then also $y = b^{-x}$. Rewrite each equation using logarithms.$\left.\right]$

   b. Use your result from part (a), together with those from Exercises 15 and 16, to sketch the graphs of $y = \log_{1/3} x$ and $y = \log_{1/5} x$.

Exercises 15–17 illustrate several graphical features of logarithmic functions $g(x) = \log_b x$. These features are the same as those exhibited by exponential functions $f(x) = b^x$, but with x and y interchanged.

- The range of $f(x) = b^x$ is $(0, \infty)$, so the domain of $g(x) = \log_b x$ is $(0, \infty)$. This means that the graph of $y = g(x)$ is entirely to the right of the y-axis.
- The domain of $f(x) = b^x$ is $(-\infty, \infty)$, so the range of $g(x) = \log_b x$ is $(-\infty, \infty)$.
- Because the graph of $f(x) = b^x$ passes through $(0, 1)$, the graph of $g(x) = \log_b x$ passes through $(1, 0)$.
- Since the x-axis is a horizontal asymptote for the graph of $f(x) = b^x$, the y-axis is a vertical asymptote for the graph of $g(x) = \log_b x$.
- Since $f(x) = b^x$ is increasing if $b > 1$, so is $g(x) = \log_b x$. (To say that a function is increasing is to say that as x increases, so does y. Interchanging x and y does not change the statement.) Similarly, $g(x)$ is decreasing if $0 < b < 1$.

In order to sketch the graph of $y = \log_b x$ quickly for any $b > 1$, plot the x-intercept at $(1, 0)$. Then plot $\left(\frac{1}{b}, -1\right)$ and $(b, 1)$, which are always on the graph. Finally, sketch a smooth curve through those points having the shape of Figure 7-17a. If $b < 1$, use the fact that $\log_{1/b} x = -\log_b x$ (see Exercise 17) to obtain a graph having the shape of Figure 7-17b.

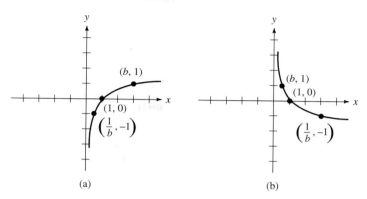

Figure 7–17

Graphical Transformations   The graphs of many logarithmic functions in the standard graphing form, $y = a \log_b(x - h) + k$, can be obtained by transforming the graph of $y = \log_b x$.

EXAMPLE 2   *Transforming the graph of* $y = \log_b x$

To sketch the graph of $y = \dfrac{1}{2} \log_3 x + 1$, begin with the basic graph $y = \log_3 x$ and make the changes

$$y = \log_3 x \rightarrow y = \frac{1}{2} \log_3 x \rightarrow y = \frac{1}{2} \log_3 x + 1$$

These changes compress the graph vertically by a factor of 2 and then shift it 1 unit up, as in Figure 7-18.

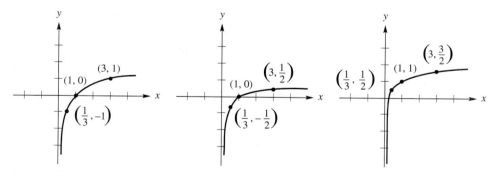

Figure 7–18

The domain of the function is $(0, \infty)$ and the range is $(-\infty, \infty)$. The function is increasing on its domain.   ∎

EXERCISES   *In Exercises 18–21, sketch the graph of each function by transforming an appropriate graph* $y = \log_b x$. *In each case,*

    a. Find the domain and range of the function.

    b. Decide whether the function is increasing or decreasing.

c. Identify the points obtained by moving $\left(\frac{1}{b}, -1\right)$, $(1, 0)$, and $(b, 1)$.

18. $y = \log_4 x - 3$ **19.** $y = 0.5 \log_2(x + 2)$

20. $y = 4 \log_7 x + 3$ **21.** $y = \log_3(x - 4) - 1$

22. (*Writing to Learn*) Describe how each transformation below affects the following features of $y = \log_b x$.

- domain and range
- asymptotes
- intercepts
- increasing/decreasing behavior

a. a horizontal stretch or compression

b. a vertical stretch or compression

c. a reflection in the x-axis

d. a reflection in the y-axis

e. a horizontal shift

f. a vertical shift

In Exercises 23–26, sketch the graph of the function and support your results with a calculator graph.

23. $y = \log x$ **24.** $y = \ln x$ (Use $e \approx 2.7$.)

25. $y = 3 \log 2x$ **26.** $y = 4 \ln x - 5$

Later in this section you will discover how to graph logarithmic functions with bases other than 10 or e on your calculator. However, you first need to learn some of the analytical properties of logarithms.

AN ANALYTICAL VIEW OF LOGARITHMIC FUNCTIONS

Basic Analytical Properties The analytical properties of logarithmic functions follow from the fact that they are inverses of exponential functions. In Section 7-3 you learned that pairs of inverse functions have the property that $f^{-1}[f(x)] = x$ for all x in the domain of f, and $f[f^{-1}(x)] = x$ for all x in the domain of f^{-1}. When stated in terms of exponential and logarithmic functions, these statements assert that

$$\log_b b^x = x \text{ for all real } x$$

and

$$b^{\log_b x} = x \text{ for all } x > 0$$

In particular, two facts you verified in Exercise 7 are worth remembering.

$$\log_b 1 = 0 \text{ for any base } b$$

and

$$\log_b b = 1 \text{ for any base } b$$

In Section 7-3 you also saw that the inverse of a function reverses the roles of the independent and dependent variables. Applied to logarithmic functions, this says that

$$y = \log_b x \quad \text{means} \quad x = b^y$$

This is another way of seeing that a logarithm is a disguised exponent. You can use this property to solve certain exponential and logarithmic equations, as in Example 3.

EXAMPLE 3   *Solving a logarithmic equation*

To solve $3 \log_5(2x + 50) = 7$

first rewrite the equation in the form

$$\log_b(\text{variable expression}) = (\text{number})$$

$$\log_5(2x + 50) = \frac{7}{3}$$

Next, convert the equation to the exponential form

$$(\text{variable expression}) = b^{(\text{number})}$$

$$2x + 50 = 5^{7/3}$$

$$2x + 50 \approx 42.75$$

$$2x \approx -7.25$$

$$x \approx -3.63$$

EXERCISES   *In Exercises 27–32, solve the equation.*

27. $4e^{5t} = 102.8$ **28.** $e^{s+2} - 27.1 = 1$ **29.** $2 \cdot 10^x + 0.42 = 3.29$

30. $\log_2 x = 3.5$ **31.** $\ln(4t - 100) = 0.57$ **32.** $5 \log(3r) = 13$

33. (*Interpreting Mathematics*) In **Crater Lake** (page 342), the charcoal sample used to determine the age of the lake contained 45% of its original carbon 14. Solving the equation $e^{-0.00012t} = 0.45$ gives the age of the lake in years. Verify that the lake is about 6700 years old.

34. a. (*Writing to Learn*) In Exercise 1 you obtained the function $y = 20 + 17(0.95)^t$ to describe the victim's body temperature in **A Murder in Ceylon**. You then used graphical methods to rewrite the function as $y = 20 + 17e^{-0.052z}$. The following analytical procedure attains the same thing. Explain why each step follows from the previous one:

$$y = 20 + 17(0.95)^t$$

$$= 20 + 17(e^{\ln 0.95})^t$$

$$\approx 20 + 17(e^{-0.052})^t$$

$$= 20 + 17e^{-0.052t}$$

b. (*Interpreting Mathematics*) In Exercise 2 you constructed the equation $27 = 20 + 17e^{-0.052t}$. Its solution represents the number of hours the victim had been dead when her body was discovered. Do the facts provide conclusive evidence of her husband's guilt or innocence? Tell why or why not.

Combining Logarithms Napier, Briggs, and other seventeenth-century mathematicians and scientists promoted the use of logarithms primarily because of their practical usefulness. Their usefulness derives from the following three distinctive properties.

If $m > 0$, $n > 0$, $b > 0$, and $b \neq 1$, then

$$\log_b(mn) = \log_b m + \log_b n$$

$$\log_b\left(\frac{m}{n}\right) = \log_b m - \log_b n$$

$$\log_b(m^a) = a \log_b m$$

The three properties can be stated in English as follows.

- To obtain the logarithm of a product, add the logarithms of the factors.
- To obtain the logarithm of a quotient, subtract the logarithms of the factors.
- To obtain the logarithm of a positive number raised to a power, multiply the exponent by the logarithm of the number.

Before discussing the practical importance of these properties, let's see why they are true. To see why $\log_b(mn) = \log_b m + \log_b n$, let's look at

$$b^{\log_b(mn)} \quad \text{and} \quad b^{(\log_b m + \log_b n)}$$

If these two powers of b are equal, then the exponents must be equal. (This is another way of saying that the function $y = b^x$ is one-to-one.)

Since $b^{\log_b x} = x$ for all $x > 0$, the first expression reduces to mn, and the second can be rewritten as $b^{\log_b m} b^{\log_b n}$, which also reduces to mn. This establishes that $\log_b(mn) = \log_b m + \log_b n$.

EXERCISE **35.** (*Making Observations*) Use arguments similar to the one just given to verify that

a. $\log_b\left(\dfrac{m}{n}\right) = \log_b m - \log_b n$

b. $\log_b(m^a) = a \log_b m$

These three properties lay at the heart of Napier's interest in logarithms. As Asimov says,

It occurred to Napier that all numbers could be expressed in exponential form. That is, 4 can be written as 2^2, while 8 can be written as 2^3, and 5, 6, and 7 can be written as 2 to some fractional power between 2 and 3. Once numbers were written in such exponential form, multiplication could be carried out by adding exponents, and division by subtracting exponents. Multiplication and division would at once become no more complicated than addition and subtraction.

In Chapter 4 you learned of Johannes Kepler's discovery that the orbits of the planets are ellipses, with the sun at one focus. This fact is the first of three famous planetary laws established by Kepler. News of the invention of logarithms reached him after he had spent almost a decade laboring over calculations based on the astronomical data of Tycho Brahe. The properties of logarithms greatly simplified his remaining work, and the development of Kepler's laws was the first significant application of logarithms. Example 4 both illustrates the properties of logarithms and indicates the manner in which they were used by Kepler to simplify computations.

EXAMPLE 4   *Using the properties of logarithms*

To evaluate an expression of the form

$$\frac{a\sqrt{b}}{c}$$

call the unknown quotient y, so that

$$y = \frac{a\sqrt{b}}{c}$$

Since logarithmic functions are one-to-one, this equation is equivalent to

$$\log y = \log \frac{a\sqrt{b}}{c}$$

The properties of logarithms tell us that multiplying and dividing numbers corresponds to adding and subtracting their logarithms.

$$\log y = \log a + \log \sqrt{b} - \log c$$

Also $\sqrt{b} = b^{1/2}$, and raising a number to a power corresponds to multiplying its logarithm by the exponent.

$$\log y = \log a + \frac{1}{2} \log b - \log c$$

In Exercise 36 you can see how Kepler might have continued from here to evaluate the expression $\frac{a\sqrt{b}}{c}$. ∎

EXERCISE   **36.** (*Making Observations*)

a. In Example 4, suppose that $a = 12.9$, $b = 535$, and $c = 8.24$. Kepler's tables of logarithms would have told him that

$$\log 12.9 \approx 1.11$$

$$\log 535 \approx 2.73$$

$$\log 8.24 \approx 0.92$$

$$\log 36.2 \approx 1.56$$

Use this information to evaluate $\frac{12.9\sqrt{535}}{8.24}$ without using your calculator.

b. How was the calculation made simpler by the use of logarithms?

In Exercises 37–40, write the expression in terms of log *a*, log *b*, *and* log *c*, *as in Example* 4.

37. $\log \dfrac{a}{bc}$ **38.** $\log \sqrt[3]{a}$

39. $\log a^2 b$ **40.** $\log \sqrt{\dfrac{b}{c}}$

41. (*Making Observations*) Interchange *x* and *y* in the equation $y = Ab^x$ and solve for *y*, and thus verify that the inverse of the function $y = Ab^x$ has the form $y = \log_b x + k$.

    With the computational power of your calculator, you do not need logarithms to simplify cumbersome arithmetic calculations. (Actually, in one sense you do need them, since your calculator does some of its work by using logarithms.) In Section 7-5, you will see an example of why the properties of logarithms are useful today.

    In an interesting aside, Asimov, who was also an accomplished science fiction writer, tells about one of Kepler's more unusual works.

> Kepler, by the way, wrote a story, "Somnium," about a man who traveled to the moon in a dream. For the first time the lunar surface was described as it really was, so that "Somnium" may be considered the first piece of authentic science fiction, as opposed to fantasy. It was published after Kepler's death.

    The properties of logarithms can be used to solve logarithmic equations more complicated than those in Exercises 27–32. When you combine logarithms in solving an equation, you must check to see whether your solutions yield undefined logarithms in the original equation. Example 5 illustrates.

EXAMPLE 5   *Solving a logarithmic equation*

To solve

$$\ln x + \ln(3x + 2) = 0$$

begin by expressing the left side as a single logarithm.

$$\ln[x(3x + 2)] = 0$$

Then rewrite the equation in an exponential form, remembering that "ln" means "$\log_e$."

$$x(3x + 2) = e^0$$
$$3x^2 + 2x = 1$$
$$3x^2 + 2x - 1 = 0$$
$$(3x - 1)(x + 1) = 0$$
$$x = \frac{1}{3} \quad \text{or} \quad x = -1$$

When $x = \dfrac{1}{3}$, the original equation becomes

$$\ln \frac{1}{3} + \ln 3 = 0$$

It is easy to verify that this is true. However, when $x = -1$, the equation becomes

$$\ln(-1) + \ln(-1) = 0$$

Since $\ln(-1)$ is not defined, the solution $x = -1$ is extraneous, and the only solution is $x = \dfrac{1}{3}$. ∎

EXERCISES   *In Exercises 42–45, solve the equation.*

42. $\log(2x - 1) - \log(x + 5) = 0$

43. $2 \log_2 x + \log_2 2x = 1$

44. $\log_x 120 - \log_x 15 = 3$

45. $\dfrac{\ln x}{\ln(x - 12)} = 2$

The Change of Base Formula   Now we are ready to return to the problem of using your calculator to graph logarithmic functions with bases other than 10 or e. Such graphs require the following **change of base formula**.

> For any bases $b > 0$, $b \neq 1$ and $c > 0$, $c \neq 1$, and any $x > 0$,
>
> $$\log_b x = \frac{\log_c x}{\log_c b}$$

Let's see why the formula is valid. Multiplying both sides of the equation by $\log_c b$, we obtain

$$(\log_c b)(\log_b x) = \log_c x$$

In words, this equation says that the power to which c is raised to obtain b, times the power to which b is raised to obtain x, equals the power to which c is raised to obtain x.

For example, suppose $c = 3$, $b = 9$, and $x = 6561$, which is both 3^8 and 9^4. Then the power to which 3 is raised to obtain 9 is 2, the power to which 9 is raised to obtain 6561 is 4, and the product $(2)(4) = 8$ is the power to which 3 is raised to obtain 6561. The statement is true in general because you multiply exponents when raising a power to a power.

In practice, c is nearly always chosen to be either 10 or e.

EXERCISES   **46.** (*Making Observations*) Evaluate $\log_2 7$ on your calculator by evaluating $\dfrac{\log 7}{\log 2}$, then by evaluating $\dfrac{\ln 7}{\ln 2}$. Confirm that the results are the same.

47. Evaluate each of the following on your calculator.

a. $\log_4 7$ **b.** $\log_7 4$

c. $\log_{2/3} 13$ **d.** $\log_{29} 29$

e. $\log_5 \sqrt{11}$ **f.** $\log_{0.4} 16$

g. $\log_3 17.6$ **h.** $\log_{12} \dfrac{16}{9}$

48. Use your calculator to graph each of the functions in Exercises 18–21.

ADDITIONAL EXERCISES

49. Make a table of values for $y = \log x$, using the x-values 1, 10, 100, 1000.

50. Make a table of values for $y = \log_3 x$, using values of x that yield integer values of y.

51. Use the observation that logarithms are disguised exponents to evaluate each of the following.

a. $\log_{36} 6$ **b.** $\log_6 36$

c. $\log_{11} \sqrt{11}$ **d.** $\log_3 \dfrac{1}{81}$

e. $\log_5 5^{-7}$ **f.** $\log_{1/2} 16$

g. $\log_{2/3} \dfrac{4}{9}$ **h.** $\log_{\sqrt{2}} 4$

In Exercises 52–55, decide whether the table fits a function $y = \log_b x + k$.

52.

x	y
-2	1
-1	2
0	4
1	8
2	16

53.

x	y
1	5
2	2
4	-1
8	-4
16	-7

54.

x	y
0.04	4
0.2	5
1	6
5	7
25	8

55.

x	y
1000	1
100	3
10	5
1	7
0.1	9

In Exercises 56–63, sketch the graph of each function by transforming an appropriate graph $y = \log_b x$. In each case,

a. Find the domain and range of the function.

b. Decide whether the function is increasing or decreasing.

c. Identify the points obtained by moving $\left(\dfrac{1}{b}, -1\right)$, $(1, 0)$, and $(b, 1)$.

Support your results with a calculator graph.

56. $y = \log(x + 2)$ **57.** $y = 4 \ln x$

58. $y = -2 \log_8(x - 4)$ **59.** $y = \dfrac{1}{3} \log_3 x + 3$

60. $y = 2 \ln x + 6$ **61.** $y = 6 \log 3x$

62. $y = 5 - \log x$ **63.** $y = \ln(x + 1) - 2$

64. **a.** (*Making Observations*) Graph each of the following pairs of functions on your calculator.

$$y = -\log_2 x \quad \text{and} \quad y = \log_{1/2} x$$

$$y = \log_2(8x) \quad \text{and} \quad y = \log_2 x + 3$$

b. (*Writing to Learn*) Explain what you observe about each pair of graphs in part (a), and support your observations analytically.

In Exercises 65–80, solve the equation.

65. $5^{3t-4} = 9$

66. $0.3^{x/2} - 1 = 2$

67. $3 \cdot 4^{v+5} - 2 = 0.21$

68. $e^{1/x} = 0.1$

69. $e^{4x+5} = 2e^{3x-1}$

70. $5^x = 10(4^x)$

71. $12 \log_8(x - 1) = 4$

72. $\log_{0.32}(5x) = 0$

73. $\ln \sqrt{t} = 5$

74. $\log x + \log(x - 3) = 1$

75. $2 \ln x = \ln(6 - x)$

76. $\log_3 x^2 - \log_3(x - 4) = 2$

77. $\log(x - 3) + \log(x^2 + 3x + 9) = \log 37$

78. $\log_x 27 + \log_x 8 = 3$

79. $\log_4(x + 5) + \log_4\left(\dfrac{1}{x}\right) = \dfrac{1}{2}$

80. $\ln x + \ln(x^2 + ex - 1) = 1$

Rewrite each expression in Exercises 81–88 to have no logarithms except $\log_4 x$.

81. $\log_4 \sqrt{x}$

82. $\log_4(16x)$

83. $\log_4 \dfrac{x}{16}$

84. $\log_4 x^4$

85. $\log_4 \dfrac{1}{x^3}$

86. $\log_4 x^3 - \log_4 x^2$

87. $\log_4(x^2 + x) - \log_4(x + 1)$

88. $\log_4 \dfrac{x^2}{y^3} + 3 \log_4 y$

89. Identify each statement as true or false.

    a. $\log_2(32x) = 5 + \log_2 x$

    b. $\log_b(x + 3) = \log_b x + \log_b 3$

    c. $\log_2(3x) = 3 \log_2 x$

    d. $\ln x^4 = 4 \ln x$

    e. $-3 \log(x - 2)^4 = \log(x - 2)^{-12}$

    f. $\log_b(bx) = \log_b x + 1$

90. (*Problem Solving*) Use the function $P = 6(1.02)^t$ from **Standing Room Only** (page 323) to predict when the world's population will reach "standing room only," one person for each square foot of land. You will need to know that the total landmass of the world is 57,280,000 square miles.

Exercises 91–93 illustrate the use of logarithmic scales to represent physical quantities.

91. (*Interpreting Mathematics*) The apparent brightness of a star is measured by the amount of light energy from the star reaching a unit area of the earth's surface. A star's apparent brightness is described by a number called its **magnitude** on a scale originated by the Greek astronomer Hipparchus (160–127 B.C.). In his textbook, *Foundations of Astronomy*, Michael Seeds says, "We must use logarithms to describe the magnitude system because Hipparchus designed it to represent the way stars look to our eyes, and our eyes work logarithmically."

The *difference* between the magnitudes M_1 and M_2 of two stars is related to the *ratio* of their apparent brightnesses I_1 and I_2. Specifically,

$$\frac{I_1}{I_2} = (2.512)^{M_2 - M_1}$$

a. Show that this equation can be rewritten

$$M_2 - M_1 = 2.5(\log I_1 - \log I_2)$$

b. How many times brighter is a star of magnitude 6 (the faintest magnitude visible to the unaided eye) than a star of magnitude 29 (the faintest magnitude visible to the Hubble space telescope)?

c. If one star is ten times as bright as another, what is the difference in their magnitudes?

92. (*Interpreting Mathematics*) The loudness of sounds is often measured in **decibels**. A sound's decibel level A is related to its intensity I by the equation

$$A = 10 \log I$$

if I is expressed in appropriate units.

a. At close range, the decibel levels of a truck and a jet plane are 100 and 120, respectively. What is the ratio of the intensity of the jet's noise to the intensity of the truck's noise?

b. In Exercise 70 of Section 7-1, you constructed the equation

$$t = 2^{21 - 0.2A}$$

to describe the maximum number of hours t that workers can safely be exposed to noise at a level of A decibels. Show that t varies approximately inversely with $I^{0.60}$.

c. Why might you want to know the information in part (b)?

93. (*Interpreting Mathematics*) The intensity of earthquakes is sometimes measured on the **Richter scale**. On this scale the magnitude M of an earthquake is related to its released energy E in ergs. Specifically,

$$M = \frac{\log E - 11.8}{1.5}$$

a. By what factor is the released energy of an earthquake multiplied by each increase of one on the Richter scale?

b. In terms of released energy, an earthquake of magnitude 8.0 is how many times as powerful as one of magnitude 4.0?

c. Two of the most famous earthquakes to occur in North America struck San Francisco in 1906 and the Alaskan peninsula in 1964. Their magnitudes were 8.25 and 8.4. How many ergs of energy were released by each earthquake?

94. (*Writing to Learn*) In the quote on page 368, Isaac Asimov says that "5, 6, and 7 can be written as 2 to some fractional power between 2 and 3." What do you think he means by the word *fractional* in this context?

SUPPLEMENTARY TOPIC
Slide Rules

A great advance in the use of logarithms was made in 1622 by William Oughtred (1574–1660). According to Asimov,

> Oughtred, who obtained his master's degree at Cambridge in 1600, was a minister and not a professional mathematician, but that makes little difference since he spent almost all the time he could spare on mathematics, even when it meant sleeping but two or three hours a night . . . .

> His greatest invention, however, came in 1622, and consisted of two rulers along which logarithmic scales were laid off. By manipulating the rulers and sliding one against the other, calculations could be performed mechanically by means of logarithms. We know it now as a slide rule, and, for centuries, engineers carried slide rules at least as lovingly as any physician ever carried his stethoscope and tongue depressor.

Figure 7-19 shows a sketch of a slide rule. The distance from the left end of a ruler to a number x on that ruler is proportional to log x.

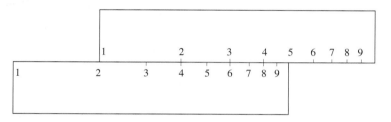

Figure 7–19

95. (*Writing to Learn*) Explain how the position of the slide rule in Figure 7-19 verifies that $2 \cdot 3 = 6$.

96. (*Extension*) Explain how a slide rule could be used to multiply numbers of any size. (*Hint*: Talk to a scientist or engineer who was educated before 1975.)

7–5   CURVE FITTING

Tables we encounter as we solve problems are seldom cooperative enough to fit one of the functions you have studied in this book. Frequently, we might want to construct a function that does not quite fit a table. For example,

- When we conduct an experiment to find the relationship between two physical quantities, we cannot measure the quantities precisely. Therefore, our table does not describe the relationship exactly.

- The table may have come from a function that is too complicated to find, or to be of practical value. We may be forced to, or prefer to, find a simpler function that fits the data approximately.

The process of constructing functions to approximate data is referred to as **curve fitting**. In this section, you will learn how to find a "best fit" for a table using linear, logarithmic, exponential, and power functions.

FITTING LINEAR FUNCTIONS TO DATA

Fitting a linear function to data is the simplest process, and it can be modified to fit logarithmic, exponential, or power functions to data. The following Mathematical Looking Glass describes a relationship that is approximately linear.

A MATHEMATICAL
LOOKING GLASS
Compasses

A navigational compass, such as you might use on a backcountry hike, works because the earth is a huge magnet. The magnetized compass needle points toward the earth's North Magnetic Pole, a point in arctic Canada about 800 miles from the geographic north pole, as shown in Figure 7-20.

Figure 7–20

By studying magnetic patterns in rocks whose ages are approximately known, scientists have concluded that the polarity (direction) of the earth's magnetic field changes about every 700,000 years. This means that a compass needle would have pointed toward the magnetic south pole 700,000 years ago and toward the magnetic north pole 700,000 years before that.

Near the time of each polarity reversal, the strength of the earth's magnetic field decreases, reaching zero at the time of the reversal. (The units used to measure field strength are the product of electrical current in amperes and area in square meters. We'll let the physicists explain why.) Using past measurements of field strength, we can predict when the next reversal will occur.

Table 33 shows the results of selected past measurements of field strength.

TABLE 33

Year	Strength
1835	8.56
1885	8.36
1910	8.25
1950	8.07
1985	7.87

Adapted from A. C. Fraser-Smith, Centered and Eccentric Geomagnetic Dipoles and Their Poles, 1600–1985, *Reviews of Geophysics*, Vol. 25, No. 1, 1987, pp. 1–16.

When you plot these data points in Exercise 1, you will see that they are almost, but not quite, collinear. Although this discrepancy is partly due to unavoidable errors in the measurements, physicists have proven that the relationship of time to field strength is not quite linear. However, a linear relationship that fits the past data closely can be expected to fit reasonably well for some time to come. In Exercises 1 and 2 you can find an appropriate linear function and use it to predict the time of the next polarity reversal. Although you will obtain a rough estimate, it will probably be accurate to within about a century.

For a few hundred years before and after the direction change, the magnetic field will be so weak as to render compasses useless, so if you happen to be out in the woods during this time, don't get lost! ■

EXERCISES

1. (*Making Observations*)

   a. Carefully plot the data from Table 33. (Suggestion: Scale both axes to intervals that do not extend very far beyond the data points.)

   b. Sketch the line that you think comes closest to passing through all the data points. (Such a line need not actually pass through any of the points.)

2. (*Problem Solving*) Estimate the coordinates of two points on your line, and use those points to find its equation. Then estimate the year in which the next direction reversal will occur.

3. (*Writing to Learn*) List several possible ways to decide what line best fits a given table. Which way makes the most sense to you? Explain why.

The most commonly accepted method of finding a line of best fit, used by your calculator, is called **linear regression**. You can use the following steps to find the **regression line** for data $(x_1, y_1), (x_2, y_2), (x_3, y_3), \ldots, (x_n, y_n)$.

- First, find the mean (the "average") of the x-values. This is $x = \dfrac{\Sigma x}{n}$, where Σx represents the sum $x_1 + x_2 + x_3 + \cdots + x_n$.
- Next, find the mean $y = \dfrac{\Sigma y}{n}$ of the y-values.
- Then, calculate the slope of the regression line.

$$m = \frac{n(\Sigma xy) - (\Sigma x)(\Sigma y)}{n(\Sigma x^2) - (\Sigma x)^2}$$

  Here $\Sigma xy = x_1 y_1 + x_2 y_2 + x_3 y_3 + \cdots + x_n y_n$, and the other Σ symbols have similar meanings.
- The equation of the regression line is

$$y - y = m(x - x)$$

EXERCISES 4. **a.** Carefully plot the following data, estimate a line of best fit, and write the equation of your line.

x	y
6	−2
8	5
9	7
10	10

   b. Find the equation of the regression line for the data. How does it compare with your result from part (a)?

5. The following table lists world record times for the mile run in selected years.

Year	Time
1875	4:24.5 (Walter Slade, England)
1895	4:17.0 (Fred Bacon, Scotland)
1915	4:12.6 (Norman Taber, United States)
1934	4:06.8 (Glen Cunningham, United States)
1954	3:59.4 (Roger Bannister, England)
1975	3:51.0 (Filbert Bayi, Tanzania)

   a. (*Creating Models*) Carefully plot the data, estimate a line of best fit, and write the equation of your line.

   b. (*Interpreting Mathematics*) Use your equation to estimate the record time for the mile run in 1985. The actual record, established that year by Steve Cram of England, was 3:46.3. How close is your estimate?

   c. (*Interpreting Mathematics*) Use your equation to estimate the record time for the mile run in the year 1, and in the year 3000.

   d. (*Writing to Learn*) For what years do you think your equation is approximately valid? Explain your answer.

Although you can estimate a line of best fit for any table, the fit is better for some relationships than for others. Furthermore, a good fit does not always indicate that the relationship is linear, but a poor fit always indicates that the relationship is not linear.

EXERCISES

6. Carefully plot each of the following tables, and assume that the values of x and y are subject to small errors. Decide whether x and y might be linearly related, and explain your answer.

a.

x	y
0.5	7
0.9	6
1.5	5
1.9	4

b.

x	y
−4	230
−2.5	250
0	350
1	460

c.

x	y
1000	10
1500	16
1800	19
2300	23

7. As parts (a)–(c) of this exercise, find the equation of the regression line for each table in Exercise 6. For which of the tables is the line reasonably close to all the data points?

FITTING LOGARITHMIC FUNCTIONS TO DATA

It is not appropriate to fit a linear function to every table. In the following Mathematical Looking Glass, patterns in the data suggest that a logarithmic function is more appropriate. Let's see how the technique of fitting a linear function to a table can be adapted to this situation.

A MATHEMATICAL LOOKING GLASS

Glottochronology

It has been said that the United States and England are two countries separated by the Atlantic Ocean and a common language. Although the English spoken by seventeenth-century American colonists was virtually identical to that spoken in England at the time, the English spoken on the two sides of the Atlantic is now distinctly different. For example, the words "lift" and "elevator" have the same meaning, but are unmistakably British and American, respectively.

Such differences illustrate that when two groups of people with a common language are separated, the language spoken by each will evolve over time. However, the fact that people in London and Denver can still understand one another shows that languages change slowly. Linguistic change is usually measured in millenia (thousands of years). Linguists need information about the rate at which languages evolve to estimate when two related languages descended from a common parent language or to determine the geographical origin of a family of languages.

Glottochronology is a mathematical method of studying the rates at which languages change. A more detailed description is given by Anthony Lo Bello and Maurice Weir in *Glottochronology*: *An Application of Calculus to Linguistics*, COMAP (1983). This analysis begins with a list of words, describing basic, time-independent concepts such as "sky," "fly," or "high,"

that were present in a language at some known benchmark date. The fundamental assumption of glottochronology, based on observations of languages
whose history is known, is that equal percentages of the words disappear
from the language over equal time intervals. This is equivalent to saying that
the number N of words remaining on the list t millenia after the benchmark
date is

$$N = Ae^{-kt}$$

In practice, N is often known but t is not. It is then convenient to regard
N as the independent variable and write the relationship in the form

$$t = m \ln N + b$$

The constants m and b are determined by curve fitting. To illustrate how
curve fitting works for approximately logarithmic data, let's begin with
Table 34. The table is based on a list of 210 Chinese words in use in the year
950 A.D. and shows the number of words remaining in the language t millenia
later. The first and last entries in the table are authentic. We made up the
others to create a simple illustration.

TABLE 34

Words (N)	Time in millenia (t)
210	0
200	0.2
186	0.5
174	0.8
167	1

The method you used for fitting linear functions to data is not practical
in this situation, because unless you are much better at graphing than we are,
you can't sketch an accurate logarithmic curve easily. Exercise 8 indicates
what to do instead.   ■

EXERCISES   8. (*Creating Models*)

a. Complete the following extension of Table 34.

N	$n = \ln N$	t
210		0
200		0.2
186		0.5
174		0.8
167		1

b. Carefully plot the data points (n, t), estimate a line of best fit, and
write the equation of your line.

c. Replace n with $\ln N$ in your equation to obtain a function $t = m \ln N + b$.

d. Graph the function in part (c) in a window large enough to include
each pair (N, t) from Table 34.

9. (*Making Observations*) Evaluate your function from Exercise 8c for
$N = 200, 186,$ and 174. Compare your results with the values of t given in
Table 34. How closely do your values fit the data?

10. **a.** Find the equation of the regression line for the data (n, t) from Table 34.

   b. Use the regression line to obtain a function $t = m \ln N + b$.

   c. Evaluate the function in part (b) for $N = 200$, 186, and 174. How closely do these values fit the data? Is the fit better or worse than what you obtained in Exercise 9?

11. Suppose that an ancient Chinese text has been found, in which half the words do not appear in the modern Chinese language. Use your function in Exercise 10 to estimate the age of the text.

FITTING EXPONENTIAL FUNCTIONS TO DATA

The dynamics of epidemics can be described by mathematical models based on differential equations. According to many such models, the growth of an epidemic during its early stages is approximately exponential. Let's see how to fit an exponential function to data on the HIV virus, which causes AIDS. (Later in this section you will decide whether a different type of function might fit the data better.)

 A MATHEMATICAL LOOKING GLASS
Growth of AIDS

During 1991, employees at the Veterans Administration Hospital in Pittsburgh were accidentally punctured on 83 occasions by needles contaminated by blood or body fluids from patients. Four of these patients were infected with the HIV virus. Although none of the employees contracted HIV, the risk of infection is a serious concern. Epidemiologist Angella Goetz gathered data in an attempt to quantify the expected degree of future risk. In making her assessment, she needed to estimate the rate at which the incidence of HIV among patients will increase over the next several years. This estimate requires a mathematical model for the spread of HIV among the entire population. Ms. Goetz consulted with one of your authors to analyze the reliability of several models. In Exercises 12–17 and 52 you will recreate two of the models and explore their reliability.

It has been difficult to collect reliable data on the HIV virus, because people with HIV exhibit no symptoms when they are first infected. However, the spread of HIV can be roughly estimated from the number of recorded deaths from AIDS. The U.S. Department of Health compiled the data in Table 35 for the years 1981–1989.

TABLE 35

Year (t)	Deaths from AIDS in each year	Total deaths from AIDS since 1981 (y)
1981	335	335
1982	950	1,285
1983	2,648	3,933
1984	5,082	9,015
1985	9,180	18,195
1986	13,257	31,452
1987	14,601	46,053
1988	9,335	55,388
1989	22,827	78,215

Fitting an exponential curve directly to the data will not work well, because it is difficult to sketch exponential graphs accurately. Once again, taking some logarithms produces an approximately linear relationship. Exercises 12 and 13 will guide you through the process.

EXERCISES

12. (*Writing to Learn*) In Table 35, let t represent time in years since 1981 and y represent the total number of deaths from AIDS since 1981. Assume that the relationship between t and y is described by an equation $y = Ae^{mt}$.

   a. Explain why the above equation is equivalent to $\ln y = \ln A + \ln e^{mt}$.

   b. Explain why the equation in part (a) is equivalent to $\ln y = \ln A + mt$.

   c. Explain why the equation in part (b) shows that t and $\ln y$ are linearly related.

13. (*Problem Solving*)

   a. Add a new column to Table 35, labeled $Y = \ln y$. Calculate the value of Y for each of the years 1981–1989.

   b. Carefully plot the data points (t, Y), estimate a line of best fit, and write the equation of your line.

   c. Your equation in part (b) has the form $Y = mt + b$. Use the fact that $Y = \ln y$ and $b = \ln A$ to obtain an equation $y = Ae^{mt}$.

   d. Graph your equation from part (c) in a window large enough to contain each point (t, y) from Table 35.

14. (*Making Observations*) Evaluate your function from Exercise 13c for each year from 1981 through 1989. Compare your results with the values of y given in Table 35. How closely do your values fit the data?

15. **a.** Find the equation of the regression line for the data (t, Y) from Table 35.

   b. Use the regression line to obtain a function $y = Ae^{mt}$.

   c. Evaluate the function in part (b) for each year from 1981 through 1989. How closely do these values fit the data? Is the fit better or worse than what you obtained in Exercise 14?

16. Use the function in Exercise 15b to predict the total number of deaths from AIDS at the end of 1991.

17. (*Writing to Learn*) The actual number of deaths from AIDS at the end of 1991 was 133,232. What conclusions can you draw by comparing this figure with your result from Exercise 16?

FYI A good fit between function and data does not always indicate that the function describes the relationship. In fact, because the spread of HIV is affected by so many complex physical and sociological factors, it is unlikely that any simple function can accurately model its long-term growth. In general, models of complex phenomena are accurate only for a short time.

FITTING POWER FUNCTIONS TO DATA

Now let's see how to fit a power function to a table, using astronomical data collected by the sixteenth-century Danish astronomer Tycho Brahe and analyzed by Johannes Kepler.

A MATHEMATICAL
LOOKING GLASS
Kepler's Third Law

In Section 7-4 you read about Kepler's three laws of planetary motion. His third law states that the square of a planet's period of revolution is proportional to the cube of its distance from the sun. Table 36 shows the distance of each planet from the sun, along with its period of revolution.

TABLE 36

Planet	Distance, in millions of kilometers	Period, in years
Mercury	57.9	0.241
Venus	108.2	0.615
Earth	149.6	1.000
Mars	227.9	1.881
Jupiter	778.3	11.86
Saturn	1427	29.46
Uranus	2870	84.01
Neptune	4497	164.8
Pluto	5900	247.7

EXERCISES

18. (*Writing to Learn*) In Table 36, let x and y represent distance and period, respectively. Explain why Kepler's third law says that x and y are related by a power function $y = Ax^m$. According to the statement of the law, what is the value of m?

19. (*Writing to Learn*)

a. Explain why the equation $y = Ax^m$ is equivalent to

$$\ln y = \ln A + \ln x^m$$

b. Explain why the equation in part (a) is equivalent to

$$\ln y = \ln A + m \ln x$$

c. Explain why the equation in part (b) shows that $\ln x$ and $\ln y$ are linearly related.

20. (*Creating Models*)

a. Use Table 36 to complete the following table.

x	$X = \ln x$	y	$Y = \ln y$
57.9		0.241	
108.2		0.615	
149.6		1.000	
227.9		1.881	
778.3		11.86	
1427		29.46	
2870		84.01	
4497		164.8	
5900		247.7	

b. Carefully plot the data points (X, Y), estimate a line of best fit, and write its equation.

c. Your equation in part (b) has the form $Y = mX + b$. Use the fact that $X = \ln x$, $Y = \ln y$ and $b = \ln A$ to obtain $y = Ax^m$.

d. Graph your equation from part (c) in a window large enough to contain each point (x, y) from Table 36.

21. (*Making Observations*) Evaluate your function from Exercise 20c for each value of x in Table 36. Compare your results with the values of y given there. How closely do your values fit the data?

22. a. Find the equation of the regression line for the data (X, Y) from Exercise 20a.

b. Use the regression line to obtain a function $y = Ax^m$.

c. Evaluate the function in part (b) for each planet. How closely do these values fit the data? Is the fit better or worse than what you obtained in Exercise 21?

23. Use the statement of Kepler's third law to write an equation relating x and y, and compare it with your equation from Exercise 22b.

CURVE FITTING ON YOUR CALCULATOR

Before proceeding any further, make sure you know how to perform curve fitting on your calculator. Consult your owner's manual for details.

If you use your calculator to construct a curve of best fit for a table, the calculator will also display a number r, called the **coefficient of linear correlation**. It is calculated by the formula

$$r = \frac{n(\Sigma xy) - (\Sigma x)(\Sigma y)}{\sqrt{n(\Sigma x^2) - (\Sigma x)^2}\,\sqrt{n(\Sigma y^2) - (\Sigma y)^2}}$$

All you need to know for now is that the value of r is always between -1 and 1 and that the best fit occurs when $|r|$ is close to 1.

EXERCISES

24. (*Creating Models*) The following table gives men's world record times for running selected distances as of September 1992.

Distance, in meters	Record time, in minutes : seconds
1,500	3:28.9 (Noureddine Morceli, Algeria)
2,000	4:50.8 (Said Aouita, Morocco)
3,000	7:29.0 (Moses Kiptanui, Kenya)
5,000	12:58.4 (Said Aouita, Morocco)
10,000	27:08.2 (Arturo Barrios, Mexico)
42,220 (marathon)	126:50 (Belayneh Densimo, Ethiopia)

Rounding all times to the nearest hundredth of a minute, use your calculator to fit each type of curve to the data. Record the value of r for each, and decide which gives the best fit.

a. a linear curve

b. a logarithmic curve

c. an exponential curve

d. a power curve

25. (*Creating Models*) Repeat Exercise 24 using the following women's world record times for the same distances.

Distance, in meters	Record time, in minutes : seconds
1,500	3:52.5 (Tatyana Kazankina, USSR)
2,000	5:28.7 (Maricica Puica, Romania)
3,000	8:22.6 (Tatyana Kazankina, USSR)
5,000	14:37.3 (Ingrid Kristiansen, Norway)
10,000	30:13.7 (Ingrid Kristiansen, Norway)
42,220 (marathon)	141:06 (Ingrid Kristiansen, Norway)

26. Use your calculator to fit the specified type of curve to the data, and compare your results with the ones you obtained in the indicated exercises.

   a. Fit a linear curve to Table 33 and compare your results with those in Exercise 2.

   b. Fit a linear curve to the table in Exercise 5 and compare your results with those in Exercise 5.

   c. Fit a logarithmic curve to Table 34 and compare your results with those in Exercise 8.

   d. Fit an exponential curve to Table 35 and compare your results with those in Exercise 13.

   e. Fit a power curve to Table 36 and compare your results with those in Exercise 20.

ADDITIONAL EXERCISES   *In Exercises 27–28, use the method of Exercises 1 and 2 to fit a linear function to the table.*

27.

x	y
1	13
2	10
3	9
4	7

28.

x	y
0.3	-120
0.5	100
0.8	360
1.0	600

29. Use the method of Exercise 8 to fit a logarithmic curve to the following table.

x	y
1	-5
7	-1
12	0
20	1

30. **a.** Find the equation of the regression line for the points (x, y) in Exercise 27.

   b. Find the equation of the regression line for the points (x, y) in Exercise 28.

   c. Find the equation of the regression line for the points $(\ln x, y)$ in Exercise 29, and use the regression line to fit a logarithmic curve to the data.

In Exercises 31–34, plot the pairs (ln x, y) *and assume that the values of x and y are subject to small errors. Decide whether x and y might be related by a logarithmic function.*

31.

x	y
2	0.6
5	1.4
10	2.0
20	2.6

32.

x	y
2	23
5	14
10	7
20	0

33.

x	y
2	0
5	1
10	10
20	50

34.

x	y
2	32
5	68
10	97
20	128

35. (*Creating Models*) During a mission in space, it is vitally important for both crew members and ground controllers to respond quickly and correctly to information displayed on control panels. For this reason, NASA has conducted studies measuring the time it takes to respond to varying numbers of choices. The results of one such study are shown in the table below, where N represents the number of choices and R is the average response time in seconds.

N	R
1	0.17
2	0.34
3	0.37
4	0.42
5	0.48
6	0.52
7	0.56
8	0.58
9	0.59

Adapted from Bernice Kastner, *Space Mathematics. A Resource for Secondary School Teachers*, NASA, 1985.

a. Carefully plot the data points (N, R), then plot the points (ln N, R) on a second set of axes. Decide whether the data looks more nearly linear or logarithmic.

b. Estimate the equation of either a line or a logarithmic curve of best fit, depending on your conclusion from part (a).

36. Use the method of Exercise 13 to estimate an exponential curve of best fit for the following table. Then use the regression line for the points $(x, \ln y)$ to find an exponential curve of best fit, and compare your results.

x	y
1	1
7	3
12	9
20	33

In Exercises 37–40, plot the pairs (x, ln y) and assume that the values of x and y are subject to small errors. Decide whether x and y might be related by an exponential function.

37.

x	y
3	8
5	9
7	10.5
9	12.5

38.

x	y
−6	1
−3	10
−1	13
0	14

39.

x	y
−1	0.5
1	2.3
2	3.9
4	14.8

40.

x	y
10	6.2
20	17.5
30	51.0
40	149.0

41. Use the method of Exercise 20 to estimate a power curve of best fit for the following data.

x	y
1	2
7	90
12	300
20	750

In Exercises 42–45, plot the pairs (ln x, ln y) on graph paper, and assume that the values of x and y are subject to small errors. Decide whether x and y might be related by a power function.

42.

x	y
1	10
2	5
3	3
4	2

43.

x	y
10	1
20	3.5
40	14.1
70	43.2

44.

x	y
3	2
5	10
6	15
8	12

45.

x	y
2	0.7
3	2.5
5	12.0
6	21.0

46. (*Writing to Learn*) Explain what goes wrong when you attempt to fit a power curve to a table when one of the entries is 0.

In Exercises 47–50, use your calculator to decide which type of curve (linear, logarithmic, exponential, or power) fits the table best.

47.

x	y
2.3	9.1
4.1	7.2
5.6	5.7
8.7	2.4

48.

x	y
5	12.1
10	13.3
15	14.6
20	16.0

49.

x	y
6	2.5
20	4.5
50	7.1
80	8.9

50.

x	y
1	20
3	85
4	180
7	1400

51. The following table gives the population of the United States as recorded in each census from 1790 through 1830.

Year	Population to the nearest thousand
1790	3,929,000
1800	5,308,000
1810	7,240,000
1820	9,564,000
1830	12,866,000

  a. (*Creating Models*) Use your calculator to decide whether the data most nearly fits a linear, logarithmic, exponential, or power function. Find the equation of the curve that fits best.

  b. (*Interpreting Mathematics*) Verify that your result from part (a) predicts the actual 1860 census figure of 31,433,000 within half a million people, but overestimates the 1870 figure of 38,558,000 by almost 3 million.

  c. (*Making Observations*) What occurred in the United States during the 1860s that would help explain the discrepancy between the predicted and actual populations for 1870?

52. (*Making Observations*) In **Growth of AIDS**, you discovered that an exponential curve of best fit for data on AIDS deaths through 1989 failed miserably to predict the number of deaths through 1991. Find a power curve of best fit for the same data, use it to predict the number of deaths through 1991, and compare your result with the actual total of 133,232.

53. (*Writing to Learn*) To find the relationship between two variables, scientists sometimes plot data on **semilog paper** (Figure 7-21a), on which the y-axis is scaled logarithmically (the scale mark for $y = y_0$ is at a distance of log y_0 units above the x-axis). They also may plot data on **log-log paper** (Figure 7-21b), on which both axes are scaled logarithmically.

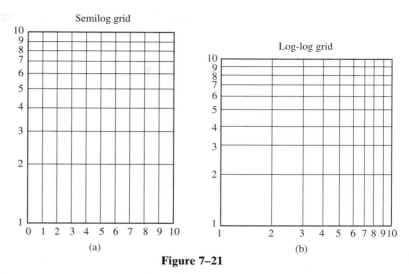

Figure 7–21

Decide what type of data would plot linearly on each type of paper, and explain why.

CHAPTER REVIEW

Complete Exercises 1–44 (Writing to Learn) before referring to the indicated pages.

WORDS AND PHRASES

In Exercises 1–11, explain the meaning of the words or phrases in your own words.

1. (pages 324, 327)   **exponential function**

2. (page 332)   **doubling time**

3. (page 332)   **geometric sequence**

4. (page 334)   **geometric series**

5. (page 334)   **sum of an infinite geometric series**

6. (page 341)   **the number e**

7. (page 341)   **continuous compounding** of interest

8. (page 346)   **inverse** of a function

9. (page 347)   **one-to-one** function

10. (page 360)   **logarithm, logarithmic function**

11. (page 376)   **curve fitting**

IDEAS

12. (page 325)   How do you decide whether a table fits an exponential function? Make up a table to illustrate your explanation.

13. (page 326)   Describe the process of fitting an exponential function to a table. Make up a table that fits an exponential function and find its equation to illustrate your explanation.

14. (page 327)   The equation $y = Ab^x$ defines an exponential function as long as $A > 0$, $b > 0$, and $b \neq 1$. For what reasons are those restrictions imposed on A and b?

15. (pages 327, 328)   How are the irrational exponents defined?

16. (pages 328, 329)   Describe the graph of a function $f(x) = Ab^x$. In particular, state its domain and range, and tell whether it is increasing or decreasing. Be sure to consider both the cases $b > 1$ and $0 < b < 1$.

17. (page 330)   Describe the process of obtaining the graph of a function $y = Ab^{x-h} + k$ from that of $y = b^x$. Illustrate your explanation with a specific function.

18. (page 333)   Explain why a geometric sequence is an exponential function. What is the difference between geometric sequences and other exponential functions?

19. (page 334)   What is the difference between a geometric sequence and a geometric series? Make up an example to illustrate the difference.

20. (pages 334, 335) How do you find the sum of a finite geometric series? Under what conditions does an infinite geometric series have a sum? How is the formula for the sum of an infinite geometric series derived from the formula for the sum of a finite geometric series?

21. (pages 340, 341) How is the formula for continuously compounded interest derived from the formula for interest compounded m times per year? What expression in n approaches a value of e if n is large? What is the value of e to the nearest tenth?

22. (pages 343, 346) Describe the process of rewriting a given exponential function to have a base of e.

23. (page 346) How does the definition of a pair of inverse functions express the idea that each function in the pair undoes the action of the other? Make up a pair of inverse functions to illustrate your response.

24. (page 347) How can you tell whether a table, a graph, or an equation describes a one-to-one function? Make up specific functions to illustrate your response.

25. (pages 347–352) Why is it that only a one-to-one function can have an inverse?

26. (page 347) How can you find the inverse of a one-to-one function that is described by a table? Illustrate the process with a specific table.

27. (page 348) How can you find the inverse of a one-to-one function that is described by a graph? Illustrate the process with a specific graph.

28. (pages 349, 350) How can you find the inverse of a one-to-one function that is described by an equation? Illustrate the process with a specific equation.

29. (pages 351, 352) When you find the inverse of a one-to-one function f analytically, why must you evaluate $f \circ f^{-1}$ and $f^{-1} \circ f$ as part of the process?

30. (pages 353–355) How can it happen that a function is not one-to-one, but becomes one-to-one when the variables are given physical meanings?

31. (pages 360–364) How do we know that a function $f(x) = b^x$ ($b > 0$, $b \neq 1$) has an inverse? Why do we need the new name *logarithm* to describe the inverse function?

32. (page 360) How can tables be used to show that logarithms are disguised exponents? Make up a table to illustrate your explanation.

33. (page 362) How can you decide whether a table fits a logarithmic function? Make up a table to illustrate your explanation.

34. (page 371) What two bases for logarithmic functions are used most often?

35. (page 364) Describe the graphical features of a function $f(x) = \log_b x$ when $b > 1$ and when $0 < b < 1$. What are the corresponding graphical features of the function $g(x) = b^x$?

36. (page 365) Describe the process of obtaining the graph of $f(x) = a \log_b(x - h) + k$ by transforming a basic graph $y = \log_b x$. Make up a specific function to illustrate your description.

37. (page 366) Explain why for any base b, $\log_b b^x = x$ for all real x, $b^{\log_b x} = x$ for $x > 0$, $\log_b 1 = 0$, and $\log_b b = 1$.

38. (page 368) In what way can we rewrite the logarithm of a product, a quotient, or a number raised to a power? How are these properties of logarithms related to the laws of exponents?

39. (page 371) What does the change of base formula say? How can it be used to evaluate logarithms on your calculator? Perform a specific calculation to illustrate your explanation.

40. (page 375) Why might we want to construct a function that does not quite fit a given table?

41. (page 377) Must a line of best fit for a given set of data pass through any of the data points? Why or why not?

42. (page 380) If the points $(\ln x, y)$ are collinear, then x and y are related by a logarithmic function. Show why this is true.

43. (page 382) If the points $(x, \ln y)$ are collinear, then x and y are related by an exponential function. Show why this is true.

44. (pages 383, 384) If the points $(\ln x, \ln y)$ are collinear, then x and y are related by a power function. Show why this is true.

CHAPTER 8

TRIGONOMETRIC FUNCTIONS

8–1 INTRODUCTION TO THE SINE AND COSINE FUNCTIONS

In Chapter 7 you learned that exponential and logarithmic functions cannot be expressed algebraically. Such functions are sometimes referred to as **transcendental functions** because they transcend, or go beyond, algebraic expression. In Chapters 8–10 you will encounter a class of transcendental functions developed more than 2000 years ago by Greek astronomers to chart the positions of stars, calculate the circumference of the earth, and measure the distance from the earth to the moon. These functions, called trigonometric functions, have also been used to determine a ship's location at sea and to represent musical sounds mathematically. Let's begin our study of trigonometric functions inside the Tevatron, the nuclear particle accelerator described in **Fermilab** (page 175).

 A MATHEMATICAL LOOKING GLASS
Fermilab 2

In **Fermilab** you learned how quarks are produced from high-speed collisions between protons and antiprotons. To attain the necessary speeds, the protons and antiprotons are accelerated in opposite directions around a circular track, represented by the unit circle $x^2 + y^2 = 1$ in Figure 8-1.

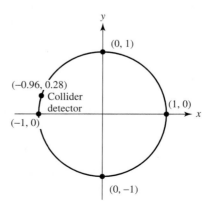

Figure 8–1

The shotmaster must make sure the collision occurs within the collider detector, at $(-0.96, 0.28)$ in Figure 8-1. For this reason, the positions of the proton and antiproton must be known at all times. To study their motion, let's assume that their speeds are a constant (and unrealistically slow) 1 km/sec.

In Figure 8-2, if a proton starting at $(1, 0)$ travels counterclockwise for t seconds, it traces out the arc PQ. If an antiproton starting at $(1, 0)$ travels clockwise for θ seconds, it traces out the arc PQ'. Both arcs are θ kilometers long. To distinguish between the paths of the two particles, it is conventional to consider the counterclockwise direction positive and the clockwise direction negative. Thus, the **directed arc** traced out in θ seconds has a **directed length** of θ kilometers for the proton and $-\theta$ kilometers for the antiproton. (For brevity, we will usually say "arc of directed length θ" to mean "directed arc of directed length θ.")

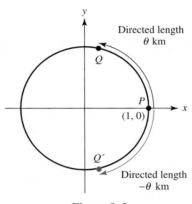

Figure 8–2   ■

TWO WAYS OF DEFINING THE SINE AND COSINE FUNCTIONS

The Unit Circle Definitions   The coordinates of the proton and antiproton in *Fermilab 2* can be described by functions $x(\theta)$ and $y(\theta)$. That is, the point $(x(\theta), y(\theta))$ represents the position of the proton after θ seconds if $\theta \geq 0$ and the position of the antiproton after $|\theta|$ seconds if $\theta < 0$.

Using more advanced mathematics, it can be shown that the functions $x(\theta)$ and $y(\theta)$ are transcendental. However, even though they cannot be expressed algebraically, they have been given names because they are so important. They are called the **cosine** and **sine** of θ, usually written

$$x(\theta) = \cos \theta$$
$$y(\theta) = \sin \theta$$

These functions can be defined in more general language. An arc of directed length θ on the unit circle is said to be in **standard position** if its **initial** (starting) **point** is $(1, 0)$, as in Figure 8-3. The cosine and sine of θ are the x- and y-coordinates at the **terminal** (ending) **point** of the arc. The names *cosine* and *sine* have not been arbitrarily chosen. You will discover their significance later in this section.

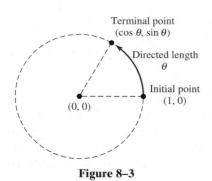

Figure 8–3

We cannot compute $\cos \theta$ and $\sin \theta$ exactly for an arbitrary value of θ, but we can compute them exactly for selected values of θ by using the fact that the circumference of the unit circle is 2π. For example, Figure 8-4 shows two arcs in standard position on the unit circle.

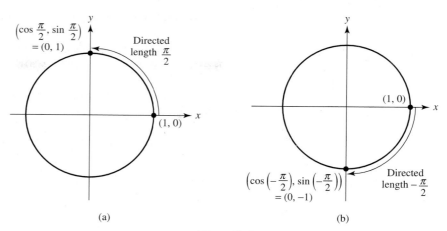

Figure 8–4

One arc has directed length $\dfrac{\pi}{2}$, and thus extends a quarter of a revolution counterclockwise from (1, 0) to (0, 1). Since its terminal point is (0, 1), we can conclude that $\cos \dfrac{\pi}{2} = 0$ and $\sin \dfrac{\pi}{2} = 1$. The other arc has directed length $-\dfrac{\pi}{2}$, and thus extends a quarter of a revolution clockwise from (1, 0) to (0, −1). Since its terminal point is (0, −1), we can conclude that $\cos\left(-\dfrac{\pi}{2}\right) = 0$ and $\sin\left(-\dfrac{\pi}{2}\right) = -1$.

EXERCISES

1. For each value of θ, sketch the unit circle and the arc with directed length θ beginning at (1, 0). Then find the point $(\cos \theta, \sin \theta)$.

    a. $\theta = \dfrac{5\pi}{2}$ **b.** $\theta = 4\pi$ **c.** $\theta = 0$ **d.** $\theta = -\pi$

    e. $\theta = 17\pi$ **f.** $\theta = -\dfrac{7\pi}{2}$

2. Evaluate each of the following:

    a. $\cos \dfrac{7\pi}{2}$ **b.** $\sin(-3\pi)$ **c.** $\sin\left(-\dfrac{\pi}{2}\right)$ **d.** $\cos 6\pi$

Your calculator can approximate $\cos \theta$ and $\sin \theta$ to 12 decimal places for any real number θ, as you will have it do in Exercise 3. Make sure your calculator is in radian mode first. (You will learn the meaning of the word *radian* later in this section.)

EXERCISE

3. For each value of θ, sketch the unit circle and the arc with directed length θ beginning at (1, 0). (Estimate the directed length of each arc from the fact that the circle's circumference is 2π, or slightly greater than 6.) Estimate the values of $\cos \theta$ and $\sin \theta$, then evaluate each on your calculator and compare.

    a. $\theta = 1$ **b.** $\theta = -\dfrac{\pi}{4}$ **c.** $\theta = 2$ **d.** $\theta = -2$ **e.** $\theta = \dfrac{\pi}{6}$ **f.** $\theta = -\dfrac{\pi}{6}$

Although the variable θ has represented the directed length of an arc throughout our discussion, it is often convenient to think of θ as a real number that may or may not represent a directed length. You will see why in Section 8-2. From this perspective, $\cos \theta$ and $\sin \theta$ are functions whose domains consist of all real numbers.

The Right Triangle Definitions The unit circle definitions of the sine and cosine functions comprise only a part of their rich history. During the past 22 centuries these functions have been defined in several seemingly unrelated ways to solve seemingly unrelated problems. For example, in Section 8-2 you will see how sines and cosines have been interpreted by astronomers as ratios of sides in right triangles. It is worthwhile to know the history of these functions for several reasons.

- It will help you to see the unifying themes in the descriptions of the functions and the problems to which they are applied.
- It will remind you that mathematics is developed to satisfy both practical needs and intellectual curiosity.
- It will allow you to appreciate the centuries of labor that have enabled you to evaluate cosines and sines instantly on your calculator.

During the second century B.C., the Greek mathematician and astronomer Hipparchus computed the length of a chord in a circle, such as the line segment AB in Figure 8-5a, from the length of the corresponding arc AB.

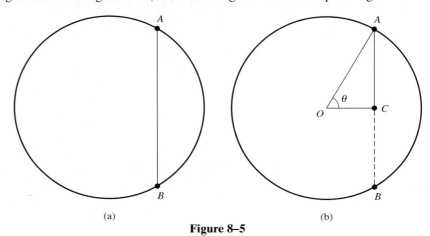

(a)   (b)

Figure 8–5

In a book, now lost, he recorded chord lengths for arcs with lengths up to half the circumference in a circle of radius 60 units. Hipparchus used these measurements to calculate the positions of stars and planets. In the sixth century A.D., the Hindu Aryabhata made a table of half-chords, such as AC in Figure 8-5b, for arcs with lengths up to a quarter of the circumference in a circle of radius 3438 units. Georg Joachim Rhäticus (1514–1576) associated the distance $\overline{AC}$ with a **central angle** (angle with its vertex at the center of the circle, the angle θ in Figure 8-5b) rather than an arc. His tables contained values of both $\overline{AC}$ and $\overline{OC}$ for angles between $0°$ and $90°$. Rhäticus was the first mathematician to define the sine and cosine functions as ratios of the sides of a right triangle.

In the right triangle OAC, if the measure of $\angle AOC$ is $\theta°$, then

$$\sin \theta° = \frac{AC}{OA} = \frac{\text{opposite side}}{\text{hypotenuse}}$$

$$\cos \theta° = \frac{OC}{OA} = \frac{\text{adjacent side}}{\text{hypotenuse}}$$

See Figure 8-6.

Figure 8–6

EXERCISE   4. (*Making Observations*)

 a. Explain why $\theta_1 = \theta_2$ in the figure below.

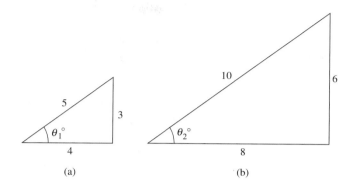

(a)   (b)

 b. Find $\sin \theta_1^\circ$ and $\cos \theta_1^\circ$ according to Rhäticus' definitions.

 c. Find $\sin \theta_2^\circ$ and $\cos \theta_2^\circ$ according to Rhäticus' definitions.

Exercise 4 illustrates that by Rhäticus' definitions, the values of $\sin \theta^\circ$ and $\cos \theta^\circ$ depend only on the value of θ, and not on the size of the right triangle.

Shortly after Rhäticus, Pitiscus (1561–1613) introduced the word **trigonometry** (tri-gono-metry, literally "three-angle-measure," or the measurement of triangles) to refer to the branch of mathematics dealing with such tables. Eventually, it became standard practice to generate these tables using a circle of radius 1. In this setting, the length $\overline{AC}$ in Figure 8-5 is the sine of θ°. The word can be traced by a tortuous path back to a Hindu word, meaning "half-chord," used by Aryabhata. By A.D. 1150, it had been translated from Hindu to Arabic to Latin (incorrectly) as *sinus*, meaning cove or bay. The length $\overline{OC}$ in Figure 8-5 is the *cosine* of θ°, which denotes the *sine* of the *c*omplementary angle $90^\circ - \theta^\circ$. In Exercise 5 you can verify that the sine of an angle is the cosine of its complement.

EXERCISE   5. (*Making Observations*) In Figure 8-7, $\phi^\circ = 90^\circ - \theta^\circ$. Explain why $\cos \theta^\circ = \sin \phi^\circ$ according to Rhäticus' definitions.

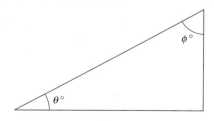

Figure 8–7

We cannot compute $\sin \theta^\circ$ and $\cos \theta^\circ$ exactly for an arbitrary value of θ between 0 and 90. However, in Exercises 6 and 7 you can calculate $\sin \theta^\circ$ and $\cos \theta^\circ$ exactly for a few selected values of θ by using the Pythagorean theorem.

EXERCISES   6. (*Writing to Learn*) Explain why Figure 8-8a shows that $\sin 45° = \dfrac{\sqrt{2}}{2}$ and $\cos 45° = \dfrac{\sqrt{2}}{2}$. (*Hint*: Explain why $\overline{OC} = \overline{AC}$. Then use the Pythagorean theorem.)

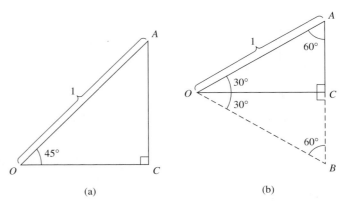

(a)   (b)

Figure 8–8

7. (*Writing to Learn*)

   a. Explain why Figure 8-8b shows that $\sin 30° = \dfrac{1}{2}$ and $\cos 30° = \dfrac{\sqrt{3}}{2}$. (*Hint*: First explain why $\overline{AC} = \overline{BC}$ and $\overline{AB} = \overline{OA}$.)

   b. Explain why Figure 8-8b shows that $\sin 60° = \dfrac{\sqrt{3}}{2}$ and $\cos 60° = \dfrac{1}{2}$. (*Hint*: See Exercise 5.)

8. Use your calculator to evaluate each of the following. Make sure your calculator is in degree mode.

   a. $\sin 23°$ **b.** $\cos 16°$

   c. $\cos 65°$ **d.** $\sin 83°$

COMPATIBILITY OF THE TWO DEFINITIONS

You have now seen two apparent differences between the unit circle and right triangle definitions of cosine and sine.

- According to the unit circle definitions, the input to the cosine and sine functions is an abstract real number, which we can think of as the directed length of an arc. According to the right triangle definitions, the input is the degree measure of an angle.

- According to the unit circle definitions, the domains of the cosine and sine functions include all real numbers. According to the right triangle definitions, the domains include only degree measures between 0° and 90°.

Directed Angles We can resolve these differences by establishing a correspondence between arcs and central angles in the unit circle. Specifically, a directed arc in standard position determines a **directed angle** in standard position, whose **initial side** extends from $(0, 0)$ to $(1, 0)$ and whose **terminal side** extends from $(0, 0)$ to the terminal point (a, b) of the arc, as in Figure 8-9.

Similarly, every directed angle in standard position determines a directed arc in standard position.

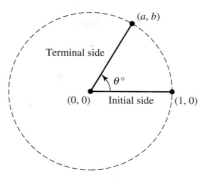

Figure 8–9

Figure 8-10 shows directed angles of 90°, 450°, and −270° in standard position.

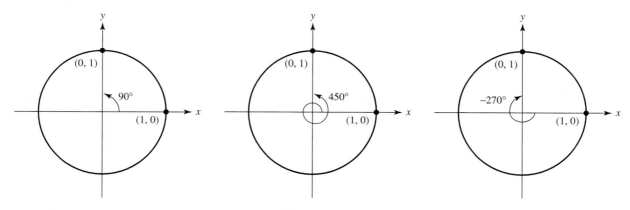

Figure 8–10

If the arc in Figure 8-9 has directed length *t* and the directed angle is θ°, we can now regard the point (*a*, *b*) as both (cos *t*, sin *t*) and (cos θ°, sin θ°). For example, Figure 8-10 shows that cos 90°, cos 450°, and cos (−270°) are all 0, and sin 90°, sin 450°, and sin (−270°) are all −1. In Exercises 11 and 12, you can discover how this point of view resolves the differences between the unit circle and right triangle definitions of the cosine and sine functions.

EXERCISES

9. For each directed angle, find the directed length of the corresponding arc in the unit circle. Then draw the angle in standard position, and find the cosine and sine of the angle.

    a. 270°   **b.** 540°

    c. −360°   **d.** −450°

10. For an arc of each directed length in the unit circle, find the corresponding directed angle. Then draw the angle in standard position, and find the sine and cosine of the angle.

    a. π **b.** 2π

    c. $-\dfrac{\pi}{2}$ **d.** $-\dfrac{7\pi}{2}$

11. (*Making Observations*) In Exercise 7a you discovered that, according to the right triangle definitions, $\cos 30° = \dfrac{\sqrt{3}}{2}$ and $\sin 30° = \dfrac{1}{2}$.

    a. Draw a directed angle of 30° in standard position in the unit circle.

    b. At what point does the unit circle intersect the terminal side of the angle? (*Hint*: Look at your solution to Exercise 7a.)

    c. Evaluate $\cos 30°$ and $\sin 30°$ according to the unit circle definitions, and compare the results with those from Exercise 7a.

12. (*Writing to Learn*) If θ is between 0 and 90, explain why $\cos \theta°$ has the same value according to both the right triangle and unit circle definitions.

Radians We have now established a close connection between directed arcs and directed angles in the unit circle. For example, an arc of directed length $\dfrac{\pi}{2}$ determines a directed angle of 90°, and vice versa. It is therefore natural to regard the directed length as a measure of the angle. The following definition formalizes this idea.

> If an arc on a circle of radius r has directed length t, we will say that the corresponding (directed) central angle has a measure of $\theta = \dfrac{t}{r}$ **radians**, as in Figure 8-11a.

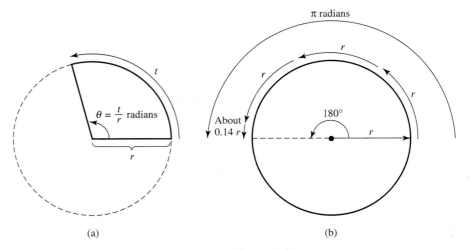

(a) (b)

Figure 8–11

The name *radian* reflects the relationship between angles and arcs. Specifically, an angle of θ radians corresponds to an arc whose directed length is θ times the radius of the circle. Thus, measuring an angle in radians is like using the radius of the circle as a measuring stick, as suggested by Figure 8-11b. In the unit circle, $r = 1$, so an angle of θ radians corresponds to an arc of directed length θ.

FYI   To obtain the radian measure of an angle you must divide one length by another. In the process the units of length cancel, so that the result is dimensionless (has no units of measurement). For this reason, angles expressed in radians are usually written as abstract numbers. For example, an arc of directed length $t = 5\pi$ meters in a circle of radius $r = 20$ meters determines an angle of $\dfrac{5\pi \text{ meters}}{20 \text{ meters}} = \dfrac{\pi}{4}\left(\text{equivalently, } \dfrac{\pi}{4} \text{ radians}\right)$.

EXERCISE   **13. a.** Find the radian measure of the angle determined by an arc with directed length 10π on a circle of radius 40.

b. On a circle with radius 8, what is the directed length of an arc that determines an angle of $\dfrac{\pi}{4}$ radians?

c. (*Making Observations*) What is the degree measure of each of the angles in parts (a) and (b)? (*Hint*: What fraction of the circumference corresponds to each arc?)

Converting Between Radians and Degrees   Throughout this book, you will encounter problems in which one form of angular measure is more convenient than the other. You therefore need to know how to convert from one system of measurement to the other. To obtain the relationship between the two, observe that the central angle corresponding to a semicircle has degree measure 180° and radian measure π as illustrated in Figure 8-11b. This means that the fractions $\dfrac{180°}{\pi}$ and $\dfrac{\pi}{180°}$ each have a value of 1. Thus, if the radian measure of an angle is θ, its degree measure is

$$\theta \cdot \frac{180°}{\pi}$$

Similarly, if the degree measure of an angle is $\theta°$, its radian measure is

$$\theta° \cdot \frac{\pi}{180°}$$

FYI   How can you remember whether to multiply by $\dfrac{180°}{\pi}$ or $\dfrac{\pi}{180°}$? Here's one way.

If an angle is initially measured in *radians*, multiply by $\dfrac{180°}{\pi}$. Since the degree symbol is in the numerator, your result is expressed in degrees.

If an angle is initially measured in *degrees*, multiply by $\dfrac{\pi}{180°}$. Since the degree symbol is in the denominator, your final result does not have a degree symbol, and is therefore expressed in radians.

EXAMPLE 1   *Converting between degree and radian measures*

a. An angle of $\dfrac{5\pi}{12}$ has degree measure

$$\frac{5\pi}{12} \cdot \frac{180°}{\pi} = 75°$$

b. An angle of $-100°$ has radian measure

$$-\frac{100°}{1} \cdot \frac{\pi}{180°} = -\frac{5\pi}{9}$$

∎

EXERCISES

14. For the given radian measure, sketch the angle in standard position, and find its degree measure:

a. 3π **b.** 6π

c. $-\dfrac{\pi}{2}$ **d.** $-\dfrac{2\pi}{3}$

e. $\dfrac{9\pi}{4}$ **f.** -3

15. For the given degree measure, sketch the angle in standard position, and find its radian measure:

a. $450°$ **b.** $900°$

c. $-135°$ **d.** $-300°$

e. $756°$ **f.** $\left(-\dfrac{180}{\pi}\right)°$

16. Find both the radian and degree measure of the central angle corresponding to an arc of directed length t on a circle of radius r, for the given values of r and t.

a. $r = 3$, $t = \pi$ **b.** $r = 100$, $t = \pi$

c. $r = 1$, $t = 3$ **d.** $r = \dfrac{1}{\pi}$, $t = \dfrac{1}{2}$

17. (*Extension*) With your calculator in parametric mode, select an undistorted viewing window (see Section B-1), centered at the origin and extending at least 1.5 units from it in all directions. Graph the equations

$$x = \cos t$$
$$y = \sin t$$

for values of t in $[0, 2\pi]$. How might you have predicted the size and shape of the graph in advance?

18. (*Extension*) Find suitable undistorted windows in which to graph each of the following sets of equations for values of t in $[0, 2\pi]$:

a. $x = 3 \cos t$ **b.** $x = 7 \cos t$
  $y = 3 \sin t$ $y = 7 \sin t$

c. $x = 20 \cos t$ **d.** $x = \dfrac{1}{2} \cos t$
  $y = 20 \sin t$ $y = \dfrac{1}{2} \sin t$

Based on your observations, describe the graph of the equations $x = r \cos t$, $y = r \sin t$ for an arbitrary value of r.

ANGULAR VELOCITY

An object traveling around a circle of radius r traces an arc whose directed length is $f(t)$ units after t seconds, corresponding to a directed angle with radian measure $\theta(t) = \dfrac{f(t)}{r}$ after t seconds. Over a time interval of length Δt the object's average **speed** is $\dfrac{\text{distance}}{\text{time}} = \dfrac{|\Delta f(t)|}{\Delta t}$ and its average **velocity** is $\dfrac{\text{directed distance}}{\text{time}} = \dfrac{\Delta f(t)}{\Delta t}$. [Recall that the symbols Δt and $\Delta f(t)$ represent changes in t and $f(t)$.] If the object travels at a constant velocity, then $\dfrac{\Delta f(t)}{\Delta t}$ is constant, so $\dfrac{\Delta \theta(t)}{\Delta t}$ is also constant. The value of $\dfrac{\Delta \theta}{\Delta t}$ is called the object's **angular velocity**.

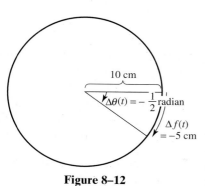

Figure 8–12

In Figure 8-12 an object travels clockwise around a circle of radius 10 cm with a speed of 5 cm/sec. Since it travels in the negative direction, its velocity is -5 cm/sec. In a time interval of $\Delta t = 1$ second, it traces an arc of directed length $\Delta f = -5$ cm, corresponding to a directed angle $\Delta \theta = \dfrac{-5 \text{ cm}}{10 \text{ cm}} = -\dfrac{1}{2}$ radian. Thus, its angular velocity is $\dfrac{\Delta \theta}{\Delta t} = -\dfrac{1}{2}$ radian per second. The object's angular velocity is equal to its velocity (-5 cm/sec) divided by the radius of the circle (10 cm). In general, if an object travels around a circle of radius r units at a velocity of v units per second, its angular velocity in radians per second is

$$\omega = \frac{v}{r}$$

EXERCISES
19. Find the angular velocity of the wheels on a car traveling at 60 mph if the radius of each tire is
    a. 15 in   **b.** 13 in

The position of an object moving in a circle is often calculated more easily from its angular velocity than from its velocity. In Exercise 20 you can develop a formula for the coordinates of an object traveling around a circle with a known angular velocity.

20. (*Writing to Learn*)
    a. If an object moves around a circle of radius r with angular velocity ω radians per second, explain why the distance it travels in t seconds is $\omega r t$.
    b. If the circle has equation $x^2 + y^2 = r^2$ and the object moves counterclockwise from $(r, 0)$, explain why its position after t seconds is $x = r \cos \omega t$, $y = r \sin \omega t$. (*Hint*: See Exercises 17 and 18.)

21. (Refer to **Fermilab 2**) For a proton that travels counterclockwise around the Tevatron at a speed of 300,000 km/sec, beginning at $(1, 0)$,
    a. Find its angular velocity.
    b. Use the angular velocity to calculate its position after 0.1 seconds.

SUMMARY

In the expressions $\cos \theta$ and $\sin \theta$, θ can be regarded as an abstract real number or the radian measure of an angle.

- If θ is an abstract real number, $(\cos\theta, \sin\theta)$ is defined as the terminal point of an arc of directed length θ in standard position.
- If θ is the radian measure of an angle, $(\cos\theta, \sin\theta)$ is defined as the point where the unit circle intersects the terminal side of the (directed) angle in standard position.

These definitions are compatible, so that $\sin\dfrac{\pi}{6} = \dfrac{1}{2}$ whether $\dfrac{\pi}{6}$ is regarded as an abstract real number or the radian measure of an angle.

In the expressions $\cos\theta°$ and $\sin\theta°$, $\theta°$ is always the degree measure of an angle.

- If θ has any real value, $(\cos\theta°, \sin\theta°)$ is defined as the point where the unit circle intersects the terminal side of the (directed) angle in standard position.
- If θ is between 0 and 90, $\sin\theta°$ can also be defined as the ratio of the opposite side to the hypotenuse and $\cos\theta°$ can also be defined as the ratio of the adjacent side to the hypotenuse in a right triangle containing an angle with degree measure $\theta°$.

These definitions are also compatible, so that $\sin 30° = \dfrac{1}{2}$ whether its value is calculated according to the unit circle definition or the right triangle definition.

Finally, all the definitions are compatible in that the values of the cosine and sine of an angle do not depend on whether the angle is expressed in degrees or radians. For example, $30° = \dfrac{\pi}{6}$ radians, and $\sin 30°$ and $\sin\dfrac{\pi}{6}$ each have a value of $\dfrac{1}{2}$.

EXERCISES

22. (*Making Observations*)

  a. With your calculator in degree mode, evaluate $\cos 20°$ and $\sin 20°$.

  b. Find the radian measure of a 20° angle.

  c. With your calculator in radian mode, evaluate the cosine and sine of the angle in part (b). Compare your answers with those you obtained in part (a).

23. (*Making Observations*)

  a. With your calculator in radian mode, evaluate $\cos 1$ and $\sin 1$.

  b. Find the degree measure of the angle whose radian measure is 1.

  c. With your calculator in degree mode, evaluate the cosine and sine of the angle in part (b). Compare your answers with those you obtained in part (a).

24. Use Figure 8-13 and the strategy of *Looking for Symmetry* to supply the following information about each directed angle whose terminal point is shown. Assume that the radian measure θ satisfies $0 \leq \theta < 2\pi$ in each case.

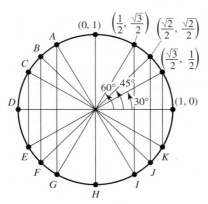

Figure 8–13

    a. the angle's degree measure

    b. the angle's radian measure

    c. $\sin \theta$

    d. $\cos \theta$

25. (*Making Observations*) Use Exercise 24 to complete Table 37.

TABLE 37

Degrees	Radians	Sine	Cosine
0°	0	0	1
30°	$\frac{\pi}{6}$	$\frac{1}{2}$	$\frac{\sqrt{3}}{2}$
45°	$\frac{\pi}{4}$	$\frac{\sqrt{2}}{2}$	$\frac{\sqrt{2}}{2}$
60°	$\frac{\pi}{3}$	$\frac{\sqrt{3}}{2}$	$\frac{1}{2}$
90°	$\frac{\pi}{2}$	1	0
120°			
135°			
150°			
		0	−1
	$\frac{7\pi}{6}$		
	$\frac{5\pi}{4}$		
	$\frac{4\pi}{3}$		
		−1	0
300°			
315°			
330°			
360°			

ADDITIONAL EXERCISES   26. Sketch the arc of directed length θ in standard position in the unit circle. Then identify the point $(\cos\theta, \sin\theta)$ without using your calculator. Use Table 37 if necessary.

a. $\theta = -\dfrac{\pi}{3}$ b. $\theta = 3\pi$

c. $\theta = \dfrac{9\pi}{4}$ d. $\theta = -\dfrac{3\pi}{2}$

e. $\theta = \dfrac{25\pi}{6}$ f. $\theta = -\dfrac{25\pi}{6}$

27. Sketch a directed angle of θ in standard position in the unit circle. Then identify the point $(\cos\theta, \sin\theta)$ without using your calculator. Use Table 37 if necessary.

a. $\theta = 135°$ b. $\theta = -30°$

c. $\theta = 180°$ d. $\theta = 420°$

e. $\theta = -405°$ f. $\theta = 510°$

28. (*Making Observations*) Describe the values of θ in the interval $[0, 2\pi]$ for which each equation is true:

a. $\cos\theta = 0$ (*Hint*: For what values of θ does the point on the unit circle have an x-coordinate of 0?)

b. $\cos\theta = 1$ c. $\cos\theta = -1$

d. $\sin\theta = 0$ e. $\sin\theta = 1$

f. $\sin\theta = -1$

29. For the given radian measure, sketch the angle in standard position, and find its degree measure:

a. $\dfrac{4\pi}{3}$ b. $-\dfrac{3\pi}{4}$

c. 10π d. $\dfrac{7\pi}{6}$

e. -1 f. $\dfrac{9\pi}{2}$

30. For the given degree measure, sketch the angle in standard position, and find its radian measure:

a. $-45°$ b. $-180°$

c. $15°$ d. $750°$

e. $-195°$ f. $1°$

31. Find both the radian and degree measure of the angle determined by an arc of directed length t on a circle of radius r.

a. $r = 12, t = \dfrac{\pi}{2}$ b. $r = 6, t = -\pi$

c. $r = \pi, t = -6$ d. $r = 20, t = \dfrac{1}{20}$

32. For each angle θ in the following figures, find $\sin\theta$ and $\cos\theta$.

a.

b.

c.

d.

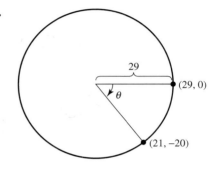

33. (*Problem Solving*)

   a. Refer to our hypothetical proton in ***Fermilab 2***, traveling counter-clockwise around the Tevatron at 1 km/sec. Find both the degree and radian measure of the angle corresponding to its position after 10 seconds, and find its coordinates at that time.

   b. Repeat for a proton traveling at the more realistic speed of 300,000 km/sec.

   c. Repeat for an antiproton traveling clockwise for 10 seconds at a speed of 300,000 km/sec.

34. Find the angular velocity of an object traveling at the given velocity around a circle with the given radius.

   a. velocity = 5 ft/sec, radius = 2 ft

   b. velocity = 50 km/hr, radius = 0.5 km

   c. velocity = 3.78 cm/min, radius = 1.26 cm

   d. velocity = 2.5 mi/hr, radius = 12.5 mi

35. Find the velocity of an object traveling with angular velocity ω around a circle with radius r.

   a. $\omega = 5$ radians/sec, $r = 7$ ft

   b. $\omega = 0.1$ radian/sec, $r = 400$ m

   c. $\omega = 1$ radian/sec, $r = 56$ cm

   d. $\omega = \pi$ radians/sec, $r = 5.5$ in

36. (*Interpreting Mathematics*) A satellite that stays above the same point on the earth is said to be in a stationary orbit. Such a satellite must make exactly one revolution around the earth's axis every 24 hours. Suppose that a satellite is in a stationary orbit 130 km above a point on the equator.

   a. Find the satellite's angular velocity.

   b. Find the satellite's velocity. (The radius of the earth is 6370 km.)

   c. If you are standing on the ground directly below the satellite, is your angular velocity around the earth's axis the same as that of the satellite? Is your velocity the same as that of the satellite? Explain your answer in each case.

8–2   THREE VIEWS OF SINE AND COSINE FUNCTIONS

The sine and cosine functions are among the most useful in all of mathematics. Although they have been defined in terms of both triangles and circles, they exhibit a surprising tendency to appear in situations seemingly unrelated to either. For example, when you turn on an electric light, the current flowing in the outlet after t seconds is a function involving $\sin t$. Because the sine and cosine appear in a variety of contexts, it is useful to regard them as functions of real numbers, which may or may not represent angles. Thus, $\sin \frac{\pi}{2} = 1$, whether $\frac{\pi}{2}$ represents a number of radians, kilometers, or seconds.

Later in this section we will explore a physical problem that illustrates the advantages of this point of view. To prepare for our exploration, let's take a closer look at the sine and cosine functions from numerical, graphical, and analytical perspectives. Throughout the discussion we will consider both functions to be defined on real numbers or, equivalently, on angles expressed in radians.

A NUMERICAL VIEW

Some important properties of the sine and cosine functions are suggested by Table 38, which you obtained in Exercise 25 of Section 8-1. You can find many of them by using the strategy of *Looking for a Pattern*.

EXERCISES   **1.** (*Making Observations*) You already know that the functions $y = \sin \theta$ and $x = \cos \theta$ each have domain $(-\infty, \infty)$. What does Table 38 suggest about their ranges?

   2. (*Making Observations*)

   a. Over what part of the interval $[0, 2\pi]$ is the function $y = \sin \theta$ increasing? Decreasing?

   b. Repeat part (a) for the function $x = \cos \theta$.

TABLE 38

Degrees	Radians	Sine	Cosine
0°	0	0	1
30°	$\dfrac{\pi}{6}$	$\dfrac{1}{2}$	$\dfrac{\sqrt{3}}{2}$
45°	$\dfrac{\pi}{4}$	$\dfrac{\sqrt{2}}{2}$	$\dfrac{\sqrt{2}}{2}$
60°	$\dfrac{\pi}{3}$	$\dfrac{\sqrt{3}}{2}$	$\dfrac{1}{2}$
90°	$\dfrac{\pi}{2}$	1	0
120°	$\dfrac{2\pi}{3}$	$\dfrac{\sqrt{3}}{2}$	$-\dfrac{1}{2}$
135°	$\dfrac{3\pi}{4}$	$\dfrac{\sqrt{2}}{2}$	$-\dfrac{\sqrt{2}}{2}$
150°	$\dfrac{5\pi}{6}$	$\dfrac{1}{2}$	$-\dfrac{\sqrt{3}}{2}$
180°	π	0	-1
210°	$\dfrac{7\pi}{6}$	$-\dfrac{1}{2}$	$-\dfrac{\sqrt{3}}{2}$
225°	$\dfrac{5\pi}{4}$	$-\dfrac{\sqrt{2}}{2}$	$-\dfrac{\sqrt{2}}{2}$
240°	$\dfrac{4\pi}{3}$	$-\dfrac{\sqrt{3}}{2}$	$-\dfrac{1}{2}$
270°	$\dfrac{3\pi}{2}$	-1	0
300°	$\dfrac{5\pi}{3}$	$-\dfrac{\sqrt{3}}{2}$	$\dfrac{1}{2}$
315°	$\dfrac{7\pi}{4}$	$-\dfrac{\sqrt{2}}{2}$	$\dfrac{\sqrt{2}}{2}$
330°	$\dfrac{11\pi}{6}$	$-\dfrac{1}{2}$	$\dfrac{\sqrt{3}}{2}$
360°	2π	0	1

   The last two columns in Table 38 also exhibit several patterns that suggest **identities** (equations that are true for all values of θ) among the values
of sin θ and cos θ. For example, the last column yields the values shown in
Table 39.

TABLE 39

Reading down from the top	Reading up from the middle
$\cos 0 = 1$	$\cos \pi = -1$
$\cos \dfrac{\pi}{6} = \dfrac{\sqrt{3}}{2}$	$\cos \dfrac{5\pi}{6} = -\dfrac{\sqrt{3}}{2}$
$\cos \dfrac{\pi}{4} = \dfrac{\sqrt{2}}{2}$	$\cos \dfrac{3\pi}{4} = -\dfrac{\sqrt{2}}{2}$
$\cos \dfrac{\pi}{3} = \dfrac{1}{2}$	$\cos \dfrac{2\pi}{3} = -\dfrac{1}{2}$

EXERCISES

3. (*Making Observations*) Explain why the data in Table 39 suggests that $\cos (\pi - \theta) = -\cos \theta$ for all θ.

4. (*Making Observations*) Construct a table similar to Table 39 to suggest the identity $\sin (\theta + \pi) = -\sin \theta$.

5. (*Making Observations*) What identity is suggested by the following table?

$\sin \dfrac{\pi}{2} = 1$	$\cos 0 = 1$
$\sin \dfrac{2\pi}{3} = \dfrac{\sqrt{3}}{2}$	$\cos \dfrac{\pi}{6} = \dfrac{\sqrt{3}}{2}$
$\sin \dfrac{3\pi}{4} = \dfrac{\sqrt{2}}{2}$	$\cos \dfrac{\pi}{4} = \dfrac{\sqrt{2}}{2}$
$\sin \dfrac{5\pi}{6} = \dfrac{1}{2}$	$\cos \dfrac{\pi}{3} = \dfrac{1}{2}$

6. (*Making Observations*) How could Table 38 be continued to suggest the identities $\sin (\theta + 2\pi) = \sin \theta$ and $\cos (\theta + 2\pi) = \cos \theta$?

You will study trigonometric identities in more detail in Section 9-2. In the meantime you can discover several more identities suggested by Table 38 in Exercises 45–48.

Because you will often need to evaluate $\sin \theta$ and $\cos \theta$ for the values of θ in Table 38, it is a good idea to memorize the table. The patterns in the last two columns can make your work easier.

AN ANALYTICAL VIEW

The properties of the sine and cosine functions in Exercises 1–6 were suggested numerically by Table 38. You can confirm them analytically by recalling that $\cos \theta$ and $\sin \theta$ are the x- and y-coordinates at the terminal point of an arc of directed length θ (or a directed angle θ) on the unit circle.

FYI For brevity, we will say "an angle θ" to mean "a directed angle of radian measure θ." However, it is important to remember that θ really represents the *measure* of the angle, not the angle itself.

EXERCISES
7. (*Writing to Learn*) As you observed in Exercise 1, Table 38 suggests that the functions $y = \sin \theta$ and $x = \cos \theta$ each have a range of $[-1, 1]$. How do the unit circle definitions of $\sin \theta$ and $\cos \theta$ confirm this?

8. (*Making Observations*) Suppose that an object travels counterclockwise around the unit circle.

    a. Over what part of the circle is the object's y-coordinate increasing? Decreasing?

    b. Over what part of the interval $[0, 2\pi]$ is the function $y = \sin \theta$ increasing? Decreasing?

    c. Over what part of the circle is the object's x-coordinate increasing? Decreasing?

    d. Over what part of the interval $[0, 2\pi]$ is the function $x = \cos \theta$ increasing? Decreasing?

    e. Do these results agree with those you obtained in Exercise 2?

Example 1 illustrates how the identities suggested by Table 38 can be confirmed analytically.

EXAMPLE 1 *Confirming a trigonometric identity analytically*

Let's confirm that $\cos (\pi - \theta) = -\cos \theta$ for all θ.

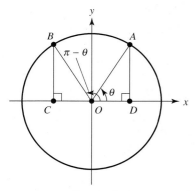

Figure 8–14

In Figure 8-14, $\overline{OA}$ is the terminal side of an angle θ and $\overline{OB}$ is the terminal side of the angle $\pi - \theta$ in the unit circle. As a first step in confirming the identity, let's establish that the triangles DOA and COB are congruent. You will need to remember that two triangles are congruent as long as any two angles and any side of one triangle have the same measure as the corresponding angles and side of the other.

Since $\angle DOB$ is supplementary to $\angle BOC$ (that is, their sum is a straight angle π), $\angle BOC = \theta$. The angles ODA and OCB are each right angles. Furthermore, since OA and OB are radii of the unit circle, $\overline{OA} = 1$ and $\overline{OB} = 1$. Therefore, the two triangles are congruent, so $\overline{OC} = \overline{OD}$. Equivalently, $|\cos (\pi - \theta)| = |\cos \theta|$.

Now let's notice that the points A and B lie on opposite sides of the y-axis for nearly all values of θ. (The exceptions occur when both A and B are on the y-axis. You can deal with this situation in Exercise 9.) This means that the x-coordinates of A and B have opposite signs, so that $\cos (\pi - \theta) = -\cos \theta$. ◼

EXERCISES
9. (*Writing to Learn*) If the points A and B in Figure 8-14 are both on the y-axis, explain why $\cos (\pi - \theta) = -\cos \theta$.

10. (*Writing to Learn*) Use a figure similar to Figure 8-14 to confirm the identity $\sin (\theta + \pi) = -\sin \theta$.

11. (*Writing to Learn*) Explain how Figure 8-15 illustrates the identity

$$\sin\left(\theta + \frac{\pi}{2}\right) = \cos \theta$$

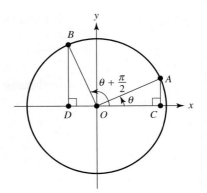

Figure 8–15

12. (*Writing to Learn*) How do the definitions of sin θ and cos θ guarantee that sin $(\theta + 2\pi) =$ sin θ and cos $(\theta + 2\pi) =$ cos θ for all θ?

FYI Because the expressions sin θ and cos θ look like products, many students assume that sin $b\theta$ and b sin θ are equal and that sin $(\theta - \phi)$ and sin $\theta -$ sin ϕ are equal. In Exercises 13 and 14 you can show that they are not. Similar statements apply to cosine functions.

EXERCISES

13. (*Making Observations*) For each value of θ, evaluate both cos 3θ and 3 cos θ, and confirm that you obtain two different values.

 a. $\theta = 0$ **b.** $\theta = \pi$ **c.** $\theta = \dfrac{\pi}{4}$ **d.** $\theta = \dfrac{2\pi}{3}$

14. (*Making Observations*) For each value of θ, evaluate both $\sin\left(\theta - \dfrac{\pi}{4}\right)$ and $\sin\theta - \sin\dfrac{\pi}{4}$, and confirm that you obtain two different values.

 a. $\theta = \pi$ **b.** $\theta = \dfrac{\pi}{2}$ **c.** $\theta = \dfrac{3\pi}{4}$ **d.** $\theta = -\dfrac{3\pi}{4}$

A GRAPHICAL VIEW

The Basic Sine and Cosine Graphs The graphs of $y =$ sin θ and $x =$ cos θ are shown in Figure 8-16.

$y = \sin\theta$

(a)

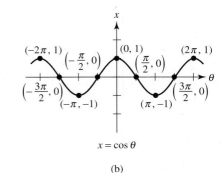

$x = \cos\theta$

(b)

Figure 8–16

In Exercises 15 and 16 you can see how you could have predicted the shapes of the graphs from your work earlier in this section.

EXERCISES **15. a.** Plot the points $(\theta, \sin \theta)$ listed in Table 38. Do your points all appear to lie on the graph of $y = \sin \theta$ as shown in Figure 8–16a?

b. Plot the points $(\theta, \cos \theta)$ listed in Table 38. Do your points all appear to lie on the graph of $x = \cos \theta$ as shown in Figure 8-16b?

16. (*Writing to Learn*) The entries in Table 38 could be used to sketch the graphs of $y = \sin \theta$ and $x = \cos \theta$ over $[0, 2\pi]$. Explain how the identities $\sin (\theta + 2\pi) = \sin \theta$ and $\cos (\theta + 2\pi) = \cos \theta$ could then be used to obtain the graphs over $[-2\pi, 0]$.

17. (*Making Observations*) Use the graphs in Figure 8-16 to answer each question, and compare your results with those in the indicated exercises.

a. What is the range of the function $y = \sin \theta$? (Exercises 1 and 7)

b. What is the range of the function $x = \cos \theta$? (Exercises 1 and 7)

c. Over what part of the interval $[0, 2\pi]$ is $y = \sin \theta$ increasing? Decreasing? (Exercises 2 and 8)

d. Over what part of the interval $[0, 2\pi]$ is $x = \cos \theta$ increasing? Decreasing? (Exercises 2 and 8)

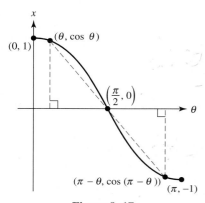

Figure 8–17

18. (*Writing to Learn*) Explain how Figure 8-17 illustrates that

$$\cos (\pi - \theta) = -\cos \theta$$

19. Use a figure similar to Figure 8-17 to illustrate the identity

$$\sin (\theta + \pi) = - \sin \theta$$

20. (*Writing to Learn*) Graph the functions

$$f(\theta) = \sin (\theta + \pi) \quad \text{and} \quad g(\theta) = - \sin \theta$$

on the same screen. How does the result support the conclusion that $\sin (\theta + \pi) = -\sin \theta$ for all θ?

21. (*Writing to Learn*) Graph the functions $f(\theta) = \sin \theta$ and $g(\theta) = \cos \theta$ on the same screen. How does the result suggest that $\sin \left(\theta + \dfrac{\pi}{2}\right) = \cos \theta$ for all θ?

Now that we have constructed the graphs of the basic sine and cosine functions, let's return to our usual convention of letting x and y represent the independent and dependent variables. In the rest of this section, we will speak of the functions $y = \sin x$ and $y = \cos x$.

Amplitude and Period The graphs of the sine and cosine functions have two especially distinctive features. First, each graph is contained between the horizontal lines $y = -1$ and $y = 1$. This is because the range of each function is $[-1, 1]$. In general, if the range of a function is an interval $[a, b]$, its graph is said to have an **amplitude** of $\dfrac{b - a}{2}$. Thus, the functions $y = \sin x$ and $y = \cos x$ each have amplitude $\dfrac{1 - (-1)}{2} = 1$. The amplitude represents the vertical distance above and below the "middle" of the graph, as shown in Figure 8-18.

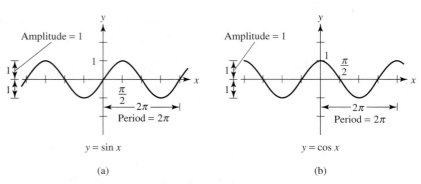

Figure 8–18

Second, each graph can be obtained by repeated horizontal translations of the segment between $x = 0$ and $x = 2\pi$. This is because $\sin(x + 2\pi) = \sin x$ and $\cos(x + 2\pi) = \cos x$. In general, if there is some positive number P for which $f(x + P) = f(x)$ for all x in the domain of f, the function f is said to be **periodic**, and the smallest such number is called the **period** of f. The portion of the graph over an interval representing a full period is called a **cycle** of the graph. Thus, the functions $y = \sin x$ and $y = \cos x$ each have a period of 2π, as shown in Figure 8-18. The portion of each graph over the interval $[0, 2\pi]$ is one cycle. The portion over $[-\pi, \pi]$ is also one cycle.

Graphical Transformations The graph of any function

$$y = A \sin b(x - h) + k \quad \text{or} \quad y = A \cos b(x - h) + k$$

can be obtained from one of the basic graphs by means of graphical transformations.

EXERCISES

22. (*Making Observations*)

   a. Refer to the graphing transformation principles listed in Section 4-5. What principle must be used to obtain the graphs of $y = 2 \sin x$ and $y = 0.5 \cos x$ from $y = \sin x$ and $y = \cos x$?

   b. Sketch at least one full cycle of each of the graphs $y = 2 \sin x$ and $y = 0.5 \cos x$. Identify the coordinates of all x-intercepts and extreme points.

   c. What is the amplitude of each function?

   d. What is the period of each function?

   e. For a function $f(x) = A \sin x$ or $f(x) = A \cos x$ with $A > 0$, express the amplitude in terms of A.

   f. Use calculator graphs of $y = 0.1 \sin x$ and $y = 3 \cos x$ to support your results in part (e).

23. (*Making Observations*)

   a. Repeat Exercise 22a for $y = -2 \sin x$ and $y = -0.5 \cos x$.

   b. Repeat Exercise 22b for $y = -2 \sin x$ and $y = -0.5 \cos x$.

   c. What is the amplitude of each function?

   d. What is the period of each function?

   e. For a function $f(x) = A \sin x$ or $f(x) = A \cos x$ with $A < 0$, express the amplitude in terms of A.

f. Use calculator graphs of $y = -0.1 \sin x$ and $y = -3 \cos x$ to support your results in part (e).

In Exercises 24–27, sketch one full cycle of the graph, and support your result with a calculator graph. Identify the coordinates of all x-intercepts and extreme points.

24. $y = 4 \sin x$ **25.** $y = -\dfrac{3}{4} \sin x$

26. $y = -50 \cos x$ **27.** $y = 1.5 \cos x$

28. (*Making Observations*)

    a. Repeat Exercise 22a for $y = \sin 2x$ and $y = \cos 0.5x$.

    b. Repeat Exercise 22b for $y = \sin 2x$ and $y = \cos 0.5x$.

    c. What is the amplitude of each function?

    d. What is the period of each function?

    e. If $b > 0$, express the period of a function

$$f(x) = \sin bx \quad \text{or} \quad f(x) = \cos bx$$

    in terms of b.

    f. Use calculator graphs of $y = \sin 0.1x$ and $y = \cos 3x$ to support your results in part (e).

In Exercises 29–32, sketch one full cycle of the graph, and support your result with a calculator graph. Identify the coordinates of all x-intercepts and extreme points.

29. $y = \sin 4x$ **30.** $y = 2 \sin \dfrac{3}{4} x$

31. $y = \cos 5x$ **32.** $y = 0.5 \cos 1.5x$

33. (*Making Observations*)

    a. Based on your results from Exercises 29–32, express the amplitude and the period of a function $f(x) = A \sin bx$ or $f(x) = A \cos bx$ with $b > 0$ in terms of A and b.

    b. If $b < 0$, then $\sin bx = -\sin |b|x$ and $\cos bx = \cos |b|x$. What does this say about the period of $f(x) = A \sin bx$ or $f(x) = A \cos bx$ when $b < 0$?

In Exercises 34–37, use the graphical transformation principles in Section 4-5 to sketch at least one full cycle of each graph, and support your results with a calculator graph. Identify the coordinates of all points obtained by moving x-intercepts and extreme points on the basic graphs.

34. $y = \sin 2x - 3$ **35.** $y = \cos 3\left(x - \dfrac{\pi}{6}\right)$

36. $y = 0.5 \cos x + 2$ **37.** $y = -3 \sin\left(x + \dfrac{\pi}{4}\right)$

38. (*Making Observations*) Based on your results from Exercises 34–37, express the amplitude and period of a function $f(x) = A \sin b(x - h) + k$ or $f(x) = A \cos b(x - h) + k$ in terms of A and b.

    A horizontal translation in a sine or cosine graph is called the **phase shift**, or **displacement**, of the graph. For example, the graphs in Exercises 35 and 37 have phase shifts of $\dfrac{\pi}{6}$ and $-\dfrac{\pi}{4}$, respectively.

To summarize, the graph of any function

$$f(x) = A \sin b(x - h) + k \quad \text{or} \quad f(x) = A \cos b(x - h) + k$$

exhibits the following features, illustrated in Figure 8-19:

- amplitude $= |A|$
- period $= \dfrac{2\pi}{|b|}$
- phase shift $= h$

$$y = A \sin b(x - h) + k$$

Figure 8–19

Because of their graphical properties, sine and cosine functions are useful models for quantitative relationships that have a well-defined amplitude and exhibit periodic behavior. The following Mathematical Looking Glass illustrates that such relationships can occur in a wide variety of contexts.

A MATHEMATICAL LOOKING GLASS
Power to the People

Duquesne Light Company provides electric power for customers in the Pittsburgh area. To estimate the cost of cooling and heating a house, the company computes the absolute difference between 65° and the average outside temperature each day. The result is multiplied by a figure that varies according to the size, shape, and amount of insulation for each house. A typical value for this multiplier might be $0.25 per day.

Duquesne Light needs a complete knowledge of average temperatures in Pittsburgh for at least two reasons:

- They need to make sure that their resources are adequate to provide sufficient power to all their customers.
- They need to compute budget bills for customers, in which monthly payments are approximately constant, based on estimated power usage over a year.

Let's see if we can create a simple function $f(t)$ to represent the average temperature in Pittsburgh on day t of the year. This means, for example, that we want $f(6)$ to be the average temperature for every January 6 since 1884, when records of daily temperatures were first kept. (That number may or may not be the average temperature on January 6 of the current year.) For simplicity, let's ignore leap years.

The domain of f is the set of integers from 1 to 365. However, if we extend the domain to all real numbers, f is periodic with a period of 365. Because t and $t + 365$ represent the same day of the year, $f(t + 365) = f(t)$. Furthermore, average temperatures fall within some finite interval, so that *f* has a well-defined amplitude. Since *f* is periodic with a fixed amplitude, it makes sense to consider making *f* a sine or cosine function. You can construct the specific function in Exercise 39. ■

EXERCISES   **39.** (*Writing to Learn*)

a. If we model the average temperature by a function

$$f(t) = A \sin b(t - h) + k$$

explain why we should choose $b = \dfrac{2\pi}{365}$.

b. The lowest average temperature in Pittsburgh is 30°, occurring near January 31 (day 31). The highest is 74°, occurring near July 24 (day 205). Explain why these facts indicate that

$$f(t) = 22 \sin \frac{2\pi}{365} (t - 118) + 52$$

is an appropriate function to use.

40. (*Writing to Learn*) The average temperature function for any city in the United States can be approximated with reasonable accuracy by a function

$$f(t) = A \sin \frac{2\pi}{365} (t - h) + k$$

A few examples are

Phoenix, AZ: $f(t) = 20 \sin \dfrac{2\pi}{365} (t - 109) + 71$

Honolulu, HI: $f(t) = 4 \sin \dfrac{2\pi}{365} (t - 141) + 75$

Bismarck, ND: $f(t) = 30 \sin \dfrac{2\pi}{365} (t - 110) + 40$

What do each of the parameters *A*, *h*, and *k* tell you about the prevailing climate in each of these cities?

HARMONIC MOTION

Sine and cosine functions occur in many natural settings. For example, every sound we hear is produced by rapid variations in air pressure. The deviation from normal pressure is a function of time that can be described by an equation

$$f(t) = a \sin \omega t \quad \text{or} \quad f(t) = a \cos \omega t$$

These equations also describe the motion of a light wave or a mass vibrating on a spring in the absence of damping forces. An object whose motion is described by such an equation is said to be in **harmonic motion**. Figure 8-20 illustrates the motion of a mass on a spring.

If *t* is measured in seconds, the period $\dfrac{2\pi}{\omega}$ is measured in seconds per cycle. In many physical contexts it is convenient to refer to the reciprocal $\dfrac{\omega}{2\pi}$, called the **frequency** of the motion, measured in cycles per second.

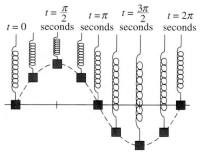

$t = 0$ $t = \dfrac{\pi}{2}$ $t = \pi$ $t = \dfrac{3\pi}{2}$ $t = 2\pi$
       seconds   seconds   seconds   seconds

Figure 8–20

EXAMPLE 2   *Frequency of harmonic motion*

The sound of a musical note is produced by an equation

$$f(t) = a \sin \omega t$$

where t is measured in seconds and $f(t)$ represents deviation from normal air pressure in lb/in². The value of the coefficient a is proportional to the loudness of the note, and the value of ω depends on its pitch (that is, how high or low the note is).

The lowest note on a standard piano has a frequency of 55 cycles per second. For that note,

$$\frac{\omega}{2\pi} = 55$$

$$\omega = 110\pi$$

The equation of the sound is $f(t) = a \sin 110\pi t$. Its period is the reciprocal of the frequency, or $\frac{1}{55}$ seconds per cycle.   ■

EXERCISES   **41.** (*Interpreting Mathematics*) The sound of middle C on a piano is described by $g(t) = a \sin 1046\pi t$. What is the frequency of middle C?

42. (*Making Observations*)

   a. When the pitch of a note is raised one octave, its frequency is doubled. What effect does this have on the equation of the sound?

   b. When the loudness of a note is doubled, what is the effect on the equation of the sound?

Mechanical devices, such as pistons in automobile engines, are often designed to exhibit harmonic motion. This is because harmonic motion minimizes wear and tear by allowing moving parts to start and stop gradually.

EXERCISES   **43.** (*Writing to Learn*) Describe in detail the motion of a piston that moves vertically and whose vertical displacement from its initial position is $f(t) = \sin t$. In particular, tell when the piston is moving up, when it is moving down, when it is moving fastest in each direction, and when it stops for an instant to change direction.

44. (*Making Observations*)

   a. For $f(t) = \sin t$, calculate the average rate of change over $\left[0, \frac{\pi}{6}\right]$, $\left[\frac{\pi}{6}, \frac{\pi}{3}\right]$, and $\left[\frac{\pi}{3}, \frac{\pi}{2}\right]$.

   b. Carefully sketch the graph of $f(t) = \sin t$ over $\left[0, \frac{\pi}{2}\right]$. Draw three secant lines whose slopes represent the average rates of change you calculated in part (a).

   c. How do your results in parts (a) and (b) support the claim that an object in harmonic motion stops gradually?

ADDITIONAL EXERCISES *In Exercises 45–48,*

    a. Use Table 38 to compare the indicated values of $\sin\theta$ and $\cos\theta$, and write an identity suggested by the table.

    b. Draw a figure in the unit circle, similar to Figure 8-14, and use it to explain why the identity is true.

    c. Draw a figure similar to Figure 8-15 that uses the graph of $y = \sin\theta$ or $x = \cos\theta$ to support your conclusions in parts (a) and (b).

45. $\sin\theta$ from 0 down to $\dfrac{\pi}{2}$; $\sin\theta$ from π up to $\dfrac{\pi}{2}$

46. $\cos\theta$ from 0 down to $\dfrac{\pi}{2}$; $\cos\theta$ from π down to $\dfrac{3\pi}{2}$

47. $\cos\theta$ from 0 down to $\dfrac{\pi}{2}$; $\cos\theta$ from 2π up to $\dfrac{3\pi}{2}$

48. $\sin\theta$ from 0 down to $\dfrac{\pi}{2}$; $\sin\theta$ from 2π up to $\dfrac{3\pi}{2}$

49. (*Making Observations*)

    a. In Section 5-3 you learned that a function is called even if its graph is symmetric about the y-axis and odd if its graph is symmetric about the origin. By observing the graphs of $y = \sin x$ and $y = \cos x$, decide whether each function is even, odd, or neither.

    b. In Section 5-3 you also learned that the graph of a function $y = f(x)$ is symmetric about the y-axis if and only if $f(-x) = f(x)$ and symmetric about the origin if and only if $f(-x) = -f(x)$. Draw a figure in the unit circle, similar to Figure 8-14, and use it to express $\sin(-x)$ and $\cos(-x)$ in terms of $\sin x$ and $\cos x$. Do your results support your conclusion in part (a)?

In Exercises 50–59,

    a. Find the domain, range, amplitude, period, and phase shift of the function.

    b. Sketch its graph over a full cycle.

    c. Identify the coordinates of all points obtained by moving x-intercepts and extreme points on the basic graphs.

50. $y = 2\sin 3x$ **51.** $y = -\sin\dfrac{x}{2}$

52. $y = -5\cos 2x$ **53.** $y = 7\cos 7x$

54. $y = \sin\left(x - \dfrac{\pi}{4}\right) - 6$ **55.** $y = -0.3\cos 2x - 1$

56. $y = \pi\cos(x + 1)$ **57.** $y = -\cos\left(4x - \dfrac{\pi}{2}\right)$

58. $y = 4\sin\dfrac{1}{3}\left(x + \dfrac{2\pi}{3}\right) + \dfrac{3}{2}$

59. $y = 8\sin\dfrac{2x + \pi}{6} - 3$

In Exercises 60–63, write the equation of a sine or cosine function with the given amplitude, period, and phase shift. (There is more than one correct answer for each.)

60. sine function;

amplitude $= 3$, period $= \pi$, phase shift $= \dfrac{\pi}{6}$

61. sine function;

amplitude $= 2$, period $= 4\pi$, phase shift $= -\dfrac{\pi}{6}$

62. cosine function;

amplitude $= 0.2$, period $= \dfrac{\pi}{3}$, phase shift $= 0$

63. cosine function;

amplitude $= 5$, period $= 2\pi$, phase shift $= \dfrac{3\pi}{4}$

The identities suggested by Exercises 64–66 will be confirmed analytically in Section 9-2.

64. (*Making Observations*)

a. Graph $y = (\cos x)^2$ in the window $[-\pi, \pi]$ by $[-1, 1]$.

b. Write the equation of a function $y = A \cos bx + k$ whose graph might coincide with the one you see.

c. Graph the two functions together.

d. Write an identity suggested by your graphs.

65. (*Making Observations*) Write the equation of a function

$$y = A \cos bx + k$$

whose graph might coincide with that of $y = (\sin x)^2$. Graph the functions together, and write an identity suggested by the graphs.

66. (*Making Observations*) Graph $y = (\sin x)^2 + (\cos x)^2$, and write an identity suggested by the graph.

67. (*Interpreting Mathematics*) The vertical displacement of a mass vibrating on a spring is $y = 5 \sin 12t$ cm after t seconds. Describe how its motion compares with that of a second mass if the equation of the second motion is

a. $y = 10 \sin 12t$ **b.** $y = 5 \sin 24t$

c. $y = 10 \sin 24t$ **d.** $y = 5 \cos 12t$

68. (*Extension*) Graph $y = \sin x$ in the window $[-2\pi, 2\pi]$ by $[-2, 2]$. Then add the following graphs to your display one at a time. (Recall that $n!$ is the product of the first n positive integers.)

a. $y = x$ **b.** $y = x - \dfrac{x^3}{3!}$

c. $y = x - \dfrac{x^3}{3!} + \dfrac{x^5}{5!}$

d. $y = x - \dfrac{x^3}{3!} + \dfrac{x^5}{5!} - \dfrac{x^7}{7!}$

How do the polynomial graphs relate to the graph of $y = \sin x$? If you study calculus, you will discover the usefulness of your observations.

69. (*Extension*) Repeat Exercise 68 for $y = \cos x$ and the following functions.

    a. $y = 1$ **b.** $y = 1 - \dfrac{x^2}{2!}$

    c. $y = 1 - \dfrac{x^2}{2!} + \dfrac{x^4}{4!}$

    d. $y = 1 - \dfrac{x^2}{2!} + \dfrac{x^4}{4!} - \dfrac{x^6}{6!}$

SUPPLEMENTARY TOPIC

Equivalent Expressions for Sine and Cosine Functions

Example 3 shows how to write several different equations to represent the same sine or cosine function. There are often practical reasons for choosing one equation over another. For example, there may be a reason for preferring sines to cosines or for using a nonnegative displacement.

EXAMPLE 3   *Expressing a function as a sine or cosine function in several ways*

Figure 8-21 shows the graph of a function having the form

$$f(x) = A \sin b(x - h) + k \quad \text{or} \quad f(x) = A \cos b(x - h) + k$$

The values of the parameters A, b, and k are determined as follows:

- The range of f is $[-1, 9]$, so $|A| = \dfrac{9 - (-1)}{2} = 5$ and $A = \pm 5$.

- One cycle begins at $x = -\dfrac{\pi}{3}$ and then ends at $x = \dfrac{2\pi}{3}$. Thus the period is
$$\frac{2\pi}{3} - \left(-\frac{\pi}{3}\right) = \pi, \text{ so } b = \frac{2\pi}{\pi} = 2.$$

- The midpoint of the range of f is $\dfrac{9 + (-1)}{2} = 4$. This is 4 units above the midpoint of the range on a basic sine or cosine graph, so $k = 4$.

There are many ways to write

$$f(x) = \pm 5 \sin 2(x - h) + 4 \quad \text{or} \quad f(x) = \pm 5 \cos 2(x - h) + 4$$

for appropriate choices of h. You can find several by comparing the given graph with those of $y = \pm 5 \sin 2x + 4$ and $y = \pm 5 \cos 2x + 4$. For example,

    a. The graph of f is that of $y = 5 \sin 2x + 4$ translated $\dfrac{\pi}{3}$ units to the left, so $f(x) = 5 \sin 2\left(x + \dfrac{\pi}{3}\right) + 4$.

    b. The graph of f is that of $y = 5 \cos 2x + 4$ translated $\dfrac{\pi}{12}$ units to the left, so $f(x) = 5 \cos 2\left(x + \dfrac{\pi}{12}\right) + 4$.

    c. The graph of f is that of $y = -5 \sin 2x + 4$ translated $\dfrac{\pi}{6}$ units to the right, so $f(x) = -5 \sin 2\left(x - \dfrac{\pi}{6}\right) + 4$.

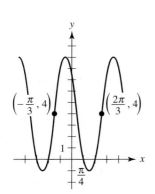

Figure 8–21

d. The graph of f is that of $y = -5 \cos 2x + 4$ translated $\dfrac{5\pi}{12}$ units to the right, so $f(x) = -5 \cos 2\left(x - \dfrac{5\pi}{12}\right) + 4$.

Another choice would be $f(x) = 5 \sin 2\left(x - \dfrac{2\pi}{3}\right) + 4$. In general, a valid representation of f can be obtained by adding any integer multiple of the period π to the value of h in any of the forms given here. ■

In Exercises 70–73,

    a. Write an expression of the form $A \sin b(x - h) + k$, with $A > 0$ and $|h|$ as small as possible, to represent the function whose graph is given.

    b. Write an expression $A \cos b(x - h) + k$ meeting the same requirements.

70.

71.

72.

73.
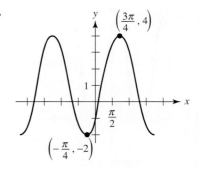

In Exercises 74–77, write an expression $A \sin b(x - h) + k$ to represent each function, with $A > 0$ and h in the indicated interval.

74. $f(x) = 40 \sin\left(x + \dfrac{5\pi}{2}\right) - 20; \ 0 \leqslant h < 2\pi$

75. $f(x) = 3 \sin (x - 373.4\pi); \ -\pi \leqslant h < \pi$

76. $f(x) = \sin 2\left(x - \dfrac{17\pi}{4}\right); \ 0 \leqslant h < \pi$

77. $f(x) = -\sin 2\left(x - \dfrac{17\pi}{4}\right); \ 0 \leqslant h < \pi$

8-3   THE SIX TRIGONOMETRIC FUNCTIONS

As you saw in Section 8-1, the sine and cosine functions can be defined for all real numbers using the unit circle. Furthermore, their unit circle definitions are both useful and compatible with their right triangle definitions. In the present section you will see how right triangles and the unit circle can both be used to define a total of six trigonometric functions.

THE DEFINITIONS OF THE TRIGONOMETRIC FUNCTIONS

The Right Triangle Definitions   As you learned in Section 8-1, Rhäticus thought of the functions $\sin \theta$ and $\cos \theta$ as ratios of side lengths in right triangles. Figure 8-22 and the table below show that in every right triangle there are a total of six ratios of side lengths, all of which were given names by Rhäticus. His names are still in use today. They follow Figure 8-22.

Figure 8–22

$$\sin \theta \text{ (the sine of } \theta) = \frac{\overline{AC}}{\overline{OA}} = \frac{\text{opposite}}{\text{hypotenuse}}$$

$$\cos \theta \text{ (the cosine of } \theta) = \frac{\overline{OC}}{\overline{OA}} = \frac{\text{adjacent}}{\text{hypotenuse}}$$

$$\tan \theta \text{ (the tangent of } \theta) = \frac{\overline{AC}}{\overline{OC}} = \frac{\text{opposite}}{\text{adjacent}}$$

$$\cot \theta \text{ (the cotangent of } \theta) = \frac{\overline{OC}}{\overline{AC}} = \frac{\text{adjacent}}{\text{opposite}}$$

$$\sec \theta \text{ (the secant of } \theta) = \frac{\overline{OA}}{\overline{OC}} = \frac{\text{hypotenuse}}{\text{adjacent}}$$

$$\csc \theta \text{ (the cosecant of } \theta) = \frac{\overline{OA}}{\overline{AC}} = \frac{\text{hypotenuse}}{\text{opposite}}$$

Like the sine and cosine, the other four ratios depend only on the angle θ, and not on the size of the triangle. Also like the sine and cosine, their definitions can be extended to apply to arbitrary angles. This will be done later in this section. All six are referred to as **trigonometric functions** of θ.

The following relationships among the six trigonometric functions, referred to as **reciprocal identities**, are immediate consequences of the definitions. You can verify them in Exercise 1.

Reciprocal Identities

$$\sin \theta = \frac{1}{\csc \theta} \qquad\qquad \csc \theta = \frac{1}{\sin \theta}$$

$$\cos \theta = \frac{1}{\sec \theta} \qquad\qquad \sec \theta = \frac{1}{\cos \theta}$$

$$\tan \theta = \frac{1}{\cot \theta} \qquad\qquad \cot \theta = \frac{1}{\tan \theta}$$

$$\tan \theta = \frac{\sin \theta}{\cos \theta} \qquad\qquad \cot \theta = \frac{\cos \theta}{\sin \theta}$$

FYI Your calculator has keys for evaluating only the sine, cosine, and tangent functions. You can use the reciprocal identities to evaluate the others. For example, to evaluate $\sec 50°$ on your calculator, evaluate $\dfrac{1}{\cos 50°}$.

Do not use the keys labeled $\sin^{-1}$, $\cos^{-1}$, and $\tan^{-1}$ to evaluate the cosecant, secant, and cotangent functions. Those keys relate to the inverse trigonometric functions. In the context of right triangles, they calculate an angle between 0° and 90° having a given sine, cosine, or tangent. In Section 8-5 you will study them in a more general context.

EXERCISES **1.** (*Writing to Learn*) Use the right triangle definitions of the trigonometric functions to explain why each of the reciprocal identities is true.

2. Use the reciprocal identities together with the exact values of $\sin 45°$ and $\cos 45°$ to find the exact value of

     a. $\tan 45°$ **b.** $\cot 45°$

     c. $\sec 45°$ **d.** $\csc 45°$

3. Repeat Exercise 2 for 30°.

4. Repeat Exercise 2 for 60°.

The Unit Circle Definitions If the angle x determines a point (a, b) on the unit circle, as in Figure 8-23, the trigonometric functions of x are defined as follows.

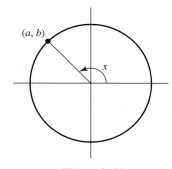

Figure 8–23

$$\sin x = b \qquad \csc x = \frac{1}{b}$$

$$\cos x = a \qquad \sec x = \frac{1}{a}$$

$$\tan x = \frac{b}{a} \qquad \cot x = \frac{a}{b}$$

The tangent and secant functions are undefined if $a = 0$, and the cotangent and cosecant functions are undefined if $b = 0$.

EXERCISES **5.** (*Writing to Learn*) If $0 < x < \dfrac{\pi}{2}$, explain why the unit circle definitions of the trigonometric functions are equivalent to their right triangle definitions.

In Exercises 6–9, evaluate all six trigonometric functions of x if x determines the given point on the unit circle.

6. $(0.6, -0.8)$ **7.** $\left(-\dfrac{5}{13}, -\dfrac{12}{13}\right)$

8. $\left(-\dfrac{\sqrt{2}}{2}, \dfrac{\sqrt{2}}{2}\right)$ **9.** $(1, 0)$

In Exercises 10–13, use the exact values of sin x and cos x to find the exact values of all six trigonometric functions of x.

10. $x = 150°$ **11.** $x = -\dfrac{\pi}{3}$

12. $x = -135°$ **13.** $x = \dfrac{5\pi}{2}$

In the context of the unit circle, the domains of the trigonometric functions are not restricted to $\left(0, \dfrac{\pi}{2}\right)$. The following Mathematical Looking Glass illustrates this advantage of the unit circle definitions in modeling physical problems.

A MATHEMATICAL LOOKING GLASS

Satellite Photography

Many satellites use scanning techniques to photograph the earth. Figure 8-24 shows a satellite employing a system of mirrors and lenses to record images as they rotate.

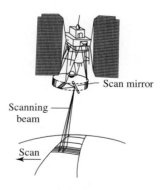

Scan mirror

Scanning beam

Scan

Figure 8–24

Adapted from Bernice Kastner,
Space Mathematics, NASA, 1985.

The recorded images are transmitted to earth as numerical data and are reconstructed as photographs with the aid of a computer. The computer must correct for several sources of distortion in the data. For example,

- An image of a curved surface (part of the earth) is being projected onto a flat surface (a piece of paper).
- If the mirrors and lenses rotate at a constant angular rate, the scanning beam does not move along the ground at a constant rate.

Let's look at some steps NASA might take in analyzing the second source of distortion. Figure 8-25 shows that when the scanning beam makes a directed angle θ with the vertical, it intersects the ground at a directed distance of x kilometers from the point directly below the satellite.

Figure 8–25

In Figure 8-25 the satellite is 200 km above the earth. The beam rotates through the angular interval $\left[-\dfrac{\pi}{4}, \dfrac{\pi}{4}\right]$ and scans the interval $[-200, 200]$ on the earth.   ■

EXERCISES   **14. a.** (*Creating Models*) Express x as a function of θ.

     b. What are the domain and range of the function in the context of the problem?

     c. (*Writing to Learn*) Explain why your model in part (a) can be created more conveniently by using the unit circle definitions of the trigonometric functions, rather than the right triangle definitions.

15. (*Making Observations*) If θ changes at a constant rate, verify that x does not. Specifically, calculate the average rate of change in x with respect to θ over each of the intervals $\left[0, \dfrac{\pi}{6}\right]$ and $\left[\dfrac{\pi}{6}, \dfrac{\pi}{4}\right]$.

Like the sine and cosine, the other four trigonometric functions can be regarded as functions of real numbers, even though their definitions are given in terms of angles. Thus, for example, $\tan \dfrac{\pi}{4} = 1$ whether $\dfrac{\pi}{4}$ is regarded as an angle in radians or an abstract real number.

Let's look at all six trigonometric functions from analytical, numerical, and graphical perspectives, just as we considered the sine and cosine functions in Section 8-2.

AN ANALYTICAL VIEW OF THE TRIGONOMETRIC FUNCTIONS

As you can verify in Exercise 16, the reciprocal identities given earlier remain valid in the unit circle.

EXERCISE   **16.** (*Making Observations*) Verify that the reciprocal identities are valid if the trigonometric functions are given their unit circle definitions.

The relationships that follow are sometimes referred to as the **Pythagorean identities**. The first is an immediate consequence of the fact that $\cos x$ and $\sin x$ are the coordinates of a point on the unit circle. You can verify the others in Exercise 18. [When a trigonometric function is raised to a power, the exponent usually follows the name of the function. For example, $(\sin x)^2$ is usually written as $\sin^2 x$.]

Pythagorean Identities

$$\cos^2 x + \sin^2 x = 1$$

$$1 + \tan^2 x = \sec^2 x$$

$$\cot^2 x + 1 = \csc^2 x$$

EXERCISES **17.** (*Writing to Learn*) By referring to the following figure, explain why the relationship $\cos^2 x + \sin^2 x = 1$ is called a Pythagorean identity.

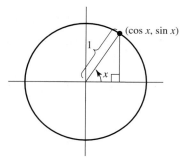

18. (*Making Observations*)

   a. Divide both sides of the identity $\cos^2 x + \sin^2 x = 1$ by $\cos^2 x$. Then use reciprocal identities to rewrite the result and thus verify that $1 + \tan^2 x = \sec^2 x$.

   b. By performing appropriate operations on the identity

$$\cos^2 x + \sin^2 x = 1$$

   verify the identity $\cot^2 x + 1 = \csc^2 x$.

   If x represents an angle in a particular quadrant, and if one trigonometric function of x is known, you can use the Pythagorean identities to find the others. Example 1 illustrates the technique.

EXAMPLE 1 *Finding all trigonometric functions of an angle when one is known*

Suppose $\tan x = -2$ and x is in quadrant IV. We can use the reciprocal identity $\cot x = \dfrac{1}{\tan x}$ to obtain $\cot x = -\dfrac{1}{2}$. Then we can use the Pythagorean identities.

$$\sec^2 x = 1 + \tan^2 x \qquad\qquad \csc^2 x = \cot^2 x + 1$$

$$\sec^2 x = 1 + (-2)^2 = 5 \qquad \csc^2 x = \left(-\frac{1}{2}\right)^2 + 1 = \frac{5}{4}$$

$$\sec x = \pm\sqrt{5} \qquad\qquad \csc x = \pm\frac{\sqrt{5}}{2}$$

Since x is in quadrant IV, it determines a point (a, b) on the unit circle with $a > 0$ and $b < 0$. Thus $\sec x > 0$ and $\csc x < 0$, so $\sec x = \sqrt{5}$ and $\csc x = -\dfrac{\sqrt{5}}{2}$.

Finally, we can use the reciprocal identities $\sin x = \dfrac{1}{\csc x}$ and $\cos x = \dfrac{1}{\sec x}$ to determine that $\sin x = -\dfrac{2\sqrt{5}}{5}$ and $\cos x = \dfrac{\sqrt{5}}{5}$.   ∎

EXERCISES   *In Exercises 19–22, evaluate all trigonometric functions of x.*

19. $\cos x = \dfrac{3}{5}$, x is in quadrant I

20. $\csc x = -3$, x is in quadrant III

21. $\tan x = 0.5$, x is in quadrant III

22. $\sin x = \dfrac{1}{3}$, x is in quadrant II

23. (*Making Observations*) How can you use the Pythagorean identities to conclude that $|\sec x| \geq 1$ and $|\csc x| \geq 1$ for all x?

24. (*Making Observations*) How can you use the reciprocal identities to conclude that $|\sec x| \geq 1$ and $|\csc x| \geq 1$ for all x?

25. (*Making Observations*) Which trigonometric functions of x are undefined when $\sin x = 0$? For what values of x are these functions undefined?

26. (*Making Observations*) Which trigonometric functions of x are undefined when $\cos x = 0$? For what values of x are these functions undefined?

A NUMERICAL VIEW OF THE TRIGONOMETRIC FUNCTIONS

In Section 8-2 you used Table 38 to discover several properties of the sine and cosine functions. In Exercises 27–32 you can extend Table 38 and use the extended table to discover several properties of the other four trigonometric functions.

EXERCISES   *Before completing Exercises 27 and 28, copy Table 38 onto a sheet of paper, and add four more columns at the right, headed* tangent, cotangent, secant, *and* cosecant.

27. (*Making Observations*) Use reciprocal identities to fill in the entries in the column headed tangent. Draw a dash in any row where the tangent function is undefined.

28. (*Making Observations*) Use reciprocal identities to fill in the entries in the remaining columns.

Use your table from Exercises 27–28 to make conjectures about some properties of the tangent, cotangent, secant, and cosecant functions in Exercises 29–32.

29. (*Making Observations*) What is the domain of each function? What is its range?

30. (*Making Observations*) For what values of x is each function positive?

31. (*Making Observations*) For what values of x is each function increasing?

32. (*Making Observations*) Since each function can be expressed in terms of the sine and cosine, each must have a period of no more than 2π. What is the period of each?

A GRAPHICAL VIEW OF THE TRIGONOMETRIC FUNCTIONS

The Graph of the Tangent Function You can use your table from Exercises 27–28 to sketch the graph of $y = \tan x$, shown in Figure 8-26.

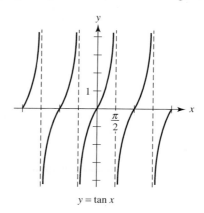

$y = \tan x$

Figure 8–26

One striking feature of the graph is that, like the graphs of the sine and cosine functions, it is periodic. However, the period of the tangent function is π, rather than 2π, so that

$$\tan (x + \pi) = \tan x$$

Exercise 33 indicates why this is so.

EXERCISE   **33.** (*Making Observations*) Suppose that a directed angle x determines a point (a, b) on the unit circle.

   a. In terms of a and b, what are the coordinates of the point determined by the directed angle $x + \pi$? (*Hint*: A picture may help.)

   b. How does your result from part (a) show that $\tan (x + \pi) = \tan x$?

Another striking feature of the graph is the presence of vertical asymptotes at odd multiples of $\dfrac{\pi}{2}$, that is, at $x = (2k + 1)\dfrac{\pi}{2}$ for each integer k. In Exercise 34, you can explain analytically why they are there.

EXERCISES   **34.** (*Making Observations*) Suppose that a directed angle x has a value near, but not equal to, $\dfrac{\pi}{2}$, so that it determines a point (a, b) in the unit circle near, but not equal to, $(0, 1)$.

   a. If x is slightly less than $\dfrac{\pi}{2}$, what can you say about the size of $\dfrac{b}{a}$? What can you say about the sign of $\dfrac{b}{a}$? How does this behavior show up on the graph of $y = \tan x$?

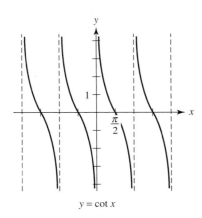

$y = \cot x$

Figure 8–27

b. If x is slightly greater than $\dfrac{\pi}{2}$, what can you say about the size of $\dfrac{b}{a}$?

What can you say about the sign of $\dfrac{b}{a}$? How does this behavior show up on the graph of $y = \tan x$?

35. (*Making Observations*) Use the graph in Figure 8-26 to find the domain and range of the tangent function.

36. (*Making Observations*) In which quadrants of the unit circle is $\tan x > 0$? How can this fact be seen on the graph of $y = \tan x$?

37. (*Making Observations*) Use the graph in Figure 8-26 to give the values of x for which $\tan x$ is increasing.

The Graph of the Cotangent Function Figure 8-27 shows the graph of $y = \cot x$, which can also be sketched from the data in your table from Exercises 27 and 28.

EXERCISE   **38.** (*Making Observations*)

a. Explain why the period of the cotangent function is π. (*Hint*: See Exercise 33.)

b. Explain how the unit circle definition of the cotangent function leads to the vertical asymptotes at $x = k\pi$ for every integer k. (*Hint*: See Exercise 34.)

c. Use the graph in Figure 8-27 to find the domain and range of the cotangent function.

d. In which quadrants of the unit circle is $\cot x > 0$? How can this fact be seen on the graph of $y = \cot x$?

e. For what values of x is $\cot x$ increasing?

The Graphs of the Secant and Cosecant Functions Figure 8-28 shows the graphs of $y = \sec x$ and $y = \csc x$. In Exercises 39–50 you can explain how they were obtained from the graphs of $y = \cos x$ and $y = \sin x$, also shown in Figure 8-28.

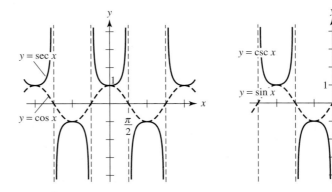

Figure 8–28

EXERCISES   **39.** (*Making Observations*) Use the fact that $\sec x = \dfrac{1}{\cos x}$ to answer the following questions.

   a. What is the value of sec x when cos $x = 1$? When cos $x = -1$?

   b. For every value of x, $|\cos x| \le 1$. What does this imply about $|\sec x|$? What is the range of the secant function?

40. (*Making Observations*) What is the period of the secant function?

41. (*Making Observations*) What can you say about sec x when cos $x = 0$? How does the graph show this behavior? What does your conclusion say about the domain of the secant function?

42. (*Making Observations*) What can you say about sec x when cos $x > 0$? When cos $x < 0$?

43. (*Making Observations*) What can you say about sec x when cos x is increasing? When cos x is decreasing?

44. (*Making Observations*) Where are the local maxima and minima for $y = \sec x$? How do these points compare with the local maxima and minima for $y = \cos x$?

45. (*Making Observations*) Use the fact that $\csc x = \dfrac{1}{\sin x}$ to answer the following questions.

   a. What is the value of csc x when sin $x = 1$? When sin $x = -1$?

   b. For every value of x, $|\sin x| \le 1$. What does this imply about $|\csc x|$? What is the range of the cosecant function?

46. (*Making Observations*) What is the period of the cosecant function?

47. (*Making Observations*) What can you say about csc x when sin $x = 0$? How does the graph show this behavior? What does your conclusion say about the domain of the secant function?

48. (*Making Observations*) What can you say about csc x when sin $x > 0$? When sin $x < 0$?

49. (*Making Observations*) What can you say about csc x when sin x is increasing? When sin x is decreasing?

50. (*Making Observations*) Where are the local maxima and minima for $y = \csc x$? How do these points compare with the local maxima and minima for $y = \sin x$?

SUMMARY

The most important fundamental properties of the six trigonometric functions are summarized in Figures 8-29 through 8-34.

Figure 8–29   Figure 8–30

Figure 8–31

Figure 8–32

Figure 8–33

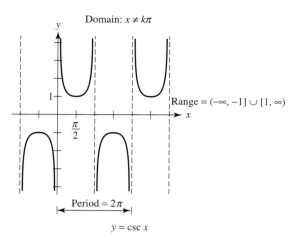

Domain: $x \neq k\pi$

Range = $(-\infty, -1] \cup [1, \infty)$

Period = 2π

$y = \csc x$

Figure 8–34

ADDITIONAL EXERCISES   *In Exercises 51–54, evaluate all six trigonometric functions of x if x determines the given point on the unit circle.*

51. $\left(-\dfrac{1}{2}, \dfrac{\sqrt{3}}{2}\right)$ **52.** $(0.28, -0.96)$

53. $\left(\dfrac{\sqrt{5}}{5}, \dfrac{2\sqrt{5}}{5}\right)$ **54.** $\left(-\dfrac{15}{17}, -\dfrac{8}{17}\right)$

In Exercises 55–62, find the exact values of all six trigonometric functions of x.

55. $x = 135°$ **56.** $x = 210°$

57. $x = -300°$ **58.** $x = 3600°$

59. $x = \dfrac{2\pi}{3}$ **60.** $x = -\dfrac{\pi}{3}$

61. $x = \dfrac{7\pi}{4}$ **62.** $x = -\dfrac{7\pi}{4}$

In Exercises 63–66, draw the angle x in the unit circle and evaluate the remaining five trigonometric functions of x.

63. $\sin x = -\dfrac{1}{4}$, x is in quadrant III

64. $\cos x = \dfrac{2}{3}$, x is in quadrant IV

65. $\cot x = -\dfrac{7}{2}$, x is in quadrant II

66. $\sec x$ is undefined, $\sin x > 0$

In Exercises 67–74,

    a. Use graphical transformations to sketch the graph through at least one full cycle.

    b. Locate all x-intercepts, vertical asymptotes, and local extrema.

    c. Indicate the domain and range of the function.

    d. Tell where the function is increasing and where it is decreasing.

67. $y = -\tan x$ **68.** $y = \cot 2x$

69. $y = 3 \sec x$ **70.** $y = \csc x - 1$

71. $y = -2 \csc\left(x - \dfrac{\pi}{4}\right)$ **72.** $y = \tan\left(-\dfrac{x}{2}\right)$

73. $y = 0.5 \cot (2x - \pi)$ **74.** $y = -4 \sec\left(3x + \dfrac{\pi}{2}\right)$

75. (*Making Observations*) Use graphical transformations to sketch the graph of $y = \cot\left(\dfrac{\pi}{2} - x\right)$, and compare the graph with that of $y = \tan x$. Write an identity suggested by the graphs.

76. (*Making Observations*) Suppose that the line segment from $(0, 0)$ to (x, y) is r units long, and makes a directed angle of θ with the positive x-axis. Verify each of the following.

a. $\sin \theta = \dfrac{y}{r}$ **b.** $\csc \theta = \dfrac{r}{y}$

c. $\cos \theta = \dfrac{x}{r}$ **d.** $\sec \theta = \dfrac{r}{x}$

e. $\tan \theta = \dfrac{y}{x}$ **f.** $\cot \theta = \dfrac{x}{y}$

8-4   THE TRIGONOMETRIC FUNCTIONS IN RIGHT TRIANGLES

HISTORICAL ORIGINS OF THE TRIGONOMETRIC FUNCTIONS

In Section 8-1 you were introduced to the sine and cosine functions in the context of Fermilab's nuclear particle accelerator. In the present section you will learn that all six trigonometric functions had their origins in a time and place far removed from Fermilab. Beginning in the third century B.C., a succession of Greek astronomers investigated several questions regarding the sun and moon. Among these were the questions of the size of the earth, moon, and sun, as well as their distances from each other. Using the strategy of *Breaking the Problem into Parts* (by default, since no individual was able to answer the questions alone), they arrived at the answers over the course of more than a century. In this section you will see how their work was related to the development of trigonometry.

What were the driving forces behind humanity's desire for astronomical knowledge, and the consequent creation of trigonometry? In *Cosmos* (Random House, 1980), Carl Sagan suggests that they were both physical and metaphysical:

> Why did people all over the world make such an effort to learn astronomy? We hunted gazelles and antelope and buffalo whose migrations ebbed and flowed with the seasons. Fruits and nuts were ready to be picked in some times but not in others. When we invented agriculture, we had to take care to plant and harvest our crops in the right season. Annual meetings of far-flung nomadic tribes were set for prescribed times. The ability to read the calendar in the skies was literally a matter of life and death. The reappearance of the crescent moon after the new moon; the return of the sun after a total eclipse; the rising of the sun in

the morning after its troublesome absence at night were noted by people around the world: these phenomena spoke to our ancestors of the possibility of surviving death. Up there in the skies was also a metaphor of immortality.

The following Mathematical Looking Glass illustrates how trigonometry can be used to obtain astronomical information.

A MATHEMATICAL LOOKING GLASS
Greek Astronomy

According to modern astronomers, the distance from the earth to the sun is about 92,900,000 miles (149,600,000 kilometers). Have you ever wondered how they know? After all, no one has ever been to the sun. Even more amazing is the fact that the Greek astronomers of antiquity, working without the advantages of modern equipment or modern mathematics, devised a valid method for measuring this distance. Many of their calculations were done before the development of trigonometry by Hipparchus in the mid-second century B.C. Let's follow their work, and see how trigonometry might have simplified some of it.

The first step toward calculating the distance to the sun was taken by Aristarchus (c. 310–230 B.C.). According to the *Encyclopaedia Brittanica*, 11th edition (1910–11), "Aristarchus was the first to maintain that the earth moves around the sun. On this account he was accused of impiety by the Stoic Cleanthes, just as Galileo, in later years, was attacked by the theologians." His book, *On the Sizes and Distances of the Sun and Moon*, contain his estimates of the relative sizes of the sun and moon, as well as their distances from the earth.

Aristarchus observed that when a half moon is directly overhead, as in Figure 8-35, the angle *OCA* is 90°. (By making observations under these special conditions, he was employing the strategy of *Examining a Special Case*.)

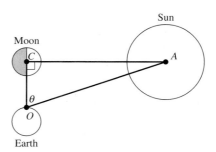

Figure 8–35

Adapted from Morris Kline, *Mathematical Thought from Ancient to Modern Times*, Oxford University Press, New York, 1972, p. 156.

Aristarchus used his observations to calculate the ratio $\dfrac{\overline{OA}}{\overline{OC}}$ and thus determine how many times farther the earth is from the sun than from the moon. Because ancient tools for measuring distances and angles had poor accuracy, his result was in error. However, his fundamental ideas were sound. You can duplicate his reasoning and calculations in Exercises 1 and 2.

EXERCISES   1. Aristarchus measured θ in Figure 8-35 as 87°. He then drew a right triangle with an 87° angle and a shorter leg of length 1, as in Figure 8-36, and measured the lengths of the longer leg and the hypotenuse as shown.

Figure 8–36

a. (*Writing to Learn*) Explain how Figure 8-36 could be used to calculate the ratio $\dfrac{\overline{OA}}{\overline{OC}}$ in Figure 8-35.

b. (*Interpreting Mathematics*) According to Aristarchus, how many times farther is the earth from the sun than from the moon?

c. What name was later given to the ratio $\dfrac{\overline{OA}}{\overline{OC}}$ by Rhäticus? (*Hint*: It is the name we still use today.)

2. According to modern measurements, θ in Figure 8-35 is about 89.857°. If $\angle QPR$ in Figure 8-36 is 89.857°, then $\overline{QR}$ is about 400.668 and $\overline{PR}$ is about 400.670. Using these values, how many times farther is the earth from the sun than from the moon?

3. Use your calculator to evaluate sec 87°. Does your result agree with the one you obtained in Exercise 1?

4. Use your calculator to evaluate sec 89.857°. Does your result agree with the one you obtained in Exercise 2?

In completing Exercise 1 you used the ratio $\dfrac{\overline{OA}}{\overline{OC}} = \sec 87°$ in Figure 8-35 to calculate a ratio of astronomical distances. From Hipparchus' time onward the Greeks made extensive use of all six trigonometric functions in the context of right triangles, although they did not formally name them.

CALCULATIONS USING TRIGONOMETRIC FUNCTIONS IN RIGHT TRIANGLES

You can follow more Greek astronomers' calculations in Exercises 33–37. In the meantime, let's observe the computational power of the trigonometric functions. First, if we know the length of one side of a right triangle, and either one of the acute angles or the length of a second side, we can find the remaining sides and angles. Second, if we know any trigonometric function of an acute angle, we can find the other five.

FYI When evaluating trigonometric functions on your calculator, you will nearly always round the results. You should round to an accuracy that depends on the precision of the angle as follows:

Precision of angle	Accuracy of trigonometric function
1° or 0.01 radians	2 significant digits
0.1° or 0.001 radians	3 significant digits
0.01° or 0.0001 radians	4 significant digits

(See Section A-1 to review the ideas of accuracy and precision.)

EXAMPLE 1   *Finding sides and angles of a right triangle*
In Figure 8-37, $A = 27.0°$ and $b = 16$.

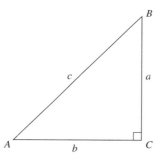

Figure 8–37

To find a, use the fact that $\frac{a}{b} = \tan A$. Thus,

$$\frac{a}{16} = \tan 27.0°$$

$$a = 16 \tan 27.0° \approx 8.15$$

To find c, use the fact that $\frac{c}{b} = \sec A$. Thus,

$$\frac{c}{16} = \sec 27.0°$$

$$c = 16 \sec 27.0° = \frac{16}{\cos 27.0} \approx 18.0$$

To find B, observe that $A + B = 90°$. Thus $B = 90° - 27.0° = 63.0°$.   ■

EXAMPLE 2   *Finding sides and angles of a right triangle*
In Figure 8-38, $a = 13.00$ and $c = 20.00$.

Figure 8–38

To find b, use the fact that by the Pythagorean theorem

$$b^2 = c^2 - a^2 = 20^2 - 13^2 = 400 - 169 = 231$$

so $b = \sqrt{231} \approx 15.20$.

To find A, use the fact that $\sin A = \frac{a}{c}$. Thus,

$$\sin A = \frac{13}{20} = 0.6500$$

$$A = \sin^{-1} 0.6500 \approx 40.54°$$

To find B, use the fact that $A + B = 90°$. Thus $B \approx 90° - 40.54° = 49.46°$.   ■

EXERCISES   *In Exercises 5–8, find all sides and angles of a right triangle ABC with sides and angles labeled as in Figure 8-38.*

5. $B = 58.3°$, $c = 50.7$ **6.** $A = 0.32$, $a = 2.5$ (*Note*: A is in radians.)

7. $a = 24.00, b = 32.00$ **8.** $b = 1.000, c = 3.000$

EXAMPLE 3   *Finding all trigonometric functions of an angle when one is known*

Suppose we know that $\cot \theta = 0.4$. Then we can place θ in any right triangle having the ratio

$$\frac{\text{side adjacent to } \theta}{\text{side opposite } \theta} = 0.4$$

One such triangle is shown in Figure 8-39.

Figure 8–39

By the Pythagorean theorem, the hypotenuse has a length of

$$\sqrt{2^2 + 5^2} = \sqrt{29}$$

Thus the trigonometric functions of θ are

$$\sin \theta = \frac{5\sqrt{29}}{29} \qquad \csc \theta = \frac{\sqrt{29}}{5}$$

$$\cos \theta = \frac{2\sqrt{29}}{29} \qquad \sec \theta = \frac{\sqrt{29}}{2}$$

$$\tan \theta = \frac{5}{2} \qquad \cot \theta = \frac{2}{5}$$ ■

EXERCISES   *In Exercises 9–12, draw a right triangle containing the angle θ, and find all six trigonometric functions of θ without using your calculator.*

9. $\tan \theta = 1$ **10.** $\csc \theta = 2$

11. $\sin \theta = \dfrac{5}{13}$ **12.** $\cos \theta = \dfrac{15}{17}$

ADDITIONAL EXERCISES   *In Exercises 13–18, find all sides and angles of a right triangle ABC with sides and angles labeled as in Figure 8-38.*

13. $A = 12°, \ b = 560$

14. $B = 1.175, \ c = 8.94$ (*Note:* B is in radians.)

15. $a = 8.64, \ c = 17.0$

16. $a = 0.390, \ b = 1.02$

17. $B = 0.56000, \ b = 1.0000$

18. $A = 85.2°, \ c = 214$

In Exercises 19–24, draw a right triangle containing the angle θ, and find all six trigonometric functions of θ without using your calculator.

19. $\sin \theta = \dfrac{4}{5}$ **20.** $\tan \theta = \sqrt{3}$

21. $\sec \theta = \dfrac{3}{2}$ **22.** $\cos \theta = 0.2$

23. $\cot \theta = \dfrac{24}{7}$ **24.** $\tan \theta = \dfrac{24}{7}$

Use the definitions of the trigonometric functions and the reciprocal identities to verify each identity in Exercises 25–28 (Making Observations) for an angle θ between 0° and 90°.

25. $\tan \theta \csc \theta = \sec \theta$

26. $\cot \theta \sec \theta = \csc \theta$

27. $\dfrac{\sin \theta}{\tan \theta} = \dfrac{1}{\sec \theta}$

28. $\tan \theta \cot \theta = \cos \theta \sec \theta$

29. (*Writing to Learn*) If θ is the acute angle between the *x*-axis and the line $y = mx + b$ with $m > 0$, explain why $m = \tan \theta$.

30. (*Problem Solving*) If a road makes an angle of θ with the horizontal, its **grade** is $|\tan \theta|$, usually expressed as a percentage. Thus a hill with a 5% grade has $|\tan \theta| = 0.05$.

Figure 8-40 shows a hillside that is to be graded in the construction of a roadbed. Points *A* and *B* are, respectively, 562 and 943 feet above sea level, and the hill from *A* to *B* will have a 3% grade. To calculate construction costs, engineers need to know the length of the road between *A* and *B*. How long will it be?

A
Elevation
562 ft

B
Elevation
943 ft

Figure 8–40

31. (*Problem Solving*) Figure 8-41a shows one end of Dave's house.

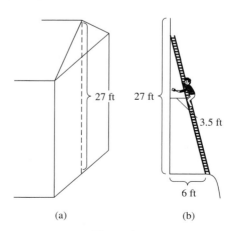

27 ft   27 ft

3.5 ft

6 ft

(a)   (b)

Figure 8–41

To paint the end wall Dave uses an extension ladder whose length can be adjusted in increments of 1 foot from 17 to 28 feet. The foot of the ladder must rest on the 6-foot strip of level ground between the wall and the hillside. Furthermore, for the sake of Dave's nerves, the ladder can make no more than a 75° angle with the ground.

a. Dave can reach no more than 5 feet above the top of the ladder. Show how he can position the ladder to reach the highest point on the end wall, 27 feet above the ground. Remember that the length of the ladder, in feet, must be an integer.

b. Dave can reach 3.5 feet in front of the ladder, as in Figure 8-41b. If the ladder is set as in part (a), what is the lowest point on the wall that he can reach?

c. Will Dave need to reset the ladder to paint the lower portion of the wall? He can paint up to a height of 8 feet without the ladder.

32. (*Problem Solving*) Allegheny Ludlum Steel Corporation produces steel from scrap metal at two locations in western Pennsylvania. One of their customers regularly orders large quantities of circular pieces, called blanks, 12 inches in diameter, to be used in manufacturing kitchenware. The blanks are cut from long strips of steel, sometimes called coils because they are rolled up like long carpets while still red hot from the furnace. Using calculus, it can be shown that the pattern in Figure 8-42 minimizes waste.

a. Explain why triangle *ABC* is equilateral.

b. How wide should the steel strip be to accommodate the pattern of blanks in Figure 8-42?

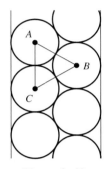

Figure 8–42

In Exercises 33–37, you can continue your explorations from **Greek Astronomy** to find the radii of the sun and moon and the distance of each from the earth. In Exercise 33 you can discover how Aristarchus expressed all four quantities in terms of a single unknown.

33. (*Problem Solving*) Figure 8-43 shows how the moon and sun appear to an observer on the earth during a solar eclipse. In another instance of *Examining a Special Case*, Aristarchus took advantage of the fact that the moon and sun appear to be the same size, and therefore occupy the same angle in the observer's field of vision. (In reality they do not appear to be exactly the same size. By assuming that they do, we introduce a small error into your final results in Exercise 37.)

Moon   Sun

Figure 8–43

a. Let *r* denote the radius of the moon ($\overline{MN}$ in Figure 8-43). Use the fact that $\angle NOP = 0.530°$ to express $\overline{OM}$ in terms of *r*.

b. In Exercise 2 you established that $\overline{OS} = 400.67\ \overline{OM}$. Use this fact to express $\overline{TS}$ and $\overline{OS}$ in terms of *r*.

In Exercises 34–37, you can follow Hipparchus' trigonometric calculations to find the value of r.

34. (*Problem Solving*) Figure 8-44 shows the earth, moon, and sun during a solar eclipse. An observer at point C sees the sun just before it begins to disappear behind the moon. During the eclipse the earth's rotation carries the observer from C to A, where the sun is again completely visible. The lines of sight from C and A intersect at some point X in space.

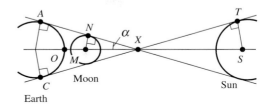

Figure 8–44

a. Explain why triangles MNX and STX are similar.

b. Justify each step in the following calculation.

$$\overline{XS} = 400.67\,\overline{MX}$$

$$\overline{OM} + \overline{MX} + \overline{XS} = 86{,}629r$$

$$\overline{MX} + \overline{XS} = 86{,}413r$$

$$401.67\,\overline{MX} = 86{,}413r$$

$$\overline{MX} \approx 215.13r$$

c. What is the value of $\sin \alpha$? Verify that $\alpha \approx 0.266°$.

35. (*Problem Solving*) In completing his calculations, Hipparchus had the advantage of knowing the size of the earth. Through an ingenious use of geometry, Eratosthenes (c. 284–192 B.C.) had calculated its radius as about 6250 kilometers, impressively close to the actual value of 6380 kilometers. (For the fascinating story of Eratosthenes' work, see Morris Kline, *Mathematical Thought from Ancient to Modern Times*, Oxford University Press, 1972.)

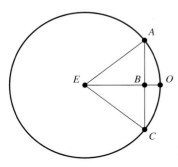

Figure 8–45

Figure 8-45 shows the arc AOC on the earth's surface from Figure 8-44. The center of the earth is at E. By timing the length of the eclipse, Hipparchus calculated the measure of $\angle AEC$. Using 6380 km for the earth's radius and 66.692° for $\angle AEC$, show that $\overline{AB} = 3506$ km, $\overline{EB} = 5328$ km, and $\overline{BO} = 1050$ km.

36. (*Writing to Learn*) Use Figure 8-46 to justify each step in the following calculation.

$$\overline{OX} = 431.34r$$

$$\cot \alpha = \frac{\overline{BO} + \overline{OX}}{\overline{BC}}$$

$$\cot 0.266° = \frac{1050 + 431.34r}{3506}$$

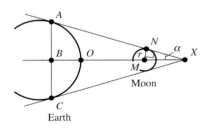

Figure 8–46

(*Hint:* See Exercise 35.)

37. (*Interpreting Mathematics*) Use your result from Exercise 36 to find *r* and calculate the radii of the sun and the moon and their distances from the earth.

Currently accepted values for the radii of the moon and sun are 1737 km and 696,000 km, respectively. The moon's distance from the earth varies between 363,000 and 406,000 km, with a mean value of 384,400 km. The sun's distance from the earth varies between 147,100,000 and 152,100,000 km, with a mean value of 149,600,000 km.

8-5   THE INVERSE TRIGONOMETRIC FUNCTIONS

In *Fermilab 2* (page 391), the shotmaster needs to know when a proton will pass through the collider detector, since this is where its collision with an antiproton must occur. In Figure 8-47 the collider detector is at $(-0.96, 0.28)$.

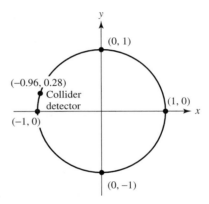

Figure 8–47

This situation is the inverse of the problem we have been considering up until now. Instead of finding outputs cos *x* and sin *x* for a given input *x*, the shotmaster must find inputs for the outputs -0.96 and 0.28. As you saw in Chapter 7, problems such as this can be solved by looking at the inverses of the given functions. Let's proceed by *Breaking the Problem into Parts*, and study the inverses of the sine and cosine functions separately.

THE INVERSE SINE FUNCTION

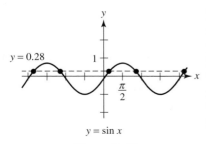

Figure 8–48

In Section 7-3 you saw that only one-to-one functions have inverses. However, Figure 8-48 makes it clear that the sine function is not one-to-one. In particular, sin *x* = 0.28 for many values of *x*. Thus the equation *y* = sin *x* does not represent *x* as a function of *y*. However, we can find an interval on which the sine function is one-to-one, and find the inverse of the function on that interval. This will give us *one* solution to the equation sin *x* = 0.28. We will then discover how to find *all* solutions to sin *x* = 0.28, and generally how to find all solutions to sin *x* = *b* for any value of *b*.

Definition of the Inverse Sine Function Figure 8-49 shows several intervals on which the function *y* = sin *x* is one-to-one.

$$\left[-\frac{3\pi}{2}, -\frac{\pi}{2}\right] \qquad \left[\frac{\pi}{2}, \frac{3\pi}{2}\right]$$

$$\left[-\frac{\pi}{2}, \frac{\pi}{2}\right]$$

$$y = \sin x$$

Figure 8–49

We could choose any of these intervals to define the inverse sine function, but it is both convenient and conventional to choose $\left[-\frac{\pi}{2}, \frac{\pi}{2}\right]$. If $-\frac{\pi}{2} \leqslant x \leqslant \frac{\pi}{2}$, the function $f(x) = \sin x$ has an inverse function $f^{-1}(x)$, denoted by $\sin^{-1}(x)$. That is,

$$y = \sin^{-1}(x) \quad \text{means} \quad x = \sin y \text{ and } -\frac{\pi}{2} \leqslant y \leqslant \frac{\pi}{2}$$

The inverse sine of x is also sometimes denoted as arcsin (x). This notation, meaning "the arc whose sine is x," recalls the historical origins of the sine function.

Because sin and $\sin^{-1}$ are inverse functions, the following relationships must hold:

$$\text{If } -\frac{\pi}{2} \leqslant x \leqslant \frac{\pi}{2}, \quad \text{then } \sin^{-1}(\sin x) = x.$$
$$\text{If } -1 \leqslant x \leqslant 1, \quad \text{then } \sin(\sin^{-1}x) = x.$$

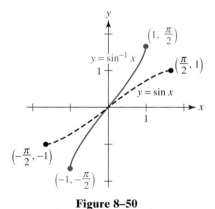

Figure 8–50

The graph of the inverse sine function can be obtained by reflecting the appropriate portion of the graph in Figure 8-49 about the line $y = x$. The result is shown in Figure 8-50.

EXERCISES

1. Use the graph in Figure 8-50 to find the domain and range of the inverse sine function.

2. Graph $y = \sin^{-1}(x)$, and trace the graph to estimate each of the following. Then evaluate each, using the $\sin^{-1}$ key on your calculator.

   a. $\sin^{-1} 0.28$ **b.** $\sin^{-1}\left(-\frac{1}{4}\right)$

   c. $\sin^{-1} 0.6$ **d.** $\sin^{-1} (-0.6)$

3. Use Table 38 from Section 8-2 to find the exact value of each of the following.

a. $\sin^{-1} 1$ **b.** $\sin^{-1}\left(\dfrac{1}{2}\right)$

c. $\sin^{-1}\left(\dfrac{\sqrt{3}}{2}\right)$ **d.** $\sin^{-1}\left(\dfrac{\sqrt{2}}{2}\right)$

4. (*Making Observations*) Use your results from Exercise 3 and the strategy of *Looking for Symmetry* in Figure 8-50 to find each of the following.

a. $\sin^{-1}(-1)$ **b.** $\sin^{-1}\left(-\dfrac{1}{2}\right)$

c. $\sin^{-1}\left(-\dfrac{\sqrt{3}}{2}\right)$ **d.** $\sin^{-1}\left(-\dfrac{\sqrt{2}}{2}\right)$

5. (*Writing to Learn*)

a. What happens when you try to evaluate $\sin^{-1} 2$ on your calculator? Explain why.

b. Evaluate $\sin^{-1}(\sin \pi)$ on your calculator. Is the result equal to π? Explain.

Solutions of sin $x = c$ As you discovered in Exercise 2, $\sin^{-1} 0.28 \approx 0.284$. Thus 0.284 is one solution of $\sin x = 0.28$. In the context of **Fermilab 2**, this means that the proton has a y-coordinate of 0.28 after 0.284 seconds. How can the shotmaster find all other times when the proton has the same y-coordinate? We can discover them by looking at the graph of the sine function.

Since the sine function is one-to-one on $\left[-\dfrac{\pi}{2}, \dfrac{\pi}{2}\right]$, the only solution of $\sin x = 0.28$ in this interval is $x \approx 0.284$.

In Exercise 45 of Section 8-2 you saw that $\sin(\pi - x) = \sin x$. Figure 8-51 shows that the only solution of $\sin x = 0.28$ in $\left[\dfrac{\pi}{2}, \dfrac{3\pi}{2}\right]$ is $x = \pi - 0.284 \approx 2.858$.

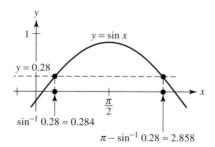

Figure 8–51

Since the period of the sine function is 2π, we can obtain all solutions to the equation $\sin x = 0.28$ by adding integer multiples of 2π to the two solutions we obtained in $\left[-\dfrac{\pi}{2}, \dfrac{3\pi}{2}\right]$. The solutions are

$$x \approx \ldots, 0.284 - 4\pi, 2.858 - 4\pi, 0.284 - 2\pi, 2.858 - 2\pi,$$
$$0.284, 2.858, 0.284 + 2\pi, 2.858 + 2\pi,$$
$$0.284 + 4\pi, 2.858 + 4\pi, \ldots$$

The exact solutions can be written concisely as

$$x = \begin{cases} \sin^{-1} 0.28 + 2k\pi \\ \pi - \sin^{-1} 0.28 + 2k\pi \end{cases}$$

where k is understood to be an integer. In **Fermilab 2**, only nonnegative values of k make sense.

Generalizing from the preceding discussion, we can make the following statement.

> For any real number c with $|c| \leqslant 1$, the solutions to the equation $\sin x = c$ are
>
> $$x = \begin{cases} \sin^{-1} c + 2k\pi \\ \pi - \sin^{-1} c + 2k\pi \end{cases}$$
>
> for all integers k.

EXERCISES

6. (*Writing to Learn*) In the preceding expressions for the solutions of $\sin x = c$, explain why the restriction $|c| \leqslant 1$ is needed.

7. Without using your calculator, find the exact value of all solutions to each equation:

    a. $\sin x = \dfrac{1}{2}$ **b.** $\sin x = -\dfrac{\sqrt{2}}{2}$

    c. $\sin x = -1$ **d.** $\sin x = 0$

    e. $\sin x = \dfrac{\sqrt{3}}{2}$ **f.** $\sin x = -3$

8. Use your calculator to find approximate values of all solutions to each equation:

    a. $\sin x = 0.37$ **b.** $\sin x = -\dfrac{2}{3}$

    c. $\sin x = \dfrac{\sqrt{5}}{5}$ **d.** $\sin x = -0.01$

THE INVERSE COSINE FUNCTION

Definition of the Inverse Cosine Function Now let's return to the problem in **Fermilab 2**. To tell when the proton is in the collider detector, the shotmaster needs to solve the equations $\sin x = 0.28$ and $\cos x = -0.96$. We have used the inverse sine function to solve the first equation. Let's see if we can define an inverse cosine function and use it to solve the second equation.

Figure 8-52 shows several intervals on which the function $g(x) = \cos x$ is one-to-one. This time the conventional choice is $[0, \pi]$. If $0 \leqslant x \leqslant \pi$, g has an inverse function $g^{-1}(x)$, usually written $\cos^{-1}(x)$ or arccos x. That is,

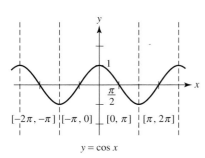

$[-2\pi, -\pi]$ $[-\pi, 0]$ $[0, \pi]$ $[\pi, 2\pi]$

$y = \cos x$

Figure 8–52

> $$y = \cos^{-1} x \quad \text{means} \quad x = \cos y \text{ and } 0 \leqslant y \leqslant \pi$$

Because cos and $\cos^{-1}$ are inverse functions, the following relationships must hold.

> If $0 \leqslant x \leqslant \pi$, then $\cos^{-1}(\cos x) = x$.
>
> If $-1 \leqslant x \leqslant 1$, then $\cos(\cos^{-1} x) = x$.

The graph of the inverse cosine function, obtained by reflecting the appropriate portion of the graph in Figure 8-52 in the line $y = x$, is shown in Figure 8-53.

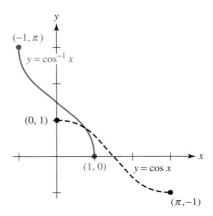

Figure 8–53

EXERCISES

9. Use the graph in Figure 8-53 to find the domain and range of the inverse cosine function.

10. Graph $y = \cos^{-1} x$, and trace the graph to estimate each of the following. Then evaluate each, using the $\cos^{-1}$ key on your calculator.

   a. $\cos^{-1}(-0.96)$ **b.** $\cos^{-1}\left(-\dfrac{1}{4}\right)$

   c. $\cos^{-1} 0.6$ **d.** $\cos^{-1}(-0.6)$

11. Use Table 38 from Section 8-2 to find the exact value of each of the following:

   a. $\cos^{-1} 1$ **b.** $\cos^{-1}\left(\dfrac{1}{2}\right)$

   c. $\cos^{-1}\left(\dfrac{\sqrt{3}}{2}\right)$ **d.** $\cos^{-1}\left(\dfrac{\sqrt{2}}{2}\right)$

12. (*Making Observations*) In Section 8-2 you showed that $\cos(\pi - x) = -\cos x$. Use this fact to find each of the following.

   a. $\cos^{-1}(-1)$ **b.** $\cos^{-1}\left(-\dfrac{1}{2}\right)$

   c. $\cos^{-1}\left(-\dfrac{\sqrt{3}}{2}\right)$ **d.** $\cos^{-1}\left(-\dfrac{\sqrt{2}}{2}\right)$

Solutions of cos $x = c$ As you discovered in Exercise 10, $\cos^{-1}(-0.96) \approx$ 2.858. Thus 2.858 is one solution of $\cos x = -0.96$. In the context of *Fermilab 2*, this means that the proton has an x-coordinate of -0.96 after 2.858 seconds. How can the shotmaster determine all other times when the proton has the same x-coordinate? We can discover them by looking at the graph of the cosine function.

Since the cosine function is one-to-one on $[0, \pi]$, the only solution of $\cos x = -0.96$ in this interval is $x \approx 2.858$.

In Exercise 51 of Section 8-2 you showed that $\cos(2\pi - x) = \cos x$. Figure 8-54 shows that the only solution of $\cos x = -0.96$ in $[\pi, 2\pi]$ is $x \approx 2\pi - 2.858 \approx 3.425$.

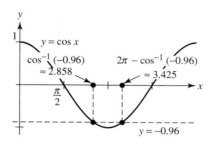

Figure 8–54

Since the period of the cosine function is 2π, we can obtain all solutions to the equation $\cos x = -0.96$ by adding integer multiples of 2π to the two solutions we obtained in $[0, 2\pi]$. The solutions are

$$x \approx \dots, 2.858 - 4\pi, 3.425 - 4\pi, 2.858 - 2\pi, 3.425 - 2\pi,$$
$$2.858, 3.425, 2.858 + 2\pi, 3.425 + 2\pi,$$
$$2.858 + 4\pi, 3.425 + 4\pi, \dots.$$

The exact solutions can be written concisely as

$$x = \begin{cases} \cos^{-1}(-0.96) + 2k\pi \\ 2\pi - \cos^{-1}(-0.96) + 2k\pi \end{cases}$$

where k is understood to be an integer. In **Fermilab 2**, only nonnegative values of k make sense.

Generalizing from the preceding discussion, we can make the following statement.

> For any real number c with $|c| \le 1$, the solutions to the equation $\cos x = c$ are
>
> $$x = \begin{cases} \cos^{-1} c + 2k\pi \\ 2\pi - \cos^{-1} c + 2k\pi \end{cases}$$
>
> for all integers k.

EXERCISES   **13.** Without using your calculator, find the exact value of all solutions to each equation:

a. $\cos x = \dfrac{1}{2}$ **b.** $\cos x = -\dfrac{\sqrt{2}}{2}$

c. $\cos x = -1$ **d.** $\cos x = 0$

e. $\cos x = \dfrac{\sqrt{3}}{2}$ **f.** $\cos x = -3$

14. Use your calculator to find approximate values of all solutions to each equation:

a. $\cos x = 0.37$ **b.** $\cos x = -\dfrac{2}{3}$

c. $\cos x = \dfrac{\sqrt{5}}{5}$ **d.** $\cos x = -0.01$

15. Find all times at which the proton in **Fermilab 2** is at the point $(-0.96, 0.28)$.

THE INVERSE TANGENT FUNCTION

Definition of the Inverse Tangent Function To find the inverse of the tangent function we can follow the same procedure we used to obtain the inverse sine and cosine functions. We will thus be using the strategy of *Examining a Related Problem*. First, let's look for intervals on which the function $h(x) = \tan x$ is one-to-one. Figure 8-55 shows several such intervals.

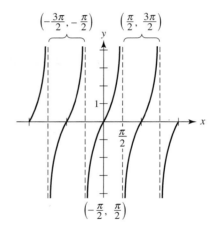

Figure 8–55

If we define $h(x) = \tan x$ on $\left(-\dfrac{\pi}{2}, \dfrac{\pi}{2}\right)$, then h has an inverse function $h^{-1}(x)$, written $\tan^{-1} x$ or $\arctan x$. That is,

$$y = \tan^{-1} x \quad \text{means} \quad x = \tan y \text{ and } -\frac{\pi}{2} < y < \frac{\pi}{2}$$

Because tan and $\tan^{-1}$ are inverse functions, the following relationships must hold.

$$\text{If } -\frac{\pi}{2} < x < \frac{\pi}{2}, \quad \text{then } \tan^{-1}(\tan x) = x.$$

If x is any real number, then $\tan(\tan^{-1} x) = x$.

The graph of the inverse tangent function, obtained by reflecting the appropriate portion of the graph in Figure 8-55 in the line $y = x$, is shown in Figure 8-56.

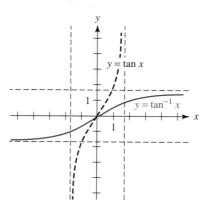

Figure 8–56

EXERCISES   **16.** (*Making Observations*) The graph in Figure 8-56 appears to have two horizontal asymptotes. What are their equations? Explain how you know.

17. What are the domain and range of the inverse tangent function?

18. Graph $y = \tan^{-1} x$ and trace along the graph to estimate each of the following. Then evaluate each, using the $\tan^{-1}$ key on your calculator.

a. $\tan^{-1} 1.52$ b. $\tan^{-1}\left(-\dfrac{1}{2}\right)$

c. $\tan^{-1} 10$ d. $\tan^{-1}(-10)$

19. Find the exact value of each of the following.

a. $\tan^{-1} 1$ b. $\tan^{-1} 0$

c. $\tan^{-1}\left(-\sqrt{3}\right)$ d. $\tan^{-1}\left(\dfrac{1}{\sqrt{3}}\right)$

Solutions of tan x = c Since the tangent function is one-to-one on $\left(-\dfrac{\pi}{2}, \dfrac{\pi}{2}\right)$, the only solution of $\tan x = c$ in this interval is $x = \tan^{-1} c$. Since the period of the tangent function is π, we can obtain all solutions to the equation $\tan x = c$ by adding integer multiples of π to the solution in $\left(-\dfrac{\pi}{2}, \dfrac{\pi}{2}\right)$. Thus,

> For any real number c, the solutions to the equation $\tan x = c$ are
> $$x = \tan^{-1} c + k\pi$$
> for all integers k.

EXERCISES   **20.** Without using your calculator, find the exact value of all solutions to each equation.

a. $\tan x = 1$ **b.** $\tan x = 0$

c. $\tan x = -\sqrt{3}$ **d.** $\tan x = \dfrac{\sqrt{3}}{3}$

21. Use your calculator to find approximate values of all solutions to each equation.

a. $\tan x = 2$ **b.** $\tan x = 0.1$

c. $\tan x = -\dfrac{1}{2}$ **d.** $\tan x = -5$

THE INVERSE COTANGENT, SECANT, AND COSECANT FUNCTIONS

Definitions of the Inverse Cotangent, Secant, and Cosecant Functions
Although inverse functions can be defined for the other three trigonometric functions, they are not encountered as often as those already discussed. The definitions most commonly given are

$$y = \cot^{-1} x \quad \text{means} \quad x = \cot y, \quad -\frac{\pi}{2} \le y \le \frac{\pi}{2}, \quad \text{and} \quad y \ne 0$$

$$y = \sec^{-1} x \quad \text{means} \quad x = \sec y, \quad 0 \le y \le \pi, \quad \text{and} \quad y \ne \frac{\pi}{2}$$

$$y = \csc^{-1} x \quad \text{means} \quad x = \csc y, \quad -\frac{\pi}{2} \le y \le \frac{\pi}{2}, \quad \text{and} \quad y \ne 0$$

The graphs of these functions are shown in Figure 8-57.

(a)   (b)

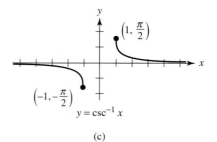

(c)

Figure 8–57

EXERCISES

22. For $y = \cot^{-1} x$, find the domain and range and the equations of the horizontal asymptotes.

23. Repeat Exercise 22 for $y = \sec^{-1} x$.

24. Repeat Exercise 22 for $y = \csc^{-1} x$.

FYI To define the inverse of the secant and cosecant functions, we need to restrict their domains to make them one-to-one. As with the other trigonometric functions, the restrictions can be made in many ways, and mathematicians do not all agree on the appropriate domain. You can explore an alternate definition of $\sec^{-1} x$ in Exercise 50. In the meantime, here is an illustration of the advantages of our definition.

To evaluate $\sec^{-1} x$ we need to find a number y such that $x = \sec y$, with y in $[0, \pi]$ and $y \neq \dfrac{\pi}{2}$. Equivalently, we must find a number y such that $\dfrac{1}{x} = \cos y$, with y in $[0, \pi]$ and $y \neq \dfrac{\pi}{2}$. That is, $y = \cos^{-1}\left(\dfrac{1}{x}\right)$. Thus you can evaluate $\sec^{-1} x$ on your calculator as $\cos^{-1}\left(\dfrac{1}{x}\right)$. Similarly, you can evaluate $\cot^{-1} x$ as $\tan^{-1}\left(\dfrac{1}{x}\right)$ and $\csc^{-1} x$ as $\sin^{-1}\left(\dfrac{1}{x}\right)$.

EXERCISES

25. Find the exact value of each of the following.

   a. $\sec^{-1} 2$ **b.** $\cot^{-1}(-1)$

   c. $\csc^{-1}\sqrt{2}$ **d.** $\sec^{-1}\left(-\dfrac{2\sqrt{3}}{3}\right)$

26. Use your calculator to approximate each of the following.

   a. $\cot^{-1} 2.5$ **b.** $\sec^{-1}(-3)$

   c. $\csc^{-1}\sqrt{6}$ **d.** $\cot^{-1} 100$

COMPOSITIONS OF TRIGONOMETRIC AND INVERSE TRIGONOMETRIC FUNCTIONS

You can sometimes express compositions of trigonometric and inverse trigonometric functions as algebraic functions. The technique is illustrated by Example 1.

EXAMPLE 1   *Finding an algebraic expression for the composition of a trigonometric function and an inverse trigonometric function*

The expression $\sin(\tan^{-1} x)$ can be rewritten in algebraic form as follows.

Denote $\tan^{-1} x$ by y. Then $x = \tan y$ and $-\dfrac{\pi}{2} < y < \dfrac{\pi}{2}$.

If $x > 0$, then $0 < y < \dfrac{\pi}{2}$, and we can indicate the relationship between x and y by drawing the right triangle in Figure 8-58, in which

$$\frac{\text{side opposite } y}{\text{side adjacent to } y} = \tan y = \frac{x}{1}$$

Figure 8–58

The hypotenuse is $\sqrt{x^2 + 1}$, so that

$$\sin(\tan^{-1} x) = \sin y = \frac{x}{\sqrt{x^2 + 1}}$$

If $x \leq 0$, it is still true that $\sin(\tan^{-1} x) = \dfrac{x}{\sqrt{x^2 + 1}}$, but the argument given here must be modified. You can supply the details in Exercise 51. ■

EXERCISES *In Exercises 27–30, write algebraic expressions equivalent to the given expressions.*

27. $\cos(\sin^{-1} x)$, $0 \leq x \leq 1$

28. $\tan(\cos^{-1} x)$, $0 \leq x \leq 1$

29. $\sec(\tan^{-1} x)$, $x > 0$

30. $\csc(\sin^{-1} x)$, $0 < x \leq 1$

ADDITIONAL EXERCISES **31.** Find the exact value of each of the following.

    a. $\sin^{-1} 0$ **b.** $\cos^{-1} 0$

    c. $\tan^{-1}(-1)$ **d.** $\cot^{-1} 0$

    e. $\sec^{-1}\left(-\sqrt{2}\right)$ **f.** $\csc^{-1} 2$

32. Use your calculator to approximate each of the following.

    a. $\sin^{-1}(-0.37)$ **b.** $\cos^{-1}(0.99)$

    c. $\tan^{-1}\left(\dfrac{5}{3}\right)$ **d.** $\cot^{-1}\left(\dfrac{3}{5}\right)$

    e. $\sec^{-1} \pi$ **f.** $\csc^{-1}(-7)$

33. Find the exact values of all solutions to each equation in parts (a)–(c), and use your calculator to find approximate values of all solutions to each equation in parts (d)–(f).

    a. $\sin x = -\dfrac{\sqrt{3}}{2}$ **b.** $\cos x = \dfrac{\sqrt{2}}{2}$

    c. $\tan x = \sqrt{3}$ **d.** $\sin x = -\dfrac{3}{8}$

    e. $\cos x = 0.41$ **f.** $\tan x = \sqrt{5}$

34. Find the exact values of all solutions to each equation in parts (a)–(c), and use your calculator to find approximate values of all solutions to each equation in parts (d)–(f). $\left(\textit{Hint:} \text{ The equations } \cot x = c, \sec x = c,\right.$

and $\csc x = c$ are equivalent to the equations $\tan x = \dfrac{1}{c}$, $\cos x = \dfrac{1}{c}$, and

$\left. \sin x = \dfrac{1}{c}.\right)$

    a. $\cot x = -\dfrac{\sqrt{3}}{3}$ **b.** $\sec x = 1$

    c. $\csc x = \sqrt{2}$ **d.** $\cot x = 0.2$

    e. $\sec x = 0.1$ **f.** $\csc x = 10$

35. Find the exact value of each of the following.

  a. $\cos\left(\tan^{-1}0\right)$ **b.** $\tan^{-1}(\cos 0)$

  c. $\sec^{-1}\!\left(\sec\dfrac{4\pi}{3}\right)$ **d.** $\sec\left(\sec^{-1}2\right)$

  e. $\csc\left(\sin^{-1}\dfrac{\sqrt{2}}{2}\right)$ **f.** $\cot^{-1}\!\left(\csc\dfrac{3\pi}{2}\right)$

In Exercises 36–41,

  a. Use graphical transformations to sketch the graph of the function.

  b. Identify its domain and range.

  c. Identify any horizontal asymptotes.

36. $y = 3\tan^{-1}x$ **37.** $y = \sin^{-1}2x$

38. $y = \sec^{-1}(x+1)$ **39.** $y = -\cos^{-1}x$

40. $y = \cot^{-1}x + \dfrac{\pi}{2}$ **41.** $y = \csc^{-1}(-x)$

In Exercises 42–45, write algebraic expressions equivalent to the given expressions.

42. $\sin\left(\tan^{-1}x\right),\ x > 0$

43. $\cot\left(\csc^{-1}x\right),\ x > 1$

44. $\sec\left(\cot^{-1}x\right),\ x > 0$

45. $\tan\left(\sin^{-1}x\right),\ 0 \leqslant x \leqslant 1$

46. (*Making Observations*) Complete the tables in parts (a) and (b).

  a.

x	$\tan^{-1}x$	$\cot^{-1}x$	$\sec^{-1}x$	$\csc^{-1}x$
1				
1,000				
1,000,000				
1,000,000,000				

  b.

x	$\tan^{-1}x$	$\cot^{-1}x$	$\sec^{-1}x$	$\csc^{-1}x$
−1				
−1,000				
−1,000,000				
−1,000,000,000				

  c. What do the entries in your tables tell you about the end behavior of $y = \tan^{-1}x$, $y = \cot^{-1}x$, $y = \sec^{-1}x$, and $y = \csc^{-1}x$?

47. (*Making Observations*)

  Explain how Figure 8-59 supports each of the following statements.

  If x is in $[\pi, 2\pi]$, then $\cos^{-1}(\cos x) = 2\pi - x$.

  If x is in $[2\pi, 3\pi]$, then $\cos^{-1}(\cos x) = x - 2\pi$.

  If x is in $[3\pi, 4\pi]$, then $\cos^{-1}(\cos x) = 4\pi - x$.

Figure 8–59

48. (*Making Observations*) Graph $y = \cos^{-1}(\cos x)$, and use the results in Exercise 47 to explain the appearance of the graph.

49. Over which of the following domains does the function $y = \sec x$ have an inverse? Over which ones does the function $y = \csc x$ have an inverse?

   a. $\left[-\dfrac{\pi}{2}, 0\right) \cup \left[\dfrac{\pi}{2}, \pi\right)$

   b. $\left[0, \dfrac{\pi}{2}\right) \cup \left[\pi, \dfrac{3\pi}{2}\right)$

   c. $\left[\dfrac{\pi}{2}, \pi\right) \cup \left(\pi, \dfrac{3\pi}{2}\right]$

   d. $\left[0, \dfrac{\pi}{2}\right) \cup \left(\dfrac{3\pi}{2}, 2\pi\right]$

50. (*Making Observations*)

   a. Use the identity $1 + \tan^2 x = \sec^2 x$ to show that
   $$\tan(\sec^{-1} x) = \pm\sqrt{x^2 - 1}$$
   for $|x| \geq 1$.

   b. If the range of the inverse secant function is $\left[0, \dfrac{\pi}{2}\right) \cup \left[\pi, \dfrac{3\pi}{2}\right)$, as in Exercise 49b, show that the $\pm$ in front of the radical can be omitted, so that $\tan(\sec^{-1} x) = \sqrt{x^2 - 1}$.

   c. If the range of the inverse secant function is $\left[0, \dfrac{\pi}{2}\right) \cup \left(\dfrac{\pi}{2}, \pi\right]$, as in our definition, explain why the $\pm$ cannot be omitted.

51. (*Making Observations*) Modify the argument in Example 1 for $x \leq 0$ as follows.

   a. If $x \leq 0$ and $y = \tan^{-1} x$, explain why $|x| = \tan|y|$.

   b. Imitate the argument in Example 1 to show that $\sin|y| = \dfrac{|x|}{\sqrt{x^2 + 1}}$.

   c. Explain why $\sin y = \dfrac{x}{\sqrt{x^2 + 1}}$.

CHAPTER REVIEW

Complete Exercises 1–49 (*Writing to Learn*) before referring to the indicated pages.

WORDS AND PHRASES

In Exercises 1–15, explain the meaning of the words or phrases in your own words.

1. (page 391)   **transcendental function**

2. (page 392)   **directed arc, directed length**

3. (page 392)   **cosine, sine** (unit circle definitions)

4. (page 392)   **standard position** for a directed arc, **initial point, terminal point**

5. (page 394)   **central angle**

6. (page 394)   **cosine, sine** (right triangle definitions)

7. (page 395)   **trigonometry**

8. (page 396)   **directed angle, initial side, terminal side**

9. (page 398)   **radian**

10. (page 421)   **trigonometric functions** (right triangle definitions)

11. (page 422)   **trigonometric functions** (unit circle definitions)

12. (pages 421, 422)   **identity**

13. (pages 411–413)   **amplitude**

14. (pages 411–413)   **periodic** function, **period**, **cycle**

15. (pages 411–413)   **phase shift** (**displacement**)

IDEAS

16. (page 392)   What is the difference between the length and the directed length of an arc? Name at least one reason for introducing the idea of directed length.

17. (page 394)   Why is it worthwhile to know the history of the trigonometric functions?

18. (pages 394, 395)   Describe how the work of Hipparchus, Aryabhata, and Rhäticus led to the right triangle definitions of the sine and cosine functions.

19. (pages 396, 397)   Identify two apparent differences between the unit circle and right triangle definitions of the cosine and sine functions, and explain how these differences are resolved by the ideas of directed angles and radian measure.

20. (page 399)   Why are the angles in radians usually expressed as abstract real numbers?

21. (page 399)   Describe the process of converting between radians and degrees. Illustrate your description with a specific example of a degree-to-radian conversion and a specific example of a radian-to-degree conversion.

22. (pages 399, 402)   How are the quantities $\cos \theta$ and $\sin \theta$ defined if θ is an abstract real number? How are they defined if θ is an angle measured in radians? What is the difference between the quantities $\cos \theta$ and $\cos \theta°$?

23. (page 402)   Why is it useful to regard the sine and cosine as functions of real numbers that may or may not represent angles?

24. (pages 406–408)   Name several properties of the sine and cosine functions that are suggested by Table 38 on page 407. (You have our permission to peek at Table 38 before answering.) Choose at least one specific property, and show how it is suggested by the table.

25. (pages 409, 410)   Name several properties of the sine and cosine functions that can be confirmed analytically by using their unit circle definitions. Choose at least one specific property, and show how the definitions can be used to suggest it.

26. (page 410)   Are the expressions $\cos b\theta$ and $b \cos \theta$ always equal? Are the expressions $\cos (\theta + \phi)$ and $\cos \theta + \cos \phi$ always equal? How can you tell?

27. (pages 411, 412)   Name several properties of the sine and cosine functions that are suggested by their graphs.

Choose at least one specific property, and show how it is suggested by the graphs.

28. (page 411)   What is the amplitude of the functions $y = \sin x$ and $y = \cos x$? How is the amplitude related to the range of each function?

29. (page 412)   What is the period of the functions $y = \sin x$ and $y = \cos x$? How is the period related to the unit circle definitions of sine and cosine?

30. (page 412)   What information do the parameters A, h, and k provide about the amplitude, period, and phase shift of a function $y = A \sin b(x - h) + k$ or $y = A \cos b(x - h) + k$? How can that information be used to sketch the graph of the function? Make up a specific function, and use it to illustrate your explanation.

31. (page 421)   How do the right triangle definitions of the trigonometric functions depend on the properties of similar triangles?

32. (page 422)   Write down the reciprocal identities for the trigonometric functions, and explain why they are called reciprocal identities.

33. (page 425)   Write down the Pythagorean identities for the trigonometric functions, and explain why they are called Pythagorean identities.

34. (pages 425, 426)   Explain how to use the reciprocal and Pythagorean identities to find all trigonometric functions of x if you know the value of one function and the quadrant containing x. Illustrate your explanation with a specific example.

35. (page 427)   Sketch the graph of $y = \tan x$. Identify the domain, range, and period, and describe how the graph illustrates each of these features.

36. (page 428)   Sketch the graph of $y = \cot x$. Identify the domain, range, and period, and describe how the graph illustrates each of these features.

37. (page 428)   Sketch the graph of $y = \sec x$. Identify the domain, range, and period, and describe how the graph illustrates each of these features.

38. (page 428)   Sketch the graph of $y = \csc x$. Identify the domain, range, and period, and describe how the graph illustrates each of these features.

39. (page 435)   Explain how trigonometric functions can be used to find sides and angles of right triangles. Illustrate your explanation with a specific calculation.

40. (pages 437–440)   Describe at least one physical problem that can be solved through the use of trigonometry. Identify all trigonometric functions used in the solution, and show how they are used.

41. (pages 440, 441)   Why must the domain of the sine function be restricted before its inverse can be defined? Describe the process by which the function $y = \sin^{-1} x$ is defined, and give its definition.

42. (page 441)   Why is $\sin(\sin^{-1} x)$ always equal to x? Why is $\sin^{-1}(\sin x)$ not always equal to x? Under what circumstances does $\sin^{-1}(\sin x) = x$?

43. (pages 442, 443)   How is the inverse sine function used to find all solutions of an equation $\sin x = c$? Make up a specific equation, and use it to illustrate your explanation.

44. (pages 443, 444)   Describe the process by which the function $y = \cos^{-1} x$ is defined, and give its definition.

45. (pages 444, 445)   How is the inverse cosine function used to find all solutions of an equation $\cos x = c$? Make up a specific equation, and use it to illustrate your explanation.

46. (pages 446, 447)   Describe the process by which the function $y = \tan^{-1} x$ is defined, and give its definition.

47. (pages 447, 448)   How is the inverse tangent function used to find all solutions of an equation $\tan x = c$? Make up a specific equation, and use it to illustrate your explanation.

48. (pages 448, 449)   Give the definitions of each of the functions $y = \cot^{-1} x$, $y = \sec^{-1} x$, and $y = \csc^{-1} x$.

49. (pages 449, 450)   How can a composition of trigonometric and inverse trigonometric functions be expressed as an algebraic function? Illustrate your explanation with a specific example.

CHAPTER 9

TRIGONOMETRIC FUNCTIONS AS ANALYTICAL TOOLS

9-1 THE LAW OF COSINES AND THE LAW OF SINES

In Section 8-3 you learned how trigonometry can be used to find all the sides and angles of a right triangle when one acute angle and the length of one side are given. In the present section you will discover how to use trigonometry to obtain information about more general triangles. A triangle is determined by any of the following combinations of side lengths and angles:

- the lengths of two sides and the measure of the included angle (side-angle-side)
- the lengths of three sides (side-side-side)
- the measure of two angles and the length of a side (angle-side-angle or angle-angle-side)

In this section you will develop two formulas to calculate the remaining sides and angles in each case. One, called the Law of Cosines, applies in the first two cases. The other, called the Law of Sines, applies in the last case. Let's begin by considering a problem whose solution requires the Law of Cosines.

THE LAW OF COSINES

Several years ago Bart Giamatti, a former English professor and President of Yale University, was chosen as President of the National Baseball League, and later as Commissioner of Baseball. In 1987 he asked physicist Robert K. Adair, also of Yale, to investigate those aspects of baseball that might be best understood through physics. Adair accepted and was given the title of "Physicist to the National League." His enthusiasm for his research eventually led him to write a book, *The Physics of Baseball* (HarperCollins, 1994), in which he states:

> To express the goals of this book, I can do no better than adopt a modification of a statement from Paul Kirkpatrick's article, "Batting the Ball": *The aim of this study is not to reform baseball, but to understand it.* As a corollary to the statement of purpose, I must emphasize that the book is not meant as a guide to a player; of all the ways to learn to throw and bat a ball better, an academic study of the mechanics of the actions must be the least useful.

455

In the same vein, the following Mathematical Looking Glass will not help you field a ground ball more efficiently, but it will use trigonometry to demonstrate some of the complex decisions a fielder must make instinctively.

A MATHEMATICAL LOOKING GLASS
Shortstop

A critical moment in the 1996 World Series occurred in the eighth inning of the third game. The Atlanta Braves had beaten the New York Yankees in the first two games, and the score of game 3 was tied 2-2. New York's Derek Jeter hit a ground ball to Atlanta shortstop Jeff Blauser and was awarded an infield single on a close play at first base. His hit started a three-run inning for the Yankees, who went on to win the series by winning four straight games.

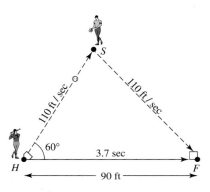

Figure 9–1

Before the ball was hit, Jeff had to position himself at some point S, as shown in Figure 9-1. If he is too shallow (that is, too close to home plate at H), the ball will bounce past him before he can react. If he is too deep, the ball will arrive too late for him to make his throw to first base (point F) in time. Ideally, he plays as deeply as he can without losing his chance to throw Derek out. His decision was complicated by the fact that the ball might be hit at any angle with any initial speed up to about 160 ft/sec. As it travels, the ball will also slow down slightly and may change direction. For simplicity, let's assume that $\angle FHS = 60°$ and the ball travels in a straight line with a constant speed of 110 ft/sec. Jeff is also aware of the following facts:

- He has a good arm. (Even if he is off balance, he can throw the ball to first base at a speed of 110 ft/sec.)
- He has a quick release. (He takes 0.9 seconds to release his throw after first picking up the ball.)
- The distance $\overline{HF}$ is 90 feet, and Derek can run from home to first base quickly (in 3.7 seconds).

From this information, he needs to find the maximum value of $\overline{HS}$. ∎

EXERCISE   **1.   a.** (*Writing to Learn*) If the ball takes t seconds to reach Jeff, explain why $\overline{HS} = 110t$.

   b. In terms of t, how many seconds will elapse from the time Jeff makes his throw until Derek reaches first base?

c. (*Writing to Learn*) If Jeff is to throw Derek out, explain why $\overline{FS} < 110(2.8 - t)$.

If $\angle HFS$ in **Shortstop** were a right angle, we could use right triangle trigonometry to calculate $\overline{HS}$ directly. Since $\angle HFS$ is unlikely to be a right angle, we need to develop a method for relating sides and angles of arbitrary triangles. In Exercises 2 and 3 you can develop a formula that can be used to solve Jeff's problem.

EXERCISES   **2.** (*Writing to Learn*) Referring to the following figure, explain why each step in the following argument is valid.

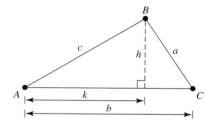

$$c^2 = h^2 + k^2$$

$$h^2 = a^2 - (b - k)^2$$

$$c^2 = a^2 - (b - k)^2 + k^2$$

$$c^2 = a^2 - b^2 + 2bk$$

$$c^2 = a^2 + b^2 - 2b(b - k)$$

$$c^2 = a^2 + b^2 - 2ab\left(\frac{b - k}{a}\right)$$

$$c^2 = a^2 + b^2 - 2ab \cos C$$

3. (*Writing to Learn*) Referring to the following figure, explain why each step in the following argument is valid.

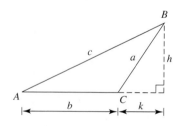

$$c^2 = h^2 + (k + b)^2$$

$$h^2 = a^2 - k^2$$

$$c^2 = a^2 - k^2 + (k + b)^2$$

$$c^2 = a^2 + b^2 + 2bk$$

$$c^2 = a^2 + b^2 + 2ab\left(\frac{k}{a}\right)$$

$$c^2 = a^2 + b^2 + 2ab \cos(\pi - C)$$

$$c^2 = a^2 + b^2 - 2ab \cos C$$

$$\left[\textit{Hint: } \cos(\pi - C) = -\cos C\right]$$

Exercises 2 and 3 demonstrate that $c^2 = a^2 + b^2 - 2ab \cos C$ in any triangle ABC. By symmetry, we can obtain the following **Law of Cosines**.

Law of Cosines

If ABC is a triangle, and if a, b, and c denote the lengths of the sides opposite angles A, B, and C, respectively, then

$$a^2 = b^2 + c^2 - 2bc \cos A$$

$$b^2 = a^2 + c^2 - 2ac \cos B$$

$$c^2 = a^2 + b^2 - 2ab \cos C$$

In words, the Law of Cosines says that the square of any side is equal to the sum of the squares of the other two sides, minus twice the product of those sides and the cosine of the angle between them.

You can use the Law of Cosines to find all sides and angles of a triangle, as long as you know either

- the lengths of any two sides and the measure of the angle between them
  or
- the lengths of all three sides.

EXAMPLE 1   *Using the Law of Cosines to find all sides and angles of a triangle*

In Figure 9-2, $a = 97.00$ m, $b = 59.00$ m, and $C = 13.00°$. Since the lengths of two sides and the measure of the angle between them are given, we can use the Law of Cosines to find c.

$$c^2 = a^2 + b^2 - 2ab \cos C$$

$$c^2 = (97)^2 + (59)^2 - 2(97)(59) \cos 13°$$

$$c^2 \approx 1737$$

$$c \approx 41.68 \text{ m}$$

$b = 59.00$ m A c

C $13.00°$

$a = 97.00$ m B

Figure 9–2

Next, let's find A:

$$a^2 = b^2 + c^2 - 2bc \cos A$$

$$(97)^2 = (59)^2 + (41.68)^2 - 2(59)(41.68) \cos A$$

$$9409 \approx 3481 + 1737 - 4918 \cos A$$

$$4918 \cos A \approx -4191$$

$$\cos A \approx -0.8522$$

$$A \approx 148.5°$$

Finally, since $A + B + C = 180°$, we have

$$B = 180° - 148.5° - 13° = 18.50°$$ ■

EXERCISES   *In Exercises 4–7, find all sides and angles of triangle ABC.*

4. $a = 3.56$ in, $b = 5.81$ in, $C = 32.5°$

5. $a = 7.2$ m, $c = 4.6$ m, $B = 47°$

6. $b = \sqrt{5.000}$, $c = \sqrt{5.000}$, $A = 110.00°$

7. $a = 35.3$ km, $b = 27.1$ km, $c = 41.5$ km

8. (*Making Observations*) In triangle ABC, suppose that $C = 90°$. Then the statement

$$c^2 = a^2 + b^2 - 2ab \cos C$$

simplifies to become what familiar theorem?

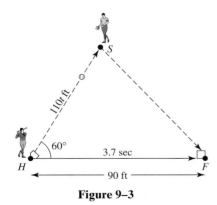

Figure 9–3

Now recall that in **Shortstop**, Jeff Blauser wants to play as deeply as he can without losing his chance to throw Derek Jeter out. This means that he wants to make side $\overline{HS}$ of triangle HFS in Figure 9-1 as long as possible. If Jeff is to throw Derek out at first base, then $\overline{FS} < 110(2.8 - t)$. The situation is summarized in Figure 9-3.

EXERCISE   **9.** (*Problem Solving*) Use the Law of Cosines to formulate an inequality in *t*. Solve the inequality to find the maximum length of $\overline{HS}$. If Jeff plays to maximize $\overline{HS}$, how long will it take the ball to reach him? When he catches it, how far will he be from home plate? From first base?

THE LAW OF SINES

If you know the measure of any two angles and the length of any side of a triangle, you cannot use the Law of Cosines alone to find the remaining sides and angles. You can develop an appropriate formula in Exercises 10 and 11.

EXERCISES   **10.** (*Writing to Learn*) Referring to the following figure, explain why each step in the following argument is valid.

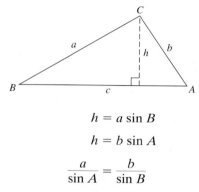

$$h = a \sin B$$

$$h = b \sin A$$

$$\frac{a}{\sin A} = \frac{b}{\sin B}$$

11. (*Writing to Learn*) Referring to the following figure, explain why $\dfrac{a}{\sin A} = \dfrac{b}{\sin B}$.

[*Hint*: $\sin(\pi - A) = \sin A$]

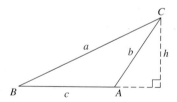

Exercises 10 and 11 demonstrate that $\dfrac{a}{\sin A} = \dfrac{b}{\sin B}$ in any triangle ABC. By symmetry, we can obtain the following **Law of Sines**:

Law of Sines

If ABC is a triangle, and if a, b, and c denote the lengths of the sides opposite angles A, B, and C, respectively, then

$$\frac{a}{\sin A} = \frac{b}{\sin B} = \frac{c}{\sin C}$$

In words, the Law of Sines says that in any triangle, the ratio of the length of each side to the sine of the opposite angle is constant.

You can use the Law of Sines to find all sides and angles of a triangle, as long as you know the measures of any two angles and the length of any side.

EXAMPLE 2   *Using the Law of Sines to find all sides and angles of a triangle*
In Figure 9-4, $a = 10$ cm, $B = 42°$, and $C = 81°$.

We can see immediately that $A = 180° - (42° + 81°) = 57°$. Knowing the measure of all three angles and the length of one side, we can use the Law of Sines to find the lengths of the other two sides. First, let's find b:

$$\frac{a}{\sin A} = \frac{b}{\sin B}$$

$$\frac{10}{\sin 57°} = \frac{b}{\sin 42°}$$

$$b = \frac{10 \sin 42°}{\sin 57°} \approx 8.0 \text{ cm}$$

Now let's find c:

$$\frac{a}{\sin A} = \frac{c}{\sin C}$$

$$\frac{10}{\sin 57°} = \frac{c}{\sin 81°}$$

$$c = \frac{10 \sin 81°}{\sin 57°} \approx 12 \text{ cm}$$ ■

Figure 9–4

EXERCISES   *In Exercises 12–15, find all sides and angles of triangle ABC.*

12. $a = 40.0$ ft, $A = 103.0°$, $B = 41.0°$
13. $b = 6.2$ m, $A = 74°$, $B = 9°$
14. $c = 15.65$ cm, $A = 50.00°$, $B = 71.57°$
15. $a = 2.94$ mi, $A = 6.4°$, $B = 132.7°$

The lengths of two sides and the measure of the angle opposite one of them may or may not determine a unique triangle. In fact, certain measurements determine two triangles, and others determine none. Figure 9-5 illustrates all possible cases when the given angle is acute.

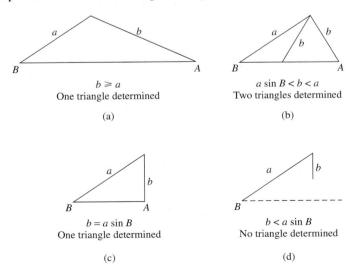

$b \geqslant a$
One triangle determined
(a)

$a \sin B < b < a$
Two triangles determined
(b)

$b = a \sin B$
One triangle determined
(c)

$b < a \sin B$
No triangle determined
(d)

Figure 9–5

If the given angle is obtuse, then one triangle is determined if $b > a$, and no triangle is determined is $b \leq a$. You can explain why in Exercise 16.

EXERCISE   **16. a.** (*Making Observations*) Sketch a figure to show that one triangle is determined if $a = 10$, $B = 150°$, and $b = 20$.

   b. (*Making Observations*) Sketch a figure to show that no triangle is determined if $a = 10$, $B = 150°$, and $b = 5$.

   c. (*Writing to Learn*) If the angle B is obtuse, explain why one triangle is determined if $b > a$ and no triangle is determined if $b \leq a$.

EXAMPLE 3   *Using the Laws of Sines and Cosines to find all sides and angles of a triangle*
In Figure 9-5a, $a = 10$, $b = 11$, and $B = 30°$. To find A,

$$\frac{a}{\sin A} = \frac{b}{\sin B}$$

$$\frac{10}{\sin A} = \frac{11}{\sin 30°}$$

$$\sin A = \frac{10 \sin 30°}{11} \approx 0.45$$

Since $0° < A < 180°$, either $A = \sin^{-1} 0.45 \approx 27°$ or $A = 180° - 27° = 153°$. Let's find C and c in each case.

If $A = 27°$, then $C = 180° - (27° + 30°) = 123°$. By the Law of Cosines,

$$c^2 = a^2 + b^2 - 2ab \cos C$$

$$= (10)^2 + (11)^2 - 2(10)(11) \cos 123° \approx 340$$

$$c \approx 18$$

If $A = 153°$, then $C = 180° - (153° + 30°) = -3°$, an impossibility, so A cannot be 153°. Thus, as indicated by Figure 9-5a, only one triangle has the given values of a, b, and B.   ■

EXAMPLE 4   *Using the Laws of Sines and Cosines to find all sides and angles of a triangle*
In Figure 9-5b, $a = 10$, $b = 8$, and $B = 30°$. To find A,

$$\frac{a}{\sin A} = \frac{b}{\sin B}$$

$$\frac{10}{\sin A} = \frac{8}{\sin 30°}$$

$$\sin A = \frac{10 \sin 30°}{8} \approx 0.63$$

Since $0° < A < 180°$, either $A = \sin^{-1} 0.63 \approx 39°$ or $A = 180° - 39° = 141°$. Let's find C and c in each case.

If $A = 39°$, then $C = 180° - (39° + 30°) = 111°$. By the Law of Cosines,

$$c^2 = a^2 + b^2 - 2ab \cos C$$

$$c^2 = (10)^2 + (8)^2 - 2(10)(8) \cos 111° \approx 220$$

$$c \approx 15$$

If $A = 141°$, then $C = 180° − (141° + 30°) = 9°$. By the Law of Cosines,

$$c^2 = a^2 + b^2 − 2ab \cos C$$

$$c^2 = (10)^2 + (8)^2 − 2(10)(8) \cos 9° \approx 6.0$$

$$c \approx 2.4$$

Thus, as indicated by Figure 9-5b, two triangles have the given values of a, b, and B. ■

EXERCISES

17. Use the Law of Sines to find all sides and angles of the triangle shown in Figure 9-5c, where $a = 10$, $b = 5$, and $B = 30°$.

18. Use the Law of Sines to verify that no triangle has sides and angles as shown in Figure 9-5d, where $a = 10$, $b = 2$, and $B = 30°$.

In Exercises 19–22, find all triangles having the given sides and angles.

19. $a = 74.2$ ft, $c = 56.9$ ft, $C = 48.0°$

20. $a = 74.2$ ft, $c = 56.9$ ft, $A = 48.0°$

21. $b = 532.4$ m, $c = 728.1$ m, $B = 142.00°$

22. $b = 532.4$ m, $c = 728.1$ m, $C = 142.00°$

A MATHEMATICAL LOOKING GLASS

Crowded Heavens

On July 21, 1994, China launched a satellite into orbit over the equator, between the U.S.-owned Rimsat-1, broadcasting TV to India, and the Japanese NTT-CS, carrying telephone traffic. The angular distance between the Chinese Apstar-1 and each of the other satellites is less than 1°, as shown in Figure 9-6.

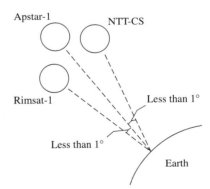

Figure 9–6

The launch immediately drew protests from both the United States and Japan, which claimed that the close proximity of Apstar-1 would interfere with the communication functions of the other two satellites. The United States was able to detect the presence of Apstar-1 because it maintains a number of tracking stations around the world to monitor satellite orbits. Two of those are located approximately on the equator, one in Ethiopia at about 78° east longitude and the other in Quito, Ecuador, at about 40° west longitude. Let's see how observations from these two stations can be coordinated to find the exact location of a satellite.

Figure 9-7 shows the earth's equator and the tracking stations at E (Ethiopia) and Q (Quito).

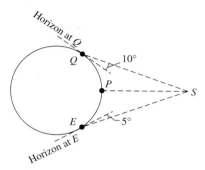

Figure 9–7

Suppose that a satellite is observed from the Ethiopian station, due west at an angle of 5° above the horizon. Its location cannot be determined from this observation alone, since it could be anywhere along the line ES. If the satellite is simultaneously observed due east from Quito at an angle of 10° above the horizon, then it must be located precisely at S. The observers would like to know the altitude of the satellite's orbit, represented by $\overline{SP}$. ■

Figure 9–8

One of the quantities in Figure 9-7 is already known to the observers. In particular, when the stations were first built, the distance $\overline{QE}$ was calculated by the following process.

In Figure 9-8, O is the center of the earth, and the radius of the earth is approximately 6380 km. The **prime meridian** is a semicircle that runs from the North Pole to the South Pole through Greenwich, England, and crosses the equator at the point indicated. Longitudes are measured east and west from the prime meridian, ranging from 0° to 180° in each direction.

EXERCISES

23. (*Writing to Learn*) Explain why the angles in triangle QOE must be as shown in Figure 9-8.

24. Use the Law of Sines to find $\overline{QE}$.

You can now calculate the sides and angles of triangle QSE by referring to Figure 9-9.

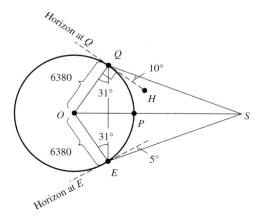

Figure 9–9

25. a. (*Writing to Learn*) Explain why $\angle OQH = 90°$.

b. Find $\angle EQS$.

c. Find $\angle QES$.

d. Use the Law of Sines and your result from Exercise 24 to find $\overline{SE}$.

26. In triangle OSE, you now know the lengths of SE and OE, and the measure of $\angle OES$.

a. Use the Law of Cosines to find $\overline{OS}$.

b. Find the altitude $\overline{PS}$ of the satellite above the earth.

ADDITIONAL EXERCISES   *In Exercises 27–46, find all sides and angles of all possible triangles ABC from the given information.*

27. $a = 24$ cm, $b = 17$ cm, $c = 29$ cm

28. $B = 72°$, $C = 49°$, $a = 5.3$ mi

29. $B = 6.7°$, $C = 102.6°$, $b = 14.3$ m

30. $A = 81°$, $b = 0.74$ in, $c = 0.56$ in

31. $C = 109.00°$, $a = 6.390$ ft, $b = 9.415$ ft

32. $A = 125°$, $a = 2.2$ mm, $b = 1.9$ mm

33. $a = 7$ ft, $b = 8$ ft, $c = 50$ ft

34. $B = 24°$, $a = 40$ m, $b = 48$ m

35. $B = 24°$, $a = 40$ m, $b = 32$ m

36. $B = 24°$, $a = 40$ m, $b = 16$ m

37. $B = 124°$, $a = 40$ m, $b = 48$ m

38. $B = 124°$, $a = 40$ m, $b = 32$ m

39. $a = 6.59$ mm, $b = 8.74$ mm, $c = 12.8$ mm

40. $a = 37$ km, $b = 35$ km, $c = 12$ km

41. $B = 7.26°$, $a = 18.85$ ft, $c = 27.08$ ft

42. $B = 117.0°$, $a = 435$ m, $c = 266$ m

43. $A = 38.1°$, $C = 78.8°$, $a = 4.82$ in

44. $A = 38.1°$, $C = 78.8°$, $c = 4.82$ in

45. $C = 120.00°$, $a = 61.94$ km, $c = 67.48$ km

46. $C = 60.00°$, $a = 61.94$ km, $c = 67.48$ km

47. (*Making Observations*)

a. In triangle ABC, suppose that sides a and b are each 10 units long. What is the length of side c if the measure of angle C is 10°? 1°? 0.1°?

b. What can you say about the length of c if C is near 0°?

c. How could you have predicted your result in part (b) without doing any calculations?

48. (*Making Observations*)

a. In triangle ABC, suppose that sides a and b are each 10 units long. What is the length of side c if the measure of angle C is 170°? 179°? 179.9°?

b. What can you say about the length of c if C is near 180°?

c. How could you have predicted your result in part (b) without doing any calculations?

9–2   TRIGONOMETRIC IDENTITIES

Recall that an equation is an **identity** if it is true whenever either side of the equation is defined. A **trigonometric identity** is an identity involving trigonometric functions.

For example, the equation $\csc x = \dfrac{1}{\sin x}$ is a trigonometric identity, because both sides are defined and equal unless x is an integer multiple of π. By contrast, $\sin x = \cos x$ is a **conditional equation**, because there are values of x (such as $x = 0$) for which both sides are defined, but are not equal.

Identities are used in problem solving to replace expressions with equivalent expressions more suited to the situation. As a simple example, you use the identity $\csc x = \dfrac{1}{\sin x}$ to evaluate the cosecant function on your calculator. In Section 9-3 you will use more sophisticated identities to study the problem of tacking a sailboat against the wind.

FUNDAMENTAL IDENTITIES

You have already seen the reciprocal and Pythagorean identities, sometimes called **fundamental identities**. They are listed again here for convenience.

Reciprocal Identities

$$\sin x = \frac{1}{\csc x} \qquad \csc x = \frac{1}{\sin x}$$

$$\cos x = \frac{1}{\sec x} \qquad \sec x = \frac{1}{\cos x}$$

$$\tan x = \frac{1}{\cot x} \qquad \cot x = \frac{1}{\tan x}$$

$$\tan x = \frac{\sin x}{\cos x} \qquad \cot x = \frac{\cos x}{\sin x}$$

Pythagorean Identities

$$\cos^2 x + \sin^2 x = 1$$

$$1 + \tan^2 x = \sec^2 x$$

$$\cot^2 x + 1 = \csc^2 x$$

REFLECTION AND ROTATION IDENTITIES

Reflection Identities If x is any directed angle in standard position, the angle $-x$ is the reflection of x in the horizontal axis, as in Figure 9-10.

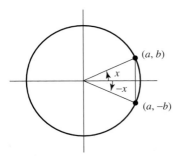

Figure 9–10

Thus if x determines the point (a, b) on the unit circle, then $-x$ determines $(a, -b)$. It is therefore not surprising that the trigonometric functions of $-x$ are related to those of x through simple equations, which we will call **reflection identities**.

The following reflection identities are consequences of Figure 9-10. We refer to them as **even-odd identities** because they express the oddness of the sine and tangent functions, and the evenness of the cosine function.

Even-Odd Identities
$\sin(-x) = -\sin x$
$\cos(-x) = \cos x$
$\tan(-x) = -\tan x$

You can use these three identities along with the reciprocal identities to derive even-odd identities for the remaining trigonometric functions in Exercise 2.

EXERCISES

1. (*Writing to Learn*) Explain how to derive the even-odd identities listed here from Figure 9-10.

2. Express $\cot(-x)$, $\sec(-x)$, and $\csc(-x)$ in terms of $\cot x$, $\sec x$, and $\csc x$.

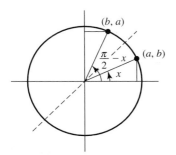

Figure 9–11

The angle $\dfrac{\pi}{2} - x$ is the reflection of the angle x in the 45° line shown in Figure 9-11. Thus the trigonometric functions of $\dfrac{\pi}{2} - x$ are related to those of x through a group of reflection identities. We refer to them as **cofunction identities** because they relate the trigonometric functions of an angle to the cofunctions of the complementary angle.

Cofunction Identities	
$\sin\left(\dfrac{\pi}{2} - x\right) = \cos x$	$\cos\left(\dfrac{\pi}{2} - x\right) = \sin x$
$\tan\left(\dfrac{\pi}{2} - x\right) = \cot x$	$\cot\left(\dfrac{\pi}{2} - x\right) = \tan x$
$\sec\left(\dfrac{\pi}{2} - x\right) = \csc x$	$\csc\left(\dfrac{\pi}{2} - x\right) = \sec x$

EXERCISES

3. (*Writing to Learn*) Explain how to derive the cofunction identities from Figure 9-11.

In Exercises 4–7, find the values of all trigonometric functions of x, $-x$, and $\dfrac{\pi}{2} - x$.

4. $\tan x = 1.25$, x is in quadrant III

5. $\sec x = -\sqrt{5}$, x is in quadrant II

6. $\sin x = -0.2$, $\cos x > 0$

7. $\sin x = -0.2$, $\cos x < 0$

In Exercises 90–91 you will discover other reflection identities for trigonometric functions of $\pi - x$ and $2\pi - x$. Because there are so many of these identities, and because they are easy to derive geometrically, it will be more efficient for you to understand the procedures for deriving them rather than memorize the identities themselves.

Rotation Identities Once again, suppose that a directed angle x determines a point (a, b) on the unit circle. Figure 9-12 shows that if the terminal side of x is rotated through a directed angle of π, the resulting angle $x + \pi$ determines the point $(-a, -b)$. Similarly, the angle $x + 2\pi$ determines the point (a, b).

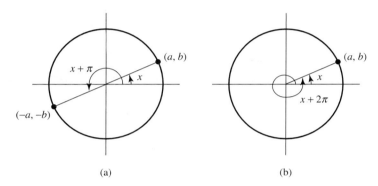

(a)   (b)

Figure 9–12

We will refer to the resulting equations relating the trigonometric functions of these angles as **rotation identities**. The following rotation identities are called **periodicity identities** because they express the periodic nature of the sine, cosine, and tangent functions.

> **Periodicity Identities**
>
> $$\sin (x + 2\pi) = \sin x$$
>
> $$\cos (x + 2\pi) = \cos x$$
>
> $$\tan (x + \pi) = \tan x$$

EXERCISES   **8.** Write rotation identities to express all other trigonometric functions of $x + \pi$ and $x + 2\pi$ in terms of trigonometric functions of x.

In Exercises 9–12, use periodicity identities to write the expression as $\sin x$ *or* $\cos x$ *with* $0 \leq x < 2\pi$.

9. $\sin \dfrac{49\pi}{8}$ **10.** $\cos \dfrac{17\pi}{5}$

11. $\cos \left(-\dfrac{9\pi}{7}\right)$ **12.** $\sin 9.1\pi$

13. Use periodicity identities to write $\tan 4.65\pi$ in two different ways as $\tan x$ with $0 \leq x < 2\pi$.

Other rotations yield similar simple relationships between the coordinates determined by x and those determined by $x + \frac{\pi}{2}$ and $x - \frac{\pi}{2}$. You will discover the resulting identities in Exercise 92.

SUM AND DIFFERENCE IDENTITIES FOR SINE AND COSINE

The following identities allow you to express $\sin (x \pm y)$ and $\cos (x \pm y)$ in terms of the sines and cosines of x and y alone. You can verify them in Exercises 14–17. Similar identities exist for the other trigonometric functions, but they are used less often. You can verify the identities for $\tan (x \pm y)$ in Exercises 93–95.

Sum and Difference Identities

$$\sin (x \pm y) = \sin x \cos y \pm \cos x \sin y$$

$$\cos (x \pm y) = \cos x \cos y \mp \sin x \sin y$$

EXERCISES   **14.** (*Making Observations*) In Figure 9-13, directed angles x and y determine points A and B on the unit circle. Angles $-y$ and $x + y$ determine the points B' and C.

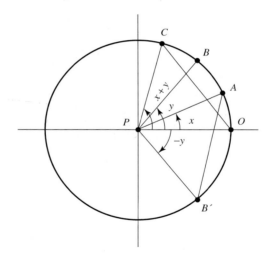

Figure 9–13

a. Explain why the coordinates of B' are $(\cos y, -\sin y)$.

b. Explain why the lengths of the line segments OC and $B'A$ are equal.

c. Use the distance formula to show that
$$\left(\overline{OC}\right)^2 = \left[1 - \cos (x + y)\right]^2 + \left[\sin (x + y)\right]^2$$

d. Show that $\left(\overline{OC}\right)^2$ can be simplified to $2 - 2 \cos (x + y)$.

e. Use the distance formula to show that
$$\left(\overline{B'A}\right)^2 = (\cos x - \cos y)^2 + (\sin x + \sin y)^2$$

f. Explain why $\left(\overline{B'A}\right)^2$ can be simplified to
$$2 - 2 \cos x \cos y + 2 \sin x \sin y$$

g. Explain how parts (a)–(f) show that
$$\cos (x + y) = \cos x \cos y - \sin x \sin y$$

15. (*Making Observations*) Replace y by $-y$ in the identity
$$\cos (x + y) = \cos x \cos y - \sin x \sin y$$
and then explain why the resulting identity is equivalent to
$$\cos (x - y) = \cos x \cos y + \sin x \sin y.$$

16. (*Making Observations*) Explain how each step follows from the previous step in deriving the identity $\sin (x + y) = \sin x \cos y + \cos x \sin y$.

$$\sin (x + y) = \cos \left[\frac{\pi}{2} - (x + y) \right]$$

$$= \cos \left[\left(\frac{\pi}{2} - x \right) - y \right]$$

$$= \cos \left(\frac{\pi}{2} - x \right) \cos y + \sin \left(\frac{\pi}{2} - x \right) \sin y$$

$$= \sin x \cos y + \cos x \sin y$$

17. (*Making Observations*) Use your result from Exercise 16 to obtain $\sin (x - y) = \sin x \cos y - \cos x \sin y$. (*Hint:* Imitate your procedure from Exercise 15.)

18. In each case, find the exact value of each side of the equation when $x = \frac{\pi}{2}$ and $y = \frac{\pi}{3}$ and verify that the two sides are equal.

a. $\sin (x + y) = \sin x \cos y + \cos x \sin y$
b. $\sin (x - y) = \sin x \cos y - \cos x \sin y$
c. $\cos (x + y) = \cos x \cos y - \sin x \sin y$
d. $\cos (x - y) = \cos x \cos y + \sin x \sin y$

19. Show that each of the following is *not* an identity by showing that the two sides are not equal when $x = \frac{\pi}{2}$ and $y = \frac{\pi}{3}$.

a. $\sin (x + y) = \sin x + \sin y$
b. $\cos (x - y) = \cos x - \cos y$

20. (*Writing to Learn*) Verify that the equation $\sin (x + y) = \sin x - \sin y$ is true for the given values of x and y.

a. $x = \frac{\pi}{3}$, $y = 0$ **b.** $x = 0$, $y = \pi$

c. $x = \frac{2\pi}{3}$, $y = 2\pi$ **d.** $x = \frac{\pi}{2}$, $y = \frac{\pi}{2}$

Does this mean that the equation is an identity? Explain your answer.

In Exercises 21–24, *use Table* 38 *from Section* 8-2 *to find the exact value of the expression.*

21. $\cos \dfrac{5\pi}{12}$ **22.** $\sin \dfrac{\pi}{12}$

$\left(Hint: \dfrac{5\pi}{12} = \dfrac{\pi}{4} + \dfrac{\pi}{6} \right)$

23. $\sin \dfrac{11\pi}{12}$ **24.** $\cos \dfrac{13\pi}{12}$

In Exercises 25–28, if directed angles x and y determine the points A and B in the unit circle, find $\sin(x + y)$, $\cos(x + y)$, $\sin(x - y)$, *and* $\cos(x - y)$.

25. $A = (0.6, -0.8)$ $B = (0.96, 0.28)$

26. $A = \left(-\dfrac{5}{13}, -\dfrac{12}{13}\right)$ $B = \left(-\dfrac{8}{17}, \dfrac{15}{17}\right)$

27. $A = \left(\dfrac{\sqrt{2}}{2}, -\dfrac{\sqrt{2}}{2}\right)$ $B = \left(\dfrac{1}{2}, \dfrac{\sqrt{3}}{2}\right)$

28. $A = \left(\dfrac{\sqrt{5}}{5}, \dfrac{2\sqrt{5}}{5}\right)$ $B = (0, -1)$

DOUBLE-ANGLE AND HALF-ANGLE IDENTITIES

Double-Angle Identities The following identities are consequences of the sum identities. You can verify them in Exercises 29–30.

> **Double-Angle Identities**
>
> $$\sin 2x = 2 \sin x \cos x$$
>
> $$\cos 2x = \cos^2 x - \sin^2 x$$
>
> $$\cos 2x = 2 \cos^2 x - 1$$
>
> $$\cos 2x = 1 - 2 \sin^2 x$$

EXERCISES

29. (*Making Observations*)

  a. Let $y = x$ in the identity for $\sin(x + y)$, and simplify to obtain $\sin 2x = 2 \sin x \cos x$.

  b. Let $y = x$ in the identity for $\cos(x + y)$, and simplify to obtain $\cos 2x = \cos^2 x - \sin^2 x$.

30. (*Making Observations*)

  a. Begin with the identity $\cos 2x = \cos^2 x - \sin^2 x$ and use a Pythagorean identity to obtain $\cos 2x = 2 \cos^2 x - 1$.

  b. Use a procedure similar to that in part (a) to obtain $\cos 2x = 1 - 2 \sin^2 x$.

31. Show that each of the double-angle identities is true for each value of x. Use the exact values of all trigonometric functions.

  a. $x = \dfrac{\pi}{2}$ **b.** $x = \dfrac{\pi}{3}$

In Exercises 32–35, find the value of $\sin 2x$ *and* $\cos 2x$.

32. $\sin x = -0.6$, $\cos x = -0.8$

33. $\sin x = \dfrac{\sqrt{5}}{5}$, $\cos x < 0$ (*Hint:* Use a Pythagorean identity to find $\cos x$.)

34. $\cos x = 0.9$, x is in quadrant IV

35. $\cos x = 0$, $\sin x < 0$

Half-Angle Identities The following identities are consequences of the double-angle identities. You can verify them in Exercises 36 and 37.

Half-Angle Identities

$$\sin^2 x = \frac{1}{2}(1 - \cos 2x)$$

$$\cos^2 x = \frac{1}{2}(1 + \cos 2x)$$

$$\sin\frac{x}{2} = \pm\sqrt{\frac{1 - \cos x}{2}}$$

$$\cos\frac{x}{2} = \pm\sqrt{\frac{1 + \cos x}{2}}$$

EXERCISES   **36.** (*Making Observations*)

a. Obtain the identity $\sin^2 x = \frac{1}{2}(1 - \cos 2x)$ by solving the double-angle identity $\cos 2x = 1 - 2\sin^2 x$ for $\sin^2 x$.

b. Obtain $\cos^2 x = \frac{1}{2}(1 + \cos 2x)$ by solving the double-angle identity $\cos 2x = 2\cos^2 x - 1$ for $\cos^2 x$.

37. (*Making Observations*)

a. Replace x by $\frac{x}{2}$ in the identity $\sin^2 x = \frac{1}{2}(1 - \cos 2x)$, and rewrite to obtain $\sin\frac{x}{2} = \pm\sqrt{\frac{1 - \cos x}{2}}$.

b. Use a similar procedure to obtain $\cos\frac{x}{2} = \pm\sqrt{\frac{1 + \cos x}{2}}$.

38. Show that each of the half-angle identities is true for each value of x. Use the exact values of all trigonometric functions.

a. $x = \dfrac{\pi}{2}$ **b.** $x = \dfrac{\pi}{3}$

Knowing the quadrant containing x is not enough to determine the sign of the radical expression for $\sin\frac{x}{2}$ or $\cos\frac{x}{2}$. Example 1 demonstrates that more specific information is needed.

EXAMPLE 1   *Finding the values of* $\sin\frac{x}{2}$ *and* $\cos\frac{x}{2}$

Suppose that $\cos x = 0.28$ and x is in quadrant I. Then

$$\sin\frac{x}{2} = \pm\sqrt{\frac{1 - 0.28}{2}} = \pm0.6$$

$$\cos\frac{x}{2} = \pm\sqrt{\frac{1 + 0.28}{2}} = \pm0.8$$

If $0 < x < \frac{\pi}{2}$ as in Figure 9-14a, then $0 < \frac{x}{2} < \frac{\pi}{4}$, so that $\frac{x}{2}$ is in quadrant I. In this case, $\sin \frac{x}{2}$ and $\cos \frac{x}{2}$ are both positive and must be equal to 0.6 and 0.8, respectively.

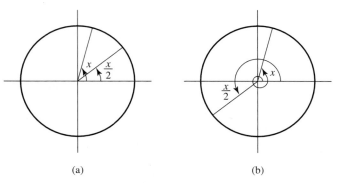

(a)   (b)

Figure 9–14

On the other hand, if $2\pi < x < \frac{5\pi}{2}$ as in Figure 9-14b, then $\pi < \frac{x}{2} < \frac{5\pi}{4}$, so that $\frac{x}{2}$ is in quadrant III. In this case, $\sin \frac{x}{2}$ and $\cos \frac{x}{2}$ are both negative and must be equal to -0.6 and -0.8, respectively.

Knowing that $\cos x = 0.28$ and x is in quadrant I is not enough to determine the values of $\sin \frac{x}{2}$ and $\cos \frac{x}{2}$, because that information is consistent with both cases shown here. ■

EXERCISES   *In Exercises 39–42, find the values of* $\sin \frac{x}{2}$ *and* $\cos \frac{x}{2}$.

39. $\cos x = -0.4, \frac{\pi}{2} < x < \pi$

40. $\cos x = \frac{1}{3}, \frac{7\pi}{2} < x < 4\pi$

41. $\cos x = \frac{1}{3}, \frac{3\pi}{2} < x < 2\pi$

42. $\cos x = -\frac{3}{4}, -\frac{\pi}{2} < x < 0$

In Exercises 43–46, find the exact value of $\sin x$ *and* $\cos x$.

43. $x = \frac{\pi}{8}$ **44.** $x = \frac{5\pi}{8}$

45. $x = \frac{7\pi}{12}$ **46.** $x = \frac{7\pi}{24}$

PRODUCT-TO-SUM IDENTITIES

The following identities are consequences of the sum and difference identities. You can verify them in Exercises 47 and 48.

Product-to-Sum Identities

$$\cos x \cos y = \frac{1}{2}[\cos (x - y) + \cos (x + y)]$$

$$\sin x \sin y = \frac{1}{2}[\cos (x - y) - \cos (x + y)]$$

$$\sin x \cos y = \frac{1}{2}[\sin (x - y) + \sin (x + y)]$$

EXERCISES   **47.** (*Making Observations*) Verify the identity

$$\cos x \cos y = \frac{1}{2}[\cos (x - y) + \cos (x + y)]$$

by using sum and difference identities to rewrite the right-hand side and simplifying.

48. (*Making Observations*) Verify the other two product-to-sum identities.

In Exercises 49–52, use a product-to-sum identity to rewrite the given expression, then verify that the result is equal to the given expression. Use exact values of all trigonometric functions.

49. $(\cos \pi)\left(\cos \dfrac{3\pi}{4}\right)$ **50.** $\left(\sin \dfrac{2\pi}{3}\right)\left(\cos \dfrac{\pi}{6}\right)$

51. $\left(\cos \dfrac{\pi}{2}\right)\left(\sin \dfrac{\pi}{3}\right)$ **52.** $\left(\sin \dfrac{\pi}{2}\right)\left(\sin \dfrac{\pi}{3}\right)$

In Exercises 53–56, rewrite the expression so that it does not involve products of trigonometric functions.

53. $\sin 3x \sin 2x$ **54.** $\cos 4x \cos x$

55. $\sin 2x \cos 7x$ **56.** $\sin 7x \cos 2x$

METHODS FOR DISTINGUISHING IDENTITIES FROM CONDITIONAL EQUATIONS

As you will see in Section 9-3, substituting equivalent trigonometric expressions for each other can help you solve problems. It is therefore useful to know whether two trigonometric expressions are equivalent. For example, you may want to know whether the equation

$$\csc x - \cot x = \frac{\tan x}{\sec x + 1}$$

is an identity. Let's see how numerical, graphical, and analytical methods can help you decide.

Numerical Methods One way to explore an equation is to evaluate both sides for several convenient values of the variables. For example, you can verify that the equation

$$\csc x - \cot x = \frac{\tan x}{\sec x + 1}$$

is true when $x = \dfrac{\pi}{6}, \dfrac{\pi}{4}$, and $\dfrac{2\pi}{3}$. This suggests that the equation is an identity.

Of course, as you saw in Exercise 20, an equation need not be an identity just because it is true for several values of x, or even several thousand. However, an equation that is false for even one value of x cannot be an identity. The strategy of using a single instance to show that a statement is not universally true is called **proof by counterexample**.

EXERCISES   *In Exercises 57–60, evaluate both sides of the equation when $x = 0, \dfrac{\pi}{3}$, and $\dfrac{3\pi}{4}$.*

Based on your results, decide whether the equation might be an identity or is definitely conditional.

57. $\sin^2 x + \tan^2 x = \sec^2 x - \cos^2 x$

58. $\sec x(\sin^2 x - 1) = \cos x$

59. $\cos^4 x - \sin^4 x = \cos 2x$

60. $2 \sin x \cos 4x = \sin 5x - \sin 3x$

Graphical Methods   Another way to explore an equation is to graph both sides.

EXERCISES   **61.** (*Making Observations*) Graph $y = \csc x - \cot x$ and $y = \dfrac{\tan x}{\sec x + 1}$ on the same screen. Do the graphs suggest that the equation

$$\csc x - \cot x = \dfrac{\tan x}{\sec x + 1}$$

is an identity? Explain why or why not.

62. Verify that the equation $\sin 2x = 2 \sin x$ is conditional by graphing $y = \sin 2x$ and $y = 2 \sin x$ on the same screen.

In Exercises 63–66, graph both sides of the equation on the same screen. Based on your results, decide whether the equation might be an identity or is definitely conditional. You worked with the same equations in Exercises 57–60.

63. $\sin^2 x + \tan^2 x = \sec^2 x - \cos^2 x$

64. $\sec x(\sin^2 x - 1) = \cos x$

65. $\cos^4 x - \sin^4 x = \cos 2x$

66. $2 \sin x \cos 4x = \sin 5x - \sin 3x$

Analytical Methods   Although numerical and graphical methods can demonstrate that certain equations are conditional, analytical methods are needed to show conclusively that an equation is an identity. You can usually perform operations on both sides of the equation to obtain a sequence of equivalent equations. If at any stage, you obtain a recognizable identity, the original equation must also be an identity. Conversely, if at any stage you obtain an equation recognizable as conditional, the original equation must also be conditional.

Examples 2–4 illustrate several techniques for operating on trigonometric equations.

EXAMPLE 2   *Deciding whether an equation is an identity*

To see whether the equation

$$\csc x - \cot x = \frac{\tan x}{\sec x + 1}$$

is an identity, let's begin by observing that the equation involves four trigonometric functions. In such a case, it is often helpful to write both sides in terms of sines and cosines.

$$\frac{1}{\sin x} - \frac{\cos x}{\sin x} = \frac{\dfrac{\sin x}{\cos x}}{\dfrac{1}{\cos x} + 1}$$

Next, let's perform the subtraction on the left, and multiply both the numerator and denominator on the right by cos x.

$$\frac{1 - \cos x}{\sin x} = \frac{\sin x}{1 + \cos x}$$

Now let's multiply both sides by $(\sin x)(1 + \cos x)$.

$$(1 - \cos x)(1 + \cos x) = \sin^2 x$$
$$1 - \cos^2 x = \sin^2 x$$

Finally, let's rewrite the left side using a Pythagorean identity.

$$\sin^2 x = \sin^2 x$$

Since the last equation is an identity, the first is also.   ■

EXAMPLE 3   *Deciding whether an equation is an identity*

To see whether the equation

$$\sin 4t = 4 \sin t \cos^3 t - 4 \sin^3 t \cos t$$

is an identity, let's begin by observing that the left side is sin $4t$, while the right side involves sin t and cos t. We can, therefore, make the two sides easier to compare by using double-angle identities. First, let's express sin $4t$ as sin $[2(2t)]$. This form allows us to use the identity for sin $2x$ on the left side, with x replaced by $2t$.

$$2 \sin 2t \cos 2t = 4 \sin t \cos^3 t - 4 \sin^3 t \cos t$$

Next, let's use the identities for sin $2x$ and cos $2x$, with x replaced by t.

$$2(2 \sin t \cos t)(\cos^2 t - \sin^2 t) = 4 \sin t \cos^3 t - 4 \sin^3 t \cos t$$

Finally, let's multiply the left side out.

$$4 \sin t \cos^3 t - 4 \sin^3 t \cos t = 4 \sin t \cos^3 t - 4 \sin^3 t \cos t$$

The original equation is, therefore, an identity.   ■

There is often more than one way to decide whether an equation is an identity. For example, in the last step of Example 3, we could have factored the right side instead of multiplying the left side out.

EXAMPLE 4   *Deciding whether an equation is an identity*

To see whether the equation

$$\tan x \sin x - \sec x = \cos x$$

is an identity, let's begin by rewriting both sides in terms of sines and cosines.

$$\left(\frac{\sin x}{\cos x}\right) \sin x - \frac{1}{\cos x} = \cos x$$

Next, let's multiply both sides by $\cos x$.

$$(\sin x)(\sin x) - 1 = \cos^2 x$$

$$\sin^2 x - 1 = \cos^2 x$$

$$-\cos^2 x = \cos^2 x$$

The last equation is false unless $\cos x = 0$, and so is conditional. The original equation is therefore also conditional.   ∎

EXERCISES   *In Exercises 67–70, decide whether the equation is an identity. You worked with the same equations in Exercises 57–60 and 63–66.*

67. $\sin^2 x + \tan^2 x = \sec^2 x - \cos^2 x$

68. $\sec x(\sin^2 x - 1) = \cos x$

69. $\cos^4 x - \sin^4 x = \cos 2x$

70. $2 \sin x \cos 4x = \sin 5x - \sin 3x$

71. Use proof by counterexample to show that the equation in Example 4 is conditional. That is, find a specific value of x for which the equation is false.

ADDITIONAL EXERCISES   *In Exercises 72–77, find each of the following.*

    a. $\sin (-x)$ **b.** $\cos (-x)$

    c. $\cot \left(\dfrac{\pi}{2} - x\right)$ **d.** $\sin \left(\dfrac{\pi}{2} - x\right)$

    e. $\tan (x - \pi)$ **f.** $\cos 2x$

    g. $\csc 2x$ **h.** $\sin \dfrac{x}{2}$

72. $x = \dfrac{5\pi}{6}$

73. $\cos x = \dfrac{5}{13}$ and $0 < x < \dfrac{\pi}{2}$

74. $\sin x = \dfrac{3}{4}$ and $\dfrac{\pi}{2} < x < \pi$

75. $\tan x = \dfrac{3}{4}$ and $\pi < x < \dfrac{3\pi}{2}$

76. $\cos x = -0.28$ and $3\pi < x < \dfrac{7\pi}{2}$

77. $\sin x = -0.4$ and $-\dfrac{5\pi}{2} < x < -2\pi$

In Exercises 78–87, decide whether the equation is an identity, and prove your answer.

78. $1 - \sec x + \tan x \sin x = 1 - \cos x$

79. $\dfrac{\sin^2 t}{1 - \cos t} = 1 - \cos t$

80. $1 - \cos \dfrac{x}{2} = 2 \sin^2 \dfrac{x}{4}$

81. $\cot A - \tan B = \dfrac{\cos (A + B)}{\sin A \cos B}$

82. $\sin \left(x + \dfrac{\pi}{2} \right) + \sin \left(x - \dfrac{\pi}{2} \right) = \cos x$

83. $\sin \left(x + \dfrac{\pi}{3} \right) + \sin \left(x - \dfrac{\pi}{3} \right) = \sin x$

84. $\dfrac{\cos (x + y)}{\cos (x - y)} = 1 - \dfrac{2 \sin x \sin y}{\cos x \cos y + \sin x \sin y}$

85. $\cos 2t + \cos 4t = 2 \cos t \cos 3t$

86. $\cos^4 x - \sin^4 x = \cos 4x$

87. $\sin x \sin 2x = 2 \cos x - 2 \cos^3 x$

88. Write a formula for $\sin 4x$ in terms of $\sin x$ and $\cos x$. (There is more than one correct answer.)

89. Write a formula for $\cos 3x$ in terms of $\sin x$ and $\cos x$. (There is more than one correct answer.)

In Exercises 90–91 you can derive additional reflection identities.

90. (*Making Observations*) The points on the unit circle determined by the directed angles x and $\pi - x$ are reflections of each other in the vertical axis. Use that fact to express each trigonometric function of $\pi - x$ in terms of a trigonometric function of x.

91. (*Making Observations*) The points on the unit circle determined by the directed angles x and $2\pi - x$ are reflections of each other in the horizontal axis. Use that fact to express each trigonometric function of $2\pi - x$ in terms of a trigonometric function of x.

In Exercise 92 you can derive additional rotation identities.

92. (*Making Observations*) Figure 9-15 shows that if a directed angle x determines the point (a, b) in the unit circle, then the directed angles $x + \dfrac{\pi}{2}$ and $x - \dfrac{\pi}{2}$ determine the points $(-b, a)$ and $(b, -a)$, respectively.

a. Use Figure 9-15 to express each trigonometric function of $x + \dfrac{\pi}{2}$ in terms of a trigonometric function of x.

b. Express each trigonometric function of $x - \dfrac{\pi}{2}$ in terms of a trigonometric function of x.

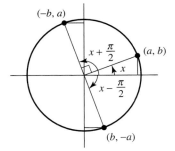

Figure 9–15

In Exercises 93–95 you can verify the following sum, difference, and double-angle identities for the tangent function.

$$\tan (x \pm y) = \frac{\tan x \pm \tan y}{1 \mp \tan x \tan y}$$

$$\tan 2x = \frac{2 \tan x}{1 - \tan^2 x}$$

93. (*Making Observations*) Verify the identity for $\tan (x + y)$ as follows.

   a. Rewrite the left side as $\dfrac{\sin (x + y)}{\cos (x + y)}$.

   b. Use sum identities to rewrite the left side again.

   c. Rewrite the right side in terms of sines and cosines, and simplify to obtain the expression in part (b).

94. (*Making Observations*) Verify the identity for $\tan (x - y)$.

95. (*Making Observations*) Verify the identity for $\tan 2x$ by letting $y = x$ in the identity for $\tan (x + y)$.

96. (*Making Observations*) If x determines the point $\left(\dfrac{\sqrt{10}}{10}, \dfrac{3\sqrt{10}}{10} \right)$ in the unit circle, find $\tan 2x$ in two different ways:

   a. Use the identity for $\tan 2x$.

   b. Use the identities for $\sin 2x$ and $\cos 2x$, and the fact that

$$\tan 2x = \frac{\sin 2x}{\cos 2x}$$

9-3   VECTORS

Many physical quantities, such as temperature and length, can be described by a single number on a scale of measurement. Such a quantity is sometimes called a **scalar quantity**, or simply a **scalar**. By contrast, other quantities must be described by two or more numbers. An example is the gravitational force that the earth exerts on you. This force has a **magnitude**, or size, that is equal to your weight. However, the effect of the force also depends on the direction in which it acts. If its direction suddenly changed from downward (Figure 9-16a) to upward (Figure 9-16b), you would shoot up into space.

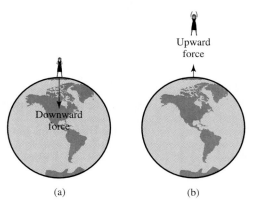

Figure 9–16

A quantity having both magnitude and direction is called a **vector quantity**, or a **vector**.

EXERCISE   **1.** Decide whether each quantity is a scalar or a vector.

    a. a distance of 3 kilometers

    b. a movement of 3 kilometers to the east

    c. an upward velocity of 50 feet per second

    d. a speed of 100 meters per second

    e. a mass of 45 kilograms

    f. a weight of 45 pounds

In this section you will discover how trigonometry can be used to analyze vectors and solve physical problems like the one in the following Mathematical Looking Glass.

A MATHEMATICAL LOOKING GLASS
Prairie Schooner

One day recently, Dave was talking to Mark Conaway, an air traffic controller who enjoys sailing. Much to Dave's surprise, Mark informed him that one of the world's best lakes for sailing is Lake Hefner, located within the limits of Oklahoma City. The lake lies at the highest point of an otherwise flat prairie extending for many miles in all directions. Because it is unprotected on all sides, there is always a strong prevailing wind.

Mark went sailing on Lake Hefner in 1993 and found the winds to be as strong as advertised. However, he wanted to sail west, directly into the wind. He therefore needed to know how the direction of the sail affects the force of the wind against it and allows him to control the motion of the boat. His setting of the sail is determined by both the direction and speed of the wind. For example, a 10-mph wind from the west must be treated differently from either a 10-mph wind from the north or a 50-mph wind from the west. ■

EXERCISE   **2.** (*Writing to Learn*) What adjustments might Mark make if the direction of the wind changed from west to north? What adjustments might he make if the speed of the wind changed from 10 mph to 50 mph?

Before solving Mark's problem, let's establish some fundamental facts about vectors.

GEOMETRIC REPRESENTATION OF VECTORS

It is convenient to represent vector quantities by means of directed line segments from a specified initial point to a specified terminal point, as in Figure 9-17. The length and direction of the directed line segment represent the magnitude and direction of the vector.

Figure 9–17

FYI In addition to its magnitude and direction, a directed line segment has a definite position, whereas a vector does not. For example, the three directed line segments in Figure 9-17 have the same magnitude and direction. Thus, they all represent the same wind, that is, the same vector. Although it is accurate to say that each directed line segment is a **representative** of the vector, it is common practice to refer to each one simply as a *vector*.

A vector is often identified by a single boldface letter or by two letters representing its initial and terminal points. The two vectors shown in Figure 9-18 are **V** = **AB** and **W** = **BA**, respectively. They are also sometimes written as $\overrightarrow{V} = \overrightarrow{AB}$ and $\overrightarrow{W} = \overrightarrow{BA}$. The points A and B are the **tail** and **head** of **V**, respectively.

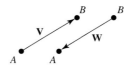

Figure 9–18

Scalar Multiplication of Vectors If **V** is any vector and c is any scalar, c**V** denotes a vector parallel to **V** whose magnitude is $|c|$ times that of **V**. The vector c**V** points in the same direction as **V** if $c > 0$ and in the opposite direction if $c < 0$. In either case, c**V** is referred to as a **scalar multiple** of **V**. For example, if the people in Figure 9-19 are pulling equally hard on their ropes, the forces being applied to the three boxes can be represented by vectors **V**, 2**V**, and -3**V**.

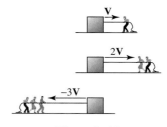

Figure 9–19

The magnitude of **V** is often denoted by $\|\mathbf{V}\|$. With this notation,

$$\|c\mathbf{V}\| = |c|\,\|\mathbf{V}\|$$

If **V** represents a force, then c**V** represents a force $|c|$ times as strong, and acting in the same or the opposite direction, according to the sign of c.

For any vector **V**, the vector 0**V** = **0** has a magnitude of 0 and undetermined direction. It is referred to as the **zero vector**.

EXERCISES   **3.** Draw a vector **V** whose length is 2 inches, and whose direction is from left to right. Then draw each of the following vectors:

    a. 1.5**V**   **b.** $-$**V**

    c. $\frac{1}{2}$**V**   **d.** $-\frac{1}{2}$**V**

4. If **W** represents a 10-pound force applied in a downward direction, represent each of the following as a scalar multiple of **W**:

   a. a 40-pound force applied downward

   b. a 40-pound force applied upward

   c. a 2-pound force applied downward

   d. a complete absence of force

Addition and Subtraction of Vectors Using physics, it can be shown that when two forces act on the same object, such as **U** and **V** in Figure 9-20a, the net effect is the same as that of a single force **W** acting along the diagonal of the parallelogram determined by **U** and **V**. For this reason, it is convenient to refer to **W** as the **vector sum** of **U** and **V** and write **W = U + V**.

(a)   (b)

Figure 9–20

An alternate geometric construction of **U** + **V** is shown in Figure 9-20b, where **V** is placed with its tail at the head of **U**. Then **U** + **V** extends from the tail of **U** to the head of **V**.

Two geometric constructions of the vector **U** − **V** = **U** + (−**V**) are shown in Figure 9-21.

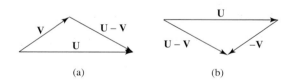

(a)   (b)

Figure 9–21

EXERCISE   **5.** For the vectors **A** and **B** below, draw each of the following.

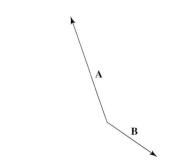

   a. A + B   **b. A − B**

   c. A + 3B   **d. 4A − 2B**

Projections Figure 9-22 illustrates that if **U** and **V** are vectors, it is possible to express **V** as the sum of two vectors, of which one is parallel to **U** and the other is perpendicular to **U**.

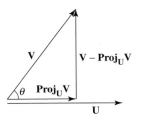

Figure 9–22

One of these is called the **projection** of **V** in the direction of **U** and is written **Proj$_U$V**. The other is called the **complementary projection** of **V** in the direction of **U** and is equal to **V** − **Proj$_U$V**. The two vectors are also referred to as the **parallel and perpendicular components** of **V** with respect to **U**.

If the angle between **U** and **V** is θ, with $0° \leq \theta \leq 180°$, then Figure 9-22 illustrates that

$$\|\mathbf{Proj_U V}\| = \|\mathbf{V}\| \, |\cos \theta| \quad \text{and} \quad \|\mathbf{V} - \mathbf{Proj_U V}\| = \|\mathbf{V}\| \, |\sin \theta|$$

EXAMPLE 1   *Finding vector projections*
In Figure 9-23,

$$\|\mathbf{Proj_U V}\| = \|\mathbf{V}\| \, |\cos 115°| \approx 20 \, |-0.4226| \approx 8.45$$

$$\|\mathbf{V} - \mathbf{Proj_U V}\| = \|\mathbf{V}\| \, |\sin 115°| \approx 20 \, |0.9063| \approx 18.13$$

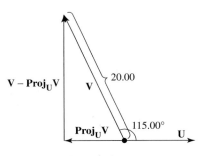

Figure 9–23   ■

EXERCISES   *In Exercises 6 and 7,*

a. Draw the vectors **U**, **V**, **Proj$_U$V**, and **V** − **Proj$_U$V**.
b. Find $\|\mathbf{Proj_U V}\|$ and $\|\mathbf{V} - \mathbf{Proj_U V}\|$.

8. (*Problem Solving*) If a 4000-pound car is parked on a hill at an angle of 10° with the horizontal, its weight is a downward force **W** with a magnitude of 4000.

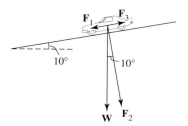

Figure 9–24

a. Figure 9-24 shows **W** as the sum of a force $\mathbf{F}_1$ acting along the road and a force $\mathbf{F}_2$ acting perpendicular to it. Find the magnitudes of $\mathbf{F}_1$ and $\mathbf{F}_2$.

b. How much force must the brakes exert in the direction of $\mathbf{F}_3$ to keep the car from rolling downhill? (*Hint:* $\|\mathbf{F}_3\| = \|\mathbf{F}_1\|$)

ANALYTICAL REPRESENTATION OF VECTORS

Two important vectors in the coordinate plane are the **unit vectors** (that is, vectors whose length is one unit) in the positive *x*- and *y*-directions. These vectors, shown in Figure 9-25a, are traditionally called **i** and **j**.

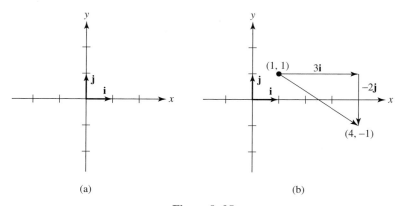

Figure 9–25

Exercises 9–14 indicate the fundamental role played by **i** and **j**.

EXERCISES   Figure 9–25b illustrates that the vector extending from (1, 1) to (4, −1) is equal to $3\mathbf{i} - 2\mathbf{j}$.

In Exercises 9–12,

a. Express the vector whose initial and terminal points are given as a sum $a\mathbf{i} + b\mathbf{j}$.

b. Draw a figure similar to Figure 9–25b to illustrate your answer to part (a).

9. $(-2, 1), (4, 5)$ **10.** $(6, -3), (2, -3)$

11. $(0, 0), (-3, 3)$ **12.** $(-3, 3), (0, 0)$

13. (*Making Observations*) If a vector **V** extends from (x_1, y_1) to (x_2, y_2), draw a figure similar to Figure 9-25b to show that **V** can be expressed as $(x_2 - x_1)\mathbf{i} + (y_2 - y_1)\mathbf{j}$.

14. (*Writing to Learn*) How do the vectors $(x_2 - x_1)\mathbf{i}$ and $(y_2 - y_1)\mathbf{j}$ in Exercise 13 express the directed changes in x and y represented by **V**?

Exercises 9–14 show that every vector **V** in the plane can be expressed as the sum of two vectors $a\mathbf{i}$ and $b\mathbf{j}$, called the components of **V**, that express directed changes in x and y. Let's see how this observation can be used to perform algebraic operations on vectors analytically.

Scalar Multiplication of Vectors If a vector **V** is placed in a coordinate plane, then for any scalar multiple $c\mathbf{V}$, the directed changes in x and y are c times those for **V**. Thus, scalar multiplication of vectors is accomplished analytically by the rule

$$c(a\mathbf{i} + b\mathbf{j}) = (ca)\mathbf{i} + (cb)\mathbf{j}$$

Addition and Subtraction of Vectors If **U** and **V** are both placed in a coordinate plane, then for the sum $\mathbf{U} + \mathbf{V}$, the directed changes in x and y are the sums of those for **U** and **V** individually. A similar statement applies to differences. Thus, addition and subtraction of vectors is accomplished analytically by the rules

$$(a_1\mathbf{i} + b_1\mathbf{j}) \pm (a_2\mathbf{i} + b_2\mathbf{j}) = (a_1 \pm a_2)\mathbf{i} + (b_1 \pm b_2)\mathbf{j}$$

EXAMPLE 2   *Performing vector operations analytically*

a. If $\mathbf{A} = -3\mathbf{i} + 7\mathbf{j}$, then $-2\mathbf{A} = -2(-3\mathbf{i} + 7\mathbf{j}) = 6\mathbf{i} - 14\mathbf{j}$.

b. If $\mathbf{A} = -3\mathbf{i} + 7\mathbf{j}$ and $\mathbf{B} = 5\mathbf{i} - 7\mathbf{j}$, then

$$\mathbf{A} + \mathbf{B} = (-3\mathbf{i} + 7\mathbf{j}) + (5\mathbf{i} - 7\mathbf{j}) = 2\mathbf{i} \qquad \blacksquare$$

EXERCISES   Draw the vectors $\mathbf{U} = 2\mathbf{i} + 4\mathbf{j}$ and $\mathbf{V} = 3\mathbf{i} - \mathbf{j}$ in a coordinate plane. Then write each of the following in terms of its components, and draw each in the coordinate plane.

15. $-\mathbf{U}$ **16.** $3\mathbf{V}$

17. $\mathbf{U} + 2\mathbf{V}$ **18.** $2\mathbf{U} - \mathbf{V}$

Finding Components from Known Magnitude and Direction Using trigonometry, we can develop a formula to find the components of a vector whose magnitude and direction are known. Specifically, suppose that $\mathbf{V} = a\mathbf{i} + b\mathbf{j}$ makes an angle of θ with the positive x-axis, as in Figure 9-26.

The terminal side of θ has a length of $\|\mathbf{V}\|$. Therefore, as you verified in Exercise 72 of Section 8-4, $\cos \theta = \dfrac{a}{\|\mathbf{V}\|}$ and $\sin \theta = \dfrac{b}{\|\mathbf{V}\|}$, so that the components of **V** are related to its magnitude and direction by the equations

$$a = \|\mathbf{V}\| \cos \theta$$

$$b = \|\mathbf{V}\| \sin \theta$$

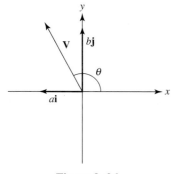

Figure 9–26

EXERCISE   **19.** (*Writing to Learn*) In Exercise 30 of Section 8-1, you observed that a circle of radius r, centered at $(0, 0)$, can be described by the equations

$x = r \cos \theta,\ y = r \sin \theta$. Use this observation to justify the equations $a = \|\mathbf{V}\| \cos \theta,\ b = \|\mathbf{V}\| \sin \theta$.

EXAMPLE 3 *Determining the components of a vector with known magnitude and direction*

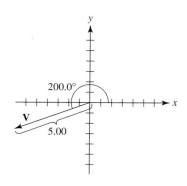

Figure 9–27

Suppose $\|\mathbf{V}\| = 5$, and $\mathbf{V}$ makes an angle of 200° with the positive x-axis, as in Figure 9-27. Then the components of $\mathbf{V}$ are $a\mathbf{i} + b\mathbf{j}$, where

$$a = 5 \cos 200° \approx -4.70$$

$$b = 5 \sin 200° \approx -1.71$$

Finding Magnitude and Direction from Known Components Figure 9-26 also allows us to express $\|\mathbf{V}\|$ and θ in terms of a and b.

$$\|\mathbf{V}\| = \sqrt{a^2 + b^2}$$

$$\tan \theta = \frac{b}{a}$$

Thus, $\|\mathbf{V}\|$ and θ can be found if the components $a\mathbf{i}$ and $b\mathbf{j}$ are known. In this case, $\tan \theta$ can be calculated directly from a and b, and some care must be taken in determining the value of θ. Example 4 shows why.

EXAMPLE 4 *Determining the magnitude and direction of a vector whose components are known*

Suppose $\mathbf{V} = -3\mathbf{i} + 3\mathbf{j}$, as in Figure 9-28. Then

$$\|\mathbf{V}\| = \sqrt{(-3)^2 + 3^2} = \sqrt{18} = 3\sqrt{2}$$

$$\tan \theta = \frac{3}{-3} = -1$$

$$\theta = \tan^{-1}(-1) + k\pi$$

or, if we want to express θ in degrees,

$$\theta = \tan^{-1}(-1) + 180k°$$

$$= -45° + 180k°$$

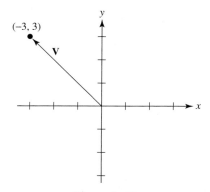

Figure 9–28

Solutions in the interval $[0°, 360°)$ are 135° and 315°. Since θ lies in the second quadrant, $\theta = 135°$.

EXERCISES   *In Exercises 20–23, find the components of* **V**.

20. $\|\mathbf{V}\| = 30.0$, $\theta = 152.0°$ **21.** $\|\mathbf{V}\| = 1001$, $\theta = 20.00°$

22. $\|\mathbf{V}\| = 0.38$, $\theta = 236°$ **23.** $\|\mathbf{V}\| = 1.0$, $\theta = 281°$

In Exercises 24–27, find $\|\mathbf{V}\|$ *and* θ.

24. $\mathbf{V} = 75\mathbf{i} - 40\mathbf{j}$ **25.** $\mathbf{V} = -12\mathbf{i} - 5\mathbf{j}$

26. $\mathbf{V} = 2\mathbf{i} + 2\sqrt{3}\,\mathbf{j}$ **27.** $\mathbf{V} = -\mathbf{i} + 7\mathbf{j}$

Now let's return to ***Prairie Schooner***, in which Mark wanted to sail west, directly against the wind. Both common sense and experience told him that he could not sail directly into the wind, as you will confirm in Exercise 33. However, he was able to make progress by means of a maneuver called **tacking** (following a zigzag course), as illustrated in Figure 9-29.

West

Figure 9–29

Figure 9–30

To tack successfully, Mark needed to understand how the motion of the boat is affected by the relative directions of the sail, wind, and boat. First, let's observe the effect of the relative directions of the sail and wind. In the overhead view in Figure 9-30, the sail makes an angle of α with the wind, represented by **W**. The force of the wind against the sail is **P**, the component of **W** perpendicular to the sail.

EXERCISE   **28.** (*Writing to Learn*) Explain why $\|\mathbf{P}\| = \|\mathbf{W}\| \sin \alpha$ as long as $0° \leq \alpha \leq 180°$.

Next, let's look at the effect of the relative directions of the sail and boat. In Figure 9-31a, the direction of the sail makes a directed angle of γ with the boat. If the force of the wind against the sail is **P**, then the boat is moved forward (Figure 9-31b) or backward (Figure 9-31c) by **M**, the component of **P** in the direction of **B**.

(a)   (b)   (c)

Figure 9–31

EXERCISE   **29.** (*Writing to Learn*) Explain why $\|\mathbf{M}\| = \|\mathbf{W}\| \sin \alpha \sin \gamma$.

Now let's look at the sail, wind, and boat together. In Figure 9-32, the sail and the boat make angles of α and β with the direction of the wind.

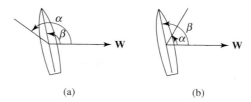

(a)   (b)

Figure 9–32

EXERCISES

30. (*Writing to Learn*) Explain why the boat moves forward if $\alpha > \beta$ as in Figure 9-32a and moves backward if $\alpha < \beta$ as in Figure 9-32b.

31. (*Writing to Learn*)

  a. Show that if $\beta > 90°$, the component of the boat's motion in the westward direction has a magnitude of $\|\mathbf{W}\| \sin \alpha \sin (\alpha - \beta) |\cos \beta|$.

  b. Show that if $\beta < 90°$, the component of the boat's motion in the eastward direction has a magnitude of $\|\mathbf{W}\| \sin \alpha \sin (\alpha - \beta) \cos \beta$.

32. In each case, the component of the wind that drives the boat west or east is a vector $k\mathbf{W}$. Use your result from Exercise 31 to find k.

  a. $\alpha = 160°, \ \beta = 110°$

  b. $\alpha = 140°, \ \beta = 100°$

  c. $\alpha = 150°, \ \beta = 120°$

  d. $\alpha = 135°, \ \beta = 90°$

33. (*Making Observations*) If $\beta = 180°$, then the boat moves backward unless $\alpha = 180°$ also. Calculate the portion of the wind's force that drives the boat westward if $\alpha = \beta = 180°$. Your result should confirm Mark's conclusion that he cannot sail directly into the wind.

  In Exercise 52 you can finish solving Mark's problem by finding the values of α and β that produce the most rapid progress toward the west.

ADDITIONAL EXERCISES

34. Draw a vector $\mathbf{W}$ with length 1.5 inches, directed upward. Then draw each of the following vectors.

  a. $2\mathbf{W}$ **b.** $-2\mathbf{W}$

  c. $-0.5\mathbf{W}$ **d.** $\dfrac{1}{3}\mathbf{W}$

35. For the vectors $\mathbf{U}$ and $\mathbf{V}$ below, draw each of the following.

  a. $2\mathbf{U} + \mathbf{V}$ **b.** $-\mathbf{U} + 2\mathbf{V}$

  c. $\dfrac{1}{2}(\mathbf{U} + \mathbf{V})$ **d.** $\mathbf{U} - \dfrac{1}{2}\mathbf{V}$

In Exercises 36 and 37,

    a. Draw the vectors **U**, **V**, **Proj$_U$V**, *and* **V − Proj$_U$V**.

    b. Find ‖**Proj$_U$V**‖ *and* ‖**V − Proj$_U$V**‖.

36.   **37.**

Draw the vectors **A** = −**i** + 0.5**j** *and* **B** = 2**j** *in a coordinate plane. Then write each vector in Exercises 38–41 in terms of its components, and draw each in the coordinate plane.*

38. 4**A**   **39.** −2**B**

40. **A** + **B**   **41.** 3(**A** − **B**)

In Exercises 42–45, find the components of **V**.

42. ‖**V**‖ = 87, θ = 32°

43. ‖**V**‖ = 3.05, θ = 146.0°

44. ‖**V**‖ = 12, θ = 189°

45. ‖**V**‖ = 312, θ = 312.0°

In Exercises 46–49, find ‖**V**‖ *and* θ.

46. **V** = −100**i** + 75**j**   **47.** **V** = −17**i** − 17**j**

48. **V** = $\sqrt{3}$ **i** + **j**   **49.** **V** = 5.10**i** − 2.76**j**

50. (*Problem Solving*) Have you ever driven along a road, or stood by the roadside, as a storm approaches the road at an angle? It is sometimes possible to see the rain advancing along the road, often at a surprisingly rapid rate. The following problem shows how a storm can advance rapidly along a road, even when the storm front is moving slowly.

Figure 9–33

Suppose you are driving from Champaign, IL, to Chicago along Interstate Highway 57, in a direction 10° east of north, as shown in Figure 9-33. A storm front is moving eastward at 10 mph. How fast must you drive to stay ahead of the storm?

51. (*Problem Solving*) A pilot flying from New Orleans to St. Louis must go directly north. However, if a strong wind is blowing from the west, the pilot must head slightly west of north, as shown in Figure 9-34. The vector **A** represents the airplane's velocity relative to the surrounding air, and **W** represents the velocity of the wind. The airplane's actual velocity relative to the ground is the vector sum **A + W**.

Figure 9–34

Suppose that the airplane's air speed $\|\mathbf{A}\|$ is 600 mph and that the wind speed is 60 mph.

a. How many degrees west of north must the pilot direct the airplane? That is, what is the degree measure of angle θ in Figure 9-34?

b. What is the airplane's ground speed $\|\mathbf{A} + \mathbf{W}\|$?

c. St. Louis is 580 air miles from New Orleans. Neglecting the time required for takeoff and landing, how long will the flight take?

Figure 9–35

52. (*Problem Solving*) Figure 9-35 relates to ***Prairie Schooner***. Mark's sail and boat make angles of α and β with the direction of the wind, $A = \pi - \alpha$ and $B = \pi - \beta$. In Exercise 31 you discovered that if $\beta > 90°$, the component of the wind driving the boat westward is proportional to $\sin \alpha \sin (\alpha - \beta)|\cos \beta|$. Mark wants to position the boat and sail so that this expression is as large as possible.

a. Explain why $\sin \alpha \sin (\alpha - \beta) |\cos \beta|$ is equal to $\sin A \sin (B - A) \cos B$.

b. Use a product-to-sum identity to rewrite $\sin A \sin (B - A)$.

c. (*Writing to Learn*) For any value of B, explain why the component of the wind's force driving the boat westward is largest when $A = \dfrac{1}{2} B$.

d. (*Making Observations*) If $A = \dfrac{1}{2} B$, show that the component of the wind's force driving the boat westward is $\dfrac{1}{2} \cos B + \dfrac{1}{2} \cos^2 B$.

e. Use graphical methods to find the value of B, in degrees with $90° < B < 180°$, for which $\dfrac{1}{2} \cos B + \dfrac{1}{2} \cos^2 B$ is largest.

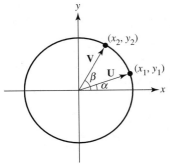

Figure 9–36

f. (*Writing to Learn*) Explain how your results from parts (a)–(e) indicate that Mark should set $A = 30°$ and $B = 60°$.

g. If $A = 30°$ and $B = 60°$, what fraction of the wind's force drives Mark's boat westward?

If $\mathbf{U} = x_1\mathbf{i} + y_1\mathbf{j}$ and $\mathbf{V} = x_2\mathbf{i} + y_2\mathbf{j}$, the quantity $x_1x_2 + y_1y_2$ is sometimes called the **scalar product**, or **dot product**, of $\mathbf{U}$ and $\mathbf{V}$, and is usually written $\mathbf{U} \cdot \mathbf{V}$. Exercises 53 and 54 show how to use the dot product to find the angle between two vectors.

53. (*Extension*) If $\|\mathbf{U}\| = 1$ and $\|\mathbf{V}\| = 1$, both $\mathbf{U}$ and $\mathbf{V}$ can be placed into the unit circle, as in Figure 9-36.

If $\mathbf{U}$ and $\mathbf{V}$ make angles α and β with the positive x-axis, then the angle between $\mathbf{U}$ and $\mathbf{V}$ is $|\alpha - \beta|$.

a. Explain why $\cos |\alpha - \beta| = \cos (\alpha - \beta)$.

b. Use a difference identity to express $\cos (\alpha - \beta)$ in terms of trigonometric functions of α and β.

c. Use the unit circle definitions of cosine and sine to express $\cos (\alpha - \beta)$ in terms of x_1, y_1, x_2, and y_2.

d. Explain why $|\alpha - \beta| = \cos^{-1}(x_1x_2 + y_1y_2)$.

54. (*Extension*) Explain why the angle between any two vectors $\mathbf{U}$ and $\mathbf{V}$ is $\cos^{-1}\left(\dfrac{\mathbf{U}}{\|\mathbf{U}\|} \cdot \dfrac{\mathbf{V}}{\|\mathbf{V}\|}\right)$. $\left(\textit{Hint:}\text{ What are the lengths of }\dfrac{\mathbf{U}}{\|\mathbf{U}\|}\text{ and }\dfrac{\mathbf{V}}{\|\mathbf{V}\|}?\right)$

It can be shown that $\dfrac{\mathbf{U}}{\|\mathbf{U}\|} \cdot \dfrac{\mathbf{V}}{\|\mathbf{V}\|}$ *is equal to* $\dfrac{\mathbf{U} \cdot \mathbf{V}}{\|\mathbf{U}\|\,\|\mathbf{V}\|}$. *In Exercises 55–58 (Extension), use this result to find the angle between $\mathbf{U}$ and $\mathbf{V}$.*

55. $\mathbf{U} = 3\mathbf{i} + 4\mathbf{j}, \mathbf{V} = 5\mathbf{i} + 12\mathbf{j}$

56. $\mathbf{U} = 3\mathbf{i} + 4\mathbf{j}, \mathbf{V} = -4\mathbf{i} + 3\mathbf{j}$

57. $\mathbf{U} = -7\mathbf{j}, \mathbf{V} = -2\mathbf{i} + 2\mathbf{j}$

58. $\mathbf{U} = -\mathbf{i} + 6\mathbf{j}, \mathbf{V} = -5\mathbf{i} - 2\mathbf{j}$

9–4   TRIGONOMETRIC EQUATIONS

Earlier in this chapter you have used trigonometry to navigate a sailboat and to find the position of a satellite orbiting the earth. Trigonometry is also needed to solve problems involving magnetic force, wave motion, and the resolution of a microscope. To solve such problems, you must often solve equations involving trigonometric expressions. Let's consider one such situation.

A MATHEMATICAL
LOOKING GLASS
Artillery

On April 12, 1861, artillerymen of the newly formed Confederate States of America fired a ten-inch mortar shell toward a garrison of 85 soldiers at Fort Sumter, South Carolina, thus beginning the U.S. Civil War. The Confederates captured the fort after bombarding it with mortar fire for 34 hours. During this time, surprisingly, there was no loss of life on either side.

The capture of Fort Sumter illustrates the effectiveness of artillery. From the time of their introduction in the fourteenth century, weapons such as cannons and mortar changed the nature of warfare by allowing armies to attack enemy positions from a distance.

Artillery also increased the importance of mathematics in military operations. Unlike a bullet, which travels in an essentially linear path, an artillery shell must be fired at a substantial angle of elevation to reach its target. Thus, an artillery weapon cannot be aimed by pointing it at its target, but instead must be "aimed" by calculating or estimating its trajectory. In past centuries, the trajectory was often controlled by varying the amount of powder used to give the shell its initial velocity. Trajectories of modern weapons are usually controlled by varying the angle of elevation θ. The crucial relationship is between θ and the distance to the target, referred to as the range of the shell. Figure 9-37 illustrates the effect of θ on the range. ∎

Figure 9–37

To construct an accurate mathematical model of the relationship between θ and the range, we would need differential equations. To construct an approximate model, let's make the simplifying assumption that once the shell is fired, it is acted on only by gravity. That is, we will ignore all other forces, most significantly air resistance.

We will also need to use a physical property of velocity. Specifically, suppose that the initial velocity of the shell is represented by a vector $\mathbf{V}$, whose horizontal and vertical components are respectively $\mathbf{V}_x$ and $\mathbf{V}_y$, with lengths V_x and V_y, as in Figure 9-38.

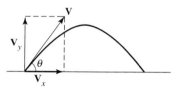

Figure 9–38

The shell's subsequent motion in the vertical direction is then identical to that of a shell fired straight up with a velocity of magnitude V_y. Similarly, its horizontal motion is identical to that of a shell fired horizontally with a velocity of magnitude V_x.

In Section 4-2 you learned that an object traveling in a vertical direction, and acted on by gravity alone, has the equation of motion

$$y = s_0 + v_0 t - 16t^2$$

In this equation t and y are expressed in seconds and feet, and s_0 and v_0 are the initial height and velocity of the object. Thus, the vertical position of a shell fired from ground level ($s_0 = 0$) with vertical velocity $v_0 = V_y$ is

$$y = V_y t - 16t^2$$

Its horizontal motion continues at a speed of V_x, since no horizontal forces are acting on the shell, so that its horizontal position is

$$x = V_x t$$

EXERCISES

1. (*Review*) If $\|\mathbf{V}\|$ is denoted by V, show that $V_x = V \cos \theta$ and $V_y = V \sin \theta$.

In Exercises 2–5,

    a. Write functions $x = V_x t$, $y = V_y t - 16t^2$ for a shell fired with an initial velocity of 2000 ft/sec at the given angle of elevation.

    b. Solve the equation $x = V_x t$ for t. Then substitute the resulting expression into the equation $y = V_y t - 16t^2$ to express y as a function of x.

    c. Find the range of the shell. (*Hint:* When the shell hits the ground, its y-coordinate is zero. Its range is the corresponding x-coordinate.)

2. $10°$ **3.** $30°$ **4.** $50°$ **5.** $70°$

In Exercises 6–9,

    a. Write functions $x = V_x t$, $y = V_y t - 16t^2$ for a shell fired with the given initial velocity at a $45°$ angle of elevation.

    b. Solve the equation $x = V_x t$ for t. Then substitute the resulting expression into the equation $y = V_y t - 16t^2$ to express y as a function of x.

    c. Find the range of the shell.

  6. 1000 ft/sec   **7.** 1500 ft/sec   **8.** 2000 ft/sec   **9.** 2500 ft/sec

10. Show that the path of a shell fired at an angle of elevation θ is described by the equation

$$y = (\tan \theta)x - \frac{16}{V^2}(\sec^2 \theta)x^2$$

To determine a shell's angle of elevation, we must solve a trigonometric equation. For example, let's fire an imaginary shell at a target 5 miles away, with an initial velocity of 1000 ft/sec. (In reality, air resistance would limit the effective range of such a shell to a much shorter distance.) We want the shell to be at ground level ($y = 0$) when $x = 5$ miles (26,400 feet). Therefore, to determine the appropriate angle, we can use the equation in Exercise 10 with $V = 1000$, $x = 26{,}400$, and $y = 0$. The resulting equation in θ is

$$26{,}400 \tan \theta - \frac{16}{(1000)^2}(26{,}400)^2 \sec^2 \theta = 0$$

Before solving this equation, let's look at some techniques for solving trigonometric equations.

THREE SPECIAL TYPES OF EQUATIONS

Most trigonometric equations you will encounter in this book can be simplified to one of the forms

$$\sin b(x - h) = c, \qquad \cos b(x - h) = c, \qquad \text{or } \tan b(x - h) = c$$

Example 1 illustrates a procedure for solving these important special types of equations.

EXAMPLE 1   *Solving an equation* $\sin b(x - h) = c$, $\cos b(x - h) = c$, *or* $\tan b(x - h) = c$

To solve $\cos 4\left(t - \dfrac{\pi}{12}\right) = \dfrac{1}{2}$, begin by recalling from Section 8-5 that the solutions to $\cos x = c$ are

$$x = \begin{cases} \cos^{-1} c + 2k\pi \\ 2\pi - \cos^{-1} c + 2k\pi \end{cases}$$

Since $\cos^{-1} \dfrac{1}{2} = \dfrac{\pi}{3}$, the possible values of $4\left(t - \dfrac{\pi}{12}\right)$ are

$$4\left(t - \frac{\pi}{12}\right) = \begin{cases} \dfrac{\pi}{3} + 2k\pi \\[2ex] \dfrac{5\pi}{3} + 2k\pi \end{cases}$$

Next divide both sides by 4.

$$t - \frac{\pi}{12} = \begin{cases} \dfrac{\pi}{12} + k\dfrac{\pi}{2} \\[2ex] \dfrac{5\pi}{12} + k\dfrac{\pi}{2} \end{cases}$$

Finally, add $\frac{\pi}{12}$ to both sides.

$$t = \begin{cases} \dfrac{\pi}{6} + k\dfrac{\pi}{2} \\[2ex] \dfrac{\pi}{2} + k\dfrac{\pi}{2} \end{cases}$$ ∎

EXERCISES   *In Exercises* 11–14, *find the exact value of all solutions to each equation.*

11. $\sin 2x = \dfrac{\sqrt{2}}{2}$ **12.** $\tan\left(x + \dfrac{\pi}{6}\right) = 1$

13. $\cos 5x = 0$ **14.** $\sin 3\left(x - \dfrac{\pi}{4}\right) = -\dfrac{\sqrt{3}}{2}$

In Exercises 15–18, *find approximate values for all solutions to each equation:*

15. $\sin 2x = 0.37$ **16.** $\tan\left(x + \dfrac{\pi}{6}\right) = 1.70$

17. $\cos 5x = 0.98$ **18.** $\sin 3\left(x - \dfrac{\pi}{4}\right) = -0.81$

OTHER TRIGONOMETRIC EQUATIONS

More complicated trigonometric equations can often be reduced to one of the forms just discussed by using one or more of the following techniques.

• Use trigonometric identities to make the argument of all trigonometric functions (the quantities to which the functions are applied) the same. For example, in an equation involving both $\sin x$ and $\sin 2x$, you might replace $\sin 2x$ by $2 \sin x \cos x$.

• Use trigonometric identities to obtain an equation involving only one trigonometric function. For example, in an equation involving both $\sin x$ and $\cos^2 x$, you might replace $\cos^2 x$ by $1 - \sin^2 x$.

• Rewrite the equation so that the right side is 0, and factor the left side.

These techniques are illustrated in Examples 2 and 3.

EXAMPLE 2   *Solving a trigonometric equation*
You can solve the equation

$$8 \sin^2 x + 6 \sin x + 1 = 0$$

by regarding it as a quadratic equation in the variable $\sin x$. Factoring the left side produces

$$(2 \sin x + 1)(4 \sin x + 1) = 0$$

$$2 \sin x + 1 = 0 \quad \text{or} \quad 4 \sin x + 1 = 0$$

$$\sin x = -\frac{1}{2} \quad \text{or} \quad \sin x = -\frac{1}{4}$$

Since $\sin^{-1}\left(-\dfrac{1}{2}\right) = -\dfrac{\pi}{6}$, the solutions are

$$x = \begin{cases} -\dfrac{\pi}{6} + 2k\pi \\[2mm] \dfrac{7\pi}{6} + 2k\pi \end{cases} \qquad x = \begin{cases} \sin^{-1}\left(-\dfrac{1}{4}\right) + 2k\pi \\[2mm] \pi - \sin^{-1}\left(-\dfrac{1}{4}\right) + 2k\pi \end{cases} \blacksquare$$

It is often useful to list solutions in the interval $[0, 2\pi)$. In Example 2, $\dfrac{7\pi}{6} + 2k\pi$ and $\pi - \sin^{-1}\left(-\dfrac{1}{4}\right) + 2k\pi$ are each in $[0, 2\pi)$ if and only if $k = 0$, and $-\dfrac{\pi}{6} + 2k\pi$ and $\sin^{-1}\left(-\dfrac{1}{4}\right) + 2k\pi$ are in $[0, 2\pi)$ if and only if $k = 1$. (Convince yourself of this by using your calculator to evaluate each expression for several values of k.) The solutions in $[0, 2\pi)$ are, therefore,

$$x = \pi - \sin^{-1}\left(-\frac{1}{4}\right) \approx 3.39 \qquad x = \frac{7\pi}{6} \approx 3.67$$

$$x = -\frac{\pi}{6} + 2\pi \approx 5.76 \qquad x = \sin^{-1}\left(-\frac{1}{4}\right) + 2\pi \approx 6.03$$

EXAMPLE 3 *Solving a trigonometric equation*
To solve

$$\sin 4x = \cos^2 2x$$

begin by using a double-angle identity, so that the argument of each trigonometric function becomes $2x$.

$$2 \sin 2x \cos 2x = \cos^2 2x$$

$$2 \sin 2x \cos 2x - \cos^2 2x = 0$$

$$\cos 2x(2 \sin 2x - \cos 2x) = 0$$

$$\cos 2x = 0 \quad \text{or} \quad 2 \sin 2x - \cos 2x = 0$$

First, let's solve $\cos 2x = 0$. Since $\cos^{-1} 0 = \dfrac{\pi}{2}$, we have

$$2x = \begin{cases} \dfrac{\pi}{2} + 2k\pi \\[3mm] \dfrac{3\pi}{2} + 2k\pi \end{cases}$$

$$x = \begin{cases} \dfrac{\pi}{4} + k\pi \\[3mm] \dfrac{3\pi}{4} + k\pi \end{cases}$$

Next, let's solve $2 \sin 2x - \cos 2x = 0$. Divide both sides by $\cos 2x$ to obtain

$$2 \tan 2x - 1 = 0$$

$$\tan 2x = \frac{1}{2}$$

$$2x = \tan^{-1}\left(\frac{1}{2}\right) + k\pi$$

$$x = \frac{1}{2}\tan^{-1}\left(\frac{1}{2}\right) + \frac{k\pi}{2}$$

The solutions in $[0, 2\pi)$ are $x = \frac{1}{2}\tan^{-1}\left(\frac{1}{2}\right)$, $\frac{\pi}{4}$, $\frac{1}{2}\tan^{-1}\left(\frac{1}{2}\right) + \frac{\pi}{2}$, $\frac{3\pi}{4}$,

$\frac{1}{2}\tan^{-1}\left(\frac{1}{2}\right) + \pi$, $\frac{5\pi}{4}$, $\frac{1}{2}\tan^{-1}\left(\frac{1}{2}\right) + \frac{3\pi}{2}$, and $\frac{7\pi}{4}$.   ■

EXERCISES

19. In solving $2 \sin 2x - \cos 2x = 0$ in Example 3, we divided both sides by $\cos 2x$. As you know, dividing both sides of an equation by a variable quantity does not always produce an equivalent equation.

    a. (*Making Observations*) Verify that $\cos 2x$ and $\sin 2x$ cannot equal 0 for the same value of x.

    b. (*Writing to Learn*) Explain how your result from part (a) guarantees that $\cos 2x \neq 0$ if x is any solution to $2 \sin 2x - \cos 2x = 0$.

    c. (*Writing to Learn*) Explain how your result from part (b) guarantees that the equations $2 \sin 2x - \cos 2x = 0$ and $2 \tan 2x - 1 = 0$ are equivalent.

Solve the trigonometric equations in Exercises 20–23, and list all solutions in $[0, 2\pi)$.

20. $2 \cos^2 x - 3 \cos x + 1 = 0$

21. $\cos 2x = \sin x$
(*Hint*: Express both sides in terms of $\sin x$.)

22. $\tan^3 x - 3 \tan x = 0$

23. $5 \sin 3x - 2 \cos^2 3x + 4 = 0$
(*Hint*: Use a Pythagorean identity.)

    Now let's return to the equation from **_Artillery_**. For a shell fired with an initial velocity of 1000 ft/sec, with a desired range of 5 miles, the angle of elevation θ must satisfy the equation

$$26{,}400 \tan \theta - \frac{16}{(1000)^2}(26{,}400)^2 \sec^2 \theta = 0$$

EXERCISES   **24.** Find approximate values of all solutions to the equation

$$26{,}400 \tan \theta - \frac{16}{(1000)^2}(26{,}400)^2 \sec^2 \theta = 0$$

in the interval $\left[0, \frac{\pi}{2}\right)$.

(*Hint*: Multiply both sides by $\cos^2 \theta$; then use a double-angle identity.)

25. (*Interpreting Mathematics*) In Exercise 24, you obtained two solutions for θ. Which, if either, would be the better solution from the perspective of a gunner? Explain your answer.

ADDITIONAL EXERCISES *In Exercises 26–29, find the exact value of each solution to each equation.*

26. $\cos 3x = -1$ **27.** $\tan 2\left(x - \dfrac{\pi}{3}\right) = -\sqrt{3}$

28. $\csc 2x = 2$ $\left(Hint: \csc x = c \text{ is equivalent to } \sin x = \dfrac{1}{c}.\right)$

29. $\cot 5\left(x + \dfrac{\pi}{4}\right) = -1$

In Exercises 30–33, find approximate values of all solutions to each equation.

30. $\cos 3x = -0.42$ **31.** $\tan 2\left(x - \dfrac{\pi}{3}\right) = -0.17$

32. $\csc 2x = 3.85$ **33.** $\cot 5\left(x + \dfrac{\pi}{4}\right) = 12.68$

Solve the trigonometric equations in Exercises 34–53, and list all solutions in $[0, 2\pi)$.

34. $(2 \sin x + 1)(3 \cos x - 1) = 0$
35. $(\tan x - 1)(\sec x - 2) = 0$
36. $3 \sin^2 x + 4 \sin x + 1 = 0$
37. $2 \cos^2 x - 5 \cos x + 2 = 0$
38. $3 \tan^3 x - \tan x = 0$
39. $\sin 4x = 2 \sin^2 4x$

40. $\sin \left(x - \dfrac{\pi}{6}\right) = -\cos \left(x - \dfrac{\pi}{6}\right)$

41. $\sin^2 \dfrac{x}{2} = 3 \cos^2 \dfrac{x}{2}$

42. $\cot x = \cot (-x)$

43. $\sec x + \csc \left(\dfrac{\pi}{2} - x\right) + 4 = 0$

44. $3 \cos x + 2 \sec x = 5$
45. $2 \sin x + 3 \csc x = 7$
46. $\sin x + \sin 2x = 0$
47. $\cos x + \cos 2x = -1$
48. $\cos^2 3x = \cos 6x$
49. $\sin 10x = \sin 5x$
50. $2 \sin^4 x - 3 \sin^2 x + 1 = 0$
51. $\cot^4 x - 4 \cot^2 x + 3 = 0$
52. $2 \cos x + \tan x = \sec x$
53. $2 \sin x \tan x - \tan x + 2 \sin x - 1 = 0$

Many trigonometric equations cannot be solved exactly. In Exercises 54–57, use graphical methods to list the approximate solutions in $[0, 2\pi)$.

54. $\cos x = \cos 0.2x$ **55.** $\sec^3 x = 2 + \cos x$

56. $\sin 2x = x$ **57.** $\tan x = x^2$

CHAPTER REVIEW

Complete Exercises 1–47 (Writing to Learn) before referring to the indicated pages.

WORDS AND PHRASES

In Exercises 1–12, explain the meaning of the words or phrases in your own words.

1. (page 465) **identity, trigonometric identity, conditional equation**
2. (page 465) **fundamental identities**
3. (page 474) **proof by counterexample**
4. (page 478) **scalar quantity, magnitude, vector quantity**
5. (page 480) **representative** of a vector
6. (page 480) **tail, head** of a vector
7. (page 480) **scalar multiple** of a vector
8. (page 480) **zero vector**
9. (page 481) **vector sum**
10. (page 482) **projection, complementary projection, parallel and perpendicular components**
11. (page 483) **unit vector**
12. (page 484) **components** of a vector

IDEAS

13. (page 455) What combinations of side lengths and angles can always be used to determine the remaining side lengths and angles of a triangle?
14. (pages 457, 458) State the Law of Cosines both symbolically and in your own words.
15. (page 458) Describe how to use the Law of Cosines to determine all side lengths and angles of a triangle when the lengths of two sides and the measure of the included angle are known. Make up an example to illustrate your explanation.
16. (page 458) Describe how to use the Law of Cosines to determine all side lengths and angles of a triangle when the lengths of all three sides sides are known. Make up an example to illustrate your explanation.
17. (page 459) State the Law of Sines both symbolically and in your own words.
18. (pages 460, 461) Describe how to use the Law of Sines to determine all side lengths and angles of a triangle when the measures of two angles and the length of the side opposite one of them are known. Make up an example to illustrate your explanation.
19. (pages 460, 461) The lengths of two sides and the measure of the angle opposite one of them sometimes determines two triangles. Make up an example to illustrate this statement, and show how to find all sides and angles of both triangles. (*Hint*: Choose an acute angle B and a value for side a; then choose b with $a \sin B < b < a$.)
20. (pages 460, 461) The lengths of two sides and the measure of the angle opposite one of them sometimes determines no triangle. Make up an example to illustrate this statement, and show that no triangle is determined. (*Hint*: Choose an acute angle B and a value for side a, then choose b with $b < a \sin B$.)
21. (pages 465–476) Give an example of a trigonometric identity, and explain how you know it is an identity. Give an example of a conditional trigonometric equation, and explain how you know it is conditional.
22. (page 473) How are trigonometric identities used in problem solving?
23. (page 466) List the even-odd identities, and explain why they are so named.
24. (page 466) List the cofunction identities, and explain why they are so named.
25. (page 466) Explain why the even-odd and cofunction identities are called reflection identities. Give a specific additional example of a reflection identity, and draw a figure to indicate why it is a reflection identity.
26. (page 467) List the periodicity identities, and explain why they are so named.
27. (page 467) Explain why the periodicity identities are called rotation identities. Give a specific additional example of a rotation identity, and draw a figure to indicate why it is a rotation identity.
28. (page 468) Write down the sum and difference identities for the sine and cosine functions.

29. (page 470)   List all the double-angle identities for the sine and cosine functions, and show how they can be derived from the sum, difference, and Pythagorean identities.

30. (page 470)   Describe how to find the value of $\sin 2x$ and $\cos 2x$ if you know the value of one of the quantities $\sin x$ or $\cos x$, and the sign of the other. Make up an example to illustrate your explanation.

31. (page 471)   List the half-angle identities for the sine and cosine functions, and explain how they can be derived from the double-angle identities.

32. (pages 471, 472)   What additional information is needed to determine the values of $\sin \frac{x}{2}$ and $\cos \frac{x}{2}$ if the value of $\sin x$ or $\cos x$ is known? Make up an example to illustrate, and find the values of $\sin \frac{x}{2}$ and $\cos \frac{x}{2}$.

33. (page 473)   List the product-to-sum identities, and explain how they can be derived from the sum and difference identities.

34. (pages 473, 474)   How can numerical methods be used to prove that an equation is conditional? Why can you not use numerical methods exclusively to prove that an equation is an identity?

35. (page 474)   How can graphs help you decide whether an equation is an identity? Can a graph prove that an equation is an identity? Why or why not?

36. (pages 474–476)   Describe several analytical strategies for deciding whether an equation is an identity. Make up a trigonometric equation, and use analytical methods to decide whether it is an identity.

37. (pages 478, 479)   What is the difference between a scalar quantity and a vector quantity? Give an example of each.

38. (page 480)   What is the difference between a vector and a representative of a vector?

39. (page 480)   Draw figures, accompanied by explanations in full sentences, to illustrate the operation of multiplying a vector by a scalar. Include both positive and negative scalars.

40. (page 481)   Draw figures, accompanied by explanations in full sentences, to illustrate two geometric methods of adding vectors and two geometric methods of subtracting vectors.

41. (page 482)   What equation relates the length of a vector to the length of its projection in the direction of a second vector? Draw a figure, accompanied by an explanation, to show that the equation is valid. Make up an example to show how the equation can be used to determine lengths of projections.

What equation relates the length of a vector to the length of its complementary projection in the direction of a second vector? Draw a figure, accompanied by an explanation, to show that the equation is valid. Make up an example to show how the equation can be used to determine lengths of projections.

42. (pages 483, 484)   What are the unit vectors **i** and **j**, and why are they important?

43. (page 483)   Describe how the vectors **i** and **j** can be used to perform the operations of scalar multiplication, addition, and subtraction. Make up examples to illustrate your explanations.

44. (pages 483, 484)   Describe the process of finding the components of a vector whose magnitude and direction are known. Make up an example to illustrate your explanation.

45. (page 484)   Describe the process of finding the magnitude and direction of a vector whose components are known. Make up an example to illustrate your explanation.

46. (pages 492, 493)   Explain the use of inverse trigonometric functions in solving the equations $\sin b(x - h) = c$, $\cos b(x - h) = c$, and $\tan b(x - h) = c$. Illustrate your explanation by solving an equation of each type.

47. (pages 493, 495)   List several analytical strategies for solving trigonometric equations. Make up an equation, and use one or more of the strategies to solve it.

CHAPTER 10

TRIGONOMETRIC FUNCTIONS AS GRAPHING TOOLS

10–1 GRAPHS OF GENERAL CONIC SECTIONS

In Chapter 9 you used trigonometry as an analytical tool, exploring physical problems by establishing relationships among the trigonometric functions and by developing techniques for solving trigonometric equations. In the present chapter you will use trigonometry as a graphical tool to describe a variety of curves in the plane and to analyze physical problems related to those curves.

Let's begin by revisiting the conic sections. In Chapter 4 you learned that conic sections with horizontal or vertical axes of symmetry have equations

$$Ax^2 + Cy^2 + Dx + Ey + F = 0$$

At that time we did not discuss equations of the more general form

$$Ax^2 + Bxy + Cy^2 + Dx + Ey + F = 0$$

These equations cannot be written in the standard graphing forms of Chapter 4 unless the coefficient B is zero. In this section you will learn that they describe conic sections whose axes of symmetry are neither horizontal nor vertical, such as the ellipse you will encounter in our next Mathematical Looking Glass. To analyze such curves, you will construct a new coordinate system whose coordinate axes have been rotated to be parallel to the graph's axes of symmetry. You will then use trigonometry to obtain an equation in standard graphing form in the new coordinate system.

COORDINATE SYSTEMS RELATED BY ROTATION OF AXES

A MATHEMATICAL LOOKING GLASS
The Comet's Tale

In August 1684, the English astronomer Edmund Halley traveled from Oxford to Cambridge to consult Isaac Newton about a troublesome mathematical question. Halley had observed a bright comet in the sky in 1682, and had attempted without success to calculate its orbit. He felt that the tools he needed might be found in Newton's recent work in the area of calculus.

According to the *Encyclopaedia Britannica*, when Halley asked Newton "what would be the curve described by a planet round the sun on the assumption that the sun's force diminished as the square of the distance, Newton replied promptly, 'an ellipse.' On being questioned by Halley as to the reason for his answer he replied, 'Why, I have calculated it.'"

Newton's calculation allowed Halley to predict that his comet would return in 1759, and in fact it was "recovered" (sighted) by an amateur astronomer on Christmas day of 1758. Historical accounts indicate that Halley's comet has been seen on 28 of its transits near the earth, dating back to May 25, 240 B.C. It was last seen in 1986 and will next be seen in 2061. In *Reading the Mind of God* (Macmillan, 1989), James Trefil comments that

> Halley's prediction depended crucially on Newtonian physics, so when it was verified it provided a dramatic proof that the new ideas really did describe the universe.
>
> . . . the story of the recovery of Halley's comet is a beautiful illustration of the classical notion of scientific method. A theory was propounded, a prediction made on the basis of the theory, and the prediction was verified by observation, lending weight and credibility to the theory.

Figure 10-1a shows the orbits of the earth and Halley's comet placed in a coordinate system. (We are assuming for simplicity that the two orbits lie in the same plane.)

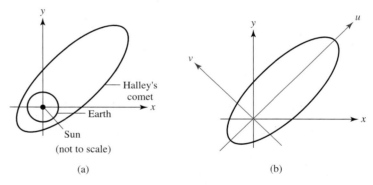

Figure 10–1

The earth's orbit is a nearly circular ellipse with the sun at one focus and its major axis on the *x*-axis. Halley's comet appears to trace out an ellipse whose major axis makes an angle of about $\frac{\pi}{4}$ with the *x*-axis. In a different coordinate system, such as the *uv*-system in Figure 10-1b, we might be able to tell whether the curve's equation describes an ellipse. ■

To find the equation of the comet's orbit in the *uv*-system, you will need to know how each point's *xy*-coordinates are related to its *uv*-coordinates. You can discover the relationship in Exercises 1–3.

EXERCISES   **1.** (*Writing to Learn*) In Figure 10-2, explain why the point *P* has *uv*-coordinates $(r \cos \phi, r \sin \phi)$ and *xy*-coordinates $(r \cos (\phi + \theta), r \sin (\phi + \theta))$.

Figure 10–2

2. (*Making Observations*) Show that the *x*-coordinate of *P* can be written as $(\cos \theta)u - (\sin \theta)v$ and the *y*-coordinate as $(\cos \theta)v + (\sin \theta)u$.
[*Hint*: Use trigonometric identities for $\cos (\phi + \theta)$ and $\sin (\phi + \theta)$.]

3. (*Making Observations*) In Exercise 2 you obtained the system of equations

$$x = (\cos \theta)u - (\sin \theta)v$$

$$y = (\cos \theta)v + (\sin \theta)u$$

Solve the system for *u* and *v* to express the *u*- and *v*-coordinates of *P* in terms of *x* and *y*.

Your results from Exercises 1–3 verify the following relationships.

Suppose that the coordinates of a point are (x, y) in the *xy*-system and (u, v) in a second system obtained by rotating the axes through a directed angle of θ. Then

$$u = x \cos \theta + y \sin \theta \qquad x = u \cos \theta - v \sin \theta$$

$$v = y \cos \theta - x \sin \theta \qquad y = v \cos \theta + u \sin \theta$$

These formulas allow you to find the *xy*-coordinates of a point whose *uv*-coordinates are given, and vice versa.

EXAMPLE 1   *Finding coordinates of a point in a system with rotated axes*

Suppose that the *xy*-system is rotated through an angle of $-\dfrac{\pi}{6}$ to obtain a *uv*-system. The point with *xy*-coordinates $(-2, 4)$ is shown in Figure 10-3.

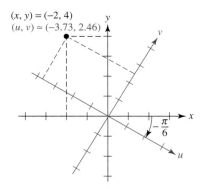

Figure 10–3

To find the point's *uv*-coordinates,

$$u = \left[\cos\left(-\frac{\pi}{6}\right)\right]x + \left[\sin\left(-\frac{\pi}{6}\right)\right]y$$

$$= \frac{\sqrt{3}}{2}(-2) + \left(-\frac{1}{2}\right)(4)$$

$$= -\sqrt{3} - 2$$

$$v = \left[\cos\left(-\frac{\pi}{6}\right)\right]y - \left[\sin\left(-\frac{\pi}{6}\right)\right]x$$

$$= \frac{\sqrt{3}}{2}(4) - \left(-\frac{1}{2}\right)(-2)$$

$$= 2\sqrt{3} - 1$$

The point has *uv*-coordinates

$$\left(-\sqrt{3} - 2, 2\sqrt{3} - 1\right) \approx (-3.73, 2.46) \qquad ■$$

EXERCISES   *In Exercises 4–7, suppose that the xy-system is rotated through a directed angle of θ to obtain a uv-system. Find the uv-coordinates of each point whose xy-coordinates are given, and vice versa. Draw both sets of coordinate axes, and plot the point.*

4. $(x, y) = (6, 0)$, $\theta = \dfrac{\pi}{4}$

5. $(x, y) = (1, \sqrt{3})$, $\theta = -\dfrac{\pi}{6}$

6. $(u, v) = (-\sqrt{3}, -1)$, $\theta = \dfrac{\pi}{6}$

7. $(u, v) = (-4, 4)$, $\theta = \dfrac{\pi}{4}$

You can use the same formulas to find the equation of a curve in a *uv*-coordinate system if you know its equation in the *xy*-system.

EXAMPLE 2   *Finding the equation of a curve in a rotated coordinate system*

Let's find the equation of the curve

$$5x^2 - 6xy + 5y^2 + 16\sqrt{2}\,x + 8 = 0$$

in a *uv*-system obtained by rotating the axes of the *xy*-system through a directed angle of $\dfrac{\pi}{4}$. We can begin by expressing x and y in terms of u and v.

$$x = \left(\cos\frac{\pi}{4}\right)u - \left(\sin\frac{\pi}{4}\right)v = \frac{\sqrt{2}}{2}(u - v)$$

$$y = \left(\cos\frac{\pi}{4}\right)v + \left(\sin\frac{\pi}{4}\right)u = \frac{\sqrt{2}}{2}(u + v)$$

Then we can replace x and y by expressions in u and v in the equation of the curve.

$$5x^2 - 6xy + 5y^2 + 16\sqrt{2}\,x + 8 = 0$$

$$5\left[\frac{\sqrt{2}}{2}(u - v)\right]^2 - 6\left[\frac{\sqrt{2}}{2}(u - v)\right]\left[\frac{\sqrt{2}}{2}(u + v)\right]$$

$$+ 5\left[\frac{\sqrt{2}}{2}(u + v)\right]^2 + 16\sqrt{2}\left[\frac{\sqrt{2}}{2}(u - v)\right] + 8 = 0$$

$$5\left(\frac{\sqrt{2}}{2}\right)^2(u - v)^2 - 6\left(\frac{\sqrt{2}}{2}\right)^2(u - v)(u + v)$$

$$+ 5\left(\frac{\sqrt{2}}{2}\right)^2(u + v)^2 + 16(u - v) + 8 = 0$$

$$\frac{5}{2}(u^2 - 2uv + v^2) - 3(u^2 - v^2) + \frac{5}{2}(u^2 + 2uv + v^2) + 16(u - v) + 8 = 0$$

$$\frac{5}{2}u^2 - 5uv + \frac{5}{2}v^2 - 3u^2 + 3v^2 + \frac{5}{2}u^2 + 5uv + \frac{5}{2}v^2 + 16u - 16v + 8 = 0$$

$$2u^2 + 8v^2 + 16u - 16v + 8 = 0$$

Since there is no *uv*-term in this equation, it can be written in standard graphing form. In Exercise 8 you can verify that the standard graphing form is

$$\frac{(u + 4)^2}{16} + \frac{(v - 1)^2}{4} = 1$$

The graph is an ellipse, as shown in Figure 10-4. ∎

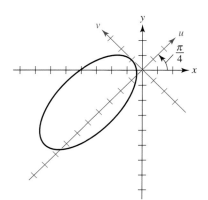

Figure 10–4

Example 2 prompts an obvious question. Is there some way to tell in advance that a rotation of $\dfrac{\pi}{4}$ produces an equation that can be written in standard graphing form? We will focus on this question after you complete Exercises 8–11.

EXERCISES   **8.** (*Review*) Verify that the equation

$$2u^2 + 8v^2 + 16u - 16v + 8 = 0$$

can be written in the standard graphing form given in Example 2.

In Exercises 9 and 10, find an equation in standard graphing form for the curve in a uv-system obtained by rotating the axes of the xy-system through a directed angle θ.

9. $xy = 4$, $\theta = \dfrac{\pi}{4}$

10. $3x^2 + 2\sqrt{3}\,xy + y^2 + 2x - 2\sqrt{3}\,y = 0$, $\theta = \dfrac{\pi}{6}$

11. a. In ***The Comet's Tale***, the orbit of Halley's comet has the equation

$$58x^2 - 102xy + 58y^2 - 42\sqrt{17}x - 42\sqrt{17}y - 147 = 0$$

Find the equation of its orbit in the *uv*-system obtained by rotating the *xy*-axes through a directed angle of $\dfrac{\pi}{4}$.

b. (*Interpreting Mathematics*) If the sun is at the origin of each coordinate system, verify that the comet's orbit is an ellipse with the sun at one focus.

CHOOSING AN APPROPRIATE ANGLE OF ROTATION

In ***The Comet's Tale***, the graph of the comet's orbit allowed us to guess that rotating the axes through an angle of $\dfrac{\pi}{4}$ would produce the equation of a conic section in standard graphing form. For other second-degree equations, there may be no obvious angle of rotation. In fact, it may not be clear that *any* rotation produces an equation in standard graphing form. In Exercises 12 and 13 you can demonstrate that a suitable angle of rotation can always be found and discover a formula for finding it.

EXERCISES   **12.** Suppose a conic section has equation

$$Ax^2 + Bxy + Cy^2 + Dx + Ey + F = 0$$

a. In a *uv*-system obtained by rotating the *xy*-system counterclockwise through an angle θ, the equation becomes

$$A(u\cos\theta - v\sin\theta)^2 + \cdots + E(v\cos\theta + u\sin\theta) + F = 0$$

Write the missing terms.

b. Combine terms involving like powers of u and v to obtain an equation

$$(A\cos^2\theta + B\cos\theta\sin\theta + C\sin^2\theta)u^2$$
$$+ \cdots + (-D\sin\theta + E\cos\theta)v + F = 0$$

Write the missing terms.

13. (*Making Observations*) The equation in Exercise 12 can be written in standard graphing form if the coefficient of *uv* is 0. Explain why this happens when $\cot 2\theta = \dfrac{A - C}{B}$.

Exercises 12 and 13 establish the following formula:

> Suppose that the *uv*-coordinate system is obtained by rotating the axes of the *xy*-system through a directed angle of θ. Then an equation
>
> $$Ax^2 + Bxy + Cy^2 + Dx + Ey + F = 0$$
>
> with $B \neq 0$ is equivalent to an equation
>
> $$A'u^2 + C'v^2 + D'u + E'v + F' = 0$$
>
> if and only if $\cot 2\theta = \dfrac{A - C}{B}$.

In Exercise 14 you can verify that $\cot 2\theta$ assumes all real values if $-\dfrac{\pi}{4} < \theta \leq \dfrac{\pi}{4}$. Thus we can always choose θ in $\left(-\dfrac{\pi}{4}, \dfrac{\pi}{4}\right]$, and it is customary to do so.

EXERCISE   **14.** (*Making Observations*) Sketch the graph of $y = \cot 2\theta$ for values of θ in $\left(-\dfrac{\pi}{4}, \dfrac{\pi}{4}\right]$. What values does y assume if θ is in $\left(-\dfrac{\pi}{4}, 0\right)$? If θ is in $\left(0, \dfrac{\pi}{4}\right]$?

EXAMPLE 3   *Simplifying the equation of a conic section through rotation of axes*
To find a *uv*-system in which

$$5x^2 - 4xy + 2y^2 + 2\sqrt{5}\,x + 4\sqrt{5}\,y + 1 = 0$$

can be written in standard graphing form, we need to rotate the axes through a directed angle of θ, where

$$\cot 2\theta = \frac{5 - 2}{-4} = -\frac{3}{4}$$

To find the equations relating x and y to u and v, we need the values of $\sin \theta$ and $\cos \theta$. We can calculate them in the following three steps.

First, recall that $\cot 2\theta = \dfrac{x}{y}$ for any point (x, y) along the terminal side of the angle 2θ in standard position. Figure 10-5 shows that the terminal side of 2θ contains the point $(3, -4)$ and that $\cos 2\theta = \dfrac{3}{5}$.

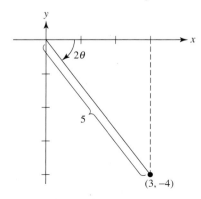

Figure 10–5

Second, use halfe-angle formulas to calculate $\sin^2 \theta = \dfrac{1 - \cos 2\theta}{2} = \dfrac{1}{5}$

and $\cos^2 \theta = \dfrac{1 + \cos 2\theta}{2} = \dfrac{4}{5}$.

Third, observe that $\cot 2\theta = -\dfrac{3}{4} < 0$, so θ is in $\left(-\dfrac{\pi}{4}, 0\right)$. Thus, $\sin \theta < 0$ and

$\cos \theta > 0$, so $\sin \theta = -\sqrt{\dfrac{1}{5}} = -\dfrac{1}{\sqrt{5}}$ and $\cos \theta = \sqrt{\dfrac{4}{5}} = \dfrac{2}{\sqrt{5}}$.

We can now express x and y in terms of u and v.

$$x = u \cos \theta - v \sin \theta = \frac{1}{\sqrt{5}}(2u + v)$$

$$y = v \cos \theta + u \sin \theta = \frac{1}{\sqrt{5}}(-u + 2v)$$

Now replace x and y by these expressions in u and v in the original equation to obtain

$$5\left[\frac{1}{5}(2u + v)^2\right] - 4\left[\frac{1}{5}(2u + v)(-u + 2v)\right] + 2\left[\frac{1}{5}(-u + 2v)^2\right]$$

$$+ 2\sqrt{5}\left(\frac{1}{\sqrt{5}}\right)(2u + v) + 4\sqrt{5}\left(\frac{1}{\sqrt{5}}\right)(-u + 2v) + 1 = 0$$

$$(4u^2 + 4uv + v^2) - \frac{4}{5}(-2u^2 + 3uv + 2v^2) + \frac{2}{5}(u^2 - 4uv + 4v^2)$$

$$+ 2(2u + v) + 4(-u + 2v) + 1 = 0 = 6u^2 + v^2 + 10v + 1 = 0$$

In standard graphing form, this becomes

$$\frac{u^2}{4} + \frac{(v + 5)^2}{24} = 1$$

To sketch the graph, we first sketch the uv-axes as in Figure 10-6, using the fact that $\tan \theta = \dfrac{\sin \theta}{\cos \theta} = -\dfrac{1}{2}$.

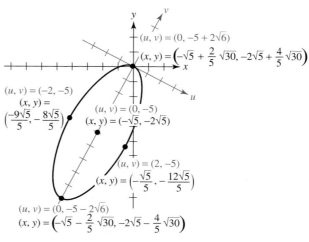

$(u, v) = (0, -5 + 2\sqrt{6})$

$(x, y) = \left(-\sqrt{5} + \dfrac{2}{5}\sqrt{30}, -2\sqrt{5} + \dfrac{4}{5}\sqrt{30}\right)$

$(u, v) = (-2, -5)$

$(x, y) = \left(\dfrac{-9\sqrt{5}}{5}, -\dfrac{8\sqrt{5}}{5}\right)$

$(u, v) = (0, -5)$

$(x, y) = (-\sqrt{5}, -2\sqrt{5})$

$(u, v) = (2, -5)$

$(x, y) = \left(-\dfrac{\sqrt{5}}{5}, -\dfrac{12\sqrt{5}}{5}\right)$

$(u, v) = (0, -5 - 2\sqrt{6})$

$(x, y) = \left(-\sqrt{5} - \dfrac{2}{5}\sqrt{30}, -2\sqrt{5} - \dfrac{4}{5}\sqrt{30}\right)$

Figure 10–6

The xy-coordinates of the important points on the graph can be obtained by substituting the uv-coordinates into the expressions $x = \dfrac{1}{\sqrt{5}}(2u + v)$ and $y = \dfrac{1}{\sqrt{5}}(-u + 2v)$. ∎

SUMMARY

Use the following steps to rewrite this equation in standard graphing form:
$Ax^2 + Bxy + Cy^2 + Dx + Ey + F = 0$.

- Use the equation $\cot 2\theta = \dfrac{A - C}{B}$ to find the value of θ in $\left(-\dfrac{\pi}{4}, \dfrac{\pi}{4}\right]$ that produces an appropriate uv-system.
- Draw the angle 2θ in standard position to find $\cos 2\theta$.
- Use half-angle identities to find $\sin^2 \theta$ and $\cos^2 \theta$.
- Take square roots, and determine the signs of $\sin \theta$ and $\cos \theta$.
- Use the equations

$$x = u \cos \theta - v \sin \theta$$
$$y = v \cos \theta + u \sin \theta$$

to express x and y in terms of u and v.
- Replace x and y by expressions in u and v in the original equation and simplify.
- Write the equation in u and v in standard graphing form.

EXERCISES

In Exercises 15–18, sketch the graph of the equation and identify all applicable important features (center, vertex or vertices, covertices, asymptotes) in terms of the xy-system.

15. $xy + 9 = 0$

16. $3x^2 - 2\sqrt{3}\,xy + y^2 + 2x + 2\sqrt{3}\,y = 0$

17. $x^2 + xy + y^2 + 2x - 2y = 0$

18. $3x^2 + 4xy - \dfrac{4}{\sqrt{5}}x + \dfrac{8}{\sqrt{5}}y - 20 = 0$

ADDITIONAL EXERCISES

In Exercises 19–26, suppose that the uv-coordinate system is obtained by rotating the axes of the xy-coordinate system through a directed angle of θ. In Exercises 19–22, find the uv-coordinates of each point whose xy-coordinates are given. Draw both sets of coordinate axes and plot the point.

19. $(x, y) = (-3, 3),\ \theta = \dfrac{\pi}{4}$

20. $(x, y) = (0, -5),\ \theta = -\dfrac{\pi}{6}$

21. $(x, y) = (2, 0),\ \theta = \dfrac{\pi}{8}$

(*Hint*: Use a half-angle identity to find $\sin \theta$ and $\cos \theta$.)

22. $(x, y) = (-8, -6),\ \theta = \sin^{-1}\dfrac{3}{5}$

In Exercises 23–26, find the xy-coordinates of each point whose uv-coordinates are given. Draw both sets of coordinate axes and plot the point.

23. $(u, v) = (4, 0),\ \theta = -\dfrac{\pi}{6}$

24. $(u, v) = (-\sqrt{2}, \sqrt{2}),\ \theta = \dfrac{\pi}{4}$

25. $(u, v) = (0, -2)$, $\theta = \dfrac{5\pi}{12}$ **26.** $(u, v) = (-8, -6)$, $\theta = \sin^{-1}\dfrac{3}{5}$

In Exercises 27–34, sketch the graph of the equation, identify the type of curve (parabola, ellipse, or hyperbola), and identify all applicable important features (center, vertex or vertices, covertices, asymptotes) in terms of the xy-system.

27. $xy + 4x + 5y + 10 = 0$

28. $\dfrac{1}{2}x^2 + xy + \dfrac{1}{2}y^2 + 4x + 5y + 10 = 0$

29. $x^2 + xy + y^2 + 4x + 5y + 10 = 0$

30. $x^2 - xy + y^2 - 12 = 0$

31. $x^2 + 10\sqrt{3}\,xy + 11y^2 - 64 = 0$

32. $x^2 + 2\sqrt{3}\,xy - y^2 + 2 = 0$

33. $9x^2 - 12xy + 4y^2 + 78x + 26y + 13 = 0$

34. $40x^2 - 36xy + 25y^2 - 8x - 12y = 0$

SUPPLEMENTARY TOPIC

The Invariance of $B^2 - 4AC$

In Section 4-4, you learned that the value of AC determines whether the graph of $Ax^2 + Cy^2 + Dx + Ey + F = 0$ is an ellipse, a parabola, or a hyperbola. The expression $B^2 - 4AC$ serves a similar purpose for equations of general conic sections.

> The graph of $Ax^2 + Bxy + Cy^2 + Dx + Ey + F = 0$ is
>
> an ellipse if $B^2 - 4AC < 0$
> a parabola if $B^2 - 4AC = 0$
> a hyperbola if $B^2 - 4AC > 0$
>
> In all cases, the graph may be a degenerate form (one or two lines, a point, or an empty graph).

Here is a brief and incomplete explanation of why the foregoing statement is true.

First, suppose that a *uv*-coordinate system is obtained by rotating the *xy*-axes and that the equation in the *uv*-system becomes

$$A'u^2 + B'uv + C'v^2 + D'u + E'v + F' = 0$$

It can be shown that $(B')^2 - 4A'C' = B^2 - 4AC$. (This fact is described by saying that the quantity $B^2 - 4AC$ is **invariant under rotation**.)

Next, observe that it is always possible to choose a *uv*-coordinate system in which $B' = 0$. In such a system, $(B')^2 - 4A'C' = -4A'C'$. If this expression is negative, then $A'C' > 0$, so the graph is an ellipse. Similarly, the graph is a parabola if $(B')^2 - 4A'C' = 0$ and a hyperbola if $(B')^2 - 4A'C' > 0$.

EXERCISE

35. (*Making Observations*)

    a. In Example 2, find the value of $B^2 - 4AC$ in the *xy*-equation, and verify that it is equal to the value of $(B')^2 - 4A'C'$ in the *uv*-equation.

    b. Repeat part (a) for Example 3.

10–2 PARAMETRIC EQUATIONS FOR CONIC SECTIONS

CIRCULAR MOTION

In Section 2-6, you learned that equations of the form $x = f(t)$, $y = g(t)$ are called **parametric equations** for a curve. You have already used parametric equations to describe lines (Section 2-6) and parabolas (Section 4-2). In this section, you will learn how to describe circles and other conic sections parametrically by using trigonometric functions of the parameter t. Those descriptions can be used to gain insight into the motion of a variety of physical objects.

A MATHEMATICAL LOOKING GLASS
The Mad Tea Party

Two of the best known children's books in the English language were written by Charles Dodgson (1832–1898), a lecturer in mathematics at Christ Church, Oxford. Dodgson enjoyed telling stories for the amusement of his friends' children, and at the urging of young Alice Liddell, wrote down an expanded version of one of his best tales. *Alice's Adventures in Wonderland* and its sequel, *Through the Looking Glass*, were written under the pen name Lewis Carroll. Alice's adventures were further popularized by Walt Disney in a 1951 animated film.

If Mr. Dodgson were alive today, he could visit Disney World in Orlando, FL, to see a ride based on one of his characters. According to Disney World's description, riders can "whirl in the Mad Hatter's giant teacups" on The Mad Tea Party. The ride consists of a large circular platform that rotates counterclockwise about its center, as shown in Figure 10-7. At the same time, five smaller circular platforms rotate clockwise about their centers. On each small platform, four teacup-shaped seats can rotate about their centers at the discretion of the riders.

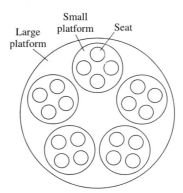

Figure 10–7

The rapid spinning and sudden changes in direction create a motion of varying velocity and acceleration along a complicated path. Engineers designing such a ride must make sure it does not pose a danger to riders. Let's see how they might begin to analyze the motion by constructing parametric equations for the rider's path. Since the platforms and the seats all rotate in circles, let's begin by constructing parametric equations to describe circular motion.

As you discovered in **Fermilab 2** (page 391), the path of an object that starts at $(1, 0)$ and travels counterclockwise at 1 radian per second around the unit circle can be described by the equations $x = \cos t, y = \sin t$. You can use this discovery and the strategy of *Examining a Related Problem* to describe other types of circular motion.

EXERCISES 1. (*Review*) Suppose that an object starts at $(1, 0)$ and travels counterclockwise around the unit circle at an angular velocity of 1 radian per second. Find the coordinates of the object after

    a. $\dfrac{\pi}{4}$ seconds   **b.** $\dfrac{2\pi}{3}$ seconds

    c. π seconds   **d.** 2.5 seconds

2. (*Review*) Explain why the path of the object in Exercise 1 can be described by the parametric equations $x = \cos t, y = \sin t$.

3. (*Review*) Suppose an object starts at $(r, 0)$ and travels counterclockwise around the circle of radius r centered at the origin with an angular velocity of 1 radian per second. Explain why its path can be described by the equations $x = r \cos t, \ y = r \sin t$. (*Hint:* See Exercise 18 of Section 8-1.)

Your observations in Exercises 1–3 can be generalized to describe motion with any uniform velocity in either direction around any circle centered at the origin. In the case where the motion begins on the positive x-axis, you can provide the details in Exercise 4, following Example 1.

EXAMPLE 1 *Parametric equations for motion around a circle centered at the origin*

    a. Suppose an object starts at $(3, 0)$ and travels around the circle of radius 3 centered at the origin with an angular velocity of 2 radians per second. After t seconds, the object has moved counterclockwise through a directed angle of $2t$ radians, as in Figure 10-8a. Its coordinates are $(3 \cos 2t, 3 \sin 2t)$, so it can be described by the equations

$$x = 3 \cos 2t$$

$$y = 3 \sin 2t$$

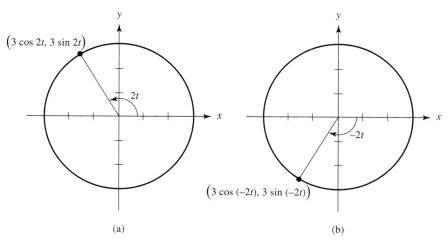

(a)   (b)

Figure 10–8

b. If the object travels with angular velocity 2 radians per second clockwise, as in Figure 10-8b, it moves through a directed angle of $-2t$ radians in t seconds. Therefore, it can be described by the equations

$$x = 3 \cos(-2t)$$

$$y = 3 \sin(-2t)$$

or, equivalently,

$$x = 3 \cos 2t$$

$$y = -3 \sin 2t$$ ∎

EXERCISES   **4.** (*Writing to Learn*) Suppose an object starts at $(r, 0)$ and travels around the circle of radius r centered at the origin.

   a. Explain why its path can be described by the equations $x = r \cos bt$, $y = r \sin bt$ if its angular velocity is b radians per second.

   b. Explain why its path can be described by the equations $x = r \cos bt$, $y = -r \sin bt$ if its angular velocity is $-b$ radians per second.

5. Write parametric equations to describe the motion of an object with angular velocity b radians per second around the circle of radius r centered at the origin, beginning at $(r, 0)$.

   a. $r = 10, b = 5$ **b.** $r = 10, b = -5$

   c. $r = 0.5, b = -\dfrac{\pi}{2}$ **d.** $r = 2, b = 2\pi$

   In Exercise 6, following Example 2, you can explain how to describe motion with a uniform velocity around a circle centered at an arbitrary point in the plane, as long as the motion begins at the rightmost point of the circle.

EXAMPLE 2   *Parametric equations for motion around an arbitrary circle*

Suppose an object begins at $(7, 3)$ and travels around the circle of radius 2 centered at $(5, 3)$ with an angular velocity of 4 radians per second. To obtain parametric equations for its motion, let's imagine a second object traveling in a similar manner around the circle of radius 2 centered at the origin, as shown in Figure 10-9.

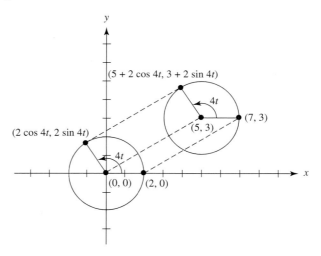

Figure 10–9

After t seconds, the second object is at $(2 \cos 4t, 2 \sin 4t)$. The first object is always 5 units farther right and 3 units farther up, so its coordinates are $(5 + 2 \cos 4t, 3 + 2 \sin 4t)$. Thus, it can be described by the equations

$$x = 5 + 2 \cos 4t$$
$$y = 3 + 2 \sin 4t$$ ∎

EXERCISES

6. (*Making Observations*) Suppose an object starts at $(x_0 + r, y_0)$ and travels around the circle of radius r centered at (x_0, y_0).

    a. Explain why its path can be described by the following equations: $x = x_0 + r \cos bt$ and $y = y_0 + r \sin bt$ if its angular velocity is b radians per second.

    b. Explain why its path can be described by the following equations: $x = x_0 + r \cos bt$, $y = y_0 - r \sin bt$ if its angular velocity is $-b$ radians per second.

7. Write parametric equations to describe the motion of an object with angular velocity b radians per second around the circle of radius r centered at (x_0, y_0), beginning at $(x_0 + r, y_0)$. Sketch a graph of the motion for $0 \le t \le 1$.

    a. $r = 4, b = \dfrac{\pi}{2},\ (x_0, y_0) = (-1, 2)$

    b. $r = 1, b = -\pi,\ (x_0, y_0) = (0, 3)$

    c. $r = 1, b = \pi,\ (x_0, y_0) = (0, 3)$

    d. $r = 3, b = \dfrac{\pi}{3},\ (x_0, y_0) = (2, -4)$

The equations in Exercise 6 can be modified to obtain parametric equations for any circular motion with a uniform velocity.

EXAMPLE 3   *Finding parametric equations to describe circular motion*

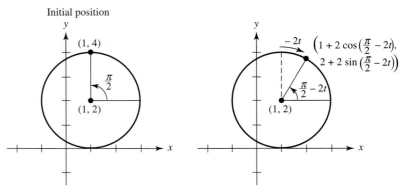

Figure 10–10

Suppose an object starts at $(1, 4)$ and travels with angular velocity -2 radians per second around the circle of radius 2 centered at $(1, 2)$, as in Figure 10-10.

After t seconds, the radius to the object makes a directed angle of $\frac{\pi}{2} - 2t$ with the positive x-direction, so its motion can be described by the equations

$$x = 1 + 2 \cos\left(\frac{\pi}{2} - 2t\right)$$

$$y = 2 + 2 \sin\left(\frac{\pi}{2} - 2t\right)$$

The identities $\cos\left(\frac{\pi}{2} - \theta\right) = \sin\theta$ and $\sin\left(\frac{\pi}{2} - \theta\right) = \cos\theta$ can be used with $\theta = 2t$ to rewrite the equations as

$$x = 1 + 2 \sin 2t$$

$$y = 2 + 2 \cos 2t \qquad \blacksquare$$

EXERCISES   **8.** Graph the equations in Example 3 for $0 \leq t \leq \pi$, and trace along the graph to estimate the position of the object after

    a. $\frac{\pi}{4}$ seconds   **b.** $\frac{2\pi}{3}$ seconds

    c. π seconds   **d.** 2.5 seconds

In Exercises 9–12, find parametric equations for the motion described.

9. at angular velocity of 1 radian per second around the circle of radius 5 centered at $(-3, 0)$, starting at $(2, 0)$

10. at angular velocity of 10 radians per second around the circle of radius 7 centered at $(1, -1)$, starting at $(-6, -1)$

11. at angular velocity of 0.5 radians per second around the circle of radius 1 centered at $(-4, -2)$, starting at $(-3, -2)$

12. at angular velocity of 3 radians per second around the circle of radius 4 centered at $(0, 0)$, starting at $(0, -4)$

If you know the parametric equations describing the circular motion of an object, you can identify its starting point, as well as the direction and speed of the motion.

EXAMPLE 4   *Identifying circular motion described by parametric equations*

The unit circle can be parametrized by the equations

$$x = \cos\left(\frac{\pi}{4} - 2t\right)$$

$$y = \sin\left(\frac{\pi}{4} - 2t\right)$$

If t represents time in seconds, what can we say about the motion described by the equations?

First, when $t = 0$, the object is at $\left(\cos\frac{\pi}{4}, \sin\frac{\pi}{4}\right) = \left(\frac{\sqrt{2}}{2}, \frac{\sqrt{2}}{2}\right)$, so the radius to the object makes an angle of $\frac{\pi}{4}$ with the positive x-direction.

Second, since $\cos\left(\dfrac{\pi}{4} - 2t\right)$ and $\sin\left(\dfrac{\pi}{4} - 2t\right)$ each have a period of $\dfrac{2\pi}{|-2|} = \pi$, the object returns to the same position every π seconds.

Third, after t seconds, the radius to the object makes a directed angle of $\dfrac{\pi}{4} - 2t$ with the positive x-direction. Since the angle decreases as t increases, the motion is clockwise. The motion is shown in Figure 10-11.

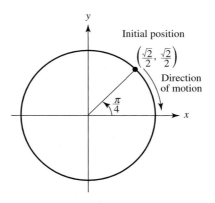

Figure 10–11 ■

EXERCISES   *Each set of equations in Exercises 13–16 describes the motion of an object around the unit circle. In each case t is measured in seconds.*

    a. Tell where the object is when $t = 0$.

    b. Tell how often the object returns to the same position.

    c. Tell whether the motion is clockwise or counterclockwise.

13. $x = \cos\left(4t - \dfrac{\pi}{3}\right)$ **14.** $x = \cos\left(\dfrac{\pi}{6} - t\right)$

    $y = \sin\left(4t - \dfrac{\pi}{3}\right)$ $y = \sin\left(\dfrac{\pi}{6} - t\right)$

15. $x = \cos \pi t$ **16.** $x = \sin 2\pi t$

    $y = -\sin \pi t$ $y = \cos 2\pi t$

Now let's return to the problem of describing the rider's motion in **The Mad Tea Party**. To simplify our task, let's consider a rider who does not rotate her teacup-shaped seat. In this case, her motion is produced by the revolution of the seat about the center of the small platform and the revolution of the small platform about the center of the large platform. Proceeding by *Breaking the Problem into Parts*, let's consider these two motions one at a time.

Suppose that the distance between the centers of the large and small platforms is R, and the distance between the center of the small platform and the seat is r. Figure 10-12a shows a coordinate system with its origin at the center of the large platform. The rider is starting at $(R + r, 0)$. In Figure 10-12b, the large platform rotates counterclockwise at a radians per second, and the small one rotates clockwise at b radians per second.

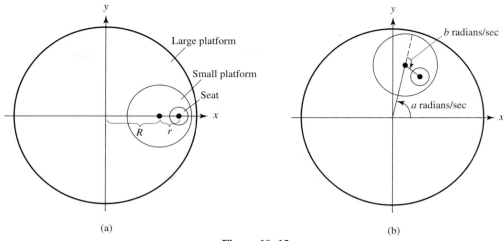

(a)   (b)

Figure 10–12

EXERCISES   **17.** (*Creating Models*) Explain why the motion of the center of the small platform can be described by the equations

$$x = R \cos at$$

$$y = R \sin at$$

18. (*Creating Models*) When the center of the small platform is at (x_0, y_0), explain why the rider's seat has coordinates $(x_0 + r \cos bt, y_0 - r \sin bt)$.

19. (*Creating Models*) Combine your results from Exercises 17 and 18 to obtain parametric equations in terms of R, r, a, and b for the position of the seat after t seconds.

20. (*Interpreting Mathematics*) Suppose $R = 50$ feet and $r = 10$ feet. For the given values for a and b, graph the rider's motion in the window $[-120, 120]$ by $[-60, 60]$ over an interval of 65 seconds, and describe the motion in words.

    a. $a = 1, b = 1$ **b.** $a = 0.1, b = 1$

    c. $a = 1, b = 0.1$

21. (*Writing to Learn*) Which combination of a and b in Exercise 20 do you think would provide the best ride? Why?

Your graph in Exercise 20a appeared to be an ellipse. Is it? Let's find out by discovering what parametric equations describe ellipses.

PARAMETRIC EQUATIONS FOR ELLIPSES

Recall that the graph of every ellipse with horizontal and vertical axes of symmetry can be obtained by performing graphical transformations on the unit circle. The equation of the ellipse can be obtained by performing corresponding operations on the equation of the unit circle. For example, to obtain the ellipse with vertices at $(\pm3, 0)$ and covertices at $(0, \pm2)$, we stretch the unit circle horizontally by a factor of 3 and vertically by a factor of 2.

In the equation of the unit circle we replace x and y by $\dfrac{x}{3}$ and $\dfrac{y}{2}$, as in Figure 10-13.

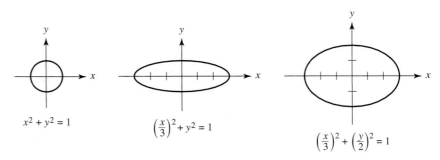

Figure 10–13

If we make the same replacements in the *parametric* equations of the unit circle, we should obtain *parametric* equations for the ellipse. You can confirm this in Exercises 22 and 23.

EXERCISES **22.** (*Making Observations*) In the equations $x = \cos t, y = \sin t$, replace x and y by $\dfrac{x}{3}$ and $\dfrac{y}{2}$, and graph the resulting equations. Does the graph appear to be the rightmost ellipse in Figure 10-13?

23. (*Making Observations*) Eliminate the parameter from your equations in Exercise 22 as follows:

    a. Solve the first equation for $\cos t$ and the second equation for $\sin t$.

    b. Use the identity $\cos^2 t + \sin^2 t = 1$ to write an equation in x and y.

    c. Verify that your equation is equivalent to that of the ellipse in Figure 10-13,

$$\frac{x^2}{9} + \frac{y^2}{4} = 1$$

Now generalize your work from Exercises 22 and 23 by completing Exercises 24 and 25.

24. (*Making Observations*) In the equations $x = \cos t, y = \sin t$, replace x and y by $\dfrac{x - h}{a}$ and $\dfrac{y - k}{b}$. Verify that the equations you obtain are equivalent to

$$x = h + a \cos t$$
$$y = k + b \sin t$$

25. (*Making Observations*) Eliminate the parameter from your equations in Exercise 24 by solving the first equation for $\cos t$ and the second for $\sin t$ and then using the identity $\cos^2 t + \sin^2 t = 1$ to obtain an equation in x and y. Verify that your equation is equivalent to

$$\frac{(x - h)^2}{a^2} + \frac{(y - k)^2}{b^2} = 1$$

which is the equation of an ellipse centered at (h, k) with vertices at $(h \pm a, k)$ and covertices at $(h, k \pm a)$.

In Exercises 22–25, you have verified the following relationship between rectangular and parametric equations for ellipses.

Equations for Ellipses

Rectangular Equation	Parametric Equations
$\dfrac{(x - h)^2}{a^2} + \dfrac{(y - k)^2}{b^2} = 1$	$x = h + a \cos t, y = k + b \sin t$
$\dfrac{(x - h)^2}{b^2} + \dfrac{(y - k)^2}{a^2} = 1$	$x = h + b \cos t, y = k + a \sin t$

This relationship allows you to find parametric equations for an ellipse if you know its rectangular equation, or vice versa.

EXAMPLE 5   *Finding parametric equations for an ellipse*

a. For the ellipse

$$\frac{(x + 5)^2}{9} + \frac{y^2}{4} = 1$$

$h = -5, k = 0, a = 3$, and $b = 2$. Thus, the ellipse has parametric equations

$$x = -5 + 3 \cos t$$

$$y = 2 \sin t$$

b. For the circle

$$(x - 8)^2 + (y - 5)^2 = 81$$

which can be rewritten as

$$\frac{(x - 8)^2}{81} + \frac{(y - 5)^2}{81} = 1$$

$h = 8, k = 5, a = 9$, and $b = 9$. Thus, the circle has parametric equations

$$x = 8 + 9 \cos t$$

$$y = 5 + 9 \sin t$$ ∎

EXAMPLE 6   *Eliminating the parameter to obtain the rectangular equation of an ellipse*
The parametric equations

$$x = -2 + 5 \cos t$$

$$y = 10 \sin t$$

are equivalent to

$$\cos t = \frac{x + 2}{5}$$

$$\sin t = \frac{y}{10}$$

The identity $\cos^2 t + \sin^2 t = 1$ can then be used to obtain the equation of an ellipse.

$$\left(\frac{x+2}{5}\right)^2 + \left(\frac{y}{10}\right)^2 = 1$$

Equivalently,

$$\frac{(x+2)^2}{25} + \frac{y^2}{100} = 1 \qquad \blacksquare$$

EXERCISES *In Exercises 26–29, find parametric equations for the ellipse. Check your answer by graphing the parametric equations on your calculator.*

26. $\dfrac{x^2}{25} + \dfrac{y^2}{64} = 1$ 　　　　　**27.** $4x^2 + 25y^2 = 100$

28. $\dfrac{(x+2)^2}{9} + (y-4)^2 = 1$

29. $\dfrac{(x-1)^2}{20} + \dfrac{(y+3)^2}{10} = 1$

In Exercises 30 and 31, state the coordinates of the center, vertices, and covertices of the ellipse, and write an equation in x and y to describe it.

30. $x = 4 + 6\cos t$ 　　**31.** $x = 57 + 31\cos t$
　　　$y = 1 + 5\sin t$ 　　　　　$y = 2.8 + 28\sin t$

32. Your equations in Exercise 20a, describing one possible motion for a rider on **The Mad Tea Party**, were

$$x = 60\cos t$$
$$y = 40\sin t$$

You now know that these equations describe an ellipse. Eliminate the parameter, and write the equation of the ellipse.

33. (*Making Observations*) In **The Mad Tea Party**, if the large and small platforms rotate at the same rate in radians per second, will the rider's motion always be elliptical? Explain why or why not.

PARAMETRIC EQUATIONS FOR HYPERBOLAS

To find parametric equations for an ellipse, we used the following identity: $\sin^2 t + \cos^2 t = 1$. To find parametric equations for the simplest hyperbola

$$x^2 - y^2 = 1$$

we might begin by trying to think of an identity involving the difference of the squares of two trigonometric functions. By *Guessing and Checking*, we might try the identity

$$\sec^2 t - \tan^2 t = 1$$

which suggests the equations

$$x = \sec t$$
$$y = \tan t$$

If we use the identity to eliminate the parameter, we obtain $x^2 - y^2 = 1$, so the parametric equations describe the hyperbola. Similarly, the hyperbola $y^2 - x^2 = 1$ can be described by the equations $x = \tan t, y = \sec t$.

The identity $\sec^2 t - \tan^2 t = 1$ can also be used to parametrize any hyperbola with a horizontal or vertical transverse axis.

EXAMPLE 7   *Finding parametric equations for a hyperbola*
To obtain parametric equations for the hyperbola

$$\frac{(x + 4)^2}{4} - \frac{(y - 1)^2}{5} = 1$$

rewrite its equation as

$$\left(\frac{x + 2}{2}\right)^2 - \left(\frac{y - 1}{\sqrt{5}}\right)^2 = 1$$

Then in the identity $\sec^2 t - \tan^2 t = 1$, make the replacements

$$\sec t = \frac{x + 2}{2}$$

$$\tan t = \frac{y - 1}{\sqrt{5}}$$

Finally, solve the two equations for x and y, respectively, to obtain

$$x = -2 + 2 \sec t$$
$$y = 1 + \sqrt{5} \tan t$$

To eliminate the parameter and obtain the equation of the hyperbola in standard graphing form, just reverse the procedure in Example 7.

EXAMPLE 8   *Eliminating the parameter to obtain the equation of a hyperbola*
To verify that the equations

$$x = 2 - 5 \tan\left(t - \frac{\pi}{6}\right)$$

$$y = 3 + 2 \sec\left(t - \frac{\pi}{6}\right)$$

describe a hyperbola, begin by solving the first equation for $\tan\left(t - \frac{\pi}{6}\right)$ and the second for $\sec\left(t - \frac{\pi}{6}\right)$.

$$\tan\left(t - \frac{\pi}{6}\right) = -\frac{x - 2}{5}$$

$$\sec\left(t - \frac{\pi}{6}\right) = \frac{y - 3}{2}$$

Then, since

$$\sec^2\left(t - \frac{\pi}{6}\right) - \tan^2\left(t - \frac{\pi}{6}\right) = 1$$

it must also be true that

$$\left(\frac{y-3}{2}\right)^2 - \left(-\frac{x-2}{5}\right)^2 = 1$$

The equation of the hyperbola in standard graphing form is

$$\frac{(y-3)^2}{4} - \frac{(x-2)^2}{25} = 1$$ ∎

EXERCISES   *In Exercises 34–37, find parametric equations for the hyperbola. Check your answer by graphing the parametric equations on your calculator.*

34. $\dfrac{x^2}{25} - \dfrac{y^2}{64} = 1$ **35.** $25y^2 - 4x^2 = 100$

36. $(y-4)^2 - \dfrac{(x+2)^2}{9} = 1$ **37.** $\dfrac{(x-1)^2}{20} - \dfrac{(y+3)^2}{10} = 1$

In Exercises 38–41, eliminate the parameter and write the equation of the hyperbola in standard graphing form.

38. $x = 2 \sec 5t$
$y = 3 \tan 5t$

39. $x = 1 + \tan\left(t - \dfrac{\pi}{4}\right)$

$y = 4 + \sec\left(t - \dfrac{\pi}{4}\right)$

40. $x = 8 - \tan 2t$ **41.** $x = 3 \sec (t^3 - 4)$
$y = 5 + 2 \sec 2t$ $y = 1 - \tan (t^3 - 4)$

ADDITIONAL EXERCISES   *In Exercises 42–45, find parametric equations for the circle.*

42. centered at $(6, 1)$, radius 5   **43.** centered at $(0, -4)$, radius 4
44. $(x + 1)^2 + y^2 = 100$ **45.** $(x - 2)^2 + (y - 7)^2 = 64$

In Exercises 46 and 47, identify the center and radius of the circle, and write an equation in x and y to describe it.

46. $x = 3 \cos t$ **47.** $x = -8 + \cos t$
$y = 3 \sin t$ $y = -3 + \sin t$

In Exercises 48–57, write parametric equations for the ellipse or hyperbola.

48. $\dfrac{x^2}{7} + \dfrac{y^2}{10} = 1$ **49.** $\dfrac{x^2}{16} - \dfrac{y^2}{16} = 1$

50. $(x - 6)^2 + (y + 6)^2 = 4$
51. $(x - 6)^2 - (y - 6)^2 = 4$
52. $\dfrac{(x - 0.2)^2}{4} - \dfrac{(y - 0.5)^2}{9} = 1$
53. $\dfrac{(x + 95)^2}{18} + \dfrac{(y + 76)^2}{50} = 1$

54. $x^2 + 9y^2 = 900$ **55.** $9x^2 + 9y^2 = 900$

56. $9x^2 - y^2 = 900$ **57.** $9x^2 - y^2 = -900$

In Exercises 58–65, eliminate the parameter and write the equation in standard graphing form.

58. $x = 4 + \cos t$ **59.** $x = 4 + 2 \cos t$
   $y = 7 + \sin t$ $y = 7 + 2 \sin t$

60. $x = 4 + \cos t$ **61.** $x = 4 + 2 \tan t$
   $y = 7 + 2 \sin t$ $y = 7 + \sec t$

62. $x = 9 - 5 \sin t$ **63.** $x = 3 \sin (t - 1)$ ·
   $y = 1 - 6 \cos t$ $y = 4 \cos (t - 1)$

64. $x = 10 - \sec (2t - \pi)$ **65.** $x = 12 \tan 3t$
   $y = 15 + \tan (2t - \pi)$ $y = 5 \sec 3t$

Each set of equations in Exercises 66–69 describes the motion of an object around the unit circle. In each case,

   a. Tell where the object is when $t = 0$.

   b. Tell how often the object returns to the same position.

   c. Tell whether the motion is clockwise or counterclockwise.

66. $x = \cos 6t$ **67.** $x = \sin 6t$
   $y = \sin 6t$ $y = \cos 6t$

68. $x = \sin (3t + \pi)$ **69.** $x = -\sin (3t + \pi)$
   $y = \cos (3t + \pi)$ $y = -\cos (3t + \pi)$

Each set of equations in Exercises 70 and 71 describes the motion of an object around an ellipse. In each case,

   a. Write the equation of the ellipse in standard graphing form.

   b. Tell where the object is when $t = 0$.

   c. Tell how often the object returns to the same position.

   d. Tell whether the motion is clockwise or counterclockwise.

70. $x = -4 + 3 \cos 2t$ **71.** $x = 5 \sin 4t$
   $y = 6 + 4 \sin 2t$ $y = 10 \cos 4t$

In Exercises 72–75, find parametric equations for the motion described.

72. at an angular velocity of 3 radians per second around the circle of radius 2 centered at $(0, 2)$, starting at $(0, 0)$

73. at an angular velocity of π radians per second around the circle of radius 5 centered at $(-3, -4)$, starting at $(-8, -4)$

74. at an angular velocity of $\dfrac{\pi}{2}$ radians per second around the circle of radius 0.4 centered at $(-2, 1)$, starting at $(-2, 1.4)$

75. at an angular velocity of 3 radians per second around the circle of radius 6 centered at $(7, 0)$, starting at $(7, 6)$

76. (*Writing to Learn*) The equations $x = \sec t$, $y = \tan t$ describe the motion of an object along the hyperbola $x^2 - y^2 = 1$. Describe the object's motion during the time interval $\left(-\dfrac{\pi}{2}, \dfrac{\pi}{2}\right)$. In particular, tell where the object is when t is near $-\dfrac{\pi}{2}$, when $-\dfrac{\pi}{2} < t < 0$, when $t = 0$, when $0 < t < \dfrac{\pi}{2}$, and when t is near $\dfrac{\pi}{2}$.

10-3   THE POLAR COORDINATE SYSTEM

Throughout this book, you have sketched graphs and written their equations using the rectangular coordinate system. You are about to learn of another system, the **polar coordinate system**, that relies on angles and straight-line distances to identify each point in the plane. The following Mathematical Looking Glass illustrates some differences between the rectangular and polar coordinate systems.

A MATHEMATICAL LOOKING GLASS
A Tale of Two Cities

Figure 10-14a shows a map of Phoenix, AZ. Phoenix occupies a site inhabited in prehistoric times by the Hohokam tribe, who grew crops in the desert by building a system of irrigation ditches. The Hohokams disappeared mysteriously by the fifteenth century, but in 1867 their canals were rebuilt by a group of settlers. Predicting that a city would rise from the ruins of the old canals, Englishman "Lord Darrell" Duppa named it after the legendary phoenix, a bird that destroyed itself by fire after a 500-year lifespan and then rose from the ashes to begin a new life cycle.

The streets of Phoenix were laid out to create simple routes between any two points in the city. The main streets are a rectangular grid, at intervals of one mile in each direction, resembling a rectangular coordinate system. For example, you could give someone directions from 59th and McDowell (the origin) to 35th and Glendale by saying, "Go 3 miles east and 5 miles north." In the language of rectangular coordinates, 35th and Glendale is at $(3, 5)$.

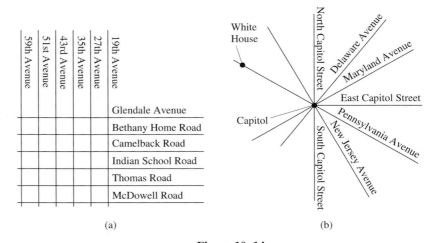

(a)

(b)

Figure 10–14

By contrast, Figure 10-14b shows a map of central Washington, DC, designed in 1791 by French military engineer Pierre-Charles L'Enfant and surveyed by Andrew Ellicott, who became U.S. surveyor-general in 1792 and was later a professor of mathematics at West Point. Although the United States consisted of only 13 states and barely a million people, L'Enfant envisioned Washington as the capital of a nation with about 50 states and 500 million people.

George Washington and Thomas Jefferson, who supervised L'Enfant, wanted the White House and the Capitol to occupy central positions in the city. Consequently, many main streets radiate in a variety of directions from each building. In Figure 10-14b, the Capitol is the origin, or **pole**, of a polar coordinate system. To give someone directions from the Capitol to the White House, you might point along Pennsylvania Avenue and say, "Go 1 mile that way." In the language of polar coordinates, the White House is at $\left(1, \dfrac{5\pi}{6}\right)$, because it is 1 mile from the Capitol along a line making a directed angle of $\dfrac{5\pi}{6}$ with the positive x-axis, or **polar axis** (East Capitol Street in Figure 10-14b).   ■

EXERCISES

1. **a.** If the origin in Figure 10-14a is at 59th and McDowell, estimate the polar coordinates of 35th and Glendale.

   b. How would you give someone directions from 59th and McDowell to 35th and Glendale by using the polar coordinate system? (To follow your directions, the person could not proceed along the streets shown.)

2. **a.** In Figure 10-14b, estimate the rectangular coordinates of the White House.

   b. How would you give someone directions from the Capitol to the White House by using the rectangular coordinate system? (To follow your directions, the person could not proceed along the streets shown.)

POLAR COORDINATES OF POINTS

Figure 10-15 shows a polar coordinate system. In this system, an ordered pair (r, θ) specifies a point at a directed distance r from the pole along a line making a directed angle θ with the polar axis.

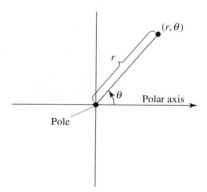

Figure 10–15

EXAMPLE 1   *Locating points identified by polar coordinates*

To locate the polar point $\left(3, \dfrac{\pi}{4}\right)$, visualize the polar axis rotated through a directed angle of $\dfrac{\pi}{4}$, as in Figure 10-16a. The point $\left(3, \dfrac{\pi}{4}\right)$ is located 3 units from the pole in the positive direction along the rotated axis.

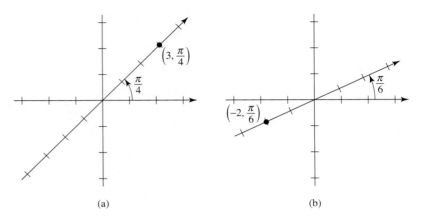

(a)   (b)

Figure 10–16

The directed distance r may be negative. For example, to locate $\left(-2, \dfrac{\pi}{6}\right)$, visualize the polar axis rotated through a directed angle of $\dfrac{\pi}{6}$, as in Figure 10-16b. The point $\left(-2, \dfrac{\pi}{6}\right)$ is located 2 units from the pole in the negative direction along the rotated axis.   ∎

Each point in the polar plane can be described by many different ordered pairs. For example, Figure 10-17 shows that the ordered pairs $\left(1, \dfrac{5\pi}{6}\right)$, $\left(1, -\dfrac{7\pi}{6}\right)$, $\left(-1, \dfrac{11\pi}{6}\right)$, and $\left(-1, -\dfrac{\pi}{6}\right)$ all describe the same point.

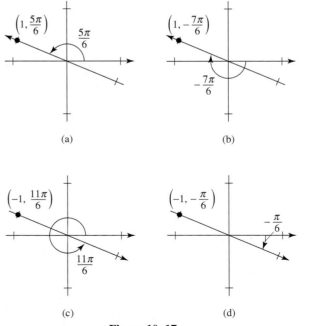

(a)   (b)

(c)   (d)

Figure 10–17

Figure 10-17a shows how one would get from the Capitol to the White House, as described in *A Tale of Two Cities*. Figure 10-17b shows a different way. Standing at the Capitol and facing east (in the direction of the polar axis), you turn through an angle of $\dfrac{7\pi}{6}$ clockwise and walk one mile straight ahead. Figure 10-17c shows how to get there by facing east, turning through an angle of $\dfrac{11\pi}{6}$ counterclockwise, and walking one mile backwards.

EXERCISES

3. Draw a figure to show that each set of polar coordinates represents the point shown in Figure 10-17.

   a. $\left(-1, \dfrac{23\pi}{6}\right)$ **b.** $\left(1, \dfrac{17\pi}{6}\right)$

   c. $\left(1, -\dfrac{19\pi}{6}\right)$ **d.** $\left(-1, -\dfrac{13\pi}{6}\right)$

4. (*Writing to Learn*) In the context of *A Tale of Two Cities*, describe the directions from the Capitol to the White House given by each set of polar coordinates in Exercise 3.

5. (*Making Observations*)

   a. Draw a figure to show that the polar coordinates $\left(5, \dfrac{\pi}{3}\right)$ and $\left(-5, \dfrac{4\pi}{3}\right)$ describe the same point.

   b. Explain why the polar coordinates (r, θ) and $(-r, \theta + \pi)$ describe the same point for all real values of r and θ.

6. (*Making Observations*)

   a. Draw a figure to show that the polar coordinates $\left(5, \dfrac{\pi}{3}\right), \left(5, \dfrac{7\pi}{3}\right)$, and $\left(5, -\dfrac{5\pi}{3}\right)$ describe the same point.

   b. Explain why the polar coordinates (r, θ) and $(r, \theta + 2k\pi)$ describe the same point for every integer k.

7. (*Making Observations*) For what values of θ do the polar coordinates $(0, \theta)$ describe the pole?

Your observations in Exercises 5–7 can be summarized as follows.

If a point P has polar coordinates (r, θ) with $r \neq 0$, then all sets of polar coordinates for P are the ordered pairs

$$(r, \theta + 2k\pi) \quad \text{and} \quad (-r, \theta + \pi + 2k\pi)$$

The pole is an exception to this rule and has coordinates $(0, \theta)$ for every value of θ.

Although each point can be represented by many sets of polar coordinates, each point except the pole has exactly one set with $r > 0$ and θ in $[0, 2\pi)$.

EXAMPLE 2   *Finding polar coordinates for a point with r > 0 and θ in [0, 2π)*

Let's find polar coordinates for $\left(-5, \dfrac{3\pi}{4}\right)$ with $r > 0$ and θ in $[0, 2\pi)$.

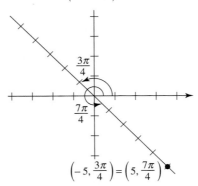

Figure 10–18

The same point has coordinates

$$\left(5, \frac{3\pi}{4} + \pi + 2k\pi\right) = \left(5, \frac{7\pi}{4} + 2k\pi\right)$$

for any integer k. As Figure 10-18 shows, we can obtain a value of θ in $[0, 2\pi)$ by choosing $\left(5, \dfrac{7\pi}{4}\right)$.   ■

EXERCISES   *In Exercises 8–11, plot the point in a polar coordinate system and find all sets of polar coordinates for it.*

8. $\left(4, \dfrac{\pi}{2}\right)$ **9.** $\left(-4, \dfrac{\pi}{2}\right)$

10. $\left(2, -\dfrac{3\pi}{4}\right)$ **11.** $(-3, -\pi)$

In Exercises 12–15, plot the point in a polar coordinate system and find a set of polar coordinates with r > 0 and θ in [0, 2π).

12. $\left(3, \dfrac{7\pi}{3}\right)$ **13.** $\left(-3, \dfrac{7\pi}{3}\right)$

14. $(-1, 5\pi)$ **15.** $\left(2, -\dfrac{\pi}{4}\right)$

RELATIONSHIPS BETWEEN RECTANGULAR AND POLAR COORDINATES

It is often useful to use both rectangular and polar coordinates to describe points in the same plane. For example, a bus driver might drive from an airport to a nearby city along an east–west road and a north–south road (that is, in a rectangular coordinate system). A pilot might fly to the same city along a straight line in a particular direction (that is, in a polar coordinate system).

Let's see how the two systems are related. Figure 10-19 indicates the following relationships for points in the first quadrant.

Figure 10–19

If a point has rectangular coordinates (x, y) and polar coordinates (r, θ), then

$$x = r \cos \theta \qquad r^2 = x^2 + y^2$$

$$y = r \sin \theta \qquad \tan \theta = \frac{y}{x} \quad (\text{if } x \neq 0)$$

EXERCISE   **16.** (*Writing to Learn*) Explain why the given relationships among x, y, r, and θ hold in all quadrants.

Obtaining Rectangular Coordinates from Polar Coordinates You can use the relationships $x = r \cos \theta$ and $y = r \sin \theta$ to find the rectangular coordinates of a point whose polar coordinates are known.

EXAMPLE 3   *Obtaining rectangular coordinates from polar coordinates*

The point with polar coordinates $\left(2, \dfrac{2\pi}{3}\right)$ has rectangular coordinates

$(x, y) = \left(2 \cos \dfrac{2\pi}{3}, 2 \sin \dfrac{2\pi}{3}\right) = (-1, \sqrt{3})$. ■

EXERCISES   *In Exercises 17–20, find the rectangular coordinates of the point whose polar coordinates are given.*

17. $\left(\sqrt{2}, -\dfrac{\pi}{4}\right)$ **18.** $(10, 3\pi)$

19. $\left(4, \dfrac{4\pi}{3}\right)$ **20.** $\left(-4, \dfrac{\pi}{3}\right)$

Obtaining Polar Coordinates from Rectangular Coordinates A little more work is required to find polar coordinates for a point whose rectangular coordinates are known. First,

$$r^2 = x^2 + y^2 \quad \text{so} \quad r = \pm\sqrt{x^2 + y^2}$$

(The positive and negative values both make sense, since r is a *directed* distance from the pole.) Next,

$$\tan \theta = \frac{y}{x} \quad (\text{if } x \neq 0) \quad \text{so} \quad \theta = \tan^{-1}\left(\frac{y}{x}\right) + k\pi$$

(If $x = 0$, then the point is on the y-axis, so $\theta = \frac{\pi}{2} + k\pi$.)

These equations show that for each point other than the pole there are two choices for r and infinitely many choices for θ. This makes sense, since we know that every point has many sets of polar coordinates. However, not all combinations of these values of r and θ yield valid polar coordinates for a given point, as Example 4 and Exercise 21 illustrate.

EXAMPLE 4   *Obtaining polar coordinates from rectangular coordinates*

The r-coordinates of the point with rectangular coordinates $(-5, -5)$ are

$$r = \pm\sqrt{(-5)^2 + (-5)^2} = \pm 5\sqrt{2}$$

The θ-coordinates are

$$\theta = \tan^{-1}\left(\frac{-5}{-5}\right) + k\pi = \tan^{-1} 1 + k\pi = \frac{\pi}{4} + k\pi$$

One way to choose a valid combination of r and θ is to choose r first. Suppose you select $r = 5\sqrt{2}$. Then, since $(-5, -5)$ is in the third quadrant, you must choose a third-quadrant angle θ. Since $\frac{\pi}{4}$ is in the first quadrant, one appropriate choice is $\theta = \frac{\pi}{4} + \pi = \frac{5\pi}{4}$. The point therefore has polar coordinates $\left(5\sqrt{2}, \frac{5\pi}{4}\right)$.

Another way to choose a valid combination of r and θ is to choose θ first. Suppose that you select $\theta = \frac{\pi}{4}$, a first-quadrant angle. Since $(-5, -5)$ is in the third quadrant, you must choose the negative r-value $r = -5\sqrt{2}$. The point therefore has polar coordinates $\left(-5\sqrt{2}, \frac{\pi}{4}\right)$.

The point also has polar coordinates $\left(5\sqrt{2}, \frac{5\pi}{4} + 2k\pi\right)$ or $\left(-5\sqrt{2}, \frac{\pi}{4} + 2k\pi\right)$ for all integer values of k.   ∎

EXERCISES   **21.** (*Writing to Learn*) In Example 3, explain why $\left(5\sqrt{2}, \frac{\pi}{4}\right)$ is not a valid set of polar coordinates for the given point.

In Exercises 22–25, the rectangular coordinates of a point are given. For each point, find at least one set of polar coordinates with $r > 0$ and one with $r < 0$.

22. $(4, -4)$ **23.** $(7, 0)$
24. $(-1, \sqrt{3})$ **25.** $(1, -\sqrt{3})$

ADDITIONAL EXERCISES   *In Exercises 26–29, plot the point in a polar coordinate system and find all sets of polar coordinates for it.*

26. $\left(5, -\dfrac{\pi}{2}\right)$ **27.** $\left(1, \dfrac{5\pi}{6}\right)$

28. $\left(-10, \dfrac{2\pi}{3}\right)$ **29.** $\left(-\sqrt{2}, -\dfrac{3\pi}{4}\right)$

In Exercises 30–33, plot the point in a polar coordinate system and find a set of polar coordinates for the point with $r > 0$ and θ in $[0, 2\pi)$.

30. $(4, 6\pi)$ **31.** $\left(-3, \dfrac{11\pi}{2}\right)$

32. $\left(-1, -\dfrac{2\pi}{3}\right)$ **33.** $\left(6, -\dfrac{5\pi}{4}\right)$

In Exercises 34–37, find the rectangular coordinates of the point whose polar coordinates are given.

34. $\left(8, \dfrac{\pi}{2}\right)$ **35.** $\left(-2, \dfrac{3\pi}{4}\right)$

36. $\left(5, -\dfrac{5\pi}{6}\right)$ **37.** $\left(0, \dfrac{\pi}{3}\right)$

In Exercises 38–41, the rectangular coordinates of a point are given. For each point, find at least one set of polar coordinates with $r > 0$ and one with $r < 0$.

38. $(\sqrt{2}, \sqrt{2})$ **39.** $(-\sqrt{2}, \sqrt{2})$

40. $(0, -9)$ **41.** $(-2\sqrt{3}, -2)$

In Exercise 42, you can derive a distance formula for the polar plane.

42. (*Extension*) The distance between (r_1, θ_1) and (r_2, θ_2) is
$$\sqrt{r_1^2 + r_2^2 - 2r_1 r_2 \cos(\theta_1 - \theta_2)}.$$

   a. Explain why this formula is valid when $r_1 > 0$ and $r_2 > 0$. (*Hint*: Use the Law of Cosines and the strategy of *Drawing a Diagram*.)

   b. Explain why the formula is still valid when one or both of the *r*-values is zero or negative.

In Exercises 43–46 (Extension), find the distance between the given points. Support your result by plotting the points and sketching the line segment between them.

43. $(6, 0), (5, \pi)$ **44.** $\left(3, \dfrac{\pi}{4}\right), \left(4, \dfrac{3\pi}{4}\right)$

45. $\left(2, \dfrac{2\pi}{3}\right), \left(-3, \dfrac{\pi}{2}\right)$ **46.** $\left(-5, \dfrac{5\pi}{6}\right), \left(-5, -\dfrac{\pi}{4}\right)$

10–4   GRAPHING IN POLAR COORDINATES

It is often helpful to use polar equations to describe curves in a plane. You are already aware that you can identify important graphical features of curves described by rectangular equations. In this section you will explore curves described by polar equations, and you will discover how their graphical features can be used in the design of an automobile engine.

POLAR AND RECTANGULAR EQUATIONS FOR THE SAME CURVE

Let's use your knowledge of rectangular curves to begin our exploration of polar curves. You can often gain insight into the graphical features of a polar equation by finding a rectangular equation for the same curve, or vice versa.

Expressing a Rectangular Equation in Polar Form By using the relationships $x = r \cos \theta$ and $y = r \sin \theta$, you can find a polar equation equivalent to a given equation in x and y. The process is sometimes referred to as expressing the equation in polar form.

EXAMPLE 1   *Expressing a rectangular equation in polar form*
A polar equation equivalent to

$$y = x^2$$

is

$$r \sin \theta = (r \cos \theta)^2$$

Thus, the equation $r \sin \theta = (r \cos \theta)^2$ describes the same parabola as $y = x^2$.

Polar equations are usually written to express r as a function of θ. As you can verify in Exercise 1, the equation just given is equivalent to $r = \sec \theta \tan \theta$. ∎

The rectangular equation in Example 1 is simpler than the polar equation for the same curve. In Exercise 2 you will encounter a polar equation that is simpler than the rectangular equation of the same curve.

EXERCISES   **1.** Verify that the equations $r \sin \theta = (r \cos \theta)^2$ and $r = \sec \theta \tan \theta$ are equivalent.
2. Show that the equation $(x^2 + y^2)^2 = x^2 - y^2$ is equivalent to $r^2 = \cos 2\theta$. (You will graph this equation in Exercise 76.)

In Exercises 3–6, convert each rectangular equation to polar form, and express r as a function of θ.

3. $y = 5$ **4.** $x = 3$
5. $x^2 - y^2 = 2x$ **6.** $x^2 + y^2 = 2x$

Expressing a Polar Equation in Rectangular Form The process of expressing a polar equation in rectangular form is more complicated. One or more of the following replacements can sometimes be made.

Replace $r \cos \theta$ by x.   Replace $r \sin \theta$ by y.

Replace r^2 by $x^2 + y^2$.   Replace $\tan \theta$ by $\dfrac{y}{x}$.

EXAMPLE 2   *Expressing a polar equation in rectangular form*
a. To express

$$r = 4 \sin \theta$$

in rectangular form, first multiply both sides by r.

$$r^2 = 4r \sin \theta$$

The left side is now r^2, which can be replaced by $x^2 + y^2$, and the right side is $4(r \sin \theta)$, which can be replaced by $4y$. Thus, an equivalent rectangular equation is

$$x^2 + y^2 = 4y$$

In rectangular form, the equation is easy to graph. You can sketch the graph in Exercise 7.

b. To express

$$\theta = \frac{\pi}{4}$$

in rectangular form, take the tangent of both sides in order to replace $\tan \theta$ by $\frac{y}{x}$.

$$\tan \theta = 1$$

$$\frac{y}{x} = 1$$

$$y = x$$

Again, the rectangular form makes graphing easy.   ■

EXERCISES   **7.** (*Review*) Sketch the graph of the equation $x^2 + y^2 = 4y$ from Example 2a.

In Exercises 8–11, write each equation in rectangular form and sketch its graph.

8. $r = 1$ **9.** $\theta = \dfrac{\pi}{3}$

10. $r = \sec \theta$ **11.** $r^2 \sin 2\theta = 2$ (*Hint*: Use a double-angle identity.)

12. (*Making Observations*) In which of Exercises 3–6 and 8–11 does the rectangular equation appear to be simpler than the polar equation? In which does the polar equation appear simpler?

13. (*Making Observations*)

a. Write the equation $\theta = \dfrac{5\pi}{4}$ in rectangular form, and compare the equation with that in Example 2b. What can you say about the graph of the equation?

b. What can you say about the graph of the equation $\theta = \dfrac{\pi}{4} + k\pi$ for any integer k?

GRAPHS OF POLAR EQUATIONS

The Graphs of $r = c$ and $\theta = c$ The simplest polar equations have the form

$$r = \text{constant} \quad \text{and} \quad \theta = \text{constant}$$

You encountered equations of each type in Exercises 8, 9, and 13. You can generalize your results in Exercises 14 and 15.

EXERCISES   14. (*Making Observations*)

  a. Write the equation $r = c$ in rectangular form.

  b. Sketch the graphs of $r = 1, r = 2, r = 3$, and $r = -3$ in the same plane.

15. (*Making Observations*)

  a. Write the equation $\theta = c$ in rectangular form.

  b. Sketch the graphs of $\theta = \dfrac{\pi}{3}$, $\theta = \dfrac{2\pi}{3}$, $\theta = \pi$, and $\theta = -2\pi$ in the same plane.

As your results from Exercises 14 and 15 suggest, the graph of the equation $r = c$ is a circle of radius $|c|$ centered at the pole, and the graph of $\theta = c$ is a straight line through the pole whose slope is $\tan c$.

EXERCISES   *Sketch the graph of each equation in Exercises 16–19 without converting to rectangular form.*

  16. $r = 4$ **17.** $r = -5$

  18. $\theta = \dfrac{3\pi}{4}$ **19.** $\theta = 5\pi$

Sketching Graphs by Plotting Points In Exercise 7 you sketched the circle $r = 4 \sin \theta$ by writing the equation in rectangular form. Let's see how the same equation might be graphed directly from its polar form by plotting points.

Since r is a function of θ, we can plot points most easily by assigning convenient values to θ and calculating r, as in Table 40. The points have been plotted in Figure 10-20a. By plotting additional points, you can obtain the graph in Figure 10-20b.

TABLE 40

θ	r	(r, θ)
0	0	$(0, 0)$
$\dfrac{\pi}{6}$	2	$\left(2, \dfrac{\pi}{6}\right)$
$\dfrac{\pi}{4}$	$2\sqrt{2}$	$\left(2\sqrt{2}, \dfrac{\pi}{4}\right)$
$\dfrac{\pi}{3}$	$2\sqrt{3}$	$\left(2\sqrt{3}, \dfrac{\pi}{3}\right)$
$\dfrac{\pi}{2}$	4	$\left(4, \dfrac{\pi}{2}\right)$

(a)

(b)

Figure 10–20

EXERCISES   *Sketch the graph of each polar equation in Exercises* 20–23.

20. $r = 4 \cos \theta$ **21.** $r = -2 \sin \theta$

22. $r = \cos \theta + \sin \theta$ **23.** $r = \theta$

24. Use your calculator to check the accuracy of your graphs in Exercises 20–23. (Consult your owner's manual for instructions about graphing in polar coordinates.)

Sketching Polar Graphs by Observing Patterns in Rectangular Graphs You can often sketch the graph of a polar equation $r = f(\theta)$ by observing patterns in the graph of the rectangular equation $y = f(x)$. Example 3 illustrates the method.

EXAMPLE 3   *Sketching a polar graph by observing a rectangular graph*

To sketch the polar graph of $r = \sin 2\theta$, let's begin by looking at the graph of $y = \sin 2x$ in Figure 10-21.

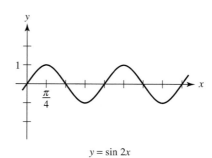

$y = \sin 2x$

Figure 10–21

On the rectangular graph, as x increases from 0 to $\dfrac{\pi}{4}$, y increases from 0 to 1.

Thus, on the polar graph, as θ increases from 0 to $\frac{\pi}{4}$, r increases from 0 to 1. That is, as the angle with the polar axis increases from 0 to $\frac{\pi}{4}$, the points on the polar graph "slide" away from the pole, as illustrated in Figure 10-22.

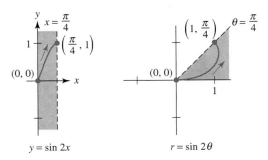

Figure 10–22

The rectangular graph also shows that as θ increases from $\frac{\pi}{4}$ to $\frac{\pi}{2}$, r decreases from 1 to 0. Thus, as the angle with the polar axis increases from $\frac{\pi}{4}$ to $\frac{\pi}{2}$, the polar graph slides back toward the pole, as in Figure 10-23.

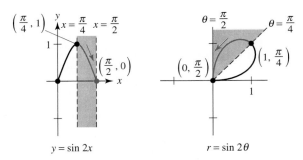

Figure 10–23

As θ increases from $\frac{\pi}{2}$ to $\frac{3\pi}{4}$, r is negative and decreases from 0 to -1. This means that the points on the polar graph are not in the shaded wedge of the second quadrant in Figure 10-24, but on the opposite side of the pole.

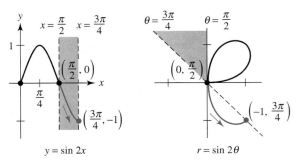

Figure 10–24

In Exercise 25 you can describe the reasoning needed to complete the polar graph, shown in Figure 10-25.

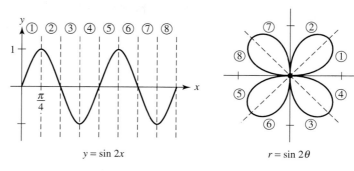

$y = \sin 2x$ $r = \sin 2\theta$

Figure 10–25

EXERCISES

25. (*Making Observations*) Describe how each segment of the rectangular graph numbered 4 through 8 in Figure 10-25 leads to the corresponding segment of the polar graph.

26. (*Making Observations*) Sketch the rectangular graph $y = 2 \cos x + 1$, and identify the segments of the graph corresponding to the numbered segments on the graph of $r = 2 \cos \theta + 1$ in Figure 10-26.

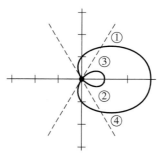

Figure 10–26

27. Sketch the rectangular graph $y = 1 - \sin x$, and use it to sketch the polar graph $r = 1 - \sin \theta$.

28. (*Making Observations*) Based on Example 3 and Exercises 25 and 26, describe the behavior of the graph of $r = f(\theta)$ when the graph of $y = f(x)$ is

    a. positive and increasing   **b.** positive and decreasing

    c. negative and decreasing   **d.** negative and increasing

Sketch each polar graph in Exercises 29–32 *by looking at a related rectangular graph.*

29. $r = 4 \sin \theta + 4$ **30.** $r = 1 - \cos \theta$

31. $r = 6 \cos 2\theta$ **32.** $r = \sin 3\theta$

Symmetry By now you have probably noticed several polar graphs that exhibit symmetry about the polar axis, the line $\theta = \dfrac{\pi}{2}$ (the y-axis in rectangular coordinates), or the pole. Let's see how such symmetry might be detected analytically. You may wish to review methods for detecting symmetry in rectangular graphs (Sections 4-5 and 5-4) before continuing.

Figure 10-27a shows that the points (r, θ) and $(r, -\theta)$ are reflections of each other in the polar axis. This means that if θ is replaced by $-\theta$ in an equation, its graph is reflected across the polar axis. In particular, if replacing θ by $-\theta$ leaves an equation unchanged, its graph is symmetric about the polar axis.

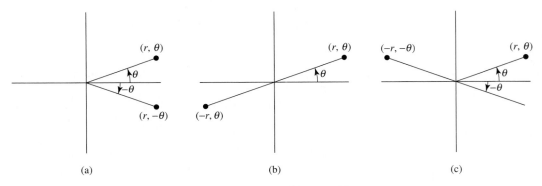

Figure 10–27

Similar observations about Figures 10-27b and 10-27c lead to the following tests for symmetry in a polar graph.

> To test for symmetry about the polar axis, replace θ by $-\theta$.
> To test for symmetry about the pole, replace r by $-r$.
>
> To test for symmetry about the line $\theta = \dfrac{\pi}{2}$, replace r by $-r$ and θ by $-\theta$.

EXAMPLE 4   *Testing a polar graph for symmetry*

To test the equation $r = \cos 3\theta$ for symmetry about the polar axis, replace θ by $-\theta$.

$$r = \cos(-3\theta)$$

Since $\cos(-t) = \cos t$ for all values of t, we have

$$r = \cos 3\theta$$

Thus the graph of the equation is symmetric about the polar axis.

To test the same equation for symmetry about the line $\theta = \dfrac{\pi}{2}$, replace r by $-r$ and θ by $-\theta$.

$$-r = \cos(-3\theta)$$

$$r = -\cos 3\theta$$

Since this equation is not equivalent to $r = \cos 3\theta$, the test does not show that the graph is symmetric about the line $\theta = \frac{\pi}{2}$.

Similarly, replacing r by $-r$ does not show that the graph is symmetric about the pole. ∎

FYI The symmetry tests given here are not the only ones possible. For example, Figure 10-28 shows that the reflection of (r, θ) in the polar axis also has coordinates $(-r, \pi - \theta)$. Thus, another test for symmetry about the polar axis is to replace r by $-r$ and θ by $\pi - \theta$.

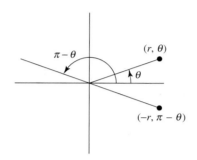

Figure 10–28

Because there are many tests for each type of symmetry, a graph may have symmetry even when a test fails to indicate it. For example, replacing θ by $-\theta$ in the equation $r = \sin 2\theta$ leads to the equation $r = -\sin 2\theta$ and thus does not indicate symmetry about the polar axis. However, the graph is symmetric about the polar axis, as you saw in Figure 10-25.

EXERCISES

33. (*Making Observations*) Show that the graph of $r = \sin 2\theta$ is symmetric about the polar axis by replacing r by $-r$ and θ by $\pi - \theta$ in the equation.

34. Find two other sets of polar coordinates for the point $(-r, \theta)$ in Figure 10-27b, and use them to write down two other tests for symmetry about the pole.

35. Find two other sets of polar coordinates for the point $(-r, -\theta)$ in Figure 10-27c, and use them to write down two other tests for symmetry about the line $\theta = \frac{\pi}{2}$.

36. Sketch the graph of $r = \cos 3\theta$. The tests applied in Example 4 did not show that the graph is symmetric about the line $\theta = \frac{\pi}{2}$ or the pole. Is it?

In Exercises 37–40,

a. Test each equation for all three types of symmetry.

b. Look at the graph from the indicated exercise, and tell whether the graph possesses any type of symmetry not indicated by your tests.

37. $r = -2 \sin \theta$ (Exercise 21)

38. $r = 4 \sin \theta + 4$ (Exercise 29)

39. $r = 6 \cos 2\theta$ (Exercise 31)

40. $r = \sin 3\theta$ (Exercise 32)

Now let's look at a physical curve whose polar equation is important to automotive engineers.

A MATHEMATICAL LOOKING GLASS

Cams

A **cam** is a machine part that rotates about an axis to cause a prescribed repetitive movement in another part, called the **follower**. Cams were the first devices used to control mechanical motions automatically and are still used in automatic packaging and typesetting, as well as in cars and analog computers. One common type, called a **plate cam**, is shown in Figure 10-29. It drives the follower up and down along a line perpendicular to its axis of rotation.

Follower

Cam

Axis of rotation

Figure 10–29

Figure 10-30 shows how a plate cam is used to control a valve in an automobile engine. As the cam rotates, the push rod moves up and down, causing the rocker arm to rock back and forth. This in turn causes the valve to open and close.

Rocker arm

Push rod (follower)

Valve

Cam

Figure 10–30

Adapted from *McGraw-Hill Encyclopedia of Science and Technology*, 1992.

Often a cam is designed to move the follower so that its displacement as a function of time is a sine or cosine function. Such a motion, desirable because it is relatively smooth and easy on the machinery, is called **harmonic motion**. A cam that produces this motion is called a **harmonic cam**. ∎

Let's see if we can find an equation to describe the outline of a harmonic cam. In Figure 10-31a, the cam and follower have been placed into a polar plane with the cam's axis of rotation at the pole.

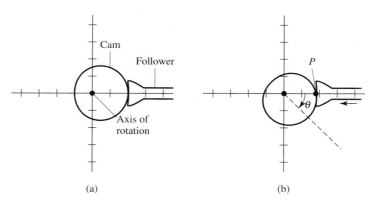

Figure 10–31

EXERCISES **41.** If the follower in Figure 10-31b is at the point $P = (3 + \cos\theta, 0)$, find the r-coordinate of the follower for each value of θ.

    a. $\dfrac{\pi}{6}$ **b.** $\dfrac{\pi}{4}$

    c. $\dfrac{\pi}{3}$ **d.** $\dfrac{\pi}{2}$

    e. $\dfrac{2\pi}{3}$ **f.** $\dfrac{3\pi}{4}$

    g. $\dfrac{5\pi}{6}$ **h.** π

42. (*Writing to Learn*) If the motion of the follower is as described in Exercise 41, explain why the follower exhibits harmonic motion between the points $(2, 0)$ and $(4, 0)$.

43. (*Writing to Learn*) Explain why the outline of the cam is described by the equation $r = 3 + \cos\theta$. (*Hint:* Why is the desired motion of the follower produced when the point P in Figure 10-31b has an r-coordinate of $3 + \cos\theta$?)

44. Use your result from Exercise 43 to sketch the outline of the cam.

45. Verify that the equation $r = 3 + \cos\theta$ is equivalent to

$$(x^2 + y^2 - x)^2 = 9(x^2 + y^2)$$

46. (*Writing to Learn*) If you were an automotive engineer, would you prefer to have the outline of a harmonic cam described by a rectangular equation or a polar equation? Explain your answer.

ADDITIONAL EXERCISES *In Exercises 47–52, convert each rectangular equation to polar form, and express r as a function of θ if possible.*

47. $y = -4$ **48.** $x = 0$

49. $y = x + 2$ **50.** $x^2 + y^2 = x + y$

51. $xy = 2x - 2y$ **52.** $y = \dfrac{x}{x - 1}$

In Exercises 53–60, write each equation in rectangular form.

53. $r = \sqrt{5}$ **54.** $\theta = 4\pi$

55. $r(\cos \theta - \sin \theta) = 6$ **56.** $r = 2 \tan \theta - \sec \theta$

57. $r^2 = \tan \theta$ **58.** $r = \cos \theta - \sin \theta$

59. $r = \dfrac{1}{\cos \theta - \sin \theta}$ **60.** $\cos 2\theta = 1$

In Exercises 61–76,

     a. Test the equation for symmetry about the polar axis, the pole, and the line $\theta = \dfrac{\pi}{2}$.

     b. Sketch the graph of the equation, and tell whether the graph possesses any type of symmetry not indicated by your tests.

61. $r = 4$ **62.** $\theta = -\dfrac{\pi}{2}$

63. $r = \csc \theta$ **64.** $r = -5 \cos \theta$

65. $r = \sec \theta + \tan \theta$ **66.** $r = 2 \sin \theta + 2 \cos \theta$

67. $r = 7 \cos 2\theta$ **68.** $r = 0.5 \sin 3\theta$

69. $r = 3 \sin 5\theta$ **70.** $r = 2 \cos 4\theta$

71. $r = 4 - 2 \cos \theta$ **72.** $r = 3 \sin \theta - 1$

73. $r = 3(\sin \theta - 1)$ **74.** $r = 3(\sin \theta - 2)$

75. $r^2 = \sin \theta$ **76.** $r^2 = \cos 2\theta$

77. (*Writing to Learn*) In what respects is the polar coordinate system easier to use than the rectangular system? In what respects is it more difficult?

78. According to an old riddle, "A hunter left his camp one morning. He walked one mile due south and one mile due east. There he shot a bear and walked back to his camp, which was one mile due north. What color was the bear?" The answer, of course, is that the bear was white because the hunter was at the North Pole. Write polar equations for each segment of his route, shown in Figure 10-32a.

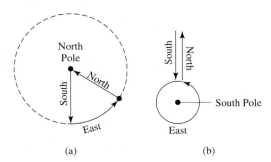

(a)   (b)

Figure 10–32

79. In addition to the North Pole, there are other places where you can walk a mile south, a mile east, and a mile north to arrive back at your starting point. All are near the South Pole. Write polar equations for each segment of the route shown in Figure 10-32b.

10-5   THE GEOMETRY OF COMPLEX NUMBERS

Before beginning this section, you may wish to review our discussion of basic operations with complex numbers in Section A-14. As you review, you may wonder whether imaginary numbers such as *i* have any practical value. Surprisingly, imaginary numbers are used in solving many physical problems. In this section you will learn how both rectangular and polar coordinates can be used to represent complex numbers geometrically. You will then discover how these geometric representations are used to simplify the analysis of the electrical circuit described in the following Mathematical Looking Glass.

A MATHEMATICAL LOOKING GLASS
Adding Voltages

The **voltage** between two points in an electrical circuit measures the tendency of current to flow between the points and is typically a function of time. Voltage provided by a wall outlet is called **ac** (alternating current) because it is a sine function of time and therefore alternates in sign, causing the current to alternate in direction. In Figure 10-33, for example, current flows from *A* to *B* if the voltage V_{AB} between *A* and *B* is positive and from *B* to *A* if V_{AB} is negative.

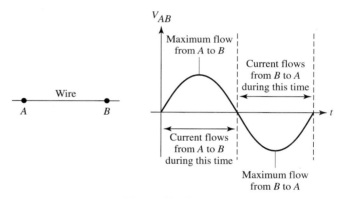

Figure 10–33

Figure 10-34 shows a portion of a wiring diagram for an electrical device.

$$A \;-\!\!/\!\backslash\!/\!\backslash\!/\!\backslash\!-\; B \;-\!000000\!-\; C$$

Figure 10–34

Only the following facts will be needed in our discussion.

- The diagram shows a **resistor** between *A* and *B*. Its presence creates a voltage V_{AB} between these points. For simplicity, let's say $V_{AB} = 16 \sin t$.
- The diagram also shows an **inductor** between *B* and *C*. Its presence creates a voltage V_{BC} between these points. It always has the same period as V_{AB}, but its graph is shifted a quarter of a cycle to the left. Let's assume that
$$V_{BC} = 12 \sin\left(t + \frac{\pi}{2}\right).$$
- The voltage between *A* and *C* is $V_{AC} = V_{AB} + V_{BC}$.

In designing the device, electrical engineers must calculate the maximum voltage between A and C. That is, they must calculate the amplitude of the voltage function V_{AC}. ■

EXERCISES

1. (*Review*)

   a. How do the equations $V_{AB} = 16 \sin t$ and $V_{BC} = 12 \sin\left(t + \frac{\pi}{2}\right)$ indicate that the two functions have the same period?

   b. What is the period of each?

   c. How do the equations indicate that the graph of V_{BC} is shifted a quarter of a cycle to the left of the graph of V_{AB}?

2. (*Making Observations*) Graph the functions V_{AB}, V_{BC}, and $V_{AC} = V_{AB} + V_{BC}$ in the same window, and estimate the amplitude of V_{AC}.

   In Exercise 23 you can calculate the exact amplitude of V_{AC}. In the meantime, let's explore some geometric properties of complex numbers that will be useful in your calculations.

THE COMPLEX PLANE

Every complex number can be written as $a + bi$ for a unique choice of real numbers a and b. Thus, if we associate the number $a + bi$ with the point (a, b), every complex number corresponds to a unique point in the coordinate plane, and vice versa. Figure 10-35 illustrates the correspondence.

When each point corresponds to a complex number in this way, the coordinate plane is called the **complex plane**.

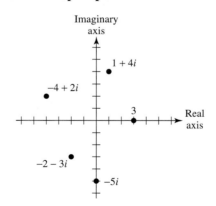

Figure 10–35

EXERCISES

3. Plot each point in the complex plane.

   a. $7 + 3i$ **b.** $-1 - 6i$

   c. $4i$ **d.** -3.5

4. Write the complex number corresponding to each point.

   a. $(1, -1)$ **b.** $(-1, 1)$

   c. $(0, \sqrt{2})$ **d.** $(0, 0)$

5. (*Writing to Learn*) The x-axis in the complex plane is called the **real axis**, and the y-axis is called the **imaginary axis**. Explain why.

6. (*Writing to Learn*) We can represent the set of real numbers geometrically on a number line. To represent the set of complex numbers geometrically, why is it more convenient to use a plane than a line?

7. (*Writing to Learn*) The **modulus** of a complex number z is its distance from the origin 0 in the complex plane and is written $|z|$. If $z = a + bi$, explain why $|z| = \sqrt{a^2 + b^2}$.

8. (*Making Observations*) Is the modulus of a real number always equal to its absolute value? Explain your answer.

9. Find the modulus of each complex number.

    a. $12 - 5i$ **b.** $-6 + 8i$

    c. $1 + 3i$ **d.** $-4i$

THE POLAR FORM OF A COMPLEX NUMBER

Every complex number can be represented by many sets of polar coordinates in the complex plane, in addition to its unique set of rectangular coordinates. For example, the number $-1 + i$ is represented by a point having rectangular coordinates $(-1, 1)$. Figure 10-36a shows that the same point also has polar coordinates $\left(\sqrt{2}, \dfrac{3\pi}{4}\right)$.

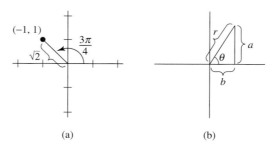

(a)   (b)

Figure 10–36

EXERCISES

10. (*Review*) Find all possible sets of polar coordinates for the point representing the number $-1 + i$.

11. (*Making Observations*)

    a. Show that the expression $\sqrt{2}\left(\cos\dfrac{3\pi}{4} + i\sin\dfrac{3\pi}{4}\right)$ can be simplified to $-1 + i$ so that both expressions represent the same number.

    b. If a point has rectangular coordinates (a, b) and polar coordinates (r, θ), explain how Figure 10-36b shows that the number $a + bi$ can also be written $r(\cos\theta + i\sin\theta)$.

Figure 10-36b shows that if a number z corresponds to a point with polar coordinates (r, θ), then $|r| = |z|$. The angle θ is called the **argument** of z. Figure 10-36 shows that the argument of $a + bi$ satisfies the equation $\tan\theta = \dfrac{b}{a}$.

The expression $r(\cos\theta + i\sin\theta)$ is called the **polar form** of z. It is customary to choose $r \geqslant 0$ and $0 \leqslant \theta < 2\pi$. Examples 1 and 2 illustrate how to obtain the rectangular form $a + bi$ from the polar form, and vice versa.

EXAMPLE 1   *Obtaining the rectangular form of a complex number from its polar form*
If a complex number has the polar form

$$z = 2\sqrt{3}\left(\cos\frac{\pi}{3} + i\sin\frac{\pi}{3}\right)$$

its rectangular form can be obtained by simplifying the expression.

$$z = 2\sqrt{3}\left(\frac{1}{2} + i\frac{\sqrt{3}}{2}\right)$$
$$= \sqrt{3} + 3i$$ ∎

EXAMPLE 2   *Obtaining the polar form of a complex number from its rectangular form*
To obtain the polar form of $z = -5 - 5i$, first observe that it corresponds to the point with rectangular coordinates $(-5, -5)$ in the complex plane. Then find a set of polar coordinates for $(-5, -5)$. This was done in Example 4 of Section 10-3, where it was found that one possible set is $\left(5\sqrt{2}, \frac{5\pi}{4}\right)$. Thus, one polar form is

$$z = 5\sqrt{2}\left(\cos\frac{5\pi}{4} + i\sin\frac{5\pi}{4}\right)$$ ∎

EXERCISES   *In Exercises 12–15, write the number in rectangular form.*

12. $8\left(\cos\frac{\pi}{4} + i\sin\frac{\pi}{4}\right)$

13. $12\left(\cos\frac{5\pi}{3} + i\sin\frac{5\pi}{3}\right)$

14. $5(\cos 2\pi + i\sin 2\pi)$

15. $\cos\left(-\frac{\pi}{2}\right) + i\sin\left(-\frac{\pi}{2}\right)$

In Exercises 16–19, write the number in polar form.

16. $6 - 6i$ **17.** $-\sqrt{3} + i$

18. -10 **19.** $10i$

ADDING AND SUBTRACTING COMPLEX NUMBERS GEOMETRICALLY

Exercises 20 and 21 indicate some similarities between the operations of adding complex numbers and adding vectors in the plane.

EXERCISES   **20.** (*Review*) Let $\mathbf{V}_1 = 3\mathbf{i} - 5\mathbf{j}$ and $\mathbf{V}_2 = 4\mathbf{i} + 2\mathbf{j}$.
   a. Add $\mathbf{V}_1$ and $\mathbf{V}_2$ analytically.
   b. Draw $\mathbf{V}_1$ and $\mathbf{V}_2$ in standard position in the plane [with their tails at $(0,0)$]. Then add $\mathbf{V}_1$ and $\mathbf{V}_2$ geometrically by drawing the diagonal of the parallelogram determined by $\mathbf{V}_1$ and $\mathbf{V}_2$. Check to see that the diagonal represents $\mathbf{V}_1 + \mathbf{V}_2$.

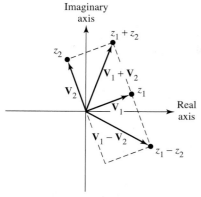

Figure 10–37

21. (*Making Observations*) Let $z_1 = 3 - 5i$ and $z_2 = 4 + 2i$.
    a. Add z_1 and z_2.
    b. Plot the points $0, z_1, z_2$, and $z_1 + z_2$ in the complex plane. They appear to lie at the vertices of a parallelogram. How does your result from Exercise 20 suggest that they actually do?

    The patterns you observed in Exercises 20 and 21 exemplify the following principle, which is also illustrated in Figure 10-37.

> If the numbers z_1 and z_2 are represented by vectors $\mathbf{V}_1$ and $\mathbf{V}_2$ in standard position in the complex plane, then the numbers $z_1 + z_2$ and $z_1 - z_2$ are represented by $\mathbf{V}_1 + \mathbf{V}_2$ and $\mathbf{V}_1 - \mathbf{V}_2$.

EXERCISE   22. Perform the indicated operations both analytically and geometrically, and check to see that your results are compatible.
    a. $2i + (-3)$ b. $5 - (-2i)$
    c. $(1 - 2i) + (-3 - i)$ d. $(-4 + 4i) - (1 + 4i)$

Now let's return to the problem of calculating the amplitude of the voltage function V_{AC} in **Adding Voltages**. Calculations such as this can be simplified through the use of a mathematical trick, which we will now explore. Specifically, let's associate every sine function $f(t) = R \sin(t + \theta)$ with the complex number whose polar coordinates are (R, θ). For example, the function $V_{AB} = 16 \sin t$ is associated with the number $z_1 = (16, 0)$, and the function $V_{BC} = 12 \sin\left(t + \dfrac{\pi}{2}\right)$ is associated with the number $z_2 = \left(12, \dfrac{\pi}{2}\right)$. In Exercises 69–72 you can show that V_{AC} is also a sine function and is associated with $z_1 + z_2$. This means that you can find the amplitude of V_{AC} by finding $|z_1 + z_2|$.

EXERCISES   23. a. Plot the polar points $z_1 = (16, 0)$ and $z_2 = \left(12, \dfrac{\pi}{2}\right)$ in the complex plane, add z_1 and z_2 geometrically, and find $|z_1 + z_2|$.
    b. Write z_1 in both polar and rectangular form.
    c. Write z_2 in both polar and rectangular form.
    d. Write $z_1 + z_2$ in rectangular form and find $|z_1 + z_2|$ analytically.
    e. What is the amplitude of the voltage function V_{AC}?

24. In **Adding Voltages**, if $V_{AB} = 12 \sin t$ and $V_{BC} = 5 \sin\left(t + \dfrac{\pi}{2}\right)$, find the amplitude of V_{AC}.

MULTIPLYING AND DIVIDING COMPLEX NUMBERS GEOMETRICALLY

Earlier in this section you saw how to add and subtract complex numbers geometrically by using their rectangular coordinates in the complex plane. The following facts allow you to multiply and divide complex numbers geometrically by using their polar coordinates.

If $z_1 = r_1(\cos\theta_1 + i\sin\theta_1)$ and $z_2 = r_2(\cos\theta_2 + i\sin\theta_2)$, then

$$z_1 z_2 = r_1 r_2\big[\cos(\theta_1 + \theta_2) + i\sin(\theta_1 + \theta_2)\big]$$

$$\frac{z_1}{z_2} = \frac{r_1}{r_2}\big[\cos(\theta_1 - \theta_2) + i\sin(\theta_1 - \theta_2)\big]$$

In words, the boxed statement says:

- To multiply complex numbers, multiply their moduli and add their arguments.
- To divide complex numbers, divide their moduli and subtract their arguments.

EXERCISES

25. (*Making Observations*)

   a. Multiply $\big(1 + \sqrt{3}\,i\big)\big(3 + \sqrt{3}\,i\big)$ using the method given in Section A-14.

   b. Multiply $\big(1 + \sqrt{3}\,i\big)\big(3 + \sqrt{3}\,i\big)$ by first writing each factor in polar form. Make sure your result agrees with that in part (a).

26. (*Writing to Learn*) Multiply $r_1(\cos\theta_1 + i\sin\theta_1)$ by $r_2(\cos\theta_2 + i\sin\theta_2)$, and simplify to show that the first statement in the preceding box is true. (*Hint*: You will need to use trigonometric identities.)

27. (*Making Observations*)

   a. Divide $\dfrac{-4}{1+i}$ using the method given in Section A–14.

   b. Divide $\dfrac{-4}{1+i}$ by first writing each factor in polar form. Make sure your result agrees with that in part (a).

28. (*Making Observations*) Divide $r_1(\cos\theta_1 + i\sin\theta_1)$ by $r_2(\cos\theta_2 + i\sin\theta_2)$, and simplify to show that the second statement in the preceding box is true. (*Hint*: You will need to use trigonometric identities.)

In Exercises 29–32,

   a. Calculate both $z_1 z_2$ and $\dfrac{z_1}{z_2}$, and write the results in both polar and rectangular form.

   b. Plot the points z_1, z_2, and $z_1 z_2$, and indicate the relationships among the arguments.

   c. Plot the points z_1, z_2, and $\dfrac{z_1}{z_2}$, and indicate the relationships among the arguments.

29. $z_1 = 6\left(\cos\dfrac{4\pi}{3} + i\sin\dfrac{4\pi}{3}\right)$, $z_2 = 2\left(\cos\dfrac{\pi}{6} + i\sin\dfrac{\pi}{6}\right)$

30. $z_1 = 5\left(\cos\dfrac{5\pi}{12} + i\sin\dfrac{5\pi}{12}\right)$, $z_2 = \cos\dfrac{7\pi}{12} + i\sin\dfrac{7\pi}{12}$

31. $z_1 = 1$, $z_2 = 4i$ **32.** $z_1 = 8 - 8i$, $z_2 = 2 - 2i$

DE MOIVRE'S THEOREM

Powers of complex numbers in polar form are also easy to calculate. The pattern you are about to discover in Exercise 33 was first noticed by the French mathematician Abraham De Moivre (1667–1754).

EXERCISE   **33.** (*Making Observations*)

    a. If $z = r(\cos \theta + i \sin \theta)$, write z^2 and z^3 in polar form.

    b. Generalize your result from part (a) to write an expression for z^n.

The observations you make in Exercise 33 can be stated as **De Moivre's theorem**.

> If $z = r(\cos \theta + i \sin \theta)$, then $z^n = r^n(\cos n\theta + i \sin n\theta)$ for every integer n.

EXERCISES   In Exercises 34–37, *perform the indicated operations and write the results in both polar and rectangular form.*

34. $\left[2\left(\cos \dfrac{2\pi}{3} + i \sin \dfrac{2\pi}{3}\right)\right]^5$

35. $\left[\sqrt{3}\left(\cos \dfrac{\pi}{4} + i \sin \dfrac{\pi}{4}\right)\right]^6$

36. $(1 - i)^8$ **37.** $\left(-6\sqrt{3} - 6i\right)^3$

38. (*Making Observations*) Let

$$u_1 = 2\left(\cos \frac{\pi}{4} + i \sin \frac{\pi}{4}\right), \qquad u_2 = 2\left(\cos \frac{3\pi}{4} + i \sin \frac{3\pi}{4}\right),$$

$$u_3 = 2\left(\cos \frac{5\pi}{4} + i \sin \frac{5\pi}{4}\right), \qquad \text{and } u_4 = 2\left(\cos \frac{7\pi}{4} + i \sin \frac{7\pi}{4}\right)$$

    a. Show that $u_1{}^4, u_2{}^4, u_3{}^4$, and $u_4{}^4$ are each equal to -16.

    b. How do the moduli of u_1, u_2, u_3, and u_4 relate to the modulus of -16? How do the arguments of u_1, u_2, u_3, and u_4 relate to the argument of -16?

    c. Draw u_1, u_2, u_3, and u_4 in the complex plane. How does your drawing indicate that $u_1{}^4, u_2{}^4, u_3{}^4$, and $u_4{}^4$ all have the same value?

    d. Do you think that $z^4 = -16$ for any other value of z? Why or why not?

Your observations in Exercise 38 can be generalized to provide a powerful tool for finding roots of complex numbers. Recall that a number u is an nth root of a number z if and only if $u^n = z$.

Every nonzero complex number $r(\cos \theta + i \sin \theta)$ has exactly n complex nth roots. They are

$$\sqrt[n]{r}\left[\cos \frac{1}{n}(\theta + 2k\pi) + i \sin \frac{1}{n}(\theta + 2k\pi)\right]$$

for $k = 0, 1, 2, \ldots, n - 1$.

EXAMPLE 3   *Finding all nth roots of a complex number*

The sixth roots of $z = 729\left(\cos \dfrac{3\pi}{4} + i \sin \dfrac{3\pi}{4}\right)$ all have modulus $\sqrt[6]{729} = 3$.

Their arguments are

$$\frac{1}{6}\left(\frac{3\pi}{4}\right) = \frac{\pi}{8}$$

$$\frac{1}{6}\left(\frac{3\pi}{4} + 2\pi\right) = \frac{11\pi}{24}$$

$$\frac{1}{6}\left(\frac{3\pi}{4} + 4\pi\right) = \frac{19\pi}{24}$$

$$\frac{1}{6}\left(\frac{3\pi}{4} + 6\pi\right) = \frac{9\pi}{8}$$

$$\frac{1}{6}\left(\frac{3\pi}{4} + 8\pi\right) = \frac{35\pi}{24}$$

$$\frac{1}{6}\left(\frac{3\pi}{4} + 10\pi\right) = \frac{43\pi}{24}$$

Thus the roots are

$$u_1 = 3\left(\cos \frac{\pi}{8} + i \sin \frac{\pi}{8}\right)$$

$$u_2 = 3\left(\cos \frac{11\pi}{24} + i \sin \frac{11\pi}{24}\right)$$

$$u_3 = 3\left(\cos \frac{19\pi}{24} + i \sin \frac{19\pi}{24}\right)$$

$$u_4 = 3\left(\cos \frac{9\pi}{8} + i \sin \frac{9\pi}{8}\right)$$

$$u_5 = 3\left(\cos \frac{35\pi}{24} + i \sin \frac{35\pi}{24}\right)$$

$$u_6 = 3\left(\cos \frac{43\pi}{24} + i \sin \frac{43\pi}{24}\right)$$

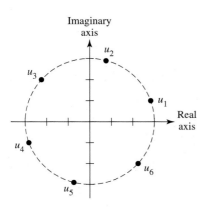

Figure 10–38

The numbers $u_1, u_2, \ldots, u_6$ are shown in the complex plane in Figure 10-38.

EXERCISES   **39.** (*Writing to Learn*) Figure 10-38 suggests that the nth roots of

$$r(\cos \theta + i \sin \theta)$$

are located on a circle centered at the origin and divide the circle into n arcs of equal length. Explain why this always happens.

40. Find all the eighth roots of 1, and plot them in the complex plane.

41. Find all the cube roots of $125\left(\cos\dfrac{4\pi}{3} + i\sin\dfrac{4\pi}{3}\right)$.

42. Find all the fifth roots of $32i$.

43. The Fundamental Theorem of Algebra guarantees that the equation $x^6 + 1 = 0$ has six complex solutions. Find them by finding all the sixth roots of -1.

ADDITIONAL EXERCISES

44. Plot each point in the complex plane.

    a. $2 + 4i$ **b.** $2 - 4i$

    c. $-6 + 3i$ **d.** $-6 - 3i$

45. Plot each point in the complex plane.

    a. $5\left(\cos\dfrac{3\pi}{2} + i\sin\dfrac{3\pi}{2}\right)$

    b. $\sqrt{2}\left(\cos\dfrac{5\pi}{4} + i\sin\dfrac{5\pi}{4}\right)$

    c. $\cos\dfrac{2\pi}{3} + i\sin\dfrac{2\pi}{3}$

    d. $4\left(\cos\dfrac{11\pi}{6} + i\sin\dfrac{11\pi}{6}\right)$

46. Find the modulus of each complex number.

    a. -7 **b.** $4 - 3i$

    c. $-8 - 15i$ **d.** $3 + 2i$

In Exercises 47–50, write both the rectangular and polar forms of the complex numbers having the given polar coordinates.

47. $(10, \pi)$ **48.** $\left(2, \dfrac{\pi}{6}\right)$

49. $\left(6\sqrt{2}, \dfrac{7\pi}{4}\right)$ **50.** $\left(4, \dfrac{4\pi}{3}\right)$

In Exercises 51–54, write both the rectangular and polar forms of the complex number having the given rectangular coordinates.

51. $\left(-\sqrt{3}, -1\right)$ **52.** $\left(-\sqrt{2}, \sqrt{2}\right)$

53. $(-5, 8)$ **54.** $(3, -2)$

In Exercises 55–62, perform the indicated operations and write the results in both polar and rectangular form.

55. $\left[7\left(\cos\dfrac{4\pi}{3} + i\sin\dfrac{4\pi}{3}\right)\right]^2$

56. $\left[\sqrt{5}\left(\cos\dfrac{\pi}{3} + i\sin\dfrac{\pi}{3}\right)\right]^6$

57. $\left(\cos\dfrac{3\pi}{4} + i\sin\dfrac{3\pi}{4}\right)^3$

58. $\left(\cos\dfrac{\pi}{4} + i\sin\dfrac{\pi}{4}\right)^9$

59. $\left(\sqrt{3} + i\right)^5$ **60.** $\left(-1 + \sqrt{3}\,i\right)^7$

61. $(2 - 2i)^4$ **62.** $(-2 - 2i)^4$

In Exercises 63–66, solve the equation by finding all the nth roots of an appropriate complex number. Check your answer by factoring and/or using the quadratic formula.

63. $x^4 - 81 = 0$ **64.** $x^2 + 25 = 0$

65. $x^3 + 8 = 0$ **66.** $x^6 - 1 = 0$

67. Perform each operation both analytically and geometrically, and check to see that your results are compatible.

    a. $(3 + 5i) + (3 - 5i)$

    b. $(3 + 5i) - (3 - 5i)$

68. (*Making Observations*) Your results in Exercise 67 suggest that if z and $\bar{z}$ are conjugates, $z + \bar{z}$ is always real and $z - \bar{z}$ is always imaginary. Is this the case? Explain your answer. (See Section A-14 for the definition of conjugate.)

Your calculations in ***Adding Voltages*** relied on a connection between the addition of sine functions and the addition of complex numbers in polar form. You can now verify that connection in Exercises 69–72.

69. (*Extension*) Suppose that the numbers z_1, z_2, and $z_1 + z_2$, shown in Figure 10-39, have polar coordinates (r_1, θ_1), (r_2, θ_2), and (r_3, θ_3) with r_1, r_2, and r_3 all positive.

Figure 10–39

    a. Use the Law of Cosines to explain why

$$r_3 = \sqrt{r_1^2 + r_2^2 - 2r_1r_2\cos\left[\pi - (\theta_2 - \theta_1)\right]}$$

    b. Explain why r_3 can be rewritten as $\sqrt{r_1^2 + r_2^2 + 2r_1r_2\cos(\theta_2 - \theta_1)}$

    c. Explain why

$$\tan\theta_3 = \frac{r_1\sin\theta_1 + r_2\sin\theta_2}{r_1\cos\theta_1 + r_2\cos\theta_2}$$

    (*Hint*: Express the horizontal and vertical sides of the two right triangles in Figure 10-39 in terms of r_1, r_2, θ_1, and θ_2.)

70. (*Extension*) Suppose that $f(t) = A_1 \sin t + A_2 \cos t$.

  a. Explain why $\left| \dfrac{A_1}{\sqrt{A_1{}^2 + A_2{}^2}} \right| \leq 1$.

  b. Explain why there is an angle θ with

$$\cos \theta = \frac{A_1}{\sqrt{A_1{}^2 + A_2{}^2}}$$

  and

$$\sin \theta = \frac{A_2}{\sqrt{A_1{}^2 + A_2{}^2}}$$

  c. If $r = \sqrt{A_1{}^2 + A_2{}^2}$, show that $f(t)$ can be written as $r \sin (t + \theta)$.
  [*Hint*: First show that $f(t) = (r \cos \theta) \sin t + (r \sin \theta) \cos t$.]

71. (*Extension*) Suppose that $f_1(t) = r_1 \sin (t + \theta_1)$ and $f_2(t) = r_2 \sin (t + \theta_2)$.

  a. Show that $(f_1 + f_2)(t) = A_1 \sin t + A_2 \cos t$, with $A_1 = r_1 \cos \theta_1 + r_2 \cos \theta_2$ and $A_2 = r_1 \sin \theta_1 + r_2 \sin \theta_2$.
  (*Hint*: Use a trigonometric identity to rewrite f_1 and f_2.)

  b. Use your result from Exercise 70 to show that

$$(f_1 + f_2)(t) = r_3 \sin (t + \theta_3)$$

  where $r_3 = \sqrt{r_1{}^2 + r_2{}^2 + 2 r_1 r_2 \cos (\theta_2 - \theta_1)}$ and

$$\tan \theta_3 = \frac{r_1 \sin \theta_1 + r_2 \sin \theta_2}{r_1 \cos \theta_1 + r_2 \cos \theta_2}$$

72. (*Extension*)

  a. Suppose that every function $f(t) = r \sin (t + \theta)$ is associated with the number z having polar coordinates (r, θ). If the functions f_1 and f_2 are associated with the numbers z_1 and z_2, how do your results from Exercises 69 and 71 show that the function $f_1 + f_2$ is associated with the number $z_1 + z_2$?

  b. How does your result from part (a) show that the amplitude of $f_1 + f_2$ is equal to $|z_1 + z_2|$?

CHAPTER REVIEW

Complete Exercises 1–33 (Writing to Learn) before referring to the indicated pages.

WORDS AND PHRASES

In Exercises 1–6, explain the meaning of the words or phrases in your own words.

  1. (page 522) **polar coordinate system**
  2. (page 523) **pole, polar axis**
  3. (page 542) **complex plane, real axis, imaginary axis**
  4. (page 543) **modulus** of a complex number
  5. (page 543) **argument** of a complex number
  6. (page 543) **polar form** of a complex number

IDEAS

  7. (page 499) In what way is the discussion of conic sections in this chapter more general than the discussion in Chapter 4?

  8. (pages 501, 502) If a uv-coordinate system is obtained from an xy-system by rotating the axes counterclockwise through an angle of θ, what equations relate x and y to u and v? Illustrate the use of the equations by choosing a specific value for θ and calculating the coordinates of a few specific points.

  9. (page 503) How can a rotation of axes simplify the process of graphing a second-degree equation in two variables?

10. (pages 503)   Describe the process of graphing a conic section by rotating axes.

11. (page 509)   Why might you want to use parametric equations to describe a conic section?

12. (pages 510–512)   What parametric equations describe the unit circle? How can these equations be modified to describe the motion of an object around any circle? Give a few specific examples of parametric equations for circular motions, and describe the motions they represent.

13. (pages 515–517)   How can the parametric equations for the unit circle be modified to describe the motion of an object around an ellipse? Give a specific example of parametric equations for motion around an ellipse, and describe the motion it represents.

14. (pages 518–520)   Why is it natural to use the secant and tangent functions to describe motion along a hyperbola? Give a specific example of parametric equations for motion along a hyperbola, and describe the motion it represents.

15. (pages 522–525)   In what respects are the polar and rectangular coordinate systems alike? In what respects are they different?

16. (pages 523–525)   Describe the process of locating a point whose polar coordinates are given. Illustrate your description with a few specific points.

17. (page 525)   Explain why every point in the plane has infinitely many sets of polar coordinates. In particular, explain why the pole has coordinates $(0, \theta)$ for every value of θ.

18. (pages 525, 526)   Explain why every point except the pole has exactly one set of polar coordinates with $r > 0$ and $0 \leqslant \theta < 2\pi$.

19. (page 527)   What equations relate the polar and rectangular coordinates of a point? Explain why the equations are valid.

20. (pages 527, 528)   Describe the processes of obtaining the rectangular coordinates of a point from its polar coordinates and vice versa. Illustrate your descriptions with specific examples.

21. (pages 530, 531)   Describe the processes of rewriting a rectangular equation in polar form, and vice versa. Make up equations to illustrate your descriptions.

22. (page 532)   What do the graphs of equations $r = c$ and $\theta = c$ look like? Use the geometric meanings of r and θ to explain the appearance of each type of graph.

23. (pages 531, 532)   When you sketch a polar graph by plotting points, why is it usually more convenient to choose values of θ and calculate the corresponding values of r?

24. (pages 533–535)   What can you say about the graph of $r = f(\theta)$ in an interval $[\theta_1, \theta_2]$ if $r > 0$? If $r < 0$? What can you say about the graph if $|r|$ is increasing? If $|r|$ is decreasing?

25. (pages 535–537)   State at least two ways of testing a polar graph for symmetry about the polar axis, the line $\theta = \dfrac{\pi}{2}$, and the pole. Explain why each test is valid.

26. (page 537)   Why might a polar graph exhibit symmetry even when an analytical test fails to detect it?

27. (page 542)   Why can every complex number be made to correspond to a unique point in the plane? Illustrate your explanation with a few specific points.

28. (pages 544, 545)   Compare the process of adding complex numbers with the process of adding vectors both analytically and geometrically. In particular, point out the similarities and illustrate with a specific example.

29. (page 544)   Describe how to obtain the rectangular form of a complex number from its polar form, and vice versa. Illustrate your description with a few specific numbers.

30. (page 546)   Explain why it is easy to calculate the modulus and argument of a product or quotient of complex numbers if you know the modulus and argument of each factor. Make up an example to illustrate your explanation.

31. (page 547)   State De Moivre's theorem and make up an example to illustrate its use.

32. (pages 547, 548)   How does De Moivre's theorem show that every complex number has exactly n complex nth roots? How does the Fundamental Theorem of Algebra (see Section 5-2) show the same thing?

33. (page 548)   Describe the method of finding all the nth roots of a complex number and make up an example to illustrate your explanation.

APPENDIX A

BASIC ALGEBRA REFERENCE

A–1 ACCURACY AND PRECISION

Numbers can represent physical quantities either exactly or approximately. For example, suppose that a friend tells you

"Out of 32 people in my algebra class, I live farthest from campus. I drive 32 miles to school."

The 32 in the first sentence is *exactly* the number of people in your friend's class. By contrast, the 32 in the second sentence is *approximately* the distance she drives to school. She probably means that the distance is closer to 32 miles than to 31 or 33 miles. That is, it is between 31.5 and 32.5 miles. As your friend's statement illustrates, you must often rely on context to decide whether a number is exact or approximate.

When a number is approximate, you often need to know how good the approximation is. Informally, you need to know how many digits you can trust. Furthermore, when you add, subtract, multiply, or divide approximate numbers, you need to know how many digits to trust in your result. The following discussion provides the needed information.

SIGNIFICANT DIGITS

If the population of the United States is reported to be 249,000,000, it is reasonable to believe that the reported number is an approximation to the nearest million, so that the actual population is between 248,500,000 and 249,500,000. Thus only the first three digits in the reported number have meaning, in the sense that their values reflect the size of the population. For this reason, the first three digits are said to be **significant**, while the last six are not.

The following rules determine which digits of an approximate number are significant.

- All nonzero digits are significant.
- All zeros between two nonzero digits are significant.
- All zeros in a string of trailing zeros are significant if the string ends to the right of the decimal point.
- No other zeros are significant.

EXAMPLE 1   *Deciding which digits of an approximate number are significant*
In the table below, the significant digits are underlined.

Approximate number	Actual value of number
3600	closer to 3600 than to 3500 or 3700, that is, between 3550 and 3650
3600.0000	closer to 3600.0000 than to 3599.9999 or 3600.0001, that is, between 3599.99995 and 3600.00005
3600.07	closer to 3600.07 than to 3600.06 or 3600.08, that is, between 3600.065 and 3600.075
0.07	closer to 0.07 than to 0.06 or 0.08, that is, between 0.065 and 0.075
0.0700	closer to 0.0700 than to 0.0699 or 0.0701, that is, between 0.06995 and 0.07005

■

The **accuracy** of an approximate number is its number of significant digits. The **precision** of the number is the decimal place of its rightmost significant digit. Thus the approximate number 3600.07 is accurate to six significant digits, and precise to the nearest hundredth.

SCIENTIFIC NOTATION

Suppose that you are writing a paper in which you want to report the population of the United States, rounded to the nearest million. Suppose, further, that the actual population is 250,341,829, and you write the rounded population as 250,000,000. In this case, the zero in the millions place has meaning in the sense that its value reflects the size of the population. However, a reader of your paper will not regard that zero as a significant digit, and will thus misinterpret your intent.

In order to overcome such ambiguities, scientists have devised a different way to represent approximate numbers. In **scientific notation**, a number is written as the product of two numbers, the first having a value between 1 and 10 and the second being an integer power of 10. *All digits in the first factor of this product are significant.* Thus the approximate population figure in the foregoing paragraph can be written as 2.50×10^8 if three significant digits are intended or as 2.5×10^8 if two significant digits are intended. Here are scientific notation representations of a few other approximate numbers:

Approximate number	Representation in scientific notation
3600	3.6×10^3
3600.0000	3.6000000×10^3
0.07	7×10^{-2}
0.0700	7.00×10^{-2}

ACCURACY OF PRODUCTS AND QUOTIENTS

How should you round off a product or quotient of two or more approximate numbers? The answer depends on the accuracy of the factors. Specifically, the accuracy of the result is equal to that of the least accurate factor.

EXAMPLE 2 *Finding the accuracy of a product*

If we multiply 432.17 by 3.089, the result is 1334.97313. However, if the factors are approximate numbers, then they are accurate to five and four significant digits, respectively. Thus the product should be rounded to four significant digits as 1335. ■

Exact numbers are more accurate than any approximation. Thus, when multiplying 432.17 (approximate) by 14 (exact), round the product 6050.38 to five significant digits as 6050.4.

PRECISION OF SUMS AND DIFFERENCES

How should you round off a sum or difference of two or more approximate numbers? The answer depends on the precision of the terms. Specifically, the precision of the result is equal to that of the least precise term.

EXAMPLE 3 *Finding the precision of a sum*

If we add 432.17 to 3.089, the result is 435.259. However, if the terms are approximate numbers, then they are precise to the nearest hundredth and the nearest thousandth (two and three decimal places), respectively. Thus the sum should be rounded to two decimal places as 435.26. ■

Exact numbers are more precise than any approximation. Thus, when adding 432.17 (approximate) to 14 (exact), round the sum 446.17 to two decimal places. That is, leave it alone.

EXERCISES *In Exercises 1–8,*

   a. Underline the significant digits in each approximate number.
   b. Find the possible actual values of the number, as in Example 1.
   c. Write the number in scientific notation.

   1. 8230   **2.** 8230.6

   3. 0.0032   **4.** 1.0032

   5. 61,400.009   **6.** 61,400.00900

   7. 0.010203   **8.** 0.010200

In Exercises 9–16, perform the indicated operations and round the result appropriately. Assume all numbers are approximate.

   9. (32.9)(0.0041)   **10.** $\dfrac{3300}{780.05}$

   11. 5.32761 + 3.5   **12.** 0.000800 − 0.00052

   13. 602.987 ÷ 7   **14.** 10,000 + 0.25

   15. $\dfrac{(4320)(1.047)}{513.92}$ **16.** 4320 + 1.047 + 513.92

In Exercises 17–20, repeat Exercises 13–16, assuming that all integers are exact.

17. $602.987 \div 7$

18. $10,000 + 0.25$

19. $\dfrac{(4320)(1.047)}{513.92}$

20. $4320 + 1.047 + 513.92$

A–2 LINEAR EQUATIONS

EQUATIONS, SOLUTIONS, AND EQUIVALENCE

An **equation** is a statement that two quantities are equal. Like other statements, an equation may be true or false. For example,

$$2 + 2 = 4$$

is a true equation, while

$$2 + 2 = 5$$

is a false equation.

Some equations involve one or more **variables**, that is, symbols which can be replaced by real numbers. A replacement that produces a true equation is a **solution** of the equation. For example, 2 is a solution of the equation

$$8x - 2 = 2(x + 5)$$

since the equation is true if x is replaced by 2. Furthermore, 2 is the only solution, since the equation is false if x is replaced by any other number.

Two equations are **equivalent** if and only if they have the same solutions. For example, the equations

$$4x - 1 = x + 5$$

$$3x = 6$$

$$x = 2$$

are all equivalent to $8x - 2 = 2(x + 5)$, since they all have 2 as their only solution.

If you perform any of the following operations on an equation, you always produce an equivalent equation.

- Replace quantities by equal quantities anywhere in the equation.
- Add or subtract equal quantities on each side of the equation.
- Multiply or divide each side of the equation by equal nonzero quantities.

SOLVING LINEAR EQUATIONS

An equation in one variable, say x, is called **linear** if the only operations performed on x are addition and subtraction, and multiplication and division by *constant* quantities. Thus $3x + 20 = 7x$ is linear, while $x^2 = 9$ is not. A linear equation in one variable can always be solved by performing only those operations listed above.

EXAMPLE 1   *Solving a linear equation in one variable*

Here is one sequence of steps which can be used to solve

$$8x - 2 = 2(x + 5)$$

Replace $2(x + 5)$ by $2x + 10$.

$$8x - 2 = 2x + 10$$

Add 2 to both sides.

$$8x = 2x + 12$$

Subtract $2x$ from both sides.

$$6x = 12$$

Divide both sides by 6.

$$x = 2$$ ■

EXERCISES   *In Exercises 1–8, solve the equation.*

1. $9 - 4x = 33$

2. $q - 13 = 5q + 7$

3. $5(t - 4) = -(7 - 2t)$

4. $0.01z = 7.86 - 2.61z$

5. $\dfrac{x}{2} + 4 = \dfrac{x}{3}$

6. $325(1 - y) = 497y$

7. $8w + 5 = 5$

8. $\sqrt{3}\, p = \dfrac{1}{\sqrt{3}}$

9. A sequence of steps to solve $8x - 2 = 2(x + 5)$, different from the sequence in the example, is shown on page A4. Identify the operation that was used to obtain each equation from the preceding one.

A-3   THE COORDINATE PLANE

A **coordinate plane** (also called a **Cartesian plane** in honor of René Descartes, whom you will meet in Chapter 7) is a plane containing two number lines at right angles to each other, as in Figure A-1.

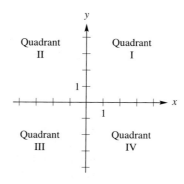

Figure A–1

The two number lines are the **horizontal axis** (*x*-**axis**) and the **vertical axis** (*y*-**axis**). Their point of intersection is the **origin**. The two axes divide the plane into four **quadrants**, numbered as in Figure A-1.

Every **ordered pair** of real numbers corresponds to a unique point in the coordinate plane, and vice versa. For example, the ordered pair $(3, -2)$ describes point A in Figure A-2, located 3 units to the right of the origin and 2 units below it. The numbers 3 and -2 are the *x*-**coordinate** and *y*-**coordinate** of A.

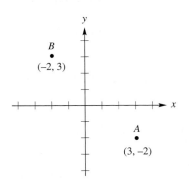

Figure A–2

Points are described by *ordered* pairs of numbers. Thus, the ordered pair $(-2, 3)$ does not describe A in Figure A-2, but instead describes B, located 2 units to the left of the origin and 3 units above it.

EXERCISES   **1.** Find the coordinates of each labeled point in the coordinate plane below.

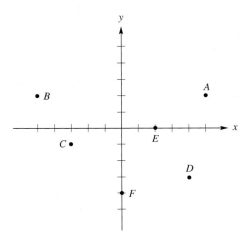

In Exercises 2–9, plot the given point in a coordinate plane, and tell which quadrant or which axis contains it.

2. $(6, 1)$ **3.** $(-1, -4)$

4. $(-2, 7)$ **5.** $(3, -1)$

6. $(5, -5)$ **7.** $(-5, 5)$

8. $(0, 2)$ **9.** $(3.5, 0)$

A–4 THE PYTHAGOREAN THEOREM AND THE DISTANCE FORMULA

THE PYTHAGOREAN THEOREM

A **right angle** is an angle having a measure of 90 degrees, as in Figure A-3a. A **right triangle** is a triangle containing a right angle, as in Figure A-3b. In a right triangle, the side opposite the 90-degree angle is called the **hypotenuse**.

Figure A–3

The **Pythagorean theorem** states:

> If the hypotenuse in a right triangle has length c and the other two sides have lengths a and b, then
>
> $$a^2 + b^2 = c^2$$

EXAMPLE 1   *Using the Pythagorean theorem to find the length of a side of a right triangle*
In Figure A-4, $a = 4$ and $c = 9$. To find b,

$$a^2 + b^2 = c^2$$
$$(4)^2 + b^2 = (9)^2$$
$$16 + b^2 = 81$$
$$b^2 = 65$$
$$b = \sqrt{65} \approx 8.06$$

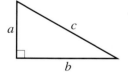

Figure A–4

THE DISTANCE FORMULA

An important consequence of the Pythagorean theorem is the **distance formula**.

> The distance between two points (x_1, y_1) and (x_2, y_2) in the coordinate plane is
>
> $$\sqrt{(x_2 - x_1)^2 + (y_2 - y_1)^2}$$

Figure A-5, on page A8, shows how the distance formula follows from the Pythagorean theorem. The distance between the points is represented by d. The horizontal distance between them is $|x_2 - x_1|$, so that the square of the horizontal distance is $(x_2 - x_1)^2$. Similarly, the square of the vertical distance is $(y_2 - y_1)^2$, so

$$d^2 = (x_2 - x_1)^2 + (y_2 - y_1)^2$$

Then

$$d = \sqrt{(x_2 - x_1)^2 + (y_2 - y_1)^2}$$

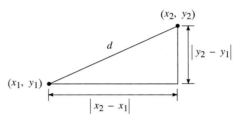

Figure A–5

EXERCISES   *In Exercises 1–8, suppose that the hypotenuse of a right triangle has length c and that the other two sides have lengths a and b. Find the length of the side not given.*

1. $a = 6$, $c = 10$ **2.** $b = 8$, $c = 17$

3. $a = 12$, $b = 5$ **4.** $a = 7$, $c = 25$

5. $b = 10$, $c = 20$ **6.** $a = 10$, $b = 10$

7. $a = 1$, $b = 3$ **8.** $a = 1$, $c = 3$

In Exercises 9–16, find the distance between the points.

9. $(4, -2)$ and $(1, 2)$ **10.** $(0, 12)$ and $(5, 0)$

11. $(-7, -3)$ and $(-9, -3)$ **12.** $(6, -3)$ and $(6, 5)$

13. $(-4, -4)$ and $(8, 3)$ **14.** $(2, 9)$ and $(7, 6)$

15. $(0, 0)$ and $(9, 40)$ **16.** $(1, 99)$ and $(-3, 97)$

A–5 BASIC GRAPHING TECHNIQUES

The equation

$$y = 2x - 1$$

describes a relationship between the variables x and y. A **solution** to this equation is an ordered pair of real numbers which produce a true equation when substituted for x and y, respectively. The **graph** of the equation consists of all points in the coordinate plane that represent solutions to the equation. The process of obtaining the graph of an equation is referred to as **graphing**. Let's graph $y = 2x - 1$.

EXAMPLE 1   *Graphing an equation*

To graph $y = 2x - 1$, begin by listing several solutions to the equation, choosing any values you please for x, and calculating the values for y.

When $x = 0$, the equation is true if $y = 2(0) - 1 = -1$, so $(0, -1)$ is a solution.

A few other solutions, calculated in the same way, are $(-1, -3)$, $(1, 1)$, $(2, 3)$, and $(3, 5)$. The points representing these solutions are shown in Figure A-6a.

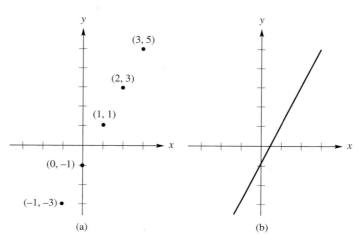

Figure A–6

Since there are infinitely many solutions, you cannot complete the graph by calculating all of them. Eventually, you must do one of two things.

• Make a guess based on the points you have plotted, or
• recall that the equation has the form $y = mx + b$, so its graph is a straight line (see Section A-6).

Figure A-6b shows the graph of $y = 2x - 1$. ■

Of course, the variables in a relationship need not be called x and y, as illustrated by the following example.

EXAMPLE 2   *Graphing an equation*
To graph

$$q = \frac{10}{p^2 + 1}$$

you must first decide whether the horizontal axis is to be the p-axis or the q-axis. Equivalently, you must decide whether the first number in an ordered pair is to represent a value of p or of q. Either choice is correct, but since the equation expresses q in terms of p, it is traditional to make the p-axis horizontal. (We will return to this idea in Section 2-1.)

Several solutions to the equation are $(-3, 1)$, $(-2, 2)$, $(-1, 5)$, $(0, 10)$, $(1, 5)$, $(2, 2)$, and $(3, 1)$, plotted in Figure A-7a (see page A10).

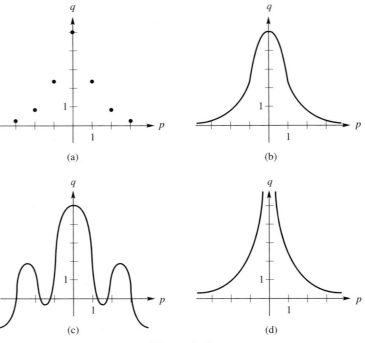

Figure A–7

Until you study equations like this one in Chapter 8, you must guess the overall shape of the graph. A good rule to follow is to sketch as simple a graph as you can, while including all of your plotted points. In particular, don't include any bends, corners, or jumps, unless your plotted points indicate that they are there. The most reasonable guess is shown in Figure A-7b, rather than Figure A-7c or A-7d. (You may need to plot a few more points to convince yourself that the graph never crosses the *p*-axis.)   ■

EXERCISES   *In Exercises 1–4, decide which points are on the graph of the equation.*

1. $y = x + 5$ $(-1, 4), (8, 3), (5, 10)$

2. $y = x^2 - 3x$ $(0, 0), (4, 4), (8, 8)$

3. $x^2 + y^2 = 25$ $(3, 2), (3, 4), (5, 0)$

4. $s = 2t$ $(2, 4), (3, 6), (100, 200)$

(Regard the first number in each ordered pair as a value of *t*.)

Sketch the graph of each equation in Exercises 5–12.

5. $y = x - 4$ **6.** $y = x^2 - 4$

7. $y = 3 - 2x$ **8.** $y = 3 - 2x^2$

9. $y = x^2 - 4x$ **10.** $y = x^2 - 4x + 4$

11. $y = x^3$ **12.** $y = x^3 - 2$

13. Sketch the graph of $z = 3w + 1$:

   a. by making the *w*-axis horizontal

   b. by making the *z*-axis horizontal

14. Sketch the graph of $v = \dfrac{u - 5}{2}$:

    a. by making the u-axis horizontal

    b. by making the v-axis horizontal

A–6   GRAPHING LINEAR EQUATIONS

An equation that can be written in the form

$$Ax + By = C$$

with A and B not both 0 is a **linear equation in two variables**.

EXAMPLE 1   *Deciding whether an equation in two variables is linear*

    a. $y = 3x - 4$ is a linear equation in two variables, since it is equivalent to $3x - y = 4$.

    b. $q - 5 = 2(p + 1)$ is a linear equation in two variables, since it is equivalent to $2p - q = -7$.

    c. $x = -8$ is a linear equation in two variables, since it is equivalent to $x + 0y = -8$.

    d. $x^2 + y^2 = 1$ is not a linear equation in two variables. ■

There are several techniques for graphing linear equations in two variables. Let's review two of the most common ones, the two-point method and the slope-intercept method.

THE TWO-POINT METHOD

You can graph a linear equation by first plotting any two points on the graph, then drawing the line through the plotted points. The **intercepts** (points where the graph intersects the coordinate axes) are convenient points to plot, especially if the equation is written in the form $Ax + By = C$. It is a good idea to plot a third point to check the accuracy of your work.

EXAMPLE 2   *Using the two-point method to graph a linear equation*

To graph

$$2x - 3y = 8$$

begin by plotting the intercepts. Since the x-intercept must have a y-coordinate of 0, you can find it by substituting 0 for y in the equation.

$$2x - 3(0) = 8$$
$$2x = 8$$
$$x = 4$$

Thus the x-intercept is at $(4, 0)$. Similarly, you can find the y-intercept at $\left(0, -\dfrac{8}{3}\right)$ by substituting 0 for x in the equation. The intercepts are shown in

Figure A-8a, along with the point $(7, 2)$, obtained by substituting 7 for x. The graph is shown in Figure A-8b.

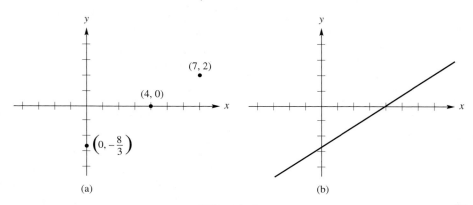

Figure A–8

EXAMPLE 3   *Using the two-point method to graph a linear equation*

To graph

$$2x - 3y = 0$$

begin by plotting the intercepts. This time, both intercepts are at $(0, 0)$, and thus provide only one point on the graph.

To obtain a second point, substitute any value except 0 for one of the variables. For example, substituting 6 for x yields

$$2(6) - 3y = 0$$
$$12 = 3y$$
$$y = 4$$

Thus, a second point on the graph is $(6, 4)$. Similarly, substituting -3 for x yields the point $(-3, -2)$. The graph is shown in Figure A-9. ∎

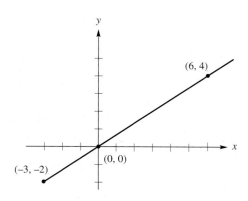

Figure A–9

THE SLOPE-INTERCEPT METHOD

An equation $Ax + By = C$ with $B \neq 0$ can be solved for y to obtain an equation $y = mx + b$. The coefficients m and b provide information about the graph of such an equation.

- The graph of $y = mx + b$ is a nonvertical straight line.
- The value of m is equal to the ratio

$$\frac{y_2 - y_1}{x_2 - x_1}$$

  for any pair of points (x_1, y_1) and (x_2, y_2) on the line. This ratio is the **slope** of the line.
- The y-intercept is $(0, b)$.

When a linear equation is written $y = mx + b$, it is said to be in **slope-intercept form**. The following example shows how to use the coefficients m and b to graph the equation.

EXAMPLE 4   *Using the slope-intercept method to graph a linear equation*
To graph

$$y = 5 - 2x$$

begin by rewriting the equation as

$$y = -2x + 5$$

so that $m = -2$ and $b = 5$.

Next, plot the y-intercept $(0, 5)$, as in Figure A-10.

The slope $-2 = \dfrac{-2}{1}$ represents the ratio of the change in y to the change in x between any two points on the line. You can therefore obtain a second point by starting at $(0, 5)$ and letting y and x change by -2 and 1, respectively. The second point is 2 units down from $(0, 5)$, and 1 unit to the right, as in Figure A-10. As in the two-point method, you can complete the graph by drawing the line through the two plotted points. ∎

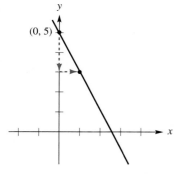

Figure A–10

HORIZONTAL AND VERTICAL LINES

Equations of the form $x = $ (constant) and $y = $ (constant) graph as vertical and horizontal lines, respectively. The slope of a horizontal line is 0, and the slope of a vertical line is undefined.

Horizontal and vertical lines are graphed most easily by remembering the geometric meanings of their equations.

EXAMPLE 5   *Graphing horizontal and vertical lines*

a. To graph

$$y = 3$$

recall that a point's y-coordinate is its directed vertical distance from the x-axis. Thus, the equation describes all points 3 units above the x-axis. Its graph is shown in Figure A-11a.

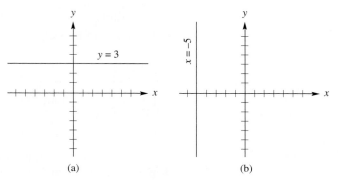

Figure A–11

b. Similarly, the equation

$$x = -5$$

describes all points 5 units to the left of the y-axis. Its graph is shown in Figure A-11b. ■

FINDING THE EQUATION OF A LINE

If a line with slope m passes through (x_1, y_1), then another point (x, y) is on the line if and only if

$$\frac{y - y_1}{x - x_1} = m$$

Multiplying both sides by $x - x_1$ produces the **point-slope form** of the equation

$$y - y_1 = m(x - x_1)$$

The following example shows how to find the equation of a line if you know either its slope and the coordinates of one of its points or the coordinates of two of its points.

EXAMPLE 6   *Finding the equation of a line*

a. The equation of the line through $(-2, 4)$ with slope 3 is

$$y - 4 = 3[x - (-2)]$$
$$y - 4 = 3(x + 2)$$

in point-slope form. Its slope-intercept form is

$$y = 3x + 10$$

b. The line through $(-1, 5)$ and $(3, 7)$ has a slope of

$$\frac{7 - 5}{3 - (-1)} = \frac{1}{2}$$

Using $(-1, 5)$ as (x_1, y_1), we find that the equation in point-slope form is

$$y - 5 = \frac{1}{2}(x + 1)$$

The slope-intercept form is

$$y = \frac{1}{2}x + \frac{11}{2}$$

[If we use $(3, 7)$ as (x_1, y_1), the point-slope equation $y - 7 = \frac{1}{2}(x - 3)$ also simplifies to $y = \frac{1}{2}x + \frac{11}{2}$.] ■

EXERCISES   *In Exercises 1–10, sketch the graph of the equation.*

1. $4x + y = 12$ **2.** $x - 3y = 6$

3. $2x + 5y = 10$ **4.** $2x + 5y = 0$

5. $y = 3x + 2$ **6.** $y = 2 - x$

7. $y = -5x$ **8.** $x = 2y - 8$

9. $x = 2$ **10.** $y = 0$

In Exercises 11–14, find the equation of the line through the given point with the given slope.

11. $(2, -2), m = 1$ **12.** $(5, 4), m = -3$

13. $(-3, -5), m = \frac{2}{3}$ **14.** $(-1, 4), m = 0$

In Exercises 15–18, find the equation of the line through the given points.

15. $(6, -1)$ and $(8, 7)$ **16.** $(-2, 9)$ and $(10, 5)$

17. $(1, -1)$ and $(0, 0)$ **18.** $(3, 0)$ and $(3, 4)$

A–7 INTERVALS

If $a < b$, the set of all numbers between a and b is an **interval**. The set of all numbers to the right of a, and the set of all numbers to the left of b, are also intervals. The intervals may or may not include their **endpoints** a and b.

Every interval can be described by an inequality. For example, the set of all numbers between -2 and 3 can be described by one of the inequalities $-2 < x < 3$, $-2 < x \le 3$, $-2 \le x < 3$, or $-2 \le x \le 3$, depending on whether the interval includes either or both of the endpoints -2 and 3.

Interval notation provides a concise way of describing intervals. Table A-1 (page A16) shows the graph of each type of interval, the inequality that describes it, and the corresponding description in interval notation.

TABLE A-1

Graph	Inequality	Interval notation
───○━━━━━○───	$a < x < b$	(a, b)
───●━━━━━●───	$a \leqslant x \leqslant b$	$[a, b]$
───●━━━━━○───	$a \leqslant x < b$	$[a, b)$
───○━━━━━●───	$a < x \leqslant b$	$(a, b]$
───○━━━━━━━━	$a < x$	(a, ∞)
───●━━━━━━━━	$a \leqslant x$	$[a, \infty)$
━━━━━━━○───	$x < b$	$(-\infty, b)$
━━━━━━━●───	$x \leqslant b$	$(-\infty, b]$

a b

The first four types of intervals are **finite**, and the last four are **semi-infinite**. The interval $(-\infty, \infty)$, which includes all real numbers, is **infinite**. A finite interval which contains both its endpoints is **closed**, and one which contains neither endpoint is **open**.

We often need to describe sets of numbers made up of two or more disjoint intervals. For example, Figure A-12 shows the set of all numbers whose absolute values are between 1 and 2, inclusive.

Figure A–12

This set is the **union** of the intervals $[-2, -1]$ and $[1, 2]$ and is written in interval notation as $[-2, -1] \cup [1, 2]$. Similarly, the set of all real numbers except π, shown in Figure A-13, can be written as $(-\infty, \pi) \cup (\pi, \infty)$.

Figure A–13

EXERCISES   *In Exercises 1–8, graph the interval described by the given inequality, and describe it in interval notation.*

1. $1 < x \leqslant 5$ **2.** $-3 \leqslant x < 0$

3. $-40 < x < 40$ **4.** $4.9 \leqslant x \leqslant 5.1$

5. $x < \sqrt{2}$ **6.** $x \geqslant -10$

7. $x \leqslant 0$ **8.** $x > 0$

In Exercises 9–16, graph the interval whose interval notation description is given, and describe it by an inequality.

9. $(-2, 2)$

10. $[-2, 2]$

11. $[0, 1.5)$

12. $[1, \infty)$

13. $(-\infty, 1)$

14. $(-10.4, -4.7]$

15. $\left[-\dfrac{1}{2}, \dfrac{2}{3}\right)$

16. $\left(-\dfrac{1}{2}, \dfrac{2}{3}\right]$

In Exercises 17–24, use both an inequality and an interval notation to describe the interval whose graph is shown.

17.

18.

19.

20.

21.

22.

23.

24.

In Exercises 25–28, graph the set of numbers described.

25. $(-4, 1) \cup [3, 5]$

26. $[2, 6) \cup (6, \infty)$

27. $(-\infty, -5] \cup (-\sqrt{3}, \sqrt{3})$

28. $[0, 1] \cup (2, 3) \cup [4, 5]$

In Exercises 29–32, use interval notation to describe the set of numbers whose graph is shown.

29.

30.

31.

32.

A–8 LINEAR INEQUALITIES

An **inequality** is a statement having one of the following forms.

$a < b$	a is less than b
$a > b$	a is greater than b
$a \le b$	a is less than or equal to b
$a \ge b$	a is greater than or equal to b

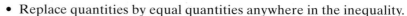

Figure A–14

Like equations, inequalities can be either true or false. For example,

$$x < 5$$

is true for some values of x, such as 1, and false for others, such as 7.

A **solution** of an inequality is a number that produces a true statement when substituted for the variable. Inequalities typically have infinitely many solutions. These can be represented graphically, as in Figure A-14.

Two inequalities are **equivalent** if they have the same solutions. If you perform any of the following operations on an inequality, you always produce an equivalent inequality.

- Replace quantities by equal quantities anywhere in the inequality.
- Add or subtract equal quantities on each side of the inequality.
- Multiply or divide each side of the inequality by equal *positive* quantities.
- Multiply or divide each side of the inequality by equal *negative* quantities, *and reverse the direction of the inequality.*

Some care must be taken in applying the last two rules. For example,

$3x < -15$ is equivalent to $x < -5$.
(Dividing both sides by 3 does not reverse the direction of the inequality.)

$-3x < -15$ is equivalent to $x > 5$.
(Dividing both sides by -3 reverses the direction of the inequality.)

An inequality in one variable, say x, is called **linear** if the corresponding equation in x is linear. Thus $3x + 20 \geq 7x$ is linear, while $x^2 \leq 9$ is not. You can always solve a linear inequality in one variable by performing only those operations listed above. The solution can be written in one of the forms $x < a, x > a, x \leq a,$ or $x \geq a$.

EXAMPLE 1   *Solving a linear inequality in one variable*

Here is one sequence of steps which can be used to solve

$$8x - 2 > 2(x + 5)$$

Replace $2(x + 5)$ by $2x + 10$.

$$8x - 2 > 2x + 10$$

Add 2 to both sides.

$$8x > 2x + 12$$

Subtract $2x$ from both sides.

$$6x > 12$$

Divide both sides by 6.

$$x > 2$$

The solution can also be represented in interval notation as $(2, \infty)$ or represented graphically, as in Figure A-15.

Figure A–15   ■

EXERCISES   *In Exercises 1–8, solve the inequality. Represent the solution as an inequality, as a graph, and in interval notation.*

1. $9 - 4x < 33$ **2.** $q - 13 \leqslant 5q + 7$

3. $5(t - 4) \geqslant -(7 - 2t)$ **4.** $0.01z > 7.86 - 2.61z$

5. $\dfrac{x}{2} + 4 \leqslant \dfrac{x}{3}$ **6.** $325(1 - y) < 497y$

7. $8w + 5 > 5$ **8.** $\sqrt{3}\, p \geqslant \dfrac{1}{\sqrt{3}}$

A–9   ABSOLUTE VALUE EQUATIONS AND INEQUALITIES

The **absolute value** of a real number x, written $|x|$, is defined as follows.

$$\text{If } x \geqslant 0, \text{ then } |x| = x.$$

$$\text{If } x < 0, \text{ then } |x| = -x.$$

According to the definition, $|x|$ is always nonnegative. For example, $|3| = 3$, and $|-4| = 4$. Thus, the absolute value of a number is found by simply eliminating the negative sign, if one is there. However, this simple rule does not apply to expressions involving more than one number or involving variables.

- It is *not* true that $|6 - 8| = 6 + 8$, because the left side is $|6 - 8| = |-2| = 2$ and the right side is $6 + 8 = 14$.
- It is *not always* true that $|-x| = x$. For example, if $x = -5$, then the left side is $|-(-5)| = |5| = 5$, and the right side is -5.

Geometrically, $|x|$ represents the undirected distance of x from 0 on the number line. Figure A-16 shows that $|-5| = 5$ and $|5| = 5$.

Figure A–16

For any constant a, the equation $|x| = a$ and the inequalities $|x| < a$, $|x| \leqslant a$, $|x| > a$, $|x| \geqslant a$ can all be written as equivalent statements without absolute values.

EXAMPLE 1   *Rewriting absolute value equations and inequalities without absolute values*

a. The equation

$$|x| = 5$$

says that x is 5 units from 0 on the number line and is therefore equivalent to the statement

$$x = -5 \quad \text{or} \quad x = 5$$

b. The inequality

$$|x| < 5$$

says that x is less than 5 units from 0. Figure A-17 shows that this is true if and only if x is in $(-5, 5)$, so the inequality is equivalent to the statement

$$-5 < x < 5$$

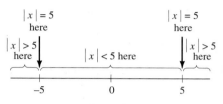

Figure A–17

c. The inequality

$$|x| > 5$$

says that x is more than 5 units from 0. Figure A-17 shows that this is true if and only if x is in $(-\infty, -5) \cup (5, \infty)$, so the inequality is equivalent to the statement

$$x < -5 \quad \text{or} \quad x > 5 \qquad \blacksquare$$

You can use the observations in Example 1 to solve equations and inequalities involving absolute values of linear expressions in x.

EXAMPLE 2   *Solving a linear absolute value equation*
The equation

$$|3 - 2x| = 5$$

is equivalent to the statement

$$3 - 2x = -5 \quad \text{or} \quad 3 - 2x = 5$$

Solving each equation yields

$$-2x = -8 \quad \text{or} \qquad -2x = 2$$
$$x = 4 \quad \text{or} \qquad x = -1$$

The equation has two solutions, namely, -1 and 4.   ∎

EXAMPLE 3   *Solving a linear absolute value inequality*
The inequality

$$|3 - 2x| < 5$$

is equivalent to the statement

$$-5 < 3 - 2x < 5$$

Solving yields

$$-8 < -2x < 2$$

$$4 > x > -1$$

Figure A–18

The solution can also be represented in interval notation as $(-1, 4)$ or graphically as in Figure A-18. ■

EXAMPLE 4   *Solving a linear absolute value inequality*
The inequality

$$|3 - 2x| \geqslant 5$$

is equivalent to the statement

$$3 - 2x \leqslant -5 \quad \text{or} \quad 3 - 2x \geqslant 5$$

Solving each equation yields

$$-2x \leqslant -8 \quad \text{or} \quad -2x \geqslant 2$$

$$x \geqslant 4 \quad \text{or} \quad x \leqslant -1$$

The solution can also be represented in interval notation as $(-\infty, -1] \cup [4, \infty)$ or graphically as in Figure A-19.

Figure A–19   ■

EXAMPLE 5   *Solving linear absolute value equations and inequalities*
a. To solve

$$|4x + 7| = -2$$

notice that the left side is an absolute value and cannot be equal to -2. Thus, the equation has no solution.

b. To solve

$$|2x - 6| > 0$$

notice that the left side must be positive unless the quantity inside the absolute value signs is 0. The inequality is true unless $2x - 6 = 0$, that is, unless $x = 3$. The solution is therefore $(-\infty, 3) \cup (3, \infty)$. ■

EXERCISES   *In Exercises 1–8,*

a. Solve the equation.
b. Replace = by $\leqslant$, and solve the inequality.
c. Replace = by $>$, and solve the inequality.

1. $|2x| = 14$

2. $|x + 6| = 0.05$

3. $|2x - 13| = 3$

4. $|13 - 2x| = 3$

5. $|3x + 5| = 13$ **6.** $|5 - 4x| = 7$

7. $|x + 3| = -1$ **8.** $|x + 3| = 0$

A–10 SYSTEMS OF LINEAR EQUATIONS

A **linear equation in n variables**, say $x_1, x_2, \ldots, x_n$, is one having the form

$$a_1 x_1 + a_2 x_2 + \cdots + a_n x_n = c$$

A **system of linear equations** is a set of one or more linear equations to be solved simultaneously. A **solution** to the system is an ordered set of n real numbers that makes every equation in the system true when substituted for $x_1, \ldots, x_n$.

A system of m linear equations in n variables is an $m \times n$ **linear system**. Let's review some commonly used methods for solving 2×2 and 3×3 linear systems.

SOLVING 2×2 LINEAR SYSTEMS

Two analytical methods of solving 2×2 linear systems are the **method of substitution** and the **method of elimination** (also called the **Gaussian elimination**). Let's review each with an example.

EXAMPLE 1   *Solving a 2×2 linear system by the method of substitution*

To solve the system

$$2x + 5y = 3$$
$$3x - 4y = 16$$

by the method of substitution, begin by solving either equation for either of the variables. For example, solve the first equation for x.

$$2x = 3 - 5y$$
$$x = \frac{3 - 5y}{2}$$

Then substitute the resulting expression for x in the other equation.

$$3x - 4y = 16$$
$$3\left(\frac{3 - 5y}{2}\right) - 4y = 16$$

Next, solve this equation for y.

$$3(3 - 5y) - 8y = 32$$
$$9 - 15y - 8y = 32$$
$$-23y = 23$$
$$y = -1$$

Finally, replace y by -1 in the expression for x.

$$x = \frac{3 - 5y}{2}$$

$$x = \frac{3 - 5(-1)}{2}$$

$$x = 4$$

The system has one solution, $(x, y) = (4, -1)$.   ■

EXAMPLE 2   *Solving a 2 × 2 linear system by the method of elimination*
To solve

$$2x + 5y = 3$$
$$3x - 4y = 16$$

by the method of elimination, begin by multiplying each equation by a constant, so that in the resulting equations, the coefficients of one variable are either equal, or opposites of each other. For example, multiply the first equation by 4 and the second by 5.

$$8x + 20y = 12$$
$$15x - 20y = 80$$

Then add corresponding sides of the two equations, and solve for x.

$$\begin{array}{rl} 8x + 20y &= 12 \\ 15x - 20y &= 80 \\ \hline 23x &= 92 \\ x &= 4 \end{array}$$

Finally, replace x by 4 in either of the original equations, and solve for y.

$$2x + 5y = 3$$
$$2(4) + 5y = 3$$
$$5y = -5$$
$$y = -1$$

The solution is $(4, -1)$.   ■

SOLVING 3 × 3 LINEAR SYSTEMS

The methods of substitution and elimination can also be used to solve 3 × 3 linear systems.

EXAMPLE 3   *Solving a 3 × 3 linear system by the method of substitution*
To solve

$$x + 2y - z = 7$$
$$2x + y + 2z = 5$$
$$x - y - 2z = -2$$

using the method of substitution, begin by solving any equation for any variable. For example, solve the last equation for x.

$$x - y - 2z = -2$$

$$x = y + 2z - 2$$

Then substitute the resulting expression for x in each of the other equations.

$$x + 2y - z = 7$$

$$2x + y + 2z = 5$$

$$(y + 2z - 2) + 2y - z = 7$$

$$2(y + 2z - 2) + y + 2z = 5$$

$$3y + z = 9$$

$$3y + 6z = 9$$

Next, solve the resulting 2×2 system. Subtracting corresponding sides of the two equations yields

$$\begin{aligned} 3y + z &= 9 \\ \underline{3y + 6z} &= 9 \\ -5z &= 0 \end{aligned}$$

$$z = 0$$

Replacing z by 0 in either equation yields

$$3y = 9$$

$$y = 3$$

Finally, replace y by 3 and z by 0 in the expression for x.

$$x = y + 2z - 2$$

$$x = (3) + 2(0) - 2$$

$$x = 1$$

The solution is $(x, y, z) = (1, 3, 0)$.   ■

EXAMPLE 4   *Solving a 3×3 linear system by the method of elimination*
To solve

$$x + 2y - z = 7$$

$$2x + y + 2z = 5$$

$$x - y - 2z = -2$$

using the method of elimination, begin by choosing any two equations, and multiply each by a constant to eliminate any variable. For example, choose the first two equations, multiply the first by 2 and subtract to eliminate x.

$$x + 2y - z = 7$$

$$2x + y + 2z = 5$$

$$2x + 4y - 2z = 14$$
$$2x + y + 2z = 5$$
$$\overline{3y - 4z = 9}$$

Then choose a different pair of equations, and eliminate the same variable. For example, eliminate x from the first and third equations by subtracting them.

$$x + 2y - z = 7$$
$$x - y - 2z = -2$$
$$\overline{3y + z = 9}$$

Next, solve the system

$$3y - 4z = 9$$
$$3y + z = 9$$

to obtain $y = 3$, $z = 0$. Finally, replace y by 3 and z by 0 in any of the original equations, and solve for x.

$$x + 2y - z = 7$$
$$x + 2(3) - (0) = 7$$
$$x = 1$$

The solution is the ordered triple $(x, y, z) = (1, 3, 0)$. ■

EXERCISES   *In Exercises 1–16, solve the linear system.*

1. $2x + 5y = 9$
   $2x - 5y = -1$

2. $s + 4t = 12$
   $2s + t = 10$

3. $4x + 3y = 6$
   $5x + 4y = 2$

4. $3x - 7y = 0.3$
   $2x - 8y = 1.5$

5. $5x - 3y = 21$
   $y = x - 5$

6. $q = 6p - 9$
   $q = 4p + 11$

7. $x = 8 - 3y$
   $y = 8 - 3x$

8. $42x - 79y = 0$
   $35x + 27y = 0$

9. $x + y + z = 0$
   $2x + y - z = -4$
   $-x + 4y + z = 7$

10. $3x - 2y + 4z = 7$
    $x + 5y - 2z = 24$
    $4x + 3y + z = 29$

11. $2x - 3y + z = 1$
    $2x + y - 6z = 2$
    $2y + z = 0$

12. $x - y = 1$
    $y - z = 2$
    $x + z = 5$

13. $a + b - 3c = 0$
    $a - 2b + c = 0$
    $2a - b - c = 0$

14. $a + b - 3c = 0$
    $a - 2b + c = 0$
    $2a - b - c = 3$

15. $x - 2y - 3z = 4$
    $x - 3y - 4z = 0$
    $x - 4y - 6z = 3$

16. $7x - 6y + 2z = 5$
    $4x + 3y - 2z = 5$
    $2x - 3y - 3z = 1$

A–11 THE LAWS OF EXPONENTS

In the expression x^p, the number p is an **exponent**. Its meaning is defined as follows.

- If p is a positive integer, then x^p is shorthand for the product $x \cdot x \cdots \cdot x$, with p factors. For example,

$$3^4 \text{ means } 3 \cdot 3 \cdot 3 \cdot 3 = 81$$

and

$$x^3 y^2 \text{ means } x \cdot x \cdot x \cdot y \cdot y$$

- If $x \neq 0$, then $x^0 = 1$. The expression 0^0 is undefined.

- If p is a positive integer, then $x^{-p} = \dfrac{1}{x^p}$. For example,

$$3^{-4} = \frac{1}{3^4} = \frac{1}{81}$$

and

$$x^3 y^{-2} = x^3 \left(\frac{1}{y^2} \right) = \frac{x^3}{y^2}$$

- If p is a rational number $\dfrac{r}{s}$ in lowest terms, then $x^p = x^{r/s}$ is defined to be either $\sqrt[s]{x^r}$ or the equivalent expression $\left(\sqrt[s]{x} \right)^r$, as long as $\sqrt[s]{x}$ is defined. For example,

$$4^{3/2} = \sqrt{4^3} = \sqrt{64} = 8$$

Equivalently,

$$4^{3/2} = \left(\sqrt{4} \right)^3 = 2^3 = 8$$

$$4^{-3/2} = \frac{1}{4^{3/2}} = \frac{1}{8}$$

$$(-4)^{3/2} = \left(\sqrt{-4} \right)^3$$

which is not defined as a real number.

When expressions involving exponents are combined, they obey the following **laws of exponents**.

1. $x^p x^q = x^{p+q}$ **2.** $\dfrac{x^p}{x^q} = x^{p-q}$

3. $(xy)^p = x^p y^p$ **4.** $\left(\dfrac{x}{y} \right)^p = \dfrac{x^p}{y^p}$

5. $(x^p)^q = x^{pq}$

In words, the laws say

- To multiply (divide) powers of the same base, add (subtract) exponents (laws 1 and 2).

- Exponents distribute over products and quotients (laws 3 and 4).
- To raise a power to a power, multiply exponents (law 5).

EXAMPLE 1   *Applying the laws of exponents*

a. $x^3(x^4 - 2x^2 + 1)$

$$= x^3x^4 - 2x^3x^2 + x^3(1)$$

$$= x^7 - 2x^5 + x^3$$

b. $\dfrac{k}{k^3} = k^{1-3} = k^{-2}$ or $\dfrac{1}{k^2}$

c. $(5y)^3 = 5^3y^3 = 125y^3$

d. $\left(\dfrac{5}{y}\right)^{-3} = \left(\dfrac{1}{5/y}\right)^3 = \left(\dfrac{y}{5}\right)^3 = \dfrac{y^3}{5^3} = \dfrac{y^3}{125}$

e. $(9t^3)^{1/2} = 9^{1/2}(t^3)^{1/2} = \sqrt{9}\ t^{3/2} = 3t^{3/2}$ ∎

Although exponents distribute over products and quotients, they do not distribute over sums and differences. For example, it is *not* correct to say that $(u^2 + v^3)^2 = u^4 + v^6$. A correct calculation is

$$(u^2 + v^3)^2 = (u^2 + v^3)(u^2 + v^3)$$

$$= u^2u^2 + v^3u^2 + u^2v^3 + v^3v^3$$

$$= u^4 + 2u^2v^3 + v^6$$

EXERCISES   *In Exercises 1–8, evaluate without using a calculator.*

1. 2^5 **2.** 6^{-2}

3. $(-1)^6$ **4.** $(-3)^{-3}$

5. 19^0 **6.** $27^{2/3}$

7. $16^{-3/4}$ **8.** $\left(\dfrac{9}{4}\right)^{-1/2}$

In Exercises 9–20, combine powers of the same base where possible, and eliminate parentheses and negative exponents.

9. k^3k^{-2} **10.** $(k^3)^{-2}$

11. $(4b^5)(2b^2)$ **12.** $(4b^5)^2$

13. $\dfrac{3z^7}{12z^3}$ **14.** $\left(\dfrac{w^2}{2w^5}\right)^3$

15. $(2xy^4)^2$ **16.** $(2x - y^4)^2$

17. $c^{-3}(c^4 + 3c^3 - 5c)$ **18.** $\left(\dfrac{x^3}{7y^2}\right)^{-2}$

19. $v^{1/2}v^{1/3}$ **20.** $(64a^{1/2})^{2/3}$

A–12   FACTORING

A **polynomial** is a sum of terms, each of which is a product of a number and a nonnegative integer power of a variable. A few examples of polynomials are

$$x^4 y^3 - 4x^2 y$$

$$3p - 6q + r$$

$$7$$

Polynomials with one, two, and three terms are called, respectively, **monomials**, **binomials**, and **trinomials**. Thus 7 is a monomial, $x^4 y^3 - 4x^2 y$ is a binomial, and $3p - 6q + r$ is a trinomial. There are no commonly used names for polynomials with four or more terms.

The **degree** of a term in a polynomial is the number of its variable factors. For example, $2x^3 = 2 \cdot x \cdot x \cdot x$ has degree 3, and the term $4x^2 y = 4 \cdot x \cdot x \cdot y$ also has degree 3. The degree of a polynomial is defined as the highest degree of any of its terms, so the three polynomials in the last paragraph have degrees, 0, 7, and 1, respectively.

To **factor** a polynomial is to express it as a product of one or more polynomial factors. Here are factorizations of several polynomials. You can check the correctness of each by performing the multiplications on the right side.

Polynomial	Factorization
$2x^3 + 8x^2$	$2x^2(x + 4)$
$x^2 - 4y^2$	$(x + 2y)(x - 2y)$
$2p^2 - 5p + 2$	$(p - 2)(2p - 1)$
$3z^5 + 3z^4 - 12z - 12$	$3(z + 1)(z^2 + 2)(z^2 - 2)$

A polynomial is **factored completely** if it is expressed as a product of polynomials that cannot be factored again. If the original polynomial has integer coefficients, then the phrase "cannot be factored again" is usually understood to mean "cannot be factored again using only integer coefficients." For example, it is not customary to write $2p - 1$ as $2\left(p - \dfrac{1}{2}\right)$ or to write $z^2 - 2$ as $\left(z + \sqrt{2}\right)\left(z - \sqrt{2}\right)$. Both of these factorizations are correct, and you will perform similar ones in Chapters 5 and 7, but in the present section, we will use only integer coefficients.

Let's review a few common factoring techniques.

REMOVAL OF COMMON FACTORS

If all terms in a polynomial have a factor in common, you can use the distributive law to write the polynomial as a product of the common factor and another polynomial. The process is referred to as the **removal of common factors**.

EXAMPLE 1   *Removing common factors*

a. $6x^2 + 12x - 3 = 3(2x^2 + 4x - 1)$

b. $x^2y^3z^6 - x^4y^3z^4 + x^2y^5z^4 = x^2y^3z^4(z^2 - x^2 + y^2)$

c. $-2x + 6 = 2(-x + 3)$ or

$-2x + 6 = -2(x - 3)$

Each factorization in part (c) is correct, and each is a complete factorization of $-2x + 6$. They differ by the removal of a common factor of -1. A factor of -1 can be removed from any completely factored polynomial to produce an equally correct complete factorization. ■

Before applying any other factoring techniques to a polynomial, it is good practice to remove all common factors first.

TECHNIQUES FOR FACTORING BINOMIALS

The following formulas are useful for factoring binomials.

$$A^2 - B^2 = (A - B)(A + B)$$

$$A^3 - B^3 = (A - B)(A^2 + AB + B^2)$$

$$A^3 + B^3 = (A + B)(A^2 - AB + B^2)$$

These are the formulas for a **difference of squares**, **difference of cubes**, and **sum of cubes**. You can check the correctness of each by performing the multiplication on the right side. In applying the formulas, you may replace A and B by any polynomial.

EXAMPLE 2   *Factoring binomials*

a. To factor

$$4r^2 - 9s^2$$

think of the polynomial as

$$(2r)^2 - (3s)^2$$

and apply the formula for the difference of squares, replacing A by $2r$ and B by $3s$.

$$A^2 - B^2 = (A - B)(A + B)$$

$$(2r)^2 - (3s)^2 = (2r - 3s)(2r + 3s)$$

b. To factor

$$2z^6 - 54$$

remove common factors first to obtain

$$2(z^6 - 27)$$

Then think of $z^6 - 27$ as $(z^2)^3 - 3^3$, and apply the formula for the difference of cubes, replacing A by z^2 and B by 3.

$$A^3 - B^3 = (A - B)(A^2 + AB + B^2)$$
$$(z^2)^3 - 3^3 = (z^2 - 3)[(z^2)^2 + (z^2)(3) + 3^2]$$
$$= (z^2 - 3)(z^4 + 3z^2 + 9)$$

The complete factorization is

$$2z^6 - 54 = 2(z^2 - 3)(z^4 + 3z^2 + 9)$$

c. To factor

$$x^9 + 1$$

think of the polynomial as

$$(x^3)^3 + 1^3$$

and apply the formula for the sum of cubes, replacing A by x^3 and B by 1.

$$A^3 + B^3 = (A + B)(A^2 - AB + B^2)$$
$$(x^3)^3 + 1^3 = (x^3 + 1)[(x^3)^2 - (x^3)(1) + 1^2]$$
$$= (x^3 + 1)(x^6 - x^3 + 1)$$

The first factor can be factored again, using the sum of cubes formula and replacing A by x and B by 1.

$$(x^3 + 1)(x^6 - x^3 + 1)$$
$$= (x + 1)(x^2 - x + 1)(x^6 - x^3 + 1) \qquad \blacksquare$$

TECHNIQUES FOR FACTORING TRINOMIALS

The most common trinomials have the form $ax^2 + bx + c$ and can sometimes be factored as

$$ax^2 + bx + c = (a_1x + s_1)(a_2x + s_2)$$

If the trinomial is **monic** ($a = 1$), then multiplying on the right confirms that $a_1a_2 = 1$, $s_1 + s_2 = b$, and $s_1s_2 = c$. These observations often allow you to factor it quickly.

EXAMPLE 3   *Factoring monic quadratic trinomials*

a. If it is possible to factor

$$x^2 + 3x - 40$$

the factorization must have the form $(x + s_1)(x + s_2)$, with $s_1 + s_2 = 3$ and $s_1s_2 = -40$. A little trial and error produces $s_1 = 8$, $s_2 = -5$ (or vice versa), so

$$x^2 + 3x - 40 = (x + 8)(x - 5)$$

b. If it is possible to factor

$$x^2 + 2x - 40$$

the factorization must have the form $(x + s_1)(x + s_2)$, with $s_1 + s_2 = 2$ and $s_1 s_2 = -40$. The only pairs of integers with a product of -40 and a positive sum are

$$-1 \text{ and } 40 \text{ (sum } = 39)$$

$$-2 \text{ and } 20 \text{ (sum } = 18)$$

$$-4 \text{ and } 10 \text{ (sum } = 6)$$

$$-5 \text{ and } 8 \text{ (sum } = 3)$$

Since no pair has a sum of 2, it is impossible to factor $x^2 + 2x - 40$ using only integer coefficients. ■

If the trinomial is not monic, a modification of the foregoing procedure can be used, as in the following example.

EXAMPLE 4   *Factoring nonmonic quadratic trinomials*

To factor a quadratic trinomial $ax^2 + bx + c$, first express the product ac as a product of two integers with a sum of b, if possible. For the particular polynomial

$$6x^2 - 19x + 15$$

this means expressing $(6)(15) = 90$ as a product of two integers with a sum of -19. The two integers are -9 and -10.

Second, rewrite the term bx using the two integers just obtained. In this case, the trinomial is rewritten as

$$6x^2 - 9x - 10x + 15$$

Third, group the four resulting terms into two pairs, and remove common factors from each pair.

$$(6x^2 - 9x) + (-10x + 15) = 3x(2x - 3) - 5(2x - 3)$$

Fourth, think of the polynomial in its present form as having two terms, $3x(2x - 3)$ and $-5(2x - 3)$, and remove the common factor $(2x - 3)$.

$$3x(2x - 3) - 5(2x - 3) = (2x - 3)(3x - 5)$$ ■

TECHNIQUES FOR FACTORING POLYNOMIALS WITH FOUR TERMS

Polynomials with four terms can sometimes be factored by the process of **grouping**, so named because it begins by factoring groups of terms within the polynomial. The technique was illustrated in the example you just read. The following example illustrates further.

EXAMPLE 5   *Factoring by grouping*

a. To factor

$$x^3 + x^2 - x - 1$$

group the terms in pairs, and remove common factors from each pair.

$$(x^3 + x^2) + (-x - 1) = x^2(x + 1) - (x + 1)$$

Then remove a common factor of $x + 1$ from the two resulting terms.

$$x^2(x + 1) - (x + 1) = (x + 1)(x^2 - 1)$$

Finally, factor $x^2 - 1$ using the formula for a difference of squares.

$$(x + 1)(x^2 - 1) = (x + 1)(x - 1)(x + 1)$$
$$= (x - 1)(x + 1)^2$$

b. To factor

$$p^3q - p^2q - pq^3 + pq^2$$

begin by removing common factors to obtain

$$pq(p^2 - p - q^2 + q)$$

Next, group the terms inside parentheses in pairs and remove common factors from each pair.

$$pq[(p^2 - p) + (-q^2 + q)] = pq[p(p - 1) - q(q - 1)]$$

At this point, the two terms $p(p - 1)$ and $q(q - 1)$ have no common factors, so you cannot continue. Nor can you leave the polynomial in its present form, since the polynomial in the brackets is not factored. Try a different grouping.

$$pq[(p^2 - q^2) + (-p + q)] = pq[(p - q)(p + q) - (p - q)]$$

Here $p^2 - q^2$ has been factored using the formula for a difference of squares. Now continue by removing a common factor of $(p - q)$.

$$pq[(p - q)(p + q) - (p - q)] = pq(p - q)[(p + q) - 1]$$
$$= pq(p - q)(p + q - 1) \qquad \blacksquare$$

EXERCISES   *In Exercises 1–20, factor the polynomial, if possible:*

1. $x^2yz + xy^2z + xyz^2$

2. $36m^2 - 54m$

3. $c^2 - 25d^2$

4. $c^2 + 25d^2$

5. $r^2 + 10r + 21$

6. $5z^3 - 40$

7. $x^2 - x - 30$

8. $10 + 3z - z^2$

9. $3x^2 + 8x - 11$

10. $ac + ad + bc + bd$

11. $2p^2 - 5p + 2$

12. $4q^2 - 17q + 4$

13. $x^4y + 27xy^4$

14. $s^2 + 7s - 44$

15. $ax^3 - 2a^2x^2 + 5a^3x$

16. $ax^3 - 6a^2x^2 + 5a^3x$

17. $x^4 - 16$

18. $y^4 - 3y^3 + 4y^2 - 12y$

19. $q^6 - 1$

20. $x^4 - 10x^2 + 9$
(*Hint*: Regard this as a quadratic trinomial in the variable x^2.)

A–13   QUADRATIC EQUATIONS

A **quadratic equation** in x is one that can be written in the form $ax^2 + bx + c = 0$ for some real numbers a, b, and c.

TECHNIQUES FOR SOLVING QUADRATIC EQUATIONS

Factoring If $ax^2 + bx + c$ can be factored, the equation $ax^2 + bx + c = 0$ can be solved using the Principle of Zero Products. This principle states that a product is equal to zero if and only if one of the factors is zero.

EXAMPLE 1   *Solving a quadratic equation by factoring*

You can solve

$$2x^2 + 3x - 2 = 0$$

by factoring the left-hand side to obtain

$$(2x - 1)(x + 2) = 0$$

By the Principle of Zero Products, this equation is equivalent to the statement

$$2x - 1 = 0 \quad \text{or} \quad x + 2 = 0$$

The solutions are $x = \dfrac{1}{2}$ and $x = -2$.   ■

The Principle of Zero Products can be used to solve a quadratic equation only when one side is zero. To solve

$$(x - 1)(x + 4) = -6,$$

it is *not* appropriate to solve $x - 1 = -6$ and $x + 4 = -6$. You must obtain zero on one side before factoring, as follows.

$$x^2 + 3x - 4 = -6$$

$$x^2 + 3x + 2 = 0$$

$$(x + 1)(x + 2) = 0$$

$$x = -1 \quad \text{or} \quad x = -2$$

Completing the Square Every expression $ax^2 + bx + c$ can be written in the form $a(x - h)^2 + k$ for some real numbers h and k. This process, called **completing the square**, can be used to solve quadratic equations.

EXAMPLE 2   *Completing the square for a quadratic expression*

To complete the square for $2x^2 + 3x - 2$,

• First, group the first and second degree terms together, and factor out the coefficient of x^2.

$$2\left(x^2 + \frac{3}{2}x\right) - 2$$

• Second, take half of the x-coefficient $\left(\text{half of } \dfrac{3}{2} \text{ is } \dfrac{3}{4}\right)$ and square it $\left[\left(\dfrac{3}{4}\right)^2 = \dfrac{9}{16}\right]$. Add and subtract this quantity inside the parentheses.

$$2\left(x^2 + \frac{3}{2}x + \frac{9}{16} - \frac{9}{16}\right) - 2$$

- Third, remove the last term from the parentheses, remembering to multiply it by the factor in front.

$$2\left(x^2 + \frac{3}{2}x + \frac{9}{16}\right) - \frac{9}{8} - 2$$

- Fourth, express the quantity inside the parentheses as the square of a binomial, and combine the constants outside the parentheses.

$$2\left(x + \frac{3}{4}\right)^2 - \frac{25}{8} \qquad \blacksquare$$

EXAMPLE 3 *Solving a quadratic equation by completing the square*

To solve

$$2x^2 + 3x - 2 = 0$$

begin by completing the square as in the last example, to obtain

$$2\left(x + \frac{3}{4}\right)^2 - \frac{25}{8} = 0$$

Continue by isolating the binomial square on one side of the equation.

$$\left(x + \frac{3}{4}\right)^2 = \frac{25}{16}$$

Then take square roots on both sides.

$$x + \frac{3}{4} = \pm\frac{5}{4}$$

Thus

$$x = -\frac{3}{4} + \frac{5}{4} = \frac{1}{2} \quad \text{or} \quad x = -\frac{3}{4} - \frac{5}{4} = -2 \qquad \blacksquare$$

The Quadratic Formula Every equation $ax^2 + bx + c = 0$ can be solved by using the **quadratic formula**, which states that the solutions are

$$x = \frac{-b \pm \sqrt{b^2 - 4ac}}{2a}$$

EXAMPLE 4 *Solving a quadratic equation by using the quadratic formula*

To solve

$$2x^2 + 3x - 2 = 0$$

replace $a, b,$ and c by 2, 3, and -2 in the quadratic formula.

$$x = \frac{-3 \pm \sqrt{3^2 - 4(2)(-2)}}{2(2)}$$

$$x = \frac{-3 + 5}{4}$$

$$x = \frac{1}{2} \quad \text{or} \quad x = -2 \qquad \blacksquare$$

COMPLEX SOLUTIONS

If i is the imaginary number $\sqrt{-1}$, then any number $a + bi$ with a and b real is a **complex number**. A complex number is real if $b = 0$ and nonreal if $b \neq 0$. The solutions to some quadratic equations are nonreal complex numbers.

EXAMPLE 5   *A quadratic equation with nonreal complex solutions*

You can solve

$$x^2 - 6x + 13 = 0$$

either by completing the square or by using the quadratic formula. Either method yields the solutions

$$x = 3 \pm \sqrt{-4}$$

which can be rewritten as

$$x = 3 \pm 2i \qquad \blacksquare$$

THE NATURE OF THE SOLUTIONS OF A QUADRATIC EQUATION

The solutions of $ax^2 + bx + c = 0$ can be described in one of three ways, depending on whether the expression $b^2 - 4ac$ in the quadratic formula is positive, zero, or negative. The three possibilities are summarized in Table A-2.

TABLE A-2

Value of $b^2 - 4ac$	Nature of the solutions
Positive	2 real
Zero	1 real
Negative	2 nonreal

Because the expression $b^2 - 4ac$ *discriminates* among the three possibilities, it is called the **discriminant** of the expression $ax^2 + bx + c$.

EXAMPLE 6   *Information provided by the discriminant*

The discriminant of $x^2 + 4x + 3$ is

$$4^2 - 4(1)(3) = 4$$

so the equation $x^2 + 4x + 3 = 0$ has 2 real solutions. (Check this by solving to obtain $x = -3$ and $x = -1$.)

The discriminant of $x^2 + 4x + 4$ is

$$4^2 - 4(1)(4) = 0$$

so the equation $x^2 + 4x + 4 = 0$ has 1 real solution. (Check this by solving to obtain $x = -2$.)

The discriminant of $x^2 + 4x + 5$ is

$$4^2 - 4(1)(5) = -4$$

so the equation $x^2 + 4x + 5 = 0$ has 2 nonreal solutions. (Check this by solving to obtain $x = -2 \pm i$.) ■

EXERCISES   *In Exercises 1–6, solve by factoring. In Exercises 7–12, solve by completing the square. In Exercises 13–18, solve by using the quadratic formula. In Exercises 19–24, solve by any method.*

1. $x^2 + 3x - 28 = 0$ **2.** $x^2 - 5x - 66 = 0$

3. $-4x^2 + x + 14 = 0$ **4.** $-2x^2 + 7x = -15$

5. $x(x - 6) = 27$ **6.** $(x - 9)(x + 13) = -117$

7. $x^2 + 5x - 9 = 0$ **8.** $2x^2 - 3x + 8 = 0$

9. $-x^2 + 8x - 14 = 0$ **10.** $x^2 - 36 = -5x$

11. $(2x - 1)^2 = 9$ **12.** $(x - 3)(x - 4) = 20$

13. $x^2 + 6x - 15 = 0$ **14.** $1 - 4x + 2x^2 = 0$

15. $7x^2 + 5 = 0$ **16.** $(1 - 2x)(3 + 4x) = 5$

17. $100x^2 - 20x + 1 = 0$ **18.** $x^2 - x - 1 = 0$

19. $x^2 + 7x + 6 = 0$ **20.** $x^2 - 5x + 4 = 0$

21. $3x^2 + 7x - 10 = 0$ **22.** $x^2 = 90 - x$

23. $x(x + 7) = 60$ **24.** $8 + 6x - 5x^2 = 0$

In Exercises 25–30, use the discriminant to determine the nature of the solutions. Check by solving, using a method of your choice.

25. $6x^2 - 10x + 8 = 0$ **26.** $4x^2 + 25 = 0$

27. $(x - 7)^2 - 79 = 0$ **28.** $3x^2 = 17x$

29. $4x^2 - 20x + 25 = 0$ **30.** $(2x + 3)(2x - 3) = -9$

A-14 OPERATIONS WITH COMPLEX NUMBERS

If i is the imaginary number $\sqrt{-1}$, then any number $a + bi$ with a and b real is a **complex number**. Thus the set of complex numbers includes all real numbers, such as -7, $\dfrac{\sqrt{3}}{2}$, and π, as well as numbers with an imaginary part, such as $3i$ and $1 - \sqrt{2}\, i$.

Operations on complex numbers are performed by treating i as an ordinary algebraic symbol, except that i^2 can be replaced by -1. The following example illustrates addition, subtraction, and multiplication of complex numbers.

EXAMPLE 1   *Adding, subtracting, and multiplying complex numbers*

a. $(5 + 7i) + (4 - i) = (5 + 4) + (7i - i) = 9 + 6i$

b. $(5 + 7i) - (4 - i) = (5 - 4) + [7i - (-i)] = 1 + 8i$

c. $(5 + 7i)(4 - i) = 20 - 5i + 28i - 7i^2$

$$= 20 + 23i - 7i^2$$

$$= 20 + 23i - 7(-1)$$

$$= 27 + 23i$$ ∎

To divide complex numbers, and express the result in the form $a + bi$, you need to make use of the fact that when a complex number $x + yi$ is multiplied by its **conjugate** $x - yi$, the result is always real. Specifically,

$$(x + yi)(x - yi) = x^2 - xyi + xyi - y^2i^2$$

$$= x^2 - y^2(-1)$$

$$= x^2 + y^2$$

The following example illustrates the use of conjugates in the division of complex numbers.

EXAMPLE 2   *Dividing complex numbers*

$$\frac{5 + 7i}{4 - i} = \frac{5 + 7i}{4 - i} \cdot \frac{4 + i}{4 + i}$$

$$= \frac{20 + 5i + 28i + 7i^2}{16 + 4i - 4i - i^2}$$

$$= \frac{20 + 33i + 7(-1)}{16 - (-1)}$$

$$= \frac{13 + 33i}{17}$$

$$= \frac{13}{17} + \frac{33}{17}i$$ ∎

By using the fact that $i^2 = -1$, you can express any integer power of i as either ± 1 or $\pm i$.

$$i^1 \qquad\qquad\qquad\qquad = i$$
$$i^2 \qquad\qquad\qquad\qquad = -1$$
$$i^3 \quad = i^2 i = (-1)i \qquad = -i$$
$$i^4 \quad = (i^2)^2 = (-1)^2 \quad = 1$$
$$i^5 \quad = i^4 i = (1)(i) \qquad = i$$
$$i^6 \quad = i^4 i^2 = (1)(-1) \quad = -1$$
$$i^7 \quad = i^4 i^3 = (1)(-i) \quad = -i$$
$$i^8 \quad = i^4 i^4 = (1)(1) \qquad = 1$$

and so on, repeating in cycles of four.

EXERCISES   *In Exercises 1–16, perform the indicated operations and express the result in the form a + bi with a and b real.*

1. $(6 + 2i) + (-4 + 5i)$
2. $-7i - (3 + 8i)$
3. $(5 + \sqrt{2}i) - (5 - \sqrt{2}i)$
4. $(5 + \sqrt{2}\,i) + (5 - \sqrt{2}\,i)$
5. $(6 + 2i)(-4 + 5i)$
6. $-7i(3 + 8i)$
7. $(5 + \sqrt{2}i)(5 - \sqrt{2}i)$
8. $(5 + \sqrt{2}\,i)^2$
9. $\dfrac{6 + 2i}{-4 + 5i}$
10. $\dfrac{-7i}{3 + 8i}$
11. $\dfrac{3 + 8i}{-7i}$
12. $\dfrac{5 + \sqrt{2}\,i}{5 - \sqrt{2}\,i}$
13. $(1 + i)^3$
14. $2i(2 - 3i)(1 + 2i)$
15. i^{100}
16. i^{99}

A–15   DIVISION AND SYNTHETIC DIVISION OF POLYNOMIALS

Division of one polynomial by another produces a quotient that can be expressed in the form

$$Q(x) + \frac{N(x)}{D(x)}$$

where Q, N, and D are polynomials and N has a smaller degree than D. The rational expression $\dfrac{N(x)}{D(x)}$ is the **remainder** of the division.

DIVISION OF MONOMIALS

You can divide a polynomial by a monomial term by term.

EXAMPLE 1   *Dividing a polynomial by a monomial*

a. $\dfrac{6x^5 + 8x - 1}{2x} = \dfrac{6x^5}{2x} + \dfrac{8x}{2x} - \dfrac{1}{2x}$

$$= 3x^4 + 4 - \frac{1}{2x}$$

The polynomial part of the quotient is $3x^4 + 4$, and the remainder is $-\dfrac{1}{2x}$.

b. $\dfrac{6x^5 + 8x - 1}{2x^2} = \dfrac{6x^5}{2x^2} + \dfrac{8x - 1}{2x^2}$

$$= 3x^3 + \frac{8x - 1}{2x^2}$$

The polynomial part of the quotient is $3x^3$, and the remainder is $\dfrac{8x - 1}{2x^2}$.

■

DIVISION BY OTHER POLYNOMIALS

When the divisor is not a monomial, polynomial division requires several steps, much like long division of numbers. The following example illustrates the process.

EXAMPLE 2   *Dividing polynomials*

To divide $3x^4 - 4x^3 + 7x - 1$ by $x^2 - 2$, begin by writing the terms of the divisor and dividend in descending order of degree.

$$x^2 - 2 \overline{\smash{\big)}3x^4 - 4x^3 \quad\quad + 7x - 1}$$

The dividend contains no term involving x^2, but your work will have better visual organization if you leave a space where that term would be.

STEP 1:   Divide the term of highest degree in the dividend by the term of highest degree in the divisor. Enter the result as the first term in the x^2 column of the quotient.

$$\begin{array}{r} 3x^2 \quad\quad\quad\quad\quad \\ x^2 - 2 \overline{\smash{\big)}3x^4 - 4x^3 \quad\quad + 7x - 1} \end{array}$$

STEP 2:   Multiply the resulting term in the quotient by the divisor. Enter the result below the dividend, keeping terms with the same degree in the same column.

$$\begin{array}{r} 3x^2 \quad\quad\quad\quad\quad \\ x^2 - 2 \overline{\smash{\big)}3x^4 - 4x^3 \quad\quad + 7x - 1} \end{array}$$

STEP 3:   Subtract.

$$\begin{array}{r} 3x^2 \quad\quad\quad\quad\quad \\ x^2 - 2 \overline{\smash{\big)}3x^4 - 4x^3 \quad\quad + 7x - 1} \\ \underline{3x^4 \quad\quad - 6x^2 \quad\quad\quad} \end{array}$$

STEP 4:   Regard the bottom line as a new dividend, and repeat steps 1–3.

$$\begin{array}{r} 3x^2 \; - 4x \quad\quad\quad\quad \\ x^2 - 2 \overline{\smash{\big)}3x^4 - 4x^3 \quad\quad + 7x - 1} \\ \underline{3x^4 \quad\quad - 6x^2 \quad\quad\quad} \\ -4x^3 + 6x^2 + 7x - 1 \\ \underline{-4x^3 \quad\quad + 8x \quad} \\ 6x^2 - x \; - 1 \end{array}$$

Continue to repeat steps 1–3 as long as the degree of the dividend is at least as large as that of the divisor.

$$
\begin{array}{r}
3x^2 - 4x + 6 \\
x^2 - 2\overline{)3x^4 - 4x^3 \qquad + 7x - 1} \\
\underline{3x^4 \qquad - 6x^2} \\
-4x^3 + 6x^2 + 7x - 1 \\
\underline{-4x^3 \qquad + 8x} \\
6x^2 - x \; - 1 \\
\underline{6x^2 \qquad - 12} \\
-x \; + 11
\end{array}
$$

The final result can be written as

$$
3x^2 - 4x + 6 + \frac{-x + 11}{x^2 - 2}
$$

∎

SYNTHETIC DIVISION

When the divisor has the form $x - a$, the long division process just described can be replaced by a streamlined process called **synthetic division**.

EXAMPLE 3   *Synthetic division of polynomials*

To divide $3x^4 - 4x^3 + 7x - 1$ by $x - 2$, begin with the following arrangement of numbers.

$$
2 \,\big|\; 3 \quad -4 \quad 0 \quad 7 \quad -1
$$

The number to the left of the vertical bar is the value of x for which the divisor is zero. The numbers to the right are the coefficients of the terms in the dividend, in order of decreasing degree.

STEP 1:   Copy the first coefficient in the dividend below the horizontal bar.

$$
2 \,\big|\; 3 \quad -4 \quad 0 \quad 7 \quad -1
$$
$$
 3
$$

STEP 2:   Multiply that number by the number to left of the vertical bar, and write the result in the next column to the right.

$$
2 \,\big|\; 3 \quad -4 \quad 0 \quad 7 \quad -1
$$
$$
 6
$$
$$
 3
$$

STEP 3:   Add the numbers in the second column.

$$
2 \,\big|\; 3 \quad -4 \quad 0 \quad 7 \quad -1
$$
$$
 6
$$
$$
 3 \quad 2
$$

STEP 4:   Repeat steps 2–3 until addition has been performed in all columns.

$$
\begin{array}{r|rrrrr}
2 & 3 & -4 & 0 & 7 & -1 \\
  &   & 6 & 4 & 8 & 30 \\
\hline
  & 3 & 2 & 4 & 15 & 29
\end{array}
$$

The quotient is represented synthetically by the numbers below the horizontal bar. The rightmost number is the numerator of the remainder. The others, from left to right, are the coefficients of the polynomial quotient, in order of decreasing degree. (The degree of the polynomial quotient is always one less than the degree of the dividend.) Thus the result of this division is

$$
3x^3 + 2x^2 + 4x + 15 + \frac{29}{x - 2} \qquad\blacksquare
$$

EXERCISES   *Perform the divisions in Exercises 1–12. Use synthetic division if the divisor has the form $x - a$.*

1. $\dfrac{x^6 - 4x^5 - 7x^3}{2x^3}$

2. $\dfrac{3x^2 + 5}{x^2}$

3. $\dfrac{2x^3 - 7x + 8}{x^2}$

4. $\dfrac{2x^3 - 7x + 8}{x^2 + 1}$

5. $\dfrac{x^7 + 4x^5 - x^2 + 3}{x^3 - 2x + 1}$

6. $\dfrac{x^5 - 6x^3 + x}{x^4 - 1}$

7. $\dfrac{x^3 - 4x^2 + 3x - 5}{x - 4}$

8. $\dfrac{2x^4 - 4x^2 + 3}{x + 2}$

9. $\dfrac{2x^2 - 5x + 2}{x - \dfrac{1}{2}}$

10. $\dfrac{x^5 + 1}{x + 1}$

11. $\dfrac{x^3 + 3x^2 - 8x + 1}{x^3 + 3x^2 + 4x - 5}$

12. $\dfrac{x^6}{x - 3}$

A-16   ALGEBRAIC FRACTIONS

An **algebraic fraction** is a quotient of two algebraic expressions. A **rational expression** is an algebraic fraction whose numerator and denominator are polynomials. Thus

$$
\frac{3y^4 + 6y^2 + 3}{y^2 - 1} \quad \text{and} \quad \frac{12}{x}
$$

are rational expressions, while

$$
\frac{a - 4}{\sqrt{a} - 2}
$$

is not, because its denominator is not a polynomial. All, however, are algebraic fractions. This section will review operations with rational expressions, but the rules given apply to all algebraic fractions.

EQUIVALENT FRACTIONS

Two algebraic fractions $\dfrac{a}{b}$ and $\dfrac{c}{d}$ are **equivalent** if and only if

- $ad = bc$

  and

- the equations $b = 0$ and $d = 0$ have the same solutions.

EXAMPLE 1   *Equivalent algebraic fractions*
a. The algebraic fractions

$$\frac{x + 3}{x + 2} \quad \text{and} \quad \frac{4x + 12}{4x + 8}$$

are equivalent, because

- Both $(x + 3)(4x + 8)$ and $(4x + 12)(x + 2)$ are equal to $4x^2 + 20x + 24$.
- The equations $x + 2 = 0$ and $4x + 8 = 0$ each have the single solution $x = -2$.

b. The fractions

$$\frac{x + 3}{x + 2} \quad \text{and} \quad \frac{(x + 3)(x - 5)}{(x + 2)(x - 5)}$$

are not equivalent if x is allowed to assume all real values, because the equations $x + 2 = 0$ and $(x + 2)(x - 5) = 0$ do not have the same solutions. However, if 5 is excluded as a replacement value for x, the two fractions are equivalent. ■

In performing operations on an algebraic fraction such as $\dfrac{x + 3}{x + 2}$, it may be convenient to write it as $\dfrac{(x + 3)(x - 5)}{(x + 2)(x - 5)}$, as though the two are equivalent. You will see why in Example 5. However, it is well to remember that the substitution is valid only if the values of x are restricted as in Example 1.

In the remainder of this section, we will assume that the values of all variables are restricted so that no denominators are zero.

REDUCTION OF FRACTIONS

The fractions $\dfrac{a}{b}$ and $\dfrac{ca}{cb}$ are equivalent (as long as $c \neq 0$), so that $\dfrac{ca}{cb}$ can be reduced by canceling a factor of c from both numerator and denominator.

EXAMPLE 2   *Reduction of algebraic fractions*

a. To reduce

$$\frac{x^2 - 9}{x^2 - 4x + 3}$$

begin by factoring the numerator and denominator.

$$\frac{(x + 3)(x - 3)}{(x - 1)(x - 3)}$$

Cancel a factor of $x - 3$ to obtain

$$\frac{x + 3}{x - 1}$$

b. To reduce

$$\frac{12 - 2x}{x^2 - 8x + 12}$$

begin by factoring

$$\frac{2(6 - x)}{(x - 2)(x - 6)}$$

Then recognize that $6 - x$ and $x - 6$ are opposites, so that $2(6 - x) = -2(x - 6)$. Thus the fraction can be reduced by rewriting it as

$$\frac{-2(x - 6)}{(x - 2)(x - 6)}$$

and canceling to obtain

$$\frac{-2}{x - 2} \quad \text{or} \quad \frac{2}{2 - x}$$

c. To reduce

$$\frac{4x - 2(x + 1)}{(x - 1)(x + 1)}$$

begin by multiplying, combining terms, and factoring the numerator.

$$\frac{4x - 2x - 2}{(x - 1)(x + 1)} = \frac{2x - 2}{(x - 1)(x + 1)}$$

$$= \frac{2(x - 1)}{(x - 1)(x + 1)}$$

Cancel to obtain

$$\frac{2}{x + 1} \qquad \blacksquare$$

Fractions can be reduced only by canceling *factors* of the numerator and denominator. Thus in part (c) of the foregoing example, you should not make the mistake of attempting to cancel the $x + 1$ immediately, because $x + 1$ is not a factor of the numerator.

MULTIPLICATION AND DIVISION OF FRACTIONS

Multiplication   Like numeric fractions, algebraic fractions are multiplied "straight across." That is, $\dfrac{a}{b} \cdot \dfrac{c}{d} = \dfrac{ac}{bd}$. For this reason, a factor in either numerator can be canceled with the same factor in either denominator.

EXAMPLE 3   *Multiplication of algebraic fractions*

$$\frac{x^3 - 8}{5x + 20} \cdot \frac{x^2 - 16}{x^3 - 6x^2 + 8x}$$

$$= \frac{(x - 2)(x^2 + 2x + 4)}{5(x + 4)} \cdot \frac{(x + 4)(x - 4)}{x(x - 4)(x - 2)}$$

$$= \frac{x^2 + 2x + 4}{5x}$$

∎

Division   Dividing by a fraction is equivalent to multiplying by its inverse its reciprocal). That is, $\dfrac{a}{b} \div \dfrac{c}{d} = \dfrac{a}{b} \cdot \dfrac{d}{c} = \dfrac{ad}{bc}$.

EXAMPLE 4   *Division of algebraic fractions*

$$\frac{a^3 y - a y^3}{ay - 4a} \div \frac{y^4 - a^4}{y^2 - 4y}$$

$$= \frac{a^3 y - a y^3}{ay - 4a} \cdot \frac{y^2 - 4y}{y^4 - a^4}$$

$$= \frac{ay(a + y)(a - y)}{a(y - 4)} \cdot \frac{y(y - 4)}{(y^2 + a^2)(y + a)(y - a)}$$

$$= \frac{-y^2}{y^2 + a^2}$$

[To cancel, regard $(a - y)$ as $(-1)(y - a)$.]   ∎

COMMON DENOMINATORS

Finding a common denominator for two algebraic fractions is the first step in performing addition or subtraction. The process is illustrated below.

EXAMPLE 5   *Finding a common denominator for two algebraic fractions*

a. To find a common denominator for

$$\frac{x + 3}{x + 2} \quad \text{and} \quad \frac{x - 1}{x - 5}$$

begin by noticing that any common denominator must contain both a factor of $x + 2$ and a factor of $x - 5$. The **least common denominator**, that is, the one containing the necessary factors and no others, is $(x + 2)(x - 5)$. The fractions equivalent to the given two, and having the required denominator, are

$$\frac{(x + 3)(x - 5)}{(x + 2)(x - 5)} \quad \text{and} \quad \frac{(x + 2)(x - 1)}{(x + 2)(x - 5)}$$

b. To find a common denominator for

$$\frac{1}{x^2 + 3x} \quad \text{and} \quad \frac{1}{x^2 - 4x}$$

begin by factoring both denominators to obtain

$$\frac{1}{x(x + 3)} \quad \text{and} \quad \frac{1}{x(x - 4)}$$

A common denominator must contain factors of $x, x + 3$ and $x - 4$. It need not contain a second factor of x. Thus, the least common denominator is $x(x + 3)(x - 4)$. The fractions equivalent to the given two, and having the required denominator, are

$$\frac{x - 4}{x(x + 3)(x - 4)} \quad \text{and} \quad \frac{x + 3}{x(x + 3)(x - 4)} \qquad ■$$

ADDITION AND SUBTRACTION OF FRACTIONS

To add or subtract algebraic fractions, you must first rewrite them with a common denominator. The numerators can then be combined. That is,

$$\frac{a}{b} \pm \frac{c}{d} = \frac{ad}{bd} \pm \frac{bc}{bd} = \frac{ad \pm bc}{bd}.$$

EXAMPLE 6   *Addition and subtraction of algebraic fractions*

a. To add

$$\frac{x + 3}{x + 2} + \frac{x - 1}{x - 5}$$

begin by finding a common denominator, as was done in the last example. The addition can then be rewritten as

$$\frac{(x + 3)(x - 5)}{(x + 2)(x - 5)} + \frac{(x + 2)(x - 1)}{(x + 2)(x - 5)}$$

The two fractions can now be added by adding their numerators.

$$= \frac{(x + 3)(x - 5) + (x + 2)(x - 1)}{(x + 2)(x - 5)}$$

$$= \frac{(x^2 - 2x - 15) + (x^2 + x - 2)}{(x + 2)(x - 5)}$$

$$= \frac{2x^2 - x - 17}{(x + 2)(x - 5)}$$

b. To subtract

$$\frac{1}{x^2 + 3x} - \frac{1}{x^2 - 4x}$$

begin by finding a common denominator, as was done in the last example. The subtraction can be rewritten as

$$\frac{x - 4}{x(x + 3)(x - 4)} - \frac{x + 3}{x(x + 3)(x - 4)}$$

The two fractions can now be subtracted by subtracting their numerators.

$$\frac{(x - 4) - (x + 3)}{x(x + 3)(x - 4)} = \frac{-7}{x(x + 3)(x - 4)}$$ ■

SIMPLIFICATION OF COMPLEX FRACTIONS

A **complex fraction** is one whose numerator and/or denominator are themselves fractions. A complex fraction can be written as a simple fraction by either of two commonly used techniques. The first makes use of the fact that a fraction denotes division, so that

$$\frac{\dfrac{a}{b}}{\dfrac{c}{d}} \quad \text{means} \quad \frac{a}{b} \div \frac{c}{d}$$

EXAMPLE 7   *Simplifying a complex fraction*

$$\frac{\dfrac{x}{x + 1}}{\dfrac{x^2}{x^2 - 1}} \quad \text{means} \quad \frac{x}{x + 1} \div \frac{x^2}{x^2 - 1}$$

$$= \frac{x}{x + 1} \cdot \frac{x^2 - 1}{x^2}$$

$$= \frac{x}{x + 1} \cdot \frac{(x + 1)(x - 1)}{x^2}$$

$$= \frac{x - 1}{x}$$ ■

A second technique, illustrated by the following example, is often effective when the numerator or denominator of a complex fraction is itself the sum or difference of two or more fractions.

EXAMPLE 8   *Simplifying a complex fraction*
To simplify

$$\frac{\dfrac{2}{x} - \dfrac{2}{y}}{\dfrac{4}{x^2} - \dfrac{4}{y^2}}$$

begin by finding a common denominator for all the terms in the numerator and denominator. The least common denominator is x^2y^2.

Next, multiply the complex fraction by another fraction whose numerator and denominator are both $\dfrac{x^2y^2}{1}$.

$$\dfrac{\dfrac{2}{x} - \dfrac{2}{y}}{\dfrac{4}{x^2} - \dfrac{4}{y^2}} \cdot \dfrac{\dfrac{x^2y^2}{1}}{\dfrac{x^2y^2}{1}}$$

$$= \dfrac{\dfrac{2}{x}\dfrac{x^2y^2}{1} - \dfrac{2}{y}\dfrac{x^2y^2}{1}}{\dfrac{4}{x^2}\dfrac{x^2y^2}{1} - \dfrac{4}{y^2}\dfrac{x^2y^2}{1}}$$

$$= \dfrac{2xy^2 - 2x^2y}{4y^2 - 4x^2}$$

$$= \dfrac{2xy(y - x)}{4(y + x)(y - x)}$$

$$= \dfrac{xy}{2(y + x)}$$ ∎

EXERCISES   *In Exercises 1–20, perform all operations and simplify.*

1. $\dfrac{4x + 20}{6x - 54} \cdot \dfrac{x - 9}{x + 5}$

2. $\dfrac{6}{x^2 - 4} \cdot \dfrac{3x + 6}{2x}$

3. $\dfrac{x - 2}{x + 3} \cdot \dfrac{x^2 - 9}{5x - 10} \cdot \dfrac{5}{x - 3}$

4. $\dfrac{x^2 + 4x + 4}{3x^4y^3} \cdot \dfrac{x^5y^2}{(2x + 4)}$

5. $\dfrac{x}{2x - 3} \div \dfrac{2x}{2x^2 - x - 3}$

6. $\dfrac{7}{x^2 + 2x + 4} \div \dfrac{14}{8 - x^3}$

7. $\dfrac{2x - 3}{x^2} \cdot \dfrac{x}{x - 4} \div (2x^2 - 5x - 12)$

8. $\dfrac{x^3 + 1}{1 - x^2} \div \dfrac{2x^2 - 2x + 2}{x + 1}$

9. $\dfrac{z}{2} + \dfrac{z}{3}$

10. $\dfrac{7}{x} - \dfrac{1}{x - 7}$

11. $\dfrac{a + 3}{a^2 - a} + \dfrac{a + 4}{a^2 - 1}$

12. $\dfrac{2q - 1}{8 - q} - \dfrac{3q + 7}{q - 8}$

13. $\dfrac{4x^2}{x - 5} - 4x$

14. $\dfrac{2x}{x - 6} + \dfrac{12}{6 - x}$

15. $\dfrac{3}{y^3 + 2y^2} + \dfrac{y - 1}{y^3 + 4y^2 + 4y}$

16. $\dfrac{x^2}{(x - 1)(x - 2)} - \dfrac{x}{(x - 2)(x - 3)} + \dfrac{1}{(x - 1)(x - 3)}$

17. $\dfrac{\dfrac{3x}{5y}}{\dfrac{12x}{15y}}$

18. $\dfrac{\dfrac{2t}{t-1}}{t^2-4t+3}$

19. $\dfrac{\dfrac{3}{x}+\dfrac{1}{2}}{\dfrac{3}{x}-\dfrac{1}{2}}$

20. $\dfrac{\dfrac{1}{x+2}-\dfrac{x}{x+5}}{\dfrac{x}{x+2}-\dfrac{1}{x+5}}$

A–17 EQUATIONS WITH ALGEBRAIC FRACTIONS

In solving an equation with algebraic fractions, the first step is often to multiply both sides by the least common denominator of all fractions in the equation. The resulting equation has no algebraic fractions, but may not be equivalent to the original one. Therefore, you need to check all solutions in the *original* equation.

EXAMPLE 1   *Solving equations with algebraic fractions*

a. To solve

$$\frac{3x+1}{6x-3} = \frac{5}{12x-6} + \frac{7}{2}$$

begin by factoring all denominators to obtain

$$\frac{3x+1}{3(2x-1)} = \frac{5}{6(2x-1)} + \frac{7}{2}$$

Next, find a common denominator for all fractional expressions in the equation. The least common denominator is $6(2x-1)$. Multiplying both sides of the equation by this quantity produces

$$\frac{3x+1}{3(2x-1)} \cdot \frac{6(2x-1)}{1} = \frac{5}{6(2x-1)} \cdot \frac{6(2x-1)}{1} + \frac{7}{2} \cdot \frac{6(2x-1)}{1}$$

$$2(3x+1) = 5 + 21(2x-1)$$

$$6x+2 = 5 + 42x - 21$$

$$18 = 36x$$

$$x = \frac{1}{2}$$

To obtain this solution, both sides of the equation were multiplied by the variable expression $6(2x-1)$, which has a value of 0 when $x = \dfrac{1}{2}$. Since multiplication of both sides of an equation by 0 does not always produce an equivalent equation, the solution may not be valid. In fact, substitution of $\dfrac{1}{2}$ for x in the original equation results in division by 0, so there is no solution.

b. To solve

$$\frac{3x + 1}{6x - 3} = \frac{-13}{12x - 6} + \frac{7}{2}$$

begin by factoring all denominators:

$$\frac{3x + 1}{3(2x - 1)} = \frac{-13}{6(2x - 1)} + \frac{7}{2}$$

As in part (a), multiply both sides by the least common denominator of $6(2x - 1)$ to obtain

$$\frac{3x + 1}{3(2x - 1)} \cdot \frac{6(2x - 1)}{1} = \frac{-13}{6(2x - 1)} \cdot \frac{6(2x - 1)}{1} + \frac{7}{2} \cdot \frac{6(2x - 1)}{1}$$

$$2(3x + 1) = -13 + 21(2x - 1)$$

$$6x + 2 = -13 + 42x - 21$$

$$36 = 36x$$

$$x = 1$$

This time, the multiplication by $6(2x - 1)$ does not invalidate the solution, since $6(2x - 1) \neq 0$ when $x = 1$. Therefore, the equation has the solution $x = 1$. ∎

In general, if both sides of an equation are multiplied by a variable expression, any resulting solution is valid *unless substituting the solution into the original equation makes one or more denominators zero.*

EXERCISES *In Exercises 1–8, solve the equation.*

1. $\dfrac{2x}{3} - \dfrac{x}{4} = \dfrac{25}{12}$

2. $\dfrac{x}{6} + \dfrac{1}{3} = \dfrac{21}{2x}$

3. $\dfrac{x}{2 - x} - 7 = \dfrac{2}{2 - x}$

4. $\dfrac{x}{2 - x} - 7 = \dfrac{10}{2 - x}$

5. $\dfrac{x}{x + 5} = \dfrac{x + 5}{x}$

6. $\dfrac{2x - 3}{x} = \dfrac{x + 7}{x - 9}$

7. $\dfrac{6}{x^2 - 9} - \dfrac{4}{x} = \dfrac{1}{x - 3}$

8. $\dfrac{x}{x + 4} + \dfrac{7}{x^2 + x - 12} = \dfrac{1}{x - 3}$

A–18 RADICALS AND RATIONAL EXPONENTS

EVEN AND ODD ROOTS

If $n \geq 2$ is an integer and $x^n = a$, then x is an **nth root** of a. The number n is the **index** of the root. It is often of interest to find all real nth roots of a given number.

EXAMPLE 1   *Real nth roots of a number*

a. Both 8 and -8 are real square roots of 64, because $8^2 = 64$ and $(-8)^2 = 64$.

b. There are no real square roots of -64, because the equation $x^2 = -64$ has no real solutions.

c. The only real cube root of 64 is 4, because $4^3 = 64$, and the equation $x^3 = 64$ has no other real solutions.

d. Both 2 and -2 are real sixth roots of 64, because $2^6 = 64$ and $(-2)^6 = 64$. ∎

The foregoing example can be generalized.

- If n is even, every positive real number a has two real nth roots, one positive and one negative. The positive root, called the **principal nth root** of a, is denoted by $\sqrt[n]{a}$. The negative root is always $-\sqrt[n]{a}$. The index n is usually omitted if $n = 2$. Thus $\sqrt{64} = 8$, $-\sqrt{64} = -8$, $\sqrt[6]{64} = 2$, and $-\sqrt[6]{64} = -2$.
- If n is even, no negative real number has a real nth root.
- If n is odd, every real number a has exactly one real nth root $\sqrt[n]{a}$. Thus $\sqrt[3]{64} = 4$ and $\sqrt[3]{-64} = -4$.
- $\sqrt[n]{0} = 0$ for every index n.

In the expression $\sqrt[n]{a}$, the symbol $\sqrt{}$ is called a **radical**, and a is the **radicand**.

SIMPLIFYING RADICALS

If n is even, then $\sqrt[n]{a^n} = |a|$ for every real number a.
If n is odd, then $\sqrt[n]{a^n} = a$ for every real number a.
For example, $\sqrt{(-5)^2} = \sqrt{25} = 5 = |-5|$, and $\sqrt[3]{(-5)^3} = \sqrt[3]{-125} = -5$.

EXAMPLE 2   *Simplifying radicals*

a. $\sqrt{25x^2} = \sqrt{(5x)^2} = |5x| = 5|x|$

b. $\sqrt{x^8} = \sqrt{(x^4)^2} = |x^4| = x^4$

c. $\sqrt[3]{8x^6} = \sqrt[3]{(2x^2)^3} = 2x^2$

d. $\sqrt[4]{x^4y^8} = \sqrt[4]{(xy^2)^4} = |xy^2| = |x|y^2$ ∎

MULTIPLYING AND DIVIDING RADICALS

If $\sqrt[n]{a}$ and $\sqrt[n]{b}$ are both real, then

$$\sqrt[n]{a} \cdot \sqrt[n]{b} = \sqrt[n]{ab}$$

$$\frac{\sqrt[n]{a}}{\sqrt[n]{b}} = \sqrt[n]{\frac{a}{b}}$$

That is, to multiply or divide radicals *of the same index*, multiply or divide the radicands. In the following examples, assume that all radical expressions represent real numbers.

EXAMPLE 3   *Multiplying and dividing radicals and simplifying*

a. $\sqrt{12x^5}\,\sqrt{3x} = \sqrt{36x^6} = \sqrt{(6x^3)^2} = 6x^3$

(We do not need to write $6|x^3|$, because $\sqrt{12x^5}$ and $\sqrt{3x}$ are not real if $x < 0$.)

b. $\dfrac{\sqrt{12x^5}}{\sqrt{3x}} = \sqrt{\dfrac{12x^5}{3x}} = \sqrt{4x^4} = 2x^2$

c. $\sqrt[3]{6x^2}\,\sqrt[3]{9x^2} = \sqrt[3]{54x^4} = \sqrt[3]{2 \cdot 27 \cdot x^3 \cdot x}$

$\qquad\qquad = \sqrt[3]{3^3}\,\sqrt[3]{x^3}\,\sqrt[3]{2x} = 3x\sqrt[3]{2x}$

d. $\dfrac{\sqrt{30x^6}}{\sqrt{10x}} = \sqrt{\dfrac{30x^6}{10x}} = \sqrt{3x^5} = \sqrt{3 \cdot (x^2)^2 \cdot x}$

$\qquad\qquad = x^2\sqrt{3x}$ ∎

ADDING AND SUBTRACTING RADICALS

If $\sqrt[n]{a}$ is real, then

$$r\sqrt[n]{a} \pm s\sqrt[n]{a} = (r \pm s)\,\sqrt[n]{a}$$

That is, to add or subtract radicals *of the same index and the same radicand*, add or subtract their coefficients.

EXAMPLE 4   *Adding and subtracting radicals and simplifying*

a. $7\sqrt{2x} - 4\sqrt{2x} = 3\sqrt{2x}$

b. $\sqrt[3]{5} + \sqrt[3]{40} = \sqrt[3]{5} + \sqrt[3]{2^3 \cdot 5} = \sqrt[3]{5} + 2\sqrt[3]{5} = 3\sqrt[3]{5}$ ∎

RATIONALIZING NUMERATORS AND DENOMINATORS

When working with an expression such as $\dfrac{\sqrt{3}}{\sqrt{5}}$, it is sometimes convenient to form an equivalent expression having no radicals in the denominator. To do so is to **rationalize the denominator**. At other times, we may want to **rationalize the numerator**.

EXAMPLE 5   *Rationalizing numerators and denominators*

a. To rationalize the denominator of $\dfrac{\sqrt{3}}{\sqrt{5}}$, multiply by $\dfrac{\sqrt{5}}{\sqrt{5}}$.

$$\frac{\sqrt{3}}{\sqrt{5}} \cdot \frac{\sqrt{5}}{\sqrt{5}} = \frac{\sqrt{15}}{\sqrt{25}} = \frac{\sqrt{15}}{5}$$

b. To rationalize the numerator of $\dfrac{\sqrt{3}}{\sqrt{5}}$, multiply by $\dfrac{\sqrt{3}}{\sqrt{3}}$.

$$\frac{\sqrt{3}}{\sqrt{5}} \cdot \frac{\sqrt{3}}{\sqrt{3}} = \frac{\sqrt{9}}{\sqrt{15}} = \frac{3}{\sqrt{15}}$$ ∎

To rationalize a numerator (denominator) consisting of a single nth root, multiply both the numerator and denominator by another nth root so that the numerator (denominator) becomes a perfect nth power.

EXAMPLE 6   *Rationalizing numerators and denominators*

a. To rationalize the numerator of $\dfrac{\sqrt[3]{x}}{\sqrt[3]{y}}$,

$$\frac{\sqrt[3]{x}}{\sqrt[3]{y}} \cdot \frac{\sqrt[3]{x^2}}{\sqrt[3]{x^2}} = \frac{\sqrt[3]{x^3}}{\sqrt[3]{x^2 y}} = \frac{x}{\sqrt[3]{x^2 y}}$$

b. To rationalize the denominator of $\dfrac{1}{\sqrt[4]{3}}$,

$$\frac{1}{\sqrt[4]{3}} \cdot \frac{\sqrt[4]{27}}{\sqrt[4]{27}} = \frac{\sqrt[4]{27}}{\sqrt[4]{81}} = \frac{\sqrt[4]{27}}{3}$$ ∎

To rationalize a numerator or denominator having the form $\sqrt{a} + \sqrt{b}$ or $\sqrt{a} - \sqrt{b}$, use the fact that

$$\left(\sqrt{a} + \sqrt{b}\right)\left(\sqrt{a} - \sqrt{b}\right) = a - b$$

EXAMPLE 7   *Rationalizing numerators and denominators*

a. To rationalize the numerator of $\dfrac{\sqrt{7} + \sqrt{3}}{\sqrt{2}}$,

$$\frac{\sqrt{7} + \sqrt{3}}{\sqrt{2}} \cdot \frac{\sqrt{7} - \sqrt{3}}{\sqrt{7} - \sqrt{3}} = \frac{7 - 3}{\sqrt{2}\left(\sqrt{7} - \sqrt{3}\right)}$$

$$= \frac{4}{\sqrt{14} - \sqrt{6}}$$

b. To rationalize the denominator of $\dfrac{3}{\sqrt{11} - \sqrt{5}}$,

$$\frac{3}{\sqrt{11} - \sqrt{5}} \cdot \frac{\sqrt{11} + \sqrt{5}}{\sqrt{11} + \sqrt{5}} = \frac{3\left(\sqrt{11} + \sqrt{5}\right)}{11 - 5}$$

$$= \frac{3\left(\sqrt{11} + \sqrt{5}\right)}{6} = \frac{\sqrt{11} + \sqrt{5}}{2}$$ ∎

RATIONAL EXPONENTS

If m and n are integers, $\dfrac{m}{n}$ is in lowest terms, and $\sqrt[n]{a}$ is real, then

$$a^{m/n} = \left(\sqrt[n]{a}\right)^m = \sqrt[n]{a^m}$$

EXAMPLE 8   *Evaluating expressions involving rational exponents*
a. $8^{5/3} = \left(\sqrt[3]{8}\right)^5 = 2^5 = 32$

b. $9^{-1/2} = \dfrac{1}{9^{1/2}} = \dfrac{1}{\sqrt{9}} = \dfrac{1}{3}$ ∎

The laws of exponents given on page A26 apply to rational exponents.

EXAMPLE 9   *Applying the laws of exponents*
   a. $x^{1/2}x^{2/3} = x^{1/2+2/3} = x^{3/6+4/6} = x^{7/6}$

   b. $(x^{1/3}y^2)^{3/2} = (x^{1/3})^{3/2}(y^2)^{3/2} = x^{1/2}y^3$ ∎

EXERCISES   *In Exercises 1–8, simplify the radical.*

   1. $\sqrt{9t^2}$ **2.** $\sqrt{100q^4}$
   3. $\sqrt{u^2v^8}$ **4.** $\sqrt{49x^4y^6}$
   5. $\sqrt[3]{27a^3}$ **6.** $\sqrt[3]{a^3b^6}$
   7. $\sqrt[4]{16y^{12}}$ **8.** $\sqrt[5]{p^5q^{20}}$

In Exercises 9–20, perform the indicated operations and simplify.

   9. $\sqrt{2}\sqrt{72}$ **10.** $\sqrt[3]{4x}\,\sqrt[3]{2x^2}$
   11. $\sqrt{10x}\sqrt{15xy}$ **12.** $\sqrt[3]{x^4y}\,\sqrt[3]{x^2y^4}$
   13. $\dfrac{\sqrt{189x^2}}{\sqrt{21}}$ **14.** $\dfrac{\sqrt[3]{250}}{\sqrt[3]{2z^3}}$
   15. $\dfrac{\sqrt{189x^3}}{\sqrt{3}}$ **16.** $\dfrac{\sqrt[3]{500z^4}}{\sqrt[3]{2z^3}}$
   17. $10\sqrt{k} + 2\sqrt{k}$ **18.** $\sqrt[4]{ab} - 5\sqrt[4]{ab}$
   19. $5\sqrt{20x} - 2\sqrt{5x^3}$ **20.** $\sqrt[3]{81} - \sqrt[3]{3}$

In Exercises 21–28,

   a. Rationalize the numerator.
   b. Rationalize the denominator.

   21. $\dfrac{\sqrt{x}}{\sqrt{6}}$ **22.** $\sqrt{\dfrac{2}{d}}$

   23. $\dfrac{\sqrt[3]{4}}{\sqrt[3]{s}}$ **24.** $\sqrt[3]{\dfrac{a}{b}}$

   25. $\dfrac{\sqrt{13} - \sqrt{5}}{\sqrt{5}}$ **26.** $\dfrac{\sqrt{x}}{\sqrt{x} + \sqrt{y}}$

   27. $\dfrac{\sqrt{2} + 1}{\sqrt{3}}$ **28.** $\dfrac{\sqrt{2} - 1}{\sqrt{3}}$

In Exercises 29–32, evaluate the expression without using a calculator.

   29. $1000^{2/3}$ **30.** $49^{-1/2}$

   31. $9^{5/2}$ **32.** $\left(\dfrac{8}{27}\right)^{1/3}$

In Exercises 33–40, simplify the expression.

   33. $a^6a^{2/3}$ **34.** $(a^6)^{2/3}$

   35. $\dfrac{a^6}{a^{2/3}}$ **36.** $(b^5c^{10})^{3/5}$

37. $\left(\dfrac{4}{x^8}\right)^{3/2}$ **38.** $\left(\dfrac{27u}{v^3}\right)^{4/3}$

39. $p^{-1/2}(p^{2/5})^{3/2}$ **40.** $\left(\dfrac{r^{5/3}}{r^{1/6}}\right)^{2/3}$

A–19 EQUATIONS WITH RADICALS

In solving an equation with square roots of variable expressions, one step is often to eliminate the radicals by squaring both sides of the equation. Since this operation does not always produce an equivalent equation, you need to check all solutions in the *original* equation.

EXAMPLE 1   *Solving an equation with radicals*

To solve

$$\sqrt{2x - 5} = x - 10$$

begin by squaring both sides of the equation.

$$\left(\sqrt{2x - 5}\right)^2 = (x - 10)^2$$
$$2x - 5 = x^2 - 20x + 100$$
$$x^2 - 22x + 105 = 0$$
$$(x - 7)(x - 15) = 0$$
$$x = 7 \quad \text{or} \quad x = 15$$

Next, substitute each solution for x in the original equation.

$\sqrt{2(7) - 5} = 7 - 10$	$\sqrt{2(15) - 5} = 15 - 10$
$\sqrt{9} = -3$	$\sqrt{25} = 5$
$3 = -3$	$5 = 5$

Substituting 7 for x produces a false equation, so 7 is not a solution. (It is sometimes called an *extraneous solution*.) Substituting 15 for x produces a true equation, so the solution is $x = 15$. ∎

EXAMPLE 2   *Solving an equation with radicals*

To solve

$$\sqrt{4x - 3} - 2 = \sqrt{2x - 5}$$

begin by squaring both sides.

$$\left(\sqrt{4x - 3} - 2\right)^2 = \left(\sqrt{2x - 5}\right)^2$$
$$(4x - 3) - 4\sqrt{4x - 3} + 4 = 2x - 5$$

The resulting equation still contains a radical. To eliminate it, isolate it on one side of the equation, then square both sides again.

$$2x + 6 = 4\sqrt{4x - 3}$$

$$(2x + 6)^2 = \left(4\sqrt{4x - 3}\right)^2$$

$$4x^2 + 24x + 36 = 16(4x - 3)$$

$$4x^2 - 40x + 84 = 0$$

$$4(x - 3)(x - 7) = 0$$

Now substitute each solution for x in the original equation.

$$\sqrt{4(3) - 3} - 2 = \sqrt{2(3) - 5}$$
$$\sqrt{9} - 2 = \sqrt{1}$$
$$3 - 2 = 1$$
$$1 = 1$$

$$\sqrt{4(7) - 3} - 2 = \sqrt{2(7) - 5}$$
$$\sqrt{25} - 2 = \sqrt{9}$$
$$5 - 2 = 3$$
$$3 = 3$$

Since substituting each solution for x produces a true equation, the solutions are $x = 3$ and $x = 7$. ∎

EXERCISES   *In Exercises 1–16, solve the equation.*

1. $\sqrt{3x + 1} = 7$

2. $\sqrt{1 - x} = 2$

3. $\sqrt{x - 5} = 0$

4. $\sqrt{x - 5} = -3$

5. $\sqrt{x + 5} = \sqrt{2x - 1}$

6. $\sqrt{3x} = \sqrt{x + 20}$

7. $\sqrt{x^2 - 9} = 4$

8. $\sqrt{x^2 + 12} = \sqrt{7x}$

9. $\sqrt{x} = x - 12$

10. $\sqrt{x + 5} = 2x$

11. $\sqrt{x^2 - 64} = x - 4$

12. $\sqrt{x^2 - 25} = 1 - x$

13. $\sqrt{x - 6} + \sqrt{x - 1} = 5$

14. $\sqrt{x} = 1 - \sqrt{x - 9}$

15. $\sqrt{x + 10} = \sqrt{x + 4} + 2$

16. $\sqrt{3x + 3} + 1 - \sqrt{6x + 7} = 0$

APPENDIX B

TIPS FOR GRAPHING FUNCTIONS WITH A CALCULATOR

PREREQUISITES BEFORE BEGINNING THIS APPENDIX, MAKE SURE YOU ARE FAMILIAR WITH:

The Keystrokes for Evaluating Functions, Graphing, Tracing, and Zooming on Your Calculator

To obtain accurate graphical information about a function from a calculator, you must do more than press keys.

- You must know what graphical information is provided by the equation of the function. You will develop that knowledge systematically beginning in Chapter 3.
- You must know how to represent graphical information accurately by controlling the appearance of the graph on your calculator screen. You will develop that knowledge in this appendix.

B-1 THE VIEWING WINDOW

When you graph a function on your calculator, the screen depicts a rectangular portion of the coordinate plane called the **viewing window**. You choose the viewing window by assigning minimum and maximum values to x and y. The appearance of a calculator graph is affected by both the *size* and the *shape* of the viewing window.

SIZE OF THE VIEWING WINDOW

The size of the viewing window affects both the extent of the graph and the amount of detail that you can see.

EXAMPLE 1   *Viewing a graph through windows of different sizes*

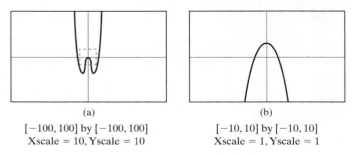

(a)

[−100, 100] by [−100, 100]
Xscale = 10, Yscale = 10

(b)

[−10, 10] by [−10, 10]
Xscale = 1, Yscale = 1

Figure B–1

Figure B-1 shows two graphs of the function $f(x) = 0.01x^4 − 1.21x^2 + 3$. The viewing window is indicated under each graph. Figure B-1b "blows up" the small dotted rectangle in Figure B-1a to fill the entire screen. That is, a portion of the coordinate plane 20 units wide and 20 units high (from −10 to 10 in each direction) now occupies the visual space previously occupied by a portion 200 units wide and 200 units high (from −100 to 100 in each direction). Thus we have **zoomed in** by a factor of 10 in each direction.

Which window is "better"? Neither is inherently better, but each brings out graphical features that the other does not. The window in Figure B-1a shows the overall shape of the graph, while the window in Figure B-1b reveals the number of *x*-intercepts. ∎

EXERCISES   *In Exercises 1–2,*

  a. Graph the function in each viewing window.

  b. Tell which window was obtained by zooming in from the other, and by what factor.

  1. $y = 4x − x^2$;   [−10, 10] by [−10, 10]
              [−1, 1] by [−1, 1]

  2. $y = 4x − x^2$;   [3, 5] by [−1, 1]
              [−5, 5] by [−5, 5]

In Exercises 3–6,

  a. Graph the function in the given window.

  b. Zoom in by a factor of 10 around the *x*-intercept shown in the window.

  c. Zoom in a second time by a factor of 10 around the *x*-intercept.

  d. Trace along the graph to approximate the *x*-intercept within 0.01.

  3. $y = x^3 − 10$; [−10, 10] by [−10, 10]

  4. $y = 7 + 2x − x^2$; [0, 10] by [−10, 10]

  5. $y = 3 − \sqrt{2x}$; [0, 10] by [−5, 5]

  6. $y = 10 − 5x^3 − x^4$; [−10, 0] by [−100, 100]

7. Match each graph with one of the viewing windows $[-10, 10]$ by $[-10, 10]$, $[-1, 1]$ by $[9, 11]$, $[9, 10]$ by $[0, 1]$. All three graphs represent the same function.

a.

b.

c.

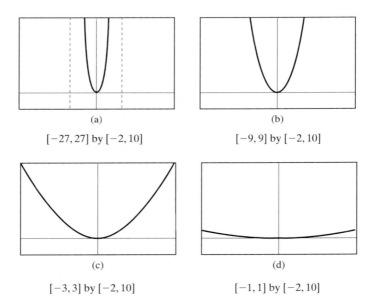

SHAPE OF THE VIEWING WINDOW

The shape of the viewing window affects the apparent steepness of a graph.

EXAMPLE 2 *Viewing a graph through windows of different shapes*

Figure B-2 shows four graphs of the function $y = x^2$. The viewing window used for each graph is indicated under it.

(a)

$[-27, 27]$ by $[-2, 10]$

(b)

$[-9, 9]$ by $[-2, 10]$

(c)

$[-3, 3]$ by $[-2, 10]$

(d)

$[-1, 1]$ by $[-2, 10]$

Figure B–2

All the graphs extend from -2 to 10 in the y-direction, and the window in Figure B-2a extends from -27 to 27 in the x-direction. In Figure B-2b we have zoomed in by a factor of 3 in the x-direction, thus expanding the dashed rectangular portion of the window in Figure B-2a to fill the entire screen. Zooming in repeatedly by a factor of 3 in the *x*-direction yields the graphs in Figures B-2c and d. Zooming in horizontally has the effect of making the graph appear horizontally stretched and, therefore, less steep.   ■

EXERCISES

8. a. Graph the function $f(x) = x^3 - x$ in each of the windows $[-3, 3]$ by $[-50, 50], [-3, 3]$ by $[-10, 10], [-3, 3]$ by $[-2, 2], [-3, 3]$ by $[-0.4, 0.4]$.

b. By what factor did you zoom in vertically to obtain each graph from the previous one?

c. How does zooming in vertically affect the apparent steepness of the graph?

Each of Exercises 9–10 shows three views of a graph. Match each graph with its viewing window. The coordinate axes have not been shown in any window.

9.

a.   **b.**

c.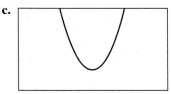

(1) $[-4, 4]$ by $[-1, 4]$
(2) $[-4, 4]$ by $[-4, 16]$
(3) $[-4, 4]$ by $[-16, 64]$

10.

a.   **b.**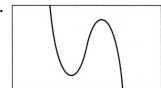

c.

(1) $[-1, 1]$ by $[-0.5, 0.5]$
(2) $[-3, 3]$ by $[-0.5, 0.5]$
(3) $[-9, 9]$ by $[-0.5, 0.5]$

UNDISTORTED WINDOWS

You have seen that some viewing windows make graphs appear steeper or flatter than they actually are. It is often necessary to choose a viewing window that depicts steepness accurately. We will refer to such a window as an **undistorted viewing window**.

EXAMPLE 3   *Distorted and undistorted viewing windows*

Figure B-3 shows the graph of $y = x$ in two windows on a Texas Instruments TI-85 calculator.

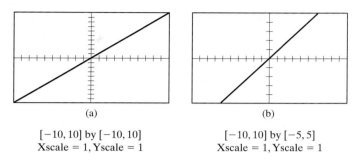

$[-10, 10]$ by $[-10, 10]$
Xscale $= 1$, Yscale $= 1$

$[-10, 10]$ by $[-5, 5]$
Xscale $= 1$, Yscale $= 1$

Figure B–3

In Figure B-3a the scale marks on the *x*-axis are farther apart than those on the *y*-axis. Thus horizontal distances are magnified in comparison with vertical distances, and the slope of the graph appears to be smaller than its actual value of 1. In Figure B-3b the scale marks are equally far apart on both axes, and the slope of the graph appears to be 1.

The screen on a TI-85 calculator is twice as wide as it is high, so that its undistorted windows show regions of the plane that are twice as wide as they are high. For example, the windows $[-20, 20]$ by $[-10, 10]$ and $[0, 100]$ by $[0, 50]$ are undistorted on a TI-85.   ■

Example 3 illustrates that the shape of your calculator screen determines whether a given window is undistorted.

EXERCISE   **11. a.** Consult your owner's manual to discover how to obtain undistorted windows on your calculator. (The manual may refer to them as "square" windows.)

   b. Find at least two undistorted windows on your calculator that are centered at $(0, 0)$, and at least one that is not.

ADJUSTING THE WINDOW

You have seen that some viewing windows make it difficult to identify important graphical features. Here are some of the most common problems, and some troubleshooting strategies that can often correct them.

Problem: The graph runs off the top or bottom of the screen and back on.
Solution: Zoom out vertically by a factor of 10.

Problem: The graph is contained in a thin horizontal strip.
Solution: Zoom in vertically by a factor of 10 around the horizontal strip. If repeated zooming yields no additional details, zoom out horizontally by a factor of 10.

Problem: The graph is contained in a thin vertical strip.
Solution: Zoom in horizontally by a factor of 10 around the vertical strip. If repeated zooming yields no additional details, zoom out vertically by a factor of 10.

Problem: The graph is contained in a small area near the center of the screen.
Solution: Zoom in by a factor of 10 around the region containing the graph.

Problem: The graph is not visible on the screen, or is visible only near one edge of the screen.
Solution: Zoom out by a factor of 10 around the center of the screen.

Problem: Any strange appearance in a graph.
Solution: Make sure you have entered the function correctly.

B–2   GRAPHING LINEAR FUNCTIONS

Although you can graph linear functions easily by hand, you may sometimes want to graph them on your calculator for several reasons. For example,

- It is easier to construct a table of values for a linear function from a calculator graph than from an equation or a hand-drawn graph.
- You must often solve problems by looking at a linear graph together with a more complicated one that would be difficult to graph manually.

Viewing windows for graphs should usually include all points of interest. For a linear function, the points of interest are the intercepts. Therefore, it is helpful to find the intercepts before graphing, especially if the coefficient of x or y is unusually large or small. A rough estimate will do.

EXAMPLE 1   *Finding a viewing window for a linear function*

To choose an appropriate viewing window for

$$y = 0.0011x - 43.5$$

begin by observing that the y-intercept is $(0, -43.5)$. The interval of y-values in the viewing window should include both -43.5 to show the y-intercept and 0 to show the x-axis. The interval $[-50, 10]$ includes both values.

Since the slope of the line is very small $\left(\text{about } \dfrac{1}{1000}\right)$, the line rises slowly, and the x-intercept has a large x-coordinate. You can obtain a rough estimate by recalling that the slope is the ratio $\dfrac{\text{change in } y}{\text{change in } x}$. From $(0, -43.5)$ to the x-intercept $(x, 0)$ the change in y is 43.5, so the change in x is about $(1000)(43.5) = 43,500$. Therefore, the interval of x-values in the viewing window should include both 0 and $43,500$.

The window $[-10,000, 50,000]$ by $[-50, 10]$ shows both intercepts. Because the window is 60,000 units long and only 60 units high, you might want to place scale marks every 10,000 units on the x-axis and every 10 units on the y-axis. The graph is shown in Figure B-4.

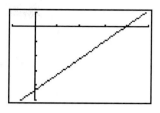

$[-10{,}000, 50{,}000]$ by $[-50, 10]$
Xscale $= 10{,}000$, Yscale $= 10$

Figure B–4

The window in Figure B-4 displays both intercepts clearly but distorts the steepness of the line. In general, when you choose a window to display some graphical feature, it may omit or distort some other feature.   ■

Example 1 illustrates a general procedure for graphing linear functions on a calculator.

- Find the y-intercept. Choose an interval containing its y-coordinate, and also containing 0. This is the interval of y-values for your viewing window.
- Find the x-intercept. Choose an interval containing its x-coordinate, and also containing 0. This is the interval of x-values for your viewing window.

- Scale the x-axis by choosing a convenient number between $\dfrac{1}{20}$ and $\dfrac{1}{5}$ of the width of the window.
- Scale the y-axis by choosing a convenient number between $\dfrac{1}{20}$ and $\dfrac{1}{5}$ of the height of the window.

EXERCISES   *In Exercises* 1–8,

    a. Find viewing windows showing both intercepts.

    b. Describe any resulting distortion in the shape of the graph.

1. $y = x + 12$ **2.** $y = 20 - 4x$

3. $y = 800 - 0.2x$

4. $y = -0.04 - 0.5x$ (Your viewing window should make it clear that the graph does not pass through the origin.)

5. $y = 0.0003x + 0.03$ **6.** $y = -7500(x + 0.4)$

7. $y = 279x + 35{,}000$ **8.** $y = 1000 - 2x$

B–3   GRAPHING QUADRATIC FUNCTIONS

For a quadratic function, the points of interest are the vertex and all intercepts. To avoid a lot of trial and error, find or estimate some of these points before graphing. On the graph of $y = ax^2 + bx + c$, the y-intercept is $(0, c)$, the x-coordinates at the x-intercepts are the solutions to $ax^2 + bx + c = 0$, and the vertex has an x-coordinate of $-\dfrac{b}{2a}$.

EXAMPLE 1   *Finding a viewing window that includes the vertex and intercepts of a quadratic graph*

To choose an appropriate viewing window for

$$y = x^2 - 92x + 180$$

begin by observing that the y-intercept is $(0, 180)$, and the vertex has an x-coordinate of $-\dfrac{-92}{2(1)} = 46$. The x-coordinates at the x-intercepts are the solutions to

$$x^2 - 92x + 180 = 0$$
$$(x - 2)(x - 90) = 0$$
$$x = 2, 90$$

To include both axes and all points of interest, the window should include the x-values 0, 2, 46, and 90, and the y-values 0 and 180. The window $[-10, 100]$ by $[-250, 250]$, shown in Figure B-5a, shows all the intercepts, but misses the vertex.

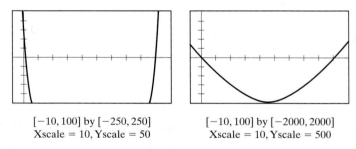

$[-10, 100]$ by $[-250, 250]$ $[-10, 100]$ by $[-2000, 2000]$
Xscale $= 10$, Yscale $= 50$ Xscale $= 10$, Yscale $= 500$

Figure B–5

Knowing the x-coordinate of the vertex, you can calculate the y-coordinate as -1936. Alternately, you can trace along the graph and observe the y-coordinates when x is near 46. Figure B-5b shows the graph in the window $[-10, 100]$ by $[-2000, 2000]$.   ■

Figure B-5 illustrates that it is often impossible to find a single ideal viewing window. The window in Figure B-5b includes all the points of interest, but it distorts the steepness of the graph. The window also makes the graph appear to pass through the origin, when in fact it does not.

Example 1 illustrates a general procedure for graphing quadratic functions on a calculator.

- Find the x-coordinates of the vertex and all x-intercepts. Choose an interval containing them, and also containing 0. This is the interval of x-values for your viewing window.
- Find the y-intercept. Choose an interval containing its y-coordinate, and also containing 0. Graph the function using this interval of y-values.
- If your window does not show the vertex, trace along the graph to find its y-coordinate and expand your interval of y-values accordingly.
- Scale the axes as you would for a linear function.

EXERCISES   *In Exercises 1–6,*

   a. Find all intercepts and the x-coordinate of the vertex analytically.
   b. Find a viewing window that shows the vertex and all intercepts.
   c. Describe any graphical features that are distorted or hidden by the viewing windows you chose.
   d. Zoom in until you can estimate the x-coordinate of the vertex and each x-intercept with an error of no more than 0.01. Make sure your results are consistent with those in part (a).

   1. $y = x^2 + x - 12$ **2.** $y = 6x - x^2$
   3. $y = x^2 - 6x + 20$ **4.** $y = -9x^2 + 37x - 38$
   5. $y = x^2 + 500x - 500{,}000$
   6. $y = 50 - 0.000001x^2$

B–4   GRAPHING QUADRATIC RELATIONS

In general, it is impossible to write equations of functions defined implicitly by a given relation. However, every quadratic relation can be represented by a pair of implicitly defined functions. On your calculator, it is more practical to work with the functions, rather than the relation, for two reasons:

- You can find the y-values for each x-value more easily.
- Most calculators graph only functions, and not more general relations.

   Example 1 demonstrates how to express a quadratic relation

$$Ax^2 + Bxy + Cy^2 + Dx + Ey + F = 0$$

as a pair of functions.

EXAMPLE 1   *Expressing a quadratic relation as a pair of functions*
Given

$$3x^2 - 6xy + y^2 + 6x - 2y + 4 = 0$$

regroup the terms to obtain

$$y^2 + (-6x - 2)y + (3x^2 + 6x + 4) = 0$$

This equation has the form $ay^2 + by + c = 0$, with

$$a = 1, \qquad b = -6x - 2, \qquad c = 3x^2 + 6x + 4$$

Using the quadratic formula $y = \dfrac{-b \pm \sqrt{b^2 - 4ac}}{2a}$, we obtain

$$y = \frac{-(-6x - 2) \pm \sqrt{(-6x - 2)^2 - 4(1)(3x^2 + 6x + 4)}}{2(1)}$$

$$= \frac{6x + 2 \pm \sqrt{24x^2 - 12}}{2}$$

$$= 3x + 1 \pm \sqrt{6x^2 - 3}$$

Thus the relation can be expressed as the pair of functions

$$f_1(x) = 3x + 1 + \sqrt{6x^2 - 3}$$

and

$$f_2(x) = 3x + 1 - \sqrt{6x^2 - 3}$$

To obtain graphs of functions defined implicitly by quadratic relations, it is often most efficient to begin with the window $[-10, 10]$ by $[-10, 10]$ and adjust it as needed by using the troubleshooting strategies listed in Section B-1. Figure B-6 shows the graphs of f_1 and f_2 in the window $[-10, 10]$ by $[-10, 10]$, along with the graph of the entire relation.

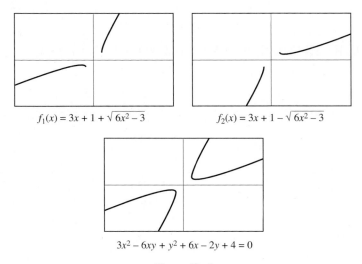

$f_1(x) = 3x + 1 + \sqrt{6x^2 - 3}$ $f_2(x) = 3x + 1 - \sqrt{6x^2 - 3}$

$3x^2 - 6xy + y^2 + 6x - 2y + 4 = 0$

Figure B–6

Example 1 illustrates a general procedure for graphing quadratic relations on a calculator.

- Express the relation as a pair of implicitly defined functions.
- Graph the functions in the window $[-10, 10]$ by $[-10, 10]$. If this window is not satisfactory, use the troubleshooting strategies in Section B-1 to adjust it.

EXERCISES   *In Exercises 1–10,*

a. Rewrite the quadratic relation by expressing y as a pair of functions of x, and graph the two functions on the same screen.

b. Trace along each graph and zoom in around each intercept until you can estimate its coordinates with an error of no more than 0.01.

1. $y^2 - 4x^2 = 16$ **2.** $4x^2 + 25y^2 = 400$

3. $2y^2 - 3x + 7y - 50 = 0$ **4.** $15 = 3x^2 - 5y^2 + 12x - 20y$

5. $x^2 - 3xy + y^2 = 0$

6. $x^2 + 6xy - 2y^2 - 10x + 10y + 25 = 0$

7. $5x - xy + 3y - 2x^2 - 2y^2 = 100$

8. $xy + 4y^2 + 6x + 4y - 9 = 0$

9. $x^2 + 5xy - 3x + 2y - 10 = 0$

10. $x^2 + 12xy = y^2 - 6x - 6y$

B–5   GRAPHING POLYNOMIAL FUNCTIONS

INCLUDING POINTS OF INTEREST

For a polynomial graph, the most important points of interest are its intercepts and turning points. Since polynomial graphs of degree greater than 2 are typically more complex than linear or quadratic graphs, you cannot always know whether a given window shows all points of interest. However, your knowledge of polynomial graphs from Chapter 7 can often help you decide.

EXAMPLE 1   *Deciding whether a given viewing window depicts a polynomial graph accurately*

A graph of $g(x) = x^4 - 64x^2 - 500$ is shown in Figure B-7.

How can you tell whether other points of interest are outside the window? The window shows two x-intercepts, which is less than the maximum for a polynomial of degree 4. However, there are three turning points, which is the maximum for a polynomial of degree 4. Since there are no other turning points outside the window, there are also no other x-intercepts. Therefore, the window shows all points of interest.

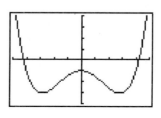

$[-10, 10]$ by $[-2000, 2000]$

Figure B–7

EXERCISES   *In Exercises 1–6,*

   a. Use your knowledge of polynomial graphs and some trial and error to find a viewing window that shows all points of interest.

   b. Trace along the graph and zoom in around each x-intercept and turning point until you can estimate its x-coordinate with an error of no more than 0.01.

   1. $f(x) = x^4 - 14x^3 + 24x^2$

   2. $f(x) = 400x - x^3$

   3. $f(x) = x^3 - 12x^2 + 49x - 50$

   4. $f(x) = x^6 - 7x^4 + 9x^2 - 12$

   5. $f(x) = x^7 - 4x^6 - 14x^5 + 56x^4 + 49x^3 - 196x^2 - 36x + 144$

   6. $f(x) = x^6 - 12x^4 + 48x^2$

The following convenient rule, popularized by textbook author John Saxon, can be used to find a window containing all intercepts and turning points of a polynomial. Calculus is needed to prove it, so we state it without proof.

> Suppose that A is the largest absolute value of any coefficient in the polynomial
>
> $$P(x) = a_n x^n + a_{n-1} x^{n-1} + a_{n-2} x^{n-2} + \cdots + a_1 x + a_0$$
>
> Then the x-values $-\dfrac{A}{|a_n|} - 1$ and $\dfrac{A}{|a_n|} + 1$ are the left and right boundaries of a viewing window containing every x-intercept and turning point of P.

EXAMPLE 2   *Applying Saxon's rule*

For $f(x) = 6x^3 - 7x^2 - 27x + 21$, $A = |-27| = 27$, and $a_n = 6$. The value of $\dfrac{A}{|a_n|} + 1$ is $\dfrac{|-27|}{|6|} + 1 = 5.5$. Therefore, some viewing window with x-values $[-5.5, 5.5]$ contains every point of interest on the graph of $y = f(x)$. To find an appropriate interval of y-values, begin with $[-10, 10]$ and adjust as needed. The window $[-5.5, 5.5]$ by $[-100, 100]$ works well for this function.   ■

EXERCISES   *In Exercises 7–10, use Saxon's rule to find a viewing window containing every point of interest on the graph of $y = f(x)$. Then trace and zoom to find the x-coordinate of each x-intercept and turning point with an error of no more than 0.01.*

   7. $f(x) = x^4 - 14x^3 + 24x^2$

   8. $f(x) = 400x - x^3$

   9. $f(x) = x^6 - 7x^4 + 9x^2 - 12$

   10. $f(x) = -2x^6 + 24x^4 - 96x^2$

It is often difficult to decide whether a given viewing window depicts a polynomial graph accurately. As with so many things in life, there is no substitute for experience. For the moment, however, here is a reasonable strategy for choosing windows.

- Choose an *x*-interval large enough to contain all points of interest. Some trial and error may be needed, but your knowledge of points of interest and end behavior, along with Saxon's rule, can make your search easier.
- View the graph in several windows, using the *x*-interval you have chosen and varying the size of the *y*-interval. Some trial and error may be needed here as well, and the best *y*-intervals are often very large. If your graph runs off the top or bottom of the screen and back on, try increasing the size of your *y*-interval by a factor of 10.

APPROXIMATING TURNING POINTS

Example 3 illustrates a method of approximating a turning point of a polynomial function with an error of less than 0.01 in both coordinates.

EXAMPLE 3

Approximating a turning point of a polynomial

Figure B-8 shows a graph of $P(x) = x^4 - 5x^3 + 6x^2$.

[−5, 5] by [−5, 5]
Xscale = 1, Yscale = 1

Figure B–8

To estimate the location of the maximum at *B*, trace along the graph near *B* until the *y*-values stop increasing and start to decrease. We obtained the following sequence of points on a Texas Instruments TI-82 calculator. Yours may yield slightly different results.

$$x = 1.0638298, \quad y = 2.0513636$$
$$x = 1.1702128, \quad y = 2.0792035$$
$$x = 1.2765957, \quad y = 2.0317722$$

Thus the local maximum value of *P* is at least 2.0792035 and occurs at some *x*-value in the interval (1.0638298, 1.2765957).

For greater accuracy, zoom in around *B*, as in Figure B-9a.

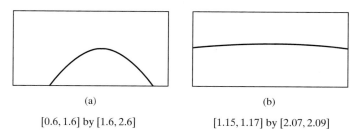

(a)

[0.6, 1.6] by [1.6, 2.6]

(b)

[1.15, 1.17] by [2.07, 2.09]

Figure B–9

Tracing again yields the following sequence of points:

$$x = 1.1425532, \quad y = 2.0791050$$
$$x = 1.1531915, \quad y = 2.0797443$$
$$x = 1.1638298, \quad y = 2.0796324$$

One more zoom, in Figure B-9b, confirms that the maximum occurs at $x \approx$ 1.16, with an error of less than 0.01.

The y-value at the maximum appears to be about 2.08. To demonstrate graphically that this estimate has an error of less than 0.01, graph $y = P(x)$ in a viewing window with $x = 1.16 \pm 0.01$ and $y = 2.08 \pm 0.01$, as in Figure B-9b.

Since the maximum is clearly in this window, it is located at $(1.16, 2.08)$, with an error of less than 0.01 in each direction.

Similarly, the local minimum at A occurs at $(0, 0)$. In Exercise 11, you can find the minimum at C.   ■

Example 3 illustrates a general procedure for locating turning points on a polynomial graph.

- Trace along the graph until the y-values stop increasing and start to decrease, or vice versa. Zoom in around this x-value and trace again. Repeat the process until you can locate the x-coordinate of the turning point with an error of no more than 0.01.
- Guess the y-coordinate at the maximum, and graph the function in a window whose y-values extend 0.01 units above and below your guess. If the turning point does not appear in the window, revise your guess and try again.

EXERCISES   **11.** Find the local minimum of $P(x) = x^4 - 5x^3 + 6x^2$ at the point C in Figure B-8.

In Exercises 12–14, *find all turning points of the polynomial function with an error of no more than* 0.01 *in each coordinate.*

12. $G(x) = 2x^4 - 3x^3 + 2x^2 - x - 3$

13. $w = 3x^3 - 10x + 6$ **14.** $y = x^7 - 3x^5 + 2x^2 - 1$

B-6   GRAPHING RATIONAL FUNCTIONS

To show the graph of a rational function accurately, a window should show all points of interest and discontinuities and indicate the graph's end behavior. It is often difficult to be certain that a window shows all points of interest, but the following strategy usually produces an accurate graph of a rational function $R(x) = \dfrac{N(x)}{D(x)}$.

- Find all discontinuities by setting $D(x) = 0$, and find all x-intercepts by setting $N(x) = 0$.
- Choose an interval of x-values that includes all x-intercepts and vertical asymptotes with some room to spare.
- Begin with a reasonable interval of y-values, such as $[-10, 10]$ by $[-10, 10]$, and adjust as needed to show turning points and vertical asymptotes clearly.
- Zoom out repeatedly until the graph's end behavior becomes apparent.

• If possible, choose a single viewing window that shows all graphical features accurately. If this is not possible, then choose two or more windows that depict the graph accurately together.

When graphing rational functions it is often helpful to put your calculator into "dot" mode, rather than "connected" mode. (See your owner's manual to learn how to do this.) Figure B-10 shows two graphs of $y = \dfrac{1}{x^2 - 4}$ to illustrate the difference between the modes. Figure B-10a, in connected mode, makes it appear that the vertical asymptotes are a part of the graph. Figure B-10b, in dot mode, makes it clear that they are not.

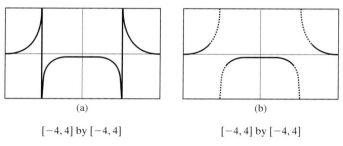

(a)   (b)

$[-4, 4]$ by $[-4, 4]$ $[-4, 4]$ by $[-4, 4]$

Figure B–10

EXAMPLE 1   *Choosing an appropriate window for a rational graph*

To graph

$$f(x) = \frac{x^2 - 4x + 3}{x + 1}$$

first observe that the denominator is zero when $x = -1$, and the numerator is zero when $x = 1$ or 3. Thus, $f(x)$ has a discontinuity at $x = -1$, and x-intercepts at $(1, 0)$ and $(3, 0)$, so the window should include x-values of -1, 1, and 3. The window in Figure B-11a uses a y-interval of $[-20, 20]$ because a smaller interval does not show the left side of the graph well. This window shows the intercepts, but does not clearly show the vertical asymptote or the end behavior.

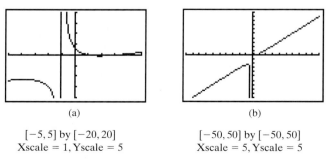

(a)   (b)

$[-5, 5]$ by $[-20, 20]$ $[-50, 50]$ by $[-50, 50]$
Xscale $= 1$, Yscale $= 5$ Xscale $= 5$, Yscale $= 5$

Figure B–11

The window in Figure B-11b was obtained by zooming out horizontally and adjusting the y-interval. This window shows both the discontinuity and the end behavior, but does not show the intercepts clearly. Both windows are needed to depict the graph accurately. ■

EXERCISES   *In Exercises* 1–4,

    a. Identify all x-intercepts and vertical asymptotes.

    b. Find one or more viewing windows that show all intercepts and vertical asymptotes and the graph's end behavior.

    c. Trace and zoom to locate all x-intercepts and vertical asymptotes with an error of no more than 0.01. Make sure your results agree with those in part (a).

1. $f(x) = \dfrac{x^2 - 4}{2x^2}$ **2.** $f(x) = \dfrac{x}{x^2 - 9}$

3. $f(x) = \dfrac{x^4}{x^3 - 25x}$ **4.** $f(x) = \dfrac{x^2 + 2}{x^4 + 4}$

ANSWERS TO
ODD-NUMBERED EXERCISES

SECTION 1–1

1. $y = -0.020t + 0.150$ **3.** 0.091

5. a. The left portion of the graph would become less steep. It would still pass through $(0.7, 0.136)$, since that point corresponds to the reading taken at 2:52. The y-intercept on the graph would be higher than in Figure 1–2.
b. Bill's friend's testimony favors the prosecution.

SECTION 1–2

1. Responses will vary.   **3.** 10.59 inches

5. The fan speed appears to be about 3700 or 3800 rpm. This result is less precise than that obtained from the equation, but is consistent with it.

7. (Sample response) The equation provides a more precise answer than the graph, but the graph provides an estimate more quickly than the equation.
The table does not provide a direct answer to either question.

9. (Sample response) I would rather have access to Table 3 because it provides reasonably precise answers to the questions that customers would ask most often.

11. Estimates will vary, but should be close to 360 parts per million (ppm) for 2000 and 420 ppm for 2050. The estimate for the year 2000 should be more trustworthy, because it assumes that the graph will show no surprising changes over a period of only 15 years beyond 1985.

13. a.

Year	CO_2 concentration (parts per million)
1960	315
1970	325
1980	335
1985	342 (Answers may vary slightly.)

b. 360 (Answers may vary.)
c. 420 (Answers may vary.)

15. a. $y = 10 - x$ **b.**

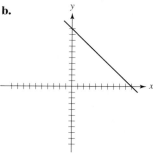

17. a. $y = \dfrac{12}{x}$ **b.**

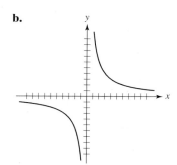

19. a.

x	y
-4	14
-2	12
0	10
2	8
4	6

b.

21. a.

x	y
-4	-6
-2	6
0	10
2	6
4	-6

b.

23. a.

x	y
-2	-3
-1	0
0	1
1	0
2	-3

b. $y \approx -24$

25. a.

x	y
-2	0
-1	-1
0	-2
1	-1
2	0

b. $y \approx 3$

27. a. (Sample response) The information is incomplete because the table does not show the life expectancy at every age.

   b. (Sample response) The information is approximate because the values of E have almost certainly been rounded.

   c. (Sample response) The information is explicit because the values of the variables can be read directly from the table.

29. a. (Sample response) The information is incomplete because it shows sea levels only for about the past 50,000 years.

   b. (Sample response) The information is approximate because sea levels must be estimated from the graph.

   c. (Sample response) The information is hidden because sea levels must be read from the graph.

31. a.

   b. about 2 years (Answers will vary.)

33. a.

D	r
200	2.0
300	3.1
400	4.2
500	5.3
600	6.4

   b.

35. Responses will vary.   **37.** Responses will vary.

SECTION 1-3

1. Responses will vary.   **3.** \$360

5. a. We cannot answer this question. We could answer if we knew the dollar amounts spent on prescription drugs in either 1990 or 1991.

   b. Physician care experienced the smallest percentage increase.

   c. We cannot answer this question. We could answer if we knew the dollar amounts spent on each category in either 1990 or 1991.

   d. We cannot answer this question. We could answer if we knew the dollar amounts spent on each category in either 1990 or 1991.

7. 105

9. Fill the larger container, pour 3 liters from it into the smaller container, then empty the smaller container. There are now 2 liters in the larger container. Pour that into the smaller container. Then once again fill the larger container. There are now 5 liters in the larger container and 2 liters in the smaller one. Pour from the larger to the smaller until the smaller one is full. There are now 4 liters in the larger container.

11. 37   **13.** 43.01 feet   **15.** 8 loaves of pumpernickel, 6 loaves of day-old bread

17. We don't know how you feel, but we would rather ride in the first elevator. The steep portion of the second graph indicates a long drop in a very short time. Our guess is that the cable on the second elevator snapped as it reached the second floor.

19. The first car has traveled farther during the hour. The two cars are going equally fast after an hour.

21. Responses will vary.

SECTION 1-4

1. The total cost of producing x shirts is the cost of the heat press, which is \$612.50, plus the cost of producing x shirts at \$3.50 for each shirt, which is $3.50x$ dollars. Therefore, the total cost is $612.50 + 3.50x$ dollars.

3. a. The total revenue is the number of shirts you sell (x), times the price you charge for each shirt $(16 - 0.005x)$, which is $16x - 0.005x^2$ dollars.

   b. Your profit in dollars is the difference between your revenue from the sale of x shirts $(16x - 0.005x^2)$ and the cost of producing x shirts $(612.50 + 3.50x)$. The difference is $-0.005x^2 + 12.50x - 612.50$.

5. 1.43

7. $-20, 120$

9. $(-\infty, 1.43)$

11. $(-20, 120)$

13. (Sample response) The x-intercepts are boundaries of intervals where the graph is above or below the x-axis. The inequality is either true in each interval where the graph is above the x-axis or true in each interval where the graph is below the x-axis. Therefore, finding the x-intercepts allows you to describe the solution to the inequality.

15. a. 16
  b. $(16, \infty)$

17. a. $-2, 2$
  b. $(-\infty, -2] \cup [0, 2]$

19. a. -2.83
  b. $[-2.83, \infty)$

21. a. $8.59, 11.41$
  b. $(8.59, 11.41)$

23. a. no solution
  b. no solution

25. a.

  b. At any particular x-value, the first graph is below the second exactly when the y-coordinate on the first graph is smaller than that on the second graph. Since $y = x^2 - 4x$ on the first graph and $y = 5$ on the second graph, the first graph is below the second at exactly those x-values for which $x^2 - 4x < 5$.
  c. $(-1, 5)$

27. a. 1.59
  b. $[1.59, \infty)$

29. a. 2
  b. $(-\infty, \infty)$

31. a. no solution
  b. no solution

SECTION 2-1

1. At any given time, Bill's blood alcohol count is a uniquely determined number.

3. At any sales level, your cost and revenue are uniquely determined. Since your profit is the difference between revenue and cost, it is also uniquely determined.

5. $H(7) = 31; H(0) = -4; H(A) = 5A - 4;$
$H(A - 1) = 5A - 9$

7. $H(7) = \dfrac{7}{8}; H(0) = 0; H(A) = \dfrac{A}{A + 1}; H(A - 1) = \dfrac{A - 1}{A}$

9. a. $y = \dfrac{9 - 4x}{2}; y$ is a function of x.

b. domain $= (-\infty, \infty)$, range $= (-\infty, \infty)$

11. a. $y = 1 - x^4; y$ is a function of x.
  b. domain $= (-\infty, \infty)$, range $= (-\infty, 1]$

13. a. $x = \dfrac{9 - 2y}{4}; x$ is a function of y.
  b. $x = 15 - 0.5y; x$ is a function of y.
  c. $x = \pm\sqrt[4]{1 - y}; x$ is not a function of y.
  d. $x = \sqrt[3]{1 - y^2}; x$ is a function of y.

15. Yes; domain $= (-\infty, \infty)$, range $= [-4, 4]$

17. No.

19. The equation $x = y^5 + y + 1$ represents y as a function of x, since no vertical line intersects its graph more than once.

21. a. y is a function of x.
  b. domain $= \{-1, 2, 5, 7, 9\}$, range $= \{0, 4, 8, 12\}$
  c. More information is needed.

23. a.

x	y
1	0.000078
10	0.00078
100	0.0077
1,000	0.068
10,000	0.30
100,000	0.47
1,000,000	0.497

  b. The table fits a function because no two rows have the same first entry.
  c. No conclusion could be drawn because the table is not complete.
  d. The table supports the conclusion that the domain is $[0, \infty)$ because it includes both small and large positive values of x. The table also supports the conclusion that the range is $\left[0, \dfrac{1}{2}\right)$ because small positive values of x correspond to values of y near 0 and large values of x correspond to values of y near $\dfrac{1}{2}$.

25. $T(50) = 2$

27. a. h represents the process of subtracting 4 times the square of the input from 100.
  b. domain $= (-\infty, \infty)$, range $= (-\infty, 100]$
  c. $36; 100; 99$
  d. $100 - 36Z^2$

29. a. h represents the process of adding the input to its reciprocal.
  b. domain $= (-\infty, 0) \cup (0, \infty)$, range $= (-\infty, -2] \cup [2, \infty)$
  c. $\dfrac{17}{4}$; undefined; -2.5
  d. $\dfrac{9Z^2 + 1}{3Z}$

31. a. h represents the process of associating each input with an output of 6.

b. domain $= (-\infty, \infty)$, range $= \{6\}$

c. 6; 6; 6

d. 6

33. $y = 2x^5 - 3x$; y is a function of x.

35. This equation cannot be solved for y.

37. y is not a function of x.

39. Yes; domain $= (-\infty, \infty)$, range $= (-\infty, \infty)$.

41. No.

43. a. In any given year, the number of metric tons of fish caught in the Pacific Ocean is uniquely determined.

  b. It appears that there are different years in which the same number of metric tons of fish were caught.

45. a. y is a function of x.

  b. domain $= \{0, -17, 2\pi, \sqrt{3}\}$, range $= \{0, -17, 2\pi, \sqrt{3}\}$

  c. More information is needed.

47. a. y is a function of x.

  b. domain $= \{2, 4, 6, 8, 10\}$, range $= \{42\}$

  c. More information is needed.

49. domain $= (0, 10.59]$; range $= (0, 3600]$

51. domain $= (0, 2000)$; range $= (0, 1,000,000)$

53. Responses will vary.

55. (Sample response) Many tables of physical relationships, such as those found in newspapers and magazines, are not complete tables of a relationship. If such a table fits a function, you need more information before concluding that it was actually generated by a function.

57. The elements "sluggish," "shiftless," and "idle" in set B all correspond to the element "lazy" in set A. Also, "evil" and "wicked" in set B correspond to "malevolent" in set A, so the correspondence is not a function from A to B.

No element in set B corresponds to both "lazy" and "malevolent," so the correspondence is a function from B to A.

59. (Sample response) The equation defines y as a function of x according to the first definition in the section because each value assumed by x determines the unique value x^2 for y. In the second definition, regard both A and B as the set of real numbers. The correspondence of a real number x in set A with its square in set B has the property that no object (number) in A corresponds to more than one object (number) in B. Therefore, the equation defines y as a function of x according to the second definition as well.

SECTION 2–2

1. $(f + g)(x) = 2x$

$(f - g)(x) = 6$

$(fg)(x) = x^2 - 9$

$\left(\dfrac{f}{g}\right)(x) = \dfrac{x + 3}{x - 3}$

3. $(f + g)(x) = \dfrac{5}{x}$

$(f - g)(x) = \dfrac{1}{x}$

$(fg)(x) = \dfrac{6}{x^2}$

$\left(\dfrac{f}{g}\right)(x) = \dfrac{3}{2}$ if $x \neq 0$

5. a. $xp(x) = x(16 - 0.005x) = 16x - 0.005x^2 = R(x)$

  b. Every company's total revenue is equal to the product of the number of items it sells and the price of each item. If x represents the number of items sold and $p(x)$ is the price of each item, then the total revenue is $xp($)

7. $\dfrac{106h - 10,000}{h - 100}$

9. $(u \circ v)(x) = 8 - 4x^2$ $\quad (v \circ u)(x) = 64 - 64x + 16x^2$

11. $(u \circ v)(x) = \dfrac{1 - 3x}{x^2}$ $\quad (v \circ u)(x) = \dfrac{1}{x^2 - 3x}$

13. $x \to x^2 \to x^2 + 3 \to 5(x^2 + 3)$

$f(x) = 5(x^2 + 3)$

15. $x \to x + 3 \to 5(x + 3) \to [5(x + 3)]^2$

$f(x) = [5(x + 3)]^2$

17. the multiplying-by-six function, the subtracting-two function, the square-root function

19. the square-root function, the multiplying-by-six function, the subtracting-two function

21. a. $(h + g)(x) = 2x - 0.1$

$(h - g)(x) = -0.1$

$(hg)(x) = x^2 - 0.1x$

$\left(\dfrac{h}{g}\right)(x) = \dfrac{x - 0.1}{x}$

  b. $(h \circ f)(x) = x + 3.9$

$(f \circ h)(x) = x + 3.9$

23. a. $(h + g)(x) = x^3 - 5x^2 + x - 6$

$(h - g)(x) = x^3 - 5x^2 - x - 6$

$(hg)(x) = x^4 - 5x^3 - 6x$

$\left(\dfrac{h}{g}\right)(x) = \dfrac{x^3 - 5x^2 - 6}{x}$

  b. $(h \circ f)(x) = x^3 + 7x^2 + 8x - 22$

$(f \circ h)(x) = x^3 - 5x^2 - 2$

25. a. $(h + g)(x) = 2x + \dfrac{1}{x}$

$(h - g)(x) = \dfrac{1}{x}$

$(hg)(x) = x^2 + 1,$ if $x \neq 0$

$\left(\dfrac{h}{g}\right)(x) = 1 + \dfrac{1}{x^2}$

  b. $(h \circ f)(x) = \dfrac{x^2 + 8x + 17}{x + 4}$

$(f \circ h)(x) = \dfrac{x^2 + 4x + 1}{x}$

27. a. $(h + g)(x) = \sqrt{4 - x} + x$
$(h - g)(x) = \sqrt{4 - x} - x$
$(hg)(x) = x\sqrt{4 - x}$
$\left(\dfrac{h}{g}\right)(x) = \dfrac{\sqrt{4 - x}}{x}$
b. $(h \circ f)(x) = \sqrt{-x}$
$(f \circ h)(x) = \sqrt{4 - x} + 4$

29. a. $(h + g)(x) = 6 + x$
$(h - g)(x) = 6 - x$
$(hg)(x) = 6x$
$\left(\dfrac{h}{g}\right)(x) = \dfrac{6}{x}$
b. $(h \circ f)(x) = 6$
$(f \circ h)(x) = 10$

31. $x \to \dfrac{x}{2} \to \dfrac{x}{2} - 5 \to \left(\dfrac{x}{2} - 5\right)^3$
$f(x) = \left(\dfrac{x}{2} - 5\right)^3$

33. $x \to x - 5 \to (x - 5)^3 \to \dfrac{(x - 5)^3}{2}$
$f(x) = \dfrac{(x - 5)^3}{2}$

35. the taking-the-reciprocal function, the multiplying-by-three function, the adding-one function

37. the adding-one function, the taking-the-reciprocal function, the multiplying-by-three function

39. a. $(I \circ N)(t)$ represents the telemarketer's income for t days.
b. $(N \circ I)(s)$ does not make sense in the physical context.

41. a. $(f + g)(2) = 10$
b. $(fg)(4) = 16$
c. $(f \circ g)(3) = 2$
d. $(g \circ f)(3) = 0$

43. a. $(g - f)(3) = -4$ **b.** $\left(\dfrac{f}{g}\right)(2) = 1$
c. $(f \circ g)(1) = 3$ **d.** $(g \circ f)(1) = 0$

45. a. $(u \circ c)(t) = 125.6$
b. $c(60) = 52$
$u(52) = 125.6$
c. $(v \circ c)(t) = 0.5t + 295.2$
d. $(v \circ c)(60) = 325.2$

47. $325.2°$

49. a. When the string fits snugly around the earth, it forms a circle whose circumference is c_0 feet. Its radius is therefore $R(c_0)$ feet. After the length of the string is increased by 10 feet, its new circumference is $c_0 + 10$ feet, and its new radius is $R(c_0 + 10)$ feet. Its distance off the surface of the earth is therefore $R(c_0 + 10) - R(c_0)$ feet.
b. 1.59 feet

SECTION 2-3

1. 0.61

3. not linear   **5.** linear

7. a. The points are collinear.

b. 0.61

c. The slope of the graph is the ratio $\dfrac{\text{change in } S}{\text{change in } T}$ between any two points on the graph. Between any two rows in the table, the value of ΔT represents a change in T and the value of ΔS represents the corresponding change in S. Therefore, the ratio $\dfrac{\Delta S}{\Delta T}$ represents $\dfrac{\text{change in } S}{\text{change in } T}$ between two points on the graph.

9. a. 2
b. $y = 2x - 3$; the slope is the coefficient of x.

11. a. 4
b. $y = 4x + 4$

13. a. 0
b. $y = -2$

15. The rate of change in the fan speed with respect to the motor pulley diameter is 340. This represents the increase in rpm corresponding to each 1-inch increase in the diameter of the pulley.

17. 0.15; each additional mile Dave drives costs him $0.15.

19. Responses will vary.

21. Table entries will vary.

23. $y = -5.6x + 175$

x	y
0	175
1	169
2	164
3	158
4	153
5	147

25. 4.46 minutes

27. a. -2.2 degrees per minute
b. 104°F
c. 15 minutes, 55 seconds

29. 15.91 hours

31. 50 mph; 60 mph; the average speed of the car was 50 mph during the first hour and 60 mph during the second hour. This suggests that the car was gradually accelerating during the two hours.

33. The average rate of change is -1.4, indicating that the coffee cooled at an average rate of $1.4°F$ per minute during the time interval $[20, 30]$.

35. The average rate of change in hours of daylight with respect to the number of days past June 21 is about -0.096, indicating that on the average, each day between June 21 and September 21 Yellowknife receives about 0.096 hours less daylight than on the previous day.

37. a. 7.45   **b.** $7.45

    c. From part (a), as your sales level increases from 500 to 510 shirts, each additional shirt sold produces about $7.45 in additional profit. At a sales level of 505 shirts, your marginal profit is $7.45.

39. Increasing your sales level from 505 to 506 shirts increases your profit more.

41. decreasing on $(-\infty, 3)$, increasing on $(3, \infty)$

43. decreasing on $(-\infty, -3)$ and $(-1, \infty)$, increasing on $(-3, -1)$

45. a. linear   **b.** 4
    c. increasing   **d.** $y = 4x + 9$

47. a. linear   **b.** -1
    c. decreasing   **d.** $y = 100 - x$

49. not a function

51. a. $A = -1800t + 36,000$

    b. The rate of change in A with respect to t is -1800. This says that the airliner descends 1800 feet each minute.

    c. $A = 360d$

    d. The rate of change in A with respect to d is 360. This says that as the airliner gets 1 mile closer to the airport, it descends 360 feet.

53. a. $L(x) = -0.1x + 0.7$

    b. $L(2) = -0.1(2) + 0.7 = 0.5 = f(2)$
    $L(5) = -0.1(5) + 0.7 = 0.2 = f(5)$

    c. $L(3) - f(3) \approx 0.067$
    $L(4) - f(4) = 0.05$

55. a. $L(x) = 13x - 10$

    b. $L(2) = 13(2) - 10 = 16 = f(2)$
    $L(5) = 13(5) - 10 = 55 = f(5)$

    c. $L(3) - f(3) = 2$
    $L(4) - f(4) = 2$

57. 100

59. -0.0005

61. a. -6.226; Increasing the length of the payment period by one year decreases the amount of the monthly payment by about $6.23. A monthly payment of $877.57 would pay the mortgage off in 30 years.

    b. 26.40 years

63. a. 6.775; The size of an average farm increased by an average of about 6.775 acres each year between 1940 and 1980.

    b. 533.075 acres

65. increasing on $(-\infty, \infty)$

67. increasing on $(-\infty, -1)$ and $(2, \infty)$, decreasing on $(-1, 2)$

69. a. The graph of g is steeper.

    b. The graph of f appears steeper.

    c. The comparative steepness of the graphs is being distorted by the difference in the viewing windows. Specifically, although the graph of g is actually 10 times as steep as the graph of f, it is being graphed in a window whose range of y-values is 50 times as great. Such zooming out in the y-direction makes the graph appear less steep.

71. a. Yes.

    b. Choices will vary. One possibility is $[-5, 5]$ by $[-5, 5]$.

73. The statement is false, since it is possible to draw, through the given points, a nonlinear curve that passes the vertical line test. The graph of $y = x^3$ is one example.

75. (Sample response) It is assumed that the airplane descends at a constant angle and that it travels at a constant speed. The first assumption is probably reasonable, but the second is probably not, since the airplane must decrease its speed to land.

77. a. milepost 286

    b. The assumption of linearity means that the rate of change in milepost numbers with respect to time remains constant. In other words, it means that Dave was driving at a constant speed.

79. a. $x \approx 15.18$ minutes   **b.** $x = 26.25$ minutes

    c. (Sample response) The interval $(92, 100)$ is much closer to 90 than is $(147, 175)$. We should therefore expect that the cooling of the coffee through the range of temperatures between $92°F$ and $100°F$ is a more accurate predictor of the time when the temperature will reach $90°F$.

SECTION 2–4

1. a.

Month	Newborns	1-Month-Olds	Adults
7	13	8	13
8	21	13	21
9	34	21	34
10	55	34	55

The number of adult female rabbits each month is the sum of the numbers in the preceding two months.

b.

Month	Adults
11	89
12	144
13	233
14	377
15	610
16	987
17	1,597
18	2,584
19	4,181
20	6,765
21	10,946
22	17,711
23	28,657
24	46,368

3. $1, \dfrac{1}{2}, \dfrac{1}{3}, \dfrac{1}{4}, \dfrac{1}{5}, \dfrac{1}{6}, \dfrac{1}{7}, \dfrac{1}{8}, \dfrac{1}{9}, \dfrac{1}{10}$

5. $\dfrac{1}{2}, \dfrac{2}{3}, \dfrac{3}{4}, \dfrac{4}{5}, \dfrac{5}{6}, \dfrac{6}{7}, \dfrac{7}{8}, \dfrac{8}{9}, \dfrac{9}{10}, \dfrac{10}{11}$

7. $1, -1, -1, 1, -1, -1, 1, -1, -1, 1$

9. $1, 3, 6, 10, 15, 21, 28, 36, 45, 55$

11. a. $3, 7, 11, 15, 19, 23, 27, 31, 35, 39$

    b. $a_n = 4n - 1, a_{1000} = 3999$

13. $3240

15. arithmetic; $a_n = 90 + 10(n - 1); 4330$

17. not arithmetic

19. $39,960

21. $-1,488,510$

23. $33,685.39$

25. 3060

27. $4, 3, 2, 1, 0, -1, -2, -3, -4, -5$

29. $1, 8, 27, 64, 125, 216, 343, 512, 729, 1000$

31. $1, 0.1, 0.01, 0.001, 0.0001, 0.00001, 0.000001, 0.0000001,$
$0.00000001, 0.000000001$

33. $1, -1, 1, -1, 1, -1, 1, -1, 1, -1$

35. a. arithmetic   **b.** $a_n = 7 + 3(n - 1)$

    c. $304; 1279$ **d.** $15,550; 273,275$

37. a. arithmetic   **b.** $a_n = 3487.6 - 0.1(n - 1)$

    c. $3477.7; 3445.2$ **d.** $348,265; 1,473,220$

39. not arithmetic

41. a. arithmetic

    b. $a_n = \pi - 2 + (\pi - 1)(n - 1)$

    c. $100\pi - 101; 425\pi - 426$

    d. $5050\pi - 5150; 90,525\pi - 90,950$

43. a. The difference between consecutive terms is 32.

    b. 1904 feet   **c.** 57,600 feet

45. a. The recursive definition of the Fibonacci sequence
states that $c_n = c_{n-1} + c_{n-2}$ for $n \geq 3$. Equivalently,
$c_{k+2} = c_{k+1} + c_k$ for $k \geq 1$. (Both equations state
that beginning with the third term, each term is the
sum of the preceding two.) Solve the second equation for c_k to obtain $c_k = c_{k+2} - c_{k+1}$.

    Now rewrite the sum $c_1 + c_2 + c_3 + \cdots + c_k$ as
$(c_3 - c_2) + (c_4 - c_3) + (c_5 - c_4) + \cdots$
$+ (c_{k+2} - c_{k+1})$. Cancel terms to obtain
$c_{k+2} - c_2 = c_{k+2} - 1$.

    b. 17,710

47. Each row is a description of what is seen in the row
above. For example, to generate the sixth row in the sequence, describe what you see in the fifth row: "three
1s, two 2s, one 1." Thus, the sixth row of the sequence is
312211.

SECTION 2-5

1. The first entry in the second column is $(12)(\$73.30) =$
$\$879.60$, representing the total cost of 12 monthly pre-
miums. Under this plan Sarah must pay all of her medical expenses up to $1000. Therefore, the other entries
in the second column represent $879.60 in premiums
plus the indicated medical expenses in the first column.

3. a. If Sarah's medical expenses are $1000 or less, the
cost of the plan is the amount of her medical expenses plus the amount of her premiums. Symbolically, this is $x + 879.60$.

    b. If Sarah's medical expenses are more than $1000,
the cost of the plan is $1879.60 plus 20% of her
medical expenses above $1000. Symbolically, this is
$1879.60 + 0.20(x - 1000) = 0.20x + 1679.60$.

5. a.   **b.**

x	y		x	y
2.5	4.0		2.95	4.90
2.6	4.2		2.96	4.92
2.7	4.4		2.97	4.94
2.8	4.6		2.98	4.96
2.9	4.8		2.99	4.98
3.0	5.0		3.00	5.00
3.1	4.9		3.01	4.99
3.2	4.8		3.02	4.98
3.3	4.7		3.03	4.97
3.4	4.6		3.04	4.96
3.5	4.5		3.05	4.95

    c. Yes, because $f(3) = 5$ and $f(x)$ is near 5 when x is
near 3.

7.

9.

11.

13. (Sample response) At an x-value where the rule defining the function changes, evaluate the two linear expressions that define the function on each side of that x-value. If the results are the same, the pieces join to form a continuous graph. Otherwise, they don't.

15. a. 0.1 cm; 0.05 cm; 0.05 cm; 0.1 cm

b. The deviation is the *size* of the difference $h - 13.90$ between actual height and designed height, without regard to the *sign* of the difference. Thus it is $|h - 13.90|$.

c. The difference between the actual capacity and the designed capacity is $\pi(4.280)^2 h - \pi(4.280)^2(13.90) = \pi(4.280)^2(h - 13.90)$. The deviation is the size of this difference, which is its absolute value $|\pi(4.280)^2(h - 13.90)| = \pi(4.280)^2|h - 13.90| \approx 57.55|h - 13.90|$.

17. $g(x) = \begin{cases} -3x + 4 & \text{if } x < 0 \\ 3x + 4 & \text{if } x \geqslant 0 \end{cases}$

19. $H(z) = \begin{cases} 5 + z & \text{if } z < 0 \\ 5 - z & \text{if } z \geqslant 0 \end{cases}$

21. a. $y = \begin{cases} -2x - 4 & \text{if } x < 0 \\ 2x - 4 & \text{if } x \geqslant 0 \end{cases}$

b. The slope of the left piece is -2. The slope of the right piece is 2. The vertex is $(0, -4)$.

c.

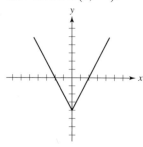

d. The axis of symmetry is the line $x = 0$ (the y-axis). The range is $[-4, \infty)$.

23. a. $y = \begin{cases} (x - 6) + 3 & \text{if } x < 6 \\ -(x - 6) + 3 & \text{if } x \geqslant 6 \end{cases}$

b. The slope of the left piece is 1. The slope of the right piece is -1. The vertex is $(6, 3)$.

c.

d. The axis of symmetry is the line $x = 6$. The range is $(-\infty, 3]$.

25. a.

x	y
2	4
3	2
4	0
5	2
6	4

b.

x	y
-2	0
-1	-2
0	-4
1	-2
2	0

c.

x	y
-7	$\frac{5}{3}$
-6	$\frac{4}{3}$
-5	1
-4	$\frac{4}{3}$
-3	$\frac{5}{3}$

d.

x	y
4	1
5	2
6	3
7	2
8	1

e. The tables indicate the slope of each piece of the graph and the location of the vertex.

27. a–d.

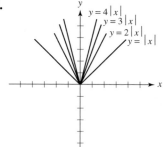

$y = 4|x|$
$y = 3|x|$
$y = 2|x|$
$y = |x|$

e. The graph must be stretched vertically by a factor of a.

29. a–d.

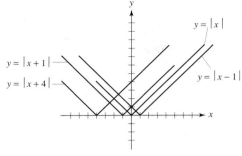

$y = |x|$

$y = |x+1|$

$y = |x+4|$

$y = |x-1|$

e. The graph must be shifted h units to the right. If $h < 0$, the actual shift is to the left.

31. a.

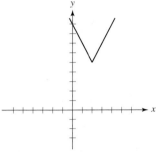

Stretch the graph of $y = |x|$ vertically by a factor of 2, then shift it 2 units to the right and 5 units up.

b.

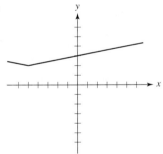

Compress the graph of $y = |x|$ vertically by a factor of 5, then shift it 5 units to the left and 2 units up.

c.

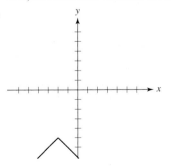

Reflect the graph of $y = |x|$ in the x-axis, then shift it 2 units to the left and 5 units down.

d.

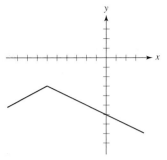

Compress the graph of $y = |x|$ vertically by a factor of 2, reflect it in the x-axis, then shift it 6 units to the left and 3 units down.

33. $t = -2, 3$ **35.** $(-\infty, -22.4] \cup [-17.6, \infty)$

37. a.

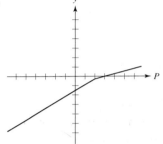

b. domain $= (-\infty, \infty)$, range $= (-\infty, \infty)$

c. increasing on $(-\infty, \infty)$

39. a.

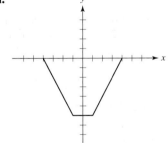

b. domain $= [-4, 4]$, range $= [-6, 0]$

c. decreasing on $[-4, -1)$, increasing on $(1, 4]$

41. a. The slope of the left piece is -6.
The slope of the right piece is 6.
The vertex is $(0, 0)$.

b.

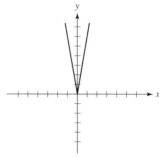

c. The axis of symmetry is the line
$x = 0$ (the y-axis).
The range is $[0, \infty)$.

d. The function is decreasing on $(-\infty, 0)$
and increasing on $(0, \infty)$.

e. The graph of $y = |x|$ must be stretched vertically by
a factor of 6.

43. a. The slope of the left piece is -1.
The slope of the right piece is 1.
The vertex is $(6, 0)$.

b.

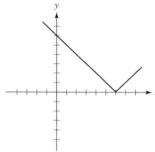

c. The axis of symmetry is the line
$x = 6$. The range is $[0, \infty)$.

d. The function is decreasing on $(-\infty, 6)$
and increasing on $(6, \infty)$.

e. The graph of $y = |x|$ must be shifted 6 units to the
right.

45. a. The slope of the left piece is -3.
The slope of the right piece is 3.
The vertex is $(-2, 0)$.

b.

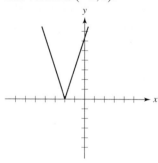

c. The axis of symmetry is the line
$x = -2$. The range is $[0, \infty)$.

d. The function is decreasing on $(-\infty, -2)$
and increasing on $(-2, \infty)$.

e. The graph of $y = |x|$ must be stretched vertically by
a factor of 3, then shifted 2 units to the left.

47. a. The slope of the left piece is -5.
The slope of the right piece is 5.
The vertex is $(-1, 7)$.

b.

c. The axis of symmetry is the line
$x = -1$. The range is $[7, \infty)$.

d. The function is decreasing on $(-\infty, -1)$ and increasing on $(-1, \infty)$.

e. The graph of $y = |x|$ must be stretched vertically by
a factor of 5, then shifted 1 unit to the left and 7
units up.

49. $x = -\dfrac{5}{3}, \dfrac{5}{3}$ **51.** $x = 15$

53. $\left(-\dfrac{26}{3}, 4\right)$ **55.** $(-\infty, 0) \cup (4, \infty)$

57. $(-\infty, \infty)$

59. For Plan 2, $g(x) = \begin{cases} x + 979.20 & \text{if } 0 \leqslant x \leqslant 500 \\ 0.20x + 1379.20 & \text{if } x > 500 \end{cases}$

For Plan 3, $h(x) = \begin{cases} x + 1065.60 & \text{if } 0 \leqslant x \leqslant 250 \\ 0.20x + 1265.60 & \text{if } x > 250 \end{cases}$

61. Responses will vary.

63. $E(t) = |t - 7756|$

65. a. $D = |M - 311|$
b. 296 or 326

67. a. $C = 0.25|T - 65|$
b. $T < 41°F$ or $T > 89°F$

69. a. $3, 7$
b. 5
c. no solution
d. A V-shaped graph can intersect the x-axis at 0, 1, or
2 points, so a linear absolute value equation has 0, 1,
or 2 solutions.

71. (Sample response) Any table can be made to fit a piecewise linear function by the process used in Exercise 70, regardless of what function generated the table. Therefore, I would not believe that the table was generated by the function in Exercise 70d without additional information.

SECTION 2-6

1. $y = \dfrac{6}{5}x - \dfrac{28}{5}$

3. $(20, 18.4)$

5.

t	x	y
0	-12	-20
1	-7	-14
2	-2	-8
3	3	-2
4	8	4
5	13	10

7. $(5.5, 1)$

9. **a.** $(5.5, 1)$; yes.

 b. 6.4 minutes; yes.

11. $y = 2x + 14$

13. $y = 8 - 2x$

15. **a.** $y = 18 - x$

 b. $x = \dfrac{10}{7}t - 4, y = -\dfrac{10}{7}t + 22$

 c. $y = 18 - x$

 d. $\left(-\dfrac{8}{7}, \dfrac{134}{7}\right)$

 e. $t = 1.4$

 f.

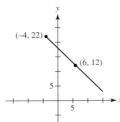

17. **a.** $y = 1 - 0.5x$

 b. $x = 6t + 2, y = -3t$

 c. $y = 1 - 0.5x$

 d. $(14, -6)$

e. $t = -\dfrac{20}{3}$

f.

19. $x = 12 - 0.35t, y = 0.35t - 7$

21. $y = 5 - x$

23. $-16 \leqslant x \leqslant 5, 0 \leqslant y \leqslant 21$

25. $y = 0.2x + 1.4$

27. $y = \dfrac{x - 7}{2}$

29. $y = 2x + 19$

31. **a.** $y = -\dfrac{2}{5}x$

 b. $x = -5t + 20, y = 2t - 8$

 c. $y = -\dfrac{2}{5}x$

33. Answers will vary.

35. **a.** $1 \leqslant x \leqslant 10, 6 \leqslant y \leqslant 15$

 b. domain $= [1, 10]$, range $= [6, 15]$

 c.

37. **a.** $0 \leqslant x \leqslant 100, 4 \leqslant y \leqslant 304$

 b. domain $= [0, 100]$, range $= [4, 304]$

 c.

39. $x = 6.19t, y = 3.09t + (9.3 \times 10^7)$

41. (Sample response) An equation in x and y describes a path in the plane, analogous to a road on a map. A set of parametric equations describes the location of an object on that path at a given time, analogous to an entry on a bus schedule.

SECTION 3–1

1.

Price per shirt	Demand	Supply
$ 3.50	2500	400
5.50	2100	600
7.50	1700	800
9.50	1300	1000
11.50	900	1200
13.50	500	1400

3. If x represents the price per shirt, in dollars, and y represents the demand for shirts, then $y = 3200 - 200x$.

5. $x = 10.50, y = 1100$

7. $(1, 5)$ **9.** $(3, 2)$

11. For the system in Exercise 8 the solution is $(-3.30, -5.15)$. For the system in Exercise 9, the solution is $(3, 2)$.

13. The terms $0.65L, 0.80M,$ and $0.65C$ represent the number of days per month Paul and his helpers will spend on landscaping, masonry, and carpentry, respectively. The equation $0.65L + 0.80M + 0.65C = 72$ expresses the condition that the total amount of time Paul and his helpers spend working should equal 72 days per month.

15. Everyone can be utilized to full capacity if the company schedules 60 days of landscaping work, 12 days of masonry, and 36 days of carpentry each month.

17.

x	y (1st equation)	y (2nd equation)
1	0	0
2	1	1
3	2	2
4	3	3

The system has infinitely many solutions. The table indicates that for every value of x, the corresponding values of y in the two equations are equal.

19. dependent   **21.** dependent

23. The system in Exercise 19 can be rewritten
$$3S - 5T = -4$$
$$9S - 15T = -12$$

Multiplying the first equation by 3 yields
$$9S - 15T = -12$$
$$9S - 15T = -12$$

Subtracting yields $0 = 0$, so the system is dependent.

25. $(4.25, 5)$ **27.** $(T, V) = (30, 4)$

29. no solution   **31.** $(x, y, z) = (10, 1, 5)$

33. $(p, q, r) = (-0.1, 0.3, 0.2)$

35. consistent and independent

37. dependent

39. a.

b.

c.

The system in part (b) is dependent.

41. a.

b.

c.

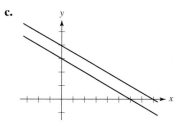

The system in part (b) is dependent.

43. \$3333.33   **45.** 18%

47. 30,000 pounds of the first alloy and 20,000 pounds of the second alloy

49. Responses will vary.

51. A system of three linear equations in two variables usually has no solutions.

53. A system of one linear equation in two variables has infinitely many solutions.

55. none   **57.** infinitely many

59. none; infinitely many; one

SECTION 3-2

1. $\begin{bmatrix} 3 & 2 & 6 \\ 1 & 7 & 5 \end{bmatrix}$ **3.** $\begin{bmatrix} 2 & 3 & 5 \\ -1 & -4 & 0 \\ 3 & 0 & 12 \\ 0 & 4 & -4 \end{bmatrix}$

5. If the variables are s and t, $\quad 23s + 6t = -2$
$\qquad\qquad\qquad\qquad\qquad 7s + 31t = 12$

7. If the variables are $x, y,$ and z, $\quad x + 3y + 9z = 5$
$$\tfrac{1}{3}x + 4y - 13z = 7$$
$$-2y + 3z = 16$$

9. The third row was multiplied by $\dfrac{1}{60}$.

11. The matrix is not in reduced row-echelon form because the third column, which contains the leading nonzero entry in the third row, contains other nonzero entries.

13. The matrix is in reduced row-echelon form.

15. $\begin{bmatrix} 1 & 0 & 2 \\ 0 & 1 & -1 \end{bmatrix}$ **17.** $\begin{bmatrix} 1 & 0 & 0 & 1.5 \\ 0 & 1 & 0 & 11 \\ 0 & 0 & 1 & 5 \end{bmatrix}$

19. $(x, y) = (2, -1)$ **21.** $(x, y, z) = (1.5, 11, 5)$

23. $0.25L + 0.15M + 0.20C = 24$
$0.60L + 0.80M + 0.70C = 56$
$0.15L + 0.05M + 0.10C = 8$

25. a–b. no solution
    c. The graph shows that the system has no solution because the graphs of the two equations in the system have no point in common. The matrix shows the same thing because the system is equivalent to one containing the self-contradictory equation $0 = -5$.

27. Here is one possible sequence of operations. Multiply row 1 by 4, add -0.60 times row 1 to row 2, add -0.15 times row 1 to row 3, interchange rows 2 and 3, multiply row 2 by -25, add -0.60 times row 2 to row 1, add -0.44 times row 2 to row 3.

29. $x = 8 - 2z_0$
$\quad y = 7 - 4z_0$
$\quad z = \qquad z_0 \quad$ for any real value of z_0

31. $w = 0.5 - 10y_0$
$\quad x = 2.5 - 20y_0$
$\quad y = \qquad y_0 \quad$ for any real value of y_0
$\quad z = 4$

33. $A = -1$
$\quad B = -1$
$\quad C = 3 - d$
$\quad D = d \quad$ for any real value of d

35. $p = 0$
$\quad q = s_0$
$\quad r = 0$
$\quad s = s_0 \quad$ for any real value of s_0

37. $\begin{bmatrix} 1 & 1 & 1 & 1 & 0 \\ 3 & -1 & 19 & 6 & \frac{1}{7} \\ 2 & 0 & 0 & 0 & 8 \end{bmatrix}$

39. $\begin{bmatrix} 2 & -3 & -4 & -5 & 0 & 0.001 \\ 0 & 2 & -3 & -4 & -5 & -0.001 \end{bmatrix}$

41. No.   **43.** Yes.

45. $\begin{bmatrix} 1 & 0 & \frac{2}{3} \\ 0 & 1 & -\frac{2}{3} \end{bmatrix}$ **47.** $\begin{bmatrix} 1 & 0 & 0 & 0 \\ 0 & 1 & 0 & 0 \\ 0 & 0 & 1 & 0 \end{bmatrix}$

49. no solution

51. $x = -\dfrac{2}{3}z_0 + \dfrac{3}{2}$
$\quad y = \dfrac{1}{3}z_0 + 2$
$\quad z = z_0 \quad$ for any real value of z_0

53. $R = 2s + 7$
$\quad S = s \quad$ for any real value of s
$\quad T = 2$

55. no solution

57. $x = -2$
$\quad y = 4 - 0.1z_0$
$\quad z = \qquad z_0$
for any real value of z_0

59. $v = 25 - 3x_0$
$\quad w = \qquad -x_0$
$\quad x = \qquad x_0$
$\quad y = 15 \qquad - 5z_0$
$\quad z = \qquad\qquad z_0$
for any real values of x_0 and z_0

61. a. $0.2P + 0.5B + 6.4S + 5.5A + 2.6M = 15$
$\qquad\qquad\qquad 40S \qquad + 10M = 50$
  b. $P = 35 \quad - 2.5B - 27.5A - 5M$
$\quad S = 1.25 \qquad\qquad\qquad - 0.25M$
for any real values of $B, A,$ and M. In the context of the problem, $P, B, S, A,$ and M must all be nonnegative.
  c. If $B = 1, A = 1,$ and $M = 1,$ then $P = 0,$ and $S = 1.$ If $B = 1, A = 0,$ and $M = 1,$ then $P = 27.5,$ and $S = 1.$ Descriptions will vary.

d. Some other solutions follow: $(P, B, S, A, M) =$ $(35, 0, 1.25, 0, 0), (7.5, 0, 1.25, 1, 0)$, and $(5, 4, 0.25, 0, 4)$.

e. (Sample response) The list of solutions in part (d) is illustrative. Most solutions to the linear system do not make reasonable meals. It would be more efficient to list portions of reasonable size and arrive at a good nutritional balance through trial and error.

63. a. $(x, y) = (2, -1)$
b. $(W, Z) = (-8, 5)$
c. $(x, y, z) = (1.5, 11, 5)$
d. $(p, q, r) = (1, 1, 1)$

SECTION 3-3

1. $x_1 + x_2 = a$
$x_1 + y_1 = b$
$y_1 + y_2 = c$
$x_2 + y_2 = d$

3. Multiply the first equation
by b_2 to obtain $a_1 b_2 x + b_1 b_2 y = c_1 b_2$
Multiply the second equation by $-b_1$ to obtain $-b_1 a_2 x - b_1 b_2 y = -b_1 c_2$
Add the equations to
obtain $(a_1 b_2 - b_1 a_2)x = c_1 b_2 - b_1 c_2$
Solve for x to obtain $x = \dfrac{c_1 b_2 - b_1 c_2}{a_1 b_2 - b_1 a_2}$

5. If $a_1 b_2 - b_1 a_2 = 0$, the expressions in Exercises 3 and 4 involve division by zero.

7. 40

9. 0

11. $x = 3, y = 2$

13. $s = 2.2, t = 2.6$

15. $D = 0$, so Cramer's Rule cannot be used. The system has no solution.

17. a. $\begin{bmatrix} -3 & 1 \\ 2 & 6 \end{bmatrix}$

b. -20

c. 20

19. a. $\begin{bmatrix} -3 & 1 \\ 0 & 5 \end{bmatrix}$

b. -15

c. -15

21. $a_{11}C_{11} + a_{12}C_{12} + a_{13}C_{13}$
$= a_{11}(-1)^{1+1}(a_{22}a_{33} - a_{23}a_{32}) + a_{12}(-1)^{1+2}(a_{21}a_{33} - a_{31}a_{23})$
$\qquad\qquad\qquad + a_{13}(-1)^{1+3}(a_{21}a_{32} - a_{31}a_{22})$
$= a_{11}(a_{22}a_{33} - a_{23}a_{32}) - a_{12}(a_{21}a_{33} - a_{31}a_{23}) +$
$a_{13}(a_{21}a_{32} - a_{31}a_{22})$
$= a_{11}a_{22}a_{33} - a_{11}a_{23}a_{32} - a_{12}a_{21}a_{33} + a_{12}a_{31}a_{23} +$
$a_{13}a_{21}a_{32} - a_{13}a_{31}a_{22}$

23. 16

25. 12

27. $x = 10, y = 1, z = 5$

29. $p = -0.1, q = 0.3, r = 0.2$

31. 25

33. 10; the minors M_{23} and M_{33} have coefficients of 0, so they do not need to be evaluated.

35. The solution to Exercise 21 shows that

$$a_{11}C_{11} + a_{12}C_{12} + a_{13}C_{13} = a_{11}a_{22}a_{33} - a_{11}a_{23}a_{32} - a_{12}a_{21}a_{33}$$
$$+ a_{12}a_{31}a_{23} + a_{13}a_{21}a_{32} - a_{13}a_{31}a_{22}$$

The terms can be rearranged to obtain

$$(a_{11}a_{22}a_{33} + a_{12}a_{23}a_{31} + a_{13}a_{21}a_{32}) - (a_{13}a_{22}a_{31}$$
$$+ a_{11}a_{23}a_{32} + a_{12}a_{21}a_{33})$$

37. -4

39. 12

41. You would need to evaluate four third-order determinants instead of two.

43. 0

45. The system does not have a unique solution.

47. One possibility is to move the counter at A to the ramp leading from A to D, as shown in the accompanying figure.

The system of equations is

$$x_2 = a$$
$$x_1 + y_1 = b$$
$$y_1 + y_2 = c$$
$$x_2 + y_2 = d$$

The system has a unique solution for any choice of a, b, c, and d.

49. 22

51. 1

53. 0

55. 0

57. 30

59. $x = -0.81, y = -0.39$

61. $p = 10, q = 51$

63. $x = 0, y = 0$

65. $s = \dfrac{3}{2}, t = \dfrac{1}{3}$

67. $x = 3, y = 5, z = 2$

69. $x = 5, y = 10, z = 1$

71. $x = 4, y = 3, z = 1$

73. $a = 5, b = 4, c = 3$

75. $w = 14$

77. $60; 360; 1,814,400; (3)(4) \cdots (n - 1)(n)$

SECTION 3-4

1. The cost of x bales of timothy at \$1.50 per bale is $1.50x$ dollars, and the cost of y bales of alfalfa at \$2.50 per bale is $2.50y$ dollars. The combined cost is $1.50x + 2.50y$ dollars, which must be equal to her expenditure of \$165.

3. (Sample response) Every linear inequality can be written in one of the forms $y > f(x), y < f(x), y \geq f(x)$, or $y \leq f(x)$. The inequality $y > f(x)$ is true at all points above the graph of $y = f(x)$. The inequality $y < f(x)$ is true at all points below the graph of $y = f(x)$. The inequality $y \geq f(x)$ is true at all points on or above the graph of $y = f(x)$. The inequality $y \leq f(x)$ is true at all points on or below the graph of $y = f(x)$.

5.

7.

9.

11.

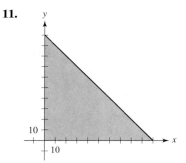

13. The feasible region represents the combinations of at most 100 bales of timothy and alfalfa Carrie can buy for no more than \$165, containing at least twice as much timothy as alfalfa.

15.

17.

19.

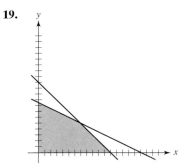

21. At $(0,0), 50 - 2x - 3y = 50$.
At $(17,0), 50 - 2x - 3y = 16$.
At $(0,12), 50 - 2x - 3y = 14$.
At $(7,5), 50 - 2x - 3y = 21$.
The minimum value is 14, and occurs at $(0,12)$.

23. -27

25.

27.

29.

31.

33.

35.

37.

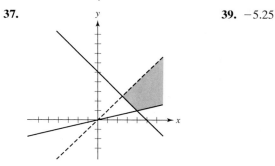

39. -5.25

41. 9   **43.** 250 newspaper ads, 10 TV spots

SECTION 4–1

1. a. At each sales level x, the profit is the y-coordinate on the graph of $y = P(x)$. The point V has the largest y-coordinate on the graph and, therefore, represents the sales level with the maximum profit.

b. Estimates will vary slightly, but should be around 1250 shirts.

3. The maximum profit appears to occur halfway between the x-intercepts.

5. a. $(0,0)$ **b.** steeper if $|a| > 1$ and flatter if $|a| < 1$

c. up if $a > 0$ and down if $a < 0$

7. a–d.

9. a–d.

11. a.

b.

c.

d.

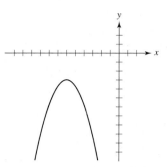

13. a. The graph of $y = x^2$ must be stretched vertically by a factor of 2, then shifted 4 units to the right and 1 unit up.

b.

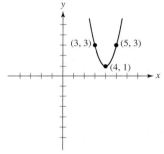

15. a. The graph of $y = x^2$ must be compressed vertically by a factor of $\dfrac{4}{3}$, reflected in the x-axis, then shifted 3 units down.

b.

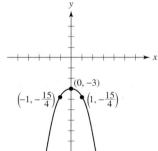

17. vertex $= (-4, -7)$; axis of symmetry $x = -4$; range $= [-7, \infty)$

19. vertex $= (0, 9)$; axis of symmetry $x = 0$; range $= (-\infty, 9]$

21. a. $x = \dfrac{1}{2}, \dfrac{9}{2}$ **b.** $x = \dfrac{5}{2}$

   c. $x = \dfrac{5}{2} \pm 2i$

23. A quadratic equation can have 0, 1, or 2 real solutions. The number of real solutions to a quadratic equation $f(x) = 0$ is equal to the number of x-intercepts on the graph of $y = f(x)$.

25. discriminant $= -16$; no x-intercepts

27. discriminant $= 0$; one x-intercept

29. $f(x) = 3x\left(x + \dfrac{8}{3}\right)$ **31.** $P(t) = 6\left(t + \dfrac{5}{3}\right)\left(t - \dfrac{1}{2}\right)$

33. $f(x) = 15\left(x + \dfrac{12}{5}\right)\left(x + \dfrac{1}{3}\right)$

35. $Q(x) = 2\left[x - \left(1 + \dfrac{\sqrt{6}}{2}\right)\right]\left[x - \left(1 - \dfrac{\sqrt{6}}{2}\right)\right]$

37. $Q(x) = 0.5\left[x - (-4 + \sqrt{6})\right]\left[x - (-4 - \sqrt{6})\right]$

39. $P(x) = -0.005(x - 50)(x - 2450);$ $(50, 0)$ and $(2450, 0)$

41. 3

43. vertex $= (0, 7)$; axis of symmetry $x = 0$;
range $= (-\infty, 7]$

45. vertex $= (1, 18)$; axis of symmetry $x = 1$;
range $= [18, \infty)$

47. a. $y = 0.5(x + 4)^2 + 1$
   b. vertex $= (-4, 1)$
   axis of symmetry $x = -4$
   range $= [1, \infty)$

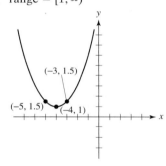

49. a. $w = -3(z - 1)^2 + 3$
   b. vertex $= (1, 3)$
   axis of symmetry $z = 1$
   range $= (-\infty, 3]$

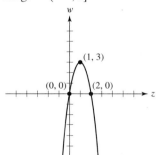

51. $(-4, 0), (-8, 0)$ **53.** $\left(-3 \pm \dfrac{\sqrt{2}}{2}, 0\right)$

55. If a and k are both positive, then $a(x - h)^2 \geqslant 0$ for all x,
so $a(x - h)^2 + k > 0$ for all x. Similarly, if a and k are
both negative, then $a(x - h)^2 + k < 0$ for all x. In each
case, $a(x - h)^2 + k$ is never 0, so the graph has no x-in-
tercepts.

57. a.

x	$y = 3x^2$	First differences	Second differences
1.0	3.00		
		3.75	
1.5	6.75		1.5
		5.25	
2.0	12.00		1.5
		6.75	
2.5	18.75		1.5
		8.25	
3.0	27.00		

b.

x	$y = ax^2$	First differences	Second differences
-1	a		
		$-a$	
0	0		$2a$
		a	
1	a		$2a$
		$3a$	
2	$4a$		$2a$
		$5a$	
3	$9a$		

c.

x	$y = ax^2$	First differences	Second differences
x_0	ax_0^2		
		$2ahx_0 + ah^2$	
$x_0 + h$	$ax_0^2 + 2ahx_0 + ah^2$		$2ah^2$
		$2ahx_0 + 3ah^2$	
$x_0 + 2h$	$ax_0^2 + 4ahx_0 + 4ah^2$		$2ah^2$
		$2ahx_0 + 5ah^2$	
$x_0 + 3h$	$ax_0^2 + 6ahx_0 + 9ah^2$		$2ah^2$
		$2ahx_0 + 7ah^2$	
$x_0 + 4h$	$ax_0^2 + 8ahx_0 + 16ah^2$		

59. quadratic   **61.** quadratic

63. a. 5.1. For speeds between 30 and 40 mph, every increase of 1 mph in speed results in an increase of
   about 5.1 feet in the required stopping distance.
   b. The average rate of change is 6.2. As your speed increases, increases of 1 mph in speed produce increasingly larger increases in the required stopping
   distance.

65. a. The graph of $y = x^2$ must be shifted 2.5 units to the
   right and 6.8 units down.
   b–c.

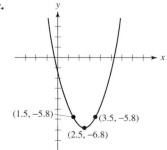

d. axis of symmetry $x = 2.5$
range $= [-6.8, \infty)$

e. decreases on $(-\infty, 2.5)$, increases on $(2.5, \infty)$

67. a. The graph of $y = r^2$ must be stretched vertically by a factor of 2π, then shifted 1 unit to the left and 2π units down.

b–c.

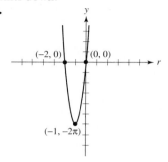

d. axis of symmetry $r = -1$
range $= [-2\pi, \infty)$

e. decreases on $(-\infty, -1)$, increases on $(-1, \infty)$

69. a. The graph of $q = p^2$ must be stretched vertically by a factor of 100, then shifted 3 units to the left and 5 units down.

b–c.

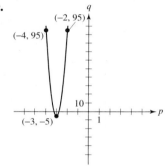

d. axis of symmetry $p = -3$
range $= [-5, \infty)$

e. decreases on $(-\infty, -3)$, increases on $(-3, \infty)$

71. a. x-intercepts $(-2, 0), (3, 0)$
vertex $= \left(\dfrac{1}{2}, \dfrac{25}{4}\right)$

b.

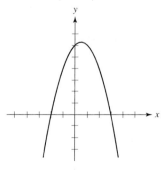

c. axis of symmetry $x = \dfrac{1}{2}$

range $= \left(-\infty, \dfrac{25}{4}\right]$

d. increases on $\left(-\infty, \dfrac{1}{2}\right)$, decreases on $\left(\dfrac{1}{2}, \infty\right)$

73. a. t-intercepts $(\pm 5, 0)$
vertex $(0, 400)$

b.

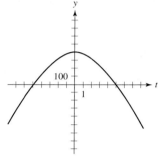

c. axis of symmetry $t = 0$
range $= (-\infty, 400]$

d. increases on $(-\infty, 0)$,
decreases on $(0, \infty)$

75. a. x-intercepts $\left(\dfrac{3 \pm \sqrt{5}}{2}, 0\right)$

vertex $\left(\dfrac{3}{2}, -\dfrac{5}{4}\right)$

b.

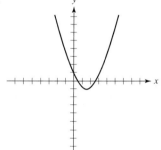

c. axis of symmetry $x = \dfrac{3}{2}$

range $= \left[-\dfrac{5}{4}, \infty\right)$

d. decreases on $\left(-\infty, \dfrac{3}{2}\right)$, increases on $\left(\dfrac{3}{2}, \infty\right)$

77. a. u-intercepts $(-6, 0), (0, 0)$
vertex $(-3, -18)$

b.

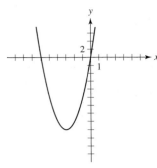

c. axis of symmetry $u = -3$
   range $= [-18, \infty)$
d. decreases on $(-\infty, -3)$, increases on $(-3, \infty)$

79. a. x-intercepts $(\pm \sqrt{5}, 0)$
   vertex $(0, -5)$
b.

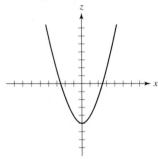

c. axis of symmetry $x = 0$
   range $= [-5, \infty)$
d. decreases on $(-\infty, 0)$, increases on $(0, \infty)$

81. a. no α-intercepts
   vertex $= (2, 80)$, other points $(1, 85)$ and $(3, 85)$
b.

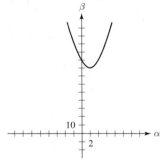

c. axis of symmetry $\alpha = 2$
   range $= [80, \infty)$

d. decreases on $(-\infty, 2)$, increases on $(2, \infty)$

83. $y = -\dfrac{2}{3}(x + 3)(x - 2)$ **85.** $y = -\dfrac{5}{16}x(x - 100)$

87. $y = 0.12(x + 4)^2$ **89.** $y = -(x + 10)^2 - 5$

91. It is the vertex.

93. a. $f(x) = ax^2 + bx + c$

$$= a\left(x^2 + \frac{b}{a}x\right) + c$$

$$= a\left(x^2 + \frac{b}{a}x + \frac{b^2}{4a^2} - \frac{b^2}{4a^2}\right) + c$$

$$= a\left(x^2 + \frac{b}{a}x + \frac{b^2}{4a^2}\right) - \frac{b^2}{4a} + c$$

$$= a\left(x + \frac{b}{2a}\right)^2 - \frac{b^2 - 4ac}{4a}$$

b. From the standard graphing form in part (a)
   the vertex is $\left(-\dfrac{b}{2a}, -\dfrac{b^2 - 4ac}{4a}\right)$.

95. a. $y = g(x)$ is steepest, $y = h(x)$ is flattest.
b. The graphs do not appear to support the conclusions in part (a). This is because the windows distort the steepness of each graph.
c. Yes.

97. $F(x) = [x - (2 + 3i)][x - (2 - 3i)]$

99. $F(x) = \left[x - \left(-2 + \dfrac{\sqrt{2}}{2}i\right)\right]\left[x - \left(-2 - \dfrac{\sqrt{2}}{2}i\right)\right]$

SECTION 4–2

1. a. $s(t) = 0$ when the height of the rocket is 0, that is, when the rocket is at ground level. Since the flight lasts 10 seconds, $s(10) = 0$.
b. $s(t) = -16t^2 + 160t$ **c.** $(5, 400)$; 400 feet

3. $(-1, 3)$; minimum   **5.** $(-6, 1176)$; maximum

7. The entire graph is above the x-axis.

9. $(-\infty, -1) \cup \left(\dfrac{1}{2}, \infty\right)$ **11.** $[-5, 20]$

13. $(a, b, c) = (0.055, 1.25, -10)$

   $0.055(30)^2 + 1.25(30) - 10 = 77$
   $0.055(40)^2 + 1.25(40) - 10 = 128$
   $0.055(50)^2 + 1.25(50) - 10 = 190$
   $0.055(60)^2 + 1.25(60) - 10 = 263$

15. $y = -0.5x^2 + 9x - 22$ **17.** $y = 4x^2 - 12x + 9$

19. $h = -2.66t^2 + 90t$

21. $y \approx 3 + x - \dfrac{31.74}{v_0^2}x^2$

23. $\sqrt{2660.87} \approx 51.58$ ft/sec

25. a. the overhand serve
   b. the overhand serve

27. $y = 2x - x^2$

29. $y = x^2 - 12x + 32$

31. 0, maximum

33. -1, minimum

35. 0.8, minimum

37. -0.25, minimum

39. $[-7, 7]$

41. $\left(-\infty, -\dfrac{7}{3}\right) \cup (2, \infty)$

43. $\{-0.5\}$

45. $(-\infty, \infty)$

47. Yes; $y = -x^2 + 11x - 12$

49. Yes; $y = -\dfrac{1}{18}x^2 - \dfrac{5}{18}x + \dfrac{340}{9}$

51. No.

53. a. 1600 shirts

    b. No, the maximum profit occurs at a sales level of 1250 shirts.

55. 78 people

57. $y = 0.00004x^2 - 0.05x + 1250$

59. a. If $M = f(v)$, then f is increasing on $(0, 45)$ and decreasing on $(45, \infty)$. Therefore, the graph of $M = f(v)$ resembles a parabola opening downward.

    b. $M = -0.015(v - 45)^2 + 30$

    c. Responses will vary, but should point out that the model predicts negative gas mileages at speeds above 90 miles per hour.

61. The condition $Q(0) = 0$ predicts a stopping distance of 0 feet from an initial velocity of 0 miles per hour, and the condition $Q(v) > 0$ whenever $v > 0$ predicts a positive stopping distance from any positive initial velocity.

SECTION 4–3

1. Any vertical line between $x = -1$ and $x = 1$ intersects the graph twice.

3. The legs of the right triangle in the diagram have lengths x and y, and the hypotenuse has length 1. By the Pythagorean theorem, $x^2 + y^2 = 1$.

5. a. Any horizontal line between $y = -1$ and $y = 1$ intersects the graph twice.

    b. The first and last rows have the same second entries and different first entries.

    c. Solving for x yields $x = \pm\sqrt{1 - y^2}$. Some values of the independent variable y generate more than one value of the dependent variable x.

7. On the upper semicircle, $y \geqslant 0$. Therefore, the equation of the upper semicircle is $y = \sqrt{1 - x^2}$. Similarly, on the lower semicircle, $y \leqslant 0$. Therefore, the equation of the lower semicircle is $y = -\sqrt{1 - x^2}$.

9. $x = [f(x)]^2$ **11.** $9x^2 + [f(x) - 5]^2 = 36$

    $x = \left(\sqrt{x}\right)^2$ $9x^2 + \left[\left(5 - 3\sqrt{4 - x^2}\right) - 5\right]^2 = 36$

    $x = x$ $9x^2 + \left(-3\sqrt{4 - x^2}\right)^2 = 36$

                       $9x^2 + 9(4 - x^2) = 36$

                       $9x^2 + 36 - 9x^2 = 36$

                       $36 = 36$

13. $y = \pm\dfrac{1}{2}\sqrt{4 - x^2}$

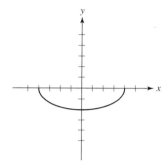

15. $y = \pm\dfrac{1}{2}\sqrt{4 + x^2}$

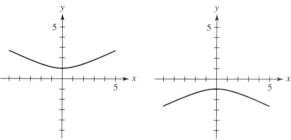

17. (Sample response) The graph of every quadratic relation consists of at most two pieces, each of which is the graph of a function. No vertical line can intersect the graph of either function more than once, so no vertical line can intersect the graph of the relation more than twice.

19. a. $y = \pm\dfrac{1}{3}\sqrt{36 - 4x^2}$ **b.**

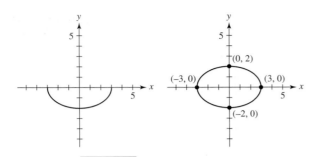

21. a. $y = \pm \sqrt{\dfrac{x - x^2 + 6}{2}}$ **b.**

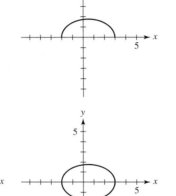

23. From Exercise 22, $\dfrac{d}{s^2} = \dfrac{1}{15}$, so $d = \dfrac{1}{15} s^2$. The value of k is $\dfrac{1}{15}$.

25. a. $s = \sqrt{15d}$

  b. As an abstract variable s can be negative, so $s = \pm\sqrt{15d}$.

27. Statement (b) best describes the speed of a tsunami. From the data in Figure 4-17, the average rate of change in speed with respect to water depth on the interval $[24{,}000, 30{,}000]$ is $\dfrac{670 - 600}{30{,}000 - 24{,}000} \approx 0.017$, indicating that the tsunami slows down about 0.017 miles per hour for each decrease of 1 foot in water depth. On the interval $[60, 600]$ the average rate of change is $\dfrac{94 - 30}{600 - 60} = 0.12$, nearly 10 times as great.

29. The graph is the upper half of the parabola $x = y^2$. To see why, solve the equation for y to obtain $y = \pm\sqrt{x}$. On the upper half of the parabola, $y \geq 0$, so $y = \sqrt{x}$.

31. The basic graph of $y = \sqrt{x}$ is stretched vertically by a factor of a if $|a| > 1$ and compressed vertically by a

factor of $\dfrac{1}{a}$ if $|a| < 1$. It is reflected in the x-axis if $a < 0$. It is shifted h units to the right (to the left if $h < 0$) and k units up (down if $k < 0$).

33.

The graph of $y = \sqrt{d}$ is stretched vertically by a factor of $\sqrt{15}$.

35. a.

  b. domain $= [-3, \infty)$, range $= (-\infty, 1]$

37. a.

  b. domain $= [0, \infty)$, range $= [-2, \infty)$

39.

41.

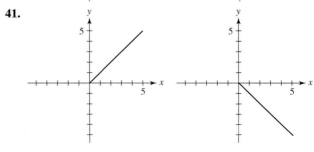

43. $9x^2 + 25[f(x)]^2 = 225$

$$9x^2 + 25\left(-\frac{3}{5}\sqrt{25 - x^2}\right)^2 = 225$$

$$9x^2 + 25\left[\frac{9}{25}(25 - x^2)\right] = 225$$

$$9x^2 + 9(25 - x^2) = 225$$
$$9x^2 + 225 - 9x^2 = 225$$
$$225 = 225$$

45. $x^2 - 4[f(x) + 1]^2 = 4$

$$x^2 - 4\left[\left(\frac{1}{2}\sqrt{x^2 - 4} - 1\right) + 1\right]^2 = 4$$

$$x^2 - 4\left(\frac{1}{2}\sqrt{x^2 - 4}\right)^2 = 4$$

$$x^2 - 4\left[\frac{1}{4}(x^2 - 4)\right] = 4$$

$$x^2 - (x^2 - 4) = 4$$
$$4 = 4$$

47. a.

b. domain = $(0, \infty]$,
range = $[0, \infty)$

49. a.

b. domain = $[1, \infty)$,
range = $[4, \infty)$

51. a.

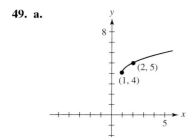

b. domain = $[-1, \infty)$,
range = $[4, \infty)$

53. a.

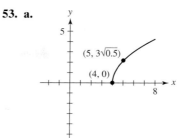

b. domain = $[4, \infty)$,
range = $[0, \infty)$

55. a. 0.0022

   b. If the tsunami is generated between 160 and 280 miles east of Honolulu, each additional mile of distance increases the arrival time by about 0.0022 hours (8 seconds).

   c. 0.0083

   d. If the tsunami is generated between 0 and 16 miles east of Honolulu, each additional mile of distance increases the arrival time by about 0.0083 hours (30 seconds). This is larger than the result in part (b). One consequence of this is that for tsunamis originating relatively close to Honolulu, estimates of the arrival time are more sensitive to errors in estimating the point of origination.

57. a. $y = -x - 5 \pm \sqrt{2x + 5}$

   b.

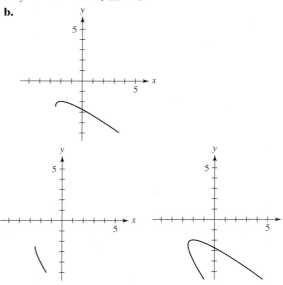

59. a. $y = \frac{1}{2}\left(x \pm \sqrt{-3x^2 + 48}\right)$

b.

c.

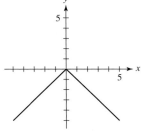

61. a.
$$x^2 - [f_1(x)]^2 = 0$$
$$x^2 - (x)^2 = 0$$
$$0 = 0$$

$$x^2 - [f_2(x)]^2 = 0$$
$$x^2 - (-x)^2 = 0$$
$$0 = 0$$

$$x^2 - [f_3(x)]^2 = 0$$
$$x^2 - (|x|)^2 = 0$$
$$0 = 0$$

$$x^2 - [f_4(x)]^2 = 0$$
$$x^2 - (-|x|)^2 = 0$$
$$0 = 0$$

b.

SECTION 4-4

1.

3.

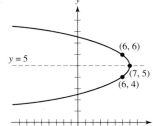

5. a. If $|x| > 3$, then $\left(\dfrac{x}{3}\right)^2 > 1$, so $\left(\dfrac{y}{2}\right)^2 < 0$ and there are no real solutions for y. This means that there are no points on the graph for which $|x| > 3$, so the entire graph lies on or between the lines $x = -3$ and $x = 3$.

b. If $|y| > 2$, then $\left(\dfrac{y}{2}\right)^2 > 1$, so $\left(\dfrac{x}{3}\right)^2 < 0$ and there are no real solutions for x. This means that there are no points on the graph for which $|y| > 2$, so the entire graph lies on or between the lines $y = -2$ and $y = 2$.

c. If $y = 0$, then $\left(\dfrac{x}{3}\right)^2 = 1$.

$$\dfrac{x}{3} = \pm 1$$

$$x = \pm 3$$

If $x = 0$, then $\left(\dfrac{y}{2}\right)^2 = 1$.

$$\dfrac{y}{2} = \pm 1$$

$$y = \pm 2$$

7. a. If $\left|\dfrac{x - 10}{2}\right| > 1$, then $\left(\dfrac{x - 10}{2}\right)^2 > 1$,

so $\left(\dfrac{y - 7}{3}\right)^2 < 0$, and there is no real solution for y.

This means that there are no points on the graph for which $\left|\dfrac{x - 10}{2}\right| > 1$. Thus all points on the

graph have $\left|\dfrac{x - 10}{2}\right| \leqslant 1$

$$-1 \leqslant \dfrac{x - 10}{2} \leqslant 1$$

$$-2 \leqslant x - 10 \leqslant 2$$

$$8 \leqslant x \leqslant 12$$

b. If $\left|\dfrac{y - 7}{3}\right| > 1$, then $\left(\dfrac{y - 7}{3}\right)^2 > 1$, so $\left(\dfrac{x - 10}{2}\right)^2 < 0$,

and there is no real solution for x. This means that there are no points on the graph for which

$\left|\dfrac{y - 7}{3}\right| > 1$. Thus all points on the graph have

$$\left|\dfrac{y - 7}{3}\right| \leqslant 1$$

$$-1 \leqslant \dfrac{y - 7}{3} \leqslant 1$$

$$-3 \leqslant y - 7 \leqslant 3$$

$$4 \leqslant y \leqslant 10$$

c. If $y = 7$, then $\left(\dfrac{x - 10}{2}\right)^2 = 1$

$$\dfrac{x - 10}{2} = \pm 1$$

$$x - 10 = \pm 2$$

$$x = 8, 12$$

If $x = 10$, then $\left(\dfrac{y - 7}{3}\right)^2 = 1$

$$\dfrac{y - 7}{3} = \pm 1$$

$$y - 7 = \pm 3$$

$$y = 4, 10$$

9. $\left(\dfrac{x}{4}\right)^2 + \left(\dfrac{y}{8}\right)^2 = 1$

center $= (0, 0)$
vertices $= (0, \pm 8)$
covertices $= (\pm 4, 0)$
major axis $x = 0$ (y-axis)
minor axis $y = 0$ (x-axis)

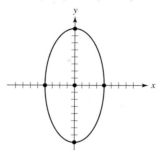

11. $\left(\dfrac{x + 4}{4}\right)^2 + \left(\dfrac{y - 2}{2}\right)^2 = 1$

center $= (-4, 2)$
vertices $= (-8, 2), (0, 2)$
covertices $= (-4, 0), (-4, 4)$
major axis $y = 2$
minor axis $x = -4$

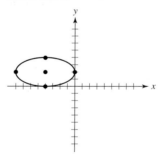

13. $\left(\dfrac{x - 30}{\sqrt{6}}\right)^2 + (y + 25)^2 = 1$

center $= (30, -25)$
vertices $= (30 \pm \sqrt{6}, -25)$
covertices $= (30, -26), (30, -24)$
major axis $y = -25$
minor axis $x = 30$

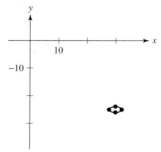

15. center = $(0, 0)$
radius = 5

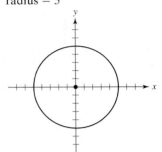

17. center = $(-6, 2)$
radius = 2

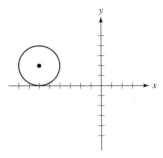

19. center = $(4, 3)$
radius = $\sqrt{2}$

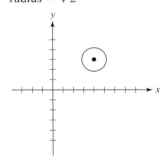

21. $x^2 - y^2 = 1$
$y^2 = x^2 - 1$
$y = \pm \sqrt{x^2 - 1}$
In the first quadrant the last equation is equivalent to $y = \sqrt{x^2 - 1}$ because $y \geqslant 0$. In general, however, y can be negative, so the equations are not equivalent.

23.

x	y
1	0
2	$\sqrt{3} \approx 1.73$
10	$\sqrt{99} \approx 9.95$
100	$\sqrt{9999} \approx 99.99$

25. If (a, b) is on the graph of $x^2 - y^2 = 1$, then $a^2 - b^2 = 1$. In that case, it is also true that $(-a)^2 + b^2 = 1$ and $a^2 + (-b)^2 = 1$, so the points $(-a, b)$ and $(a, -b)$ are also on the graph.

27. center = $(-5, -6)$
vertices = $(-8, -6), (-2, -6)$
transverse axis $y = -6$
conjugate axis $x = -5$
asymptotes $y + 6 = \pm \dfrac{4}{3}(x + 5)$

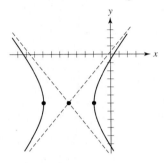

29. center = $(-129, 85)$
vertices = $(-129, 50), (-129, 120)$
transverse axis $x = -129$
conjugate axis $y = 85$
asymptotes $y - 85 = \pm \dfrac{7}{10}(x + 129)$

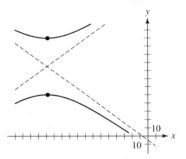

31. The graph is a single point at the origin.
33. The graph consists of the two lines $y = x$ and $y = -x$.
35. The graph is empty.
37. The graph consists of the lines $y = \pm \dfrac{1}{2}(x - 3)$.
39. $x = 3y^2$
vertex = $(0, 0)$
axis of symmetry $y = 0$ (x-axis)

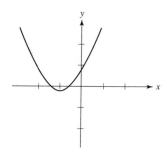

41. $\left(\dfrac{x}{\sqrt{14}}\right)^2 + \left(\dfrac{y}{2}\right)^2 = 1$

center $= (0, 0)$
vertices $= (\pm \sqrt{14}, 0)$
covertices $= (0, \pm 2)$
major axis $y = 0$ (x-axis)
minor axis $x = 0$ (y-axis)

47. $\left(\dfrac{x - 7}{2}\right)^2 + \left(\dfrac{y}{\sqrt{20}}\right)^2 = 1$

center $= (7, 0)$
vertices $= (7, \pm 2\sqrt{5})$
covertices $= (5, 0), (9, 0)$
major axis $x = 7$
minor axis $y = 0$ (x-axis)

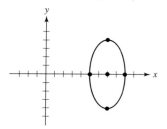

43. $(x + 1)^2 + (y - 2)^2 = \dfrac{9}{4}$

center $= (-1, 2)$
radius $= \dfrac{3}{2}$

49. $y = 2(x - 3)^2 - 8$
vertex $= (3, -8)$
axis of symmetry $x = 3$

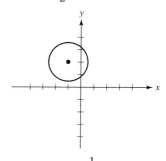

51. The graph is a pair of intersecting lines, with equations $y + 4 = \pm 2(x + 3)$.

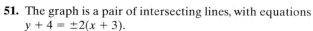

45. $y = (x + 1)^2 - \dfrac{1}{4}$

vertex $\left(-1, -\dfrac{1}{4}\right)$
axis of symmetry $x = -1$

53. $\left(\dfrac{x+0.1}{10}\right)^2 + \left(\dfrac{y-0.6}{5}\right)^2 = 1$

center $= (-0.1, 0.6)$
vertices $= (-10.1, 0.6), (9.9, 0.6)$
covertices $= (-0.1, -4.4), (-0.1, 5.6)$
major axis $y = 0.6$
minor axis $x = -0.1$

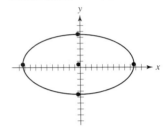

55. $\left(\dfrac{x+3}{5}\right)^2 + \left(\dfrac{y-1}{2}\right)^2 = 1$

center $= (-3, 1)$
vertices $= (-8, 1), (2, 1)$
covertices $= (-3, -1), (-3, 3)$
major axis $y = 1$
minor axis $x = -3$

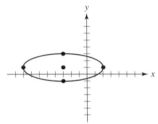

57. The graph is empty.

59. $\left(\dfrac{x+3}{5}\right)^2 - \left(\dfrac{y-1}{2}\right)^2 = 1$

center $= (-3, 1)$
vertices $= (-8, 1), (2, 1)$
transverse axis $y = 1$
conjugate axis $x = -3$

asymptotes $\dfrac{y-1}{2} = \pm\dfrac{x+3}{5}$

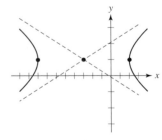

61. $(y+2)^2 - \left(\dfrac{x+3}{\sqrt{5}}\right)^2 = 1$

center $= (-3, -2)$
vertices $= (-3, -3), (-3, -1)$
transverse axis $x = -3$
conjugate axis $y = -2$

asymptotes $y + 2 = \pm\dfrac{x+3}{\sqrt{5}}$

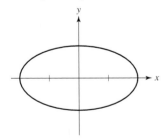

63. $x = 0.365$

SECTION 4-5

1. a. $\left(\dfrac{-2}{2}\right)^2 + (0)^2 = 1; \left(\dfrac{0}{2}\right)^2 + (1)^2 = 1;$

$\left(\dfrac{0}{2}\right)^2 + (-1)^2 = 1; \left(\dfrac{2}{2}\right)^2 + (0)^2 = 1$

b. (Sample response) The expression $\dfrac{x}{2}$ in the new equation plays the same role as x in the unit circle. Therefore, if an ordered pair (a, b) satisfies the equation of the unit circle and (c, b) satisfies the new equation, then $\dfrac{c}{2} = a$ or, equivalently, $c = 2a$.

c. Each x-coordinate on the new graph is twice the corresponding x-coordinate on the unit circle.

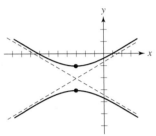

d. The circle is stretched horizontally by a factor of 2.

e. $y = \pm\sqrt{1 - \left(\dfrac{x}{2}\right)^2}$

3. a.

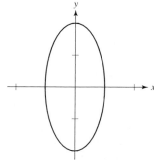

b. The circle was stretched vertically by a factor of 2.

c. $y = \pm 2\sqrt{1-x^2}$

5. The graph is compressed horizontally by a factor of 3.

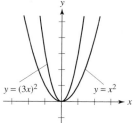

7. The graph is stretched horizontally by a factor of 2.

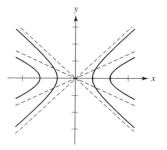

9. The graph is stretched horizontally by a factor of 3 and vertically by a factor of 4.

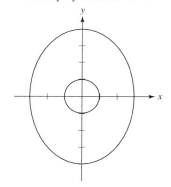

11. a. $(2-3)^2 + (0)^2 = 1; (3-3)^2 + (1)^2 = 1;$
$(3-3)^2 + (-1)^2 = 1; (4-3)^2 + (0)^2 = 1$

    b. (Sample response) The expression $x - 3$ in the new equation plays the same role as x in the unit circle. Therefore, if an ordered pair (a, b) satisfies the equation of the unit circle and (c, b) satisfies the new equation, then $c - 3 = a$ or, equivalently, $c = a + 3$.

    c. Each x-coordinate on the new graph is three more than the corresponding x-coordinate on the unit circle.

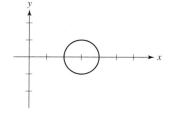

d. The circle is shifted 3 units to the right.

e. $y = \pm \sqrt{1 - (x - 3)^2}$

13. a.

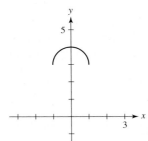

b. The circle is shifted 3 units up.

c. $y = 3 \pm \sqrt{1 - x^2}$

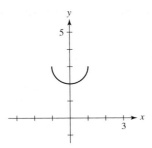

15. The graph is shifted 3 units to the right.

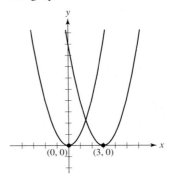

17. The graph is shifted 2 units to the left.

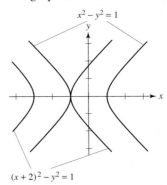

19. The graph is shifted 3 units to the left and 4 units down.

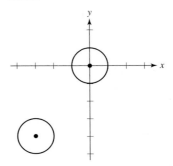

21. $x^2 + y^2 = 1; x^2 + y^2 = 1$. The graph of each equation is the unit circle.

23. a. $x^2 + (-y - 2)^2 = 1$
$x^2 + [-(y + 2)]^2 = 1$
$x^2 + (y + 2)^2 = 1$

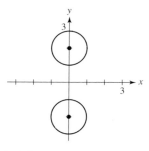

b. The second graph is the reflection of the first graph in the x-axis.

25. symmetric about the x-axis

27. symmetric about both axes

29. a.

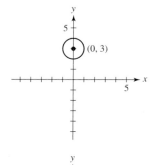

b. $x^2 + \left(\dfrac{y}{2} - 3\right)^2 = 1$

31. center $(-129, 85)$
vertices $= (-179, 85), (-79, 85)$
covertices $= (-129, 50), (-129, 120)$
major axis $y = 85$
minor axis $x = -129$

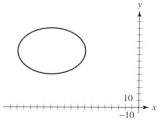

33. center $= (0.5, 1.5)$
vertices $= (0.5, -0.5), (0.5, 3.5)$
covertices $= (-0.5, 1.5), (1.5, 1.5)$
major axis $x = 0.5$
minor axis $y = 1.5$

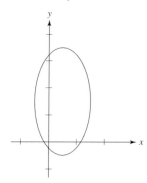

35. center $= (-129, 85)$
vertices $= (-179, 85), (-79, 85)$
transverse axis $y = 85$
conjugate axis $x = -129$
asymptotes $\dfrac{y - 85}{35} = \pm \dfrac{x + 129}{50}$

37. center $= (0.5, 1.5)$
vertices $= (0.5, -0.5), (0.5, 3.5)$
transverse axis $x = 0.5$

conjugate axis $y = 1.5$

asymptotes $\dfrac{y - 1.5}{2} = \pm\,(x - 0.5)$

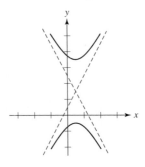

39. The graph of $x = y^2$ is reflected in the y-axis and stretched horizontally by a factor of 5.
vertex $= (0, 0)$
axis of symmetry $y = 0$ (x-axis)

41. The graph of $x^2 + y^2 = 1$ is stretched horizontally by a factor of $\sqrt{6}$ and vertically by a factor of $\sqrt{5}$.
center $= (0, 0)$
vertices $= (\pm\sqrt{6}, 0)$
covertices $= (0, \pm\sqrt{5})$
major axis $y = 0$ (x-axis)
minor axis $x = 0$ (y-axis)

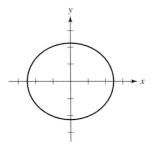

43. The graph of $x^2 + y^2 = 1$ is stretched both horizontally and vertically by a factor of 5, then shifted 10 units to the left and 20 units up.
center $= (-10, 20)$
radius $= 5$

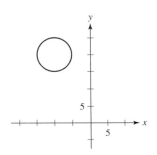

45. The graph of $y = x^2$ is shifted 10 units to the left and 5 units down.
vertex $= (-10, -5)$
axis of symmetry $x = -10$

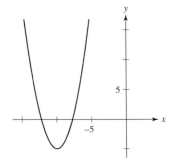

47. The graph of $x^2 + y^2 = 1$ is stretched horizontally by a factor of 4 and vertically by a factor of 6, then shifted 2 units to the left and 3 units down.
center $= (-2, -3)$
vertices $= (-2, -9), (-2, 3)$
covertices $= (-6, -3), (2, -3)$
major axis $x = -2$
minor axis $y = -3$

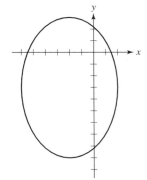

49. The graph of $x^2 - y^2 = 1$ is stretched horizontally by a factor of 4 and vertically by a factor of 6, then shifted 2 units to the left and 3 units down.

center $= (-2, -3)$
vertices $= (-6, -3), (2, -3)$
transverse axis $y = -3$
conjugate axis $x = -2$
asymptotes $\dfrac{y + 3}{6} = \pm\dfrac{x + 2}{4}$

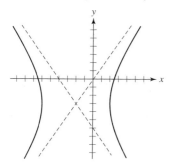

51. symmetric about the y-axis
53. not symmetric about either axis
55. symmetric about the x-axis
57. a. $(x - 5)^2 + (y + 2)^2 = 1$ **b.** $\left(\dfrac{x}{2}\right)^2 + (2y)^2 = 1$

c. $\left(\dfrac{x + 1}{3}\right)^2 + y^2 = 1$ **d.** $\left(\dfrac{x}{3} + 1\right)^2 + y^2 = 1$

59.

61.

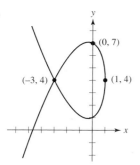

63. $y = (-2.4 \times 10^{-5})x^2 - (3.2 \times 10^{-9})x^3$

65. a–b.

c. $\dfrac{y + 2}{3} = |x - 4|$

$y + 2 = 3|x - 4|$
$y = 3|x - 4| - 2$

67.

69.

71. a.

b.

c.

d.

d.

73. a.

b.

c.

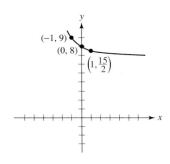

SECTION 4–6

1. Folding the paper as shown in Figure 4-39 superimposes point D on point F, while leaving point P stationary. Thus the line segment PD is superimposed on PF, so the two segments must have the same length.

3. Since both sides of the equation in Exercise 2 are positive, squaring both sides produces an equivalent equation:

$$x^2 + (y - p)^2 = (y + p)^2$$
$$x^2 + y^2 - 2py + p^2 = y^2 + 2py + p^2$$
$$x^2 = 4py$$
$$y = \frac{1}{4p}x^2$$

5. focus $= (0, 1)$
    directrix $x = 4$

7. focus $= (0.4, 0.08)$
    directrix $y = -0.02$

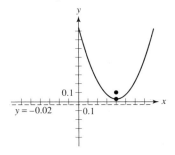

9. about 4.17 feet

11. The points P, F_1, and F_2 have coordinates (x, y), $(-c, 0)$, and $(c, 0)$, respectively. By the distance formula,

$$PF_1 = \sqrt{[x - (-c)]^2 + (y - 0)^2}$$

and

$$PF_2 = \sqrt{(x - c)^2 + (y - 0)^2}$$

Since $PF_1 + PF_2 = 2a$, we have

$$\sqrt{(x + c)^2 + y^2} + \sqrt{(x - c)^2 + y^2} = 2a$$

13. a. It is more nearly circular.

   b. It would be a circle.

15. foci $= (\pm 3, 0)$

eccentricity $= \dfrac{3}{5}$

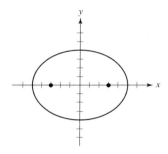

17. foci $= \left(-1, -3 \pm 2\sqrt{5}\right)$

eccentricity $= \dfrac{2\sqrt{5}}{6}$

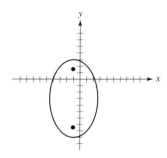

19. at the foci, about 7.07 feet from the center

21. The points P, F_1, and F_2 have coordinates (x, y), $(-c, 0)$, and $(c, 0)$, respectively. By the distance formula,

$$PF_1 = \sqrt{[x - (-c)]^2 + (y - 0)^2}$$

and

$$PF_2 = \sqrt{(x - c)^2 + (y - 0)^2}$$

Since $PF_1 - PF_2 = 2a$, we have

$$\sqrt{(x + c)^2 + y^2} - \sqrt{(x - c)^2 + y^2} = 2a$$

23. foci $= (\pm\sqrt{41}, 0)$

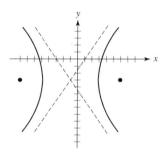

25. foci $= (-1 \pm 2\sqrt{13}, -3)$

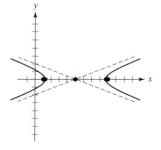

27. The primary shock wave took 0.5 seconds longer to reach Petaluma than to reach Sacramento. Since the shock wave travels at 25 km/sec, this means that the epicenter was 12.5 km closer to Sacramento than to Petaluma. The points satisfying that condition lie on one branch of a hyperbola with foci at Sacramento and Petaluma. The value of c is half the distance between the two cities, or 55. The value of a is half the difference in the distances from the epicenter, or 6.25. This value could also be calculated as 12.5 times the difference of 0.5 seconds in arrival times.

29. center $= (5, 0)$

vertices $= \left(5 \pm \sqrt{14}, 0\right)$

foci $= (1, 0), (9, 0)$

asymptotes $\dfrac{x - 5}{\sqrt{14}} = \pm \dfrac{y}{\sqrt{2}}$

31. center $= (5, 0)$

vertices $= \left(5 \pm \sqrt{14}, 0\right)$

covertices $= \left(5, \pm \sqrt{2}\right)$

foci $= \left(5 \pm 2\sqrt{3}, 0\right)$

33.

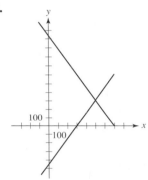

35. center $= \left(\dfrac{1}{2}, -\dfrac{1}{2}\right)$

vertices $= \left(\dfrac{1}{2} \pm \dfrac{9\sqrt{2}}{2}, -\dfrac{1}{2}\right)$

covertices $= \left(\dfrac{1}{2}, -\dfrac{1}{2} \pm \dfrac{3\sqrt{2}}{2}\right)$

foci $= (-5.5, -0.5), (6.5, -0.5)$

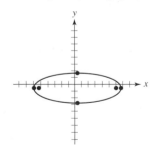

37. center $= (69.5, 46.5)$

radius $= \sqrt{6992.5}$

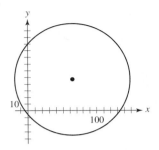

39. a. If the bulb is at the focus of the parabola and the axis of symmetry points straight ahead of the car, then all light rays coming from the bulb will strike the reflective surface and will be reflected straight ahead of the car, resulting in maximum illumination.

b. about 0.68 inches in front of the vertex

41. a. 0.0090

b. $a = 30.06, c = 0.27$

c. $(x - 0.27)^2 + y^2 = 30.06^2$

43.

The graphs intersect at approximately $(-28.80, \pm 7.64)$. Estimates of the fraction of Pluto's orbit inside Neptune's orbit will vary.

45. Since PD is vertical, its length is $y - (k - p)$. By the distance formula, $\overline{PF} = \sqrt{(x - h)^2 + [y - (k + p)]^2}$. Therefore,

$$\sqrt{(x - h)^2 + [y - (k + p)]^2} = y - (k - p)$$

Since both sides of the equation represent positive distances, squaring both sides produces an equivalent equation:

$(x - h)^2 + [y - (k + p)]^2 = y^2 - 2(k - p)y + (k - p)^2$

$(x - h)^2 + y^2 - 2(k + p)y + (k + p)^2$
$\quad = y^2 - 2(k - p)y + (k - p)^2$

$(x - h)^2 + y^2 - 2ky - 2py + k^2 + 2pk + p^2$
$\quad = y^2 - 2ky + 2py + k^2 - 2pk + p^2$

$(x - h)^2 = 4py - 4pk$

$4py = (x - h)^2 + 4pk$

$y = \dfrac{1}{4p}(x - h)^2 + k$

47. By the distance formula,

$\overline{PF_1} = \sqrt{[x - (h - c)]^2 + (y - k)^2}$

$\qquad = \sqrt{[(x - h) + c]^2 + (y - k)^2}$

and

$$\overline{PF_2} = \sqrt{[x - (h + c)]^2 + (y - k)^2}$$
$$= \sqrt{[(x - h) - c]^2 + (y - k)^2}.$$

Thus, $\sqrt{[(x - h) + c]^2 + (y - k)^2}$

$$- \sqrt{[(x - h) - c]^2 + (y - k)^2} = 2a$$

$$\sqrt{[(x - h) + c]^2 + (y - k)^2} = 2a$$
$$+ \sqrt{[(x - h) - c]^2 + (y - k)^2}$$

$$\left\{\sqrt{[(x - h) + c]^2 + (y - k)^2}\right\}^2$$
$$= \left\{2a + \sqrt{[(x - h) - c]^2 + (y - k)^2}\right\}^2$$

$$[(x - h) + c]^2 + (y - k)^2$$
$$= 4a^2 + 4a\sqrt{[(x - h) - c]^2 + (y - k)^2}$$
$$+ [(x - h) - c]^2 + (y - k)^2$$

$$4a\sqrt{[(x - h) - c]^2 + (y - k)^2}$$
$$= -4a^2 - [(x - h) - c]^2 + [(x - h) + c]^2$$

$$4a\sqrt{[(x - h) - c]^2 + (y - k)^2}$$
$$= -4a^2 - (x - h)^2 + 2c(x - h) - c^2 + (x - h)^2$$
$$+ 2c(x - h) + c^2$$

$$4a\sqrt{[(x - h) - c]^2 + (y - k)^2} = -4a^2 + 4c(x - h)$$
$$a\sqrt{[(x - h) - c]^2 + (y - k)^2} = -a^2 + c(x - h)$$

$$\left\{a\sqrt{[(x - h) - c]^2 + (y - k)^2}\right\}^2 = [-a^2 + c(x - h)]^2$$
$$a^2\{[(x - h) - c]^2 + (y - k)^2\}$$
$$= a^4 - 2a^2c(x - h) + c^2(x - h)^2$$

$$a^2[(x - h)^2 - 2c(x - h) + c^2 + (y - k)^2]$$
$$= a^4 - 2a^2c(x - h) + c^2(x - h)^2$$

$$a^2(x - h)^2 - 2a^2c(x - h) + a^2c^2 + a^2(y - k)^2$$
$$= a^4 - 2a^2c(x - h) + c^2(x - h)^2$$

$$a^2c^2 - a^4 = c^2(x - h)^2 - a^2(x - h)^2 - a^2(y - k)^2$$
$$(c^2 - a^2)(x - h)^2 - a^2(y - k)^2 = a^2(c^2 - a^2)$$

$$\frac{(x - h)^2}{a^2} - \frac{(y - k)^2}{c^2 - a^2} = 1$$

SECTION 4-7

1. $\dfrac{(y + 55)^2}{50^2} - \dfrac{x^2}{22.91^2} = 1$; lower branch

3. $(0, 1)$ **5.** $(-4, -6), (4, 6)$

7. $(-\sqrt{8}, -\sqrt{11}), (-\sqrt{8}, \sqrt{11}), (\sqrt{8}, -\sqrt{11}),$
$(\sqrt{8}, \sqrt{11})$

9. $(-\sqrt{5}, -\sqrt{3}), (-\sqrt{5}, \sqrt{3}), (\sqrt{5}, -\sqrt{3}), (\sqrt{5}, \sqrt{3})$

11. $(-\sqrt{13}, -2\sqrt{3}), (-2\sqrt{3}, -\sqrt{13}), (2\sqrt{3}, \sqrt{13}),$
$(\sqrt{13}, 2\sqrt{3})$

13. $(-6, 1), (-2, 3), (2, -3), (6, -1)$

15. a.

0 solutions   1 solution

2 solutions   3 solutions

4 solutions

b.

0 solutions   1 solution

2 solutions   3 solutions

4 solutions

17. $(-4.48, 0.52), (0.73, 5.73)$ **19.** no real solutions

21. a–b. The ship lies on or inside the circle
$x^2 + (y + 40)^2 = 400$.

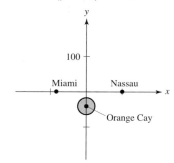

23. No. The region in Exercise 22 does not include the entire region in Exercise 21.

25.

27.

29.

31.

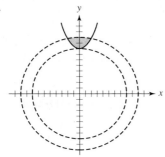

33. $x^2 + (y + 40)^2 \le 400$
$(x - 100)^2 + y^2 \le 10{,}000$

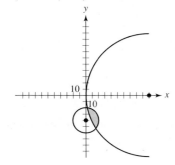

35. Responses will vary.

37. $(-2\sqrt{5}, 0.4\sqrt{5}), (-2, 2), (2, -2), (2\sqrt{5}, -0.4\sqrt{5})$

39. no real solutions

41. $(-3.32, 5), (0, -6), (3.32, 5)$ **43.** $(0, -2), (4, 0)$

45.

47.

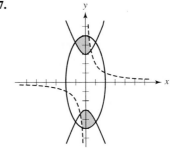

49. a. No.
   b. twice, at about $(-0.47, 0.89)$ and $(0.91, -0.48)$

51. a. The time for the batted ball to reach the fielder
   plus the time for the thrown ball to reach first base
   must be no more than 3 seconds. Since the ball always travels at 120 feet per second, the total distance traveled by the ball must be 360 feet. Thus
   $\overline{AB} + \overline{BC} = 360$. This means that the fielder is on an
   ellipse with foci at A and C and with $a = 180$.

   b. $\left(\dfrac{x - 45}{180}\right)^2 + \left(\dfrac{y}{174.28}\right)^2 \le 1$
   $x > 0$
   $y > 0$

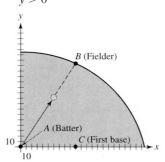

SECTION 5–1

1. The graphs all pass through $(-1, -1), (0, 0),$ and $(1, 1)$.
   The left tail of each graph points down, and the right
   tail points up. For larger powers, the graph is flatter
   near the origin, and steeper when $|x| > 1$.

3.

5.

7.

9. a.

b. $y = (-x)^3$

11. a.

b. $y = 4x^3 + 2$

13. Yes; $y = 0.1x^5$

15. No.

17. 65

19. In Exercise 13, $y = 320{,}000$.
In Exercise 14, $y = 16{,}000$.
In Exercise 16, $y = 80{,}000$.

21. a. (Sample response) The volume of a regular geometric solid is proportional to the cube of any linear dimension. For example, a cube with side length s has a volume of s^3 and a sphere with radius r has a volume of $\frac{4}{3}\pi r^3$. It is reasonable to expect that the volume V of a *Brachiosaurus* (or a scale model of one) is proportional to the cube of any linear measurement, say, its length L. That is, $V = kL^3$ for some constant k.

If the length of the scale model is L_0, then

$$728 = k(L_0)^3$$
$$k = \frac{728}{(L_0)^3}$$
$$V = \frac{728}{(L_0)^3} L^3$$

The length of a real *Brachiosaurus* is $40L_0$, so its volume is

$$V = \frac{728}{(L_0)^3}(40L_0)^3$$
$$= (40)^3(728)$$

b. 102,500 pounds; this is greater than the estimate in Exercise 20, but within the range of other researchers' estimates.

23. a.

b. domain $= (-\infty, \infty)$,
range $= (-\infty, \infty)$

25. a.

b. domain $= (-\infty, \infty)$,
range $= [4, \infty)$

45. $27.32°$

47. 46%

SECTION 5-2

1. a. $x = -3, 2$ **b.** $x = 0, 2, 3$
  c. $x = \pm 1, -2$ **d.** $x = \pm 1, \pm 3$

3. a. $x = 4$ **b.** $x = 0, \pm 3$
  c. $x = 1, \pm i$ **d.** $x = \pm \sqrt{2}$

5. $-3, -1, 2$ **7.** $-2, \dfrac{1 \pm \sqrt{5}}{2}$

9. a.

27. a.

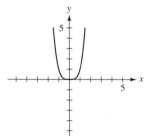

b. $y = x^4$

29. a.

b. $y = (x + 1)^4$

b. Each real zero must correspond to an x-intercept on the graph. Since the graph has no x-intercepts, the function has no real zeros.
  c. $x = \pm i$

11. In Exercise 3a, degree $= 2$, number of zeros $= 2$.
In Exercise 3b, degree $= 5$, number of zeros $= 5$.
In Exercise 3c, degree $= 3$, number of zeros $= 3$.
In Exercise 3d, degree $= 4$, number of zeros $= 4$.

13. $0, -7, 6$ **15.** $3, \pm \sqrt{2}$

17. a. $g(-1) = 0; -1$ is a zero of g.
  b. $g(x) = x^3(x + 1)(x - 1)(x^2 + 2); x = 0, \pm 1, \pm \sqrt{2}\, i$

19. a. $g(-1) = 9; -1$ is not a zero of g.
  b. $g(x) = (2x^2 + 1)(x^2 + 2); x = \pm \dfrac{\sqrt{2}}{2} i, \pm \sqrt{2}\, i$

31. a.

b. $y = 3x^4 + 2$

21. $b = 4$, multiplicity 3   **23.** $r = -2, 4$, multiplicity 3 each
  $b = -\pi$, multiplicity 5 $r = -5, 3$, multiplicity 2 each

25. $z = \pm \sqrt{2}$, multiplicity 2 each
  $z = \pm \sqrt{2}\, i$, multiplicity 1 each

27. $x = -1 \pm 2i$; degree $= 2$, number of zeros $= 2$

29. $r = 0, \pm \sqrt{3}\, i$; degree $= 5$, number of zeros $= 5$

33. Yes; $y = 30{,}000 x^4$

35. Yes; $y = -0.02 x^2$

37. In Exercise 33, $y = 30{,}000$.
In Exercise 34, $y = 15$.
In Exercise 35, $y = -0.02$.

39. $\dfrac{9}{5}$

41. 1

43. a. $C = 0.01275 H$
  b. $G = 80 N$
  c. The relationship is not one of direct variation.
  d. $W = 0.0031 N$

31. a. The degree of the polynomial is 3, and we have found only one solution. The number of solutions counted according to multiplicity must be 3.
  b. $x = 11 \pm \sqrt{23}\, i$

33. a. $P(x) = \left(x - \dfrac{1}{2}\right)^2 (x^2 - 3)$

$$= \left(x^2 - x + \dfrac{1}{4}\right)(x^2 - 3)$$

$$= x^4 - x^3 - \dfrac{11}{4} x^2 + 3x - \dfrac{3}{4}$$

b.

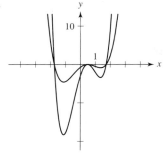

The two graphs have the same x-intercepts.

c. $P(x) = 0$ if and only if $4P(x) = 0$.

d. If the zeros of Q are the same as those of P with the same multiplicities, then Q also has factors of $\left(x - \frac{1}{2}\right)^2, \left(x + \sqrt{3}\right),$ and $\left(x - \sqrt{3}\right)$. Since Q has no other zeros, it can have no other linear factors. Thus Q must be a constant multiple of $\left(x - \frac{1}{2}\right)^2 \left(x + \sqrt{3}\right)\left(x - \sqrt{3}\right).$

35. $P(x) = (x - 3)^2(2x - 1)$ **37.** $P(x) = (x + 5)(2x + 3)^2$

39. $P(x) = (x^2 - 2)^2$ **41.** $4, \dfrac{-3 \pm \sqrt{29}}{2}$

43. $-\dfrac{7}{3}, \dfrac{1}{2}, 3$ **45.** $4, -1 \pm \sqrt{5}$

47. $-3, 1.80$

SECTION 5-3

1. a. $x = -4$, multiplicity 1; $x = -1$, multiplicity 1; $x = 3$, multiplicity 3
$(-4, 0), (-1, 0), (3, 0)$

   b. $x = 0$, multiplicity 1
$(0, 0)$

   c. $x = 0$, multiplicity 2; $x = \pm 2\sqrt{7}$, multiplicity 1 each
$(0, 0), (-2\sqrt{7}, 0), (2\sqrt{7}, 0)$

   d. $x = \pm 2$, multiplicity 1 each
$(-2, 0), (2, 0)$

3. a. $P(x) = (x + 2)^2(x - 1)^2(x - 3)^2; (-2, 0), (1, 0), (3, 0)$

   b. $P(x) = x^3 - 6x; (0, 0), (\pm \sqrt{6}, 0)$

   c. $P(x) = (2x - 3)^2(3x + 1)(3x - 1); \left(\pm \frac{1}{3}, 0\right), \left(\frac{3}{2}, 0\right)$

   d. $P(x) = (x^2 - 2)(x^2 - 3); (\pm \sqrt{2}, 0), (\pm \sqrt{3}, 0)$

5. $f(x) = (x + 4)^3(x - 1)^2(x - 3)$

7. $f(x) = x^2(x + 4)^2(x - 4)^2$

9. local minimum at $(-2, -16)$, local maximum at $(2, 16)$, decreasing on $(-\infty, -2)$ and $(2, \infty)$, increasing on $(-2, 2)$

11. local maximum at about $(-1.05, 13.03)$, local minimum at about $(1.05, -1.03)$, increasing on $(-\infty, -1.05)$ and $(1.05, \infty)$, decreasing on $(-1.05, 1.05)$

13. The second and third graphs can be obtained by shifting the first graph vertically.

15.

Value of x	Value of x^4	Value of $-100x^3$	Value of $P(x)$
$-10,000$	10^{16}	10^{14}	1.01×10^{16}
$-1,000$	10^{12}	10^{11}	1.1×10^{12}
-100	10^8	10^8	2×10^8
0	0	0	0
100	10^8	-10^8	0
$1,000$	10^{12}	-10^{11}	9×10^{11}
$10,000$	10^{16}	-10^{14}	9.9×10^{15}

17. both tails up

19. both tails down

21. Figure 5-17a

23. Figure 5-17c

25. (Sample response) If n is odd, the left tail of the graph of $y = x^n$ points down, and the right tail points up. The graph of $y = ax^n$ is obtained by stretching or compressing the basic graph $y = x^n$ vertically and reflecting it in the x-axis if $a < 0$. Stretching or compressing the graph vertically leaves the direction of the tails unchanged, and reflecting the graph in the x-axis reverses them. Therefore, if $a < 0$, the left tail of the graph of $y = ax^n$ points up, and the right tail points down. Similarly, if n is even, both tails of the graph of $y = x^n$ point up. If $a > 0$, so do both tails of the graph of $y = ax^n$. If $a < 0$, both tails point down.

27. symmetric about neither the y-axis nor the origin

29. symmetric about the origin

31. symmetric about neither the y-axis nor the origin

33. A graph that is symmetric about the x-axis contains the point $(a, -b)$ whenever it contains (a, b). If $b \neq 0$, these are two distinct points lying on the same vertical line. Thus, the graph of a function $y = f(x)$ cannot be symmetric about the x-axis unless $f(x) = 0$ for all values of x.

35. $y = x(x + 5)(x - 3)$

37. $y = (x + \sqrt{2})(x - \sqrt{2})(x - 4)$

39. $y = (x + 5)^2(x + 3)(x - 3)^2(x - 5)$

41. $y = (x + 4)^2(x + 2)(x - 4)^3$

43. a. $0, \dfrac{5}{4}$

   b. local maximum at $(0, 0)$, local minimum at $(1, -1)$

   c. left tail down, right tail up

   d. Answers will vary. The window $[-3, 3]$ by $[-3, 3]$ works well.

   e. increasing over $(-\infty, 0)$ and $(1, \infty)$, decreasing over $(0, 1)$

45. a. about 4.38

   b. local minimum at about $(-1.37, 5.24)$, local maximum at about $(2.69, 9.45)$

   c. left tail up, right tail down

d. Answers will vary. The window $[-10, 10]$ by $[-50, 50]$ works well.

e. decreasing over $(-\infty, -1.37)$ and $(2.69, \infty)$, increasing over $(-1.37, 2.69)$

47. a. about $-6.16, -2.24, 0.16, 2.24, 6.00$

b. local maxima at about $(-4.91, 1324.19)$ and $(1.31, 132.09)$, local minima at about $(-1.20, -173.27)$ and $(4.81, -1100.60)$

c. left tail down, right tail up

d. Answers will vary. The window $[-10, 10]$ by $[-1500, 1500]$ works well, although it does not show clearly that the graph does not pass through $(0, 0)$.

e. increasing over $(-\infty, -4.91)$, $(-1.20, 1.31)$, and $(4.81, \infty)$, decreasing over $(-4.91, -1.20)$ and $(1.31, 4.81)$

49. odd

51. even

53. (Sample response) If the degree of a polynomial is odd, the tails of its graph point in opposite directions. The tail that points up includes arbitrarily large y-values, and the tail that points down includes arbitrarily small (large in the negative direction) y-values. Since polynomials are continuous, the range includes all y-values in between as well and is thus $(-\infty, \infty)$.

If the degree is even, both tails of the graph point in the same direction, say, up for the sake of illustration. In this case, the graph must have an absolute minimum somewhere between the two tails, and the range includes only those y-values that are at least equal to the absolute minimum.

55. domain $= (-\infty, \infty)$, range $\approx [-24.05, \infty)$

57. domain $= (-\infty, \infty)$, range $\approx [-1706.35, \infty)$

59. domain $= (-\infty, \infty)$, range $= (-\infty, \infty)$

61. (Sample response) As you trace along the graph of a polynomial function near a local maximum, the y-coordinates change very slowly in comparison with the x-coordinates. This is another way of saying that the average rate of change in y with respect to x is small near a maximum, or that the graph is relatively flat there.

SECTION 5-4

1. a.

x	$f(x)$	First differences	Second differences	Third differences
1	3			
		7.125		
1.5	10.125		6.75	
		13.875		2.25
2	24		9	
		22.875		2.25
2.5	46.875		11.25	
		34.125		2.25
3	81		13.5	
		47.625		
3.5	128.625			

b.

x	$f(x)$	First differences	Second differences	Third differences
-1	$-a$			
		a		
0	0		0	
		a		$6a$
1	a		$6a$	
		$7a$		$6a$
2	$8a$		$12a$	
		$19a$		$6a$
3	$27a$		$18a$	
		$37a$		
4	$64a$			

c.

x	$f(x)$	First differences	Second differences	Third differences
x_0	ax_0^3			
		$3ax_0^2h + 3ax_0h^2 + ah^3$		
$x_0 + h$	$ax_0^3 + 3ax_0^2h + 3ax_0h^2 + ah^3$		$6ax_0h^2 + 6ah^3$	
		$3ax_0^2h + 9ax_0h^2 + 7ah^3$		$6ah^3$
$x_0 + 2h$	$ax_0^3 + 6ax_0^2h + 12ax_0h^2 + 8ah^3$		$6ax_0h^2 + 12ah^3$	
		$3ax_0^2h + 15ax_0h^2 + 19ah^3$		$6ah^3$
$x_0 + 3h$	$ax_0^3 + 9ax_0^2h + 27ax_0h^2 + 27ah^3$		$6ax_0h^2 + 18ah^3$	
		$3ax_0^2h + 21ax_0h^2 + 37ah^3$		$6ah^3$
$x_0 + 4h$	$ax_0^3 + 12ax_0^2h + 48ax_0h^2 + 64ah^3$		$6ax_0h^2 + 24ah^3$	
		$3ax_0^2h + 27ax_0h^2 + 61ah^3$		
$x_0 + 5h$	$ax_0^3 + 15ax_0^2h + 75ax_0h^2 + 125ah^3$			

3. 3   **5.** 2

7. (Sample response) A table containing four data points with equally spaced x-values has only one third difference. Thus the third differences are constant by default, so the table must fit a cubic polynomial function.

9. Some possibilities are $(4, 26)$, $(5, 61)$, $(6, 120)$, $(7, 209)$, $(8, 334)$, $(9, 501)$.

11. $f(x) = \dfrac{4}{3}x^3 - 6x^2 + \dfrac{20}{3}x + 1$

13. $f(x) = x^4 - 2x^3 - 3x + 1$

15. 1.37   **17.** 1.62

19. 1, 3   **21.** 1.84, 1.20, 1.52

23. 6   **25.** 2

27. $f(x) = \dfrac{25}{6}x^3 - \dfrac{41}{2}x^2 + \dfrac{79}{3}x$

29. $f(x) = \dfrac{1}{2}x^2 - \dfrac{13}{2}x + 18$

31. 2.28   **33.** 4.38

35. 1.34   **37.** -1.27

SECTION 5-5

1. $(1.20, 2.55) \cup (4.25, \infty)$ **3.** $\{-1, 3\}$

5. (Sample response) According to the new cost function, $C(x)$ represents the cost of producing $100x$ shirts. According to the former cost function, the cost of producing $100x$ shirts is
$C(100x) = 3.50(100x) + 612.50 = 350x + 612.50$. Thus the new cost function should be $C(x) = 350x + 612.50$.

Similarly, according to the former revenue function, the revenue produced by the sale of $100x$ shirts is
$R(100x) = -0.005(100x)^2 + 16(100x) = -50x^2 + 1600x$.

7. a. $P(x) = -0.08x^3 - 44x^2 + 1250x - 612.50$
   b. $0.50 \leq x \leq 26.59$; between 50 and 2659 shirts
   c. The third x-intercept is about $(-577.10, 0)$ and the graph is above the x-axis to the left of this intercept. The solution is $(-\infty, -577.10] \cup [0.50, 26.60]$. This differs from the solution in part (b) because the negative values of x do not make sense in the physical context.

9. $(-\infty, -2.83] \cup \{0\} \cup [2.83, \infty)$ **11.** $[0, \infty)$

13. $(-\infty, -7] \cup \left[-\dfrac{4}{3}, \infty\right)$ **15.** $\left(1, \dfrac{7}{2}\right) \cup \left(\dfrac{7}{2}, 5\right) \cup (5, \infty)$

17. Responses will vary.   **19.** $(-\infty, -5) \cup (0, 3)$

21. $[-2, 2]$ **23.** $(-\infty, -5) \cup (0, 3)$

25. $[-2, -1] \cup [1, 2]$ **27.** $\left(-\infty, -\dfrac{5}{2}\right) \cup \left(\dfrac{1}{3}, 7\right)$

29. $(-\infty, -4) \cup (12, \infty)$

31.

33.

35. (Sample response) The sign pattern in Exercise 15 indicates that the graph is above the x-axis when $x < 1$ and below it when $x > 1$, except at the x-intercepts $\left(\dfrac{7}{2}, 0\right)$ and $(5, 0)$. Conversely, the graph in Exercise 33 indicates that $y < 0$ in $\left(1, \dfrac{7}{2}\right) \cup \left(\dfrac{7}{2}, 5\right) \cup (5, \infty)$.

37.

39.

SECTION 5-6

1. a. There are six terms.
   b. The degree of each term is 5.
   c. The first term involves p^5, and the power of p decreases by 1 in each successive term. The first term involves q^0, and the power of q increases by 1 in each successive term.

3. a. The expanded form of $(p + q)^4$ is $p^4 + 4p^3q + 6p^2q^2 + 4pq^3 + q^4$. This expression must be multiplied by $p + q$ in order to obtain $(p + q)^5$. Multiplying by p yields terms involving $p^5, p^4q, p^3q^2, p^2q^3$, and pq^4 with coefficients of 1, 4, 6, 4, and 1, respectively. Multiplying by q yields terms involving p^4q, p^3q^2, p^2q^3, pq^4, and q^5 with the same coefficients. Aligning like terms and adding yields the following pattern of coefficients.

1	4	6	4	1	
	1	4	6	4	1
1	5	10	10	5	1

  b. $p^5 + 5p^4q + 10p^3q^2 + 10p^2q^3 + 5pq^4 + q^5$

5. a. $n + 1$
  b. n
  c. The first term involves p^n, and the power of p decreases by 1 in each successive term. The first term involves q^0, and the power of q increases by 1 in each successive term.

7. Write the coefficients in the expanded form of $(p + q)^{n-1}$ in a row. Then write them again in a second row, shifted one place to the right. That is, write the first number in the second row under the second number in the first row, and so on. Add the numbers in each column to obtain the coefficients in the expanded form of $(p + q)^n$.

9. $p^7 + 7p^6q + 21p^5q^2 + 35p^4q^3 + 35p^3q^4 + 21p^2q^5 + 7pq^6 + q^7$

11. $p^9 + 9p^8q + 36p^7q^2 + 84p^6q^3 + 126p^5q^4 + 126p^4q^5 + 84p^3q^6 + 36p^2q^7 + 9pq^8 + q^9$

13. $16a^4 - 96a^3b + 216a^2b^2 - 216ab^3 + 81b^4$

15. $32a^{10} + 80a^8b^5 + 80a^6b^{10} + 40a^4b^{15} + 10a^2b^{20} + b^{25}$

17. a. Each of the 20 strings in Table 28 has three possible positions for the third p.
  b. To distinguish among the three p's, use p_1 to represent the first p placed in the string, p_2 to represent the second p, and p_3 to represent the third p. The string ppp _ _ would appear in six different places in the table as $p_1p_2p_3$ _ _ , $p_1p_3p_2$ _ _ , $p_2p_1p_3$ _ _ , $p_2p_3p_1$ _ _ , $p_3p_1p_2$ _ _ , and $p_3p_2p_1$ _ _ .
  c. Each string has p's in three fixed positions. If we distinguish among the p's according to the order in which they are chosen, as in part (b), there are six different ways to arrange the three p's among the three positions.
  d. $10; ppp$ _ _ , pp _ p _ , pp _ _ p, p _ pp _ , p _ p _ p, p _ _ pp, _ ppp _ , _ pp _ p, _ p _ pp, and _ _ ppp.

19. a. 720
  b. 5040
  c. 3,628,800
  d. about 1.55×10^{25}

21. 15

23. 56

25. $p^6 + 6p^5q + 15p^4q^2 + 20p^3q^3 + 15p^2q^4 + 6pq^5 + q^6$

27. $125w^3 + 150w^2x + 60wx^2 + 8x^3$

29. a. You can choose any 6 of the 33 numbers. Each choice corresponds to a different string of 6 p's and 27 q's, and vice versa.
  b. 1,107,568

31. a. 1
  b. 15
  c. 120
  d. 12

33. a–b. $p^5 + 5p^4q + 10p^3q^2 + 10p^2q^3 + 5pq^4 + q^5$

35. a–b. $216y^3 + 540y^2z + 450yz^2 + 125z^3$

37. $a^6 - 18a^5b + 225a^4b^2 - 540a^3b^3 + 1215a^2b^4 - 1458ab^5 + 729b^6$

39. $10,000u^4 + 4000u^3v^2 + 600u^2v^4 + 40uv^6 + v^8$

41. $32x^{10} + 240x^8y^3 + 720x^6y^6 + 1080x^4y^9 + 810x^2y^{12} + 243y^{15}$

43. In an n-letter string consisting of p's and q's, the first letter can be chosen in two different ways (p or q). For each of those choices, the second letter can also be either p or q. Thus the first two letters can be chosen in $2^2 = 4$ different ways ($pp, pq, qp,$ or qq). Similarly, the first three letters can be chosen in $2^3 = 8$ different ways. Continuing, the n letters can be chosen in 2^n different ways.

45. a. 210
  b. $\dfrac{210}{1024} \approx 0.21$

47. $\dfrac{1}{1024} \approx 0.00098$; this question differs from that in Exercise 45b in that the positions of the four heads in the sequence was specified in this exercise.

49. $\dfrac{20}{64} \approx 0.31$

SECTION 6–1

1. $g(s) = \dfrac{450}{s}$

3. a. about 1 hour and 8 minutes
  b. You save about 48 minutes at a cost of about $3.13 per minute.

5. a.

x	$f(x)$
1	1
10	0.1
100	0.01
1000	0.001

b.

x	$f(x)$
-1	-1
-10	-0.1
-100	-0.01
-1000	-0.001

c.

x	$f(x)$
1	1
0.1	10
0.01	100
0.001	1000

d.

x	$f(x)$
-1	-1
-0.1	-10
-0.01	-100
-0.001	-1000

7. The tables illustrate that when $|x|$ is large, both $|f(x)|$ and $|g(x)|$ are small. They also illustrate that when $|x|$ is small, both $|f(x)|$ and $|g(x)|$ are large. Finally, they illustrate that $g(x)$ is always positive, but $f(x)$ can be either positive or negative.

9. a. 45 minutes   **b.** 2.7 minutes
   c. 0.0024 seconds

11. a. 7.5 days   **b.** 5.14 years
   c. 4500 years

13. Yes; $y = \dfrac{84}{x}$.   **15.** Yes; $y = \dfrac{900}{x^2}$.

17. 12   **19.** 2

21. The graph of $y = \dfrac{1}{x}$ has a vertical asymptote at $x = 0$ (on the y-axis). When the graph is shifted 1 unit to the left, the asymptote shifts 1 unit to the left, to the line $x = -1$. Similarly, when the graph is shifted 2 units up, the horizontal asymptote shifts from $y = 0$ to $y = 2$.

23.

vertical asymptote $x = -3$
horizontal asymptote $y = -2$
domain $= (-\infty, -3) \cup (-3, \infty)$
range $= (-\infty, -2) \cup (-2, \infty)$

25.

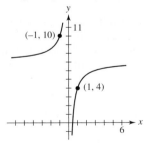

vertical asymptote $x = 0$ (y-axis)
horizontal asymptote $y = 7$
domain $= (-\infty, 0) \cup (0, \infty)$
range $= (-\infty, 7) \cup (7, \infty)$

27.

vertical asymptote $x = -1$
horizontal asymptote $y = 3$

29.

vertical asymptote $x = 2$
horizontal asymptote $y = -1$

31. Yes; $y = \dfrac{0.1}{x^2}$.   **33.** Yes; $y = \dfrac{4096}{x^2}$.

35. 18   **37.** $\dfrac{1}{2}$

39. a. 60 lb/in^2 **b.** $\dfrac{1}{5}$

41. 0.45 newtons

43.

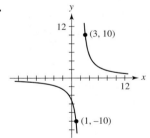

vertical asymptote $x = 2$
horizontal asymptote $y = 0$ (x-axis)
domain $= (-\infty, 2) \cup (2, \infty)$
range $= (-\infty, 0) \cup (0, \infty)$

45.

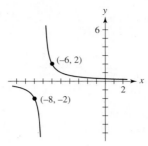

vertical asymptote $x = -7$
horizontal asymptote $y = 0$ (x-axis)
domain $= (-\infty, -7) \cup (-7, \infty)$
range $= (-\infty, 0) \cup (0, \infty)$

47.

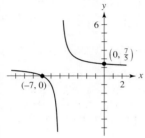

vertical asymptote $x = -5$
horizontal asymptote $y = 1$

49.

vertical asymptote $x = 0$ (y-axis)
horizontal asymptote $y = 2$

51.

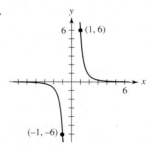

vertical asymptote $x = 0$ (y-axis)
horizontal asymptote $y = 0$ (x-axis)
domain $= (-\infty, 0) \cup (0, \infty)$
range $= (-\infty, 0) \cup (0, \infty)$

53.

vertical asymptote $x = 0$ (u-axis)
horizontal asymptote $u = 3$
domain $= (-\infty, 0) \cup (0, \infty)$
range $= (-\infty, 3)$

SECTION 6-2

1. a. The equation $x^2 + 1 = 0$ has only the solutions $x = \pm i$, so $x^2 + 1$ is not 0 for any real value of x.
   b. The graph has no x-intercepts.

3. a.

x	$h(x)$
0	-1
0.9	-10
0.99	-100
0.999	-1000
1	undefined
1.001	1000
1.01	100
1.1	10
2	1

   b. As the values of x become close to 1, $|h(x)|$ becomes arbitrarily large.

5. (Sample response) If x is near 1, then the value of $x - 1$ is near 0. The reciprocal of a number near 0 has a large absolute value, so $|h(x)| = \left| \dfrac{1}{x - 1} \right|$ is large.

7. vertical asymptote at $x = -4$

9. missing point at $x = 5$

11. (Sample response) As the level of production increases, the given description indicates that the cost of advertising increases slowly at first, then more rapidly. Thus the cost function does not exhibit a constant rate of change. As x increases, the graph of the cost function becomes steeper, suggesting a parabolic shape.

13. a.

x	y
100	11.50
10	73.60
1	703.51
0.1	7003.50

Each row shows the average cost per shirt at a given level of production.

b. When the production level is low, the average cost per shirt is very high. On the graph of $y = A(x)$, x-values near 0 correspond to large y-values. This suggests that the graph has a vertical asymptote at $x = 0$.

15. a. 9.50

   b. If you increase production, each additional shirt will cost about $9.50 to produce. Since you are selling shirts for only $9.00 each, you will be losing money on the additional shirts. Therefore, you should not increase production.

17. a. $\dfrac{x-6}{x-1} = 1 - \dfrac{5}{x-1}$

   b.

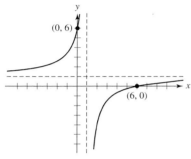

   horizontal asymptote $y = 1$

19. a. (Sample response) When $|x|$ is large, then $|x - 1|$ is also large. Since the reciprocal of a large number is near 0, any expression $\left|\dfrac{\text{constant}}{x-1}\right|$ is small when $|x|$ is large.

   b. (Sample response) Since $h(x)$ is the sum of the expressions $x + 2$ and $-\dfrac{4}{x-1}$, and since the second expression is near 0 when $|x|$ is large, $h(x)$ must be near $x + 2$ when $|x|$ is large.

21.

x	$g(x)$ (approximate)
10	0.44
100	0.95
1,000	0.995
10,000	0.9995

The table suggests that if $|x|$ is large, $g(x)$ is near 1.

23. Yes; $y = \dfrac{1}{3}$ **25.** No; both tails down.

27. a. $P(t) = 2000t$ **b.** $Q(t) = 500 + 50t$

   c. $f(t) = \dfrac{2000t}{500 + 50t}$. This represents the ratio of open-edition pieces to original and limited-edition pieces in Robert Arnold's work.

29. a–b. vertical asymptote at $x = 4$ **c.** Yes; $y = 1$

31. a–b. missing point at $x = 0$

   c. No; left tail down, right tail up

33. a–b. vertical asymptote at $x = -3$ **c.** Yes; $y = -1$

35. a–b. vertical asymptotes at $x = \pm\sqrt{3}$

   c. Yes; $y = 0$

37. a–b. missing point at $x = 1$ **c.** Yes; $y = 0$

   vertical asymptote at $x = 7$

39. a–b. missing point at $x = -2$

   c. No; left tail down, right tail up

41. a–b. vertical asymptotes at $x = \pm\sqrt[4]{2}$

   c. Yes; $y = -1$

43. a–b. no discontinuities   **c.** Yes; $y = 0$

45. The discontinuity occurs when $50t + 500 = 0$, that is, when $t = -10$. The discontinuity may be interpreted as follows. Robert Arnold had produced 500 original and limited-edition pieces prior to 1990. If he had produced them at his current rate of 50 per year, he would have begun in 1980, at $t = -10$.

47. a. $\dfrac{21.7}{\pi r^2}$ **b.** $S(r) = 2\pi r^2 + 2\pi r\left(\dfrac{21.7}{\pi r^2}\right)$

   c. S is discontinuous when $r = 0$, and its graph has a vertical asymptote there. This says that if the radius of the can is near 0, the surface area is extremely large (because the height must be large to produce a volume of 21.7 in^3).

   d. The end behavior of S agrees with that of $S = r^2$, so the right tail of the graph points up. This says that if the radius of the can is large, the surface area is also large.

49. a. The function $f(x)$ is the sum of the expressions $x^2 + 2$ and $\dfrac{3x - 2}{2x^2 + x - 4}$. Exercise 48 shows that the second expression is near 0 if $|x|$ is large, so that $f(x)$ is near $x^2 + 2$.

   b. The function $f(x)$ is the sum of $P(x)$ and the remainder from the division. The remainder is a rational function in which the denominator has a larger degree than the numerator. By the Highest Degree Theorem, the remainder is near 0 when $|x|$ is large, so $f(x)$ is near $P(x)$.

51. $y = x$ **53.** $y = 3x + 4.5$

55. $y = 2x + 2$ **57.** $y = 0.5x$

SECTION 6-3

1. 4   **3.** no solution

5. a. 200 and 350

   b. If your production level is either 200 or 350 shirts per month, the average cost per shirt is $9.

7. $[-3, 2) \cup [7, \infty)$ **9.** $(1, \infty)$

11. a. $(-\infty, -3) \cup [-1, 3)$ **b.** $[-3, 2) \cup [7, \infty)$

   c. $(-\infty, -5) \cup (-3, \infty)$ **d.** $(1, \infty)$

13. a. $(-\infty, -3) \cup [-1, 3)$ **b.** $[-3, 2) \cup [7, \infty)$

   c. $(-\infty, -5) \cup (-3, \infty)$ **d.** $(1, \infty)$

15. $(-3, 0) \cup (4, \infty)$ **17.** $[-3, -2) \cup [1, \infty)$

19. $(0, 1)$ **21.** $[-\sqrt{6}, 0) \cup [\sqrt{6}, \infty)$

23. $(-\infty, -20) \cup (-20, -10) \cup (10, 30) \cup (30, \infty)$

25. $[0, \infty)$ **27.** $(-\infty, 0) \cup \left[\frac{1}{2}, \infty\right)$

29. $(-\infty, 1) \cup (7, \infty)$ **31.** $(-\infty, -2) \cup \{0\}$

33. a. $0.44 < r < 3.75$ **b.** 1.51

  c. (Sample response) A can of the optimal shape is difficult to hold and drink from. Furthermore, cans of a nonstandard shape would be hard to recognize in a supermarket, and would not fit most vending machines. Finally, the cost of retooling a canning plant would probably outweigh any savings in materials.

SECTION 7–1

1. a. (Sample response) Each entry in the second column is obtained by multiplying the previous entry by 1.02. Thus the entry opposite $t = 4$ can be obtained by starting with the entry of 6 opposite $t = 0$ and multiplying by 1.02 four times. The final result is $6(1.02)^4$.

  b. (Sample response) If Table 29 is continued, the entry opposite any value of t can be obtained by starting with the entry of 6 opposite $t = 0$ and multiplying by 1.02 t times. The final result is $6(1.02)^t$.

3. Yes; 3.   **5.** No.

7. a.

x	y
5	243
6	729
7	2,187
8	6,561
9	19,683
10	59,049

When x is large, y is extremely large.

b.

x	y (approx.)
-1	0.33
-2	0.11
-3	0.037
-4	0.012
-5	0.0041
-6	0.0014
-7	0.00046
-8	0.00015
-9	0.000051
-10	0.000017

When $x < 0$ and $|x|$ is large, y is extremely close to 0.

9. a. $(0, \infty)$

  b. The left tail points up, and the right tail approaches the x-axis.

  c. The left tail approaches the x-axis, and the right tail points up.

11. In Exercise 3, $y = \frac{2}{3} 3^x$. In Exercise 4, $y = 64 \left(\frac{1}{2}\right)^x$.

In Exercise 6, $y = \left(-\frac{16}{3}\right)\left(-\frac{3}{2}\right)^x$.

13. (Sample response) Whenever x is increased by 1, the power of b is increased by 1, so the value of the function is multiplied by b. The ratio of the new value of f to the previous value is b.

15. a. 14.75   **b.** 36.46

17. a. (Sample response) Raising $\frac{1}{2}$ to a large positive power is equivalent to taking the reciprocal of 2 to a large positive power. Since the reciprocal of a large positive number is a small positive number, the value of $g(x)$ is positive and extremely near 0 when x is large.

  b. (Sample response) Raising $\frac{1}{2}$ to a large negative power is equivalent to raising 2 to a large positive power. Thus when $x < 0$ and $|x|$ is large, $g(x)$ is extremely large.

  c. (Sample response) The results taken together suggest that as the exponent x increases, the value of b^x decreases if $0 < b < 1$, so that $g(x) = b^x$ is a decreasing function, becoming arbitrarily large when $x < 0$ and $|x|$ is large and becoming arbitrarily close to 0 when x is large. The results also suggest that $b^x > 0$ for all values of x, so that the range of g is $(0, \infty)$.

19. a.

  b.

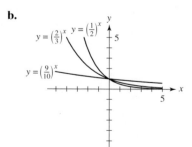

  c. domain $= (-\infty, \infty)$, range $= (0, \infty)$

  d. y-intercept $= (0, 1)$, no x-intercepts

  e. $f(1) = b, f(-1) = \frac{1}{b}$

f. horizontal asymptote on the right side of the x-axis
 g. decreasing on $(-\infty, \infty)$

21. a. domain $= (-\infty, \infty)$, range $= (-2, \infty)$
 b. decreasing
 c.

23. a. domain $= (-\infty, \infty)$, range $= (-\infty, 0)$
 b. increasing
 c.

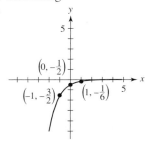

25. $f(0) = 1, f(1.71) \approx 2, f(3.42) \approx 4, f(5.13) \approx 8; 1.71$

27. $f(-2.32) \approx 1, f(-1.32) \approx 2, f(-0.32) \approx 4, f(0.68) \approx 8;$
 1.00

29. $s = 2^n$

31. (Sample response) One reason is that by asking 20 yes/no questions, a child can distinguish among more than a million objects. Another reason is that most children choose familiar objects for their opponents to identify, so that the pool of likely objects is usually much smaller than a million.

33. $a = \left\{2, \ -\dfrac{2}{3}, \dfrac{2}{9}, -\dfrac{2}{27}, \dfrac{2}{81}, \cdots \right\}$; yes.

35. $a = \left\{\dfrac{1}{5}, \dfrac{2}{5}, \dfrac{4}{5}, \dfrac{8}{5}, \dfrac{16}{5}, \cdots \right\}$; yes.

37. No.   **39.** Yes; 2.59.

41. a. $Ar + Ar^2 + Ar^3 + \cdots + Ar^n + Ar^{n+1}$
 b. $A(1 - r^{n+1})$ **c.** $S_n = \dfrac{A(1 - r^{n+1})}{1 - r}$

43. 12.80, 12.80, 12.80 **45.** 12.21, 31.87, 114.55

47. In Exercise 42, $\dfrac{A}{1 - r} = 9$. In Exercise 43, $\dfrac{A}{1 - r} = 12.8$.

49. Yes; $y = (0.0021)(9^x)$. **51.** No.

53. $y = 7\left(\sqrt{7}\right)^x$ **55.** $(47.54)(0.94)^x$

57. a. domain $= (-\infty, \infty)$, range $= (1, \infty)$
 b. increasing
 c.

59. a. domain $= (-\infty, \infty)$, range $= (-\infty, 8)$
 b. decreasing
 c.

61. a.   **b.** 64

63. a. Yes.   **b.** $-(0.9^n); -0.43$
 c. -10

65. a. Yes.   **b.** $10 \cdot 5^n; 3{,}906{,}250$
 c. No finite sum exists.

67. a. Yes.   **b.** $\left(-\dfrac{3}{4}\right)^n; 0.10$
 c. $\dfrac{4}{7}$

69. No.   **71.** $f(n) = 220\left(\sqrt[12]{2}\right)^n$

73. 29 days

SECTION 7-2

1. a. After 1 year you owe the original $10,000, plus interest amounting to 5.9% of $10,000. The total amount you owe, in dollars, is $10{,}000 + 0.059(10{,}000) = 10{,}000(1 + 0.059)$.

b.

Interest periods	Amount owed
0	10,000.00
1	10,590.00
2	11,214.81
3	11,876.48
4	12,577.20
5	13,319.25

c. $10,000(1.059)^t$

3. a. After one-third of a year you owe the original $10,000, plus interest amounting to one-third of 5.9% of $10,000. The total amount you owe, in dollars, is $10,000\left(1 + \dfrac{0.059}{3}\right)$.

b.

Interest periods	Amount owed
0	10,000.00
1	10,196.67
2	10,397.20
3	10,601.68
4	10,810.18
5	11,022.78
6	11,239.56
7	11,460.61
8	11,686.00
9	11,915.82
10	12,150.17
11	12,389.12
12	12,632.77
13	12,881.22
14	13,134.55
15	13,392.86

c. $10,000\left(1 + \dfrac{0.059}{3}\right)^{3t}$

5. $13,431

7.

n	$\left(1 + \dfrac{1}{n}\right)^n$
1,000	2.717
1,000,000	2.718280
1,000,000,000	2.718281827

9. a. annually, $6067.64; monthly, $6237.27; continuously, $6254.46
b. annually, $594,107.13; monthly, $653,813.37; continuously, $659,856.91
c. annually, $866.03; monthly, $1260.81; continuously, $1359.14

11.

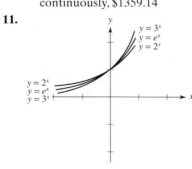

13. (Sample response) If the independent variable is time, then the average rate of change in f over any time interval is the average rate at which f grows over that interval. If the average rate of change is always equal to the value of f at some point within the interval, then f grows at a rate equal to its size.

15. 44%   **17.** $y \approx e^{1.10t}$

19. $y \approx 3e^{-0.69t}$

21.

23.

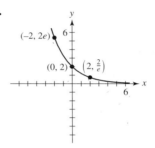

25. The graph in Exercise 21 is reflected in the y-axis to generate the graph in Exercise 22. That graph is stretched vertically by a factor of 2 to generate the graph in Exercise 23. That graph is shifted 3 units down to generate the graph in Exercise 24.

27. a. $f(t) = (10,114.32)e^{0.0485t/12}$
b. $10,270.86

29. a. $10.31   **b.** $21,124.14

31. a. $f(t) = 2^{(-2.22 \times 10^{-10})t}$ **b.** 97%
c. about 2 billion years

33. $6! = 720; 7! = 5040; 10! = 3,628,800; 25! \approx 1.55 \times 10^{25}$

SECTION 7-3

1. $(F \circ G)(x) = F[G(x)] = F(x - 2) = (x - 2) + 2 = x$
$(G \circ F)(x) = G[F(x)] = G(x + 2) = (x + 2) - 2 = x$

3. No.   **5.** No.

7. Yes.   **9.** No.

11. Yes.   **13.** Yes.

15. $(f \circ g)(x) = x; (g \circ f)(x) = x;$ both results can be simplified to x for all real values of x.

17. a.

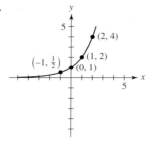

b. The points $\left(-1, \dfrac{1}{2}\right), (0, 1), (1, 2)$, and $(2, 4)$ make up a table generated by $f(x) = 2^x$. The corresponding table generated by $y = f^{-1}(x)$ contains those points with their coordinates reversed.

c. The plotted points suggest that the points (a, b) and (b, a) are always reflections of each other in the line $y = x$.

d.

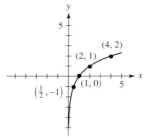

19. a. $(-\infty, 1) \cup (1, \infty)$

b. $f[g(x)] = f\left(-\dfrac{2x}{x - 1}\right)$

$$= \dfrac{-\dfrac{2x}{x - 1}}{-\dfrac{2x}{x - 1} + 2} = \dfrac{-\dfrac{2x}{x - 1}}{-\dfrac{2x}{x - 1} + 2} \cdot \dfrac{x - 1}{x - 1}$$

$$= \dfrac{-2x}{-2x + 2(x - 1)} \text{ (if } x \neq 1) = \dfrac{-2x}{-2} = x$$

for all x in the domain of g

21. a.

b. $f^{-1}(x) = \dfrac{6 - x}{4}$

23. a.

b. $f^{-1}(x) = \dfrac{x + 1}{2x}$

25. a.

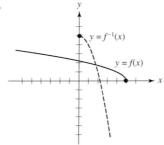

b. $f^{-1}(x) = 5 - x^2$ for $x \geq 0$

27. a.

b. $f^{-1}(x) = x^4$ for $x \geq 0$

29. a. $v = \dfrac{-20 \pm 5\sqrt{16 + 6d}}{3}$

b. The function $d = 0.06v^2 + 0.8v$ is one-to-one on $[0, \infty)$. Its inverse is found by reversing the roles of the independent and dependent variables, as was done in part (a).

31. $f_1(x) = 10 - 2x^2$ for $x \leq 0$; $f_2(x) = 10 - 2x^2$ for $x \geq 0$

$$f_1^{-1}(x) = -\sqrt{\dfrac{10 - x}{2}}; \, f_2^{-1}(x) = \sqrt{\dfrac{10 - x}{2}}$$

33. $f_1(x) = 3(x - 1)^2 + 2$ for $x \leq 1$; $f_2(x) = 3(x - 1)^2 + 2$ for $x \geq 1$

$$f_1^{-1}(x) = 1 - \sqrt{\dfrac{x - 2}{3}}; \, f_2^{-1}(x) = 1 + \sqrt{\dfrac{x - 2}{3}}$$

35. Answers will vary.

37. Yes.   **39.** Yes.

41.

x	$f^{-1}(x)$
$\sqrt{3}$	1
12,946	2.8
0	7π
e	$\dfrac{2}{3}$

$f^{-1}[f(1)] = 1$ $f[f^{-1}(\sqrt{3})] = \sqrt{3}$

$f^{-1}[f(2.8)] = 2.8$ $f[f^{-1}(12{,}946)] = 12{,}946$

$f^{-1}[f(7\pi)] = 7\pi$ $f[f^{-1}(0)] = 0$

$f^{-1}\left[f\left(\dfrac{2}{3}\right)\right] = \dfrac{2}{3}$ $f[f^{-1}(e)] = e$

43. No.

45. Yes.

47. Yes.

49. Yes.

51. No.

53. Yes.

55. Yes.   **57.** Yes.

59. No.   **61.** No.

63. $f^{-1}(x) = 5x + 4$

65. $f^{-1}(x) = \sqrt[3]{3x - 8}$

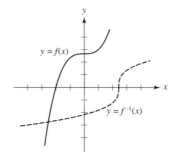

67. $f^{-1}(x) = \dfrac{5x + 5}{x}$

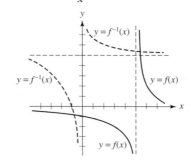

69. $f^{-1}(x) = 10 - x^2$ for $x \geqslant 0$

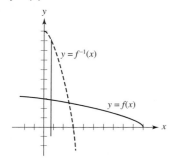

71. $f_1(x) = 7 - 3x^2$ for $x \leqslant 0$; $f_2(x) = 7 - 3x^2$ for $x \geqslant 0$

$f_1^{-1}(x) = -\sqrt{\dfrac{7-x}{3}}$; $f_2^{-1}(x) = \sqrt{\dfrac{7-x}{3}}$

73. $f_1(x) = (x + 2)^2 + 1$ for $x \leqslant -2$; $f_2(x) = (x + 2)^2 + 1$ for $x \geqslant -2$

$f_1^{-1}(x) = -\sqrt{x - 1} - 2$; $f_2^{-1}(x) = \sqrt{x - 1} - 2$

75. $(-\infty, -1), (-1, 0), (0, 1), (1, \infty)$

77. $(-\infty, -3), (-3, 0), (0, 3), (3, \infty)$

79.

81.

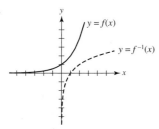

SECTION 7-4

1. a. $y = 20 + 17(0.95)^t$

   b. $y = 20 + 17e^{-0.052t}$

3. a. No horizontal line intersects the graph of $y = 2^x$ more than once.

   b. (Sample response) Solving $2^x = 7$ is equivalent to finding the point $(x_0, 7)$ on the graph of $y = g(x)$. This, in turn, is equivalent to finding the point $(7, x_0)$ on the graph of $y = g^{-1}(x)$, so that $x_0 = g^{-1}(7)$.

5.

x	$y = \log_5 x$
5	1
25	2
125	3
625	4

7. It must be true that $\log_b 1 = 0$ for any base $b > 0$ because $b^0 = 1$ for any $b > 0$. Similarly, $\log_b b = 1$ because $b^1 = b$ for any $b > 0$.

9. Yes.   **11.** Yes.

13. a–b.

x	$h(x)$	x	$h^{-1}(x)$
−2	1	1	−2
−1	3	3	−1
0	9	9	0
1	27	27	1
2	81	81	2

   c. The table for the inverse of $h(x) = 9 \cdot 3^x$ in Exercise 13b and the table for $y = \log_3 x - 2$ in Exercise 12c are identical.

15.

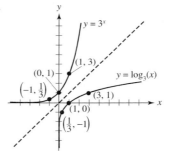

17. a. If $c = \log_{1/b} x$, then $x = \left(\dfrac{1}{b}\right)^c$, so it is also true that $x = b^{-c}$, which means that $-c = \log_b x$. Thus $\log_{1/b} x = -\log_b x$.

   b.

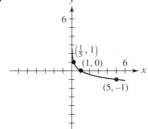

19. a. domain $= (-2, \infty)$, range $= (-\infty, \infty)$
  b. increasing
  c.

21. a. domain $= (4, \infty)$, range $= (-\infty, \infty)$
  b. increasing
  c.

23.

25.

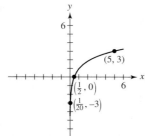

27. $\dfrac{\ln 25.7}{5}$

29. $\log\left(\dfrac{2.87}{2}\right)$

31. $\dfrac{e^{0.57} + 100}{4}$

33. $-\dfrac{\ln 0.45}{0.00012} \approx 6654$

35. a. To see why $\log_b\left(\dfrac{m}{n}\right) = \log_b m - \log_b n$, let's look at

the expressions $b^{\log_b (m/n)}$ and $b^{(\log_b m - \log_b n)}$.

If these two powers of b are equal, then the exponents must be equal. (This is another way of saying that the function $y = b^x$ is one-to-one.)

The first expression reduces to $\dfrac{m}{n}$, and the second

can be rewritten as $\dfrac{b^{\log_b m}}{b^{\log_b n}}$, which also reduces to

$\dfrac{m}{n}$. This establishes that $\log_b\left(\dfrac{m}{n}\right) = \log_b m - \log_b n$.

  b. Look at the expressions $b^{\log_b(m^a)}$ and $b^{a \log_b m}$.
    The first expression reduces to m^a, and the second
    can be rewritten as $(b^{\log_b m})^a$, which also reduces to
    m^a. This establishes that $\log_b(m^a) = a \log_b m$.

37. $\log a - (\log b + \log c)$ **39.** $2 \log a + \log b$

41. $x = Ab^y$
    $\log_b x = \log_b(Ab^y)$
    $\log_b x = \log_b A + \log_b b^y$
    $\log_b x = \log_b A + y$
    $y = \log_b x - \log_b A$
    This equation has the form $y = \log_b x + k$, with
    $k = -\log_b A$.

43. 1   **45.** 16

47. a. 1.40   **b.** 0.71
  c. -6.33 **d.** 1
  e. 0.74   **f.** -3.03
  g. 2.61   **h.** 0.23

49.

x	y
1	0
10	1
100	2
1000	3

51. a. $\dfrac{1}{2}$ **b.** 2

  c. $\dfrac{1}{2}$ **d.** -4

  e. -7 **f.** -4
  g. 2   **h.** 4

53. Yes.   **55.** Yes.

57. a. domain $= (0, \infty)$, range $= (-\infty, \infty)$
  b. increasing
  c.

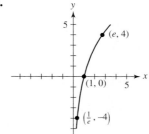

59. a. domain $= (0, \infty)$, range $= (-\infty, \infty)$
   b. increasing
   c.

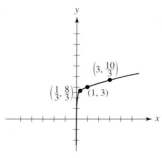

61. a. domain $= (0, \infty)$, range $= (-\infty, \infty)$
   b. increasing
   c.

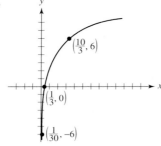

63. a. domain $= (-1, \infty)$, range $= (-\infty, \infty)$
   b. increasing
   c.

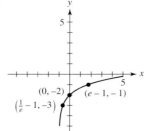

65. $\dfrac{\log_5 9 + 4}{3} \approx 1.79$ **67.** $\log_4\left(\dfrac{2.21}{3}\right) - 5 \approx -5.22$

69. $\ln 2 - 6 \approx -5.31$

71. 3   **73.** $e^{10} \approx 22{,}026.47$

75. 2   **77.** 4

79. 5

81. $\dfrac{1}{2} \log_4 x$ **83.** $\log_4 x - 2$

85. $-3 \log_4 x$ **87.** $\log_4 x$

89. a. true   **b.** false
   c. false   **d.** true
   e. true   **f.** true

91. a. $\dfrac{I_1}{I_2} = (2.512)^{M_2 - M_1}$

$$\log\left(\dfrac{I_1}{I_2}\right) = \log(2.512)^{M_2 - M_1}$$

$$\log I_1 - \log I_2 = (M_2 - M_1)\log 2.512$$

$$M_2 - M_1 = \dfrac{1}{\log 2.512}(\log I_1 - \log I_2)$$

$$M_2 - M_1 \approx 2.50(\log I_1 - \log I_2)$$

   b. about 1.59 billion   **c.** about 2.50

93. a. about 31.62   **b.** 1,000,000
   c. about 1.51×10^{24}; about 2.51×10^{24}

95. The 2 on the lower ruler and the 3 on the upper ruler represent lengths of log 2 and log 3, respectively. The position of the slide rule demonstrates that the sum of these two lengths is log 6, since the 3 on the upper ruler is opposite the 6 on the lower ruler. Thus $\log 6 = \log 2 + \log 3 = \log(2 \cdot 3)$, so $6 = 2 \cdot 3$.

SECTION 7-5

1. a.

   b. Lines will vary.

3. Answers will vary.

5. a.

Lines and equations will vary.
   b. Estimates will vary, but should be around 3:47.
   c. Answers will vary, but should be unrealistically large for the year 1 and negative for the year 3000.
   d. Answers will vary, but the line interval should exclude the years 1 and 3000.

7. a. $y = -2.07x + 7.98$ **b.** $y = 43.90x + 382.87$
   c. $y = 0.01x + 0.5$

9. Answers will vary.   **11.** Estimates will vary but should be around 3000 years.

13. a.

t	y	$Y = \ln y$
0	335	5.81
1	1,285	7.16
2	3,933	8.28
3	9,015	9.11
4	18,195	9.81
5	31,452	10.36
6	46,053	10.74
7	55,388	10.92
8	78,215	11.27

b. Equations will vary, but should be close to $\ln y = 6 + 0.6t$.

c. Equations will vary, but should be close to $y = 400\,e^{0.6t}$.

d. Graphs will vary.

15. a. $Y = 0.65t + 6.67$ **b.** $y = 788.40e^{0.65t}$

c.

t	y
0	788
1	1,510
2	2,893
3	5,541
4	10,615
5	20,333
6	38,949
7	74,608
8	142,915

17. The exponential model constructed in Exercise 15 was not an accurate predictor of the number of deaths from AIDS in 1991. This indicates that the actual relationship between t and y may not be exponential.

19. a. If $y = Ax^m$, then $\ln y = \ln(Ax^m) = \ln A + \ln x^m$.

b. Because $\ln x^m = m \ln x$, the equation in part (a) can be written $\ln y = \ln A + m \ln x$.

c. The equation is $\ln y = m \ln x + \ln A$. This has the form $\ln y = (\text{constant})\ln x + (\text{constant})$, with m and $\ln A$ as the two constants.

21. Tables will vary.   **23.** $y \approx 0.00055x^{3/2}$

25. a. $y \approx 0.0034x - 1.94, r \approx 0.99986$

b. $y \approx 39.78x - 306.17, r \approx 0.92$

c. $y \approx 6.59(1.000077)^x, r \approx 0.92$

d. $y \approx 0.0016x^{1.071}, r \approx 0.99994$; this gives the best fit.

27. Answers will vary.   **29.** Equations will vary, but should be close to $y = 2 \ln x - 5$.

31.

Yes.

33.

No.

35. a.

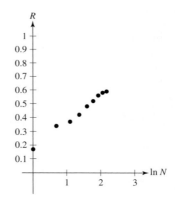

The data looks more nearly logarithmic.

b. Equations will vary, but should be close to $R = 0.2 + 0.2 \ln N$.

37.

Yes.

39.

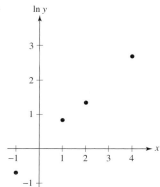

Yes.

41. Equations will vary, but should be close to $y = 2x^2$.

43.

Yes.

45.

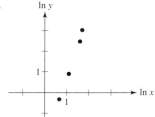

Yes.

47. linear ($y \approx 11.51 - 1.05x$), $r \approx -0.99994$

49. power ($y \approx 1.04x^{0.49}$), $r \approx -0.999988$

51. a. $y \approx 3.77(10^{-17})(1.03)^t$

b. The equation predicts an 1860 population of about 31,392,000 people and an 1870 population of about 42,211,000 people.

c. The Civil War would account for some of the short-fall.

53. a. The horizontal and vertical distances on semilog paper are proportional to x and log y, respectively. The plot is linear if the relationship between x and log y is linear, that is, if x and y are related by an exponential function.

b. The horizontal and vertical distances on log-log paper are proportional to log x and log y, respectively. The plot is linear if the relationship between log x and log y is linear, that is, if x and y are related by a power function.

SECTION 8–1

1. a. $(\cos \theta, \sin \theta) = (0, 1)$ **b.** $(\cos \theta, \sin \theta) = (1, 0)$

c. $(\cos \theta, \sin \theta) = (1, 0)$ **d.** $(\cos \theta, \sin \theta) = (-1, 0)$

e. $(\cos \theta, \sin \theta) = (-1, 0)$ **f.** $(\cos \theta, \sin \theta) = (0, 1)$

3. a. $(\cos \theta, \sin \theta) \approx (0.54, 0.84)$

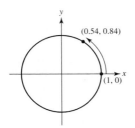

b. $(\cos \theta, \sin \theta) \approx (0.71, -0.71)$

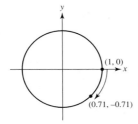

c. $(\cos \theta, \sin \theta) \approx (-0.42, 0.91)$

d. $(\cos \theta, \sin \theta) \approx (-0.42, -0.91)$

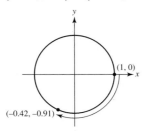

e. $(\cos \theta, \sin \theta) \approx (0.87, 0.50)$

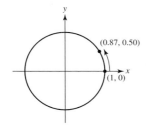

f. $(\cos \theta, \sin \theta) \approx (0.87, -0.50)$

5. The side of the triangle adjacent to the angle of $\theta°$ is the side opposite the angle of $\phi°$, so the ratio $\dfrac{\text{adjacent side}}{\text{hypotenuse}}$ for $\theta°$ is the same as the ratio $\dfrac{\text{opposite side}}{\text{hypotenuse}}$ for $\phi°$. By definition, this means that $\cos \theta° = \sin \phi°$.

7. a. The upper and lower triangles each have a 30° angle at O and a 90° angle at C, and they share the side OC. Therefore, the triangles are congruent, so $\overline{AC} = \overline{BC}$. Furthermore, the angles at A and B must each be 60°, so triangle OAB is equilateral. Thus $\overline{AB} = \overline{OA}$, so $\overline{AC}$ and $\overline{BC}$ are each $\frac{1}{2}\,\overline{OA}$. By the Pythagorean theorem, $\overline{OC} = \sqrt{(\overline{OA})^2 - (\overline{AC})^2} =$

$$\sqrt{(\overline{OA})^2 - \left(\frac{1}{2}\,\overline{OA}\right)^2} = \sqrt{\frac{3}{4}\,(\overline{OA})^2} = \frac{\sqrt{3}}{2}\,\overline{OA}.$$

Therefore, $\sin 30° = \dfrac{\overline{AC}}{\overline{OA}} = \dfrac{1}{2}$ and

$\cos 30° = \dfrac{\overline{OC}}{\overline{OA}} = \dfrac{\sqrt{3}}{2}$.

b. From Exercise 5, the sine of an angle is the cosine of its complement, and vice versa. Therefore, $\sin 60° = \cos 30° = \dfrac{\sqrt{3}}{2}$ and $\cos 60° = \sin 30° = \dfrac{1}{2}$.

9. a. $\dfrac{3\pi}{2}$: $(\cos 270°, \sin 270°) = (0, -1)$

b. 3π: $(\cos 540°, \sin 540°) = (-1, 0)$

c. -2π; $(\cos(-360°), \sin(-360°)) = (1, 0)$

d. $-\dfrac{5\pi}{2}$; $(\cos(-450°), \sin(-450°)) = (0, -1)$

11. a.

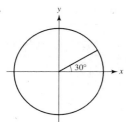

b. $\left(\dfrac{\sqrt{3}}{2}, \dfrac{1}{2}\right)$

c. $\cos 30° = \dfrac{\sqrt{3}}{2}$, $\sin 30° = \dfrac{1}{2}$

13. a. $\dfrac{\pi}{4}$ **b.** 2π

c. $45°$

15. a. $\dfrac{5\pi}{2}$ **b.** 5π

c. $-\dfrac{3\pi}{4}$ **d.** $-\dfrac{5\pi}{3}$

e. $\dfrac{21\pi}{5}$ **f.** -1

17.

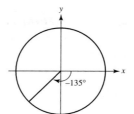

For every value of t, the point $(x, y) = (\cos t, \sin t)$ is the terminal point of some directed arc on the unit circle.

19. a. 70.4 radians per second

  b. 81.23 radians per second

21. a. 300,000 radians per second **b.** $(-0.60, -0.80)$

23. a. $\cos 1 \approx 0.54$, $\sin 1 \approx 0.84$ **b.** $57.30°$

  c. $\cos 57.30° \approx 0.54$, $\sin 57.30° \approx 0.84$

25.

Degrees	Radians	Sine	Cosine
0°	0	0	1
30°	$\dfrac{\pi}{6}$	$\dfrac{1}{2}$	$\dfrac{\sqrt{3}}{2}$
45°	$\dfrac{\pi}{4}$	$\dfrac{\sqrt{2}}{2}$	$\dfrac{\sqrt{2}}{2}$
60°	$\dfrac{\pi}{3}$	$\dfrac{\sqrt{3}}{2}$	$\dfrac{1}{2}$
90°	$\dfrac{\pi}{2}$	1	0
120°	$\dfrac{2\pi}{3}$	$\dfrac{\sqrt{3}}{2}$	$-\dfrac{1}{2}$
135°	$\dfrac{3\pi}{4}$	$\dfrac{\sqrt{2}}{2}$	$-\dfrac{\sqrt{2}}{2}$
150°	$\dfrac{5\pi}{6}$	$\dfrac{1}{2}$	$-\dfrac{\sqrt{3}}{2}$
180°	π	0	-1
210°	$\dfrac{7\pi}{6}$	$-\dfrac{1}{2}$	$-\dfrac{\sqrt{3}}{2}$
225°	$\dfrac{5\pi}{4}$	$-\dfrac{\sqrt{2}}{2}$	$-\dfrac{\sqrt{2}}{2}$
240°	$\dfrac{4\pi}{3}$	$-\dfrac{\sqrt{3}}{2}$	$-\dfrac{1}{2}$
270°	$\dfrac{3\pi}{2}$	-1	0
300°	$\dfrac{5\pi}{3}$	$-\dfrac{\sqrt{3}}{2}$	$\dfrac{1}{2}$
315°	$\dfrac{7\pi}{4}$	$-\dfrac{\sqrt{2}}{2}$	$\dfrac{\sqrt{2}}{2}$
330°	$\dfrac{11\pi}{6}$	$-\dfrac{1}{2}$	$\dfrac{\sqrt{3}}{2}$
360°	2π	0	1

27. a. $(\cos\theta, \sin\theta) = \left(-\dfrac{\sqrt{2}}{2}, \dfrac{\sqrt{2}}{2}\right)$

b. $(\cos\theta, \sin\theta) = \left(\dfrac{\sqrt{3}}{2}, -\dfrac{1}{2}\right)$

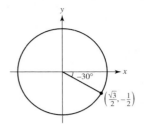

c. $(\cos\theta, \sin\theta) = (-1, 0)$

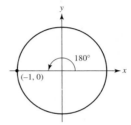

d. $(\cos\theta, \sin\theta) = \left(\dfrac{1}{2}, \dfrac{\sqrt{3}}{2}\right)$

e. $(\cos\theta, \sin\theta) = \left(\dfrac{\sqrt{2}}{2}, -\dfrac{\sqrt{2}}{2}\right)$

f. $(\cos\theta, \sin\theta) = \left(-\dfrac{\sqrt{3}}{2}, \dfrac{1}{2}\right)$

29. a. 240°

b. −135°

c. 1800°

d. 210°

e. $-\dfrac{180°}{\pi}$

f. 810°

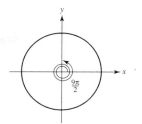

31. a. $\dfrac{\pi}{24}$, 7.5°

b. $-\dfrac{\pi}{6}$, −30°

c. $-\dfrac{6}{\pi}$, $-\left(\dfrac{1080}{\pi^2}\right)^{\circ}$

d. $\dfrac{1}{400}$, 0.45°

33. a. 572.96°; 10, (−0.84, −0.54)

b. $(1.72 \times 10^8)°$, 3×10^6; (0.48, −0.88)

c. $(-1.72 \times 10^8)°$, -3×10^6, (0.48, 0.88)

35. a. 35 ft/sec

b. 40 m/sec

c. 56 cm/sec

d. 5.5π in/sec

SECTION 8-2

1. The range of each function is $[-1, 1]$.

3. In each row of Table 38, the two angles have a sum of π, and their cosines are the opposites of each other.

5. $\sin \theta = \cos\left(\theta - \dfrac{\pi}{2}\right)$

7. For every real value of θ, the values of $\sin \theta$ and $\cos \theta$ are the y-coordinate and x-coordinate of a point on the unit circle. Since the unit circle is centered at $(0, 0)$ and has a radius of 1, the coordinates of every point on the circle must be between -1 and 1.

9. If the points A and B are on the y-axis, then $\cos \theta$ and $\cos (\pi - \theta)$ are both zero, so $\cos (\pi - \theta) = -\cos \theta$.

11. In triangle OCA the angle at O is θ, so the angle at A is $\dfrac{\pi}{2} - \theta$. In triangle BDO the angle at O is

$$\pi - \left(\theta + \dfrac{\pi}{2}\right) = \dfrac{\pi}{2} - \theta,$$ so the angle at B is θ. Furthermore, the hypotenuse of each triangle is a radius of the unit circle and therefore has a length of 1. Thus the two triangles are congruent. Now

$$\sin\left(\theta + \dfrac{\pi}{2}\right) = \overline{DB} = \overline{OC} = \cos \theta$$

13. a. $\cos 3\theta = 1, 3\cos\theta = 3$ **b.** $\cos 3\theta = -1, 3\cos\theta = -3$

   c. $\cos 3\theta = -\dfrac{\sqrt{2}}{2}, 3\cos\theta = \dfrac{3\sqrt{2}}{2}$

   d. $\cos 3\theta = 1, 3\cos\theta = -\dfrac{3}{2}$

15. a.

   Yes.

   b.

   Yes.

17. a. $[-1, 1]$ **b.** $[-1, 1]$

   c. increasing over $\left(0, \dfrac{\pi}{2}\right)$ and $\left(\dfrac{3\pi}{2}, 2\pi\right)$, decreasing

   over $\left(\dfrac{\pi}{2}, \dfrac{3\pi}{2}\right)$

   d. increasing over $(\pi, 2\pi)$, decreasing over $(0, \pi)$

19.

21.

   The graph of $y = \cos\theta$ is the graph of $y = \sin\theta$ shifted

   $\dfrac{\pi}{2}$ units to the left.

23. a. Multiplying a function by -1 reflects its graph in
   the x-axis.

   b.

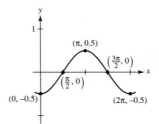

   c. $2, 0.5$ **d.** $2\pi, 2\pi$

   e. $|A|$

   f.

25.

27.

29.

31.

33. a. amplitude $= |A|$, period $= \dfrac{2\pi}{b}$

　　b. The period is $\dfrac{2\pi}{|b|}$.

35.

37.

39. a. The period of f must be 365, so $\dfrac{2\pi}{b} = 365$. Solving

　　for b yields $b = \dfrac{2\pi}{365}$.

b. The amplitude of f is half the difference between its maximum and minimum values, so $A = \dfrac{1}{2}(74 - 30) = 22$. A cycle of f should begin halfway between day 31 and day 205, at $t = \dfrac{1}{2}(31 + 205) = 118$. The average temperature on that day is probably near the average of the minimum and maximum; that is, $\dfrac{1}{2}(30 + 74) = 52°$. Thus a cycle of f should begin at the point $(h, k) = (118, 52)$.

41. 523 cycles per second.

43. The piston moves up when $f(t)$ is increasing and down when $f(t)$ is decreasing. In the interval $0 \leqslant t \leqslant 2\pi$, the piston moves up for $0 \leqslant t < \dfrac{\pi}{2}$ and $\dfrac{3\pi}{2} < t \leqslant 2\pi$ and down for $\dfrac{\pi}{2} < t < \dfrac{3\pi}{2}$. It is moving fastest when the graph of $f(t)$ is steepest, at $t = 0$, π, and 2π. It stops for an instant to change direction when $f(t)$ changes from increasing to decreasing, or vice versa. This happens at the extreme points on the graph, when $t = \dfrac{\pi}{2}$ and $t = \dfrac{3\pi}{2}$.

45. a. $\sin(\pi - \theta) = \sin\theta$

　　b.

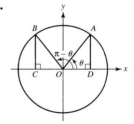

The argument in Example 1 of Section 8-2 can be used to show that triangles DOA and COB are congruent. Therefore, $\overline{AD} = \overline{BC}$, so $|\sin(\pi - \theta)| = |\sin\theta|$. Furthermore, A and B are on the same side of the x-axis, so $\sin(\pi - \theta)$ and $\sin\theta$ have the same sign. Therefore, $\sin(\pi - \theta) = \sin\theta$.

　　c.

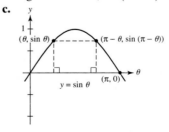

47. a. $\cos(2\pi - \theta) = \cos\theta$

b.

$|\cos(2\pi - \theta)| = OC = |\cos\theta|$, and A and B are on the same side of the y-axis, so $\cos\theta$ and $\cos(2\pi - \theta)$ have the same sign. Therefore, $\cos(2\pi - \theta) = \cos\theta$.

c.

49. a. $y\sin x$ is odd, $y = \cos x$ is even

b.

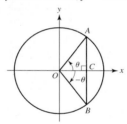

$\sin(-x) = -\sin x; \cos(-x) = -\cos x$; yes

51. a. domain $= (-\infty, \infty)$, range $= [-1, 1]$, amplitude $= 1$, period $= 4\pi$, phase shift $= 0$

b–c.

53. a. domain $= (-\infty, \infty)$, range $= [-7, 7]$, amplitude $= 7$, period $= \dfrac{2\pi}{7}$, phase shift $= 0$

b–c.

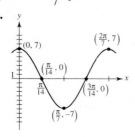

55. a. domain $= (-\infty, \infty)$, range $= [-1.3, -0.7]$, amplitude $= 0.3$, period $= \pi$, phase shift $= 0$

b–c.

57. a. domain $= (-\infty, \infty)$, range $= [-1, 1]$, amplitude $= 1$, period $= \dfrac{\pi}{2}$, phase shift $= \dfrac{\pi}{8}$

b–c.

59. a. domain $= (-\infty, \infty)$, range $= [-11, 5]$, amplitude $= 8$, period $= 6\pi$, phase shift $= -\dfrac{\pi}{2}$

b–c.

61. One possible equation is $y = 2\sin\dfrac{1}{2}\left(x + \dfrac{\pi}{6}\right)$.

63. One possible equation is $y = 5\cos\left(x - \dfrac{3\pi}{4}\right)$.

65.

$y = -\dfrac{1}{2}\cos 2x + \dfrac{1}{2}; (\sin x)^2 = \dfrac{1}{2}(-\cos 2x + 1)$

67. a. The amplitude is twice that of the first motion, so the maximum displacement of the second spring is twice that of the first.

b. The period is half that of the first motion, so the second spring vibrates twice as rapidly as the first.

c. The second spring vibrates twice as rapidly as the first, and its maximum displacement is twice as great.

d. The graph of $y = 5 \cos 12t$ can be obtained by shifting the graph of $y = 5 \sin 12t, \frac{\pi}{24}$ units to the left. Therefore, the second spring occupies a given position $\frac{\pi}{24}$ seconds before the first.

69. a–d. $y = 1 - \frac{x^2}{2!} + \frac{x^4}{4!}$

$y = 1 - \frac{x^2}{2!} + \frac{x^4}{4!} - \frac{x^6}{6!}$ $y = 1 - \frac{x^2}{2!}$

Each successive polynomial graph approximates the graph of $y = \cos x$ more closely for values of x near 0.

71. a. $f(x) = 2 \sin \frac{1}{2}\left(x + \frac{\pi}{2}\right)$

b. $f(x) = 2 \cos \frac{1}{2}\left(x - \frac{\pi}{2}\right)$

73. a. $f(x) = 3 \sin\left(x - \frac{\pi}{4}\right)$

b. $f(x) = 3 \cos\left(x - \frac{3\pi}{4}\right)$

75. $f(x) = 3 \sin(x + 0.6\pi)$

77. $f(x) = \sin 2\left(x - \frac{3\pi}{4}\right)$

SECTION 8-3

1. According to the right triangle definitions,

$\sin \theta = \dfrac{\text{opposite side}}{\text{hypotenuse}}$ and $\csc \theta = \dfrac{\text{hypotenuse}}{\text{opposite side}}$, so by definition $\sin \theta$ and $\csc \theta$ are reciprocals. Similarly, $\cos \theta$ and $\sec \theta$ are reciprocals and $\tan \theta$ and $\cot \theta$ are reciprocals. the quotient

$$\dfrac{\sin \theta}{\cos \theta} = \dfrac{\frac{\text{opposite side}}{\text{hypotenuse}}}{\frac{\text{adjacent side}}{\text{hypotenuse}}} = \left(\dfrac{\text{opposite side}}{\text{hypotenuse}}\right)\left(\dfrac{\text{hypotenuse}}{\text{adjacent side}}\right)$$

$$= \dfrac{\text{opposite side}}{\text{adjacent side}} = \tan \theta.$$ Since $\cot \theta$ is the reciprocal of

$\tan \theta$, it is also the reciprocal of $\dfrac{\sin \theta}{\cos \theta}$, which is $\dfrac{\cos \theta}{\sin \theta}$.

3. a. $\tan 30° = \dfrac{\sqrt{3}}{3}$

b. $\cot 30° = \sqrt{3}$

c. $\sec 30° = \dfrac{2\sqrt{3}}{3}$

d. $\csc 30° = 2$

5. If $0 < x < \dfrac{\pi}{2}$, the angle x determines a right triangle in the first quadrant whose horizontal and vertical sides have lengths a and b and whose hypotenuse has a length of 1, as in the accompanying figure.

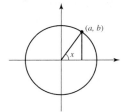

According to the right triangle definitions of the trigonometric functions,

$\sin x = \dfrac{\text{opposite side}}{\text{hypotenuse}} = \dfrac{b}{1}$ $\cos x = \dfrac{\text{adjacent side}}{\text{hypotenuse}} = \dfrac{a}{1}$

$\tan x = \dfrac{\text{opposite side}}{\text{adjacent side}} = \dfrac{b}{a}$ $\cot x = \dfrac{\text{adjacent side}}{\text{opposite side}} = \dfrac{a}{b}$

$\sec x = \dfrac{\text{hypotenuse}}{\text{adjacent side}} = \dfrac{1}{a}$ $\csc x = \dfrac{\text{hypotenuse}}{\text{opposite side}} = \dfrac{1}{b}$

7. $\sin x = -\dfrac{12}{13}$ $\cos x = -\dfrac{5}{13}$ $\tan x = \dfrac{12}{5}$

$\cot x = \dfrac{5}{12}$ $\sec x = -\dfrac{13}{5}$ $\csc x = -\dfrac{13}{12}$

9. $\sin x = 0$ $\cos x = 1$ $\tan x = 0$

$\cot x = \dfrac{1}{0}$ (undefined) $\sec x = 1$

$\csc x = \dfrac{1}{0}$ (undefined)

11. $\sin\left(-\dfrac{\pi}{3}\right) = -\dfrac{\sqrt{3}}{2}$ $\cos\left(-\dfrac{\pi}{3}\right) = \dfrac{1}{2}$ $\tan\left(-\dfrac{\pi}{3}\right) = -\sqrt{3}$

$\cot\left(-\dfrac{\pi}{3}\right) = -\dfrac{\sqrt{3}}{3}$ $\sec\left(-\dfrac{\pi}{3}\right) = 2$ $\csc\left(-\dfrac{\pi}{3}\right) = -\dfrac{2\sqrt{3}}{3}$

13. $\sin \dfrac{5\pi}{2} = 1$ $\cos \dfrac{5\pi}{2} = 0$ $\tan \dfrac{5\pi}{2} = \dfrac{1}{0}$ (undefined)

$\cot \dfrac{5\pi}{2} = 0$ $\sec \dfrac{5\pi}{2} = \dfrac{1}{0}$ (undefined) $\csc \dfrac{5\pi}{2} = 1$

15. $221; 323$

17. The lengths of the horizontal and vertical sides of the right triangle in the figure are $\cos x$ and $\sin x$. By the Pythagorean theorem, $(\cos x)^2 + (\sin x)^2 = 1^2$, or $\cos^2 x + \sin^2 x = 1$.

19. $\sin x = \dfrac{4}{5}$ $\cos x = \dfrac{3}{5}$ $\tan x = \dfrac{4}{3}$

$\cot x = \dfrac{3}{4}$ $\sec x = \dfrac{5}{3}$ $\csc x = \dfrac{5}{4}$

21. $\sin x = -\dfrac{\sqrt{5}}{5}$ $\cos x = -\dfrac{2\sqrt{5}}{5}$ $\tan x = \dfrac{1}{2}$

$\cot x = 2$ $\sec x = -\dfrac{\sqrt{5}}{2}$ $\csc x = -\sqrt{5}$

23. Since $\tan^2 x = (\tan x)^2 \geqslant 0$ for all values of x, $\sec^2 x = 1 + \tan^2 x \geqslant 1$ for all values of x. Thus $|\sec x| = \sqrt{\sec^2 x} \geqslant 1$ for all x. Similarly, since $\cot^2 x = (\cot x)^2 \geqslant 0$ for all x, $\csc^2 x = 1 + \cot^2 x \geqslant 1$ for all x. Thus $|\csc x| = \sqrt{\csc^2 x} \geqslant 1$ for all x.

25. The functions $\cot x = \dfrac{\cos x}{\sin x}$ and $\csc x = \dfrac{1}{\sin x}$ are undefined when $\sin x = 0$. This occurs when $x = k\pi$ for any integer k.

27.

Degrees	Radians	Sine	Cosine	Tangent
0°	0	0	1	0
30°	$\dfrac{\pi}{6}$	$\dfrac{1}{2}$	$\dfrac{\sqrt{3}}{2}$	$\dfrac{\sqrt{3}}{3}$
45°	$\dfrac{\pi}{4}$	$\dfrac{\sqrt{2}}{2}$	$\dfrac{\sqrt{2}}{2}$	1
60°	$\dfrac{\pi}{3}$	$\dfrac{\sqrt{3}}{2}$	$\dfrac{1}{2}$	$\sqrt{3}$
90°	$\dfrac{\pi}{2}$	1	0	—
120°	$\dfrac{2\pi}{3}$	$\dfrac{\sqrt{3}}{2}$	$-\dfrac{1}{2}$	$-\sqrt{3}$
135°	$\dfrac{3\pi}{4}$	$\dfrac{\sqrt{2}}{2}$	$-\dfrac{\sqrt{2}}{2}$	-1
150°	$\dfrac{5\pi}{6}$	$\dfrac{1}{2}$	$-\dfrac{\sqrt{3}}{2}$	$-\dfrac{\sqrt{3}}{3}$
180°	π	0	-1	0
210°	$\dfrac{7\pi}{6}$	$-\dfrac{1}{2}$	$-\dfrac{\sqrt{3}}{2}$	$\dfrac{\sqrt{3}}{3}$
225°	$\dfrac{5\pi}{4}$	$-\dfrac{\sqrt{2}}{2}$	$-\dfrac{\sqrt{2}}{2}$	1
240°	$\dfrac{4\pi}{3}$	$-\dfrac{\sqrt{3}}{2}$	$-\dfrac{1}{2}$	$\sqrt{3}$
270°	$\dfrac{3\pi}{2}$	-1	0	—
300°	$\dfrac{5\pi}{3}$	$-\dfrac{\sqrt{3}}{2}$	$\dfrac{1}{2}$	$-\sqrt{3}$
315°	$\dfrac{7\pi}{4}$	$-\dfrac{\sqrt{2}}{2}$	$\dfrac{\sqrt{2}}{2}$	-1
330°	$\dfrac{11\pi}{6}$	$-\dfrac{1}{2}$	$\dfrac{\sqrt{3}}{2}$	$-\dfrac{\sqrt{3}}{3}$
360°	2π	0	1	0

29. The domain of the tangent and secant functions includes all real numbers except odd multiples of $\dfrac{\pi}{2}$. The domain of the cotangent and cosecant functions includes all real numbers except integer multiples of π. The range of the tangent and cotangent functions is $(-\infty, \infty)$. The range of the secant and cosecant functions is $(-\infty, -1] \cup [1, \infty)$.

31. The tangent function is increasing wherever it is defined. The cotangent function is never increasing. The secant function is increasing for values of x corresponding to angles in the first and second quadrants. The cosecant function is increasing for values of x corresponding to angles in the second and third quadrants.

33. a. $(-a, -b)$ **b.** $\tan(x + \pi) = \dfrac{-b}{-a} = \dfrac{b}{a} = \tan x$

35. The domain includes all real numbers except odd multiples of $\dfrac{\pi}{2}$. The range is $(-\infty, \infty)$.

37. $\tan x$ is increasing wherever it is defined.

39. a. When $\cos x = \pm 1$, $\sec x$ has the same value.
   b. $|\sec x| \geqslant 1$ for all x. The range is $(-\infty, -1] \cup [1, \infty)$.

41. When $\cos x = 0$, $\sec x = \dfrac{1}{\cos x}$ is undefined. This fact appears in the graph of $y = \sec x$ as a vertical asymptote when $x = \dfrac{\pi}{2} + k\pi$ for every integer k. The domain of the secant function excludes these values of x.

43. When $\cos x$ is increasing, $\sec x = \dfrac{1}{\cos x}$ is decreasing, and vice versa.

45. a. When $\sin x = \pm 1$, $\csc x$ has the same value.
   b. $|\csc x| \geqslant 1$ for all x. The range is $(-\infty, -1] \cup [1, \infty)$.

47. When $\sin x = 0$, $\csc x = \dfrac{1}{\sin x}$ is undefined. This fact appears in the graph of $y = \csc x$ as a vertical asymptote when $x = k\pi$ for every integer k. The domain of the cosecant function excludes these values of x.

49. When $\sin x$ is increasing, $\csc x = \dfrac{1}{\sin x}$ is decreasing, and vice versa.

51. $\sin x = \dfrac{\sqrt{3}}{2}$ $\cos x = -\dfrac{1}{2}$ $\tan x = -\sqrt{3}$

$\cot x = -\dfrac{\sqrt{3}}{3}$ $\sec x = -2$ $\csc x = \dfrac{2\sqrt{3}}{3}$

53. $\sin x = \dfrac{2\sqrt{5}}{5}$ $\cos x = \dfrac{\sqrt{5}}{5}$ $\tan x = 2$

$\cot x = \dfrac{1}{2}$ $\sec x = \sqrt{5}$ $\csc x = \dfrac{\sqrt{5}}{2}$

55. $\sin 135° = \dfrac{\sqrt{2}}{2}$ $\cos 135° = -\dfrac{\sqrt{2}}{2}$ $\tan 135° = -1$

$\cot 135° = -1$ $\sec 135° = -\sqrt{2}$ $\csc 135° = \sqrt{2}$

57. $\sin(-300°) = \dfrac{\sqrt{3}}{2}$ $\cos(-300°) = \dfrac{1}{2}$

     $\tan(-300°) = \sqrt{3}$ $\cot(-300°) = \dfrac{\sqrt{3}}{3}$

     $\sec(-300°) = 2$ $\csc(-300°) = \dfrac{2\sqrt{3}}{3}$

59. $\sin\dfrac{2\pi}{3} = \dfrac{\sqrt{3}}{2}$ $\cos\dfrac{2\pi}{3} = -\dfrac{1}{2}$ $\tan\dfrac{2\pi}{3} = -\sqrt{3}$

   $\cot\dfrac{2\pi}{3} = -\dfrac{\sqrt{3}}{3}$ $\sec\dfrac{2\pi}{3} = -2$ $\csc\dfrac{2\pi}{3} = \dfrac{2\sqrt{3}}{3}$

61. $\sin\dfrac{7\pi}{4} = -\dfrac{\sqrt{2}}{2}$ $\cos\dfrac{7\pi}{4} = \dfrac{\sqrt{2}}{2}$ $\tan\dfrac{7\pi}{4} = -1$

   $\cot\dfrac{7\pi}{4} = -1$ $\sec\dfrac{7\pi}{4} = \sqrt{2}$ $\csc\dfrac{7\pi}{4} = -\sqrt{2}$

63.

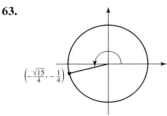

$\sin x = -\dfrac{1}{4}$ $\cos x = -\dfrac{\sqrt{15}}{4}$ $\tan x = \dfrac{\sqrt{15}}{15}$

$\cot x = \sqrt{15}$ $\sec x = -\dfrac{4\sqrt{15}}{15}$ $\csc x = -4$

65.

$\sin x = \dfrac{2\sqrt{53}}{53}$ $\cos x = -\dfrac{7\sqrt{53}}{53}$ $\tan x = -\dfrac{2}{7}$

$\cot x = -\dfrac{7}{2}$ $\sec x = -\dfrac{\sqrt{53}}{7}$ $\csc x = \dfrac{\sqrt{53}}{2}$

67. a–b.

c. The domain includes all real numbers except odd multiples of $\dfrac{\pi}{2}$. The range is $(-\infty, \infty)$.

d. The function is decreasing on each interval where it is defined.

69. a–b.

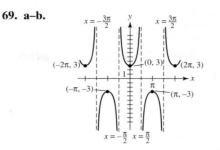

c. The domain includes all real numbers except odd multiples of $\dfrac{\pi}{2}$. The range is $(-\infty, -3] \cup [3, \infty)$.

d. In the window shown, the function is increasing on $\left(-2\pi, -\dfrac{3\pi}{2}\right), \left(-\dfrac{3\pi}{2}, -\pi\right), \left(0, \dfrac{\pi}{2}\right),$ and $\left(\dfrac{\pi}{2}, \pi\right)$ and de-

creasing on $\left(-\pi, -\dfrac{\pi}{2}\right), \left(-\dfrac{\pi}{2}, 0\right), \left(\pi, \dfrac{3\pi}{2}\right),$ and $\left(\dfrac{3\pi}{2}, 2\pi\right)$.

71. a–b.

c. The domain includes all real numbers except $\dfrac{\pi}{4} + k\pi$ for integer values of k. The range is $(-\infty, -2] \cup [2, \infty)$.

d. In the window shown, the function is increasing on $\left(\dfrac{-7\pi}{4}, \dfrac{-5\pi}{4}\right), \left(\dfrac{-\pi}{4}, \dfrac{\pi}{4}\right), \left(\dfrac{\pi}{4}, \dfrac{3\pi}{4}\right),$ and $\left(\dfrac{7\pi}{4}, \dfrac{9\pi}{4}\right)$ and

decreasing on $\left(\dfrac{-5\pi}{4}, \dfrac{-3\pi}{4}\right), \left(\dfrac{-3\pi}{4}, \dfrac{-\pi}{4}\right),$

$\left(\dfrac{3\pi}{4}, \dfrac{5\pi}{4}\right),$ and $\left(\dfrac{5\pi}{4}, \dfrac{7\pi}{4}\right)$.

73. a–b.

c. The domain includes all real numbers except integer multiples of $\frac{\pi}{2}$. The range is $(-\infty, \infty)$.

d. The function is decreasing on each interval where it is defined.

75.

$y = \cot x$

$y = \cot(-x)$

$y = \cot\left(\frac{\pi}{2} - x\right)$

$$\cot\left(\frac{\pi}{2} - x\right) = \tan x$$

SECTION 8-4

1. a. The triangle in Figure 8-36 is similar to triangle OCA in Figure 8-35. Therefore, the ratio $\frac{OA}{OC}$ in Figure 8-35 is equal to the ratio $\frac{PR}{PQ}$ in Figure 8-36, or 19.11.

b. 19.11

c. the secant of θ

3. 19.11; yes

5. $a \approx 26.6, b \approx 43.1, A = 31.7°$

7. $c = 40, A \approx 36.87°, B \approx 53.13°$

9.

$\sin \theta = \dfrac{\sqrt{2}}{2}, \cos \theta = \dfrac{\sqrt{2}}{2}, \tan \theta = 1, \cot \theta = 1,$

$\sec \theta = \sqrt{2}, \csc \theta = \sqrt{2}$

11.

$\sin \theta = \dfrac{5}{13}, \cos \theta = \dfrac{12}{13}, \tan \theta = \dfrac{5}{12}, \cot \theta = \dfrac{12}{5},$

$\sec \theta = \dfrac{13}{12}, \csc \theta = \dfrac{13}{5}$

13. $a \approx 120$
$c \approx 570$
$B = 78°$

15. $b \approx 14.6$
$A \approx 30.5°$
$B \approx 59.5°$

17. $a \approx 1.5950$
$c \approx 1.8826$
$A \approx 1.01080 \approx 57.914°$

19.

$\sin \theta = \dfrac{4}{5}, \cos \theta = \dfrac{3}{5}, \tan \theta = \dfrac{4}{3}, \cot \theta = \dfrac{3}{4}, \sec \theta = \dfrac{5}{3},$

$\csc \theta = \dfrac{5}{4}$

21.

$\sin \theta = \dfrac{\sqrt{5}}{3}, \cos \theta = \dfrac{2}{3}, \tan \theta = \dfrac{\sqrt{5}}{2}, \cot \theta = \dfrac{2\sqrt{5}}{5},$

$\sec \theta = \dfrac{3}{2}, \csc \theta = \dfrac{3\sqrt{5}}{5}$

23.

$\sin \theta = \dfrac{7}{25}, \cos \theta = \dfrac{24}{25}, \tan \theta = \dfrac{7}{24}, \cot \theta = \dfrac{24}{7},$

$\sec \theta = \dfrac{25}{24}, \csc \theta = \dfrac{25}{7}$

25. $(\tan \theta)(\csc \theta) = \sec \theta$

$\left(\dfrac{\sin \theta}{\cos \theta}\right)\left(\dfrac{1}{\sin \theta}\right) = \left(\dfrac{1}{\cos \theta}\right)$

$\dfrac{1}{\cos \theta} = \dfrac{1}{\cos \theta}$

27. $\dfrac{\sin \theta}{\tan \theta} = \dfrac{1}{\sec \theta}$

$\dfrac{\sin \theta}{\dfrac{\sin \theta}{\cos \theta}} = \dfrac{1}{\dfrac{1}{\cos \theta}}$

$$(\sin \theta)\left(\frac{\cos \theta}{\sin \theta}\right) = \cos \theta$$

$$\cos \theta = \cos \theta$$

29.

If P is the x-intercept and Q is any point on the line above the x-axis, then $\tan \theta$ is the ratio $\dfrac{\Delta y}{\Delta x}$ of the changes in y and x between P and Q. This ratio is also the slope m.

31.

a. In the triangle shown at the left, $b = 6$ feet and a must be at least $27 - 5 = 22$ feet. If $a = 22$ feet, then $c = \sqrt{6^2 + 22^2} \approx 22.80$ feet. Since c must be an integer, c must be at least 23 feet. If $c = 23$ feet, then $\cos A = \dfrac{6}{23}$, so $A = \cos^{-1} \dfrac{6}{23} \approx 74.88°$. (You may verify that if $c = 24$ feet, then $A > 75°$, which makes Dave too nervous to paint.)

b. about 9 feet off the ground

c. Yes.

33. a. $OM \approx 216.21r$

b. $OS \approx 86,629r, ST \approx 400.67r$

35. First, $\angle AEB = \dfrac{1}{2} \angle AEC = 33.346°$. Therefore,

$\dfrac{\overline{AB}}{\overline{AE}} = \sin 33.346°$, so $\overline{AB} = \overline{AE} \sin 33.346° =$

$6378 \ \sin 33.346° \approx 3506$ km. Similarly, $\overline{BC} = 3506$ km.

Next, $\dfrac{\overline{EB}}{\overline{AE}} = \cos 33.346°$, so $\overline{EB} = \overline{AE} \cos 33.346° =$

$6378 \cos 33.346° \approx 5328$ km. Therefore, $\overline{BO} =$
$6378 - 5328 = 1050$ km.

37. $r \approx 1748$

The radius of the moon is $r \approx 1748$ km, and its distance from the earth is $216.21r \approx 377,900$ km. The radius of the sun is $400.67r \approx 700,400$ km, and its distance from the earth is $86,629r \approx 151,400,000$ km.

SECTION 8-5

1. domain $= [-1, 1]$, range $= \left[-\dfrac{\pi}{2}, \dfrac{\pi}{2}\right]$

3. a. $\dfrac{\pi}{2}$ **b.** $\dfrac{\pi}{6}$

   c. $\dfrac{\pi}{3}$ **d.** $\dfrac{\pi}{4}$

5. a. You will get an error message. There is no value of x for which $\sin x = 2$, so $\sin^{-1} 2$ is undefined.

   b. The number π is not in $\left[-\dfrac{\pi}{2}, \dfrac{\pi}{2}\right]$, the range of the inverse sine function, so $\sin^{-1}(\sin \pi)$ cannot be equal to π. In fact, $\sin^{-1}(\sin \pi) = \sin^{-1} 0 = 0$.

7. a. $x = \begin{cases} \dfrac{\pi}{6} + 2k\pi \\[2ex] \dfrac{5\pi}{6} + 2k\pi \end{cases}$

   b. $x = \begin{cases} -\dfrac{\pi}{4} + 2k\pi \\[2ex] \dfrac{5\pi}{4} + 2k\pi \end{cases}$

   c. $x = \dfrac{3\pi}{2} + 2k\pi$

   d. $x = \begin{cases} 2k\pi \\ \pi + 2k\pi \end{cases}$

   e. $x = \begin{cases} \dfrac{\pi}{3} + 2k\pi \\[2ex] \dfrac{2\pi}{3} + 2k\pi \end{cases}$

   f. no solutions

9. domain $= [-1, 1]$, range $= [0, \pi]$

11. a. 0 **b.** $\dfrac{\pi}{3}$

   c. $\dfrac{\pi}{6}$ **d.** $\dfrac{\pi}{4}$

13. a. $x = \begin{cases} \dfrac{\pi}{3} + 2k\pi \\[2ex] \dfrac{5\pi}{3} + 2k\pi \end{cases}$

   b. $x = \begin{cases} \dfrac{3\pi}{4} + 2k\pi \\[2ex] \dfrac{5\pi}{4} + 2k\pi \end{cases}$

   c. $x = \pi + 2k\pi$

   d. $x = \begin{cases} \dfrac{\pi}{2} + 2k\pi \\[2ex] \dfrac{3\pi}{2} + 2k\pi \end{cases}$

e. $x = \begin{cases} \dfrac{\pi}{6} + 2k\pi \\[2mm] \dfrac{11\pi}{6} + 2k\pi \end{cases}$

f. no solutions

15. $2.858 + 2k\pi$ seconds

17. domain $= (-\infty, \infty)$, range $= \left(-\dfrac{\pi}{2}, \dfrac{\pi}{2}\right)$

19. a. $\dfrac{\pi}{4}$ **b.** 0 **c.** $-\dfrac{\pi}{3}$ **d.** $\dfrac{\pi}{6}$

21. a. $1.11 + k\pi$ **b.** $0.10 + k\pi$
    c. $-0.46 + k\pi$ **d.** $-1.37 + k\pi$

23. domain $= (-\infty, -1] \cup [1, \infty)$, range $=$

$\left[0, \dfrac{\pi}{2}\right) \cup \left(\dfrac{\pi}{2}, \pi\right]$, horizontal asymptote at $y = \dfrac{\pi}{2}$

25. a. $\dfrac{\pi}{3}$ **b.** $-\dfrac{\pi}{4}$ **c.** $\dfrac{\pi}{4}$ **d.** $\dfrac{5\pi}{6}$

27. $\sqrt{1 - x^2}$ **29.** $\sqrt{1 + x^2}$

31. a. 0 **b.** $\dfrac{\pi}{2}$ **c.** $-\dfrac{\pi}{4}$
    d. $\dfrac{\pi}{2}$ **e.** $\dfrac{3\pi}{4}$ **f.** $\dfrac{\pi}{6}$

33. a. $x = \begin{cases} -\dfrac{\pi}{3} + 2k\pi \\[2mm] \dfrac{4\pi}{3} + 2k\pi \end{cases}$ **b.** $x = \begin{cases} \dfrac{\pi}{4} + 2k\pi \\[2mm] \dfrac{7\pi}{4} + 2k\pi \end{cases}$

c. $x = \dfrac{\pi}{3} + k\pi$

d. $x = \begin{cases} -0.38 + 2k\pi \\[2mm] 3.53 + 2k\pi \end{cases}$ **e.** $x = \begin{cases} 1.15 + 2k\pi \\[2mm] 5.13 + 2k\pi \end{cases}$

f. $x = 1.15 + k\pi$

35. a. 1 **b.** $\dfrac{\pi}{4}$ **c.** $\dfrac{2\pi}{3}$
    d. 2 **e.** $\sqrt{2}$ **f.** $-\dfrac{\pi}{4}$

37. a.

b. domain $= \left[-\dfrac{1}{2}, \dfrac{1}{2}\right]$, range $= \left[-\dfrac{\pi}{2}, \dfrac{\pi}{2}\right]$

c. None

39. a.

b. domain $= [-1, 1]$, range $= [-\pi, 0]$
c. none

41. a.

b. domain $= (-\infty, -1] \cup [1, \infty)$,

   range $= \left[-\dfrac{\pi}{2}, 0\right) \cup \left(0, \dfrac{\pi}{2}\right]$

c. $y = 0$

43. $\sqrt{x^2 - 1}$ **45.** $\dfrac{x}{\sqrt{1 - x^2}}$

47. Denote the x-coordinates of the four points on the graph in Figure 8-59 (page 451) by x_1, x_2, x_3, and x_4. Then x_1 can be expressed in terms of each of the others as follows. First, x_1 and x_2 have a sum of 2π, so $x_1 = 2\pi - x_2$. Also, x_1 and x_3 are one cycle apart on the graph, so $x_1 = x_3 - 2\pi$. Finally, x_2 and x_4 are one cycle apart on the graph, so $x_2 = x_4 - 2\pi$. Therefore, $x_1 = 2\pi - x_2 = 2\pi - (x_4 - 2\pi) = 4\pi - x_4$.

The graph indicates that $\cos x_1, \cos x_2, \cos x_3$, and $\cos x_4$ all have the same value. Thus

$$\cos^{-1}(\cos x_2) = \cos^{-1}(\cos x_1) = x_1 = 2\pi - x_2$$
$$\cos^{-1}(\cos x_3) = \cos^{-1}(\cos x_1) = x_1 = x_3 - 2\pi$$
$$\cos^{-1}(\cos x_4) = \cos^{-1}(\cos x_1) = x_1 = 4\pi - x_4$$

49. a. Both have inverses.   **b.** Both have inverses.
    c. $y = \csc x$ has an inverse. **d.** $y = \csc x$ has an inverse.

51. a. If $x \leqslant 0$ and $y = \tan^{-1} x$, then $y \leqslant 0$ also. Thus $\tan|y| = \tan(-y) = -\tan y = -x = |x|$.

b. In Figure 8-58 from Example 1, the angle is $|y|$ and the side opposite y is $|x|$. Thus $\sin|y| = \dfrac{|x|}{\sqrt{x^2 + 1}}$.

c. Since $|x| = -x$ and $|y| = -y$, $\sin y = -\sin|y| = -\dfrac{|x|}{\sqrt{x^2 + 1}} = \dfrac{x}{\sqrt{x^2 + 1}}$.

SECTION 9-1

1. a. The ball travels at 110 feet per second for t seconds, so it travels $110t$ feet.

b. $2.8 - t$

c. If Jeff is to throw Derek out, the ball must travel from Jeff's position at S to first base at F in less than $2.8 - t$ seconds at 110 feet per second. Therefore, the distance $\overline{FS}$ must be less than $110(2.8 - t)$ feet.

3. By applying the Pythagorean theorem to the larger right triangle, we obtain $c^2 = h^2 + (k + b)^2$. Then by applying the Pythagorean theorem to the smaller right triangle, $h^2 = a^2 - k^2$. Substituting $a^2 - k^2$ for h^2 in the first equation yields $c^2 = a^2 - k^2 + (k + b)^2$. Since $(k + b)^2 = k^2 + 2bk + b^2$, the equation can be written $c^2 = a^2 - k^2 + (k^2 + 2bk + b^2)$, or $c^2 = a^2 + b^2 + 2bk$. Multiplying and dividing the last term by a, we obtain $c^2 = a^2 + b^2 + 2ab\left(\dfrac{k}{a}\right)$. In the smaller right triangle, $\dfrac{k}{a} = \cos(\pi - C) = -\cos C$, so the equation can be rewritten as $c^2 = a^2 + b^2 - 2ab\cos C$.

5. $b \approx 5.3$ m, $A \approx 93°$, $C \approx 40°$

7. $A \approx 57.4°$, $B \approx 40.3°$, $C \approx 82.3°$

9. 1.50 seconds, 165 feet, 143 feet

11. In the larger right triangle, $\sin B = \dfrac{h}{a}$, so $h = a \sin B$.

In the smaller right triangle, $\sin(\pi - A) = \dfrac{h}{b}$, so $h = b \sin(\pi - A)$. Since $\sin(\pi - A) = \sin A$, we have $h = b \sin A$. Now $a \sin B = b \sin A$, since both expressions are equal to h. Dividing both sides by $(\sin A)(\sin B)$ yields $\dfrac{a}{\sin A} = \dfrac{b}{\sin B}$.

13. $C = 97°$, $a \approx 38$ m, $c \approx 39$ m

15. $C = 40.9°$, $b \approx 19.4$ mi, $c \approx 17.3$ mi

17. $A = 90°$, $C = 60°$, $c \approx 8.66$

19. $A = 75.7°$, $B = 56.3°$, $b \approx 63.7$ ft
$A = 104.3°$, $B = 27.7°$, $b \approx 35.6$ ft

21. No triangle is determined.

23. In the circle, $\angle QOE$ is a central angle subtending an arc of 118°, so its measure must be 118°. Since OQ and OE are each radii, the triangle QOE is isosceles, so $\angle OQE = \angle OEQ$, and the sum of their measures is $180° - 118° = 62°$. Therefore, the measure of each is 31°.

25. a. The line QH is the horizon at Q, so it is horizontal from the perspective of a person standing at Q. The line OQ is a radius of the earth, so it is vertical from the perspective of a person standing at Q. Therefore, the lines are perpendicular, and $\angle OQH = 90°$.

b. 69°　　　　**c.** 64°　　　　**d.** 13,900 km

27. $A \approx 56°$, $B \approx 36°$, $C \approx 88°$

29. $A = 70.7°$, $a \approx 116$ m, $c \approx 120$ m

31. $c \approx 12.99$ ft, $A \approx 27.72°$, $B \approx 43.28°$

33. No triangle is determined.

35. $A = 31°$, $C = 125°$, $c \approx 64$ m
$A = 149°$, $C = 7°$, $c \approx 9.1$ m

37. $A \approx 44°$, $C \approx 12°$, $c \approx 12$ m

39. $A \approx 28.4°$, $B \approx 39.1°$, $C \approx 112.5°$

41. $b \approx 8.713$ ft, $A \approx 15.87°$, $C \approx 156.87°$
$b \approx 8.713$ ft, $A \approx 164.13°$, $C \approx 8.61°$

43. $B = 63.1°$, $b \approx 6.97$ in, $c \approx 7.66$ in

45. $A \approx 52.65°$, $B \approx 7.35°$, $b \approx 9.969$ km

47. a. 1.74, 0.175, 0.0175

b. If C is near 0°, then c is near 0.

c. As C decreases toward 0°, sides a and b are brought closer together, so side c becomes very small.

SECTION 9-2

1. In Figure 9-10, $\sin(-x) = -b = -\sin x$
$\cos(-x) = a = \cos x$
$\tan(-x) = \dfrac{-b}{a} = -\dfrac{b}{a} = -\tan x$

3. In Figure 9-11, $\sin\left(\dfrac{\pi}{2} - x\right) = a = \cos x$
$\cos\left(\dfrac{\pi}{2} - x\right) = b = \sin x$
$\tan\left(\dfrac{\pi}{2} - x\right) = \dfrac{a}{b} = \cot x$
$\cot\left(\dfrac{\pi}{2} - x\right) = \dfrac{b}{a} = \tan x$
$\sec\left(\dfrac{\pi}{2} - x\right) = \dfrac{1}{b} = \csc x$
$\csc\left(\dfrac{\pi}{2} - x\right) = \dfrac{1}{a} = \sec x$

5. $\sin x = \dfrac{2\sqrt{5}}{5}$
$\cos x = -\dfrac{\sqrt{5}}{5}$
$\tan x = -2$
$\cot x = -\dfrac{1}{2}$
$\sec x = -\sqrt{5}$
$\csc x = \dfrac{\sqrt{5}}{2}$
$\sin(-x) = -\dfrac{2\sqrt{5}}{5}$　　$\sin\left(\dfrac{\pi}{2} - x\right) = -\dfrac{\sqrt{5}}{5}$
$\cos(-x) = -\dfrac{\sqrt{5}}{5}$　　$\cos\left(\dfrac{\pi}{2} - x\right) = \dfrac{2\sqrt{5}}{5}$

$\tan(-x) = 2 \qquad \tan\left(\dfrac{\pi}{2} - x\right) = -\dfrac{1}{2}$

$\cot(-x) = \dfrac{1}{2} \qquad \cot\left(\dfrac{\pi}{2} - x\right) = -2$

$\sec(-x) = -\sqrt{5} \qquad \sec\left(\dfrac{\pi}{2} - x\right) = \dfrac{\sqrt{5}}{2}$

$\csc(-x) = -\dfrac{\sqrt{5}}{2} \qquad \csc\left(\dfrac{\pi}{2} - x\right) = -\sqrt{5}$

7. $\sin x = -\dfrac{1}{5}$

$\cos x = -\dfrac{2\sqrt{6}}{5}$

$\tan x = \dfrac{\sqrt{6}}{12}$

$\cot x = 2\sqrt{6}$

$\sec x = -\dfrac{5\sqrt{6}}{12}$

$\csc x = -5$

$\sin(-x) = \dfrac{1}{5} \qquad \sin\left(\dfrac{\pi}{2} - x\right) = -\dfrac{2\sqrt{6}}{5}$

$\cos(-x) = -\dfrac{2\sqrt{6}}{5} \qquad \cos\left(\dfrac{\pi}{2} - x\right) = -\dfrac{1}{5}$

$\tan(-x) = -\dfrac{\sqrt{6}}{12} \qquad \tan\left(\dfrac{\pi}{2} - x\right) = 2\sqrt{6}$

$\cot(-x) = -2\sqrt{6} \qquad \cot\left(\dfrac{\pi}{2} - x\right) = \dfrac{\sqrt{6}}{12}$

$\sec(-x) = -\dfrac{5\sqrt{6}}{12} \qquad \sec\left(\dfrac{\pi}{2} - x\right) = -5$

$\csc(-x) = 5 \qquad \csc\left(\dfrac{\pi}{2} - x\right) = -\dfrac{5\sqrt{6}}{12}$

9. $\sin\dfrac{\pi}{8}$ **11.** $\cos\dfrac{5\pi}{7}$

13. $\tan 0.65\pi, \tan 1.65\pi$

15. From Exercise 14, $\cos(x - y) = \cos[x + (-y)] = \cos x \cos(-y) - \sin x \sin(-y)$. Even–odd identities can be used to rewrite $\cos(-y)$ as $\cos y$ and $\sin(-y)$ as $-\sin y$. Therefore, $\cos(x - y) = \cos x \cos y + \sin x \sin y$.

17. From Exercise 16, $\sin(x - y) = \sin[x + (-y)] = \sin x \cos(-y) + \cos x \sin(-y)$. Even–odd identities can be used to rewrite $\cos(-y)$ as $\cos y$ and $\sin(-y)$ as $-\sin y$. Therefore, $\sin(x - y) = \sin x \cos y - \cos x \sin y$.

19. a. $\sin\left(\dfrac{\pi}{2} + \dfrac{\pi}{3}\right) = \dfrac{1}{2}$; $\sin\dfrac{\pi}{2} + \sin\dfrac{\pi}{3} = 1 + \dfrac{\sqrt{3}}{2}$

b. $\cos\left(\dfrac{\pi}{2} - \dfrac{\pi}{3}\right) = \dfrac{\sqrt{3}}{2}$; $\cos\dfrac{\pi}{2} - \cos\dfrac{\pi}{3} = -\dfrac{1}{2}$

21. $\dfrac{\sqrt{6} - \sqrt{2}}{4}$ **23.** $\dfrac{\sqrt{6} - \sqrt{2}}{4}$

25. $\sin(x + y) = -0.6$
$\cos(x + y) = 0.8$
$\sin(x - y) = -0.936$
$\cos(x - y) = 0.352$

27. $\sin(x + y) = \dfrac{\sqrt{6} - \sqrt{2}}{4}$

$\cos(x + y) = \dfrac{\sqrt{6} + \sqrt{2}}{4}$

$\sin(x - y) = \dfrac{-\sqrt{6} - \sqrt{2}}{4}$

$\cos(x - y) = \dfrac{-\sqrt{6} + \sqrt{2}}{4}$

29. a. $\sin(x + x) = \sin x \cos x + \cos x \sin x = 2\sin x \cos x$
b. $\cos(x + x) = \cos x \cos x - \sin x \sin x = \cos^2 x - \sin^2 x$

31. a. $\sin 2\left(\dfrac{\pi}{2}\right) = 0$; $2\sin\dfrac{\pi}{2}\cos\dfrac{\pi}{2} = 0$;

$\cos 2\left(\dfrac{\pi}{2}\right) = -1$; $\cos^2\dfrac{\pi}{2} - \sin^2\dfrac{\pi}{2} = -1$;

$2\cos^2\dfrac{\pi}{2} - 1 = -1$; $1 - 2\sin^2\dfrac{\pi}{2} = -1$

b. $\sin 2\left(\dfrac{\pi}{3}\right) = \dfrac{\sqrt{3}}{2}$; $2\sin\dfrac{\pi}{3}\cos\dfrac{\pi}{3} = \dfrac{\sqrt{3}}{2}$;

$\cos 2\left(\dfrac{\pi}{3}\right) = -\dfrac{1}{2}$; $\cos^2\left(\dfrac{\pi}{3}\right) - \sin^2\left(\dfrac{\pi}{3}\right) = -\dfrac{1}{2}$;

$2\cos^2\left(\dfrac{\pi}{3}\right) - 1 = -\dfrac{1}{2}$; $1 - 2\sin^2\left(\dfrac{\pi}{3}\right) = -\dfrac{1}{2}$

33. $-\dfrac{4}{5}, \dfrac{3}{5}$

35. $0, -1$

37. a. $\sin^2\dfrac{x}{2} = \dfrac{1}{2}\left[1 - \cos 2\left(\dfrac{x}{2}\right)\right]$

$\sin^2\dfrac{x}{2} = \dfrac{1 - \cos x}{2}$

$\sin\dfrac{x}{2} = \pm\sqrt{\dfrac{1 - \cos x}{2}}$

b. $\cos^2\dfrac{x}{2} = \dfrac{1}{2}\left[1 + \cos 2\left(\dfrac{x}{2}\right)\right]$

$\cos^2\dfrac{x}{2} = \dfrac{1 + \cos x}{2}$

$\cos\dfrac{x}{2} = \pm\sqrt{\dfrac{1 + \cos x}{2}}$

39. $\sqrt{0.7}, \sqrt{0.3}$

41. $\sqrt{\dfrac{1}{3}}, -\sqrt{\dfrac{2}{3}}$

43. $\dfrac{\sqrt{2 - \sqrt{2}}}{2}, \dfrac{\sqrt{2 + \sqrt{2}}}{2}$

45. $\dfrac{\sqrt{2 + \sqrt{3}}}{2}, -\dfrac{\sqrt{2 - \sqrt{3}}}{2}$

47. $\frac{1}{2}[\cos(x - y) + \cos(x + y)]$

$= \frac{1}{2}[(\cos x \cos y + \sin x \sin y) +$

$(\cos x \cos y - \sin x \sin y)]$

$= \frac{1}{2}(2 \cos x \cos y)$

$= \cos x \cos y$

49. $(\cos \pi)\left(\cos \frac{3\pi}{4}\right) = \frac{\sqrt{2}}{2}$; $\frac{1}{2}\left(\cos \frac{\pi}{4} + \cos \frac{7\pi}{4}\right) = \frac{\sqrt{2}}{2}$

51. $\left(\sin \frac{\pi}{3}\right)\left(\cos \frac{\pi}{2}\right) = 0$; $\frac{1}{2}\left[\sin\left(-\frac{\pi}{6}\right) + \sin \frac{5\pi}{6}\right] = 0$

53. $\frac{1}{2}(\cos x - \cos 5x)$ **55.** $\frac{1}{2}[\sin 9x - \sin 5x]$

57. $\sin^2 0 + \tan^2 0 = 0$; $\sec^2 0 - \cos^2 0 = 0$

$\sin^2 \frac{\pi}{3} + \tan^2 \frac{\pi}{3} = \frac{15}{4}$; $\sec^2 \frac{\pi}{3} - \cos^2 \frac{\pi}{3} = \frac{15}{4}$

$\sin^2 \frac{3\pi}{4} + \tan^2 \frac{3\pi}{4} = \frac{3}{2}$; $\sec^2 \frac{3\pi}{4} - \cos^2 \frac{3\pi}{4} = \frac{3}{2}$

Based on these results, the equation might be an identity.

59. $\cos^4 0 - \sin^4 0 = 1$; $\cos 2(0) = 1$

$\cos^4 \frac{\pi}{3} - \sin^4 \frac{\pi}{3} = -\frac{1}{2}$; $\cos 2\left(\frac{\pi}{3}\right) = -\frac{1}{2}$

$\cos^4 \frac{3\pi}{4} - \sin^4 \frac{3\pi}{4} = 0$; $\cos 2\left(\frac{3\pi}{4}\right) = 0$

Based on these results, the equation might be an identity.

61.

For each value of x, the values of y on the two graphs represent $\csc x - \cot x$ and $\frac{\tan x}{\sec x + 1}$, respectively. The graphs appear to coincide, suggesting that

$\csc x - \cot x = \frac{\tan x}{\sec x + 1}$ for all values of x.

63.

The equation might be an identity.

65.

The equation might be an identity.

67. $\sin^2 x + \tan^2 x = \sec^2 x - \cos^2 x$

$\sin^2 x + \cos^2 x = \sec^2 x - \tan^2 x$

$1 = 1$

Since the last equation is an identity, so is the first one.

69. $\cos^4 x - \sin^4 x = \cos 2x$

$(\cos^2 x + \sin^2 x)(\cos^2 x - \sin^2 x) = \cos 2x$

$(1)(\cos^2 x - \sin^2 x) = \cos 2x$

$\cos 2x = \cos 2x$

Since the last equation is an identity, so is the first one.

71. Since the equation is false unless $\cos x = 0$, choose any value of x with $\cos x \neq 0$. For example, when $x = 0$, the left side of the equation is $\tan 0 \sin 0 - \sec 0 = -1$, and the right side is $\cos 0 = 1$. Since $-1 \neq 1$, the equation is false when $x = 0$ and is therefore conditional.

73. **a.** $-\frac{12}{13}$ **b.** $\frac{5}{13}$ **c.** $\frac{12}{5}$ **d.** $\frac{5}{13}$

      e. $\frac{12}{5}$ **f.** $-\frac{119}{169}$ **g.** $\frac{169}{120}$ **h.** $\frac{2\sqrt{13}}{13}$

75. **a.** $\frac{3}{5}$ **b.** $-\frac{4}{5}$ **c.** $\frac{3}{4}$ **d.** $-\frac{4}{5}$

      e. $\frac{3}{4}$ **f.** $\frac{7}{25}$ **g.** $\frac{25}{24}$ **h.** $\frac{3\sqrt{10}}{10}$

77. **a.** $\frac{2}{5}$ **b.** $\frac{\sqrt{21}}{5}$ **c.** $-\frac{2\sqrt{21}}{21}$ **d.** $\frac{\sqrt{21}}{5}$

      e. $-\frac{2\sqrt{21}}{21}$ **f.** $\frac{17}{25}$ **g.** $-\frac{25\sqrt{21}}{84}$

      h. $\sqrt{\frac{5 - \sqrt{21}}{10}}$

79. $\frac{\sin^2 t}{1 - \cos t} = 1 - \cos t$

The equation is false if $t = \pi$: $\frac{\sin^2 \pi}{1 - \cos \pi} = 1 - \cos \pi$

$\frac{(0)^2}{1 - (-1)} = 1 - (-1)$

$0 = 2$

81. $\cot A - \tan B = \frac{\cos(A + B)}{\sin A \cos B}$

$\frac{\cos A}{\sin A} - \frac{\sin B}{\cos B} = \frac{\cos(A + B)}{\sin A \cos B}$

$\cos A \cos B - \sin A \sin B = \cos(A + B)$

$\cos(A + B) = \cos(A + B)$

The equation is an identity.

83. $\sin\left(x + \dfrac{\pi}{3}\right) + \sin\left(x - \dfrac{\pi}{3}\right) = \sin x$

$\left(\sin x \cos \dfrac{\pi}{3} + \cos x \sin \dfrac{\pi}{3}\right) +$

$\left(\sin x \cos \dfrac{\pi}{3} - \cos x \sin \dfrac{\pi}{3}\right) = \sin x$

$2 \sin x \cos \dfrac{\pi}{3} = \sin x$

$2 \sin x \left(\dfrac{1}{2}\right) = \sin x$

$\sin x = \sin x$

The equation is an identity.

85. $\cos 2t + \cos 4t = 2 \cos t \cos 3t$

$\cos 2t + \cos 4t = 2\left\{\dfrac{1}{2}[\cos (t + 3t) + \cos (t - 3t)]\right\}$

$\cos 2t + \cos 4t = \cos 4t + \cos (-2t)$

$\cos 4t + \cos 2t = \cos 4t + \cos 2t$

The equation is an identity.

87. $\sin x \sin 2x = 2 \cos x - 2 \cos^3 x$

$\sin x (2 \sin x \cos x) = 2 \cos x (1 - \cos^2 x)$

$2 \sin^2 x \cos x = 2 \cos x \sin^2 x$

$2 \sin^2 x \cos x = 2 \sin^2 x \cos x$

The equation is an identity.

89. One formula is $\cos 3x = \cos^3 x - 3 \sin^2 x \cos x$.

91. $\sin (2\pi - x) = -\sin x$

$\cos (2\pi - x) = \cos x$

$\tan (2\pi - x) = -\tan x$

$\cot (2\pi - x) = -\cot x$

$\sec (2\pi - x) = \sec x$

$\csc (2\pi - x) = -\csc x$

93. a. $\dfrac{\sin (x + y)}{\cos (x + y)} = \dfrac{\tan x + \tan y}{1 - \tan x \tan y}$

b. $\dfrac{\sin x \cos y + \cos x \sin y}{\cos x \cos y - \sin x \sin y} = \dfrac{\tan x + \tan y}{1 - \tan x \tan y}$

c. $\dfrac{\sin x \cos y + \cos x \sin y}{\cos x \cos y - \sin x \sin y} = \dfrac{\dfrac{\sin x}{\cos x} + \dfrac{\sin y}{\cos y}}{1 - \left(\dfrac{\sin x}{\cos x}\right)\left(\dfrac{\sin y}{\cos y}\right)}$

$\dfrac{\sin x \cos y + \cos x \sin y}{\cos x \cos y - \sin x \sin y} = \dfrac{\dfrac{\sin x}{\cos x} + \dfrac{\sin y}{\cos y}}{1 - \left(\dfrac{\sin x}{\cos x}\right)\left(\dfrac{\sin y}{\cos y}\right)}$

$\cdot \dfrac{\cos x \cos y}{\cos x \cos y}$

$\dfrac{\sin x \cos y + \cos x \sin y}{\cos x \cos y - \sin x \sin y} = \dfrac{\sin x \cos y + \cos x \sin y}{\cos x \cos y - \sin x \sin y}$

95. $\tan (x + x) = \dfrac{\tan x + \tan x}{1 - \tan x \tan x} = \dfrac{2 \tan x}{1 - \tan^2 x}$

SECTION 9-3

1. a. scalar   **b.** vector   **c.** vector

     d. scalar   **e.** scalar   **f.** vector

3.

5. a.   **b.**

c.

d.

7. a.

b. $\|\mathbf{Proj_U V}\| \approx 215.3,\ \ \|\mathbf{V} - \mathbf{Proj_U V}\| \approx 186.7$

9. a. $6\mathbf{i} + 4\mathbf{j}$ **b.**

11. a. $-3\mathbf{i} + 3\mathbf{j}$ **b.**

13.

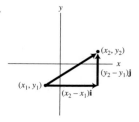

15. $-\mathbf{U} = -2\mathbf{i} - 4\mathbf{j}$

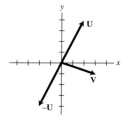

17. $\mathbf{U} + 2\mathbf{V} = 8\mathbf{i} + 2\mathbf{j}$

19. If the vector $\mathbf{V}$ is placed in standard position, its terminal point is on a circle of radius $\|\mathbf{V}\|$ centered at $(0, 0)$. The coordinates of the terminal point are therefore $(\|\mathbf{V}\| \cos \theta, \|\mathbf{V}\| \sin \theta)$, so $\mathbf{V} = (\|\mathbf{V}\| \cos \theta)\mathbf{i} + (\|\mathbf{V}\| \sin \theta)\mathbf{j}$.

21. $940.6\mathbf{i}, 342.4\mathbf{j}$ **23.** $0.19\mathbf{i}, -0.98\mathbf{j}$

25. $\|\mathbf{V}\| = 13, \theta = 202.62°$ **27.** $\|\mathbf{V}\| = \sqrt{50}, \theta = 98.13°$

29. The angle between $\mathbf{P}$ and $\mathbf{B}$ is $90° - \gamma$. Since $\mathbf{M}$ is the component of $\mathbf{P}$ in the direction of $\mathbf{B}$, $\mathbf{M} = \mathbf{Proj}_{\mathbf{B}}\mathbf{P}$, so $\|\mathbf{M}\| = \|\mathbf{P}\| \cos (90° - \gamma)$. This is equal to $\|\mathbf{P}\| \sin \gamma$ because $\cos (90° - \gamma) = \sin \gamma$. From Exercise 28, $\|\mathbf{P}\| = \|\mathbf{W}\| \sin \alpha$, so $\|\mathbf{M}\| = \|\mathbf{W}\| \sin \alpha \sin \gamma$.

31. a. The boat moves in the direction of $\mathbf{M}$. The westward component of the motion is $\mathbf{Proj}_{\mathbf{W}}\mathbf{M}$. Since the angle between $\mathbf{W}$ and $\mathbf{M}$ is β, the magnitude of $\mathbf{Proj}_{\mathbf{W}}\mathbf{M}$ is $\|\mathbf{M}\| |\cos \beta|$. From Exercise 29, $\|\mathbf{M}\| = \|\mathbf{W}\| \sin \alpha \sin \gamma$, and $\gamma = \alpha - \beta$. Therefore, $\|\mathbf{Proj}_{\mathbf{W}}\mathbf{M}\| = \|\mathbf{W}\| \sin \alpha \sin (\alpha - \beta) |\cos \beta|$.
 b. If $\beta < 90°$, $\mathbf{Proj}_{\mathbf{W}}\mathbf{M}$ represents the eastward component of the boat's motion. Its magnitude is still $\|\mathbf{W}\| \sin \alpha \sin (\alpha - \beta) |\cos \beta|$, as in part (a). Since $\beta < 90°$, $\cos \beta > 0$, so $\|\mathbf{Proj}_{\mathbf{W}}\mathbf{M}\| = \|\mathbf{W}\| \sin \alpha \sin (\alpha - \beta) \cos \beta$.

33. 0

35. a.

b.

c.

d.

37. a.

 b. $\|\mathbf{Proj}_{\mathbf{U}}\mathbf{V}\| = 0$, $\|\mathbf{V} - \mathbf{Proj}_{\mathbf{U}}\mathbf{V}\| = 25$

39. $-4\mathbf{j}$

41. $-3\mathbf{i} - 4.5\mathbf{j}$

43. $-2.53\mathbf{i}, 1.71\mathbf{j}$ **45.** $209\mathbf{i}, -232\mathbf{j}$

47. $\|\mathbf{V}\| \approx 24.04, \theta = 225°$ **49.** $\|\mathbf{V}\| \approx 5.80, \theta = 331.58°$

51. a. $5.74°$

   b. 597.0 mph

   c. 0.97 hours (58.3 minutes)

53. a. Depending on the directions of $\mathbf{U}$ and $\mathbf{V}$, $|\alpha - \beta| = \pm (\alpha - \beta)$. Since $\cos [-(\alpha - \beta)] = \cos (\alpha - \beta)$, $\cos |\alpha - \beta| = \cos (\alpha - \beta)$ in either case.

   b. $\cos (\alpha - \beta) = \cos \alpha \cos \beta + \sin \alpha \sin \beta$

   c. The terminal point of $\mathbf{U}$ has coordinates $(x_1, y_1) = (\cos \alpha, \sin \alpha)$, and the terminal point of $\mathbf{V}$ has coordinates $(x_2, y_2) = (\cos \beta, \sin \beta)$. Therefore, $\cos |\alpha - \beta| = \cos \alpha \cos \beta + \sin \alpha \sin \beta = x_1 x_2 + y_1 y_2$.

   d. Since $|\alpha - \beta|$ is an undirected angle, it lies between $0°$ and $180°$. The cosine function is one-to-one on this domain, and $\cos|\alpha - \beta| = x_1 x_2 + y_1 y_2$, so $|\alpha - \beta| = \cos^{-1} (x_1 x_2 + y_1 y_2)$.

55. $14.25°$ **57.** $135°$

SECTION 9-4

1. The quantity V_x is the magnitude of the horizontal component of the shell's initial velocity, so it is equal to $\|\mathbf{V}\| \cos \theta = V \cos \theta$. Similarly, V_y is the magnitude of the vertical component of the velocity, so it is equal to $\|\mathbf{V}\| \sin \theta = V \sin \theta$.

3. a. $x \approx 1732.05t$

      $y = 1000t - 16t^2$

   b. $y \approx 0.58x - 0.0000053x^2$

   c. $109,000$ feet (about 20.7 miles)

5. a. $x \approx 684.04t$

      $y \approx 1879.39t - 16t^2$

   b. $y \approx 2.75x - 0.000034x^2$

   c. $81,000$ feet (about 15.3 miles)

7. a. $x \approx 1060.66t$

      $y \approx 1060.66t - 16t^2$

   b. $y \approx x - 0.000014x^2$

   c. $70,000$ feet (about 13.3 miles)

9. a. $x \approx 1767.77t$

      $y \approx 1767.77t - 16t^2$

   b. $y \approx x - 0.0000051x^2$

   c. $195,000$ feet (about 37.0 miles)

11. $\dfrac{\pi}{8} + k\pi, \dfrac{3\pi}{8} + k\pi$

13. $\dfrac{\pi}{10} + \dfrac{2k\pi}{5}, \dfrac{3\pi}{10} + \dfrac{2k\pi}{5}$

15. $0.19 + k\pi, 1.38 + k\pi$

17. $0.04 + \dfrac{2k\pi}{5}, 1.22 + \dfrac{2k\pi}{5}$

19. a. $\cos 2x = 0$ when $2x = \dfrac{\pi}{2} + 2k\pi$ or $2x = \dfrac{3\pi}{2} + 2k\pi$

     $\sin 2x = 0$ when $2x = 0 + 2k\pi$ or $2x = \pi + 2k\pi$
     The equations have no solutions in common.

   b. If $\cos 2x = 0$, then $\sin 2x \neq 0$, so the equation $2 \sin 2x - \cos 2x = 0$ cannot be true.

   c. The second equation is obtained from the first by dividing both sides by $\cos 2x$. If the first equation is true, then $\cos 2x \neq 0$, so the division is a valid operation. Therefore, the second equation is equivalent to the first.

21. $\dfrac{\pi}{6}, \dfrac{5\pi}{6}, \dfrac{3\pi}{2}$ **23.** $\dfrac{7\pi}{18}, \dfrac{11\pi}{18}, \dfrac{19\pi}{18}, \dfrac{23\pi}{18}, \dfrac{31\pi}{18}, \dfrac{35\pi}{18}$

25. (Sample response) The gunner might prefer the smaller value of θ because it produces a flatter trajectory. The shell can then hit a tall target, such as a building, even if there is a small error in the calculation of the range.

27. $\dfrac{\pi}{6} + \dfrac{k\pi}{2}$ **29.** $-\dfrac{3\pi}{10} + \dfrac{k\pi}{5}$

31. $0.96 + \dfrac{k\pi}{2}$ **33.** $-0.77 + \dfrac{k\pi}{5}$

35. $\dfrac{\pi}{4}, \dfrac{\pi}{3}, \dfrac{5\pi}{4}, \dfrac{5\pi}{3}$ **37.** $\dfrac{\pi}{3}, \dfrac{5\pi}{3}$

39. $0, \dfrac{\pi}{24}, \dfrac{5\pi}{24}, \dfrac{\pi}{4}, \dfrac{\pi}{2}, \dfrac{13\pi}{24}, \dfrac{17\pi}{24}, \dfrac{3\pi}{4}, \pi, \dfrac{25\pi}{24}, \dfrac{29\pi}{24}, \dfrac{5\pi}{4}, \dfrac{3\pi}{2}, \dfrac{37\pi}{24}, \dfrac{41\pi}{24}, \dfrac{7\pi}{4}$

41. $\dfrac{2\pi}{3}, \dfrac{4\pi}{3}$ **43.** $\dfrac{2\pi}{3}, \dfrac{4\pi}{3}$

45. $\dfrac{\pi}{6}, \dfrac{5\pi}{6}$ **47.** $\dfrac{\pi}{2}, \dfrac{2\pi}{3}, \dfrac{4\pi}{3}, \dfrac{3\pi}{2}$

49. $0, \dfrac{\pi}{15}, \dfrac{\pi}{5}, \dfrac{\pi}{3}, \dfrac{2\pi}{5}, \dfrac{7\pi}{15}, \dfrac{3\pi}{5}, \dfrac{11\pi}{15}, \dfrac{4\pi}{5}, \dfrac{13\pi}{15}, \pi, \dfrac{17\pi}{15}, \dfrac{6\pi}{5}, \dfrac{19\pi}{15}, \dfrac{7\pi}{5}, \dfrac{23\pi}{15}, \dfrac{8\pi}{5}, \dfrac{5\pi}{3}, \dfrac{9\pi}{5}, \dfrac{29\pi}{15}$

51. $\dfrac{\pi}{6}, \dfrac{\pi}{4}, \dfrac{3\pi}{4}, \dfrac{5\pi}{6}, \dfrac{7\pi}{6}, \dfrac{5\pi}{4}, \dfrac{7\pi}{4}, \dfrac{11\pi}{6}$

53. $\dfrac{\pi}{6}, \dfrac{3\pi}{4}, \dfrac{5\pi}{6}, \dfrac{7\pi}{4}$

55. $0.77, 5.51$ **57.** $0, 4.67$

SECTION 10-1

1. Figure 10-2 shows a right triangle with legs parallel to the u- and v-axes. The length of the leg parallel to the u-axis is $r \cos \phi$, and this length is the u-coordinate of P. The length of the leg parallel to the v-axis is $r \sin \phi$, and this length is the v-coordinate of P.

Figure 10-2 also shows a right triangle with legs parallel to the x- and y-axes. The length of the leg parallel to the x-axis is $r \cos (\theta + \phi)$, and this length is the x-coordinate of P. The length of the leg parallel to the y-axis is $r \sin (\theta + \phi)$, and this length is the y-coordinate of P.

3. $u = (\cos \theta)x + (\sin \theta)y$
$v = (\cos \theta)y - (\sin \theta)x$

5. $(0, 2)$

7. $\left(-4\sqrt{2}, 0\right)$

9. $\dfrac{u^2}{8} - \dfrac{v^2}{8} = 1$

11. a. $\dfrac{(u - 3\sqrt{34})^2}{327} + \dfrac{v^2}{21} = 1$

   b. In the uv-system, the center of the ellipse is $\left(3\sqrt{34}, 0\right)$. The foci are c units to the left and right of the center, where $c = \sqrt{a^2 - b^2} = \sqrt{327 - 21} = \sqrt{306} = 3\sqrt{34}$. Therefore, the foci are at $(0, 0)$ and $\left(6\sqrt{34}, 0\right)$.

13. The coefficient of uv can be written as

$$(C - A)(2 \cos \theta \sin \theta) + B(\cos^2 \theta - \sin^2 \theta)$$

Double-angle identities can be used to rewrite this as $(C - A) \sin 2\theta + B \cos 2\theta$. The coefficient of uv is 0 when

$$\begin{aligned}
(C - A) \sin 2\theta + B \cos 2\theta &= 0 \\
B \cos 2\theta &= -(C - A) \sin 2\theta \\
\frac{\cos 2\theta}{\sin 2\theta} &= -\frac{C - A}{B} \\
\cot 2\theta &= \frac{A - C}{B}
\end{aligned}$$

15.

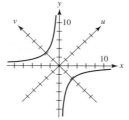

	(u, v)	(x, y)
Center	$(0, 0)$	$(0, 0)$
Vertices	$\left(0, \pm 3\sqrt{2}\right)$	$(3, -3), (-3, 3)$
Asymptotes	$v = \pm u$	$x = 0, y = 0$

17.

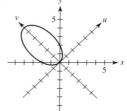

	(u, v)	(x, y)
Center	$\left(0, 2\sqrt{2}\right)$	$(-2, 2)$
Vertices	$(0, 0), \left(0, 4\sqrt{2}\right)$	$(0, 0), (-4, 4)$
Covertices	$\left(\pm\dfrac{2\sqrt{6}}{3}, 2\sqrt{2}\right)$	$\left(\dfrac{2\sqrt{3}}{3} - 2, \dfrac{2\sqrt{3}}{3} + 2\right),$ $\left(-\dfrac{2\sqrt{3}}{3} - 2, -\dfrac{2\sqrt{3}}{3} + 2\right)$

19. $\left(0, 3\sqrt{2}\right)$

21. $\left(\sqrt{2 + \sqrt{2}}, -\sqrt{2 - \sqrt{2}}\right)$

23. $\left(2\sqrt{3}, -2\right)$

25. $\left(\sqrt{2+\sqrt{3}}, -\sqrt{2-\sqrt{3}}\right)$

27.

	(u,v)	(x,y)
Center	$\left(-\dfrac{9\sqrt{2}}{2}, \dfrac{\sqrt{2}}{2}\right)$	$(-5,-4)$
Vertices	$\left(-\dfrac{9\sqrt{2}}{2} \pm \sqrt{20}, \dfrac{\sqrt{2}}{2}\right)$	$(-5 \pm \sqrt{10}, -4 \pm \sqrt{10})$
Asymptotes	$v = u + 5\sqrt{2}$	$x = -5$
	$v = -u - 4\sqrt{2}$	$y = -4$

29. The graph is empty.

31.

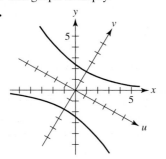

	(u,v)	(x,y)
Center	$(0,0)$	$(0,0)$
Vertices	$(0, \pm 2)$	$\left(\pm 1, \pm\sqrt{3}\right)$
Asymptotes	$v = \pm\dfrac{1}{2}u$	$y = \dfrac{-2 \pm \sqrt{3}}{2\sqrt{3} \pm 1} x$

33.

	(u,v)	(x,y)
Vertex	$\left(-\dfrac{7\sqrt{13}}{13}, \dfrac{2\sqrt{13}}{13}\right)$	$\left(-\dfrac{17}{13}, \dfrac{20}{13}\right)$

35. a. $B^2 - 4AC = -64, (B')^2 - 4A'C' = -64$
b. $B^2 - 4AC = -24, (B')^2 - 4A'C' = -24$

SECTION 10-2

1. a. $\left(\dfrac{\sqrt{2}}{2}, \dfrac{\sqrt{2}}{2}\right)$

b. $\left(-\dfrac{1}{2}, \dfrac{\sqrt{3}}{2}\right)$

c. $(-1,0)$

d. $(-0.80, 0.60)$

3. After t seconds, the ray from the origin to the object makes a directed angle of t radians with the positive x-axis. That ray intersects the unit circle at $(\cos t, \sin t)$ and intersects the circle of radius r at $(r \cos t, r \sin t)$. Thus the position of the object always satisfies the equations $x = r \cos t, y = r \sin t$.

5. a. $x = 10 \cos 5t, y = 10 \sin 5t$

b. $x = 10 \cos(-5t), y = 10 \sin(-5t)$
Equivalently, $x = 10 \cos 5t, y = -10 \sin 5t$.

c. $x = 0.5 \cos\left(-\dfrac{\pi t}{2}\right), y = 0.5 \sin\left(-\dfrac{\pi t}{2}\right)$

Equivalently, $x = 0.5 \cos\dfrac{\pi t}{2}, y = -0.5 \sin\dfrac{\pi t}{2}$

d. $x = 2 \cos 2\pi t, y = 2 \sin 2\pi t$

7. a. $x = -1 + 4\cos\dfrac{\pi t}{2}, y = 2 + 4\sin\dfrac{\pi t}{2}$

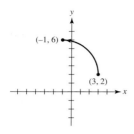

b. $x = \cos \pi t, y = 3 - \sin \pi t$

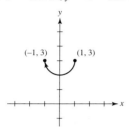

c. $x = \cos \pi t, y = 3 + \sin \pi t$

d. $x = 2 + 3\cos\dfrac{\pi t}{3}, y = -4 + 3\sin\dfrac{\pi t}{3}$

9. $x = -3 + 5\cos t, y = 5\sin t$

11. $x = -4 + \cos 0.5t, y = -2 - \sin 0.5t$

13. a. $\left(\dfrac{1}{2}, -\dfrac{\sqrt{3}}{2}\right)$

    b. every $\dfrac{\pi}{2}$ seconds

    c. counterclockwise

15. a. $(1, 0)$

    b. every 2 seconds

    c. clockwise

17. The center of the small platform moves counterclockwise at a radians per second around a circle of radius R centered at the origin, beginning at $(R, 0)$. Such a motion can be described by the equations $x = R\cos at$, $y = R\sin at$.

19. $x = R\cos at + r\cos bt, y = R\sin at - r\sin bt$

21. Answers will vary.

23. a. $\cos t = \dfrac{x}{3}, \sin t = \dfrac{y}{2}$

    b–c. $\cos^2 t + \sin^2 t = 1$

$$\left(\dfrac{x}{3}\right)^2 + \left(\dfrac{y}{2}\right)^2 = 1$$

$$\dfrac{x^2}{9} + \dfrac{y^2}{4} = 1$$

25. $\cos t = \dfrac{x - h}{a}$ and $\sin t = \dfrac{y - k}{b}$.

$$\cos^2 t + \sin^2 t = 1$$

$$\left(\dfrac{x - h}{a}\right)^2 + \left(\dfrac{y - k}{b}\right)^2 = 1$$

$$\dfrac{(x - h)^2}{a^2} + \dfrac{(y - k)^2}{b^2} = 1$$

27. $x = 5\cos t, y = 2\sin t$

29. $x = 1 + 2\sqrt{5}\cos t, y = -3 + \sqrt{10}\sin t$

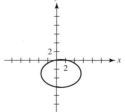

31. $\dfrac{(x - 57)^2}{31^2} + \dfrac{(y - 2.8)^2}{28^2} = 1$; center $= (57, 2.8)$,
vertices $= (57 \pm 31, 2.8) = (26, 2.8), (88, 2.8)$,
covertices $= (57, 2.8 \pm 28) = (57, -25.2), (27, 30.8)$

33. If both platforms rotate at the rate of a radians per second, the rider's seat can be described by the equations

$$x = R\cos at + r\cos at$$

$$y = R\sin at - r\sin at$$

Equivalently,

$$x = (R + r) \cos at$$

$$y = (R - r) \sin at$$

These equations describe the ellipse

$$\frac{x^2}{(R + r)^2} + \frac{y^2}{(R - r)^2} = 1.$$

35. In standard graphing form, the rectangular equation is $\frac{y^2}{4} - \frac{x^2}{25} = 1$. The parametric equations are $x = 5 \tan t$, $y = 2 \sec t$.

37. $x = 1 + 2\sqrt{5} \sec t, y = -3 + \sqrt{10} \tan t$

39. $(y - 4)^2 - (x - 1)^2 = 1$

41. $\frac{x^2}{9} - (y - 1)^2 = 1$

43. $x = 4 \cos t, y = -4 + 4 \sin t$

45. $x = 2 + 8 \cos t, y = 7 + 8 \sin t$

47. center $= (-8, -3)$, radius $= 1$; $(x + 8)^2 + (y + 3)^2 = 1$

49. $x = 4 \sec t, y = 4 \tan t$

51. $x = 6 + 2 \sec t, y = 6 - 2 \tan t$

53. $x = -95 + 3\sqrt{2} \cos t, y = -76 + 5\sqrt{2} \sin t$

55. $x = 10 \cos t, y = 10 \sin t$

57. $x = 10 \tan t, y = 30 \sec t$

59. $(x - 4)^2 + (y - 7)^2 = 4$

61. $(y - 7)^2 - \frac{(x - 4)^2}{4} = 1$

63. $\frac{x^2}{9} + \frac{y^2}{16} = 1$

65. $\frac{y^2}{25} - \frac{x^2}{144} = 1$

67. a. $(0, 1)$

    b. every $\frac{\pi}{3}$ seconds

    c. clockwise

69. a. $(0, 1)$

    b. every $\frac{2\pi}{3}$ seconds

    c. clockwise

71. a. $\frac{x^2}{25} + \frac{y^2}{100} = 1$

    b. $(0, 10)$

    c. every $\frac{\pi}{2}$ seconds

    d. clockwise

73. $x = -3 - 5 \cos \pi t, y = -4 + 5 \sin \pi t$

75. $x = 7 - 6 \sin 3t, y = 6 \cos 3t$

SECTION 10-3

1. a. $\left(6, \frac{\pi}{3}\right)$ (Answers may vary.)

    b. Face east, looking along McDowell Avenue. Then turn about 60° to the left, and walk straight ahead for about 6 miles.

3. a.   **b.**

c.   **d.**

5. a.

    b. The angles θ and $\theta + \pi$ are 180° apart. Therefore, going r units in the positive direction on the terminal side of θ is the same as going r units in the negative direction on the terminal side of $\theta + \pi$.

7. all values of θ

9.

$$\left(-4, \frac{\pi}{2} + 2k\pi\right), \left(4, \frac{3\pi}{2} + 2k\pi\right)$$

11.

$$(-3, -\pi + 2k\pi), (3, 2k\pi)$$

13.

$$\left(3, \frac{4\pi}{3}\right)$$

15.

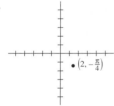

$$\left(2, \frac{7\pi}{4}\right)$$

17. $(1, -1)$ **19.** $\left(-2, -2\sqrt{3}\right)$

21. The angle $\dfrac{\pi}{4}$ is in the first quadrant, so the polar point $\left(5\sqrt{2}, \dfrac{\pi}{4}\right)$ is also in the first quadrant. These polar coordinates cannot represent the rectangular point $(-5, -5)$, which is in the third quadrant.

23. $(7, 0), (-7, \pi)$ **25.** $\left(2, -\dfrac{\pi}{3}\right), \left(-2, \dfrac{2\pi}{3}\right)$

27.

$$\left(1, \frac{5\pi}{6} + 2k\pi\right), \left(-1, \frac{11\pi}{6} + 2k\pi\right)$$

29.

$$\left(-\sqrt{2}, -\frac{3\pi}{4} + 2k\pi\right), \left(\sqrt{2}, \frac{\pi}{4} + 2k\pi\right)$$

31.

$$\left(3, \frac{\pi}{2}\right)$$

33.

$$\left(6, \frac{3\pi}{4}\right)$$

35. $(\sqrt{2}, -\sqrt{2})$ **37.** $(0, 0)$

39. $\left(2, \dfrac{3\pi}{4}\right), \left(-2, \dfrac{7\pi}{4}\right)$ **41.** $\left(4, \dfrac{7\pi}{6}\right), \left(-4, \dfrac{\pi}{6}\right)$

43. 11   **45.** $\sqrt{13 + 6\sqrt{3}} \approx 4.84$

SECTION 10-4

1. $r \sin \theta = (r \cos \theta)^2$

$r \sin \theta = r^2 \cos^2 \theta$

$\sin \theta = r \cos^2 \theta$

$r = \dfrac{\sin \theta}{\cos^2 \theta}$

$r = \left(\dfrac{1}{\cos \theta}\right)\left(\dfrac{\sin \theta}{\cos \theta}\right)$

$r = \sec \theta \tan \theta$

In the above process, we have divided both sides of the equation by r, so we must make sure that the equations are equivalent when $r = 0$, that is, at the pole. Since both the first and last equations are true when $r = 0$ and $\theta = 0$, the equations are equivalent.

3. $r = 5 \csc \theta$ **5.** $r = \dfrac{2 \cos \theta}{\cos 2\theta}$

7.

9. $y = \sqrt{3}\, x$

11. $y = \dfrac{1}{x}$

13. a. $y = x$

The graph is the same as the graph of $\theta = \dfrac{\pi}{4}$.

    b. It is the line $y = x$.

15. a. $y = (\tan c)x$

    b.

17.

19.

21.

23.

25. As θ increases from $\dfrac{3\pi}{4}$ to π, r is negative and increases from -1 to 0. Although the values of θ are in the second quadrant, the points on the graph are in the fourth quadrant and moving in toward the pole, as in the segment numbered 4 on the polar graph.

As θ increases from π to $\dfrac{5\pi}{4}$, r is positive and increases from 0 to 1. Since the values of θ are in the third quadrant, the points on the graph are in the third quadrant and moving outward from the pole, as in the segment numbered 5 on the polar graph.

As θ increases from $\dfrac{5\pi}{4}$ to $\dfrac{3\pi}{2}$, r is positive and decreases from 1 to 0. Since the values of θ are in the third quadrant, the points on the graph are in the third quadrant and moving in toward the pole, as in the segment numbered 6 on the polar graph.

As θ increases from $\dfrac{3\pi}{2}$ to $\dfrac{7\pi}{4}$, r is negative and decreases from 0 to -1. Although the values of θ are in the fourth quadrant, the points on the graph are in the second quadrant and moving outward from the pole, as in the segment numbered 7 on the polar graph.

As θ increases from $\dfrac{7\pi}{4}$ to 2π, r is negative and increases from -1 to 0. Although the values of θ are in the fourth quadrant, the points on the graph are in the second quadrant and moving in toward the pole, as in the segment numbered 8 on the polar graph.

27.

29.

31.

33. The replacements lead to the equation
$$-r = \sin 2\,(\pi - \theta)$$
$$-r = \sin (2\pi - 2\theta)$$
$$-r = \sin (-2\theta)$$
$$-r = -\sin 2\theta$$
$$r = \sin 2\theta$$

35. $(-r, -\theta) = (r, \pi - \theta)$
$(-r, -\theta) = (-r, 2\pi - \theta)$
Replace θ by $\pi - \theta$.
Replace r by $-r$ and θ by $2\pi - \theta$.

37. a. The tests show that the graph is symmetric about the line $\theta = \dfrac{\pi}{2}$.
    b. The graph does not possess any type of symmetry not shown by the tests.

39. a. The tests show that the graph is symmetric about the polar axis.
    b. The graph is also symmetric about the pole and about the line $\theta = \dfrac{\pi}{2}$.

41. a. 3.87   **b.** 3.71   **c.** 3.5   **d.** 3
    e. 2.5   **f.** 2.29   **g.** 2.13   **h.** 2

43. After the follower has turned clockwise through an angle of θ, the circumference of the cam must intersect the polar axis at the point $(3 + \cos\theta, 0)$ in order to place the follower at the proper position. Before the cam begins to turn, this point on the circumference must have coordinates $(3 + \cos\theta, \theta)$, so it must satisfy the equation $r = 3 + \cos\theta$.

45. $r = 3 + \cos\theta$
$r^2 = 3r + r\cos\theta$
$r^2 - r\cos\theta = 3r$
$x^2 + y^2 - x = 3r$
$(x^2 + y^2 - x)^2 = 9r^2$
$(x^2 + y^2 - x)^2 = 9(x^2 + y^2)$

47. $r = -4\csc\theta$ **49.** $r = \dfrac{2}{\sin\theta - \cos\theta}$

51. $r = \dfrac{4\cos\theta - 4\sin\theta}{\sin 2\theta}$ **53.** $x^2 + y^2 = 5$

55. $x - y = 6$ **57.** $x^2 + y^2 = \dfrac{y}{x}$ **59.** $x - y = 1$

61. a. The tests show that the graph is symmetric about the polar axis.
    b. The graph is also symmetric about the pole and the line $\theta = \dfrac{\pi}{2}$.

63. a. The tests show that the graph is symmetric about the line $\theta = \dfrac{\pi}{2}$.
    b. The graph does not possess any type of symmetry not shown by the tests.

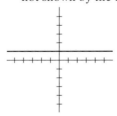

65. a. The tests do not show that the graph possesses any type of symmetry.
    b. The graph is symmetric about the polar axis.

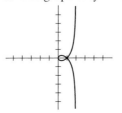

67. a. The tests show that the graph is symmetric about the polar axis.
b. The graph is also symmetric about the pole and the line $\theta = \dfrac{\pi}{2}$.

69. a. The tests show that the graph is symmetric about the line $\theta = \dfrac{\pi}{2}$.
b. The graph does not possess any type of symmetry not shown by the tests.

71. a. The tests show that the graph is symmetric about the polar axis.
b. The graph does not possess any type of symmetry not shown by the tests.

73. a. The tests do not show that the graph possesses any type of symmetry.
b. The graph is symmetric about the line $\theta = \dfrac{\pi}{2}$.

75. a. The tests show that the graph is symmetric about the pole.
b. The graph is also symmetric about the polar axis and the line $\theta = \dfrac{\pi}{2}$.

77. (Sample response) Some curves, such as circles centered at the origin and circles tangent to one of the coordinate axes at the origin, have polar equations that are simpler than their rectangular equations. On the other hand, the fact that every point has many sets of polar coordinates makes the polar system more difficult to work with in some respects. For example, a graph may have symmetry about an axis or about the pole even though an analytical test fails to detect it.

79. The first segment lies along the line $\theta = \dfrac{\pi}{2}$ for $\dfrac{1}{2\pi} \le r \le \dfrac{1}{2\pi} + 1$. The second segment is the circle $r = \dfrac{1}{2\pi}$. The third segment coincides with the first.

SECTION 10–5

1. a. The coefficient of t is the same in each equation.
b. 2π
c. A quarter of a cycle of each graph is $\dfrac{\pi}{2}$ units. In the second equation, t has been replaced by $t + \dfrac{\pi}{2}$, and this replacement shifts the graph $\dfrac{\pi}{2}$ units to the left.

3. a–d.

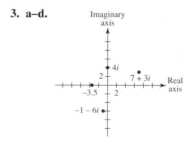

5. Every real number is represented by a point on the x-axis, and every imaginary number is represented by a point on the y-axis.

7. The distance of z from 0 is the distance between $(0, 0)$ and (a, b) in the plane. By the distance formula, the distance is $\sqrt{(a - 0)^2 + (b - 0)^2} = \sqrt{a^2 + b^2}$.

9. a. 13   **b.** 10   **c.** $\sqrt{10}$ **d.** 4

11. a. $\sqrt{2}\left(\cos \dfrac{3\pi}{4} + i \sin \dfrac{3\pi}{4}\right) = \sqrt{2}\left(-\dfrac{\sqrt{2}}{2} + i \dfrac{\sqrt{2}}{2}\right) = -\dfrac{2}{2} + i\dfrac{2}{2} = -1 + i$

b. The rectangular coordinates (a, b) are related to the polar coordinates (r, θ) by the equations $a = r \cos \theta$, $b = r \sin \theta$. Therefore, the number $a + bi$ can be written $(r \cos \theta) + (r \sin \theta)i = r(\cos \theta + i \sin \theta)$.

13. $6 - 6\sqrt{3}\,i$ **15.** $-i$

17. $2\left(\cos \dfrac{5\pi}{6} + i \sin \dfrac{5\pi}{6}\right)$

19. $10\left(\cos \dfrac{\pi}{2} + i \sin \dfrac{\pi}{2}\right)$

21. a. $7 - 3i$ **b.**

In Section 9-3 it was established that the points corresponding to the vectors $\mathbf{0}, \mathbf{V}_1, \mathbf{V}_2,$ and $\mathbf{V}_1 + \mathbf{V}_2$ lie at the vertices of a parallelogram. The same points correspond to the numbers $0, z_1, z_2,$ and $z_1 + z_2$.

23. a.

$|z_1 + z_2| = 20$

b. $16(\cos 0 + i \sin 0); 16$

c. $12\left(\cos \dfrac{\pi}{2} + i \sin \dfrac{\pi}{2}\right); 2i$

d. $16 + 12i; 20$

e. 20

25. a. $4\sqrt{3}\,i$ **b.** $4\sqrt{3}\,i$

27. a. $-2 + 2i$ **b.** $-2 + 2i$

29. a. $z_1 z_2 = 12\left(\cos \dfrac{3\pi}{2} + i \sin \dfrac{3\pi}{2}\right) = -12i$

$\dfrac{z_1}{z_2} = 3\left(\cos \dfrac{7\pi}{6} + i \sin \dfrac{7\pi}{6}\right) = -\dfrac{3\sqrt{3}}{2} - \dfrac{3}{2}i$

b.

c.

31. a. $z_1 z_2 = 4\left(\cos \dfrac{\pi}{2} + i \sin \dfrac{\pi}{2}\right) = 4i$

$\dfrac{z_1}{z_2} = \dfrac{1}{4}\left(\cos \dfrac{3\pi}{2} + i \sin \dfrac{3\pi}{2}\right) = -\dfrac{1}{4}i$

b.

c.

33. a. $z^2 = r^2(\cos 2\theta + i \sin 2\theta),\ z^3 = r^3(\cos 3\theta + i \sin 3\theta)$

    b. $z^n = r^n(\cos n\theta + i \sin n\theta)$

35. $27\left(\cos \dfrac{3\pi}{2} + i \sin \dfrac{3\pi}{2}\right), -27i$

37. $1728\left(\cos \dfrac{3\pi}{2} + i \sin \dfrac{3\pi}{2}\right), -1728i$

39. The nth roots of $r(\cos \theta + i \sin \theta)$ each have modulus $\sqrt[n]{r}$, so they are all located on the circle of radius $\sqrt[n]{r}$ centered at the origin. Furthermore, their arguments are spaced at intervals of $\dfrac{2\pi}{n}$, so they divide the circle

into n equal arcs, each subtended by an angle of $\dfrac{2\pi}{n}$.

41. $5\left(\cos\dfrac{4\pi}{9} + i\sin\dfrac{4\pi}{9}\right),$

$5\left(\cos\dfrac{10\pi}{9} + i\sin\dfrac{10\pi}{9}\right),$

$5\left(\cos\dfrac{16\pi}{9} + i\sin\dfrac{16\pi}{9}\right)$

43. $\cos\dfrac{\pi}{6} + i\sin\dfrac{\pi}{6}$ $\cos\dfrac{7\pi}{6} + i\sin\dfrac{7\pi}{6}$

$\cos\dfrac{\pi}{2} + i\sin\dfrac{\pi}{2}$ $\cos\dfrac{3\pi}{2} + i\sin\dfrac{3\pi}{2}$

$\cos\dfrac{5\pi}{6} + i\sin\dfrac{5\pi}{6}$ $\cos\dfrac{11\pi}{6} + i\sin\dfrac{11\pi}{6}$

45.

47. $-10, 10(\cos\pi + i\sin\pi)$

49. $6 - 6i, 6\sqrt{2}\left(\cos\dfrac{7\pi}{4} + i\sin\dfrac{7\pi}{4}\right)$

51. $-\sqrt{3} - i, 2\left(\cos\dfrac{7\pi}{6} + i\sin\dfrac{7\pi}{6}\right)$

53. $-5 + 8i, \sqrt{89}\,(\cos 2.13 + i\sin 2.13)$

55. $49\left(\cos\dfrac{2\pi}{3} + i\sin\dfrac{2\pi}{3}\right), -\dfrac{49}{2} + \dfrac{49\sqrt{3}}{2}i$

57. $\cos\dfrac{\pi}{4} + i\sin\dfrac{\pi}{4}, \dfrac{\sqrt{2}}{2} + \dfrac{\sqrt{2}}{2}i$

59. $32\left(\cos\dfrac{5\pi}{6} + i\sin\dfrac{5\pi}{6}\right), -16\sqrt{3} + 16i$

61. $64(\cos\pi + i\sin\pi), -64$

63. $\pm 3, \pm 3i$ **65.** $-2, 1 \pm \sqrt{3}\,i$

67. a. 6   **b.** $10i$

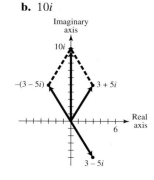

69. a. In the parallelogram, the angle at the origin is $\theta_2 - \theta_1$, so the angle at z_1 is $\pi - (\theta_2 - \theta_1)$. In the solid triangle, the sides adjacent to z_1 have lengths r_1 and r_2, and the third side has length r_3. By the Law of Cosines,

$$r_3 = \sqrt{r_1^2 + r_2^2 - 2r_1 r_2 \cos[\pi - (\theta_2 - \theta_1)]}$$

b. Since $\cos[\pi - (\theta_2 - \theta_1)] = -\cos(\theta_2 - \theta_1)$,

$$r_3 = \sqrt{r_1^2 + r_2^2 + 2r_1 r_2 \cos(\theta_2 - \theta_1)}$$

c. In polar form, $z_1 = r_1(\cos\theta_1 + i\sin\theta_1)$ and $z_2 = r_2(\cos\theta_2 + i\sin\theta_2)$, so the x-coordinate of $z_1 + z_2$ is $x_3 = r_1\cos\theta_1 + r_2\cos\theta_2$ and the y-coordinate is $y_3 = r_1\sin\theta_1 + r_2\sin\theta_2$. Therefore,

$$\tan\theta_3 = \dfrac{y_3}{x_3} = \dfrac{r_1\sin\theta_1 + r_2\sin\theta_2}{r_1\cos\theta_1 + r_2\cos\theta_2}$$

71. a. $(f_1 + f_2)(t) = r_1\sin(t + \theta_1) + r_2\sin(t + \theta_2)$
$= r_1(\cos\theta_1\sin t + \sin\theta_1\cos t) + r_2(\cos\theta_2\sin t + \sin\theta_2\cos t)$
$= (r_1\cos\theta_1 + r_2\cos\theta_2)\sin t + (r_1\sin\theta_1 + r_2\sin\theta_2)\cos t$

b. Let $A_1 = r_1\cos\theta_1 + r_2\cos\theta_2$ and $A_2 = r_1\sin\theta_1 + r_2\sin\theta_2$, so $(f_1 + f_2)(t) = A_1\sin t + A_2\cos t$. By the result of Exercise 70, we can rewrite $(f_1 + f_2)(t)$ as $r_3\sin(t + \theta_3)$. Furthermore,

$$r_3 = \sqrt{A_1^2 + A_2^2}$$
$$= \sqrt{(r_1\cos\theta_1 + r_2\cos\theta_2)^2 + (r_1\sin\theta_1 + r_2\sin\theta_2)^2}$$

$$= \sqrt{\begin{array}{l} r_1^2\cos^2\theta_1 + 2r_1 r_2\cos\theta_1\cos\theta_2 + r_2^2\cos^2\theta_2 + \\ r_1^2\sin^2\theta_1 + 2r_1 r_2\sin\theta_1\sin\theta_2 + r_2^2\sin^2\theta_2 \end{array}}$$

$$= \sqrt{\begin{array}{l} r_1^2(\cos^2\theta_1 + \sin^2\theta_1) + 2r_1 r_2(\cos\theta_1\cos\theta_2 + \\ \sin\theta_1\sin\theta_2) + r_2^2(\cos^2\theta_2 + \sin^2\theta_2) \end{array}}$$

$$= \sqrt{r_1^2 + r_2^2 + 2r_1 r_2\cos(\theta_2 - \theta_1)}$$

and $\tan\theta_3 = \dfrac{\sin\theta_3}{\cos\theta_3}$

$$= \dfrac{\dfrac{A_2}{\sqrt{A_1^2 + A_2^2}}}{\dfrac{A_1}{\sqrt{A_1^2 + A_2^2}}}$$

$$= \dfrac{A_2}{A_1} = \dfrac{r_1\sin\theta_1 + r_2\sin\theta_2}{r_1\cos\theta_1 + r_2\cos\theta_2}$$

SECTION A-1

1. a. 8230   **b.** between 8225 and 8235
c. 8.23×10^3

3. a. 0.0032   **b.** between 0.00315 and 0.00325
c. 3.2×10^{-3}

5. a. <u>61,400.009</u>
  b. between 61,400.0085 and 61,400.0095
  c. 6.1400009×10^4

7. a. 0.0<u>10203</u>
  b. between 0.0102025 and 0.0102035
  c. 1.0203×10^{-2}

9. 0.13   **11.** 8.8

13. 90   **15.** 8.80

17. 86.141   **19.** 8.801

SECTION A-2

1. -6 **3.** $\dfrac{13}{3}$

5. -24 **7.** 0

9. Both sides of the equation were divided by 2. Then 1 was added to both sides of the equation and x was subtracted from both sides of the equation. Both sides of the equation were divided by 3.

SECTION A-3

1. $A = (5, 2)$ $B = (-5, 2)$ $C = (-3, -1)$
  $D = (4, -3)$ $E = (2, 0)$ $F = (0, -4)$

3. quadrant III   **5.** quadrant IV

7. quadrant II   **9.** x-axis

SECTION A-4

1. 8   **3.** 13

5. $\sqrt{300}$ **7.** $\sqrt{10}$

9. 5   **11.** 2

13. $\sqrt{193}$ **15.** 41

SECTION A-5

1. $(-1, 4), (5, 10)$ **3.** $(3, 4), (5, 0)$

5.

7.

9.

11.

13. a.

13. b.

SECTION A-6

1.

3.

5.

7.

9.

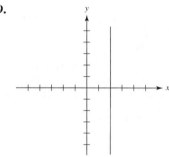

11. $y = x - 4$ **13.** $y = \frac{2}{3}x - 3$

15. $y = 4x - 25$ **17.** $y = -x$

SECTION A-7

1. $(1, 5]$

3. $(-40, 40)$

5. $(-\infty, \sqrt{2})$

7. $(-\infty, 0]$

9. $-2 < x < 2$

11. $0 \leq x < 1.5$

13. $x < 1$

15. $-\frac{1}{2} \leq x < \frac{2}{3}$

17. $-2 < x \leq 3$ $(-2, 3]$ **19.** $-2 \leq x \leq 3$ $[-2, 3]$

21. $x > 3$ $(3, \infty)$ **23.** $x \geq -2$ $[-2, \infty)$

25.

27.

29. $(-\infty, -2] \cup [3, \infty)$ **31.** $(-\infty, -2] \cup (3, \infty)$

SECTION A-8

1. $(-6, \infty)$

3. $\left[\dfrac{13}{3}, \infty\right)$

5. $(-\infty, -24]$

7. $(0, \infty)$

SECTION A-9

1. a. ± 7 **b.** $[-7, 7]$
   c. $(\infty, -7) \cup (7, \infty)$

3. a. $5, 8$ **b.** $[5, 8]$
   c. $(-\infty, 5) \cup (8, \infty)$

5. a. $-6, \dfrac{8}{3}$ **b.** $\left[-6, \dfrac{8}{3}\right]$

   c. $(-\infty, -6) \cup \left(\dfrac{8}{3}, \infty\right)$

7. a. no solution   **b.** no solution
   c. $(-\infty, \infty)$

SECTION A-10

1. $(x, y) = (2, 1)$ **3.** $(x, y) = (18, -22)$

5. $(x, y) = (3, -2)$ **7.** $(x, y) = (2, 2)$

9. $(x, y, z) = (-2, 1, 1)$

11. $(x, y, z) = \left(\dfrac{23}{36}, \dfrac{1}{18}, -\dfrac{1}{9}\right)$

13. $(a, b, c) = (0, 0, 0)$ **15.** $(x, y, z) = (5, 11, -7)$

SECTION A-11

1. 32 **3.** 1

5. 1 **7.** $\dfrac{1}{8}$

9. k **11.** $8b^7$

13. $\dfrac{z^4}{4}$ **15.** $4x^2 y^8$

17. $c + 3 - \dfrac{5}{c^2}$ **19.** $v^{5/6}$

SECTION A-12

1. $xyz(x + y + z)$ **3.** $(c + 5d)(c - 5d)$

5. $(r + 7)(r + 3)$ **7.** $(x + 5)(x - 6)$

9. $(3x + 11)(x - 1)$ **11.** $(2p - 1)(p - 2)$

13. $xy(x + 3y)(x^2 - 3xy + 9y^2)$

15. $ax(x^2 - 2ax + 5a^2)$

17. $(x^2 + 4)(x + 2)(x - 2)$

19. $(q + 1)(q^2 - q + 1)(q - 1)(q^2 + q + 1)$

SECTION A-13

1. $-7, 4$ **3.** $-\dfrac{7}{4}, 2$

5. $-3, 9$ **7.** $-\dfrac{5}{2} \pm \dfrac{\sqrt{61}}{2}$

9. $4 \pm \sqrt{2}$ **11.** $-1, 2$

13. $-3 \pm 2\sqrt{6}$ **15.** $\pm \dfrac{\sqrt{35}}{7} i$

17. 0.1 **19.** $-6, -1$

21. $-\dfrac{10}{3}, 1$ **23.** $-12, 5$

25. no real solutions; $\dfrac{5 \pm \sqrt{23}\, i}{6}$

27. two real solutions; $7 \pm \sqrt{79}$

29. one real solution; $\dfrac{5}{2}$

SECTION A-14

1. $2 + 7i$ **3.** $2\sqrt{2}\, i$

5. $-34 + 22i$ **7.** 27

9. $-\dfrac{14}{41} - \dfrac{38}{41} i$ **11.** $-\dfrac{8}{7} + \dfrac{3}{7} i$

13. $-2 + 2i$ **15.** 1

SECTION A-15

1. $\dfrac{1}{2}x^3 - 2x^2 - \dfrac{7}{2}$ **3.** $2x - \dfrac{7}{x} + \dfrac{8}{x^2}$

5. $x^4 + 6x^2 - x + 12 + \dfrac{-7x^2 + 22x - 8}{x^3 - 2x + 1}$

7. $x^2 + 3 + \dfrac{7}{x - 4}$

9. $2x - 4$ **11.** $1 + \dfrac{-12x + 6}{x^3 + 3x^2 + 4x - 5}$

SECTION A-16

1. $\dfrac{2}{3}$ **3.** 1

5. $\dfrac{x + 1}{2}$ **7.** $\dfrac{2x - 3}{x(2x + 3)(x - 4)^2}$

9. $\dfrac{5z}{6}$

11. $\dfrac{2a^2 + 8a + 3}{a(a + 1)(a - 1)}$

13. $\dfrac{20x}{x - 5}$

15. $\dfrac{y^2 + 2y + 6}{y^2(y + 2)^2}$

17. $\dfrac{3}{4}$

19. $\dfrac{6 + x}{6 - x}$

SECTION A–17

1. 5

3. no solution

5. $-\dfrac{5}{2}$

7. $\dfrac{-12}{5}$

SECTION A–18

1. $3|t|$

3. $|u|v^4$

5. $3a$

7. $2|y|^3$

9. 12

11. $5x\sqrt{6y}$

13. $3|x|$

15. $3x\sqrt{7x}$

17. $12\sqrt{k}$

19. $(10 - 2x)\sqrt{5x}$

21. a. $\dfrac{x}{\sqrt{6x}}$ **b.** $\dfrac{\sqrt{6x}}{6}$

23. a. $\dfrac{2}{\sqrt[3]{2s}}$ **b.** $\dfrac{\sqrt[3]{4s^2}}{s}$

25. a. $\dfrac{8}{\sqrt{65} + 5}$ **b.** $\dfrac{\sqrt{65} - 5}{5}$

27. a. $\dfrac{1}{\sqrt{6} - \sqrt{3}}$ **b.** $\dfrac{\sqrt{6} + \sqrt{3}}{3}$

29. 100

31. 243

33. $a^{20/3}$

35. $a^{16/3}$

37. $\dfrac{8}{x^{12}}$

39. $p^{1/10}$

SECTION A–19

1. 16

3. 5

5. 6

7. ± 5

9. 16

11. 10

13. 10

15. $-\dfrac{15}{4}$

SECTION B–1

1. a.

b. The second window was obtained by zooming in from the first window by a factor of 10.

3. a.

b–d. $(2.15, 0)$

5. a.

b–d. $(4.50, 0)$

7. a. $[-1, 1]$ by $[9, 11]$ **b.** $[-10, 10]$ by $[-10, 10]$
   c. $[9, 10]$ by $[0, 1]$

9. a. $[-4, 4]$ by $[-4, 16]$ **b.** $[-4, 4]$ by $[-16, 64]$
   c. $[-4, 4]$ by $[-1, 4]$

11. Responses will vary.

SECTION B–2

1. a. Choices will vary.   **b.** Responses will vary.

3. a. Choices will vary.   **b.** Responses will vary.

5. a. Choices will vary.
   b. The graph appears steeper than it really is.

7. a. Choices will vary.
   b. The graph appears flatter than it really is.

SECTION B–3

1. a. y-intercept $= (0, -12)$; vertex at $x = -\dfrac{1}{2}$;
   x-intercepts at $x = -4, 3$
   b–c. Responses will vary.   **d.** See part (a).

3. a. y-intercept $= (0, 20)$; vertex at $x = 3$; no x-intercepts
   b–c. Responses will vary.   **d.** See part (a).

5. a. y-intercept $= (0, -500{,}000)$; vertex at $x = -250$;
   x-intercepts at $x = -1000, 500$
   b–c. Responses will vary.   **d.** See part (a).

SECTION B-4

1. a. $y = \pm \sqrt{4x^2 + 16}$

b. y-intercepts $\approx (0, \pm 4.00)$
no x-intercepts

3. a. $y = -\dfrac{7}{4} \pm \dfrac{\sqrt{24x + 449}}{4}$

b. y-intercepts $\approx (0, -7.05), (0, 3.55)$
x-intercepts $\approx (-16.67, 0)$

5. a. $y = \dfrac{3}{2}x \pm \dfrac{\sqrt{5}}{2}|x|$

b. intercept $= (0, 0)$

7. a. $y = \dfrac{3}{4} - \dfrac{1}{4}x \pm \dfrac{\sqrt{-15x^2 + 34x - 791}}{4}$

b. no intercepts

9. a. $y = \dfrac{-x^2 + 3x + 10}{5x + 2}$

b. y-intercept $\approx (0, 5.00)$
x-intercepts $\approx (-2.00, 0), (5.00, 0)$

SECTION B-5

1. a. Choices will vary.
b. $(0, 0), (2, 0), (12, 0)$

3. a. Choices will vary.
b. $(1.51, 0)$

5. a. Choices will vary.
b. $(-3, 0), (-2, 0), (-1, 0), (1, 0), (2, 0), (3, 0), (4, 0)$

7. $[-25, 25]; x$-intercepts $(0, 0), (2, 0), (12, 0)$;
turning points $(0, 0), (1.31, 12.66), (9.19, -1706.35)$

9. $[-13, 13]; x$-intercepts $(\pm 2.41, 0)$;
turning points $(0, -12), (\pm 0.88, -8.76), (\pm 1.97, -24.05)$

11. $(2.59, -1.62)$

13. local maximum $\approx (-1.05, 13.03)$
local minimum $\approx (1.05, -1.03)$

SECTION B-6

1. a. x-intercepts $(\pm 2, 0)$; vertical asymptote $x = 0$
b. Choices will vary.

3. a. no x-intercepts; vertical asymptotes $x = \pm 5$
b. Choices will vary.

INDEX